FIBONACCI AND LUCAS NUMBERS WITH APPLICATIONS

PURE AND APPLIED MATHEMATICS

A Wiley Series of Texts, Monographs, and Tracts

Founded by RICHARD COURANT
Editors Emeriti: MYRON B. ALLEN III, PETER HILTON, HARRY HOCHSTADT, ERWIN KREYSZIG, PETER LAX, JOHN TOLAND

A complete list of the titles in this series appears at the end of this volume.

FIBONACCI AND LUCAS NUMBERS WITH APPLICATIONS

Volume Two

THOMAS KOSHY
Framingham State University

WILEY

This edition first published 2019

Registered Offices
John Wiley & Sons, Inc., 111 River Street, Hoboken, NJ 07030, USA

Editorial Office
111 River Street, Hoboken, NJ 07030, USA

For details of our global editorial offices, customer services, and more information about Wiley products, visit us at www.wiley.com.

Wiley also publishes its books in a variety of electronic formats and by print-on-demand. Some content that appears in standard print versions of this book may not be available in other formats.

Library of Congress Cataloging-in-Publication Data

Names: Koshy, Thomas.
Title: Fibonacci and Lucas numbers with applications / Thomas Koshy, Framingham State University.
Description: Second edition. | Hoboken, New Jersey : John Wiley & Sons, Inc., [2019]- | Series: Pure and applied mathematics: a Wiley series of texts, monographs, and tracts | Includes bibliographical references and index.
Identifiers: LCCN 2016018243 | ISBN 9781118742082 (cloth : v. 2)
Subjects: LCSH: Fibonacci numbers. | Lucas numbers.
Classification: LCC QA246.5 .K67 2019 | DDC 512.7/2–dc23 LC record available at https://lccn.loc.gov/2016018243

Cover image: © NDogan/Shutterstock
Cover design by Wiley

Set in 10/12pt, TimesNewRomanMTStd by SPi Global, Chennai, India

Printed in the United States of America

V10006580_120618

Dedicated to
the loving memory of
Dr. Kolathu Mathew Alexander
(1930–2017)

CONTENTS

LIST OF SYMBOLS

Symbol	Meaning
⇐ or ⇒	marginal symbol for alerting the change in notation
?	unsolved problem
■	end of a proof or solution; end of a lemma, theorem, or corollary when it does not end in a proof
$\mathbb{C}$	set of complex numbers
$(a_1, a_2, \ldots, a_n)$	greatest common divisor (gcd) of the positive integers $a_1, a_2, \ldots, a_n$
$[a_1, a_2, \ldots, a_n]$	least common multiple (lcm) of the positive integers $a_1, a_2, \ldots, a_n$
Δ	$\sqrt{x^2+4}$
$\alpha(x)$	$\frac{x+\Delta}{2}$
$\beta(x)$	$\frac{x-\Delta}{2}$
D	$\sqrt{x^2+1}$
$\gamma(x)$	$x+D$
$\delta(x)$	$x-D$
$a(x) \bmod b(x)$	remainder when $a(x)$ is divided by $b(x)$
$a(x) \equiv b(x) \pmod{c(x)}$	$a(x)$ is congruent to $b(x)$ modulo $c(x)$

Symbol	Meaning
$[a_0; \overline{a_1, \ldots, a_n}]$	infinite simple continued fraction
$w(\text{tile})$	weight of tile
$\mu(x)$	characteristic of the gibonacci family
F_n^*	$F_n F_{n-1} \cdots F_1$, where $F_0^* = 1$
$\begin{bmatrix} n \\ r \end{bmatrix}$	fibonomial coefficient $\dfrac{F_n^*}{F_r^* F_{n-r}^*}$
f_n^*	$f_n f_{n-1} \cdots f_1$, where $f_0^* = 1$
$\left[\begin{bmatrix} n \\ r \end{bmatrix}\right]$	gibonomial coefficient $\dfrac{f_n^*}{f_r^* f_{n-r}^*}$
$\begin{Bmatrix} n \\ r \end{Bmatrix}_q$	q-binomial coefficient $\dfrac{1-q^m}{1-q} \cdot \dfrac{1-q^{m-1}}{1-q^2} \cdots \dfrac{1-q^{m-r+1}}{1-q^r}$
$\Delta(x, y)$	$\sqrt{x^2 + 4y}$
$\boxed{!}$	switching variables

PREFACE

> Man has the faculty of becoming completely absorbed in one subject, no matter how trivial, and no subject is so trivial that it will not assume infinite proportions if one's entire attention is devoted to it.
>
> –Tolstoy, *War and Peace*

THE TWIN SHINING STARS REVISITED

The main focus of Volume One was to showcase the beauty, applications, and ubiquity of Fibonacci and Lucas numbers in many areas of human endeavor. Although these numbers have been investigated for centuries, they continue to charm both creative amateurs and mathematicians alike, and provide exciting new tools for expanding the frontiers of mathematical study. In addition to being great fun, they also stimulate our curiosity and sharpen mathematical skills such as pattern recognition, conjecturing, proof techniques, and problem-solving. The area is still so fertile that growth opportunities appear to be endless.

EXTENDED GIBONACCI FAMILY

The gibonacci numbers in Chapter 7 provide a unified approach to Fibonacci and Lucas numbers. In a similar way, we can extend these twin numeric families to twin polynomial families. For the first time, the present volume extends the gibonacci polynomial family even further. Besides Fibonacci and Lucas polynomials and their numeric counterparts, the extended gibonacci family includes Pell, Pell–Lucas, Jacobsthal, Jacobsthal–Lucas, Chebyshev, and

Vieta polynomials, and their numeric counterparts as subfamilies. This unified approach gives a comprehensive view of a very large family of polynomial functions, and the fascinating relationships among the subfamilies. The present volume provides the largest and most extensive study of this spectacular area of discrete mathematics to date.

Over the years, I have had the privilege of hearing from many Fibonacci enthusiasts around the world. Their interest gave me the strength and courage to embark on this massive task.

AUDIENCE

The present volume, which is a continuation of Volume One, is intended for a wide audience, including professional mathematicians, physicists, engineers, and creative amateurs. It provides numerous delightful opportunities for proposing and solving problems, as well as material for talks, seminars, group discussions, essays, applications, and extending known facts.

This volume is the result of extensive research using over 520 references, which are listed in the bibliography. It should serve as an invaluable resource for Fibonacci enthusiasts in many fields. It is my sincere hope that this volume will aid them in exploring this exciting field, and in advancing the boundaries of our current knowledge with great enthusiasm and satisfaction.

PREREQUISITES

A familiarity with the fundamental properties of Fibonacci and Lucas numbers, as in Volume One, is an indispensable prerequisite. So is a basic knowledge of combinatorics, generating functions, graph theory, linear algebra, number theory, recursion, techniques of solving recurrences, and trigonometry.

ORGANIZATION

The book is divided into 19 chapters of manageable size. Chapters 31 and 32 present an extensive study of Fibonacci and Lucas polynomials, including a continuing discussion of Pell and Pell–Lucas polynomials. They are followed by combinatorial and graph-theoretic models for them in Chapters 33 and 34. Chapters 35–39 offer additional properties of gibonacci polynomials, followed in Chapter 40 by a blend of trigonometry and gibonacci polynomials. Chapters 41 and 42 deal with a short introduction to Chebyshev polynomials and combinatorial models for them. Chapters 44 and 45 are two delightful studies of Jacobsthal and Jacobsthal–Lucas polynomials, and their numeric counterparts. Chapters 43, 46, and 48 contain a short discussion of bivariate gibonacci polynomials and their combinatorial models. Chapter 47 gives a brief discourse on Vieta polynomials, combinatorial models, and the relationships among the gibonacci subfamilies. Chapter 49 presents tribonacci numbers and polynomials; it also highlights their combinatorial and graph-theoretic models.

SALIENT FEATURES

This volume, like Volume One, emphasizes a user-friendly and historical approach; it includes a wealth of applications, examples, and exercises; numerous identities of varying degrees of sophistication; current applications and examples; combinatorial and graph-theoretic models; geometric interpretations; and links among and applications of gibonacci subfamilies.

HISTORICAL PERSPECTIVE

As in Volume One, I have made every attempt to present the material in a historical context, including the name and affiliation of every contributor, and the year of the contribution; indirectly, this puts a human face behind each discovery. I have also included photographs of some mathematicians who have made significant contributions to this ever-growing field.

Again, my apologies to those contributors whose names or affiliations are missing; I would be grateful to hear about any omissions.

EXERCISES AND SOLUTIONS

The book features over 1,230 exercises of varying degrees of difficulty. I encourage students and Fibonacci enthusiasts to have fun with them; they may open new avenues for further exploration. Abbreviated solutions to all odd-numbered exercises are given at the end of the book.

ABBREVIATIONS AND SYMBOLS INDEXES

An updated list of symbols, standard and nonstandard, appears in the front of the book. In addition, I have used a number of abbreviations in the interest of brevity; they are listed at the end of the book.

APPENDIX

The Appendix contains four tables: the first 100 Fibonacci and Lucas numbers; the first 100 Pell and Pell–Lucas numbers; the first 100 Jacobsthal and Jacobsthal–Lucas numbers; and a table of 100 tribonacci numbers. These should be useful for hand computations.

ACKNOWLEDGMENTS

A massive project such as this is not possible without constructive input from a number of sources. I am grateful to all those who played a significant role in enhancing the quality of the manuscript with their thoughts, suggestions, and comments.

My gratitude also goes to George E. Andrews, Marjorie Bicknell-Johnson, Ralph P. Grimaldi, R.S. Melham, and M.N.S. Swamy for sharing their brief biographies and photographs; to Margarite Landry for her superb editorial assistance; to Zhenguang Gao for preparing the tables in the Appendix; and to the staff at John Wiley & Sons, especially Susanne Steitz (former mathematics editor), Kathleen Pagliaro, and Jon Gurstelle for their enthusiasm and confidence in this huge endeavor.

Finally, I would be grateful to hear from readers about any inadvertent errors or typos, and especially delighted to hear from anyone who has discovered new properties or applications.

Thomas Koshy
tkoshy@emeriti.framingham.edu
Framingham, Massachusetts
August, 2018

If I have been able to see farther, it was only
because I stood on the shoulders of giants.
–Sir Isaac Newton (1643–1727)

31

FIBONACCI AND LUCAS POLYNOMIALS I

A man may die,
nations may rise and fall,
but an idea lives on.
–John F. Kennedy (1917–1963)

The celebrated Fibonacci polynomials $f_n(x)$ were originally studied beginning in 1883 by the Belgian mathematician Eugene C. Catalan, and later by the German mathematician Ernst Jacobsthal (1882–1965). They were further investigated by M.N.S. Swamy at the University of Saskatchewan, Canada. The equally famous Lucas polynomials $l_n(x)$ were studied beginning in 1970 by Marjorie Bicknell of Santa Clara, California [37].

Fibonacci and Lucas Numbers with Applications, Volume Two. Thomas Koshy.

Eugène Charles Catalan (1814–1894) was born in Bruges, Belgium, and received his Doctor of Science from the École Polytechnique in Paris. After working briefly at the Department of Bridges and Highways, he became professor of mathematics at Collège de Chalons-sur-Marne, and then at Collège Charlemagne. Catalan went on to teach at Lycée Saint Louis. In 1865, he became professor of analysis at the University of Liège. He published *Éléments de Géométrie* (1843) and *Notions d'astronomie* (1860), as well as many articles on multiple integrals, the theory of surfaces, mathematical analysis, calculus of probability, and geometry. Catalan is well known for extensive research on spherical harmonics, analysis of differential equations, transformation of variables in multiple integrals, continued fractions, series, and infinite products.

M.N.S. Swamy was born in Karnataka, India. He received his B.Sc. (Hons) in Mathematics from Mysore University in 1954; Diploma in Electrical Engineering from the Indian Institute of Science, Bangalore, in 1957; and M.Sc. (1960) and Ph.D. (1963) in Electrical Engineering from the University of Saskatchewan, Canada.

A former Chair of the Department of Electrical Engineering and Dean of Engineering and Computer Science at Concordia University, Canada, Swamy is currently a Research Professor and the Director of the Center for Signal Processing and Communications. He has also taught at the Technical University of Nova Scotia, and the Universities of Calgary and Saskatchewan.

Swamy is a prolific problem-proposer and problem-solver well known to the Fibonacci audience. He has published extensively in number theory, circuits, systems, and signal processing and has written three books. He is the editor-in-chief of *Circuits, Systems, and Signal Processing*, and an associate editor of *The Fibonacci Quarterly*, and a sustaining member of the Fibonacci Association.

Swamy received the Commemorative Medal for the 125th Anniversary of the Confederation of Canada in 1993 in recognition of his significant contributions to Canada. In 2001, he was awarded D.Sc. in Engineering by Ansted University, British Virgin Islands, "in recognition of his exemplary contributions to the research in Electrical and Computer Engineering and to Engineering Education, as well as his dedication to the promotion of Signal Processing and Communications Applications."

Marjorie Bicknell-Johnson was born in Santa Rosa, California. She received her B.S. (1962) and M.A. (1964) in Mathematics from San Jose State University, California, where she wrote her Master's thesis, *The Lambda Number of a Matrix*, under the guidance of V.E. Hoggatt, Jr.

The concept of the lambda number of a matrix first appears in the unpublished notes of Fenton S. Stancliff (1895–1962) of Meadville, Pennsylvania. (He died in Springfield, Ohio in 1962.) His extensive notes are pages of numerical examples without proofs or coherent definitions, that provided material for further study. Bicknell developed the mathematics of the lambda function in her thesis [40].

A charter member of the Fibonacci Association, Bicknell-Johnson has been a member of its Board of Directors since 1967, as well as Secretary (1965–2010) and Treasurer (1981–1999). In 2012, she wrote a history of the first 50 years of the Association [39].

Bicknell-Johnson has been a passionate and enthusiastic contributor to the world of Fibonacci and Lucas numbers, as author or co-author of F_{11} research papers, 32 of them written with Hoggatt. Her 1980 obituary of Hoggatt remains a fine testimonial to their productive association [38].

31.1 FIBONACCI AND LUCAS POLYNOMIALS

As we might expect, they satisfy the same polynomial recurrence $g_n(x) = xg_{n-1}(x) + g_{n-2}(x)$, where $n \geq 2$. When $g_0(x) = 0$ and $g_1(x) = 1$, $g_n(x) = f_n(x)$; and when $g_0(x) = 2$ and $g_1(x) = x$, $g_n(x) = l_n(x)$. Table 31.1 gives the first ten Fibonacci and Lucas polynomials in x. Clearly, $f_n(1) = F_n$ and $l_n(1) = L_n$.

In the interest of brevity and clarity, we drop the argument in the functional notation, when such deletions do *not* cause any confusion. Thus g_n will mean $g_n(x)$, although g_n is technically a functional name and *not* an output value. ⇐

TABLE 31.1. First 10 Fibonacci and Lucas Polynomials

n	$f_n(x)$	$l_n(x)$
1	1	x
2	x	x^2+2
3	x^2+1	x^3+3x
4	x^3+2x	x^4+4x^2+2
5	x^4+3x^2+1	x^5+5x^3+5x
6	x^5+4x^3+3x	$x^6+6x^4+9x^2+2$
7	$x^6+5x^4+6x^2+1$	$x^7+7x^5+14x^3+7x$
8	$x^7+6x^5+10x^3+4x$	$x^8+8x^6+20x^4+16x^2+2$
9	$x^8+7x^6+15x^4+10x^2+1$	$x^9+9x^7+27x^5+30x^3+9x$
10	$x^9+8x^7+21x^5+20x^3+5x$	$x^{10}+10x^8+35x^6+50x^4+25x^2+2$

For the curious-minded, we add that f_n is an even function when n is odd, and an odd function when n is even; and l_n is an odd function when n is odd, and even when n is even.

TABLE 31.2. Triangular Array A

n \ k	0	1	2		Row Sums
1	1				1
2	1				1
3	1	1			2
4	1	2			3
5	1	3	1		5
6	1	4	3		8
7	1	5	6	1	13
8	1	6	(10)	4	21
			↑ t_n		↑ F_n

Table 31.1 contains some hidden treasures. To see them, we arrange the nonzero coefficients of the Fibonacci polynomials in a left-justified array A; see Table 31.2. Column 2 of the array consists of the triangular numbers $t_n = n(n+1)/2$, and the nth row sum is F_n.

Let $a_{n,k}$ denote the element in row n and column k of the array. Clearly, $a_{n,k}$ is the coefficient of x^{n-2k-1} in f_n; so $a_{n,k} = \binom{n-k-1}{k}$. Recall that $\sum_{k\geq 0}\binom{n-k-1}{k} = F_n$ [287].

Consequently, it can be defined recursively:

$$a_{1,0} = 1 = a_{2,0}$$
$$a_{n,k} = a_{n-1,k} + a_{n-2,k-1},$$

where $n \geq 3$ and $k \geq 1$; see the arrows in Table 31.2. This can be confirmed; see Exercise 31.1.

Let d_n denote the nth rising diagonal sum. The sequence $\{d_n\}$ shows an interesting pattern: 1, 1, 1, 2, 3, 4, ⑥, 9, 13, ...; see Figure 31.1. We can also define d_n recursively:

$$d_1 = d_2 = d_3 = 1$$
$$d_n = d_{n-1} + d_{n-3},$$

where $n \geq 4$.

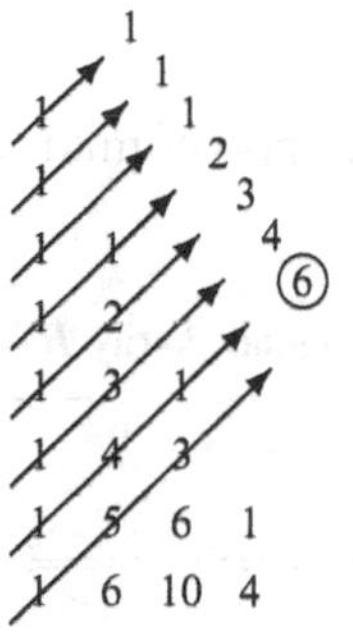

Figure 31.1.

Since $a_{n,k} = \binom{n-k-1}{k}$, it follows that

$$d_n = \sum_{k=0}^{\lfloor (n-1)/3 \rfloor} a_{n-k,k}$$
$$= \sum_{k=0}^{\lfloor (n-1)/3 \rfloor} \binom{n-2k-1}{k}.$$

For example, $d_8 = \sum_{k=0}^{2} \binom{7-2k}{k} = \binom{7}{0} + \binom{5}{1} + \binom{3}{2} = 9$.

The falling diagonal sums also exhibit an interesting pattern: 1, 2, 4, 8, 16, ...; see Figure 31.2. This is so, since the nth such sum is given by

$$\sum_{k=0}^{n-1} a_{n+k,k} = \sum_{k=0}^{n-1} \binom{n-1}{k}$$
$$= 2^{n-1},$$

where $n \geq 1$.

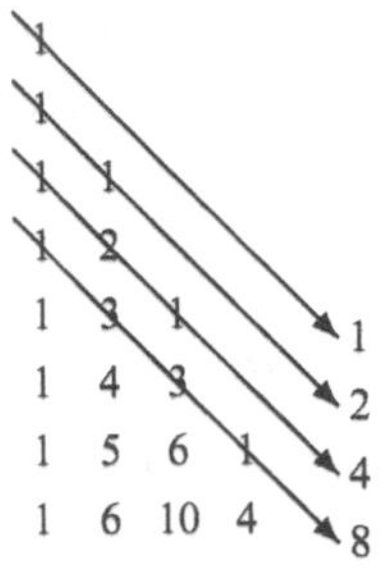

Figure 31.2.

The nonzero elements of Lucas polynomials also manifest interesting properties; see array B in Table 31.3.

TABLE 31.3. Triangular Array B

n \ k	0	1	2	3	4	Row Sums
1	1					1
2	1	2				3
3	1	3				4
4	1	4	2			7
5	1	5	5			11
6	1	6	9	2		18
7	1	7	14	7		29
8	1	8	20	16	2	47

↑ L_n

Let $b_{n,k}$ denote the element in row n and column k, where $n \geq 1$ and $k \geq 0$. Then

1) $\sum_{k=0}^{\lfloor n/2 \rfloor} b_{n,k} = L_n$.
2) $b_{n,k} = b_{n-1,k} + b_{n-2,k-1}$, where $b_{1,0} = 1 = b_{2,0}$, $b_{2,1} = 2$, $n \geq 3$, and $k \geq 0$.
3) Let x_n denote the nth rising diagonal sum. Then $x_1 = 1 = x_2$, $x_3 = 3$, and $x_n = x_{n-1} + x_{n-3}$, where $n \geq 4$.
4) $x_n = \sum_{k=0}^{\lfloor (n-1)/3 \rfloor} \frac{n-k}{n-2k}\binom{n-2k}{k}$.

For example, $x_7 = \sum_{k=0}^{2} \frac{7-k}{7-2k}\binom{7-2k}{k} = \frac{7}{7}\binom{7}{0} + \frac{6}{5}\binom{5}{1} + \frac{5}{3}\binom{3}{2} = 12.$
In the interest of brevity, we omit their proofs; see Exercises 31.2–31.5.
Next we construct a graph-theoretic model for Fibonacci polynomials.

Weighted Fibonacci Trees

Recall from Chapter 4 that the nth *Fibonacci tree* T_n is a (rooted) binary tree [287] such that

1) both T_1 and T_2 consist of exactly one vertex; and
2) T_n is a binary tree whose left subtree is T_{n-1} and right subtree is T_{n-2}, where $n \geq 3$. It has $2F_n - 1$ vertices, F_n leaves, $F_n - 1$ internal vertices, and $2F_n - 2$ edges.

Figure 31.3 shows the first five Fibonacci trees.

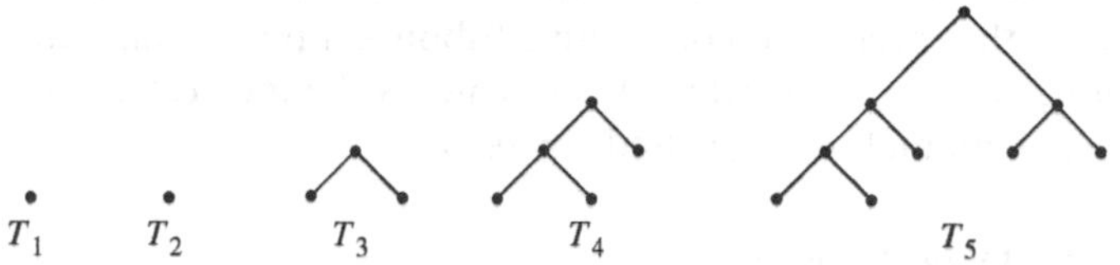

Figure 31.3.

We now assign a *weight* to T_n recursively. The weight of T_1 is 1 and that of T_2 is x. Then the weight $w(T_n)$ of T_n is defined by $w(T_n) = x \cdot w(T_{n-1}) + w(T_{n-2})$, where $n \geq 3$.

For example, $w(T_3) = x \cdot w(T_2) + w(T_1) = x^2 + 1$; and $w(T_4) = x \cdot w(T_3) + w(T_2) = x^3 + 2x$.

Since $w(T_1) = f_1$, and $w(T_2) = f_2$, it follows by the recursive definition that $w(T_n) = f_n$, where $n \geq 1$. Clearly, $w(T_n)$ gives the number of leaves of T_n when $x = 1$.

Binet-like Formulas

Using the recurrence $g_n = xg_{n-1} + g_{n-2}$ and the initial conditions, we can derive explicit formulas for both f_n and l_n; see Exercises 31.6 and 31.7:

$$f_n = \frac{\alpha^n - \beta^n}{\alpha - \beta} \quad \text{and} \quad l_n = \alpha^n + \beta^n,$$

where $\alpha = \alpha(x) = \dfrac{x+\Delta}{2}$ and $\beta = \beta(x) = \dfrac{x-\Delta}{2}$ are the solutions of the equation $t^2 - xt - 1 = 0$ and $\Delta = \Delta(x) = \sqrt{x^2+4}$. Notice that $\alpha + \beta = x$, $\alpha - \beta = \Delta$, and $\alpha\beta = -1$.

Since $\alpha = \alpha f_1 + f_0$ and $\alpha^2 = \alpha x + 1$, it follows by the principle of mathematical induction (PMI) that $\alpha^n = \alpha f_n + f_{n-1}$, where $n \geq 1$; see Exercise 31.8. Similarly $\beta^n = \beta f_n + f_{n-1}$.

Using the Binet-like formulas, we can extend the definitions of Fibonacci and Lucas polynomials to negative subscripts: $f_{-n} = (-1)^{n-1} f_n$ and $l_{-n} = (-1)^n l_n$.

Using the Binet-like formulas, we can also extract a plethora of properties of Fibonacci and Lucas polynomials; see Exercises 31.14–31.97. For example, it is fairly easy to establish that

$$f_n l_n = f_{2n};$$

$$f_{n+1} + f_{n-1} = l_n; \tag{31.1}$$

$$x f_{n-1} + l_{n-1} = 2f_n; \tag{31.2}$$

$$l_{2n} + 2(-1)^n = l_n^2;$$

$$f_{n+1} f_{n-1} - f_n^2 = (-1)^n;$$

$$l_{n+1} l_{n-1} - l_n^2 = (-1)^{n-1}(x^2+4).$$

The last two identities are Cassini-like formulas. It follows from the Cassini-like formula for f_n that every two consecutive Fibonacci polynomials are relatively prime; that is, $(f_n, f_{n-1}) = 1$, where (a, b) denotes the greatest common divisor (gcd) of the polynomials $a = a(x)$ and $b = b(x)$.

Cassini-like Formulas Revisited

Since $l_n(2i) = 2i^n$, it follows that $(x \pm 2i) \nmid l_n$, where $i = \sqrt{-1}$. Consequently, by the Cassini-like formula for l_n, every two consecutive Lucas polynomials are relatively prime, that is, $(l_n, l_{n+1}) = 1$.

The Cassini-like formulas have added dividends. For instance, $(f_{n+4k} + f_n, f_{n+4k-1} + f_{n-1}) = l_{2k}$. To see this, we have

$$\begin{aligned} \Delta(f_{n+4k} + f_n) &= \left(\alpha^{n+4k} - \beta^{n+4k}\right) + (\alpha^n - \beta^n) \\ &= \left(\alpha^{n+2k} - \beta^{n+2k}\right)(\alpha^{2k} + \beta^{2k}) \\ f_{n+4k} + f_n &= f_{n+2k} l_{2k}. \end{aligned}$$

Replacing n with $n-1$, this implies $f_{n+4k-1} + f_{n-1} = f_{n+2k-1} l_{2k}$. Thus

$$\begin{aligned}(f_{n+4k} + f_n, f_{n+4k-1} + f_{n-1}) &= l_{2k} \cdot (f_{n+2k}, f_{n+2k-1}) \\ &= l_{2k} \cdot 1 \\ &= l_{2k}. \qquad (31.3)\end{aligned}$$

Similarly,

$$(l_{n+4k} + l_n, l_{n+4k-1} + l_{n-1}) = l_{2k}; \qquad (31.4)$$

see Exercise 31.102.

It follows from properties (31.3) and (31.4) that

$$(F_{n+4k} + F_n, F_{n+4k-1} + F_{n-1}) = L_{2k};$$

$$(L_{n+4k} + L_n, L_{n+4k-1} + L_{n-1}) = L_{2k}.$$

For example, $(L_{23} + L_7, L_{22} + L_6) = (64079 + 29, 39603 + 18) = 47 = L_8$.

Pythagorean Triples

The identities $l_{n+1} + l_{n-1} = \Delta^2 f_n$ and $l_{2n} = \Delta^2 f_n^2 + 2(-1)^n$ (see Exercises 31.32 and 31.49) can be employed to construct Pythagorean triples (a, b, c). To see this, let $c = \Delta^2 f_{2n+3}$ and $a = x l_{2n+3} - 4(-1)^n$. We now find b such that (a, b, c) is a Pythagorean triple.

Since $c = l_{2n+4} + l_{2n+2}$, we have

$$\begin{aligned}c + a &= l_{2n+4} + (x l_{2n+3} + l_{2n+2}) - 4(-1)^n \\ &= 2[l_{2n+4} - 2(-1)^{n+2}] \\ &= 2\Delta^2 f_{n+2}^2; \\ c - a &= (l_{2n+4} - x l_{2n+3}) + l_{2n+2} + 4(-1)^n \\ &= 2[l_{2n+2} - 2(-1)^{n+1}] \\ &= 2\Delta^2 f_{n+1}^2.\end{aligned}$$

Therefore, $b^2 = c^2 - a^2 = (2\Delta^2 f_{n+2}^2)(2\Delta^2 f_{n+1}^2) = 4\Delta^4 f_{n+2}^2 f_{n+1}^2$; so we obtain $b = 2\Delta^2 f_{n+2} f_{n+1}$.

Thus $(a, b, c) = (x l_{2n+3} - 4(-1)^n, 2\Delta^2 f_{n+2} f_{n+1}, \Delta^2 f_{2n+3})$ is a Pythagorean triple.

Clearly, $\Delta^2 | b$ and $\Delta^2 | c$; so $\Delta^4 | (c^2 - b^2)$. Consequently, $\Delta^4 | a^2$ and hence $\Delta^2 | a$. Thus (a, b, c) is *not* a primitive Pythagorean triple.

H.T. Freitag (1908–2005) of Roanoke, Virginia, studied the Pythagorean triple for the special case $x = 1$ in 1991 [168].

Recall from Chapter 16 that $\lim_{n\to\infty} \frac{F_{n+1}}{F_n} = \alpha$. So what can we say about $\lim_{n\to\infty} \frac{f_{n+1}}{f_n}$? Next we investigate this.

Suppose $x > 0$. Then $0 < x/\Delta < 1$. Since $\frac{\beta}{\alpha} = \frac{x-\Delta}{x+\Delta} = -\frac{1-x/\Delta}{1+x/\Delta}$, $|\beta/\alpha| < 1$. Consequently,

$$\begin{aligned}
\frac{f_{n+1}}{f_n} &= \frac{\alpha^{n+1} - \beta^{n+1}}{\alpha^n - \beta^n}\\
&= \frac{\alpha^{n+1}}{\alpha^n} \cdot \frac{1-(\beta/\alpha)^{n+1}}{1-(\beta/\alpha)^n}\\
\lim_{n\to\infty} \frac{f_{n+1}}{f_n} &= \alpha \cdot \frac{1-0}{1-0}\\
&= \alpha.
\end{aligned}$$

Similarly, $\lim_{n\to\infty} \frac{l_{n+1}}{l_n} = \alpha$. Thus

$$\lim_{n\to\infty} \frac{f_{n+1}}{f_n} = \alpha = \lim_{n\to\infty} \frac{l_{n+1}}{l_n}, \tag{31.5}$$

where $x > 0$.

For the curious-minded, we add that

$$\frac{f_{n+1}(0)}{f_n(0)} = \begin{cases} 0 & \text{if } n \text{ is odd}\\ \text{undefined} & \text{otherwise;}\end{cases}$$

$$\frac{l_{n+1}(0)}{l_n(0)} = \begin{cases} \text{undefined} & \text{if } n \text{ is odd}\\ 0 & \text{otherwise.}\end{cases}$$

It follows by the recursive definition that $\deg(f_n) = n - 1$ and $\deg(l_n) = n$, where $\deg(h_n)$ denotes the degree of the polynomial $h_n(x)$ and $n \geq 1$. Suppose $a, b \geq 2$. Then $(a-1)(b-1) \geq 1$; consequently, $ab > a + b - 1$. Suppose also that $x \geq 1$. Since $\deg(f_a f_b) = \deg(f_a) + \deg(f_b) = a + b - 2$, it follows that $f_{ab} > f_a f_b$. Likewise, $l_{ab} > l_a l_b$.

The facts that $2\alpha = x + \Delta$, $2\beta = x - \Delta$, and $\Delta = \sqrt{x^2+4}$ can be used to develop two interesting identities, one involving Fibonacci polynomials and the other involving Lucas polynomials.

To begin, we have

$$(x^2+4)^n = (2\alpha - x)^{2n}$$
$$= \sum_{k=0}^{2n} \binom{2n}{k} (2\alpha)^k (-x)^{2n-k}. \tag{31.6}$$

Similarly,

$$(x^2+4)^n = \sum_{k=0}^{2n} \binom{2n}{k} (2\beta)^k (-x)^{2n-k}. \tag{31.7}$$

It follows by equations (31.6) and (31.7) that

$$2(x^2+4)^n = \sum_{k=0}^{2n} \binom{2n}{k} (-2)^k l_k x^{2n-k} \tag{31.8}$$

$$0 = \sum_{k=0}^{2n} \binom{2n}{k} (-2)^k f_k x^{2n-k}; \tag{31.9}$$

see Exercise 31.71.

For example,

$$\sum_{k=0}^{4} \binom{4}{k} (-2)^k l_k x^{4-k} = l_0 x^4 - 8l_1 x^3 + 24 l_2 x^2 - 32 l_3 x + 16 l_4$$
$$= 2(x^4 + 8x^2 + 16)$$
$$= 2(x^2+4)^2.$$

Identity (31.8), in particular, yields

$$\sum_{k=0}^{2n} \binom{2n}{k} (-2)^k L_k = 2 \cdot 5^n.$$

J.L. Brown of Pennsylvania State University found this result in 1965 [59].

The next example is an interesting application of identity (31.1).

Example 31.1. *Prove that*

$$f_{n+1} = \begin{cases} l_n - l_{n-2} + l_{n-4} - \cdots - l_2 + 1 & \text{if } n \equiv 0 \pmod 4 \\ l_n - l_{n-2} + l_{n-4} - \cdots - l_3 + x & \text{if } n \equiv 1 \pmod 4 \\ l_n - l_{n-2} + l_{n-4} - \cdots - l_2 - 1 & \text{if } n \equiv 2 \pmod 4 \\ l_n - l_{n-2} + l_{n-4} - \cdots - l_3 - x & \text{otherwise.} \end{cases}$$

Proof. Suppose n is even. Let $S_n = l_n - l_{n-2} + l_{n-4} - \cdots + (-1)^{(n-2)/2}$. Using identity (31.1), we can rewrite S_n as a telescoping sum:

$$\begin{aligned} S_n &= (f_{n+1} + f_{n-1}) - (f_{n-1} + f_{n-3}) + \cdots + (-1)^{(n-2)/2}(f_3 + f_1) \\ &= f_{n+1} + (-1)^{(n-2)/2} f_1 \\ &= \begin{cases} f_{n+1} - 1 & \text{if } n \equiv 0 \pmod 4 \\ f_{n+1} + 1 & \text{if } n \equiv 2 \pmod 4. \end{cases} \end{aligned}$$

This yields the desired formulas when n is even.

The formulas when n is odd follow similarly; see Exercise 31.72. ∎

In 1996, R. Euler of Northwest Missouri State University studied this example for the case $x = 1$ [152].

In particular, let $n = 7$. Then

$$\begin{aligned} l_7 - l_5 + l_3 - x &= (x^7 + 7x^5 + 14x^3 + 7x) - (x^5 + 5x^3 + 5x) + (x^3 + 3x) - x \\ &= x^7 + 6x^5 + 10x^3 + 4x \\ &= f_8. \end{aligned}$$

Generalized Cassini-like Formulas

The Cassini-like formulas can be generalized as follows:

$$f_m f_{m+n+k} - f_{m+k} f_{m+n} = (-1)^{m+1} f_n f_k; \tag{31.10}$$

$$l_m l_{m+n+k} - l_{m+k} l_{m+n} = (-1)^m (x^2 + 4) f_n f_k; \tag{31.11}$$

see Exercises 31.73 and 31.74.

It follows that both $f_m f_{m+n+k} - f_{m+k} f_{m+n}$ and $l_m l_{m+n+k} - l_{m+k} l_{m+n}$ are divisible by $f_n f_k$. In particular, both $f_m f_{m+2n} - f_{m+n}^2$ and $l_m l_{m+2n} - l_{m+n}^2$ are divisible by f_n^2. It also follows that $F_m F_{m+n+k} - F_{m+k} F_{m+n} = (-1)^{m+1} F_n F_k$ and $L_m L_{m+n+k} - L_{m+k} L_{m+n} = (-1)^m 5 L_n L_k$.

It follows from identities (31.10) and (31.11) that

$$f_m f_{n+1} - f_{m+1} f_n = (-1)^n f_{m-n}; \tag{31.12}$$

$$l_m l_{n+1} - l_{m+1} l_n = (-1)^n (x^2 + 4) f_{m-n}. \tag{31.13}$$

Identity (31.12) is a generalization of the *d'Ocagne identity* $F_m F_{n+1} - F_{m+1} F_n = (-1)^n F_{m-n}$, named after the French mathematician Philbert Maurice d'Ocagne (1862–1938).

It also follows from identities (31.10) and (31.11) that

$$f_{n+k}f_{n-k} - f_n^2 = (-1)^{n+k+1} f_k^2; \tag{31.14}$$

$$l_{n+k}l_{n-k} - l_n^2 = (-1)^{n+k}(x^2+4)f_k^2; \tag{31.15}$$

see Exercises 31.76 and 31.77.

These two identities imply that

$$f_{2n+1} = f_{n+1}^2 + f_n^2; \tag{31.16}$$

$$xl_{2n+1} = l_{n+1}^2 - (x^2+4)f_n^2 \tag{31.17}$$

$$= (x^2+4)f_{n+1}^2 - l_n^2; \tag{31.18}$$

see Exercises 31.78–31.80. Consequently, $l_{n+1}^2 + l_n^2 = (x^2+4)f_{2n+1}$.

The next example features a neat application of identity (31.10). It was originally studied in 1969 by Swamy [489].

Example 31.2. *Prove that*

$$\left(1 + \sum_{k=1}^{n} \frac{1}{f_{2k-1}f_{2k+1}}\right)\left(1 - \sum_{k=1}^{n} \frac{x^2}{f_{2k}f_{2k+2}}\right) = 1.$$

Proof. It follows from identity (31.10) that

$$f_{a+1}f_{a-2} - f_a f_{a-1} = (-1)^{a+1}x. \tag{31.19}$$

Consequently, we have

$$\frac{x}{f_{2k-1}f_{2k+1}} = \frac{f_{2k+2}}{f_{2k+1}} - \frac{f_{2k}}{f_{2k-1}}$$

$$\sum_{k=1}^{n} \frac{x}{f_{2k-1}f_{2k+1}} = \frac{f_{2n+2}}{f_{2n+1}} - \frac{f_2}{f_1}$$

$$= \frac{f_{2n+2}}{f_{2n+1}} - x$$

$$1 + \sum_{k=1}^{n} \frac{1}{f_{2k-1}f_{2k+1}} = \frac{f_{2n+2}}{xf_{2n+1}}. \tag{31.20}$$

It also follows by identity (31.10) that

$$-\frac{x}{f_{2k}f_{2k+2}} = \frac{f_{2k+3}}{f_{2k+2}} - \frac{f_{2k+1}}{f_{2k}}$$

$$-\sum_{k=1}^{n} \frac{x}{f_{2k}f_{2k+2}} = \frac{f_{2n+3}}{f_{2n+2}} - \frac{f_3}{f_2}$$

$$= \frac{xf_{2n+2} + f_{2n+1}}{f_{2n+2}} - \frac{x^2+1}{x}$$

$$= \frac{f_{2n+1}}{f_{2n+2}} - \frac{1}{x}$$

$$1 - \sum_{k=1}^{n} \frac{x^2}{f_{2k}f_{2k+2}} = x\frac{f_{2n+1}}{f_{2n+2}}. \qquad (31.21)$$

The given result now follows by equations (31.20) and (31.21). ■

The formula in Example 31.2 has a Lucas counterpart:

$$\left[x^2+2-\sum_{k=1}^{n} \frac{x^2(x^2+4)}{l_{2k-1}l_{2k+1}}\right]\left[\frac{1}{x^2+2}+\sum_{k=1}^{n} \frac{x^2+4}{l_{2k}l_{2k+2}}\right] = 1;$$

see Exercise 31.148.

A quick look at Table 31.1 reveals that the constant term in f_n is 1 if n is odd, and 0 if n is even; and the constant term in l_n is 0 if n is odd, and 2 if n is even. We now confirm these observations.

Ends of the Polynomials f_n and l_n

Since $\Delta(0) = 2$, $\alpha(0) = 1 = -\beta(0)$. Therefore, by the Binet-like formula for f_n, $f_n(0) = \dfrac{\alpha^n(0) - \beta^n(0)}{\alpha(0) - \beta(0)} = \dfrac{1-(-1)^n}{2}$. So f_n ends in 1 if n is odd, and 0 if n is even.

On the other hand, let

$$\kappa_n = \begin{cases} 0 & \text{if } n \text{ is odd} \\ 2 & \text{otherwise.} \end{cases}$$

Then $l_n(0) = 1 + (-1)^n = \kappa_n$. So l_n ends in 0 if n is odd, and 2 otherwise.

Next we develop two bridges linking f_n and l_n, by employing a bit of differential and integral calculus.

Links Between f_n and l_n

Since $\Delta' = x/\Delta$, it follows that $\alpha' = \alpha/\Delta$ and $\beta' = -\beta/\Delta$, where the prime denotes differentiation with respect to x. By the Binet-like formula for l_n, we then have

$$\begin{aligned} l_n' &= n\alpha^{n-1} \cdot \frac{\alpha}{\Delta} - n\beta^{n-1} \cdot \frac{\beta}{\Delta} \\ &= nf_n. \end{aligned} \tag{31.22}$$

It follows from identity (31.22) that $l_n'(1) = nF_n$.

For example, $l_6 = x^6 + 6x^4 + 9x^2 + 2$; so $l_6' = 6(x^5 + 4x^3 + 3x) = 6f_6$; and $l_6'(1) = 6 \cdot 8 = 6F_6$.

To see a related link, property (31.22) implies that we can recover l_n from f_n by integrating both sides from 0 to x:

$$\begin{aligned} \int_0^x l_n'(y)dy &= n\int_0^x f_n(y)dy \\ l_n - l_n(0) &= n\int_0^x f_n(y)dy \\ l_n &= \kappa_n + n\int_0^x f_n(y)dy. \end{aligned} \tag{31.23}$$

It follows from (31.23) that

$$\int_0^1 f_n dx = \frac{1}{n}(L_n - \kappa_n). \tag{31.24}$$

For example, $l_5 = 0 + 5\int_0^x f_5(y)dy = 5\int_0^x (y^4 + 3y^2 + 1)dy = x^5 + 5x^3 + 5x$; and $l_6 = 2 + 6\int_0^x (y^5 + 4y^3 + 3y)dy = x^6 + 6x^4 + 9x^2 + 2$. Clearly, $\int_0^1 f_6 dx = \int_0^1 (x^5 + 4x^3 + 3x)dx = \frac{8}{3} = \frac{1}{6}(L_6 - 2)$.

We can use the Binet-like formula for f_n, coupled with property (31.22), to develop a second-order differential equation for l_n.

A Differential Equation for l_n

By the Binet-like formula, we have

$$f_n' = \frac{n\Delta\left(\frac{\alpha^n}{\Delta} + \frac{\beta^n}{\Delta}\right) - (\alpha^n - \beta^n)\frac{x}{\Delta}}{\Delta^2}$$

$$\frac{1}{n}l_n'' = \frac{nl_n - xf_n}{x^2 + 4} \tag{31.25}$$

$$(x^2+4)l_n'' = n\left(nl_n - \frac{x}{n}l_n'\right)$$

$$(x^2+4)l_n'' + xl_n' - n^2 l_n = 0. \tag{31.26}$$

It follows from equation (31.26) that $l_n''(1) = n(nL_n - F_n)/5$.

For example, $l_6 = x^6 + 6x^4 + 9x^2 + 2$; $l_6' = 6x^5 + 24x^3 + 18x$; $l_6'' = 30x^4 + 72x^2 + 18$. Then $(x^2+4)l_6'' + xl_6' - 36l_6 = 0$; and $l_6''(1) = 6(6 \cdot 18 - 8)/5 = 120$.

Alternate Explicit Formulas

Fibonacci and Lucas polynomials can be defined explicitly in alternate ways:

$$f_{n+1} = \sum_{k=0}^{\lfloor n/2 \rfloor} \binom{n-k}{k} x^{n-2k}; \tag{31.27}$$

$$l_n = \sum_{k=0}^{\lfloor n/2 \rfloor} \frac{n}{n-k}\binom{n-k}{k} x^{n-2k}. \tag{31.28}$$

Both can be confirmed using PMI; see Exercises 31.96 and 31.97.

We now establish both, using different techniques.

Alternate Methods

To establish the Lucas-like formula (31.27), we employ a bit of operator theory [284, 498]. To this end, let

$$S_n = S_n(x) = \sum_{k=0}^{\lfloor n/2 \rfloor} \binom{n-k}{k} x^{n-2k}.$$

Let $D(S_n) = S_{n+1} - xS_n$. Then

$$\begin{aligned}
D(S_n) &= \sum_{k\ge 0}\left[\binom{n+1-k}{k} - \binom{n-k}{k}\right] x^{n-2k+1}\\
&= \sum_{k\ge 0}\binom{n-k}{k-1} x^{n-2k+1}\\
&= \sum_{j\ge 0}\binom{n-j-1}{j} x^{n-2j-1};\\
D^2(S_n) &= D(D(S_n))\\
&= D(S_{n+1} - xS_n)\\
&= D(S_{n+1}) - xD(S_n)
\end{aligned}$$

$$= \sum_{k\geq 0} \left[\binom{n+1-k}{k-1} - \binom{n-k}{k-1} \right] x^{n-2k+2}$$

$$= \sum_{k\geq 0} \binom{n-k}{k-2} x^{n-2k+2}$$

$$= \sum_{j\geq 0} \binom{n-j-1}{j-1} x^{n-2j}.$$

Consequently,

$$D^2(S_n) + xD(S_n) = \sum_{j\geq 0} \left[\binom{n-j-1}{j} + \binom{n-j-1}{j-1} \right] x^{n-2j}$$

$$S_{n+2} - xS_{n+1} = \sum_{j\geq 0} \binom{n-j}{j} x^{n-2j}$$

$$= S_n.$$

Thus S_n satisfies the Fibonacci polynomial recurrence. Since $S_0 = 1 = f_1$ and $S_1 = x = f_2$, this implies $S_n = f_{n+1}$. This yields the Lucas-like formula.

Formula (31.28) follows by a similar argument; see Exercise 31.98.

Interestingly, we can recover formula (31.28) from (31.27) by using equation (31.23):

$$l_n = \kappa_n + n \int_0^x f_n(y)dy$$

$$= \kappa_n + n \int_0^x \sum_{k=0}^{\lfloor (n-1)/2 \rfloor} \binom{n-k-1}{k} y^{n-2k-1} dy$$

$$= \kappa_n + \sum_{k=0}^{\lfloor (n-1)/2 \rfloor} \frac{n}{n-2k} \binom{n-k-1}{k} x^{n-2k}$$

$$= \kappa_n + \sum_{k=0}^{\lfloor (n-1)/2 \rfloor} \frac{n}{n-k} \binom{n-k}{k} x^{n-2k}$$

$$= \sum_{k=0}^{\lfloor n/2 \rfloor} \frac{n}{n-k} \binom{n-k}{k} x^{n-2k},$$

as desired.

It follows from formula (31.27) that the coefficient of x^{n-5} in f_n is $\binom{n-3}{2} = t_{n-4}$, where t_k denotes the kth triangular number and $n \geq 5$. For example, the coefficient of x^4 in f_9 is $t_5 = 15$; see Table 31.1. Likewise, the coefficient of x^{n-3} in f_n is $n-2$, where $n \geq 3$.

Formulas (31.24) and (31.27) Revisited

It follows by the Lucas-like formula (31.27) that

$$\int_0^1 f_n(x)dx = \sum_{k=0}^{\lfloor (n-1)/2 \rfloor} \frac{1}{n-2k}\binom{n-k-1}{k}.$$

This, coupled with formula (31.24), yields the summation formula

$$\sum_{k=0}^{\lfloor (n-1)/2 \rfloor} \frac{1}{n-2k}\binom{n-k-1}{k} = \frac{1}{n}(L_n - \kappa_n).$$

For example, $\sum_{k=0}^{2} \frac{1}{6-2k}\binom{5-k}{k} = \frac{1}{6}\binom{5}{0} + \frac{1}{4}\binom{4}{1} + \frac{1}{2}\binom{3}{2} = \frac{16}{6} = \frac{1}{6}(L_6 - 2).$
Similarly, $\sum_{k=0}^{3} \frac{1}{7-2k}\binom{6-k}{k} = \frac{29}{7} = \frac{1}{7}(L_7 - 0).$

31.2 PASCAL'S TRIANGLE

Fibonacci Polynomials

It follows by (31.27) that f_{n+1} can be found by adding up the binomial coefficients $\binom{n-k}{k}$ along the northeast diagonal n, with weights x^{n-2k}, where $0 \le k \le \lfloor n/2 \rfloor$.

For example,

$$\begin{aligned} f_5 &= \binom{4}{0}x^4 + \binom{3}{1}x^2 + \binom{2}{2}x^0 \\ &= \textcircled{1}x^4 + \textcircled{3}x^2 + \textcircled{1}1; \end{aligned}$$

see Figure 31.4. Clearly, $\textcircled{1} + \textcircled{3} + \textcircled{1} = 5 = F_5$.

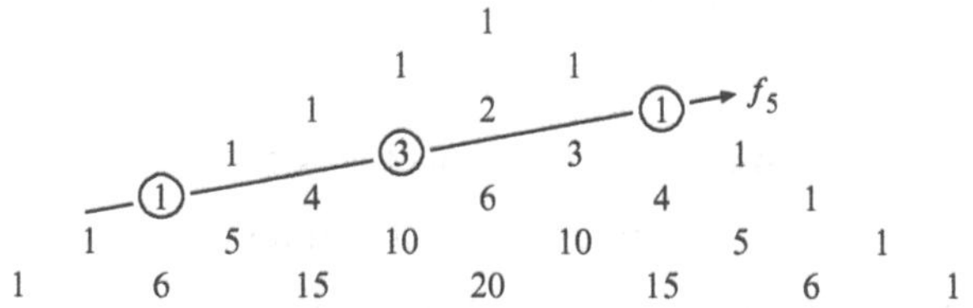

Figure 31.4. Pascal's Triangle.

This method of finding Fibonacci polynomials can be re-phrased slightly differently. This time, we construct a left-justified triangular array A. Its nth row consists of the terms in the binomial expansion of $(x+1)^n = \sum_{r=0}^{n} \binom{n}{r} x^{n-r}$; see Figure 31.5.

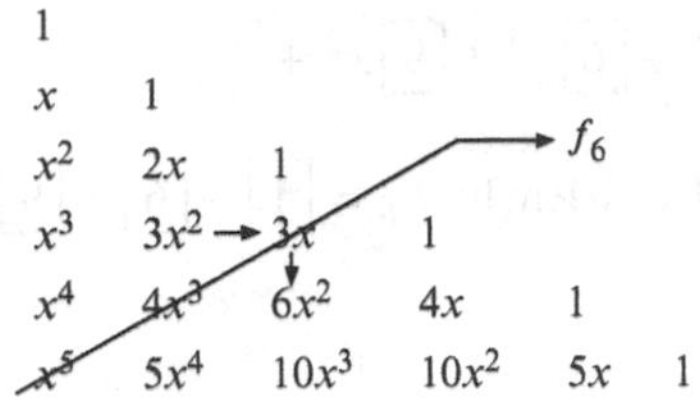

Figure 31.5.

Let $A(n,r)$ denote the element in row n and column r of array A. Then $A(n,r) = \binom{n}{r} x^{n-r}$, where $0 \le r \le n$.

Array A can be defined recursively as well:

$$A(0,0) = 1, \quad A(1,0) = x$$
$$A(n,r) = xA(n-1,r) + A(n-1,r-1),$$

where $0 \le r \le n$ and $n \ge 1$; see the arrows in Figure 31.5.

The nth rising diagonal sum $d_n(x)$ of the array is given by

$$\begin{aligned} d_n(x) &= \sum_{r=0}^{\lfloor n/2 \rfloor} A(n-r,r) \\ &= \sum_{r=0}^{\lfloor n/2 \rfloor} \binom{n-r}{r} x^{n-2r} \\ &= f_{n+1}, \end{aligned}$$

where $n \ge 0$.

For example, $d_5(x) = \sum_{r=0}^{2} \binom{5-r}{r} x^{5-2r} = x^5 + 4x^3 + 3x = f_6$; see Figure 31.5.

Lucas Polynomials

Since $\frac{n}{n-k}\binom{n-k}{k} = \binom{n-k}{k} + \binom{n-k-1}{k-1}$, it follows by formula (31.28) that

$$l_n = \sum_{k=0}^{\lfloor n/2 \rfloor} \left[\binom{n-k}{k} + \binom{n-k-1}{k-1} \right] x^{n-2k}.$$

Consequently, we can find the coefficients in l_n by adding the adjacent entries on the alternate diagonals $n-1$ and $n-3$. Multiplying the sums with x^{n-2k} and then adding the products yields l_n.

For example,

$$\begin{aligned} l_6 &= (1+0)x^6 + (5+1)x^4 + (6+3)x^2 + (1+1)x^0 \\ &= \boxed{1}\,x^6 + \boxed{6}\,x^4 + \boxed{9}\,x^2 + \boxed{2}; \end{aligned}$$

see the loops in Figure 31.6. Clearly, $L_6 = \boxed{1} + \boxed{6} + \boxed{9} + \boxed{2} = 18$.

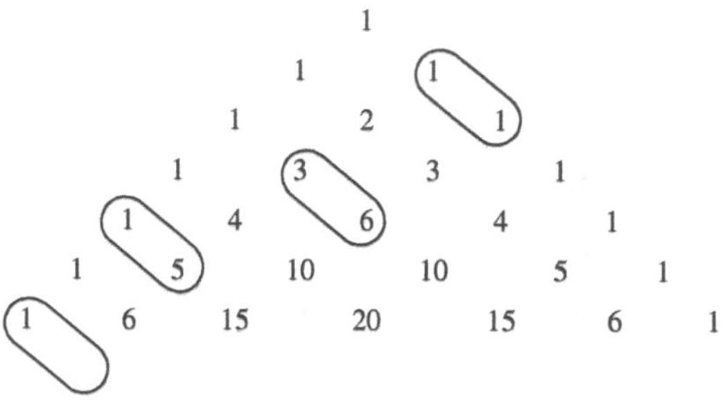

Figure 31.6. Pascal's Triangle.

As we saw earlier, this technique can be re-phrased a bit differently. To this end, we construct a new triangular array B using array A. Let $B(0,0) = 2$. We obtain the elements $B(n,r)$ of array B in row n by adding the corresponding elements in rows $n-1$ and n of array A, where $n \geq 1$. Thus row n of array B consists of the (distinct) terms in the sum $(x+1)^{n-1} + (x+1)^n = (x+2)(x+1)^{n-1}$; thus $B(n,r) = \left[\binom{n}{r} + \binom{n-1}{r-1}\right] x^{n-r}$; see Figure 31.7.

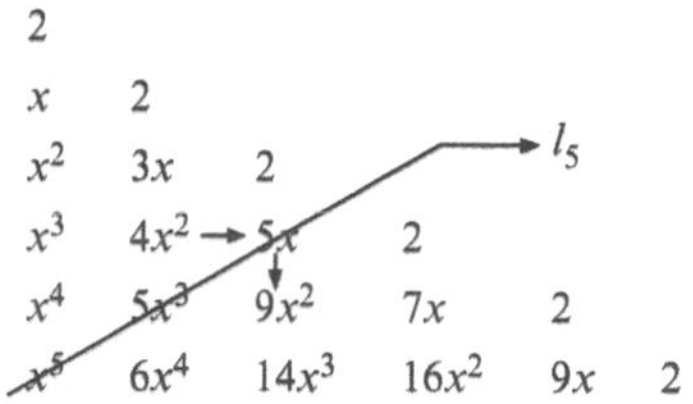

Figure 31.7.

The nth rising diagonal sum $b_n(x)$ of array B is given by

$$\begin{aligned} b_n(x) &= \sum_{r=0}^{\lfloor n/2 \rfloor} B(n-r, r) \\ &= \sum_{r=0}^{\lfloor n/2 \rfloor} \left[\binom{n}{r} + \binom{n-1}{r-1} \right] x^{n-2r} \\ &= l_n, \end{aligned}$$

where $n \geq 0$.

Next we find formula (31.28) for l_n in yet another way.

Lockwood's Identity

In 1967, E.H. Lockwood developed the identity

$$u^n + v^n = (u+v)^n + \sum_{k=1}^{\lfloor n/2 \rfloor} (-1)^k \left[\binom{n-k}{k} + \binom{n-k-1}{k-1} \right] (uv)^k (u+v)^{n-2k}$$

from the binomial theorem [333]. This can be rewritten as

$$u^n + v^n = \sum_{k=0}^{\lfloor n/2 \rfloor} (-1)^k \frac{n}{n-k} \binom{n-k}{k} (uv)^k (u+v)^{n-2k}. \tag{31.29}$$

Letting $u = \alpha$ and $v = \beta$, this yields the Lucas-like formula

$$l_n = \sum_{k=0}^{\lfloor n/2 \rfloor} \frac{n}{n-k} \binom{n-k}{k} x^{n-2k}.$$

Lockwood's identity has an added byproduct. To see this, we let $u = \alpha$ and $v = -\beta$, but change n to $2n+1$. Then equation (31.29) yields an interesting formula for f_{2n+1}:

$$\Delta f_{2n+1} = \sum_{k=0}^{n} (-1)^k \frac{2n+1}{2n-k+1} \binom{2n-k+1}{k} \Delta^{n-2k+1}$$

$$f_{2n+1} = \sum_{k=0}^{n} (-1)^k \frac{2n+1}{2n-k+1} \binom{2n-k+1}{k} (x^2+4)^{n-k}. \tag{31.30}$$

For example, $f_7 = \sum_{k=0}^{3}(-1)^k \frac{7}{7-k}\binom{7-k}{k}(x^2+4)^{7-2k} = x^6 + 5x^4 + 6x^2 + 1.$

It follows from formula (31.30) that

$$F_{2n+1} = \sum_{k=0}^{n}(-1)^k \frac{2n+1}{2n-k+1}\binom{2n-k+1}{k}5^{n-k}.$$

31.3 ADDITIONAL EXPLICIT FORMULAS

The Binet-like formulas, coupled with the binomial theorem, can be used to develop explicit formulas for f_n and l_n (see Exercises 31.99 and 31.100):

$$f_n = \frac{1}{2^{n-1}} \sum_{k=0}^{\lfloor (n-1)/2 \rfloor} \binom{n}{2k+1}(x^2+4)^k x^{n-2k-1}; \tag{31.31}$$

$$l_n = \frac{1}{2^{n-1}} \sum_{k=0}^{\lfloor n/2 \rfloor} \binom{n}{2k}(x^2+4)^k x^{n-2k}. \tag{31.32}$$

For example, $f_5 = \frac{1}{16}\sum_{k=0}^{2}\binom{5}{2k+1}(x^2+4)^k x^{4-2k} = x^4 + 3x^2 + 1$; and $l_5 = \frac{1}{16}\sum_{k=0}^{2}\binom{5}{2k}(x^2+4)^k x^{5-2k} = x^5 + 5x^3 + 5x.$

It follows from formulas (31.31) and (31.32) that

$$F_n = \frac{1}{2^{n-1}} \sum_{k=0}^{\lfloor (n-1)/2 \rfloor} \binom{n}{2k+1}5^k;$$

$$L_n = \frac{1}{2^{n-1}} \sum_{k=0}^{\lfloor n/2 \rfloor} \binom{n}{2k}5^k.$$

Consequently, $2^{n-1}F_n \equiv n \pmod 5$ and $2^{n-1}L_n \equiv 1 \pmod 5$.

Since $2^{n-1} = \sum_{k=0}^{\lfloor (n-1)/2 \rfloor}\binom{n}{2k+1} = \sum_{k=0}^{\lfloor n/2 \rfloor}\binom{n}{2k}$, formulas (31.31) and (31.32) can be rewritten with different looks:

$$f_n = \frac{\sum_{k=0}^{\lfloor (n-1)/2 \rfloor}\binom{n}{2k+1}(x^2+4)^k x^{n-2k-1}}{\sum_{k=0}^{\lfloor (n-1)/2 \rfloor}\binom{n}{2k+1}};$$

$$l_n = \frac{\sum_{k=0}^{\lfloor n/2 \rfloor} \binom{n}{2k} (x^2+4)^k x^{n-2k}}{\sum_{k=0}^{\lfloor n/2 \rfloor} \binom{n}{2k}}.$$

Using the binomial theorem, we can develop formulas for $l_{2n} \pm l_n$, as the next example shows.

Example 31.3. *Prove that*

$$l_{2n} + l_n = \sum_{k=0}^{\lfloor n/2 \rfloor} \binom{n}{2k} l_{2k} \left(x^{2k} + x^{n-2k}\right) + \sum_{k=0}^{\lfloor (n-1)/2 \rfloor} \binom{n}{2k+1} l_{2k+1} \left(x^{2k+1} - x^{n-2k-1}\right); \tag{31.33}$$

$$l_{2n} - l_n = \sum_{k=0}^{\lfloor n/2 \rfloor} \binom{n}{2k} l_{2k} \left(x^{2k} - x^{n-2k}\right) + \sum_{k=0}^{\lfloor (n-1)/2 \rfloor} \binom{n}{2k+1} l_{2k+1} \left(x^{2k+1} + x^{n-2k-1}\right). \tag{31.34}$$

Proof. By the binomial theorem, we have

$$\begin{aligned} (u+v)^n &= \sum_{k=0}^{n} \binom{n}{k} u^{n-k} v^k \\ &= \sum_{k=0}^{\lfloor n/2 \rfloor} \binom{n}{2k} u^{n-2k} v^{2k} + \sum_{k=0}^{\lfloor (n-1)/2 \rfloor} \binom{n}{2k+1} u^{n-2k-1} v^{2k+1}. \end{aligned} \tag{31.35}$$

Recall that $1 + \alpha x = \alpha^2$. Letting $u = 1$ and $v = \alpha x$, this yields

$$\alpha^{2n} = \sum_{k=0}^{\lfloor n/2 \rfloor} \binom{n}{2k} (\alpha x)^{2k} + \sum_{k=0}^{\lfloor (n-1)/2 \rfloor} \binom{n}{2k+1} (\alpha x)^{2k+1}.$$

Similarly,

$$\beta^{2n} = \sum_{k=0}^{\lfloor n/2 \rfloor} \binom{n}{2k} (\beta x)^{2k} + \sum_{k=0}^{\lfloor (n-1)/2 \rfloor} \binom{n}{2k+1} (\beta x)^{2k+1}.$$

Adding these two equations, we get

$$l_{2n} = \sum_{k=0}^{\lfloor n/2 \rfloor} \binom{n}{2k} l_{2k} x^{2k} + \sum_{k=0}^{\lfloor (n-1)/2 \rfloor} \binom{n}{2k+1} l_{2k+1} x^{2k+1}. \tag{31.36}$$

Since $\alpha + \beta = x$, letting $u = x$ and $v = -\alpha$ in equation (31.35), we get

$$\beta^n = \sum_{k=0}^{\lfloor n/2 \rfloor} \binom{n}{2k} \alpha^{2k} x^{n-2k} - \sum_{k=0}^{\lfloor (n-1)/2 \rfloor} \binom{n}{2k+1} \alpha^{2k+1} x^{n-2k-1}. \quad (31.37)$$

Similarly,

$$\alpha^n = \sum_{k=0}^{\lfloor n/2 \rfloor} \binom{n}{2k} \beta^{2k} x^{n-2k} - \sum_{k=0}^{\lfloor (n-1)/2 \rfloor} \binom{n}{2k+1} \beta^{2k+1} x^{n-2k-1}. \quad (31.38)$$

Adding equations (31.37) and (31.38), we get

$$l_n = \sum_{k=0}^{\lfloor n/2 \rfloor} \binom{n}{2k} l_{2k} x^{n-2k} - \sum_{k=0}^{\lfloor (n-1)/2 \rfloor} \binom{n}{2k+1} l_{2k+1} x^{n-2k-1}. \quad (31.39)$$

The given identities now follow by combining equations (31.36) and (31.39). ■

Identities (31.33) and (31.34) have Fibonacci counterparts:

$$f_{2n} + f_n = \sum_{k=0}^{\lfloor n/2 \rfloor} \binom{n}{2k} f_{2k} \left(x^{2k} - x^{n-2k}\right) + \sum_{k=0}^{\lfloor (n-1)/2 \rfloor} \binom{n}{2k+1} f_{2k+1} \left(x^{2k+1} + x^{n-2k-1}\right);$$

$$f_{2n} - f_n = \sum_{k=0}^{\lfloor n/2 \rfloor} \binom{n}{2k} f_{2k} \left(x^{2k} + x^{n-2k}\right) + \sum_{k=0}^{\lfloor (n-1)/2 \rfloor} \binom{n}{2k+1} f_{2k+1} \left(x^{2k+1} - x^{n-2k-1}\right). \quad (31.40)$$

In the interest of brevity, we omit their proofs; see Exercise 31.101.

We can similarly show that

$$f_{3n} - A + B + C - D = \sum_{k=0}^{\lfloor (n-1)/2 \rfloor} \binom{n}{2k+1} l_{2(n-2k)} f_{2k} \left(1 - x^{2k}\right) +$$
$$\sum_{k=0}^{\lfloor (n-1)/2 \rfloor} \binom{n}{2k+1} l_{2(n-2k-1)} f_{2k+1} \left(1 + x^{2k+1}\right), \quad (31.41)$$

where $\Delta A = [(x^2-1)\alpha + 2x]^n$, $\Delta B = [(x^2-1)\beta + 2x]^n$, $\Delta C = (2\alpha x - x^2 + 1)^n$, and $\Delta D = (2\beta x - x^2 + 1)^n$; see Exercise 31.177.

It follows from equations (31.33), (31.34), and (31.41) that

$$L_{2n} + L_n = 2\sum_{k=0}^{\lfloor n/2\rfloor} \binom{n}{2k} L_{2k};$$

$$L_{2n} - L_n = 2\sum_{k=0}^{\lfloor (n-1)/2\rfloor} \binom{n}{2k+1} L_{2k+1};$$

$$F_{3n} + 2^n F_n = 2\sum_{k=0}^{\lfloor (n-1)/2\rfloor} \binom{n}{2k+1} L_{2(n-2k-1)} F_{2k+1}.$$

H.T. Leonard, Jr. and Hoggatt developed these three special cases in 1968 [319].

Next we show how the Lucas recurrence and a suitable modulus can be effectively used to find the ends of the numbers l_n, where x is a positive integer such that $x^2 + s = A \cdot 10^t$ for some positive integers s, t, and A.

31.4 ENDS OF THE NUMBERS l_n

Suppose we would like to compute the units digit in l_n, where x is a positive integer. We choose $x^2 + 1$ as the modulus. Since $l_0 = 2$ and $l_1 = x$, it follows by the Lucas recurrence that the sequence $\{l_n \pmod{x^2+1}\}_{n\geq 0}$ is periodic with period 12; see Table 31.4. Consequently, $l_{12n+r} \equiv l_r \pmod{x^2+1}$, where $0 \leq r < 12$. We can confirm this using induction.

TABLE 31.4.

n	0	1	2	3	4	5	6	7	8	9	10	11
$l_n \pmod{x^2+1}$	2	x	1	$2x$	-1	x	-2	$-x$	-1	$-2x$	1	$-x$

For example, $l_{2035} = l_{12\cdot 169+7} \equiv l_7 \equiv -x \pmod{x^2+1}$. Consequently, $l_{2035}(3) \equiv -3 \equiv 7 \pmod{10}$; thus $l_{2035}(3)$ ends in 7.

On the other hand, suppose we would like to determine the last three digits in l_n. Since $114^2 + 4 = 13{,}000$, we will choose $\Delta^2 = x^2 + 4$ as the new modulus. Then the sequence $\{l_n \pmod{\Delta^2}\}_{n\geq 0}$ is periodic with period 4; see Table 31.5. Consequently, $l_{4n+r} \equiv l_r \pmod{\Delta^2}$, where $0 \leq r < 4$. This can be confirmed using PMI.

TABLE 31.5.

n	0	1	2	3	4	5	6	7
$l_n \pmod{\Delta^2}$	2	x	-2	$-x$	2	x	-2	$-x$

Thus

$$l_n = \begin{cases} 2 \pmod{\Delta^2} & \text{if } n \equiv 0 \pmod 4 \\ x \pmod{\Delta^2} & \text{if } n \equiv 1 \pmod 4 \\ -2 \pmod{\Delta^2} & \text{if } n \equiv 2 \pmod 4 \\ -x \pmod{\Delta^2} & \text{otherwise.} \end{cases}$$

As an example, $l_{2779} = l_{4\cdot694+3} \equiv l_3 \equiv -x \pmod{\Delta^2}$. In particular, $l_{2779} \equiv -114 \equiv 12{,}886 \pmod{13{,}000}$; so $l_{2779}(114)$ ends in 886.

An Important Observation

It follows from Table 31.5 that $l_n \not\equiv 0 \pmod{\Delta^2}$; so *no* Lucas polynomial is divisible by $x^2 + 4$. Consequently, *no* Lucas number ends in 0 or 5.

Next we find generating functions for the sequences $\{f_n\}$ and $\{l_n\}$.

31.5 GENERATING FUNCTIONS

Let $g(z) = \sum_{n=0}^{\infty} g_n z^n$, where $g_n = g_n(x)$. Then, by the Fibonacci recurrence, we have

$$(1 - xz - z^2)g(z) = g_0 + (g_1 - xg_0)z$$

$$g(z) = \frac{g_0 + (g_1 - xg_0)z}{1 - xz - z^2}.$$

Thus

$$\frac{g_0 + (g_1 - xg_0)z}{1 - xz - z^2} = \sum_{n=0}^{\infty} g_n z^n.$$

In particular,

$$\frac{z}{1 - xz - z^2} = \sum_{n=0}^{\infty} f_n z^n; \tag{31.42}$$

$$\frac{2 - xz}{1 - xz - z^2} = \sum_{n=0}^{\infty} l_n z^n.$$

The generating function (31.42), together with differentiation, can be used to develop a summation formula for f_n', as Swamy did in 1965 [475].

A Summation Formula for f_n'

Differentiating both sides of (31.42) with respect to x, we get

$$\begin{aligned}\sum_{n=0}^{\infty} f_n' z^n &= \left(\frac{z}{1-xz-z^2}\right)^2\\ &= \left(\sum_{n=0}^{\infty} f_n z^n\right)^2\\ &= \sum_{n=0}^{\infty}\left(\sum_{k=0}^{n} f_k f_{n-k}\right) z^n.\end{aligned}$$

Equating the coefficients of z^n from both sides, we get the desired formula:

$$f_n' = \sum_{k=1}^{n-1} f_k f_{n-k}.$$

For example, $\sum_{k=1}^{4} f_k f_{5-k} = f_1 f_4 + f_2 f_3 + f_3 f_2 + f_4 f_1 = 2[(x^3+2x) + x(x^2+1)] = 4x^3 + 6x = f_5'$.

Now we introduce briefly two interesting subfamilies of the Fibonacci–Lucas family [285].

31.6 PELL AND PELL–LUCAS POLYNOMIALS

Pell polynomials $p_n(x)$ and *Pell–Lucas polynomials* $q_n(x)$ are defined by $p_n(x) = f_n(2x)$ and $q_n(x) = l_n(2x)$, respectively. Both satisfy the same second-order recurrence $g_n(x) = 2xg_{n-1}(x) + g_{n-2}(x)$, where $n \geq 2$. When $g_0(x) = 0$ and $g_1(x) = 1$, $g_n(x) = p_n(x)$; and when $g_0(x) = 2$ and $g_1(x) = 2x$, $g_n(x) = q_n$. Again, ⇒ we delete the argument from the functional notation when such a notational switch causes *no* confusion. Table 31.6 gives the first ten Pell and Pell–Lucas polynomials.

Correspondingly, the *Pell numbers* P_n and *Pell–Lucas numbers* Q_n are given by $P_n = p_n(1) = f_n(2)$ and $2Q_n = q_n(1) = l_n(2)$. Table 31.7 gives the first ten Pell and Pell–Lucas numbers. Table A.2 in the Appendix gives the first 100 Pell and Pell–Lucas numbers. Clearly, $p_n(1/2) = F_n$ and $q_n(1/2) = L_n$.

Pell polynomials were mistakenly named after the English mathematician John Pell (1611–1685). Although Pell numbers occur in the study of the (Pell's) equation $u^2 - 2v^2 = (-1)^n$, the attribution of Pell's name to this equation, and hence to Pell numbers, is due to an innocent error by the great Swiss mathematician L. Euler.

TABLE 31.6. First 10 Pell and Pell–Lucas Polynomials

n	$p_n(x)$	$q_n(x)$
1	1	$2x$
2	$2x$	$4x^2+2$
3	$4x^2+1$	$8x^3+6x$
4	$8x^3+4x$	$16x^4+16x^2+2$
5	$16x^4+12x^2+1$	$32x^5+40x^3+10x$
6	$32x^5+32x^3+6x$	$64x^6+96x^4+36x^2+2$
7	$64x^6+80x^4+24x^2+1$	$128x^7+224x^5+112x^3+14x$
8	$128x^7+192x^5+80x^3+8x$	$256x^8+512x^6+320x^4+64x^2+2$
9	$256x^8+448x^6+240x^4+40x^2+1$	$512x^9+1152x^7+864x^5+240x^3+18x$
10	$512x^9+1024x^7+672x^5+160x^3+10x$	$1024x^{10}+2560x^8+2240x^6+800x^4+100x^2+2$

TABLE 31.7. First 10 Pell and Pell–Lucas Numbers

n	1	2	3	4	5	6	7	8	9	10
P_n	1	2	5	12	29	70	169	408	985	2378
Q_n	1	3	7	17	41	99	239	577	1393	3363

Leonhard Euler (1707–1783) was born in Basel, Switzerland. His father, a Calvinist pastor and a mathematician, wanted him to become a pastor. Although young Euler had his own ideas, he followed his father's wishes and studied Hebrew and theology at the University of Basel. His exceptional mathematical ability brought him to the attention of the well-known mathematician Johann Bernoulli (1667–1748). Recognizing the young Euler's remarkable talents, Bernoulli succeeded in persuading the father to change his mind, and Euler pursued his passion for mathematics.

At age 19, Euler published his first paper. Although it failed to win the prestigious Paris Prize in 1727, he won it twelve times in later years.

In 1727, Euler became the chair of mathematics at St. Petersburg Academy, founded by Peter the Great. Fourteen years later, he accepted the invitation of Frederick the Great to run the Prussian Academy in Berlin.

Euler was undoubtedly one of the most prolific mathematicians in history, making significant contributions to every branch of mathematics. He is known as the father of *graph theory*. With his phenomenal memory, he had every formula at his finger tips, and his genius enabled him to work anywhere and under any conditions. His productivity did not diminish when he became totally blind in 1768. Among mathematicians, Euler belongs in a class by himself.

Pell–Lucas numbers are named after both Pell and Lucas, although, like Pell, Lucas had nothing to do with them. Like Fibonacci and Lucas numbers, Pell and Pell–Lucas numbers behave like number-theoretic twins, and bring a great deal of joy and excitement to the mathematical community.

Binet-like Formulas for the Pell Family

Pell and Pell–Lucas polynomials, and the corresponding numbers, also can be defined by the Binet-like formulas

$$p_n = \frac{\gamma^n(x) - \delta^n(x)}{\gamma(x) - \delta(x)} \quad \text{and} \quad q_n = \gamma^n(x) + \delta^n(x)$$
$$P_n = \frac{\gamma^n - \delta^n}{\gamma - \delta} \quad \text{and} \quad Q_n = \frac{\gamma^n + \delta^n}{2},$$

where $\gamma(x) = x + D$, $\delta(x) = x - D$, $D = \sqrt{x^2+1}$, $\gamma = 1 + \sqrt{2}$, and $\delta = 1 - \sqrt{2}$. Notice that $\gamma(x) + \delta(x) = 2x$, $\gamma(x) - \delta(x) = 2D$, and $\gamma(x)\delta(x) = -1$.

Since $\alpha(2x) = \gamma(x)$ and $\beta(2x) = \delta(x)$, it follows from equation (31.5) that

$$\lim_{n\to\infty} \frac{p_{n+1}}{p_n} = \gamma(x) = \lim_{n\to\infty} \frac{q_{n+1}}{q_n},$$

where $|\gamma(x)| > |\delta(x)|$. In particular,

$$\lim_{n\to\infty} \frac{P_{n+1}}{P_n} = \gamma = \lim_{n\to\infty} \frac{Q_{n+1}}{Q_n}.$$

Pell Counterparts

Since $p_n(x) = f_n(2x)$ and $q_n(x) = l_n(2x)$, Fibonacci and Lucas properties have Pell counterparts.

For instance, Example 31.1 implies that

$$p_{n+1} = \begin{cases} q_n - q_{n-2} + q_{n-4} - \cdots - q_2 + 1 & \text{if } n \equiv 0 \pmod 4 \\ q_n - q_{n-2} + q_{n-4} - \cdots - q_3 + 2x & \text{if } n \equiv 1 \pmod 4 \\ q_n - q_{n-2} + q_{n-4} - \cdots - q_2 - 1 & \text{if } n \equiv 2 \pmod 4 \\ q_n - q_{n-2} + q_{n-4} - \cdots - q_3 - 2x & \text{otherwise.} \end{cases}$$

In particular,

$$P_{n+1} = \begin{cases} 2(Q_n - Q_{n-2} + Q_{n-4} - \cdots - Q_2) + 1 & \text{if } n \equiv 0 \pmod 4 \\ 2(Q_n - Q_{n-2} + Q_{n-4} - \cdots - Q_3) + 2 & \text{if } n \equiv 1 \pmod 4 \\ 2(Q_n - Q_{n-2} + Q_{n-4} - \cdots - Q_2) - 1 & \text{if } n \equiv 2 \pmod 4 \\ 2(Q_n - Q_{n-2} + Q_{n-4} - \cdots - Q_3) - 2 & \text{otherwise.} \end{cases}$$

For example,

$$\begin{aligned} q_7 - q_5 + q_3 - 2x &= (128x^7 + 224x^5 + 112x^3 + 14x) - (32x^5 + 40x^3 + 10x) \\ &\quad + (8x^3 + 6x) - 2x \\ &= 128x^7 + 192x^5 + 80x^3 + 8x \\ &= p_8; \end{aligned}$$

and $2(Q_7 - Q_5 + Q_3) - 2 = 2(239 - 41 + 7) - 2 = 408 = P_8$.

Cassini-like Formulas for the Pell Family

It follows from identities (31.10) and (31.11) that

$$\begin{aligned} p_m p_{m+n+k} - p_{m+k} p_{m+n} &= (-1)^{m+1} p_n p_k; \\ q_m q_{m+n+k} - q_{m+k} q_{m+n} &= 4(-1)^m (x^2+1) p_n p_k; \\ P_m P_{m+n+k} - P_{m+k} P_{m+n} &= (-1)^{m+1} P_n P_k; \\ Q_m Q_{m+n+k} - Q_{m+k} Q_{m+n} &= 2(-1)^m (x^2+1) P_n P_k. \end{aligned}$$

Links Between p_n and q_n

Since $q_n(x) = l_n(2x)$, it follows from equation (31.22) that $q_n'(x) = 2l_n'(2x) = 2nf_n(2x) = 2np_n(x)$; and hence $q_n'(1) = 2nP_n$.

As an example, $q_5 = 32x^5 + 40x^3 + 10x$; so $q_5' = 10(16x^4 + 12x^2 + 1) = 10p_5$; and hence $q_5'(1) = 290 = 10P_5$.

It follows from property (31.23) that

$$\begin{aligned} q_n &= \kappa_n + n\int_0^x f_n(2t)d(2t) \\ &= \kappa_n + 2n\int_0^x p_n(t)dt \qquad (31.43) \end{aligned}$$

$$2Q_n = \kappa_n + 2n\int_0^1 p_n(t)dt. \qquad (31.44)$$

For example, $q_5 = 0 + 10\int_0^x p_5(t)d(t) = 10\int_0^x (16t^4 + 12t^2 + 1)dt = 32x^5 + 40x^3 + 10x$; so $Q_5 = 82/2 = 41$. Similarly, $q_6 = 2 + 12\int_0^x (32t^5 + 32t^3 + 6t)dt = 64x^6 + 96x^4 + 36x^2 + 2$, and $Q_6 = 198/2 = 99$; see Tables 31.6 and 31.7.

Since $q_n'(x) = 2l_n'(2x)$ and $q_n''(x) = 4l_n''(2x)$, it follows from equation (31.26) that q_n satisfies the second-order differential equation

$$(x^2+1)q_n'' + xq_n' - n^2 q_n = 0. \qquad (31.45)$$

For example, $(x^2+1)q_5''+xq_5'-25q_5=(x^2+1)(640x^3+240x)+x(160x^4+120x^2+10)-25(32x^5+40x^3+10x)=0.$

Additional Formulas for the Pell Family

It follows from formulas (31.1), (31.3), (31.31), and (31.32) that

$$p_{n+1}=\sum_{k=0}^{\lfloor n/2\rfloor}\binom{n-k}{k}(2x)^{n-2k};$$

$$q_n=\sum_{k=0}^{\lfloor n/2\rfloor}\frac{n}{n-k}\binom{n-k}{k}(2x)^{n-2k};$$

$$p_{n+1}=\sum_{k=0}^{\lfloor n/2\rfloor}\binom{n+1}{2k+1}(x^2+1)^k x^{n-2k};$$

$$q_n=2\sum_{k=0}^{\lfloor n/2\rfloor}\binom{n}{2k}(x^2+1)^k x^{n-2k};$$

$$P_{n+1}=\sum_{k=0}^{\lfloor n/2\rfloor}\binom{n-k}{k}2^{n-2k};$$

$$Q_n=\sum_{k=0}^{\lfloor n/2\rfloor}\frac{n}{n-k}\binom{n-k}{k}2^{n-2k-1}$$

$$P_{n+1}=\sum_{k=0}^{\lfloor n/2\rfloor}\binom{n+1}{2k+1}2^k;$$

$$Q_n=\sum_{k=0}^{\lfloor n/2\rfloor}\binom{n}{2k}2^k.$$

From formula (31.30), we can extract a summation formula for odd-numbered Pell polynomials:

$$p_{2n+1}=\sum_{k=0}^{n}(-1)^k\frac{(2n+1)4^{n-k}}{2n-k+1}\binom{2n-k+1}{k}(x^2+1)^{n-k}. \qquad (31.46)$$

For example,

$$\begin{aligned}p_5&=\sum_{k=0}^{2}(-1)^k\frac{5\cdot 4^{2-k}}{5-k}\binom{5-k}{k}(x^2+1)^{2-k}\\&=16x^4+12x^2+1,\end{aligned}$$

as expected.

In particular, formula (31.46) yields

$$P_{2n+1} = \sum_{k=0}^{n} (-1)^k \frac{2n+1}{2n-k+1} \binom{2n-k+1}{k} 8^{n-k}.$$

Consequently,

$$\begin{aligned} P_7 &= \sum_{k=0}^{3} (-1)^k \frac{7}{7-k} \binom{7-k}{k} 8^{3-k} \\ &= \frac{7}{7}\binom{7}{0}8^3 - \frac{7}{6}\binom{6}{1}8^2 + \frac{7}{5}\binom{5}{2}8^1 - \frac{7}{4}\binom{4}{3}8^0 \\ &= 169. \end{aligned}$$

Pell and Pell–Lucas polynomials can be extended to negative subscripts also: $p_{-n} = (-1)^{n-1}p_n$ and $q_{-n} = (-1)^n q_n$; see Exercises 31.107 and 31.108. They too have their own versions of Cassini-like formulas:

$$\begin{aligned} p_{n+1}p_{n-1} - p_n^2 &= (-1)^n; \\ q_{n+1}q_{n-1} - q_n^2 &= (-1)^{n-1}(x^2+1); \end{aligned}$$

see Exercises 31.109 and 31.110.

Like Fibonacci and Lucas polynomials, Pell and Pell–Lucas polynomials also can be found using the northeast diagonals in Pascal's triangle, with weights $(2x)^{n-2k-1}$. For example,

$$\begin{aligned} p_5 &= \boxed{1}(2x)^4 + \boxed{3}(2x)^2 + \boxed{1}(2x)^0 \\ &= 16x^4 + 12x^2 + 1; \end{aligned}$$

$$\begin{aligned} q_6 &= \boxed{1}(2x)^6 + \boxed{6}(2x)^4 + \boxed{9}(2x)^2 + \boxed{2}(2x)^0 \\ &= 64x^6 + 96x^4 + 36x^2 + 2. \end{aligned}$$

Clearly, $P_5 = 16 + 12 + 1 = 29$ and $Q_6 = (64 + 96 + 36 + 2)/2 = 99$; see Tables 31.6 and 31.7.

It follows from properties (31.3) and (31.4) that

$$\begin{aligned} (p_{n+4k} + p_n, p_{n+4k-1} + p_{n-1}) &= q_{2k}; \\ (q_{n+4k} + q_n, q_{n+4k-1} + q_{n-1}) &= q_{2k}; \\ (P_{n+4k} + P_n, P_{n+4k-1} + P_{n-1}) &= 2Q_{2k}; \\ (Q_{n+4k} + Q_n, Q_{n+4k-1} + Q_{n-1}) &= 2Q_{2k}. \end{aligned}$$

For example, let $n = 7$ and $k = 4$. Then

$$(P_{23} + P_7, P_{22} + P_6) = (225058681 + 169, 93222358 + 70) = 1154 = 2Q_8;$$
$$(Q_{23} + Q_7, Q_{22} + Q_6) = (318281039 + 239, 131836323 + 99) = 1154 = 2Q_8.$$

Next we study an interesting Fibonacci inequality.

Swamy's Inequality

In 1966, Swamy discovered that $f_n^2 \leq (x^2+1)^2(x^2+2)^{n-3}$, where x is a real number ≥ 1 and $n \geq 3$ [478]. The proof is fairly straightforward and follows by PMI; see Exercise 31.131. Since $f_3 = x^2 + 1$ and $f_4 = x^3 + 2x$, the inequality can be rewritten as $x^{n-3} f_n^2 \leq f_3^2 f_4^{n-3}$. In particular, $4x^{n-3}p_n^2 \leq p_3^2 p_4^{n-3}$. Thus $F_n^2 \leq 4 \cdot 3^{n-3}$ and $4P_n^2 \leq 25 \cdot 12^{n-3}$.

For example,

$$\begin{aligned} x^2 f_5^2 &= x^2(x^4 + 3x^2 + 1)^2 \\ &= x^2(x^8 + 6x^6 + 11x^4 + 6x^2 + 1) \\ &\leq (x^2+1)^2(x^3+2x)^3 \\ &= f_3^2 f_4^2. \end{aligned}$$

As we might expect, Swamy's inequality has a Lucas counterpart: $l_n^2 \leq l_3^2 l_4^{n-3}$, where x is a real number ≥ 1 and $n \geq 3$; see Exercise 31.132. Consequently, $q_n^2 \leq q_3^2 q_4^{n-3}$, $L_n^2 \leq 16 \cdot 7^{n-3}$, and $Q_n^2 \leq 49 \cdot 34^{n-3}$.

31.7 COMPOSITION OF LUCAS POLYNOMIALS

It is well known that the composition operation, in general, is *not* commutative. But in the next example, we show that composition is commutative on a special family of Lucas polynomials. G.W. Smith of Brunswick, Maine, found this surprising property in 1998 [467]. Our proof invokes PMI [68]; see [68] for a trigonometric proof.

Example 31.4. *Let m and n be odd positive integers. Prove that $l_m \circ l_n = l_n \circ l_m$, where $\circ$ denotes the composition operation.*

Proof. Let $y = l_m(x)$. Then, by the Lucas recurrence, we have

$$\begin{aligned} l_n(y) &= y l_{n-1}(y) + l_{n-2}(y) \\ l_n(l_m(x)) &= l_m(x) l_{n-1}(l_m(x)) + l_{n-2}(l_m(x)). \end{aligned}$$

Clearly, we have

$$\begin{aligned} l_{1\cdot n}(x) &= xl_{n-1}(x) + (-1)^{1-1}l_{n-2}(x); \\ l_{2\cdot n}(x) &= x[xl_{2n-2}(x) + l_{2n-3}(x)] + l_{2n-2}(x) \\ &= (x^2+1)l_{2n-2}(x) + [l_{2n-2}(x) - l_{2n-4}(x)] \\ &= (x^2+2)l_{2n-2}(x) - l_{2n-4}(x) \\ &= l_2(x)l_{2(n-1)}(x) + (-1)^{2-1}l_{2(n-2)}(x). \end{aligned}$$

More generally, it follows by the strong version of PMI that

$$l_{nm}(x) = l_m(x)l_{(n-1)m}(x) + (-1)^{m-1}l_{(n-2)m}(x).$$

Since m is odd, we then have $l_n(l_m(x)) = l_{nm}(x)$. Similarly, when n is odd, $l_m(l_n(x)) = l_{mn}(x)$.

Thus $l_m(l_n(x)) = l_{mn}(x) = l_n(l_m(x))$; that is, $l_m \circ l_n = l_n \circ l_m$, as desired. ■

It follows by this example that $l_m(L_n) = L_{mn} = l_n(L_m)$, where mn is odd.

For example,

$$\begin{aligned} (l_5 \circ l_3)(x) &= l_5(x^3+3x) \\ &= (x^3+3x)^5 + 5(x^3+3x)^3 + 5(x^3+3x) \\ &= x^{15} + 15x^{13} + 90x^{11} + 275x^9 + 450x^7 + 378x^5 + 140x^3 + 15x \\ &= (x^5+5x^3+5x)^3 + 3(x^5+5x^3+5x) \\ &= (l_3 \circ l_5)(x); \end{aligned}$$

and $(l_5 \circ l_3)(1) = 1364 = 4^5 + 5\cdot 4^3 + 5\cdot 4 = l_5(4) = l_5(L_3)$.

It follows by Example 31.4 that $l_m(q_n(x)) = l_n(q_m(x))$ and hence $l_m(2Q_n) = 2Q_{mn}$, where mn is odd.

For example,

$$\begin{aligned} l_5(q_3(x)) &= l_5(8x^3+6x) \\ &= (8x^3+6x)^5 + 5(8x^3+6x)^3 + 5(8x^3+6x) \\ &= (32x^5+40x^3+10x)^3 + 3(32x^5+40x^3+10x) \\ &= l_3(q_5(x)); \end{aligned}$$

and $l_5(2Q_3) = l_5(14) = 551{,}614 = l_3(82) = l_3(2Q_5) = 2Q_{15}$.

Next we use the Binet-like formulas to develop results that remind us of the well known De Moivre's theorem in trigonometry, named after the French mathematician Abraham De Moivre (1667–1754) [287].

31.8 DE MOIVRE-LIKE FORMULAS

Since $\Delta f_n = \alpha^n - \beta^n$ and $l_n = \alpha^n + \beta^n$, $\frac{l_n + \Delta f_n}{2} = \alpha^n$ and $\frac{l_n - \Delta f_n}{2} = \beta^n$, where $\Delta = \Delta(x) = \alpha - \beta = \sqrt{x^2+4}$. Then

$$\left(\frac{l_n + \Delta f_n}{2}\right)^m = \alpha^{nm} = \frac{l_{nm} + \Delta f_{nm}}{2}; \tag{31.47}$$

$$\left(\frac{l_n - \Delta f_n}{2}\right)^m = \beta^{nm} = \frac{l_{nm} - \Delta f_{nm}}{2}. \tag{31.48}$$

These two De Moivre-like formulas can be used to develop interesting expansions of l_{nm} and f_{nm}. By the binomial theorem, formula (31.47) yields

$$\frac{1}{2^m}\sum_{r=0}^{m}\binom{m}{r} l_n^{m-r}\Delta^r f_n^r = \frac{l_{nm} + \Delta f_{nm}}{2}.$$

Equating the rational parts from both sides, we get

$$\begin{aligned} l_{nm} &= \frac{1}{2^{m-1}} \sum_{\substack{0\le r\le m \\ r \text{ even}}} \binom{m}{r} l_n^{m-r}\Delta^r f_n^r \\ &= \frac{1}{2^{m-1}} \sum_{r=0}^{\lfloor m/2\rfloor} \binom{m}{2r} (x^2+4)^r l_n^{m-2r} f_n^{2r}. \end{aligned} \tag{31.49}$$

Similarly, we get

$$f_{nm} = \frac{1}{2^{m-1}} \sum_{r=0}^{\lfloor (m-1)/2\rfloor} \binom{m}{2r+1} (x^2+4)^r l_n^{m-2r-1} f_n^{2r+1}. \tag{31.50}$$

Letting $m = 3$ in formulas (31.48) and (31.49), we get

$$\begin{aligned} 4f_{3n} &= 3l_n^2 f_n + (x^2+4)f_n^3 \\ 4l_{3n} &= l_n^3 + 3(x^2+4)l_n f_n^2. \end{aligned}$$

In particular, $f_6 = x^5 + 4x^3 + 3x$ and $l_6 = x^6 + 6x^4 + 9x^2 + 2$; see Table 31.1.

Formula (31.50) has an interesting byproduct. It implies that $f_n | f_{nm}$, where $n \neq 0$.

The De Moivre-like formulas have another delightful byproduct. It follows by formulas (31.47) and (31.48) that

$$l_{nm}^2 - \Delta^2 f_{nm}^2 = 4(-1)^{nm}.$$

In particular, this yields

$$l_n^2 - (x^2+4)f_n^2 = 4(-1)^n. \tag{31.51}$$

For example, $l_4^2 - (x^2+4)f_4^2 = (x^4+4x^2+2)^2 - (x^2+4)(x^3+2x)^2 = 4$; and $l_5^2 - (x^2+4)f_5^2 = (x^5+5x^3+5x)^2 - (x^2+4)(x^4+3x^2+1)^2 = -4$.

Identity (31.51) has a number of interesting applications. We will see some shortly, and also in Chapter 32.

Next we pursue some Fibonacci–Lucas bridges with several implications.

31.9 FIBONACCI–LUCAS BRIDGES

Recall that $\alpha\beta = -1$ and $\Delta = \sqrt{x^2+4}$. By the Binet-like formulas, we have

$$\begin{aligned}
\Delta^2 f_m f_{n-m} &= (\alpha^m - \beta^m)(\alpha^{n-m} - \beta^{n-m}) \\
(x^2+4) f_m f_{n-m} &= (\alpha^m + \beta^m) - \alpha^{n-m}(-1/\alpha)^m - \beta^{n-m}(-1/\beta)^m \\
&= l_n - (-1)^m l_{n-2m}.
\end{aligned} \tag{31.52}$$

For example,

$$\begin{aligned}
(x^2+4)f_3 f_4 &= (x^2+4)(x^2+1)(x^3+2x) \\
&= (x^7+7x^5+14x^3+7x)+x \\
&= l_7 + l_1.
\end{aligned}$$

Identity (31.52) has an interesting consequence. Replacing n with $n+m$, it yields

$$l_{n+m} - (-1)^m l_{n-m} = (x^2+4) f_m f_n. \tag{31.53}$$

Consequently, $l_{n+m} - (-1)^m l_{n-m} \equiv 0 \pmod{x^2+4}$, where $n \ge m$.

For example, let $n = 5$ and $m = 3$. Then

$$\begin{aligned}
l_8 + l_2 &= (x^8+8x^6+20x^4+16x^2+2) + (x^2+2) \\
&= x^8+8x^6+20x^4+17x^2+4 \\
&\equiv (-4)^4 + 8(-4)^3 + 20(-4)^2 + 17(-4) + 4 \pmod{x^2+4} \\
&\equiv 0 \pmod{x^2+4},
\end{aligned}$$

as expected.

Now replace n with $n-m$ and m with $m+1$ in (31.53). Also replace n with $n-m-1$ in the identity. These two operations yield

$$l_{n+1} + (-1)^{m+1} l_{n-2m-1} = (x^2+4) f_{m+1} f_{n-m}$$
$$l_{n-1} - (-1)^{m} l_{n-2m-1} = (x^2+4) f_{m} f_{n-m-1}.$$

Subtracting we get

$$l_{n+1} - l_{n-1} + 2(-1)^{m} l_{n-2m-1} = (x^2+4)(f_{m+1} f_{n-m} - f_m f_{n-m-1})$$
$$x l_n + 2(-1)^{m} l_{n-2m-1} = (x^2+4)(f_{m+1} f_{n-m} - f_m f_{n-m-1}). \quad (31.54)$$

Consequently, $x l_n + 2(-1)^m l_{n-2m-1} \equiv 0 \pmod{x^2+4}$. In particular, $L_n + 2(-1)^m L_{n-2m-1} \equiv 0 \pmod 5$. M.N. Deshpande of the Institute of Science, Nagpur, India, found this special case in 2001 [125].

As an added dividend, it follows from identity (31.53) that $L_{n+m} - (-1)^m L_{n-m} = 5F_m F_n$, and hence $L_{n+5} + L_{n-5} \equiv 0 \pmod{25}$.

Identities (31.52)–(31.54) have additional consequences:

$$4(x^2+1) p_m p_{n-m} = q_n - (-1)^m q_{n-2m};$$
$$(x^2+4) f_n^2 = l_{2n} - 2(-1)^n; \quad (31.55)$$
$$4(x^2+1) p_n^2 = q_{2n} - 2(-1)^n;$$
$$5F_m F_{n-m} = L_n - (-1)^m L_{n-2m};$$
$$4P_m P_{n-m} = Q_n - (-1)^m Q_{n-2m};$$
$$5F_n^2 = L_{2n} - 2(-1)^n;$$
$$4P_n^2 = Q_{2n} - (-1)^n.$$

Next we study some applications of identity (31.51).

31.10 APPLICATIONS OF IDENTITY (31.51)

Since $l_{2n} = l_n^2 - 2(-1)^n$ (see Exercise 31.48), it follows from identity (31.55) that

$$l_n^2 - (x^2+4) f_n^2 = 4(-1)^n,$$

which is identity (31.51).

It implies that

$$q_n^2 - 4(x^2+1) p_n^2 = 4(-1)^n; \quad (31.56)$$
$$L_n^2 - 5F_n^2 = 4(-1)^n; \quad (31.57)$$
$$Q_n^2 - 2P_n^2 = (-1)^n. \quad (31.58)$$

It follows from identity (31.51) that $[l_n^2 - (x^2+4)f_n^2]^2 - 16 = 0$. This equation shows the existence of a polynomial $f(x, y)$ with integral coefficients such that $f(F_n, L_n) = 0$, namely, $f(x, y) = (5x^2 - y^2)^2 - 16 = 25x^4 - 10x^2y^2 + y^4 - 16$. Likewise, there is a polynomial such that $g(P_n, Q_n) = 0$, namely, $g(x, y) = 4x^4 - 4x^2y^2 + y^4 - 1$.

It follows from identities (31.57) and (31.58) that the Pell's equations $x^2 - 5y^2 = 4(-1)^n$ and $x^2 - 2y^2 = (-1)^n$ are solvable [285, 287]. The solutions are the lattice points (L_n, F_n) and (Q_n, P_n), respectively; they lie on the hyperbolas $x^2 - 5y^2 = \pm 4$ and $x^2 - 2y^2 = \pm 1$, respectively; see Figure 31.8.

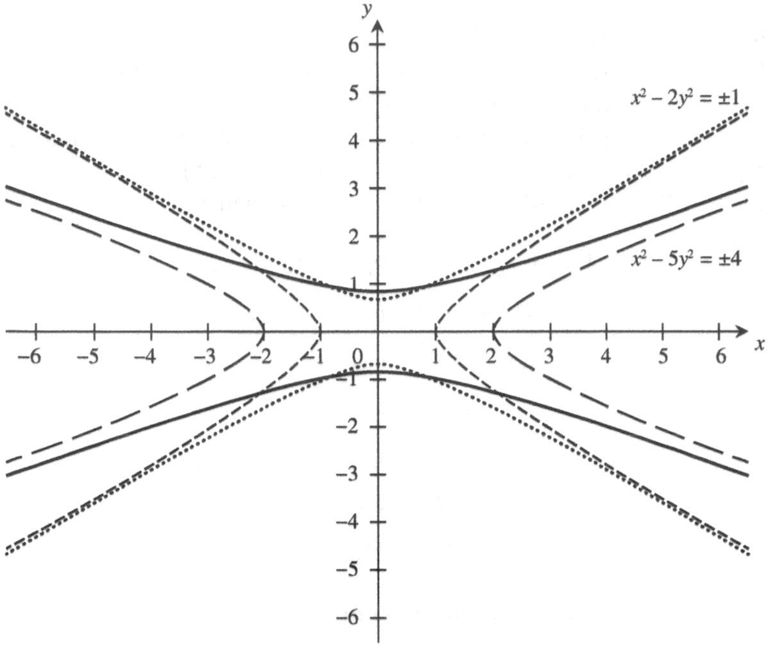

Figure 31.8.

To enjoy a simple application of the equation $x^2 - 5y^2 = -4$, consider the diophantine equation $x^2 + y^2 + 1 = 3xy$, studied in 1970 by G. Ledin, Jr. of the Institute of Chemical Biology, University of San Francisco, California [318]. It can be rewritten as the Pell's equation $(2x - 3y)^2 - 5y^2 = -4$. Its solutions are given by $(2x_n - 3y_n, y_n) = (L_{2n+1}, F_{2n+1})$. Then

$$\begin{aligned} 2x_n &= 3y_n + L_{2n+1} \\ &= 3F_{2n+1} + (F_{2n+2} + F_{2n}) \\ &= 2F_{2n+3}. \end{aligned}$$

Thus the solutions of the diophantine equation $x^2 + y^2 + 1 = 3xy$ are given by $(x_n, y_n) = (F_{2n+3}, F_{2n+1})$. (Notice that $F_{2n+3}^2 + F_{2n+1}^2 + 1 = 3F_{2n+3}F_{2n+1}$.)

Identity (31.57) has an additional application [11]. To see this, let n be an integer such that $L_n \equiv 2 \pmod{p}$, where p is an odd prime. Since n is even, identity (31.57) implies that $5F_n^2 \equiv 0 \pmod{p}$; so $5F_n \equiv 0 \pmod{p}$. Then

$$\begin{aligned} 2L_{n+1} &= (L_{n+1} + L_{n-1}) + L_n \\ &= 5F_n + L_n \\ &\equiv 0 + 2 \pmod{p}. \end{aligned}$$

This implies that $L_{n+1} \equiv 1 \pmod{p}$.

For example, let $n = 18$ and $p = 19$. Then $L_{18} = 5778 \equiv 2 \pmod{19}$ and $L_{19} = 9349 \equiv 1 \pmod{19}$. Similarly, $L_{14} = 843 \equiv 2 \pmod{23}$ and $L_{15} = 1364 \equiv 1 \pmod{23}$.

The next example is a fine application of identity (31.57). H. Ohtsuka of Saitama, Japan, studied it in 2012 [369, 370].

Example 31.5. *Evaluate the quotient* $\frac{A}{B}$, *where* $A = \sum_{k=1}^{\infty} \frac{1}{F_{4k}^2} - \sum_{k=1}^{\infty} \frac{1}{L_{4k}^2} + \sum_{k=1}^{\infty} \frac{1}{L_{2k}^2}$ *and* $B = \sum_{k=1}^{\infty} \frac{1}{F_{2k}^2}$.

Solution. The infinite series in both A and B are convergent; so the quotient $\frac{A}{B}$ is defined.

It follows from identity (31.57) that when $n \neq 0$, $\frac{5}{L_n^2} = \frac{1}{F_n^2} - \frac{4(-1)^n}{F_{2n}^2}$. Then

$$\begin{aligned} 5A - B &= \sum_{k=1}^{\infty} \left(\frac{5}{F_{4k}^2} - \frac{5}{L_{4k}^2} \right) + \sum_{k=1}^{\infty} \left(\frac{5}{L_{2k}^2} - \frac{1}{F_{2k}^2} \right) \\ &= \sum_{k=1}^{\infty} \left(\frac{5}{F_{4k}^2} - \frac{1}{F_{4k}^2} + \frac{4}{F_{2^{2k+1}}^2} \right) - \sum_{k=1}^{\infty} \frac{4}{F_{2^{k+1}}^2} \\ &= 4\sum_{k=1}^{\infty} \left(\frac{1}{F_{2^{2k}}^2} + \frac{1}{F_{2^{2k+1}}^2} \right) - \sum_{k=1}^{\infty} \frac{4}{F_{2^{k+1}}^2} \\ &= 4\sum_{k=2}^{\infty} \frac{1}{F_{2^k}^2} - 4\sum_{k=2}^{\infty} \frac{1}{F_{2^k}^2} \\ &= 0. \end{aligned}$$

So $\frac{A}{B} = \frac{1}{5}$. ■

We now study a few additional applications of identity (31.51).

Example 31.6. *It follows from identity (31.51) that* $l_n^2 \equiv 4(-1)^n \pmod{x^2+4}$. *For example,* $l_8 \equiv (-4)^4 + 8(-4)^3 + 20(-4)^2 + 16(-4) + 2 \equiv 2 \pmod{x^2+4}$*; so* $l_8^2 \equiv 4 \pmod{x^2+4}$.

More generally, let m and n be any positive integers. Then

$$l_{2mn}^2 \equiv 4 \pmod{(x^2+4)f_{2mn}^2}; \tag{31.59}$$

see Exercise 31.119. In particular, $l_{2mn}^2(114) \equiv 4 \pmod{13{,}000 f_{2mn}^2(114)}$ *and hence* $L_{2mn}^2(114) \equiv 4 \pmod{13{,}000}$. ■

It follows from congruence (31.59) that

$$\begin{aligned} q_{2mn}^2 &\equiv 4 \pmod{4(x^2+1)p_{2mn}^2}; \\ L_{2mn}^2 &\equiv 4 \pmod{5F_{2mn}^2}; \\ Q_{2mn}^2 &\equiv 1 \pmod{2P_{2mn}^2}. \end{aligned} \tag{31.60}$$

Congruence (31.60) implies that $L_{10n}^2 \equiv 4 \pmod{5F_{10n}^2}$. Since $F_{10}|F_{10n}$, $55|F_{10n}$. So $5^2 \cdot 55^2 | 5F_{10n}^2$. Consequently, $L_{10n}^2 \equiv 4 \pmod{15{,}125}$.

For example, $L_{30} = 1{,}860{,}498 \equiv 123 \pmod{15{,}125}$; so $L_{30}^2 \equiv 123^2 \equiv 4 \pmod{15{,}125}$, as expected.

Example 31.7. *Let m be an odd positive integer and* $\Delta = \sqrt{x^2+4}$. *Then*

$$\sum_{k=0}^{m} l_{n+k}^2 = \Delta^2 \sum_{k=0}^{m} f_{n+k}^2. \tag{31.61}$$

Proof. We have

$$\begin{aligned} l_n^2 + l_{n+1}^2 &= \Delta^2(f_n^2 + f_{n+1}^2) \\ l_{n+2}^2 + l_{n+3}^2 &= \Delta^2(f_{n+2}^2 + f_{n+3}^2) \\ &\vdots \\ l_{n+m-1}^2 + l_{n+m}^2 &= \Delta^2(f_{n+m-1}^2 + f_{n+m}^2). \end{aligned}$$

The given result now follows by adding these equations. ■

For example,

$$\begin{aligned} \sum_{k=0}^{3} l_{2+k}^2 &= l_2^2 + l_3^2 + l_4^2 + l_5^2 \\ &= (x^2+2)^2 + (x^3+3x)^2 + (x^4+4x^2+2)^2 + (x^5+5x^3+5x)^2 \end{aligned}$$

$$= x^{10} + 11x^8 + 44x^6 + 77x^4 + 54x^2 + 8$$
$$= (x^2+4)[x^2 + (x^2+1)^2 + (x^3+2x)^2 + (x^4+3x^2+1)^2]$$
$$= (x^2+4)\sum_{k=0}^{3} f_{2+k}^2.$$

It follows from identity (31.61) that

$$\sum_{k=0}^{m} q_{n+k}^2 = 4(x^2+1)\sum_{k=0}^{m} p_{n+k}^2;$$
$$\sum_{k=0}^{m} L_{n+k}^2 = 5\sum_{k=0}^{m} F_{n+k}^2; \qquad (31.62)$$
$$\sum_{k=0}^{m} Q_{n+k}^2 = 2\sum_{k=0}^{m} P_{n+k}^2,$$

where m is odd. Swamy found identity (31.62) in 1995 [491].

Identity (31.62) has a companion result when m is even:

$$\sum_{k=0}^{m} l_{n+k}^2 = (x^2+4)\sum_{k=0}^{m} f_{n+k}^2 + 4(-1)^n;$$

see Exercises 31.120–31.123.

In the next example, we study a function with a very special property.

Example 31.8. *Consider the function f, where $f(w,x,y,z) = (w^2+x^2)^k + (y^2+z^2)^k$. Prove that $f(l_n, l_{n+1}, l_{n+2}, l_{n+3}) = \Delta^{2k} f(f_n, f_{n+1}, f_{n+2}, f_{n+3})$.*

Proof. Recall that $f_n^2 + f_{n+1}^2 = f_{2n+1}$, $l_n^2 = \Delta^2 f_n^2 + 4(-1)^n$, and $l_n^2 + l_{n+1}^2 = \Delta^2(f_n^2 + f_{n+1}^2)$. Consequently,

$$f(l_n, l_{n+1}, l_{n+2}, l_{n+3}) = (l_n^2 + l_{n+1}^2)^k + (l_{n+2}^2 + l_{n+3}^2)^k$$
$$= \Delta^{2k}[(f_n^2 + f_{n+1}^2)^k + (f_{n+2}^2 + f_{n+3}^2)^k]$$
$$= \Delta^{2k} f(f_n, f_{n+1}, f_{n+2}, f_{n+3}),$$

as desired. ■

H.-J. Seiffert studied the function f in 1994 [425]. The functions $a(w,x,y,z) = (w^2+x^2+y^2+z^2)^k$, $b(w,x,y,z) = (w^2+x^2)^k(y^2+z^2)^k$, and $c(w,x,y,z) = (w^2-y^2)^k(x^2-z^2)^k$ also enjoy the same remarkable property; see Exercises 31.124–31.126. A.N. ’t Woord, L.A.G. Dresel, and J. Suck found these three functions, respectively [425].

In the next application, we show that $(x^2+4)(f_m^2 - f_{m-n}^2) + 4(-1)^m$ can be expressed as the product of two Lucas polynomials when n is odd, and $f_m^2 - f_{m-n}^2$ as the product of two Fibonacci polynomials when n is even.

Example 31.9. *Let m be any integer. Prove that $(x^2+4)(f_m^2 - f_{m-n}^2) + 4(-1)^m = l_{2m-n}l_n$ if n is odd, and $f_m^2 - f_{m-n}^2 = f_{2m-n}f_n$ if n is even.*

Proof. First, it follows by identity (31.51) that

$$\begin{aligned}(x^2+4)(f_m^2 - f_{m-n}^2) &= (l_m^2 - l_{m-n}^2) - 4(-1)^m[1-(-1)^n]\\ &= \begin{cases} l_m^2 - l_{m-n}^2 - 8(-1)^m & \text{if } n \text{ is odd}\\ l_m^2 - l_{m-n}^2 & \text{otherwise.}\end{cases}\end{aligned} \tag{31.63}$$

Since $\alpha\beta = -1$, $\alpha^n\beta^{2m-n} = (-1)^n\beta^{2m-2n}$ and $\beta^n\alpha^{2m-n} = (-1)^n\alpha^{2m-2n}$. Let n be odd. By the Binet-like formula for l_k, we then have

$$\begin{aligned}l_m^2 - l_{m-n}^2 &= \alpha^{2m} + \beta^{2m} - \alpha^{2m-2n} - \beta^{2m-2n} + 2(-1)^m - 2(-1)^{m-n}\\ &= \alpha^{2m} + \beta^{2m} + \beta^n\alpha^{2m-n} + \alpha^n\beta^{2m-n} + 4(-1)^m\\ &= (\alpha^n + \beta^n)(\alpha^{2m-n} + \beta^{2m-n}) + 4(-1)^m\\ &= l_{2m-n}l_n + 4(-1)^m.\end{aligned}$$

Consequently,

$$\begin{aligned}(x^2+4)(f_m^2 - f_{m-n}^2) + 4(-1)^m &= l_m^2 - l_{m-n}^2 - 4(-1)^m\\ &= l_{2m-n}l_n,\end{aligned}$$

as claimed.

We leave the other half as an exercise; see Exercise 31.127. ∎

It follows from this example that

$$\begin{aligned}4(x^2+1)(p_m^2 - p_{m-n}^2) + 4(-1)^m &= q_{2m-n}q_n;\\ 5(F_m^2 - F_{m-n}^2) + 4(-1)^m &= L_{2m-n}L_n;\\ 2(P_m^2 - P_{m-n}^2) + 4(-1)^m &= Q_{2m-n}Q_n,\end{aligned}$$

where n is odd; and

$$\begin{aligned}p_m^2 - p_{m-n}^2 &= p_{2m-n}p_n;\\ F_m^2 - F_{m-n}^2 &= F_{2m-n}F_n;\\ P_m^2 - P_{m-n}^2 &= P_{2m-n}P_n,\end{aligned}$$

where n is even. S. Edwards of Southern Polytechnic University in Georgia, developed the two special cases for Fibonacci numbers in 2004 [147].

For a specific example, let $m = 5$ and $n = 3$. Then

$$\begin{aligned}(x^2+4)(f_5^2 - f_2^2) - 4 &= (x^2+4)[(x^4+3x^2+1)^2 - x^2] - 4\\ &= x^{10} + 10x^8 + 35x^6 + 49x^4 + 21x^2\\ &= (x^7 + 7x^5 + 14x^3 + 7x)(x^3 + 3x)\\ &= l_7 l_3.\end{aligned}$$

The next example is still another application of identity (31.51).

Example 31.10. *Prove that* $l_{n+1}^4 - (x^2+4)^2 f_n f_{n+1}^2 f_{n+2} = (-1)^n (x^2-4) l_{n+1}^2 + 4.$

Proof. We let the Cassini-like formula for f_n and determinants do the job for us.

$$\begin{aligned}\text{LHS} &= \begin{vmatrix} l_{n+1}^2 & \Delta^2 f_n f_{n+2}\\ \Delta^2 f_{n+1}^2 & l_{n+1}^2\end{vmatrix}\\ &= \begin{vmatrix} l_{n+1}^2 + 0 & \Delta^2[f_{n+1}^2 + (-1)^{n+1}]\\ \Delta^2 f_{n+1}^2 & l_{n+1}^2\end{vmatrix}\\ &= \begin{vmatrix} l_{n+1}^2 & \Delta^2 f_{n+1}^2\\ \Delta^2 f_{n+1}^2 & l_{n+1}^2\end{vmatrix} + \begin{vmatrix} 0 & \Delta^2(-1)^{n+1}\\ \Delta^2 f_{n+1}^2 & l_{n+1}^2\end{vmatrix}\\ &= (l_{n+1}^4 - \Delta^4 f_{n+1}^4) - \Delta^4(-1)^{n+1} f_{n+1}^2\\ &= (l_{n+1}^2 - \Delta^2 f_{n+1}^2)(l_{n+1}^2 + \Delta^2 f_{n+1}^2) - \Delta^4(-1)^{n+1} f_{n+1}^2\\ &= 4(-1)^{n+1}(l_{n+1}^2 + \Delta^2 f_{n+1}^2) - \Delta^4(-1)^{n+1} f_{n+1}^2\\ &= 4(-1)^{n+1} l_{n+1}^2 - x^2\Delta^2(-1)^{n+1} f_{n+1}^2\\ &= (-1)^{n+1}[4l_{n+1}^2 - x^2(x^2+4) f_{n+1}^2]\\ &= (-1)^n[x^2(x^2+4) f_{n+1}^2 - 4l_{n+1}^2]\\ &= (-1)^n\{x^2[l_{n+1}^2 - 4(-1)^{n+1}] - 4l_{n+1}^2\}\\ &= (-1)^n[(x^2-4) l_{n+1}^2 + 4(-1)^n]\\ &= (-1)^n x^2(x^2-4) l_{n+1}^2 + 4,\end{aligned}$$

as desired. ■

As additional applications, we can use identity (31.51), coupled with the identities $2f_{n+1} - xf_n = l_n$ (see Exercise 31.18) and $2l_{n+1} - xl_n = (x^2+4)f_n$ (see Exercise 31.26), to express f_{n+1} and l_{n+1} as functions of f_n and l_n, respectively:

$$f_{n+1} = \frac{xf_n + \sqrt{(x^2+4)f_n^2 + 4(-1)^n}}{2};$$

$$l_{n+1} = \frac{xl_n + \sqrt{(x^2+4)[l_n^2 - 4(-1)^n]}}{2}.$$

For example,

$$\begin{aligned} f_4 &= \frac{xf_3 + \sqrt{(x^2+4)f_3^2 - 4}}{2} \\ &= \frac{x(x^2+1) + \sqrt{(x^2+4)(x^2+1)^2 - 4}}{2} \\ &= x^3 + 2x. \end{aligned}$$

Next we find two close links between triangular numbers $t_n = n(n+1)/2$ and identity (31.51).

Triangular Numbers

Example 31.11. *Prove that* $\dfrac{(x^2+4)f_{6n}^2}{32} = \dfrac{\lambda(\lambda+1)}{2}$, *where* $n \geq 1$ *and*

$$\lambda = \begin{cases} \dfrac{\Delta^2 f_{3n}^2}{4} - 1 & \text{if } n \text{ is odd} \\ \dfrac{\Delta^2 f_{3n}^2}{4} & \text{otherwise.} \end{cases}$$

Proof. We have

$$\begin{aligned} f_{6n}^2 &= f_{3n}^2 l_{3n}^2 \\ &= f_{3n}^2 \left[\Delta^2 f_{3n}^2 + 4(-1)^{3n}\right] \\ \frac{\Delta^2 f_{6n}^2}{32} &= \frac{1}{2}\left(\frac{\Delta^2 f_{3n}^2}{4}\right)\left[\frac{\Delta^2 f_{3n}^2}{4} + (-1)^n\right]. \end{aligned} \tag{31.64}$$

Consequently, $\dfrac{(x^2+4)f_{6n}^2}{32} = \dfrac{\lambda(\lambda+1)}{2}$, as desired. ■

This example has a delightful Fibonacci byproduct. To see this, we let $x = 1$. Then

$$\lambda = \begin{cases} \dfrac{5F_{3n}^2}{4} - 1 & \text{if } n \text{ is odd} \\[2ex] \dfrac{5F_{3n}^2}{4} & \text{otherwise.} \end{cases}$$

Since $2|F_{3n}$, λ is an integer.

Thus $\dfrac{5F_{6n}^2}{32} = t_\lambda$; that is, $\dfrac{5F_{6n}^2}{32}$ is a triangular number. J. Pla of Paris, France, discovered this result in 2003 [385].

In particular, let $n = 3$. Then $\lambda = \dfrac{5F_9^2}{4} - 1 = 1{,}444$ and $t_\lambda = t_{1444} = 1{,}044{,}735 = \dfrac{5 \cdot 2584^2}{32} = \dfrac{5F_{18}^2}{32}$. Similarly, when $n = 4$, $\lambda = 25{,}920$ and $t_\lambda = 335{,}936{,}160 = \dfrac{5F_{24}^2}{32}$.

Identity (31.64) has a Pell consequence also: $\dfrac{(x^2+1)p_{6n}^2}{8} = \dfrac{\nu(\nu+1)}{2}$, where

$$\nu = \nu(x) = \begin{cases} (x^2+1)p_{3n}^2 - 1 & \text{if } n \text{ is odd} \\[2ex] (x^2+1)p_{3n}^2 & \text{otherwise.} \end{cases}$$

Consequently, $\dfrac{P_{6n}^2}{4} = t_\nu$.

For example, when $n = 1$, $\nu = 2P_3^2 - 1 = 49$ and $t_{49} = 1{,}225 = \dfrac{70^2}{4} = \dfrac{P_6^2}{4}$; and when $n = 2$, $\nu = 2P_6^2 = 9{,}800$ and $t_{9800} = 48{,}024{,}900 = \dfrac{P_{12}^2}{4}$.

Example 31.12. *Let k be an arbitrary positive integer. Then $t_{k-2} + t_{k-1} + t_k + t_{k+1} = 2(k^2+1)$. Consequently,*

$$t_{g_n-2} + t_{g_n-1} + t_{g_n} + t_{g_n+1} = 2(g_n^2 + 1), \tag{31.65}$$

where $g_k = g_k(x)$ is an integer-valued function satisfying the gibonacci recurrence. Thus $2(g_n^2+1)$ can be expressed as the sum of four consecutive triangular numbers. ■

For example,

$$\begin{aligned}
2(f_5^2+1) &= 2[(x^4+3x^2+1)^2+1] \\
&= 2x^8+12x^6+22x^4+12x^2+4 \\
&= \frac{(x^4+3x^2-1)(x^4+3x^2)}{2} + \frac{(x^4+3x^2)(x^4+3x^2+1)}{2} \\
&\quad + \frac{(x^4+3x^2+1)(x^4+3x^2+2)}{2} + \frac{(x^4+3x^2+2)(x^4+3x^2+3)}{2} \\
&= t_{x^4+3x^2-1} + t_{x^4+3x^2} + t_{x^4+3x^2+1} + t_{x^4+3x^2+2} \\
&= t_{f_5-2} + t_{f_5-1} + t_{f_5} + t_{f_5+1}.
\end{aligned}$$

Since $l_n^2 = \Delta^2 f_n^2 + 4(-1)^n$, it follows from (31.65) that

$$t_{\frac{1}{2}l_{2n+1}-2} + t_{\frac{1}{2}l_{2n+1}-1} + t_{\frac{1}{2}l_{2n+1}} + t_{\frac{1}{2}l_{2n+1}+1} = \frac{1}{2}\left(l_{2n+1}^2+4\right).$$

Consequently,

$$t_{\frac{1}{2}q_{2n+1}-2} + t_{\frac{1}{2}q_{2n+1}-1} + t_{\frac{1}{2}q_{2n+1}} + t_{\frac{1}{2}q_{2n+1}+1} = \frac{\Delta^2}{2} f_{2n+1}^2;$$

$$t_{Q_{2n+1}-2} + t_{Q_{2n+1}-1} + t_{Q_{2n+1}} + t_{Q_{2n+1}+1} = 4P_{2n+1}^2. \tag{31.66}$$

K.S. Bhanu and M.N. Deshpande discovered identity (31.66) in 2008 [33].

For example,

$$\begin{aligned}
t_{Q_7-2} + t_{Q_7-1} + t_{Q_7} + t_{Q_7+1} &= t_{237} + t_{238} + t_{239} + t_{240} \\
&= 114{,}244 = 4 \cdot 169^2 \\
&= 4P_7^2.
\end{aligned}$$

We now study a few additional implications of identity (31.51).

Additional Implications

The identity $l_n^2 - (x^2+4)f_n^2 = 4(-1)^n$ can be generalized in several directions:

$$l_{n+c}^2 - (x^2+4)f_{n+c}^2 = 4(-1)^{n+c};$$

$$l_n^2 - (x^2+4)f_{n+c}^2 = \begin{cases} l_{-c}l_{2n+c} & \text{if } c \text{ is odd} \\ (x^2+4)f_{-c}f_{2n+c} + 4(-1)^n & \text{otherwise;} \end{cases} \tag{31.67}$$

$$l_m l_n - (x^2+4) f_{m+c} f_{n+c} = \begin{cases} l_{-c} l_{m+n+c} & \text{if } c \text{ is odd} \\ (x^2+4) f_{-c} f_{m+n+c} + 2(-1)^n l_{m-n} & \text{otherwise;} \end{cases}$$

$$l_m l_{n+c} - (x^2+4) f_m f_{n+c} = 2(-1)^m l_{n-m+c};$$

$$l_m l_n - (x^2+4) f_{m-c} f_{n+c} = 2(-1)^m l_{-c} l_{n-m+c};$$

$$l_{m+c} l_{n+d} - (x^2+4) f_{m+c} f_{n+d} = 2(-1)^{n+c} l_{m-n}.$$

In the interest of brevity, we will prove identity (31.67) and leave the others as straightforward applications of the Binet-like formulas; see Exercises 31.142–31.147.

Proof.

$$\begin{aligned} \text{LHS} &= (\alpha^n + \beta^n)^2 - (\alpha^{n+c} + \beta^{n+c})^2 \\ &= \alpha^{2n} + \beta^{2n} - \alpha^{2n+c}\alpha^c - \beta^{2n+c}\beta^c + 2(-1)^n[1 + (-1)^c]. \end{aligned}$$

Case 1. Suppose c is odd. Then

$$\begin{aligned} \text{LHS} &= \alpha^{2n} + \beta^{2n} + \alpha^{2n+c}\beta^{-c} + \beta^{2n+c}\alpha^{-c} \\ &= (\alpha^{-c} + \beta^{-c})(\alpha^{2n+c} + \beta^{2n+c}) \\ &= l_{-c} l_{2n+c}. \end{aligned}$$

Case 2. On the other hand, suppose c is even. Then

$$\begin{aligned} \text{LHS} &= \alpha^{2n} + \beta^{2n} - \alpha^{2n+c}\beta^{-c} - \beta^{2n+c}\alpha^{-c} + 4(-1)^n \\ &= (\alpha^{-c} - \beta^{-c})(\alpha^{2n+c} - \beta^{2n+c}) + 4(-1)^n \\ &= (x^2+4) f_{-c} f_{2n+c} + 4(-1)^n. \end{aligned}$$

Combining the two cases, we get the desired result. ■

Identity (31.57) Revisited

Suppose we would like to solve the diophantine equation $x^2 + y^2 - 3xy = 4(-1)^n$. It can be rewritten as a Pell's equation: $(x - 3y/2)^2 - 5(y/2)^2 = 4(-1)^n$. Its solutions are given by $(x_n - 3y_n/2, y_n/2) = (L_n, F_n)$. Then $y_n = 2F_n$ and hence $x_n = 3F_n + L_n = 2F_{n+2}$. So the desired solutions are given by $(x_n, y_n) = (2F_{n+2}, 2F_n)$. Consequently, $F_{n+2}^2 + F_n^2 - 3F_{n+2}F_n = (-1)^n$.

Here is a related problem: Solve the diophantine equation $x^2 + y^2 - 3xy = (-1)^n$. Its solutions are given by $(x_n - 3y_n/2, y_n/2) = (L_n/2, F_n/2)$; so

$(x_n, y_n) = (F_{n+2}, F_n)$. G. Ledin, Jr. of San Francisco, California, studied the equation in 1967 with n even [315].

Next we pursue an application of identity (31.58), studied in 1967 by G.L. Alexanderson of the University of Santa Clara, California [3].

Identity (31.58) Revisited

Suppose we would like to solve the diophantine equation $x^2 + (x+1)^2 = z^2$. Letting $w = 2x + 1$, it becomes the Pell's equation $w^2 - 2z^2 = -1$. Its positive solutions are given by $(w_n, z_n) = (Q_{2n-1}, P_{2n-1})$ [285]. Then $x_n = \dfrac{Q_{2n-1} - 1}{2}$. The odd parity of Pell–Lucas numbers guarantees that x_n is always an integer. Thus $(x_n, z_n) = ((Q_{2n-1} - 1)/2, P_{2n-1})$, where $n \geq 1$. Consequently, $Q_{2n-1}^2 + 1 = 2P_{2n-1}^2$.

Similarly, the positive solutions of the diophantine equation $x^2 + (x-1)^2 = z^2$ are given by $(x_n, z_n) = ((Q_{2n-1} + 1)/2, P_{2n-1})$. (This follows by replacing x with $x - 1$ in the preceding diophantine equation.)

Next we study two infinite products involving Fibonacci polynomials. As we might predict, they have Pell implications.

31.11 INFINITE PRODUCTS

The proof of the infinite product in the next example uses the following identities:

$$f_{2n} - f_{n-1}l_{n+1} = (-1)^{n+1}x;$$

$$f_{2n} - f_{n+1}l_{n-1} = (-1)^{n}x;$$

see Exercises 31.60 and 31.61. In addition, we will use the facts that $\lim_{n\to\infty} \dfrac{f_{n+1}}{f_n} = \alpha = \lim_{n\to\infty} \dfrac{l_{n+1}}{l_n}$.

Example 31.13. *Let x be a real number ≥ 1. Prove that* $\displaystyle\prod_{k=2}^{\infty} \frac{f_{2k} + x}{f_{2k} - x} = \frac{x^2 + 2}{x^2}$.

Proof. Suppose n is even, say $n = 2m$. Then

$$\begin{aligned}\prod_{k=2}^{n} \frac{f_{2k} + x}{f_{2k} - x} &= \prod_{k=2}^{2m} \frac{f_{2k} + x}{f_{2k} - x} \\ &= \prod_{\substack{k=2 \\ k \text{ even}}}^{2m} \frac{f_{2k} + x}{f_{2k} - x} \cdot \prod_{\substack{k=3 \\ k \text{ odd}}}^{2m-1} \frac{f_{2k} + x}{f_{2k} - x}\end{aligned}$$

$$= \prod_{\substack{k=2 \\ k \text{ even}}}^{2m} \frac{f_{k-1}l_{k+1}}{f_{k+1}l_{k-1}} \cdot \prod_{\substack{k=3 \\ k \text{ odd}}}^{2m-1} \frac{f_{k+1}l_{k-1}}{f_{k-1}l_{k+1}}$$

$$= \frac{f_1 l_{2m+1}}{f_{2m+1}l_1} \cdot \frac{f_{2m}l_2}{f_2 l_{2m}}$$

$$= \frac{f_1 l_2}{f_2 l_1} \cdot \frac{f_n}{f_{n+1}} \cdot \frac{l_{n+1}}{l_n}$$

$$\prod_{k=2}^{\infty} \frac{f_{2k}+x}{f_{2k}-x} = \frac{x^2+2}{x^2} \cdot \lim_{n\to\infty} \frac{f_n}{f_{n+1}} \cdot \lim_{n\to\infty} \frac{l_{n+1}}{l_n}$$

$$= \frac{x^2+2}{x^2} \cdot \frac{1}{\alpha} \cdot \alpha$$

$$= \frac{x^2+2}{x^2}.$$

On the other hand, suppose n is odd, say, $n = 2m + 1$. Then, as before, we have

$$\prod_{k=2}^{n} \frac{f_{2k}+x}{f_{2k}-x} = \frac{f_1 l_2}{f_2 l_1} \cdot \frac{f_{2m+2}}{f_{2m+1}} \cdot \frac{l_{2m+1}}{l_{2m+2}}$$

$$\prod_{k=2}^{\infty} \frac{f_{2k}+x}{f_{2k}-x} = \frac{x^2+2}{x^2} \cdot \lim_{n\to\infty} \frac{f_{n+1}}{f_n} \cdot \lim_{n\to\infty} \frac{l_n}{l_{n+1}}$$

$$= \frac{x^2+2}{x^2} \cdot \alpha \cdot \frac{1}{\alpha}$$

$$= \frac{x^2+2}{x^2}.$$

Thus, in both cases, the infinite product converges to the same limit, as claimed. ■

It follows from this example that

$$\prod_{k=2}^{\infty} \frac{p_{2k}+2x}{p_{2k}-2x} = \frac{2x^2+1}{2x^2};$$

$$\prod_{k=2}^{\infty} \frac{F_{2k}+1}{F_{2k}-1} = 3;$$

$$\prod_{k=2}^{\infty} \frac{P_{2k}+2}{P_{2k}-2} = \frac{3}{2}.$$

In 2006, M. Goldberg and M. Kaplan of Baltimore Polytechnic Institute in Maryland, studied the special case of the infinite product for the case $x = 1$ [187].

Example 31.14. *Let x be a real number* ≥ 1. *Evaluate the product*

$$\prod_{k=1}^{\infty}\left(1+\frac{2}{\sqrt{x^2+4f_{2^k}^2+4}}\right).$$

Solution. Let $A_k = A_k(x)$ denote the expression in the infinite product. Since $\beta^2 = \beta x + 1$ and $1 - \beta^2 = -\beta x = x/\alpha$, by the Binet-like formula for f_k, we then have

$$\begin{aligned}
A_k &= 1 + \frac{2}{\sqrt{(\alpha^{2^k} - \beta^{2^k}) + 4}}\\
&= 1 + \frac{2}{\alpha^{2^k} + \beta^{2^k}}\\
&= 1 + \frac{2\alpha^{-2^k}}{1 + (\beta/\alpha)^{2^k}}\\
&= 1 + \frac{2\beta^{2^k}}{1 + (-\beta^2)^{2^k}}\\
&= 1 + \frac{2\beta^{2^k}}{1 + \beta^{2^{k+1}}}\\
&= \frac{(1 + \beta^{2^k})^2}{1 + \beta^{2^{k+1}}};\\
\prod_{k=1}^{n} A_k &= \frac{(1+\beta^2)^2(1+\beta^4)^2\cdots(1+\beta^{2^n})^2}{(1+\beta^4)(1+\beta^8)\cdots(1+\beta^{2^n})(1+\beta^{2^{n+1}})}\\
&= \frac{(1+\beta^2)^2}{1+\beta^{2^{n+1}}}\cdot(1+\beta^4)(1+\beta^8)\cdots(1+\beta^{2^n})\\
&= \frac{1+\beta^2}{1+\beta^{2^{n+1}}}\cdot\frac{1-\beta^{2n+1}}{1-\beta^2};\\
\prod_{k=1}^{\infty} A_k &= \frac{1+\beta^2}{1+\beta^2}\cdot\lim_{n\to\infty}\frac{1-\beta^{2n+1}}{1+\beta^{2n+1}}\\
&= \frac{1+\beta^2}{1-\beta^2}\\
&= \frac{2+\beta x}{x/\alpha}\\
&= \frac{2\alpha - x}{x}.
\end{aligned}$$

■

In particular,

$$\prod_{k=1}^{\infty}\left(1+\frac{2}{\sqrt{(x^2+1)p_{2^k}^2+1}}\right)=\frac{\gamma-x}{x};$$

$$\prod_{k=1}^{\infty}\left(1+\frac{2}{\sqrt{5F_{2^k}^2+4}}\right)=\sqrt{5};$$

$$\prod_{k=1}^{\infty}\left(1+\frac{2}{\sqrt{2P_{2^k}^2+1}}\right)=\sqrt{2}. \quad (31.68)$$

M. Catalani of the University of Torino, Italy, studied the special case (31.68) in 2004 [94].

31.12 PUTNAM DELIGHT REVISITED

In Example 5.2, we studied a charming problem from the 68th Putnam Mathematical Competition [287, 505]:

Solve the recurrence $x_{n+1} = 3x_n + \lfloor\sqrt{5}x_n\rfloor$, where x is a positive integer, $x_0 = 1$, and $n \geq 0$.

We established that $x_n = 2^{n-1}F_{2n+3}$, where $n \geq 0$.

This delightful Putnam problem has an equally charming Lucas counterpart [295].

Lucas Counterpart

Solve the recurrence $x_{n+1} = 3x_n + \lceil\sqrt{5}x_n\rceil$, where x is a positive integer, $x_0 = 2$, and $n \geq 0$.

The first few values of x_n exhibit a clear pattern:

$$\begin{aligned}
x_0 &= \quad 2 = 2^{-1}\cdot 4\\
x_1 &= \quad 11 = 2^{0}\cdot 11\\
x_2 &= \quad 58 = 2^{1}\cdot 29\\
x_3 &= \quad 304 = 2^{2}\cdot 76\\
x_4 &= 1592 = 2^{3}\cdot 199\\
&\ \ \vdots \qquad\qquad\quad \uparrow\\
&\qquad\qquad \text{Lucas numbers}
\end{aligned}$$

So we conjecture that $x_n = 2^{n-1}L_{2n+3}$, where $n \geq 0$. The proof is quite similar, so in the interest of brevity, we omit it; see Exercise 31.178.

As we can predict, the Fibonacci and Lucas versions have Pell and Pell–Lucas counterparts [295]:

Consider the sequence $\{x_n\}$, where $x_{n+1} = 3x_n + \lceil 2\sqrt{2}x_n \rceil$, x is a positive integer, $n \geq 0$, and $x_0 = 2$. Then $x_n = P_{2n+2}$.

On the other hand, if $x_{n+1} = 3x_n + \lfloor 2\sqrt{2}x_n \rfloor$ and $x_0 = 3$, then $x_n = Q_{2n+2}$.

See Exercises 31.179 and 31.180.

Polynomial Extensions

Next we study the Fibonacci, Lucas, Pell, and Pell–Lucas polynomial extensions of the Putnam delight [297].

Fibonacci Extension

Solve the recurrence $x_{n+1} = (x^2+2)x_n + \lfloor x\Delta x_n \rfloor$, $x_0 = \dfrac{x^2+1}{2}$, $\Delta = \Delta(x) = \sqrt{x^2+4}$, $n \geq 0$, and x is a positive integer.

We establish that $x_n = 2^{n-1}f_{2n+3}$.

Proof. We confirm this statement using PMI. It is clearly true when $n = 0$. Now assume it is true for an arbitrary nonnegative integer n. Then

$$\begin{aligned}
x_{n+1} &= (x^2+2)x_n + \lfloor x\Delta x_n \rfloor \\
&= \lfloor (x^2 + x\Delta + 2)x_n \rfloor \\
&= \lfloor 2\alpha^2 x_n \rfloor \\
&= \lfloor 2\alpha^2 \cdot 2^{n-1} f_{2n+3} \rfloor \\
&= \left\lfloor 2^n\alpha^2 \cdot \frac{\alpha^{2n+3} - \beta^{2n+3}}{\alpha - \beta} \right\rfloor \\
&= \left\lfloor \frac{2^n}{\alpha - \beta}\left[(\alpha^{2n+5} - \beta^{2n+5}) + (\beta^{2n+5} - \beta^{2n+1})\right] \right\rfloor \\
&= 2^n f_{2n+5} + \left\lfloor \frac{2^n\beta^{2n+3}}{\alpha - \beta}(\beta^2 - \alpha^2) \right\rfloor \\
&= 2^n f_{2n+5} + \left\lfloor -2^n x\beta^{2n+3} \right\rfloor.
\end{aligned}$$

We now show that $\left\lfloor -2^n x\beta^{2n+3} \right\rfloor = 0$.

Since

$$\begin{aligned} 2\alpha^2 &= x^2 + x\Delta + 2 \\ &> 1 + 1 + 2 \\ &= 4, \end{aligned}$$

$\alpha^2 > 2$. Consequently, $0 < 2\beta^2 = \dfrac{2}{\alpha^2} < 1$, and hence $0 < \beta^2 < 1$.

Since

$$\begin{aligned} -2\beta &= \Delta - x \\ &= \frac{\Delta^2 - x^2}{\Delta + x} \\ &= \frac{4}{\Delta + x} \\ &< \frac{4}{x + x} \\ &= \frac{2}{x}, \end{aligned}$$

$0 < -\beta x < 1$.

Then $0 < (2\beta^2)^n \beta^2(-\beta x) < 1$; so $\lfloor -2^n x \beta^{2n+3} \rfloor = 0$.

Thus $x_{n+1} = 2^n f_{2n+5}$, so the formula works for $n + 1$ also. Consequently, the formula works for all $n \geq 0$, as desired. ■

For example, $x_5 = (x^2 + 2)x_4 + \lfloor x\Delta x_4 \rfloor = 2^4 f_{13} = 16(x^{12} + 11x^{10} + 45x^8 + 84x^6 + 70x^4 + 21x^2 + 1)$.

Clearly, when $x = 1$, the formula yields the solution to the original Putnam problem [295, 505].

Next we study the Lucas extension of the problem [297].

Lucas Extension

Solve the recurrence $x_{n+1} = (x^2 + 2)x_n + \lceil x\Delta x_n \rceil$, $x_0 = \dfrac{x^3 + 3x}{2}$, $\Delta = \Delta(x) = \sqrt{x^2 + 4}$, $n \geq 0$, *x is a positive integer.*

We claim that $x_n = 2^{n-1} l_{2n+3}$. The proof follows as above; see Exercise 31.181.

For example, $x_4 = (x^2 + 2)x_n + \lceil x\Delta x_n \rceil = 2^3 l_{11} = 8(x^{11} + 11x^9 + 44x^7 + 77x^5 + 55x^3 + 11x)$.

In particular, consider the recurrence $x_{n+1} = 3x_n + \lceil \sqrt{5}x_n \rceil$, where $x_0 = 2$ and $n \geq 0$. Then $x_n = 2^{n-1} L_{2n+3}$.

Finally, we present the Pell and Pell–Lucas extensions [297].

Pell and Pell–Lucas Extensions

Solve the recurrence $x_{n+1} = (2x^2 + 1)x_n + \lceil 2x\sqrt{x^2 + 1}x_n \rceil$, *$x$ is a positive integer,* $n \geq 0$, *and* $x_0 = 2x$. *Then* $x_n = p_{2n+2}$.

On the other hand, if $x_{n+1} = (2x^2 + 1)x_n + \lfloor 2x\sqrt{x^2 + 1}x_n \rfloor$ *and* $x_0 = 4x^2 + 2$, *then* $x_n = q_{2n+2}$.

The proofs follow using similar steps, so we omit them in the interest of brevity. But we encourage gibonacci enthusiasts to confirm both formulas; see Exercises 31.182 and 31.183.

In particular, let $x = 1$. Then the recurrence $x_{n+1} = 3x_n + \lceil 2\sqrt{2}x_n \rceil$ with the initial condition $x_0 = 2$, implies $x_n = P_{2n+2}$; likewise, the recurrence $x_{n+1} = 3x_n + \lfloor 2\sqrt{2}x_n \rfloor$, coupled with the initial condition $x_0 = 6$, yields $x_n = Q_{2n+2}$ [297].

For example, consider the latter recurrence with $x_0 = 6$. Then

$$\begin{aligned} x_{10} &= 3x_9 + \lceil 2\sqrt{2}x_9 \rceil \\ &= 3(2Q_{20}) + \lceil 2\sqrt{2}(2Q_{20}) \rceil \\ &= 24 \cdot 22{,}619{,}537 + \lceil 4\sqrt{2} \cdot 22{,}619{,}537 \rceil \\ &= 2 \cdot 131{,}836{,}323 \\ &= 2Q_{22}, \end{aligned}$$

as desired.

31.13 INFINITE SIMPLE CONTINUED FRACTION

In the next example, we study an infinite simple continued fraction (ISCF) which has interesting implications. Seiffert investigated it in 1992 [420].

Example 31.15. *Show that the ISCF* $z = [a; \overline{b, a}]$ *converges and then find the limit, where* $a = a(x)$, $b = b(x)$, *and x are integers.*

Proof. Let $\dfrac{c_n}{d_n}$ denote the nth convergent of the ISCF, where $n \geq 0$. Then

$$\begin{aligned} c_0 &= a, \quad d_0 = 1, \\ c_1 &= ab + 1, \quad d_1 = b, \\ c_{2n} &= ac_{2n-1} + c_{2n-2}, \quad d_{2n} = ad_{2n-1} + d_{2n-2}, \\ c_{2n+1} &= ac_{2n} + c_{2n-1}, \quad d_{2n+1} = ad_{2n} + d_{2n-1}. \end{aligned}$$

Consequently, we have

$$\begin{aligned} c_{2n} &= a(bc_{2n-2} + c_{2n-3}) + c_{2n-2} \\ &= (ab + 1)c_{2n-2} + (c_{2n-2} - c_{2n-4}) \\ &= (ab + 2)c_{2n-2} - c_{2n-4}. \end{aligned}$$

Similarly, $d_{2n} = (ab + 2)d_{2n-2} - d_{2n-4}$.

These two second-order recurrences have the same characteristic equation $t^2 - (ab+2)t + 1 = 0$, with solutions $r = \frac{ab+2+D}{2}$ and $s = \frac{ab+2-D}{2}$, where $D = \sqrt{ab(ab+4)}$. So both c_{2n} and d_{2n} have the same general solution $t = Ar^n + Bs^n$. This, coupled with the initial conditions $c_0 = a$ and $c_2 = a(ab+2)$, and $d_0 = 1$ and $d_2 = ab + 1$, yield the explicit formulas

$$c_{2n} = \frac{a}{D}\left(r^{n+1} - s^{n+1}\right)$$

$$d_{2n} = \frac{1}{D}\left[(r-1)^n - (s-1)^n\right],$$

where $n \geq 0$.

Since $r > s > 0$, we have

$$\lim_{n\to\infty} \frac{c_{2n}}{d_{2n}} = \lim_{n\to\infty} \frac{a[1 - (s/r)^{n+1}]}{(r-1)(1/r) - \frac{s-1}{r}(s/r)^n}$$

$$= \frac{ar}{r-1}.$$

Similarly,

$$\lim_{n\to\infty} \frac{c_{2n+1}}{d_{2n+1}} = \frac{ar}{r-1}.$$

Thus, in both cases, the limits exist and are the same. Consequently, $\lim_{n\to\infty} \frac{c_n}{d_n}$ exists and equals $\frac{ar}{r-1}$; in other words, $z = \frac{ar}{r-1}$. ■

Since we have confirmed that the ISCF converges, we can take a different route to compute the limit z. Since $z > 0$ and $z = a + \cfrac{1}{b + \cfrac{1}{z}}$, it follows that

$$z = \frac{ab}{2}\left[1 + \sqrt{1 + \frac{4}{ab}}\right].$$

In particular,

$$[f_n; \overline{f_n, l_n}] = \frac{f_{2n}}{2}\left[1 + \sqrt{1 + \frac{4}{f_{2n}}}\right]$$

$$[f_n; \overline{f_{n+1}, f_n}] = \frac{f_{n+1}f_n}{2}\left[1 + \sqrt{1 + \frac{4}{f_{n+1}f_n}}\right].$$

L. Kupiers of Sierre, Switzerland, studied these two continued fractions in 1991 [306, 420].

An interesting observation: Suppose $n \geq 2$ and $x \geq 1$. Since $l_n = f_{n+1} + f_{n-1}$, it follows that $l_n > f_{n+1}$; so $f_{2n} > f_{n+1}f_n$. Consequently, $[f_n; \overline{f_n, l_n}] > [f_n; \overline{f_{n+1}, f_n}]$.

EXERCISES 31

Prove each, where $g_k = g_k(x)$, and t_k denotes the kth triangular number.

Let $a_{n,k}$ denote the element in row n and column k of array A in Table 31.2. Then

1. $a_{n,k} = a_{n-1,k} + a_{n-2,k-1}$, where $a_{1,0} = 1 = a_{2,0}$, $n \geq 3$, and $k \geq 1$.

Let $b_{n,k}$ denote the element in row n and column k of array B in Table 31.3. Then

2. $\sum_{k=0}^{\lfloor n/2 \rfloor} b_{n,k} = L_n$.
3. $b_{n,k} = b_{n-1,k} + b_{n-2,k-1}$, where $b_{1,0} = 1 = b_{2,0}, b_{2,1} = 2, n \geq 3$, and $k \geq 1$.

Let x_n denote the nth rising diagonal sum of array B in Table 31.3. Then

4. $x_n = x_{n-1} + x_{n-3}$, where $n \geq 4$.
5. $x_n = \sum_{k=0}^{\lfloor (n-1)/3 \rfloor} \frac{n-k}{n-2k}\binom{n-2k}{k}$.
6. $f_n = \frac{\alpha^n - \beta^n}{\alpha - \beta}$.
7. $l_n = \alpha^n + \beta^n$.
8. $\alpha^n = \alpha f_n + f_{n-1}$.
9. $\Delta\alpha^n = \alpha l_n + l_{n-1}$.
10. $\alpha^m(\alpha f_n + f_{n-1}) = \alpha^n(\alpha f_n + f_{n-1})$.
11. $\alpha^m(\alpha l_n + l_{n-1}) = \alpha^n(\alpha l_m + l_{m-1})$.
12. $\alpha(l_n + xf_n) + (l_{n-1} + xf_{n-1}) = 2\alpha^{n+1}$.
13. $\alpha(l_n - xf_n) + (l_{n-1} - xf_{n-1}) = -2\alpha^{n-1}$.
14. $f_n l_n = f_{2n}$.
15. $f_{n+1} + f_{n-1} = l_n$.
16. $f_{n+1}^2 - f_{n-1}^2 = xf_{2n}$.
17. $xf_n + 2f_{n-1} = l_n$.
18. $xf_n + l_n = 2f_{n+1}$.
19. $l_n^2 - x^2 f_n^2 = 4f_{n+1}f_{n-1}$.

20. $f_{n+2} + f_{n-2} = (x^2+2)f_n$.
21. $f_{n+2} - f_{n-2} = xl_n$.
22. $f_{n+2}^2 - f_{n-2}^2 = x(x^2+2)f_{2n}$.
23. $l_{n+2}^2 - l_{n-2}^2 = x(x^2+2)(x^2+4)f_{2n}$.
24. $(x^2+4)(f_{n+3}^2 - f_n^2) = (x^3+3x)l_{2n+3} + 4(-1)^n$.
25. $l_{n+3}^2 - l_n^2 = (x^3+3x)l_{2n+3} - 4(-1)^n$.
26. $(x^2+4)f_{n-1} + xl_{n-1} = 2l_n$.
27. $f_{n+1}^2 + f_n^2 = f_{2n+1}$.
28. $f_{n+3}^2 + f_n^2 = (x^2+1)f_{2n+3}$.
29. $l_{n+3}^2 + l_n^2 = (x^2+1)(x^2+4)f_{2n+3}$.
30. $(x^2+1)f_n + xf_{n-1} - f_{n-2} = xl_n$.
31. $(x^2+1)l_n + xl_{n-1} - l_{n-2} = x(x^2+4)f_n$.
32. $l_{n+1} + l_{n-1} = (x^2+4)f_n$.
33. $l_{n+1}^2 - l_{n-1}^2 = x(x^2+4)f_{2n}$.
34. $l_{n+2} + l_{n-2} = (x^2+2)l_n$.
35. $l_{n+2} - l_{n-2} = x(x^2+4)f_n$.
36. $f_{n+k} + f_{n-k} = \begin{cases} f_k l_n & \text{if } k \text{ is odd} \\ f_n l_k & \text{otherwise.} \end{cases}$
37. $f_{n+k} - f_{n-k} = \begin{cases} f_n l_k & \text{if } k \text{ is odd} \\ f_k l_n & \text{otherwise.} \end{cases}$
38. $f_{n+k}^2 - f_{n-k}^2 = f_{2n}f_{2k}$.
39. $l_{n+k} + l_{n-k} = \begin{cases} (x^2+4)f_k l_n & \text{if } k \text{ is odd} \\ f_n l_k & \text{otherwise.} \end{cases}$
40. $l_{n+k} - l_{n-k} = \begin{cases} f_n l_k & \text{if } k \text{ is odd} \\ (x^2+4)f_k l_n & \text{otherwise.} \end{cases}$
41. $l_{n+k}^2 - l_{n-k}^2 = (x^2+4)f_{2n}f_{2k}$.
42. $l_{n+2} + (x^2+2)l_{n-1} = (x^2+4)f_n$.
43. $(x^2+4)f_n + xl_n = 2l_{n+1}$.
44. $xl_{n-2} + (x^2+2)l_{n-1} = (x^2+4)f_n$.
45. Let $R_n = \dfrac{f_n}{l_n}$. Then $R_{n+1} = \dfrac{xR_n + 1}{(x^2+4)R_n + 1}$.
46. $l_{n+1}^2 + l_n^2 = (x^2+4)f_{2n+1}$.

47. $l_{n+1}^2 - l_n^2 = xl_{2n+1} - 4(-1)^n$.
48. $l_{2n} = l_n^2 - 2(-1)^n$.
49. $l_{2n} = (x^2+4)f_n^2 + 2(-1)^n$.
50. $2l_{2n} = l_n^2 + (x^2+4)f_n^2$.
51. $l_n^2 + l_{2n} = 2[l_{2n} + (-1)^n]$.
52. $(x^2+4)f_n f_{n+1} = l_{2n+1} - (-1)^n x$.
53. $l_n l_{n+1} = l_{2n+1} + (-1)^n x$.
54. $l_{n+2}l_{n-1} = l_{2n+1} - (-1)^n l_3$.
55. $(x^2+4)f_{n+2}f_{n-1} = l_{2n+1} + (-1)^n l_3$.
56. $l_n^2 - (x^2+4)f_n^2 = 4(-1)^n$.
57. $(x^2+4)f_n^4 = f_{2n}^2 - 4(-1)^n f_n^2$.
58. $f_{2n+1} - f_n l_{n+1} = (-1)^n$.
59. $f_{2n+1} - f_{n+1}l_n = (-1)^{n+1}$.
60. $f_{2n} - f_{n-1}l_{n+1} = (-1)^{n+1}x$.
61. $f_{2n} - f_{n+1}l_{n-1} = (-1)^n x$.
62. $f_{n+1}f_{n-1} - f_n^2 = (-1)^n$.
63. $l_{n+1}l_{n-1} - l_n^2 = (-1)^{n+1}(x^2+4)$.
64. $l_n f_{n-1} - f_n l_{n-1} = 2(-1)^n$.
65. $l_n f_{n-1} + f_n l_{n-1} = 2f_{2n-1}$.
66. $f_{n+1}l_{n+2} - xf_{n+2}l_n = f_{2n+1} - (x-1)(-1)^n$.
67. $l_{2n}l_{2n+2} - (x^2+4)f_{2n+1}^2 = x^2$.
68. $4f_{n+1}^4 + f_{n-1}^4 = (f_{n+1}^2 + l_n^2)(f_{n+1}^2 + x^2 f_n^2)$.
69. $4l_{n+1}^4 + l_{n-1}^4 = [l_{n+1}^2 + (x^2+4)f_n^2](l_{n+1}^2 + x^2 l_n^2)$.
70. $\left(\dfrac{l_m - \Delta f_m}{2}\right)^n = \dfrac{l_{mn} - \Delta f_{mn}}{2}$.
71. $\displaystyle\sum_{k=0}^{2n} \binom{2n}{k}(-2)^k x^{2n-k} f_k = 0$.
72. Formula in Example 31.1, where n is odd and $n \geq 3$.
73. $f_m f_{m+n+k} - f_{m+k}f_{m+n} = (-1)^{m+1} f_n f_k$.
74. $l_m l_{m+n+k} - l_{m+k}l_{m+n} = (-1)^m (x^2+4) f_n f_k$.
75. $l_{n+1}l_{n+3} - (x^2+4)f_n f_{n+4} = (-1)^n l_1 l_3$.
76. $f_{n+k}f_{n-k} - f_n^2 = (-1)^{n+k+1} f_k^2$.
77. $l_{n+k}l_{n-k} - l_n^2 = (-1)^{n+k}(x^2+4)f_k^2$.

78. $f_{2n+1} = f_{n+1}^2 + f_n^2$, using Exercise 31.73.
79. $xl_{2n+1} = l_{n+1}^2 - (x^2+4)f_n^2$, using Exercise 31.74.
80. $xl_{2n+1} = (x^2+4)f_{n+1}^2 - l_n^2$.
81. $f_{3n} = f_{2n}l_n - (-1)^n f_n$.
82. $f_{3n} = (x^2+4)f_n^3 + 3(-1)^n f_n$.
83. $f_{5n} = f_n\left[\Delta^4 f_n^4 + 5\Delta^2(-1)^n f_n^2 + 5\right]$.
84. $l_{3n} = l_n^3 - 3(-1)^n l_n$.
85. $l_{4n} = l_n^4 - 4(-1)^n l_n^2 + 2$.
86. $l_{5n} = l_n\left[l_n^4 - 5(-1)^n l_n^2 + 5\right]$.
87. $l_{6n} = l_n^6 - 6(-1)^n l_n^4 + 9l_n^2 - 2(-1)^n$.
88. $\Delta^4 f_n^4 = l_{4n} - 4(-1)^n l_{2n} + 6$.
89. $\Delta^4 f_n^4 = l_n^4 - 6(-1)^n l_n^2 + 16$.
90. $\Delta^2(f_{n+1}^2 + f_{n-1}^2) = (x^2+2)l_{2n} + 4(-1)^n$.
91. $\Delta^2(f_{n+1}^2 - f_{n-1}^2) = xf_{2n}\left[(x^2+2)l_{2n} + 4(-1)^n\right]$.
92. $l_{n+1}^2 + l_{n-1}^2 = (x^2+2)l_{2n} - 4(-1)^n$.
93. $l_{n+1}^4 - l_{n-1}^4 = x\Delta^2 f_{2n}\left[(x^2+2)l_{2n} - 4(-1)^n\right]$.
94. $f_{2m+2n}^2 - f_{2n}^2 = f_{2m}f_{2m+4n}$.
95. Let $(x+\Delta)^n = a_n + \Delta b_n$. Then $a_n = 2^{n-1}l_n$ and $b_n = 2^{n-1}f_n$.
96. $$f_{n+1} = \sum_{k=0}^{\lfloor n/2\rfloor} \binom{n-k}{k} x^{n-2k}.$$
97. $$l_n = \sum_{k=0}^{\lfloor n/2\rfloor} \frac{n}{n-k}\binom{n-k}{k} x^{n-2k}.$$
98. Formula (31.28), using the operator D.
99. Formula (31.31).
100. Formula (31.32).
101. Formulas (31.40).
102. $(l_{n+4k} + l_n, l_{n+4k-1} + l_{n-1}) = l_{2k}$.
103. $$P_n = \begin{cases} \left\lceil \dfrac{\gamma^n}{2\sqrt{2}} \right\rceil & \text{if } n \text{ is odd} \\ \left\lfloor \dfrac{\gamma^n}{2\sqrt{2}} \right\rfloor & \text{otherwise.} \end{cases}$$

104. $Q_n = \begin{cases} \left\lfloor \dfrac{\gamma^n}{2} \right\rfloor & \text{if } n \text{ is odd} \\ \left\lceil \dfrac{\gamma^n}{2} \right\rceil & \text{otherwise.} \end{cases}$

105. The number of digits in P_n equals $\lceil n \log \gamma - 1.5 \log 2 \rceil$.

106. The number of digits in Q_n equals $\lceil n \log \gamma - \log 2 \rceil$.

107. $p_{-n} = (-1)^{n-1} p_n$.

108. $q_{-n} = (-1)^n q_n$.

109. $p_{n+1}p_{n-1} - p_n^2 = (-1)^n$.

110. $q_{n+1}q_{n-1} - q_n^2 = (-1)^{n-1}(x^2 + 1)$.

111. $(x^2 + 4)f_n^3 = f_{3n} - 3(-1)^n f_n$.

112. $l_n^3 = l_{3n} + 3(-1)^n l_n$.

113. $4(x^2 + 1)p_n^3 = p_{3n} - 3(-1)^n p_n$.

114. $4(x^2 + 1)q_n^3 = q_{3n} - 3(-1)^n q_n$.

115. $\Delta^{2m} f_n^{2m+1} = \sum_{k=0}^{m} (-1)^{(n+1)k} \binom{2m+1}{k} f_{2n(m-k)+n}$.

116. $l_n^{2m+1} = \sum_{k=0}^{m} (-1)^{nk} \binom{2m+1}{k} l_{2n(m-k)+n}$.

117. Identity (31.53).

118. Identity (31.54).

119. $l_{2mn}^2 \equiv 4 \pmod{(x^2 + 4) f_{mn}^2}$.

120. $\sum_{k=0}^{m} l_{n+k}^2 = (x^2 + 4) \sum_{k=0}^{m} f_{n+k}^2 + 4(-1)^n$, where m is even.

121. $\sum_{k=0}^{m} q_{n+k}^2 = 4(x^2 + 1) \sum_{k=0}^{m} p_{n+k}^2 + 4(-1)^n$, where m is even.

122. $\sum_{k=0}^{m} L_{n+k}^2 = 5 \sum_{k=0}^{m} f_{n+k}^2 + 4(-1)^n$, where m is even.

123. $\sum_{k=0}^{m} Q_{n+k}^2 = 2 \sum_{k=0}^{m} P_{n+k}^2 + (-1)^n$, where m is even.

124. $a(l_n, l_{n+1}, l_{n+2}, l_{n+3}) = \Delta^{4k} a(f_n, f_{n+1}, f_{n+2}, f_{n+3})$, where $a(w, x, y, z) = (w^2 + x^2 + y^2 + z^2)^k$ (Woord, [425]).

125. $b(l_n, l_{n+1}, l_{n+2}, l_{n+3}) = \Delta^{4k} b(f_n, f_{n+1}, f_{n+2}, f_{n+3})$, where $b(w, x, y, z) = (w^2 + x^2)^k (y^2 + z^2)^k$ (Dresel, [425]).

126. $c(l_n, l_{n+1}, l_{n+2}, l_{n+3}) = \Delta^{4k} c(f_n, f_{n+1}, f_{n+2}, f_{n+3})$, where $c(w, x, y, z) = (w^2 - x^2)^k (y^2 - z^2)^k$ (Suck, [425]).
127. Let m be any integer and n an even integer. Then $f_m^2 - f_{m-n}^2 = f_{2m-n} f_n$.
128. $(l_n l_{n+3})^2 + 4(l_{n+1} l_{n+2})^2 = 5l_{2n+3}^2 + 2(-1)^n (x^3 - x) l_{2n+3} + (x^6 + 6x^4 + 13x^2)$.
129. Let $x \geq 1$ and $n \geq 4$. Then $x f_{n+1} < x l_n < f_{n+2}$.
130. Let $x \geq 1$ and $n \geq 4$. Then $f_{2n+1} < x l_n^2 < f_{2n+2}$.
131. $x^{n-3} f_n^2 \leq f_3^2 f_4^{n-3}$, where $n \geq 3$ (Swamy, [478]).
132. $l_n^2 \leq l_3^2 l_4^{n-3}$, where $n \geq 3$.
133. $t_{g_n} + t_{g_n - 1} = g_n^2$.
134. $t_{t_{g_n}} + t_{t_{g_n} - 1} = t_{g_n}^2$.
135. $t_{g_n}^2 + t_{g_n - 1}^2 = t_{g_n^2}$.
136. $t_{g_n}^2 - t_{g_n - 1}^2 = g_n^3$.
137. $x \sum_{k=1}^{n} g_k^2 = g_n g_{n+1} - g_0 g_1$.
138. $x \sum_{k=1}^{n} f_k^2 = f_n f_{n+1}$.
139. $x \sum_{k=1}^{n} l_k^2 = l_n l_{n+1} - 2x$.
140. $\sum_{r=0}^{m} (-1)^r \binom{m}{r} (x^2 + 4)^r l_n^{2m-2r} f_n^{2r} = (-1)^{mn} 4^m$.
141. $\sum_{r=0}^{m} (-1)^r \binom{m}{r} 4^r (x^2 + 1)^r q_n^{2m-2r} p_n^{2r} = (-1)^{mn} 4^m$.
142. $l_{n+c}^2 - (x^2 + 4) f_{n+c}^2 = 4(-1)^{n+c}$.
143. $l_m l_n - (x^2 + 4) f_{m+c} f_{n+c} = \begin{cases} l_{-c} l_{m+n+c} & \text{if } c \text{ is odd} \\ (x^2 + 4) f_{-c} f_{m+n+c} + 2(-1)^n l_{m-n} & \text{otherwise.} \end{cases}$
144. $l_m l_{n+c} - (x^2 + 4) f_m f_{n+c} = 2(-1)^m l_{n-m+c}$.
145. $l_m l_n - (x^2 + 4) f_{m-c} f_{n+c} = (-1)^m l_{-c} l_{n-m+c}$.
146. $l_{m+c} l_{n+d} - (x^2 + 4) f_{m+c} f_{n+d} = 2(-1)^{n+c} l_{m-n}$.
147. $l_m l_n - (x^2 + 4) f_m f_n = 2(-1)^m l_{n-m}$.
148. $\left[x^2 + 2 - \sum_{k=1}^{n} \frac{x^2(x^2 + 4)}{l_{2k-1} l_{2k+1}} \right] \left[\frac{1}{x^2 + 2} + \sum_{k=1}^{n} \frac{x^2 + 4}{l_{2k} l_{2k+2}} \right] = 1.$

149. $\left(\frac{1}{g_1} - \sum_{k=1}^{n} \frac{xg_{2k}}{g_{2k+1}g_{2k-1}}\right)\left(\frac{1}{g_2} - \sum_{k=1}^{n} \frac{xg_{2k+1}}{g_{2k+2}g_{2k}}\right) = \frac{1}{g_{2n+2}g_{2n+1}}.$

150. $\left(1 - \sum_{k=1}^{n} \frac{xf_{2k}}{f_{2k+1}f_{2k-1}}\right)\left(1 - \sum_{k=1}^{n} \frac{x^2 f_{2k+1}}{f_{2k+2}f_{2k}}\right) = \frac{x}{f_{2n+2}f_{2n+1}}.$

151. $\left(1 - \sum_{k=1}^{n} \frac{x^2 l_{2k}}{l_{2k+1}l_{2k-1}}\right)\left(1 - \sum_{k=1}^{n} \frac{x(x^2+2)l_{2k+1}}{l_{2k+2}l_{2k}}\right) = \frac{x(x^2+2)}{l_{2n+2}l_{2n+1}}.$

152. Derive a generating function for $p_n(z)$.

153. Derive a generating function for $q_n(z)$.

Define each recursively.

154. $z_n = 3f_{2n-1} + x + 1.$

155. $z_n = 3l_{2n-1} + x + 1.$

156. $z_n = \frac{1}{2}f_{3^n}.$

157. $z_n = \frac{1}{2}l_{3^n}.$

158. $z_k = \frac{f_{kn}}{f_n}$, where n is a fixed nonzero integer.

159. $z_k = l_{kn}.$

160. $z_n = f_{3^n}.$

161. $z_n = l_{3^n}.$

162. $z_n = f_{5^n}.$

163. $z_n = l_{2^n}.$

164. $z_n = l_{4^n}.$

165. $z_n = l_{6^n}.$

Evaluate each sum.

166. $\frac{f_{n+1}}{\alpha^{n-1}} + \frac{f_n}{\alpha^n}.$

167. $\frac{l_{n+1}}{\alpha^{n-1}} + \frac{l_n}{\alpha^n}.$

168. $x + \sum_{i=1}^{n} \frac{(-1)^{i+1}}{f_i f_{i+1}}.$

169. $x + \sum_{i=1}^{\infty} \frac{(-1)^{i+1}}{f_i f_{i+1}}.$

Evaluate each product, where x is a real number ≥ 1.

170. $\displaystyle\prod_{k=1}^{n} \frac{xg_{2k}g_{2k+2} + g_{2k-1}g_{2k+2}}{xg_{2k}g_{2k+2} + g_{2k}g_{2k+1}}$.

171. $\displaystyle\prod_{k=1}^{n} \left(x + \frac{g_{k-1}}{g_k}\right)$.

172. $\displaystyle\prod_{k=1}^{\infty} \frac{xf_{2k}f_{2k+2} + f_{2k-1}f_{2k+2}}{xf_{2k}f_{2k+2} + f_{2k}f_{2k+1}}$.

173. $\displaystyle\prod_{k=1}^{\infty} \frac{xl_{2k}l_{2k+2} + l_{2k-1}l_{2k+2}}{xl_{2k}l_{2k+2} + l_{2k}l_{2k+1}}$.

174. $\displaystyle\prod_{k=1}^{\infty} \left(1 + \frac{2}{\sqrt{(x^2+1)p_{2^k}^2 + 1}}\right)$.

175. $\displaystyle\prod_{k=1}^{\infty} \left(1 + \frac{2}{\sqrt{5F_{2^k}^2 + 4}}\right)$.

176. $\displaystyle\prod_{k=1}^{\infty} \left(1 + \frac{2}{\sqrt{2P_{2^k}^2 + 1}}\right)$ (Catalani, [94]).

177. Identity (31.41).

Prove each, where x is a positive integer and $n \geq 0$.

178. Let $x_{n+1} = 3x_n + \lceil\sqrt{5}x_n\rceil$, where $x_0 = 2$. Then $x_n = 2^{n-1}L_{2n+3}$.

179. Let $x_{n+1} = 3x_n + \lceil 2\sqrt{2}x_n\rceil$, where $x_0 = 2$. Then $x_n = P_{2n+2}$.

180. Let $x_{n+1} = 3x_n + \lfloor 2\sqrt{2}x_n\rfloor$, where $x_0 = 3$. Then $x_n = Q_{2n+2}$.

181. Let $x_{n+1} = (x^2+2)x_n + \lceil x\Delta x_n\rceil$, $x_0 = \dfrac{x^3+3x}{2}$, $\Delta = \sqrt{x^2+4}$. Then $x_n = 2^{n-1}l_{2n+3}$.

182. Let $x_{n+1} = (2x^2+1)x_n + \lceil 2x\sqrt{x^2+1}x_n\rceil$, where $x_0 = 2x$. Then $x_n = p_{2n+2}$.

183. Let $x_{n+1} = (2x^2+1)x_n + \lfloor 2x\sqrt{x^2+1}x_n\rfloor$, where $x_0 = 4x^2+2$. Then $x_n = q_{2n+2}$.

32

FIBONACCI AND LUCAS POLYNOMIALS II

> Imagination disposes everything; it creates
> beauty, justice, and happiness,
> which are everything in this world.
> –Blaise Pascal (1623–1662)

The preceding chapter laid the foundation for our investigation of Fibonacci, Lucas, Pell, and Pell–Lucas polynomials. We now build upon this foundation to study additional properties. We begin our journey with an addition formula for Fibonacci polynomials. Although we can derive it using the Binet-like formula, we use matrices instead, to minimize the amount of basic algebra needed.

32.1 *Q*-MATRIX

Let

$$Q = Q(x) = \begin{bmatrix} x & 1 \\ 1 & 0 \end{bmatrix}.$$

It then follows by PMI that

$$Q^n = \begin{bmatrix} f_{n+1} & f_n \\ f_n & f_{n-1} \end{bmatrix},$$

where $n \geq 1$; see Exercise 32.1.

Using Q^n, we can easily confirm the Cassini-like formula for both f_n and l_n.

Fibonacci and Lucas Numbers with Applications, Volume Two. Thomas Koshy.

Cassini-like Formulas Revisited

Since $|Q^n| = |Q|^n = (-1)^n$, the Cassini-like formula for f_n follows from $|Q^n|$, where $|M|$ denotes the determinant of the square matrix M.

To establish its counterpart for l_n, we let I denote the 2×2 identity matrix. Then

$$\begin{aligned} Q^2 + I &= \begin{bmatrix} x^2+1 & x \\ x & 1 \end{bmatrix} + \begin{bmatrix} 1 & 0 \\ 0 & 1 \end{bmatrix} \\ &= \begin{bmatrix} x^2+2 & x \\ x & 2 \end{bmatrix} \end{aligned}$$

and $|Q^2 + I| = x^2 + 4$.

We also have

$$\begin{aligned} Q^{n+1} + Q^{n-1} &= \begin{bmatrix} f_{n+2} & f_{n+1} \\ f_{n+1} & f_n \end{bmatrix} + \begin{bmatrix} f_n & f_{n-1} \\ f_{n-1} & f_{n-2} \end{bmatrix} \\ Q^{n-1}(Q^2 + I) &= \begin{bmatrix} l_{n+1} & l_n \\ l_n & l_{n-1} \end{bmatrix} \\ |Q^{n-1}(Q^2 + I)| &= l_{n+1}l_{n-1} - l_n^2. \end{aligned}$$

Since $|Q^{n-1}(Q^2 + I)| = |Q^{n-1}||(Q^2 + I)| = (f_n f_{n-2} - f_{n-1}^2)(x^2 + 4) =$ $(-1)^{n-1}(x^2 + 4)$, this yields the Cassini-like formula for l_n, as desired.

Interestingly, we can establish the identity $xf_{n-1} + l_{n-1} = 2f_n$ also using Q^n. We accomplish this by computing the sum $xQ^n + Q^{n+1} + Q^{n-1}$ in two different ways.

To begin with,

$$\begin{aligned} xQ^n + Q^{n+1} + Q^{n-1} &= (xQ^n + Q^{n+1}) + Q^{n-1} \\ &= \begin{bmatrix} * & * \\ * & (xf_{n-1} + f_{n-2}) + f_n \end{bmatrix} \\ &= \begin{bmatrix} * & * \\ * & 2f_n \end{bmatrix}, \end{aligned} \tag{32.1}$$

where an asterisk indicates some element of the matrix.

We also have

$$\begin{aligned} xQ^n + Q^{n+1} + Q^{n-1} &= xQ^n + (Q^{n+1} + Q^{n-1}) \\ &= x\begin{bmatrix} f_{n+1} & f_n \\ f_n & f_{n-1} \end{bmatrix} + \begin{bmatrix} l_{n+1} & l_n \\ l_n & l_{n-1} \end{bmatrix} \\ &= \begin{bmatrix} xf_{n+1} + l_{n+1} & xf_n + l_n \\ xf_n + l_n & xf_{n-1} + l_{n-1} \end{bmatrix}. \end{aligned} \tag{32.2}$$

The desired identity now follows by equating the elements in the lower right-hand corners of the matrices in equations (32.1) and (32.2).

An Interesting Byproduct

Equation (32.2) yields an interesting byproduct. To see this, we have

$$\begin{aligned} Q^2 + xQ + I &= \begin{bmatrix} x^2+1 & x \\ x & 1 \end{bmatrix} + \begin{bmatrix} x^2 & x \\ x & 0 \end{bmatrix} + \begin{bmatrix} 1 & 0 \\ 0 & 1 \end{bmatrix} \\ &= \begin{bmatrix} 2x^2+2 & 2x \\ 2x & 2 \end{bmatrix} \\ |Q^2 + xQ + I| &= 4. \end{aligned}$$

Consequently, by equation (32.2), we have

$$\begin{aligned} |xQ^n + Q^{n+1} + Q^{n-1}| &= |Q^{n-1}(Q^2 + xQ + I)| \\ (xf_{n+1} + l_{n+1})(xf_{n-1} + l_{n-1}) - (xf_n + l_n)^2 &= |Q^{n-1}| \cdot |Q^2 + xQ + I| \\ &= 4(f_n f_{n-2} - f_{n-1}^2) \\ &= 4(-1)^{n-1}. \end{aligned} \tag{32.3}$$

For example,

$$\begin{aligned} (xf_4 + l_4)(xf_2 + l_2) - (xf_3 + l_3)^2 &= (2x^4 + 6x^2 + 2)(2x^2 + 2) - (2x^3 + 4x)^2 \\ &= 4 = 4(-1)^{3-1}. \end{aligned}$$

Notice that property (32.3) follows fairly quickly, since $xf_{k-1} + l_{k-1} = 2f_k$.

32.2 SUMMATION FORMULAS

The Q-matrix has additional consequences. For example, we can use it to extract summation formulas. To begin with, we now employ the matrix Q^n and the Cassini-like formula for Fibonacci polynomials to establish the summation formula

$$x\sum_{k=1}^{n} f_k = f_{n+1} + f_n - 1. \tag{32.4}$$

Notice that $|Q - I| \neq 0$; so $Q - I$ is invertible and $(Q - I)^{-1} = \begin{bmatrix} 0 & 1 \\ 1 & 1-x \end{bmatrix}$.

Since $(Q-I)(I+Q+Q^2+\cdots+Q^n)=Q^{n+1}-I$, it follows that

$$\begin{aligned}
I+Q+Q^2+\cdots+Q^n &= (Q^{n+1}-I)(Q-I)^{-1}\\
&= \begin{bmatrix} f_{n+2}-1 & f_{n+1}\\ f_{n+1} & f_n-1\end{bmatrix}\begin{bmatrix}0 & 1\\ 1 & 1-x\end{bmatrix}\\
&= \begin{bmatrix} f_{n+1} & f_{n+1}+f_n-1\\ f_n-1 & f_n+f_{n-1}-1\end{bmatrix}.
\end{aligned}$$

Equating the upper right-hand elements from both sides, we get the desired formula.

Using the identity $f_{k-1}+f_{k+1}=l_n$ and the summation formula (32.4), we can find a formula for $\sum\limits_{k=0}^{n} l_k$:

$$\begin{aligned}
x\sum_{k=0}^{n} l_k &= x\sum_{k=0}^{n}(f_{k-1}+f_{k+1})\\
&= (f_n+f_{n-1}+x-1)+(f_{n+2}+f_{n+1}-1)\\
&= (f_n+f_{n+2})+(f_{n-1}+f_{n+1})+x-2\\
&= l_{n+1}+l_n+x-2. \qquad (32.5)
\end{aligned}$$

For example,

$$\begin{aligned}
x\sum_{k=0}^{3} l_k &= x(l_0+l_1+l_2+l_3)\\
&= x[2+x+(x^2+2)+(x^3+3x)]\\
&= x^4+x^3+4x^2+4x\\
&= (x^4+4x^2+2)+(x^3+3x)+x-2\\
&= l_4+l_3+x-2.
\end{aligned}$$

Next we use the Q-matrix to develop recursive summation formulas for Fibonacci and Lucas polynomials.

Recursive Summation Formulas

The characteristic polynomial of the Q-matrix is

$$|Q-\lambda I| = \begin{vmatrix} x-\lambda & 1\\ 1 & -\lambda\end{vmatrix} = \lambda^2 - x\lambda - 1.$$

Consequently, its eigenvalues are $\lambda = \dfrac{x\pm\sqrt{x^2+4}}{2}$, that is, $\lambda=\alpha,\beta$.

By the well-known Cayley–Hamilton theorem in linear algebra [13], every square matrix satisfies its characteristic equation; so $Q^2 = xQ + I$. Then

$$Q^{2n} = (xQ + I)^n = \sum_{k=0}^{n} \binom{n}{k} Q^k x^k.$$

So

$$Q^{2n} \cdot Q^m = \sum_{k=0}^{n} \binom{n}{k} Q^k \cdot Q^m x^k$$

$$Q^{2n+m} = \sum_{k=0}^{n} \binom{n}{k} Q^{k+m} x^k.$$

This implies

$$f_{2n+m} = \sum_{k=0}^{n} \binom{n}{k} f_{k+m} x^k. \tag{32.6}$$

In particular,

$$f_{2n} = \sum_{k=0}^{n} \binom{n}{k} f_k x^k;$$

$$f_{2n+1} = \sum_{k=0}^{n} \binom{n}{k} f_{k+1} x^k;$$

$$f_{2n-1} = \sum_{k=0}^{n} \binom{n}{k} f_{k-1} x^k.$$

It follows by the last two formulas that

$$l_{2n} = \sum_{k=0}^{n} \binom{n}{k} l_k x^k. \tag{32.7}$$

For example,

$$l_6 = \sum_{k=0}^{3} \binom{3}{k} l_k x^k = l_0 + 3l_1 x + 3l_2 x^2 + l_3 x^3 = x^6 + 6x^4 + 9x^2 + 2.$$

Interesting Byproducts

To see another application of the Q-matrix, we have

$$\begin{aligned} f_nQ + f_{n-1}I &= f_n \begin{bmatrix} x & 1 \\ 1 & 0 \end{bmatrix} + f_{n-1} \begin{bmatrix} 1 & 0 \\ 0 & 1 \end{bmatrix} \\ &= \begin{bmatrix} f_{n+1} & f_n \\ f_n & f_{n-1} \end{bmatrix} \\ &= Q^n. \end{aligned}$$

Then, by the binomial theorem,

$$\begin{aligned} Q^{mn} &= (f_nQ + f_{n-1}I)^m \\ &= \sum_{i=0}^{m} \binom{m}{i} Q^i f_n^i f_{n-1}^{m-i} \\ Q^{mn+k} &= \sum_{i=0}^{m} \binom{m}{i} Q^{i+k} f_n^i f_{n-1}^{m-i}. \end{aligned}$$

This implies

$$f_{mn+k} = \sum_{i=0}^{m} \binom{m}{i} f_{i+k} f_n^i f_{n-1}^{m-i}, \tag{32.8}$$

where $m \geq 0$.

In particular, this yields

$$F_{mn+k} = \sum_{i=0}^{m} \binom{m}{i} F_{i+k} F_n^i F_{n-1}^{m-i}.$$

Hoggatt and I.D. Ruggles of then San Jose State College discovered this formula in 1963 [228].

The matrix Q has an added byproduct. To see this, notice that $Q \begin{bmatrix} f_{k+1} \\ f_k \end{bmatrix} = \begin{bmatrix} f_{k+2} \\ f_{k+1} \end{bmatrix}$. More generally, $Q^n \begin{bmatrix} f_{k+1} \\ f_k \end{bmatrix} = \begin{bmatrix} f_{n+k+1} \\ f_{n+k} \end{bmatrix}$. Letting $k = 0$, this implies $Q \begin{bmatrix} f_1 \\ f_0 \end{bmatrix} = \begin{bmatrix} f_{n+1} \\ f_n \end{bmatrix}$. Thus

$$Q^n \begin{bmatrix} f_{k+1} & f_1 \\ f_k & f_0 \end{bmatrix} = \begin{bmatrix} f_{n+k+1} & f_{n+1} \\ f_{n+k} & f_n \end{bmatrix}.$$

Since $|Q| = -1$ and $|AB| = |A| \cdot |B|$, this implies $f_{n+k+1}f_n - f_{n+k}f_{n+1} = (-1)^{n+1}f_k$, where A and B are square matrices of the same order. Clearly, this identity is a generalization of the Cassini-like identity for Fibonacci polynomials. See Exercise 31.73 for a generalization of this identity.

32.3 ADDITION FORMULAS

Since $Q^{m+n} = Q^m \cdot Q^n$, it follows by equating corresponding elements that

$$f_{m+n} = f_{m+1}f_n + f_m f_{n-1}. \tag{32.9}$$

An immediate consequence of this *addition formula* is the property that $f_{mn} > f_m f_n$, where $x \geq 1$ and $m, n \geq 2$; see Exercise 32.9. It also follows that $f_{2n} = f_n l_n$ and $f_{2n+1} = f_{n+1}^2 + f_n^2$. In addition, it implies that

$$\begin{aligned} f_{-(m-n)} &= f_{-(m-1)}f_n + f_{-m}f_{n-1} \\ f_{m-n} &= (-1)^n \left(f_m f_{n-1} - f_{m-1}f_n\right). \end{aligned} \tag{32.10}$$

We can arrive at identity (32.10) by computing Q^{-n} also. Since $|Q| \neq 0$, Q is invertible, and $Q^{-1} = -\begin{bmatrix} 0 & -1 \\ -1 & x \end{bmatrix}$. Then, by Cassini's formula and PMI, $Q^{-n} = (-1)^n \begin{bmatrix} f_{n-1} & -f_n \\ -f_n & f_{n+1} \end{bmatrix}$. So

$$\begin{aligned} Q^{m-n} &= Q^m \cdot Q^{-n} \\ \begin{bmatrix} f_{m-n+1} & f_{m-n} \\ f_{m-n} & f_{m-n-1} \end{bmatrix} &= (-1)^n \begin{bmatrix} f_{m+1} & f_m \\ f_m & f_{m-1} \end{bmatrix} \begin{bmatrix} f_{n-1} & -f_n \\ -f_n & f_{n+1} \end{bmatrix} \\ &= (-1)^n \begin{bmatrix} f_{m+1}f_{n-1} - f_m f_n & f_m f_{n+1} - f_{m+1}f_n \\ f_m f_{n-1} - f_{m-1}f_n & f_{m-1}f_{n+1} - f_m f_n \end{bmatrix}. \end{aligned}$$

This yields the desired identity.

It follows by identities (32.9) and (32.10) that

$$\begin{aligned} f_{m+n} + f_{m-n} &= \begin{cases} l_m f_n & \text{if } n \text{ is odd} \\ f_m l_n & \text{otherwise;} \end{cases} \\ f_{m+n} - f_{m-n} &= \begin{cases} f_m l_n & \text{if } n \text{ is odd} \\ l_m f_n & \text{otherwise.} \end{cases} \end{aligned} \tag{32.11}$$

Consequently,

$$f_{m+n}^2 - f_{m-n}^2 = f_{2m}f_{2n}; \tag{32.12}$$

$$p_{m+n}^2 - p_{m-n}^2 = p_{2m}p_{2n}; \tag{32.13}$$

and hence $F_{m+n}^2 - F_{m-n}^2 = F_{2m}F_{2n}$ and $P_{m+n}^2 - P_{m-n}^2 = P_{2m}P_{2n}$.

It follows from identity (32.9) that

$$\begin{aligned}
f_{a+b} &= f_{a+1}f_b + f_af_{b-1}\\
&= f_af_{b+1} + f_{a-1}f_b\\
2f_{a+b} &= f_a(f_{b+1} + f_{b-1}) + f_b(f_{a+1} + f_{a-1})\\
&= f_al_b + f_bl_a, \qquad (32.14)
\end{aligned}$$

where a and b are any integers.

Consequently, $2f_{a-b} = (-1)^b\left(f_al_b - f_bl_a\right)$. Thus

$$f_{a+b} + (-1)^b f_{a-b} = f_al_b. \qquad (32.15)$$

For example, let $a = 7$ and $b = 3$. Then

$$\begin{aligned}
f_{10} - f_4 &= (x^9 + 8x^7 + 21x^5 + 20x^3 + 5x) - (x^3 + 2x)\\
&= x^9 + 8x^7 + 21x^5 + 19x^3 + 3x\\
&= (x^6 + 5x^4 + 6x^2 + 1)(x^3 + 3x)\\
&= f_7l_3.
\end{aligned}$$

Similarly, $f_{10} + f_2 = (x^5 + 4x^3 + 3x)(x^4 + 4x^2 + 2) = f_6l_4$.

We can show likewise that

$$\begin{aligned}
l_{a+b} &= f_{a+1}l_b + f_al_{b-1};\\
l_{a-b} &= (-1)^b\left(f_{a+1}l_b - f_al_{b+1}\right);
\end{aligned}$$

see Exercises 32.14 and 32.15. So

$$l_{a+b} + (-1)^b l_{a-b} = 2f_{a+1}l_b. \qquad (32.16)$$

Identity (32.14) has an analogous result for Lucas polynomials:

$$2l_{a+b} = l_al_b + (x^2 + 4)f_af_b; \qquad (32.17)$$

see Exercise 32.18.

This result has an interesting consequence. To see this, let $i + j = h + k$. Then it follows by identity (32.17) that $2l_{i+j} \equiv l_il_j \pmod{x^2 + 4}$ and $2l_{h+k} \equiv l_hl_k \pmod{x^2 + 4}$. Since $i + j = h + k$, it follows from these two congruences that $l_il_j \equiv l_hl_k \pmod{x^2 + 4}$.

For example,

$$\begin{aligned} l_3 l_5 &= (x^3 + 3x)(x^5 + 5x^3 + 5x) \\ &\equiv (-x)x \equiv (-2)(-2) \pmod{x^2 + 4} \\ &\equiv (x^2 + 2)(x^6 + 6x^4 + 9x^2 + 2) \pmod{x^2 + 4} \\ &\equiv l_2 l_6 \pmod{x^2 + 4}. \end{aligned}$$

The addition formula $xf_{m+n} = f_{m+1}f_{n+1} - f_{m-1}f_{n-1}$ (see Exercise 32.10), coupled with a telescoping sum, has an interesting application; it can be used to find a formula for $x \sum_{k=1}^{n} f_{4k+3}$.

To see this, we have $xf_{4k+3} = xf_{(2k+3)+(2k)} = f_{2k+4}f_{2k+1} - f_{2k+2}f_{2k-1}$. Consequently,

$$\begin{aligned} x\sum_{k=1}^{n} f_{4k+3} &= \sum_{k=1}^{n} (f_{2k+4}f_{2k+1} - f_{2k+2}f_{2k-1}) \\ &= f_{2n+4}f_{2n+1} - f_4 f_1 \\ &= f_{2n+4}f_{2n+1} - (x^3 + 2x). \end{aligned} \tag{32.18}$$

It follows from formula (32.18) that

$$\sum_{k=1}^{n} F_{4k+3} = F_{2n+4}F_{2n+1} - 3; \tag{32.19}$$

$$2x\sum_{k=1}^{n} p_{4k+3} = p_{2n+4}p_{2n+1} - (8x^3 + 4x);$$

$$2\sum_{k=1}^{n} P_{4k+3} = P_{2n+4}P_{2n+1} - 12.$$

For example, $2 \sum_{k=1}^{4} P_{4k+3} = 13{,}652{,}088 = 13{,}860 \cdot 985 - 12 = P_{12}P_9 - 12.$

M.A. Khan of Lucknow, India, found formula (32.18) in 2015 [267].

Similarly, we can show (see Exercise 32.38) that

$$x\sum_{k=1}^{n} l_{4k+3} = f_{2n+4}l_{2n+1} - (x^3 + 2x)x. \tag{32.20}$$

This implies

$$\sum_{k=1}^{n} L_{4k+3} = F_{2n+4}L_{2n+1} - 3;$$

$$2x \sum_{k=1}^{n} q_{4k+3} = p_{2n+4}q_{2n+1} - 2x(8x^3 + 4x);$$

$$2 \sum_{k=1}^{n} Q_{4k+3} = P_{2n+4}Q_{2n+1} - 12.$$

For example, $2 \sum_{k=1}^{4} Q_{4k+3} = 568{,}330 = 2378 \cdot 239 - 12 = P_{10}Q_7 - 12.$

Next we evaluate an interesting infinite product.

Example 32.1. *Evaluate the infinite product*

$$P = x\left(1 + \frac{1}{f_3}\right)\left(1 + \frac{1}{f_7}\right)\left(1 + \frac{1}{f_{15}}\right)\cdots.$$

Solution. Using the identities $f_{2n} = f_n l_n$, $f_{n+1} + f_{n-1} = l_n$, and $f_{n+1}f_{n-1} - f_n^2 = (-1)^n$, we can show that $f_{2^{n+1}-1} + 1 = f_{2^n-1}l_{2^n}$; see Exercise 32.29. In addition, using PMI, we can confirm that $x \prod_{k=1}^{n} l_{2^k} = f_{2^{n+1}}$; see Exercise 32.30.

We then have

$$\begin{aligned}
x\prod_{n=1}^{m}\left(1 + \frac{1}{f_{2^{n+1}-1}}\right) &= x\prod_{n=1}^{m}\frac{f_{2^{n+1}-1}+1}{f_{2^{n+1}-1}}\\
&= x\prod_{n=1}^{m}\frac{f_{2^n-1}l_{2^n}}{f_{2^{n+1}-1}}\\
&= \prod_{n=1}^{m}\frac{f_{2^n-1}}{f_{2^{n+1}-1}} \cdot f_{2^{m+1}}\\
&= \frac{f_{2^{m+1}}}{f_{2^{m+1}-1}}\\
P &= \lim_{m\to\infty}\frac{f_{2^{m+1}}}{f_{2^{m+1}-1}}\\
&= \alpha.
\end{aligned}$$

■

J. Shallit of Palo Alto, California, studied this infinite product in 1980 for the case $x = 1$ [457].

If we return to identity (32.16), we can show that

$$l_{a+b} + l_{a-b} = \begin{cases} (x^2+4)f_a f_b & \text{if } b \text{ is odd} \\ l_a l_b & \text{otherwise;} \end{cases}$$

see Exercise 32.19.

The next example is a simple and direct application of this result.

Example 32.2. *Let $S_{k,n} = l_{2n} + l_{4n} + l_{6n} + \cdots + l_{2kn}$, where k is even. Prove that*

$$S_{k,n} = \begin{cases} 0 \pmod{\Delta^2 f_n} & \text{if } n \text{ is odd} \\ 0 \pmod{l_n} & \text{otherwise.} \end{cases}$$

Proof. By pairing every two adjacent summands, we can rewrite the sum as

$$S_{k,n} = \sum_{j=1}^{k/2} \left(l_{(4j-1)n+n} + l_{(4j-1)n-n} \right)$$

$$= \begin{cases} \Delta^2 f_n \sum_{j=1}^{k/2} l_{(4j-1)n} & \text{if } n \text{ is odd} \\ l_n \sum_{j=1}^{k/2} l_{(4j-1)n} & \text{otherwise.} \end{cases}$$

This yields the desired result. (Freitag studied this example in 1993 for the case $x = 1$ [174].) ■

It follows by this example that $S_{k,n} \equiv 0 \pmod{l_n}$, when both n and and k are even. But we can make this congruence stronger: $S_{k,n} \equiv 0 \pmod{l_n^2}$, when both n and and k are even.

To establish this, we need to lay some groundwork:

1) $l_n | l_{jn}$, where j is odd.
2) $l_n | f_{(j+1)n}$, where j is odd.
3) Let j be odd, and n even. As in Example 32.2, if we pair the adjacent terms, we get

$$\begin{aligned} l_{2jn} + l_{2(j+1)n} &= \left(\alpha^{2jn} + \beta^{2jn}\right) + \left[\alpha^{2(j+1)n} + \beta^{2(j+1)n}\right] \\ &= \left(\alpha^{jn} + \beta^{jn}\right)^2 + \left[\alpha^{(j+1)n} - \beta^{(j+1)n}\right]^2 \\ &= l_{jn}^2 + \Delta^2 f_{(j+1)n}^2. \end{aligned}$$

Since $l_n | l_{jn}$ and $l_n | f_{(j+1)n}$, it follows that $l_{2jn} + l_{2(j+1)n} \equiv 0 \pmod{l_n^2}$.

Consequently, $S_{k,n} \equiv 0 \pmod{l_n^2}$, as desired.

For a specific case, we let $n = 2$ and $k = 4$. We have

$$\begin{aligned} l_2^2 &= x^4 + 2x^2 + 4 \\ l_4 &= x^4 + 4x^2 + 2 \equiv -2 \pmod{l_2^2} \\ l_8 &= x^8 + 8x^6 + 20x^4 + 16x^2 + 2 \equiv 2 \pmod{l_2^2} \\ l_{12} &= x^{12} + 12x^{10} + 54x^8 + 112x^6 + 105x^4 + 36x^2 + 2 \equiv -2 \pmod{l_2^2} \\ l_{16} &= x^{16} + 16x^{14} + 104x^{12} + 352x^{10} + 660x^8 + 672x^6 \\ &\quad + 336x^4 + 64x^2 + 2 \equiv 2 \pmod{l_2^2}. \end{aligned}$$

So $l_4 + l_8 + l_{12} + l_{16} \equiv -2 + 2 - 2 + 2 \equiv 0 \pmod{l_2^2}$, as expected.

Identity (32.15) has a delightful consequence. The identity can be used to develop an interesting recurrence for f_{n^2}. To this end, we need the following hybrid identity:

$$xl_{m+4} = (x^4 + 4x^2 + 2)l_{m+1} - (x^2 + 1)(x^2 + 4)f_m; \tag{32.21}$$

see Exercise 32.34.

32.4 A RECURRENCE FOR f_{n^2}

Let $R_n = f_{n^2}$. Letting $a = 2n$ and $b = n^2 + 1$, identity (32.15) gives

$$\begin{aligned} f_{(n+1)^2} + (-1)^{n^2+1} f_{-(n-1)^2} &= f_{2n} l_{n^2+1} \\ f_{(n+1)^2} - f_{(n-1)^2} &= f_{2n} l_{n^2+1} \\ R_{n+1} - R_{n-1} &= f_{2n} l_{n^2+1}. \end{aligned} \tag{32.22}$$

With $a = 4n$ and $b = n^2 + 4$, identity (32.15) yields

$$\begin{aligned} f_{(n+2)^2} + (-1)^{n^2+4} f_{-(n-2)^2} &= f_{4n} l_{n^2+4} \\ f_{(n+2)^2} - f_{(n-2)^2} &= f_{4n} l_{n^2+4} \\ R_{n+2} - R_{n-2} &= f_{4n} l_{n^2+4}. \end{aligned} \tag{32.23}$$

With $m = n^2$, identity (32.21) becomes

$$xl_{n^2+4} = (x^4 + 4x^2 + 2)l_{n^2+1} - (x^2 + 1)(x^2 + 4)f_{n^2}.$$

This, coupled with (32.22) and (32.23), yields

$$\begin{aligned} x(R_{n+2} - R_{n-2}) &- (x^4 + 4x^2 + 2)l_{2n}(R_{n+1} - R_{n-1}) \\ &= xf_{4n} l_{n^2+4} - (x^4 + 4x^2 + 2)l_{2n}(f_{2n} l_{n^2+1}) \\ &= f_{4n}\left[xl_{n^2+4} - (x^4 + 4x^2 + 2)l_{n^2+1}\right] \\ &= f_{4n}\left[-(x^2 + 1)(x^2 + 4)R_n\right]. \end{aligned}$$

This gives a fourth-order recurrence for $R_n = f_{n^2}$:

$$\begin{aligned} xR_{n+4} = (x^4 + 4x^2 + 2)l_{2n+4}R_{n+3} &- (x^2 + 1)(x^2 + 4)f_{4n+8}R_{n+2} \\ &- (x^4 + 4x^2 + 2)l_{2n+2}R_{n+1} + xR_n. \end{aligned} \tag{32.24}$$

It follows from this recurrence that

$$\begin{aligned}
xS_{n+4} &= (8x^4+8x^2+1)q_{2n+4}S_{n+3} - 2(x^2+1)(4x^2+1)p_{4n+8}S_{n+2} \\
&\quad - (8x^4+8x^2+1)q_{2n+2}S_{n+1} + xS_n; \\
F_{(n+4)^2} &= 7L_{2n+4}F_{(n+3)^2} - 10F_{4n+8}F_{(n+2)^2} - 7L_{2n+2}F_{(n+1)^2} + F_{n^2}; \qquad (32.25) \\
P_{(n+4)^2} &= 34Q_{2n+4}P_{(n+3)^2} - 20P_{4n+8}P_{(n+2)^2} - 34Q_{2n+2}P_{(n+1)^2} + P_{n^2},
\end{aligned}$$

where $S_k = S_k(x) = p_{k^2}$. I. Strazdins developed the recurrence (32.25) in 2000 [472].

Identity (32.14) Revisited

Identity (32.14) can be used to develop another addition formula for Lucas polynomials. Using the identities $f_{n+1} + f_{n-1} = l_n$ and $l_{n+1} + l_{n-1} = (x^2+4)f_n$, identity (32.14) yields

$$\begin{aligned}
2f_{a+b} &= f_a l_b + f_b l_a \\
2f_{a+b+2} &= f_{a+2} l_b + f_b l_{a+2} \\
2(f_{a+b} + f_{a+b+2}) &= (f_a + f_{a+2}) l_b + f_b(l_a + l_{a+2}) \\
2l_{a+b+1} &= l_{a+1} l_b + (x^2+4) f_{a+1} f_b \\
2l_{a+b} &= l_a l_b + (x^2+4) f_a f_b.
\end{aligned}$$

Consequently,

$$\begin{aligned}
2l_{a-b} &= l_a l_{-b} + (x^2+4) f_a f_{-b} \\
&= (-1)^b[l_a l_b - (x^2+4) f_a f_b].
\end{aligned}$$

Thus

$$l_{a+b} + (-1)^b l_{a-b} = l_a l_b. \qquad (32.26)$$

Similarly,

$$l_{a+b} - (-1)^b l_{a-b} = \Delta^2 f_a f_b. \qquad (32.27)$$

This implies

$$l_{2m} - (-1)^{m+n} l_{2n} = \Delta^2 f_{m+n} f_{m-n}. \qquad (32.28)$$

The next two examples highlight the beauty and power of identities (32.26) and (32.27). In addition, Example 32.3 uses the identity $f_{2n} = f_n l_n$ and Example 32.4 uses $l_{2n} = l_n^2 - 2(-1)^n$.

Example 32.3. *Establish each formula, where $x \geq 1$.*

1) $\displaystyle\sum_{k=1}^{\infty} \frac{l_{2^{k+1}}}{f_{3\cdot 2^k}} = \frac{x^2+4}{x(x^2+3)}$.

2) $\displaystyle\sum_{k=1}^{\infty} \frac{f_{2^{k-1}}^2}{l_{2^k}^2 - 1} = \frac{x^2+5}{(x+1)(x+3)(x^2+4)}$.

Proof. It follows by identities (32.26) and (32.27) that $l_{2^{k+1}} + l_{2^k} = l_{3\cdot 2^{k-1}} l_{2^{k-1}}$, $l_{2^k} + 2 = l_{2^{k-1}}^2$, and $l_{2^k} - 2 = \Delta^2 f_{2^{k-1}}^2$, where $k \geq 2$. Since

$$\begin{aligned}
\frac{l_{2^{k+1}}}{f_{3\cdot 2^k}} &= \frac{l_{3\cdot 2^{k-1}} l_{2^{k-1}}}{f_{3\cdot 2^{k-1}} l_{3\cdot 2^{k-1}}} - \frac{l_{2^k}}{f_{3\cdot 2^k}} \\
&= \frac{l_{2^{k-1}}}{f_{3\cdot 2^{k-1}}} - \frac{l_{2^k}}{f_{3\cdot 2^k}},
\end{aligned}$$

$$\begin{aligned}
\sum_{k=1}^{\infty} \frac{l_{2^{k+1}}}{f_{3\cdot 2^k}} &= \frac{l_4}{f_6} + \sum_{k=2}^{\infty} \left(\frac{l_{2^{k-1}}}{f_{3\cdot 2^{k-1}}} - \frac{l_{2^k}}{f_{3\cdot 2^k}} \right) \\
&= \frac{l_4}{f_6} + \frac{l_2}{f_6} \\
&= \frac{x^4 + 5x^2 + 4}{x^5 + 4x^3 + 3x} \\
&= \frac{x^2+4}{x(x^2+3)}.
\end{aligned}$$

We can similarly establish the second formula; see Exercise 32.39. ■

It follows from this example that

$$\sum_{k=1}^{\infty} \frac{L_{2^{k+1}}}{F_{3\cdot 2^k}} = \frac{5}{4}; \tag{32.29}$$

$$\sum_{k=1}^{\infty} \frac{q_{2^{k+1}}}{p_{3\cdot 2^k}} = \frac{2(x^2+1)}{x(4x^2+3)};$$

$$\sum_{k=1}^{\infty} \frac{Q_{2^{k+1}}}{P_{3\cdot 2^k}} = \frac{2}{7};$$

$$\sum_{k=1}^{\infty} \frac{F_{2^{k-1}}^2}{L_{2^k}^2 - 1} = \frac{3}{20}; \tag{32.30}$$

$$\sum_{k=1}^{\infty} \frac{p_{2k-1}^2}{q_{2k}^2 - 1} = \frac{4x^2+5}{4(2x+1)(2x+3)(x^2+1)};$$

$$\sum_{k=1}^{\infty} \frac{P_{2k-1}^2}{4Q_{2k}^2 - 1} = \frac{1}{24}.$$

Ohtsuka found formulas (32.29) and (32.30) in 2014 [313, 371].

Example 32.4. *Prove that* $\Delta^4 f_{a+b+c} f_{a+b-c} f_{b+c-a} f_{c+a-b} = l_{2a} l_{2b} l_{2c} - \nu\left(l_{2a}^2 + l_{2b}^2 + l_{2c}^2\right) - 4\nu$, *where* $\nu = (-1)^{a+b+c}$.

Proof. We have

$$\begin{aligned}
\text{LHS} &= (l_{2a+2b} - \nu\, l_{2c})(l_{2c} - \nu\, l_{2a-2b}) \\
&= l_{2c}(l_{2a+2b} + l_{2a-2b}) - \nu\, l_{2c}^2 - \nu\, l_{2a+2b} l_{2a-2b} \\
&= l_{2c}(l_{2a} l_{2b}) - \nu\, l_{2c}^2 - \nu(l_{4a} + l_{4b}) \\
&= l_{2a} l_{2b} l_{2c} - \nu\, l_{2c}^2 - \nu\left[l_{2a}^2 + l_{2b}^2 - 2(-1)^{2a} - 2(-1)^{2b}\right] \\
&= l_{2a} l_{2b} l_{2c} - \nu\left(l_{2a}^2 + l_{2b}^2 + l_{2c}^2\right) + 4\nu,
\end{aligned}$$

as desired. ■

Notice that

$$\Delta^4 f_{a+b+c} f_{a+b-c} f_{b+c-a} f_{c+a-b}$$
$$= \begin{cases} l_{2a} l_{2b} l_{2c} + \left(l_{2a}^2 + l_{2b}^2 + l_{2c}^2\right) - 4 & \text{if } a+b+c \text{ is odd} \\ l_{2a} l_{2b} l_{2c} - \left(l_{2a}^2 + l_{2b}^2 + l_{2c}^2\right) + 4 & \text{otherwise.} \end{cases}$$

G. Wulczyn of Lewisburg, Pennsylvania, studied this example in 1990 for the case $x = 1$ [507, 508].

The formula in Example 32.4 has a Lucas counterpart:

$$l_{a+b+c} l_{a+b-c} l_{b+c-a} l_{c+a-b} = l_{2a} l_{2b} l_{2c} + \nu\left(l_{2a}^2 + l_{2b}^2 + l_{2c}^2\right) - 4\nu;$$

see Exercise 32.35.

Next we pursue an application of the identity $l_n^2 - (x^2+4) f_n^2 = 4(-1)^n$ to Pythagorean triples.

Pythagorean Triples

We can use identity (32.26), together with the identity $l_n^2 - (x^2+4)f_n^2 = 4(-1)^n$, to develop Pythagorean triples. To this end, we have

$$\begin{aligned}[l_{2n+3} + 4(-1)^n]^2 + 4[l_{2n+3} - (-1)^n]^2 &= 5l_{2n+3}^2 + 20 \\ &= 5(x^2+4)f_{2n+3}^2. \qquad (32.31)\end{aligned}$$

Identity (32.31) implies that

$$[L_{2n+3} + 4(-1)^n]^2 + 4[L_{2n+3} - (-1)^n]^2 = 25F_{2n+3}^2. \qquad (32.32)$$

So $(L_{2n+3} + 4(-1)^n, 2L_{2n+3} - 2(-1)^n, 5F_{2n+3})$ is a Pythagorean triple for all values of n.

In particular, let $n = 5$. Then $(L_{13} - 4, 2L_{13} - 2, 5F_{13}) = (521 - 4, 2 \cdot 521 + 2, 5 \cdot 233) = (517, 1044, 1165)$: $517^2 + 1044^2 = 1165^2$.

Identity (32.32) is basically the same as the identity $(L_nL_{n+3})^2 + (2L_{n+1}L_{n+2})^2 = (L_{2n+5} - L_{2n+1})^2$, discovered by M.J. Karameh of Jerusalem, Israel, in 2004 [263]. This is in fact a slight variation of the identity $(L_nL_{n+3})^2 + (2L_{n+1}L_{n+2})^2 = (L_{n+2}L_{n+3} - L_nL_{n+1})^2$, discovered by H.L. Umansky and M.H. Tallman in 1966 [497].

It also follows from identity (32.31) that

$$[q_{2n+3} + 4(-1)^n]^2 + 4[q_{2n+3} - (-1)^n]^2 = 20(x^2+1)p_{2n+3}^2$$
$$[q_{2n+3}(2) + 4(-1)^n]^2 + 4[q_{2n+3}(2) - (-1)^n]^2 = 100p_{2n+3}^2(2).$$

So $(q_{2n+3}(2) + 4(-1)^n, 2q_{2n+3}(2) - 2(-1)^n, 10p_{2n+3}(2))$ is also a Pythagorean triple.

For example, let $n = 1$. Then $q_5(2) = 32 \cdot 2^5 + 40 \cdot 2^3 + 10 \cdot 2 = 1364$; $p_5(2) = 16 \cdot 2^4 + 12 \cdot 2^2 + 1 = 305$; and $(q_5(2) - 4, 2q_5(2) + 2, 10p_5(2)) = (1360, 2730, 3050)$ is a Pythagorean triple: $1360^2 + 2730^2 = 3050^2$. On the other hand, let $n = 2$. Then $(q_7(2) - 4, 2q_7(2) + 2, 10p_7(2)) = (24480, 48950, 54730)$ is a Pythagorean triple: $24480^2 + 48950^2 = 54730^2$.

Addition Formulas Revisited

Interestingly, the addition formulas $2f_{a+b} = f_al_b + f_bl_a$ and $2l_{a+b} = l_al_b + \Delta^2 f_af_b$ can be combined into a matrix equation

$$2\begin{bmatrix} f_{a+b} \\ l_{a+b} \end{bmatrix} = \begin{bmatrix} l_b & f_b \\ \Delta^2 f_b & l_b \end{bmatrix} \begin{bmatrix} f_a \\ l_a \end{bmatrix}.$$

This matrix approach pays an interesting dividend. To see this, we let $a = (k-1)n$, $b = n$, and $k \geq 1$. Then

$$\begin{bmatrix} f_{kn} \\ l_{kn} \end{bmatrix} = \frac{1}{2}\begin{bmatrix} l_n & f_n \\ \Delta^2 f_n & l_n \end{bmatrix} \begin{bmatrix} f_{(k-1)n} \\ l_{(k-1)n} \end{bmatrix}.$$

Using iteration, this yields formulas for f_{kn} and l_{kn}:

$$\begin{aligned}\begin{bmatrix} f_{kn} \\ l_{kn} \end{bmatrix} &= \frac{1}{2^k} \begin{bmatrix} l_n & f_n \\ \Delta^2 f_n & l_n \end{bmatrix}^k \begin{bmatrix} f_0 \\ l_0 \end{bmatrix} \\ &= \frac{1}{2^k} \begin{bmatrix} l_n & f_n \\ \Delta^2 f_n & l_n \end{bmatrix}^k \begin{bmatrix} 0 \\ 2 \end{bmatrix}. \end{aligned} \tag{32.33}$$

In particular,

$$\begin{aligned}\begin{bmatrix} f_{2n} \\ l_{2n} \end{bmatrix} &= \frac{1}{4} \begin{bmatrix} l_n^2 + \Delta^2 f_n^2 & 2f_n l_n \\ \Delta^2 f_n l_n & l_n^2 + \Delta^2 f_n^2 \end{bmatrix} \begin{bmatrix} 0 \\ 2 \end{bmatrix} \\ &= \frac{1}{2} \begin{bmatrix} 2f_n l_n \\ l_n^2 + \Delta^2 f_n^2 \end{bmatrix}. \end{aligned}$$

Consequently, $f_{2n} = f_n l_n$ and $2l_{2n} = l_n^2 + (x^2+4)^2 f_n^2$.

Likewise,

$$\begin{aligned}\begin{bmatrix} f_{3n} \\ l_{3n} \end{bmatrix} &= \frac{1}{2} \begin{bmatrix} l_n & f_n \\ \Delta^2 f_n & l_n \end{bmatrix} \cdot \frac{1}{2} \begin{bmatrix} 2f_n l_n \\ l_n^2 + \Delta^2 f_n^2 \end{bmatrix} \\ &= \frac{1}{4} \begin{bmatrix} \Delta^2 f_n^3 + 3f_n l_n^2 \\ l_n^3 + 3\Delta^2 l_n f_n^2 \end{bmatrix}. \end{aligned}$$

Thus $4f_{3n} = (x^2+4)f_n^3 + f_n l_n^2$ and $4l_{3n} = l_n^3 + 3(x^2+4)l_n f_n^2$.

Similarly, $2f_{4n} = (x^2+4)f_n^3 l_n + f_n l_n^3$ and $8l_{4n} = l_n^4 + 6(x^2+4)l_n^2 f_n^2 + (x^2+4)^2 f_n^4$; see Exercises 32.31 and 32.32.

Identity (32.11) has an interesting byproduct, as the next example shows. To this end, we need the summation formula $x \sum_{j=1}^{n} f_{2j} = f_{2n+1} - 1$; see Exercise 32.5.

Example 32.5. *Prove that* $x \sum_{j=1}^{n} f_{2j} = f_{n+[n \pmod 2]} l_{n+1-[n \pmod 2]}$.

Proof. In light of the summation formula, it suffices to show that $f_{n+[n \pmod 2]} l_{n+1-[n \pmod 2]} = f_{2n+1} - 1$. By identity (32.11), we have

$$\begin{aligned} f_{n+[n \pmod 2]} l_{n+1-[n \pmod 2]} &= \begin{cases} f_{n+1} l_n & \text{if } n \text{ is odd} \\ f_n l_{n+1} & \text{otherwise} \end{cases} \\ &= \begin{cases} f_{(n+1)+n} - f_{(n+1)-n} & \text{if } n \text{ is odd} \\ f_{n+(n+1)} - f_{n-(n+1)} & \text{otherwise} \end{cases} \end{aligned}$$

$$= \begin{cases} f_{2n+1} - f_1 & \text{if } n \text{ is odd} \\ f_{2n+1} - f_{-1} & \text{otherwise} \end{cases}$$
$$= f_{2n+1} - 1,$$

as desired. ∎

It follows from this example that $F_{n+[n \pmod 2)]} L_{n+1-[n \pmod 2)]} = \sum_{j=1}^{n} F_{2j} = F_{2n+1} - 1$. R. Euler found this special case in 2001 [153].

Identity (32.12) has a delightful application. To see this, let $a + b$ be even. Then $a - b$ is also even. Let $m = (a + b)/2$ and $n = (a - b)/2$. Then identity (32.11) implies that $f_a^2 - f_b^2 = f_{a+b} f_{a-b}$; that is, $(f_a + f_b)(f_a - f_b) = f_{a+b} f_{a-b}$.

Identity (32.12) also has a Lucas counterpart:

$$l_{m+n}^2 - l_{m-n}^2 = (x^2 + 4) f_{2m} f_{2n}; \tag{32.34}$$

see Exercise 32.21.

In addition, we can also show that $l_a^2 - l_b^2 = l_{a+b} l_{a-b} + 4(-1)^a$, where a is odd and b is even; see Exercise 32.25. Consequently, $(l_a + l_b)(l_a - l_b) = l_{a+b} l_{a-b} + 4(-1)^a$ and $L_a^2 - L_b^2 = L_{a+b} L_{a-b} + 4(-1)^a$, where a is odd and b is even.

Next we study some interesting and useful divisibility properties that are extensions of the familiar divisibility properties of Fibonacci numbers.

32.5 DIVISIBILITY PROPERTIES

In 1998, S. Rabinowitz of Westford, Massachusetts, established that *no* Lucas polynomial l_n is divisible by $x - 1$ [392]. This result, an application of the well-known factor theorem in basic algebra, is a special case of the following theorem; it was discovered independently by L. Somer of The Catholic University of America, Washington, D.C. [468].

Theorem 32.1 (Somer, 1998 [468]). *Let a be a nonzero real number. Then $(x - a) \nmid l_n$, where $n \geq 0$; and $(x - a) \nmid f_n$, where $n \geq 1$.*

Proof. By the factor theorem, $(x - a) | l_n$ if and only if $l_n(a) = 0$; likewise, $(x - a) | f_n$ if and only if $f_n(a) = 0$. Since $l_n(-a) = (-1)^n l_n(a)$ and $f_n(-a) = (-1)^{n-1} f_n(a)$, it follows that $|l_n(-a)| = l_n(a)$ and $|f_n(-a)| = f_n(a)$. Consequently, it suffices to show that $l_n(a) > 0$ when $n \geq 0$; and $f_n(a) > 0$ when $n \geq 1$, where $a > 0$.

Suppose $a > 0$. Since $l_0(a) = 2$ and $l_1(a) = a$, $l_n(a) > 0$ when $n = 0$ or 1. Likewise, $f_0(a) = 0$ and $f_1(a) = 1$; so $f_1(a) > 0$ when $n = 1$.

Since both $l_n(a)$ and $f_n(a)$ satisfy the same recurrence $g_{n+2} = a g_{n+1} + g_n$, it follows by the strong version of PMI that $l_n(a) > 0$ when $n \geq 0$; and $f_n(a) > 0$ when $n \geq 1$.

Thus the desired results follow. ■

Let m and n be any positive integers. Clearly, $f_m | f_m$ and $f_m | f_{2m}$. Suppose $f_m | f_{mn}$, where $n \geq 2$. By the Fibonacci addition formula (32.9), $f_{m(n+1)} = f_{mn+m} = f_{mn+1} f_m + f_{mn} f_{m-1}$. This implies $f_m | f_{m(n+1)}$. Thus, by PMI, $f_m | f_{mn}$ for every $n \geq 1$. This gives the next result.

Theorem 32.2. *If $m|n$, then $f_m | f_n$.* ■

Consequently, if $m|n$, then $p_m | p_n$, $F_m | F_n$, and $P_m | P_n$.

Conversely, we can establish that if $f_m | f_n$, then $m|n$. To this end, we need the following lemma.

Lemma 32.1. *If $f_m | f_n$, then $f_m | f_{n-qm}$, where $0 \leq q \leq \lfloor n/m \rfloor$.*

Proof. Clearly, the result is true when $q = 0$. Suppose it is true for an arbitrary nonnegative integer $< q$. Since $f_n = f_{(n-qm)+qm} = f_{n-qm+1} f_{qm} + f_{n-qm} f_{qm-1}$, it follows that $f_m | f_{n-qm} f_{qm-1}$.

By the Cassini-like formula, $(f_{qm}, f_{qm-1}) = 1$. Since $f_m | f_{qm}$ by Theorem 32.2, it follows that $(f_m, f_{qm-1}) = 1$. Consequently, $f_m | f_{n-qm}$.

Thus, by PMI, the result is true for every q, where $0 \leq q \leq \lfloor n/m \rfloor$. ■

We are now ready to establish the aforementioned statement.

Theorem 32.3. *If $f_m | f_n$, then $m|n$.*

Proof. The statement is clearly true when $m = 1$; so assume $m \geq 2$. By the division algorithm, let $n = mq + r$, where $0 \leq r < m$. Since $f_m | f_n$, by Lemma 32.1, $f_m | f_{n-qm}$; that is, $f_m | f_r$, where $\deg f_r < \deg f_m$[†]. This is impossible, unless $r = 0$. Then $n = mq$ and hence $m|n$, as claimed. ■

It follows by Theorems 32.2 and 32.3 that $f_m | f_n$ if and only if $m|n$. Consequently, $p_m | p_n$ if and only if $m|n$.

Next we show that $(f_m, f_n) = f_{(m,n)}$. To this end, we need the following result.

Lemma 32.2. *Let $n = mq + r$, where $0 \leq r < m$. Then $(f_n, f_m) = (f_m, f_r)$.*

Proof. Since $(f_{qm-1}, f_m) = 1$, by the Fibonacci addition formula, we then have

$$\begin{aligned}
(f_n, f_m) &= (f_{qm+r}, f_m) \\
&= (f_{mq} f_{r+1} + f_{mq-1} f_r, f_m) \\
&= (f_{mq-1} f_r, f_m) \\
&= (f_r, f_m),
\end{aligned}$$

as desired. ■

[†] $\deg f_k$ denotes the degree of the polynomial f_k.

For example, let $n = 8$ and $m = 6$. Then $r = 2$ and $(f_8, f_6) = (f_2, f_6) = (x, x^5 + 4x^3 + 3x) = x$.

We are now ready for the next elegant and beautiful result.

Theorem 32.4. $(f_n, f_m) = f_{(n,m)}$.

Proof. Without any loss of generality, assume that $n \geq m$. Using the traditional Euclidean algorithm with n as the dividend and m as the divisor, we get the following array of equations:

$$\begin{aligned}
n &= q_0 m + r_1, & 0 &\leq r_1 < m\\
m &= q_1 r_1 + r_2, & 0 &\leq r_2 < r_1\\
r_1 &= q_2 r_2 + r_3, & 0 &\leq r_3 < r_2\\
&\vdots\\
r_{n-2} &= q_{n-1} r_{n-1} + r_n, & 0 &\leq r_n < r_{n-1}\\
r_{n-1} &= q_n r_n.
\end{aligned}$$

So $r_n = (n, m)$.

It follows by Lemma 32.2 that $(f_n, f_m) = (f_m, f_{r_1}) = f_{r_1}, f_{r_2}) = \cdots = (f_{r_{n-1}}, f_{r_n})$. Since $r_n | r_{r_{n-1}}$, it follows by Theorem 32.3 that $(f_{r_{n-1}}, f_{r_n}) = f_{r_n} = f_{(n,m)}$. Thus $(f_n, f_m) = f_{(n,m)}$, as claimed. ■

For example, $(f_6, f_3) = f_{(6,3)} = f_3 = x^2 + 1$; and $(f_{10}, f_6) = f_{(10,6)} = f_2 = x$. It follows from Theorem 32.4 that

1) $(p_n, p_m) = p_{(n,m)}$, $(F_n, F_m) = F_{(n,m)}$, and $(P_n, P_m) = P_{(n,m)}$.
   For example, $(P_{15}, P_6) = (195025, 70) = 5 = P_3 = P_{(15,6)}$.
2) $(f_n, f_m) = 1$ and only if $(n, m) = 1$.
3) If $(n, m) = 1$, then $f_n f_m | f_{nm}$; see Exercise 32.36.

For example, $f_2 | f_6$ and $f_3 | f_6$. Since $(f_2, f_3) = 1$, it follows that $f_2 f_3 | f_6$. Notice that $f_6 = x^5 + 4x^3 + 3x = x(x^2 + 1)(x^2 + 3) = (x^2 + 3) f_2 f_6$.

EXERCISES 32

Prove each, where $\nu = (-1)^{a+b+c}$.

1. Let $Q = \begin{bmatrix} x & 1\\ 1 & 0\end{bmatrix}$. Then $Q^n = \begin{bmatrix} f_{n+1} & f_n\\ f_n & f_{n-1}\end{bmatrix}$, where $n \geq 1$.
2. Let $M = \begin{bmatrix} 2x & 1\\ 1 & 0\end{bmatrix}$. Then $M^n = \begin{bmatrix} p_{n+1} & p_n\\ p_n & p_{n-1}\end{bmatrix}$, where $n \geq 1$.

3. $p_{n+1}p_{n-1} - p_n^2 = (-1)^n$.
4. $q_{n+1}q_{n-1} - q_n^2 = (-1)^{n-1}(x^2+1)$.
5. $x\sum_{j=1}^{n} f_{2j} = f_{2n+1} - 1$.
6. $2x\sum_{j=1}^{n} p_{2j} = p_{2n+1} - 1$.
7. $f_n = f_{k+1}f_{n-k} + f_k f_{n-k-1}$.
8. $l_n = f_{k+1}l_{n-k} + f_k l_{n-k-1}$.
9. $f_{mn} > f_m f_n$, where $m, n \geq 2$ and $x \geq 1$.
10. $xf_{m+n} = f_{m+1}f_{n+1} - f_{m-1}f_{n-1}$.
11. $xl_{m+n} = f_{m+1}l_{n+1} - f_{m-1}l_{n-1}$.
12. $(x^2+4)f_{m+n} = l_{m+1}l_n + l_m l_{n-1}$.
13. $f_{m+n} - (-1)^n f_{m-n} = l_m f_n$.
14. $l_{m+n} = f_{m+1}l_n + f_m l_{n-1}$.
15. $l_{m-n} = (-1)^n(f_{m+1}l_n - f_m l_{n+1})$.
16. $l_{m+n} - (-1)^n l_{m-n} = (x^2+4)f_m f_n$.
17. $l_{n+4k} - l_n = (x^2+4)f_{2k}f_{n+2k}$.
18. $2l_{m+n} = l_m l_n + (x^2+4)f_m f_n$.
19. $l_{m+n} + l_{m-n} = \begin{cases} (x^2+4)f_m f_n & \text{if } n \text{ is odd} \\ l_m l_n & \text{otherwise.} \end{cases}$
20. $l_{m+n} - l_{m-n} = \begin{cases} l_m l_n & \text{if } n \text{ is odd} \\ (x^2+4)f_m f_n & \text{otherwise.} \end{cases}$
21. $l_{m+n}^2 - l_{m-n}^2 = (x^2+4)f_{2m}f_{2n}$.
22. $L_{m+n}^2 - L_{m-n}^2 = 5F_{2m}F_{2n}$.
23. $l_{m+n}^2 + l_{m-n}^2 = l_{2m}l_{2n} + 4(-1)^{m+n}$.
24. $l_{m+n}^2 + (x^2+4)f_{m-n}^2 = l_{2m}l_{2n}$.
25. $l_a^2 - l_b^2 = l_{a+b}l_{a-b} + 4(-1)^a$, where a is odd and b is even.
26. $q_{m+n}^2 - q_{m-n}^2 = 4(x^2+1)p_{2m}p_{2n}$.
27. $Q_{m+n}^2 - Q_{m-n}^2 = 2P_{2m}P_{2n}$.
28. $(x^2+4)(f_{n+1}^2 - f_n^2) = xl_{2n+1} + 4(-1)^n$.
29. $f_{2^{n+1}-1} + 1 = f_{2^n-1}l_{2^n}$.
30. $x\prod_{k=1}^{n} l_{2^k} = f_{2^{n+1}}$.

31. $2f_{4n} = (x^2+4)f_n^3 l_n + f_n l_n^3$.
32. $8l_{4n} = l_n^4 + 6(x^2+4)l_n^2 f_n^2 + (x^2+4)^2 f_n^4$.
33. $xl_n + 2(-1)^m l_{n-2m-1} = (x^2+4)(f_{n-m}f_{m+1} - f_{n-m-1}f_m)$.
34. $xl_{m+4} = (x^4+4x^2+2)l_{m+1} - (x^2+1)(x^2+4)f_m$.
35. $l_{a+b+c}l_{a+b-c}l_{b+c-a}l_{c+a-b} = l_{2a}l_{2b}l_{2c} + \nu\left(l_{2a}^2 + l_{2b}^2 + l_{2c}^2\right) - 4\nu$.
36. If $(n,m)=1$, then $f_n f_m | f_{nm}$.
37. Suppose the sequence $\{u_n\}$ is defined recursively by $au_n + bxu_{n-1} + cu_{n-2} = 0$, where $u_0 = 0$, $u_1 = 1$, and $u_n = u_n(x)$. Then $au_{m+n+1} = bu_{m+1}u_{n+1} - cu_m u_n$, where $m, n \geq 0$. (This generalizes the Fibonacci addition formula.)
38. $x\sum_{k=1}^{n} l_{4k+3} = f_{2n+4}l_{2n+1} - (x^3+2x)x$.
39. $\sum_{k=1}^{\infty} \frac{f_{2k-1}^2}{l_{2k}^2 - 1} = \frac{x^2+5}{(x+1)(x+3)(x^2+4)}$.

Evaluate each sum S.

40. $\sum_{k=1}^{\infty}\left(\frac{1}{f_{2k}f_{2k+1}} + \frac{1}{f_{2k}f_{2k+2}} - \frac{1}{f_{2k+1}f_{2k+2}}\right)$.
41. $\sum_{k=1}^{\infty}\left(\frac{1}{l_{2k}l_{2k+1}} + \frac{1}{l_{2k}l_{2k+2}} - \frac{1}{l_{2k+1}l_{2k+2}}\right)$.

33

COMBINATORIAL MODELS II

You don't really understand something
if you only understand it one way.
–Marvin Minsky (1927–2016)

Recall that the number of compositions of a positive integer n with summands 1 and 2 is F_{n+1}, where $n \geq 1$ [287]. An immediate consequence of this fundamental fact is that a $1 \times n$ board can be tiled with square tiles and dominoes in F_{n+1} distinct ways. Likewise, a circular board with n labeled cells can be tiled with (curved) square tiles and (curved) dominoes in L_n different ways.

We now extend these ideas for constructing combinatorial models for Fibonacci and Lucas polynomials. Again, we *omit* the argument from the functional notation for brevity and convenience, where the omission causes *no* confusion. $\boxed{\Leftarrow}$

Once again, the fundamental tool we employ throughout our discussion is the well-known *Fubini's principle*, named after the Italian mathematician Guido Fubini (1879–1943): *Counting the objects in a set in two different ways yields the same result*.

33.1 A MODEL FOR FIBONACCI POLYNOMIALS

To begin with, we introduce the concept of the *weight* of a tile. We define the *weight* of a square (tile) to be x: $\boxed{x}$; and that of a domino to be 1: $\boxed{\quad 1 \quad}$. The

Fibonacci and Lucas Numbers with Applications, Volume Two. Thomas Koshy.

weight of a tiling is the product of the weights of all tiles in the tiling. For instance, the weight of the tiling $\boxed{x\,|\,x\,|\,x\,|\quad 1\quad}$ is x^3; and that of the tiling $\boxed{\quad 1\quad|\quad 1\quad}$ is 1. We define the *weight* of the empty tiling to be one.

Figure 33.1 shows the tilings and the sum of their weights of a $1 \times n$ board, where $0 \leq n \leq 5$. Using the experimental data, we conjecture that the sum of the weights is f_{n+1}. The next theorem confirms this observation.

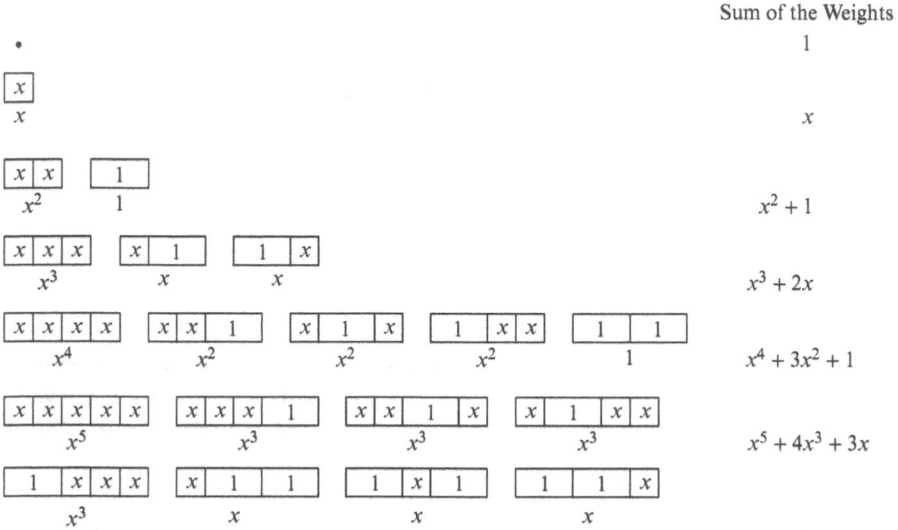

Figure 33.1. Tilings of a $1 \times n$ board, where $0 \leq n \leq 5$.

Theorem 33.1. *The sum of the weights of tilings of a* $1 \times n$ *board is* f_{n+1}, *where* $n \geq 0$.

Proof. Let S_n denote the sum of the weights of tilings of the board. Clearly, $S_0 = 1 = f_1$ and $S_1 = x = f_2$.

We now show that S_n satisfies the Fibonacci polynomial recurrence. To this end, consider an arbitrary tiling of the board. Suppose it ends in a square: $\underbrace{\text{subtiling}}_{\text{length } n-1}\boxed{x}$. The sum of the weights of such tilings is xS_{n-1}.

On the other hand, suppose the tiling ends in a domino: $\underbrace{\text{subtiling}}_{\text{length } n-2}\boxed{\quad 1\quad}$. The sum of the weights of such tilings is $1 \cdot S_{n-2} = S_{n-2}$.

Consequently, the sum of the weights of all tilings of the board is $xS_{n-1} + S_{n-2}$, where $n \geq 2$. So $S_n = xS_{n-1} + S_{n-2}$. Thus S_n satisfies the Fibonacci recurrence; so $S_n = f_{n+1}$, as desired. ■

The next two results follow from this theorem.

Corollary 33.1. *The number of tilings of a* $1 \times n$ *board is* F_{n+1}, *where* $n \geq 0$. ■

Corollary 33.2. *Suppose the weight of a square is* $2x$ *and that of a domino is* 1. *Then the sum of the weights of tilings of a* $1 \times n$ *board is* p_{n+1}, *where* $n \geq 0$. ■

Suppose a tiling of a $1 \times n$ board contains exactly k dominoes. Then it contains $n - 2k$ squares, and a total of $k + (n - 2k) = n - k$ tiles. The weight of such a tiling is $x^{n-2k} \cdot 1^k = x^{n-2k}$. Since the k dominoes can be placed among the $n - k$ tiles in $\binom{n-k}{k}$ different ways, there are $\binom{n-k}{k}$ such tilings. So the sum of the weights of tilings with exactly k dominoes is $\binom{n-k}{k}x^{n-2k}$, where $0 \leq k \leq \lfloor n/2 \rfloor$. Thus the sum of the weights of all tilings of length n is $\sum_{k=0}^{\lfloor n/2 \rfloor} \binom{n-k}{k}x^{n-2k}$. This, coupled with Theorem 33.1, gives the explicit formula (31.27) for f_{n+1}.

Example 33.1. *Let* $n \geq 0$. *Then* $f_{n+1} = \sum_{k=0}^{\lfloor n/2 \rfloor} \binom{n-k}{k}x^{n-2k}$. ■

For example, consider the eight tilings of a 1×5 board; see Figure 33.1. There is one tiling with no dominoes; its weight is x^5. There are four tilings with exactly one domino each; the sum of their weights is $4x^3$; and there are three tilings with exactly two dominoes each; the sum of their weights is $3x$. Thus the sum of the weights of the tilings is $x^5 + 4x^3 + 3x = f_6$.

In the next example, we pursue another summation formula. This time, the strategy is slightly different.

Example 33.2. *Confirm combinatorially the identity* $\sum_{k=1}^{n} \binom{n}{k} f_k x^k = f_{2n}$.

Proof. Consider a $1 \times (2n - 1)$ board. By Theorem 33.1, the sum of the weights of its tilings is f_{2n}.

We now compute the sum in a different way. Since the length of the board is odd, each tiling must have an odd number of squares. So each tiling must contain at least $(n - 1) + 1 = n$ tiles.

Let T_k be a tiling with k squares (and hence $n - k$ dominoes) among the first n tiles. The k squares can be placed among the n tiles in $\binom{n}{k}$ ways; so there are $\binom{n}{k}$ such tilings T_k.

The first n tiles partition T_k into two subtilings, one of length $k + 2(n - k) = 2n - k$ and the other of length $k - 1$: $\underbrace{\text{subtiling}}_{k \text{ squares}}\underbrace{\text{subtiling}}_{\text{length } k-1}$. Since each square has weight x and the sum of the weights of a tiling of length $k-1$ is f_k, the weight of tiling T_k is $f_k x^k$. Since there are $\binom{n}{k}$ tilings T_k, the sum of the weights of all such tilings equals $\binom{n}{k} f_k x^k$.

Consequently, the sum of the weights of all tilings of the board equals $\sum_{k=0}^{n} \binom{n}{k} f_k x^k$. Since $f_0 = 0$, this, coupled with the initial result, yields the formula. ■

Next we establish a double-summation formula for even-numbered Fibonacci polynomials f_{2n+2}, using a tiling board of odd length. To this end, first we make an important observation. Since the length of the board is odd, each tiling must contain an odd number of squares. So the tiling must contain a special square M that has an equal number of squares on either side. We call it the *median square*.

For example, consider the tiling of the 1×11 board in Figure 33.2. Its median square occurs in cell $m = 8$.

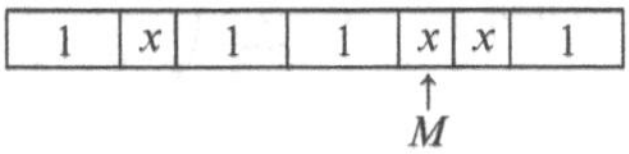

Figure 33.2.

We are now ready to present the summation formula for f_{2n+2}. Its proof is short, but still beautiful.

Example 33.3. *Establish the identity*

$$f_{2n+2} = \sum_{\substack{i,j \geq 0 \\ i+j \leq n}} \binom{n-i}{j} \binom{n-j}{i} x^{2n-2i-2j+1}.$$

Proof. Consider a $1 \times (2n+1)$ board. The sum of the weights of its tilings is f_{2n+2}.

Consider an arbitrary tiling of the board. Suppose there are i dominoes to the left of its median square M and j dominoes to its right: $\underbrace{i \text{ dominoes}}_{\text{on left}}\ \underset{\overset{\uparrow}{M}}{\boxed{x}}\ \underbrace{j \text{ dominoes}}_{\text{on right}}$. Then there are $2n+1-2i-2j$ squares in the tiling; so there are $n-i-j$ squares on either side of M. Consequently, there are $(n-i-j)+i = n-j$ tiles to the left of M. So the i dominoes can be placed to the left of M in $\binom{n-j}{i}$ different ways. The weight of the corresponding subtiling is $\binom{n-j}{i} x^{n-i-j}$.

Similarly, the weight of the subtiling to the right of M is $\binom{n-i}{j}x^{n-i-j}$.
As a result, the weight of the tiling is $\binom{n-j}{i}x^{n-i-j}\cdot x\cdot\binom{n-i}{j}x^{n-i-j} = \binom{n-i}{j}\binom{n-i}{j}x^{2n-2i-2j+1}$, where $0 \le i+j \le n$.

Thus the sum of the weights of all tilings of the board is $\sum_{i,j\ge 0}\binom{n-i}{j}\binom{n-j}{i}x^{2n-2i-2j+1}$. This, together with the earlier sum, gives the desired result. ■

For a specific example, let $n = 2$. Figure 33.3 shows the tilings of a 1×5 board, where the up arrows point to the median square M in each tiling.

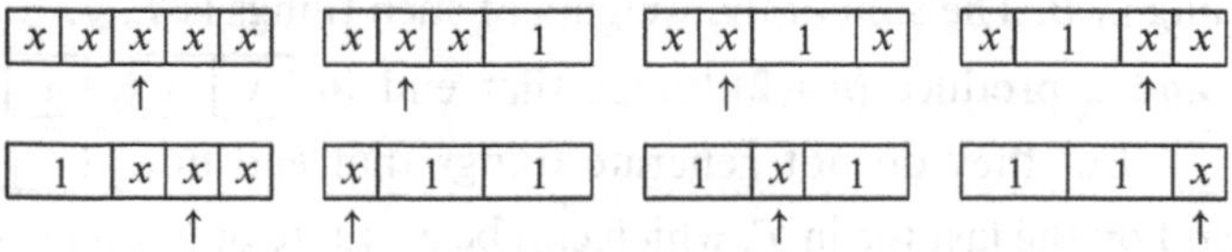

Figure 33.3.

Let m denote the location of M. Table 33.1 shows the possible values of m, the corresponding values of i and j, and the weight(s) of the corresponding tiling(s). It follows from the table that the sum of the weights of all tilings of the board is

$$x^5 + 4x^3 + 3x = f_6 = \sum_{0\le i+j\le 2}\binom{2-i}{j}\binom{2-j}{i}x^{5-2i-2j}.$$

TABLE 33.1.

m	i	j	Sum of the Weight(s) of Tiling(s)
1	0	2	x
2	0	1	$2x^3$
3	0	0	x^5
	1	1	x
4	1	0	$2x^3$
5	2	0	x

↑
Cumulative sum = $x^5 + 4x^3 + 3x$

In the next example, we establish combinatorially the identity $f_{n+2} + f_{n-2} = (x^2+2)f_n$; equivalently, $f_{n+3} + f_{n-1} = (x^2+2)f_{n+1}$. We accomplish this by establishing a *one-to-three correspondence* between two appropriate sets of tilings.

Example 33.4. *Prove combinatorially that* $f_{n+3} + f_{n-1} = (x^2+2)f_{n+1}$.

Proof. Let A, B, and C denote the sets of tilings of a board of length $n, n+2$, and $n-2$, respectively. Clearly, the sums of the weights of their tilings are f_{n+1}, f_{n+3}, and f_{n-1}, respectively. So the sum of the weights of the tilings in $B \cup C$ is $f_{n+3} + f_{n-1}$, where $X \cup Y$ denotes the *union* of sets X and Y.

We now develop a three-step algorithm to construct a one-to-three correspondence between A, and $B \cup C$. To this end, consider an arbitrary tiling T in A.

Step 1. Append two squares at the end of T: $\underbrace{\text{tiling } T}_{\text{length } n}\boxed{x \mid x}$. This creates a unique tiling (of length $n+2$) in B. The sum of the weights of such tilings is $x^2 f_{n+1}$.

Step 2. Append a domino at the end of T: $\underbrace{\text{tiling } T}_{\text{length } n}\boxed{\quad 1 \quad}$. This also produces a unique tiling in B. The sum of the weights of such tilings is f_{n+1}.

Steps 1 and 2 produce $(n+2)$-tilings that end in $\boxed{x \mid x}$, $\boxed{x \mid \quad 1 \quad}$, or $\boxed{\quad 1 \quad \mid \quad 1 \quad}$. But they do not generate tilings that end in $\boxed{\quad 1 \quad \mid x}$. Such tilings depend on the last tile in T, which can be a square or a domino. This hole takes us to Step 3, which has two parts.

Step 3A. Suppose T ends in a square: $\underbrace{\text{subtiling}}_{\text{length } n-1}\boxed{x}$. Then insert a domino immediately to the left of the square; this produces exactly one $(n+2)$-tiling that ends in $\boxed{\quad 1 \quad \mid x}$: $\underbrace{\text{subtiling}}_{\text{length } n-1}\boxed{\quad 1 \quad \mid x}$. The sum of the weights of such tilings is $x \cdot 1 \cdot f_n = x f_n$.

Step 3B. On the other hand, suppose T ends in a domino: $\underbrace{\text{subtiling}}_{\text{length } n-2}\boxed{\quad 1 \quad}$. Deleting the domino yields a single tiling in C: $\underbrace{\text{subtiling}}_{\text{length } n-2}$. The sum of the weights of such the tilings is f_{n-1}.

Clearly, every step produces a single element in $B \cup C$. Thus, by Steps 1–3, the sum of the weights of the tilings in $B \cup C$ equals $x^2 f_{n+1} + f_{n+1} + (x f_n + f_{n-1}) = (x^2 + 2) f_{n+1}$.

Is this algorithm reversible? In other words, does every tiling T in B or S in C generate a unique element in A in the reverse order? The answer is yes. To see this, suppose T ends in $\boxed{x \mid x}$; deleting this pair creates a single element in A. If T ends in $\boxed{\quad 1 \quad}$, then dropping it yields exactly one element in A. If T ends in $\boxed{\quad 1 \quad \mid x}$, then removing the domino also creates a single element in A. Finally, appending a domino at the end of each S gives a unique element in A. Thus the algorithm is reversible.

Consequently, the two sums of weights are equal; that is, $f_{n+3} + f_{n-1} = (x^2 + 2)f_{n+1}$, as claimed. ■

We now illustrate Steps 1–3 for the case $n = 3$. Figure 33.4 shows the tilings of boards of length 3, 5, and 1, and the sums of their weights.

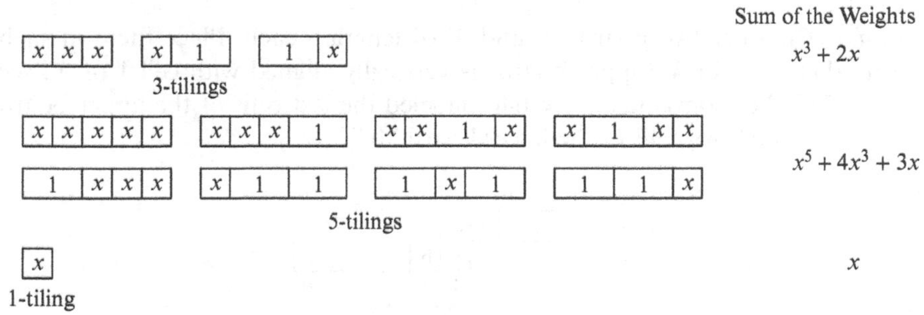

Figure 33.4.

Step 1. Appending two squares at the end of each 3-tiling yields three 5-tilings; see Figure 33.5. The sum of their weights is $x^5 + 2x^3 = x^4 f_4$.

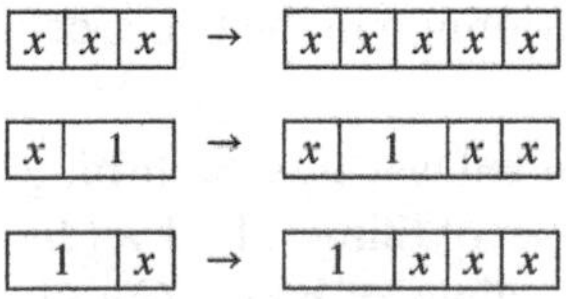

Figure 33.5.

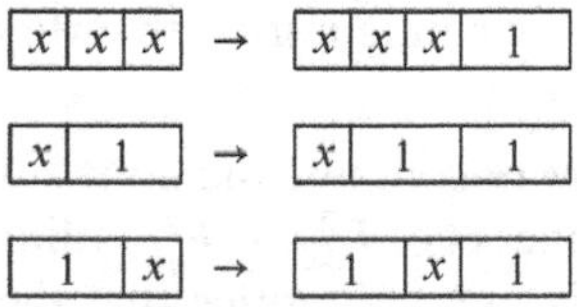

Figure 33.6.

Step 2. Appending a domino at the end of each 3-tiling produces three more 5-tilings; see Figure 33.6. The sum of their weights is $x^3 + 2x = f_4$.
Step 3A. Two 3-tilings end in a square. Inserting a domino immediately to the left of the last square creates two additional 5-tilings; see Figure 33.7. The sum of their weights is $x^3 + x = xf_3$.

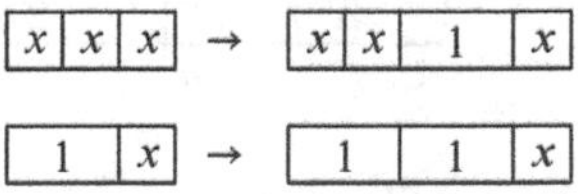

Figure 33.7.

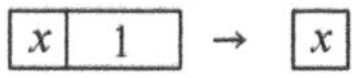

Figure 33.8.

Step 3B. Exactly one 3-tiling ends in a domino. Deleting the domino yields the one 1-tiling; see Figure 33.8. Its weight is $x = f_2$.

The sum of the weights of all 5- or 1-tilings equals $f_6 + f_2 = x^5 + 4x^3 + 4x = (x^2 + 2)(x^3 + 3x) = (x^2 + 2)f_4$, as expected.

In the next example, we establish combinatorially the Lucas-like identity $f_{n+1}^2 - f_{n-1}^2 = xf_{2n}$. Although the algebraic proof takes only a few seconds, the combinatorial proof is a bit complicated, yet still elegant. Our strategy involves employing two boards of length n.

Example 33.5. *Prove that* $f_{n+1}^2 - f_{n-1}^2 = xf_{2n}$.

Proof. Consider two boards X and Y of length n each. Place them in such a way that cell n of X (upper board) is vertically aligned with cell 1 of Y; see Figure 33.9. For convenience, we have labeled the last cells of the upper board "a" and the first two cells of the lower board "b."

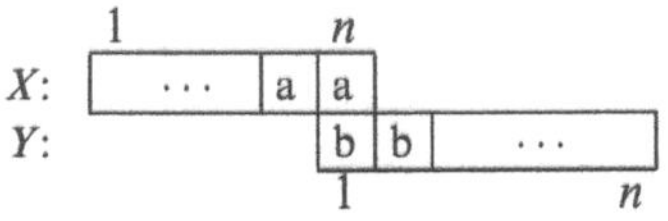

Figure 33.9.

Clearly, the configuration (X, Y) in Figure 33.9 can be tiled in F_{n+1}^2 ways; the sum of the weights of those tilings is f_{n+1}^2. Interestingly, F_{n-1}^2 of them have a domino occupying the two a-cells and the two b-cells; the sum of the weights of such tilings is f_{n-1}^2. Consequently, the sum of the weights of the $F_{n+1}^2 - F_{n-1}^2$ tilings of the configuration that do *not* have a domino in the two a-cells or b-cells is $f_{n+1}^2 - f_{n-1}^2$.

We now re-compute the same sum in a different way by establishing a bijection between the set of $F_{n+1}^2 - F_{n-1}^2$ tilings and the set of tilings of a $1 \times (2n-1)$ board C. To this end, first notice that there are F_{2n} tilings of board C and the sum of their weights is f_{2n}.

We devise an algorithm to accomplish this in three steps, where S denotes a square and D a domino:

Step 1. Suppose a square occupies both cell n of X and cell 1 of Y. Then move the tiles of X into the first n cells of board C, and place the tiles in the last $n-1$ cells of Y into those of C:

Step 2. Suppose a domino occupies the two a-cells of X, and a square occupies cell 1 of Y. Then place the tiles of X into cells 1 through n of C, and the tiles of cells 2 through n of Y into cells $n+1$ through $2n-1$ of C:

Step 3. Suppose a square occupies cell n of X, and a domino occupies the two b-cells of Y. Then place the tiles of cells 1 through $n-1$ of X into cells 1 through $n-1$ of C, and the tiles in Y into the remaining n cells of C:

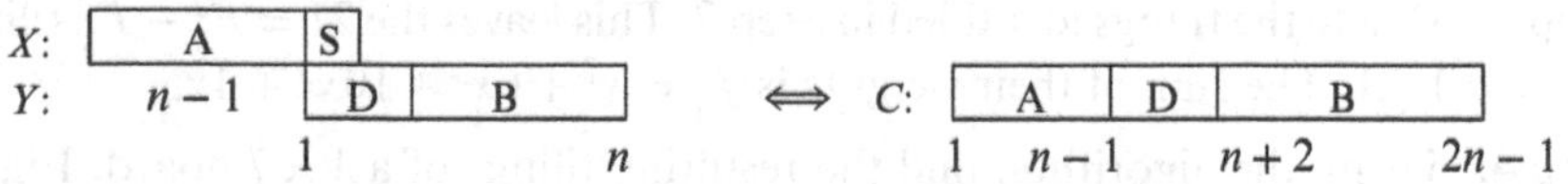

Steps 1–3 of this algorithm establish a unique match C for each of the $F_{n+1}^2 - F_{n-1}^2$ tilings of the configuration (X, Y) in Figure 33.9. Fortunately, the matching works in the reverse order as well; that is, given a tiling C, we can easily recover the corresponding pair (X, Y). Thus the algorithm is reversible; so the matching between the set of $F_{n+1}^2 - F_{n-1}^2$ tilings and the set of tilings of board C is a bijection.

Unfortunately, this bijection does *not* imply that the sums of their weights are equal; note that $f_{n+1}^2 - f_{n-1}^2 \neq f_{2n}$. We need to make a small correction. When a tiling of C is created from the configuration (X, Y), a square is lost in the process in each case. Consequently, in order to recover the weight $f_{n+1}^2 - f_{n-1}^2$ from the sum of the weights of tilings of C, we need to account for the missing square by multiplying f_{2n} with x. Thus $f_{n+1}^2 - f_{n-1}^2 = xf_{2n}$, as desired. ■

We now demonstrate the steps using a specific example. Let $n = 4$.

Step 1. Find the different tilings of the following configuration:

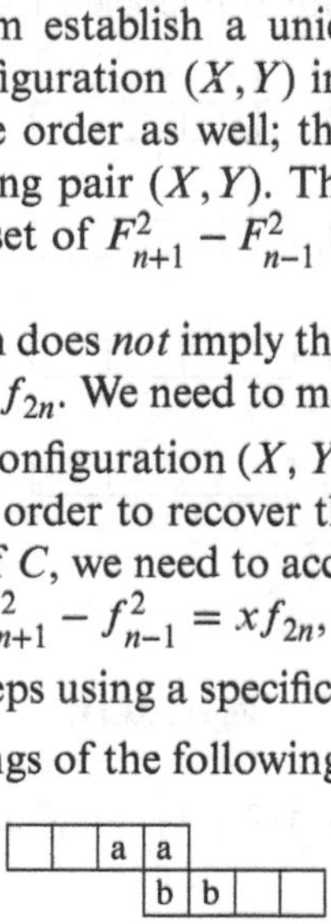

Figure 33.10 shows the $25 = F_5^2$ tilings of the configuration. For convenience, 21 of them are labeled 1–21.

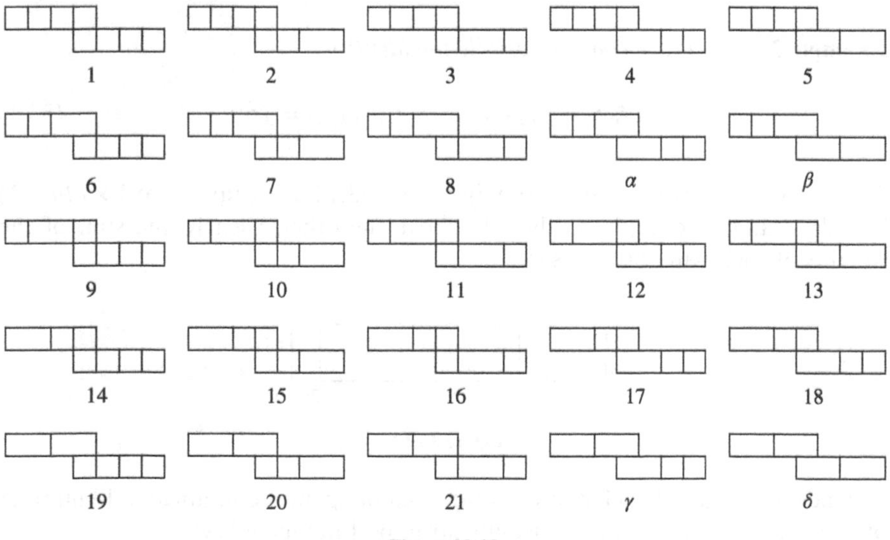

Figure 33.10.

Step 2. Identify those tilings with a domino in the two a-cells and the two b-cells. There are $4 = F_3^2$ such tilings. For easy identification, we have labeled them α, β, γ, and δ in Figure 33.10. The sum of their weights is $(x^2+1)^2$.

Step 3. Delete the tilings identified in Step 2. This leaves the $21 = F_5^2 - F_3^2$ tilings marked 1–21. The sum of their weights is $f_8 = x^7 + 6x^5 + 10x^3 + 4x$.

Step 4. Using the algorithm, find the resulting tilings of a 1×7 board. Figure 33.11 shows such tilings; they are also labeled 1–21 for easy matching.

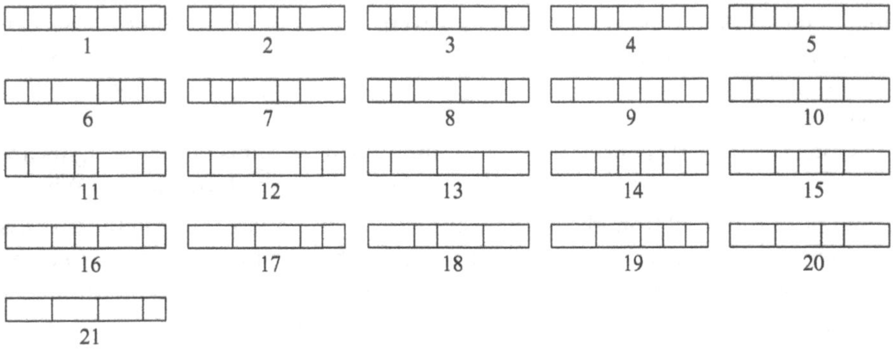

Figure 33.11.

Step 5. Then $f_5^2 - f_3^2 = (x^4 + 3x^2 + 1)^2 - (x^2+1)^2 = x^8 + 6x^6 + 10x^4 + 4x^2 = xf_8$, as expected.

The combinatorial proof of the next summation formula employs a different strategy. It involves a pair of boards of length $2n-1$ each. The proof is quite beautiful.

Example 33.6. *Prove combinatorially the identity*

$$x(f_0 f_1 + f_1 f_2 + \cdots + f_{2n-1} f_{2n}) = f_{2n}^2. \tag{33.1}$$

Proof. Consider the configuration in Figure 33.12 made up of two $1 \times (2n-1)$ boards A and B, one placed directly above the other. Clearly, the sum of the weights of such pairs of tilings is f_{2n}^2.

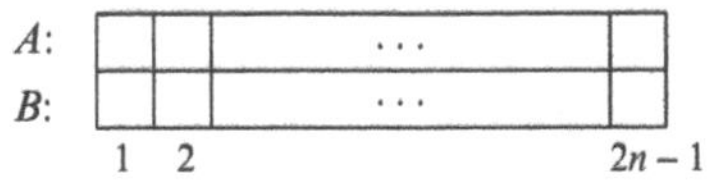

Figure 33.12.

Since the length of each board is odd, each tiling must contain an odd number of squares. In addition, they must appear in odd-numbered cells.

Let $2k - 1$ be the label of the first column containing a square. Suppose it occurs in board A; see Figure 33.13. The subtilings in cells 1 through $2k - 2$ must be made up of dominoes in either board. The weights of the subtilings of lengths $2n - 2k$ and $2n - 2k + 1$ are $f_{2n-2k+1}$ and $f_{2n-2k+2}$, respectively. Consequently, the weight of the corresponding tiling of the configuration is $xf_{2n-2k+1}f_{2n-2k+2}$.

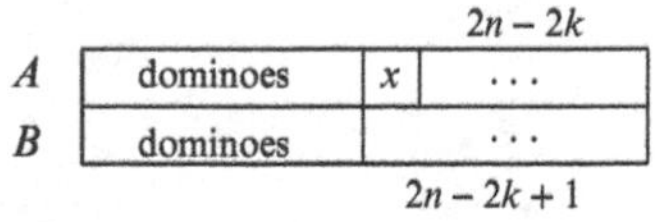

Figure 33.13.

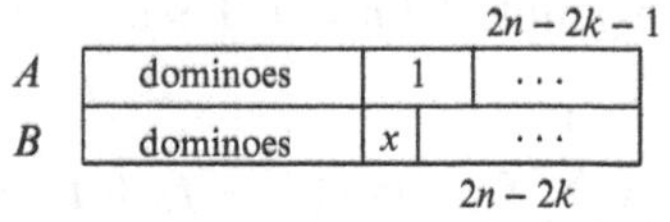

Figure 33.14.

On the other hand, suppose the first square is *not* in A. Then a domino must occupy cells $2k - 1$ and $2k$ of A; see Figure 33.14. So cell $2k - 1$ of B must contain a square. So the weight of the corresponding tiling is $xf_{2n-2k}f_{2n-2k+1}$.

Consequently, the sum of the weights of such tilings of the configuration is $xf_{2n-2k+1}f_{2n-2k+2} + xf_{2n-2k}f_{2n-2k+1}$, where $1 \le k \le n$. Thus the sum of the weights of all pairs of tilings equals

$$x\sum_{k=1}^{n}\left(f_{2n-2k}f_{2n-2k+1} + f_{2n-2k+1}f_{2n-2k+2}\right) = x\sum_{k=0}^{n-1}\left(f_{2k}f_{2k+1} + f_{2k+1}f_{2k+2}\right).$$

This sum, coupled with the initial one, yields the desired result. ■

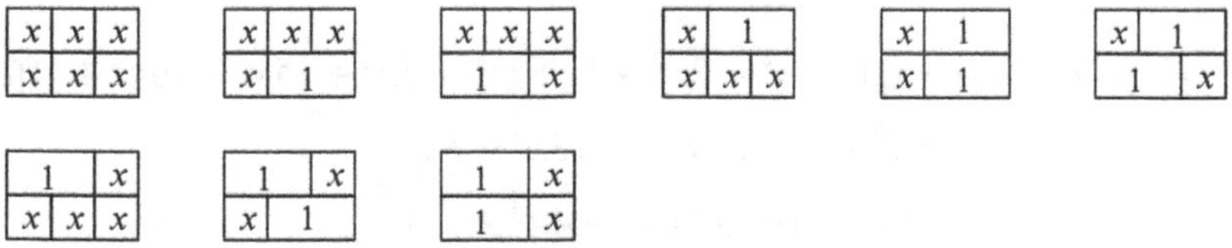

Figure 33.15.

For example, consider the tilings in Figure 33.15 of the configuration with $n = 2$. The sum of the weights of all tilings is given by

$$\begin{aligned} x\sum_{k=0}^{1}\left(f_{2k}f_{2k+1} + f_{2k+1}f_{2k+2}\right) &= x\left(f_0f_1 + f_1f_2 + f_2f_3 + f_3f_4\right) \\ &= x[0 + x + x(x^2+1) + (x^2+1)(x^3+2x)] \\ &= (x^3+2x)^2 \\ &= f_4^2. \end{aligned}$$

It follows from formula (33.1) that

$$F_0F_1 + F_1F_2 + \cdots + F_{2n-1}F_{2n} = F_{2n}^2 \tag{33.2}$$

$$x(f_2 + f_6 + f_{10} + \cdots + f_{4n-2}) = f_{2n}^2; \tag{33.3}$$

see Exercise 33.1.

For example, let $n = 4$. Then

$$\begin{aligned} x\sum_{k=1}^{4} f_{4k-2} &= x(f_2 + f_6 + f_{10} + f_{14}) \\ &= x^{14} + 12x^{12} + 56x^{10} + 128x^8 + 148x^6 + 80x^4 + 16x^2 \\ &= (x^7 + 6x^5 + 10x^3 + 4x)^2 \\ &= f_8^2. \end{aligned}$$

In particular, formula (33.3) implies that

$$F_2 + F_6 + F_{10} + \cdots + F_{4n-2} = F_{2n}^2. \tag{33.4}$$

We can easily confirm algebraically that

$$x(f_1 + f_5 + f_9 + \cdots + f_{4n-3}) = f_{2n-1}f_{2n}; \tag{33.5}$$

see Exercise 33.2.

For a specific case, we have

$$\begin{aligned} x(f_1 + f_5 + f_9) &= x[1 + (x^4 + 3x^2 + 1) + (x^8 + 7x^6 + 15x^4 + 10x^2 + 1)] \\ &= x^9 + 7x^7 + 16x^5 + 13x^3 + 3x \\ &= (x^4 + 3x^2 + 1)(x^5 + 4x^3 + 3x) \\ &= f_5f_6. \end{aligned}$$

It follows from (33.5) that

$$F_1 + F_5 + F_9 + \cdots + F_{4n-3} = F_{2n-1}F_{2n}. \tag{33.6}$$

The summation formulas (33.4) and (33.6) then yield the following results:

$$F_3 + F_7 + F_{11} + \cdots + F_{4n-1} = F_{2n}F_{2n+1}; \tag{33.7}$$

$$L_2 + L_6 + L_{10} + \cdots + L_{4n-2} = F_{4n}; \tag{33.8}$$

see Exercises 33.3 and 33.4.

Next we introduce the concept of breakability.

33.2 BREAKABILITY

A tiling is *unbreakable* at cell i if a domino occupies cells i and $i+1$; otherwise, it is *breakable* at cell i. For example, the tiling in Figure 33.16 is breakable at cells 1, 2, 4, 6, 7, 9, and 10, but not at cells 3, 5, and 8.

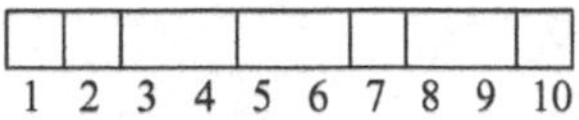

Figure 33.16.

The concept of breakability comes in handy in combinatorially establishing the Fibonacci addition formula, as the following theorem shows.

Example 33.7 (Addition formula). $f_{m+n} = f_{m+1}f_n + f_m f_{n-1}$.

Proof. Consider the tilings of a $1 \times (m+n-1)$ board. By Theorem 33.1, the sum of the weights of its tilings is f_{m+n}.

Now consider an arbitrary tiling of the board. Suppose it is breakable at cell m. This cell partitions the tiling into two disjoint subtilings, one of length m and the other of length $n-1$: $\underbrace{\text{subtiling}}_{\text{length } m}\underbrace{\text{subtiling}}_{\text{length } n-1}$. The sums of the weights of the subtilings are f_{m+1} and f_n, respectively. So the sum of the weights of tilings breakable at cell m is $f_{m+1}f_n$.

On the other hand, suppose the tiling is *not* breakable at cell m. Then it has a domino at cells m and $m+1$. The domino partitions the tiling into three disjoint subtilings: $\underbrace{\text{subtiling}}_{\text{length } m-1}\boxed{\;1\;}\underbrace{\text{subtiling}}_{\text{length } n-2}$. The sum of the weights of such tilings is $f_m f_{n-1}$.

Thus the sum of the weights of all tilings of the board is $f_{m+1}f_n + f_m f_{n-1}$. Combining the two sums, we get the desired formula. ■

The addition formula implies the following identities; see Exercises 33.5 and 33.6.

- $f_{n+1}^2 + f_n^2 = f_{2n+1}$.
- $f_{2n} = f_n l_n$.

The addition formula has an added byproduct. Since $f_{mn} = f_{m+(n-1)m} = f_{m+1}f_{(n-1)m} + f_m f_{(n-1)m-1}$, it follows by PMI that $f_m | f_{mn}$, where $m \geq 1$ and $n \geq 0$. This can be proved independently; see Exercise 33.10.

The next example gives a combinatorial proof of Cassini's formula. The proof involves tilings of two pairs of linear boards, PMI, and a bijection between sets of tilings from each pair.

Example 33.8 (Cassini's formula). *Let $n \geq 1$. Then $f_{n+1}f_{n-1} - f_n^2 = (-1)^n$.*

Proof. Clearly, the formula works when $n = 1$ and $n = 2$. Assume it works for an arbitrary integer $n \geq 2$.

Consider the two pairs of configurations A and B of boards in Figure 33.17. The boards in A are of length $n+1$ and $n-1$, whereas those in B are of the same length n. By Theorem 33.1, the sums of the weights of the tilings in A and B are $f_n f_{n+2}$ and f_{n+1}^2, respectively.

Figure 33.17.

Suppose we move cell 1 from the lower board of A to the beginning of the upper board. Such a shift creates a tiling of configuration B. On the other hand, suppose a domino occupies cells 1 and 2. Shifting cells 1 and 2 to the upper board does *not* yield a tiling of configuration B. So the only tilings of A that do not have a matching tile of B are the tilings of A that have a domino occupying cell 1 in the lower board of A. The sum of the weights of such unmatchable tilings is f_n^2. Consequently, the sum of the weights of tilings of A with matchable tilings of B is $f_n f_{n+2} - f_n^2$.

We now reverse the order. Shifting cell 1 from the upper board of B to the beginning of the lower board gives a tiling of A. But moving the domino in cells 1 and 2 from the upper board of B to the lower board does *not* create a tiling of A; the sum of the weights of such unmatchable tilings equals $f_{n-1}f_{n+1}$. So the sum of the weights of matchable tilings of B equals $f_{n+1}^2 - f_{n-1}f_{n+1}$, which equals $f_{n+1}^2 - [f_n^2 + (-1)^n]$, by the inductive hypothesis.

Since there is a bijection between the two sets of matchable tilings in A and B, the sums of their weights must be equal; that is, $f_n f_{n+2} - f_n^2 = f_{n+1}^2 - [f_n^2 + (-1)^n]$. This implies that $f_n f_{n+2} - f_{n+1}^2 = (-1)^{n+1}$. Consequently, the formula works for $n+1$.

Thus, by PMI, it works for all positive integers n. ∎

We now illustrate the various steps in the proof with $n = 3$. Figure 33.18 shows the corresponding configurations.

Figure 33.18.

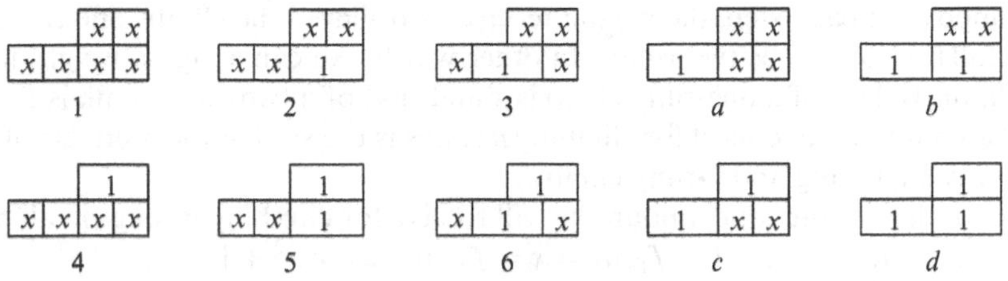

Figure 33.19.

The sum of the weights of tilings of A equals $x^2(x^4+3x^2+1)+1\cdot(x^4+3x^2+1)=(x^2+1)(x^4+3x^2+1)=f_3f_5$ (see Figure 33.19); and that of the tilings of B equals $x^3(x^3+2x)+x(2x^3+4x)=(x^3+2x)^2=f_4^2$ (see Figure 33.20).

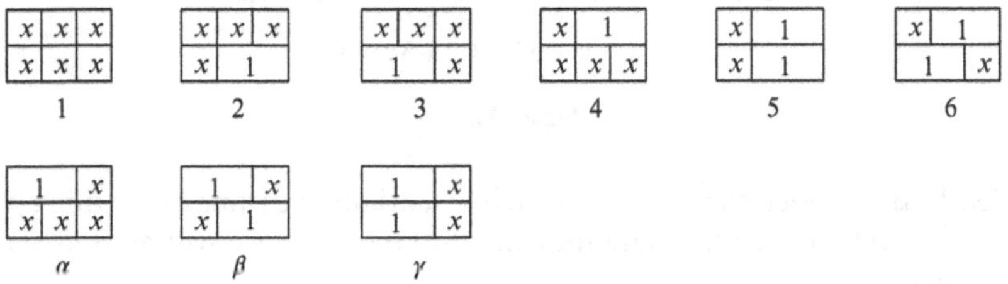

Figure 33.20.

The matchable tilings in Figures 33.19 and 33.20 are numbered 1 through 6; and the unmatchable ones in Figure 33.19 by a through d, and the ones in Figure 33.20 by α, β, and γ. The sum of the weights of the unmatchable tilings a through d equals $x^4+2x^2+1=(x^2+1)^2=f_3^2$, and hence that of the matchable tilings equals $(x^2+1)(x^4+3x^2+1)-(x^2+1)^2=x^2(x^2+1)(x^2+2)=f_3f_5-f_3^2$. Similarly, the sum of the weights of tilings of B equals $(x^3+2x)^2-(x^4+2x^2)=x^2(x^2+1)(x^2+2)=f_4^2-f_2f_3=f_4^2-[f_3^2+(-1)^3]$.

Thus $f_3f_5-f_3^2=f_4^2-[f_3^2+(-1)^3]$. This implies that $f_3f_5-f_4^2=(-1)^4$, as expected.

Next we present a related combinatorial model for Fibonacci polynomials.

33.3 A LADDER MODEL

Suppose a person decides to climb up a ladder with n rungs. At each step, he can climb either one or two rungs. It is well known that there are F_{n+1} different ways of accomplishing this task [106, 287].

Interestingly, this combinatorial problem can be extended to construct Fibonacci polynomials by simply modifying the linear model in Section 33.1.

Suppose, at each step, the person receives x dollars if he climbs one rung, or one dollar if he climbs two rungs. In other words, we can assign a weight for a climb; the weight of a one-rung climb is x and that of a two-rung climb is 1. The total amount he can collect for climbing n rungs is the sum of the products of the amounts for 1-rung and 2-rung climbs.

Let a_n denote the total amount he will receive for climbing n rungs in all possible ways. Clearly, $a_0 = 1 = f_1$, $a_1 = x = f_2$, and $a_2 = x^2 + 1 = f_3$.

Figure 33.21 shows the possible climbs and the sum of all amounts (weights) when $n = 5$.

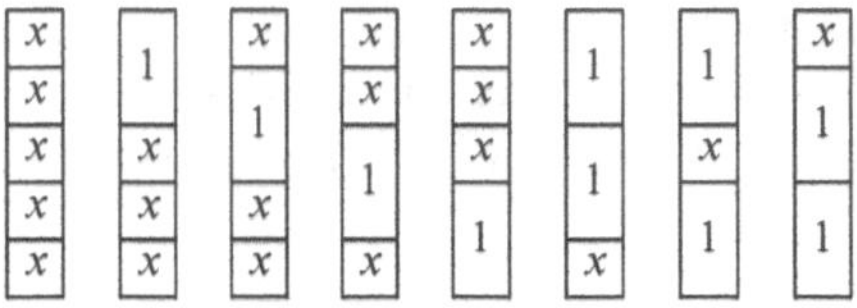

Total Sum $= x^5 + 4x^3 + 3x = f_6$

Figure 33.21.

Clearly, this ladder-climbing problem is essentially the same as the tiling problem of a $1 \times n$ board with square tiles and dominoes. So we will not pursue this any further.

Next we present a combinatorial model for Pell–Lucas polynomials.

33.4 A MODEL FOR PELL–LUCAS POLYNOMIALS: LINEAR BOARDS

In the Fibonacci tiling model in Theorem 33.1, we let the weight of a square tile be x. This time we let the weight be $2x$, except for the first square in the tiling. The weight of the domino remains 1.

Figure 33.22 shows such tilings of a $1 \times n$ board and the sum of their weights, where $0 \le n \le 4$. Based on these data, we conjecture that the sum of the weights of tilings of a board of length n is $\frac{1}{2}q_n$, where $n \ge 0$. The following theorem confirms it; we omit its proof for the sake of brevity; see Exercise 33.11.

Theorem 33.2. *The sum of the weights of tilings of a* $1 \times n$ *board is* $\frac{1}{2}q_n$, *where the weight of a domino is* 1 *and that of a square is* $2x$ *with one exception: If a tiling begins with a square, its weight is* x. ■

Theorem 33.2, coupled with breakability, can be used to prove the addition formula for Pell–Lucas polynomials; see Exercise 33.12.

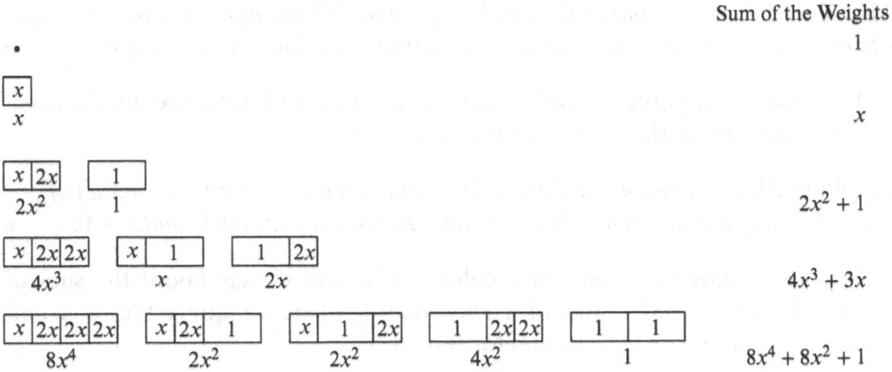

Figure 33.22.

Next we introduce colored tilings. They yield us a surprising dividend.

33.5 COLORED TILINGS

Suppose square tiles come in two colors, black and white. As before, each square has weight x and each domino 1. Again, we define the weight of the empty tiling to be one.

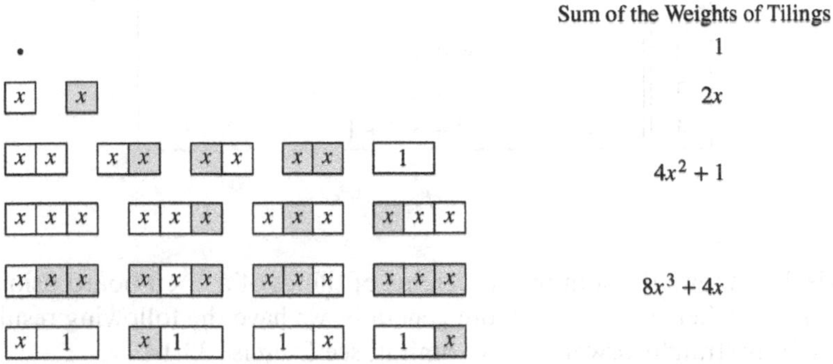

Figure 33.23.

Figure 33.23 shows the sum of the weights of such colored tilings of a $1 \times n$ board, where $0 \leq n \leq 3$. The sum of the weights of a board of length n is p_{n+1}, as the next theorem reveals. Again, we omit its proof in the interest of brevity; see Exercise 33.13.

Theorem 33.3. *The sum of the weights of colored tilings of a* $1 \times n$ *board is* p_{n+1}, *where the weight of a square tile is* x *and that of a domino is* 1, *and* $n \geq 0$. ■

In particular, suppose the weight of a square tile is $x/2$. It follows by Theorem 33.3 that the sum of the weights of the tilings is f_{n+1}.

Corollary 33.3. *The sum of the weights of colored tilings of a* $1 \times n$ *board is* f_{n+1}, *where the weight of a square tile is* $x/2$ *and that of a domino is* 1, *and* $n \geq 0$. ■

Suppose square tiles come in k colors. What can we say about the sum of the weights of colored tilings of a $1 \times n$ board, where w(square tile) = x and w(domino) = 1? Some possibilities for thoughts.

33.6 A NEW TILING SCHEME

Suppose square tiles are available in x colors, and both squares and dominoes have the same weight 1. The weight of the empty tiling is one. So the weight of each tiling is 1 and the sum of the weights of tilings of a $1 \times n$ board is precisely the number of such tilings.

TABLE 33.2.

n	Sum of the Weights of Tilings of Length n
0	1
1	x
2	$x^2 + 1$
3	$x^3 + 2x$
4	$x^4 + 3x^2 + 1$

↑
f_{n+1}

Table 33.2 shows the sum of the weights of tilings of a $1 \times n$ board using this tiling scheme, where $0 \leq n \leq 4$. More generally, we have the following result. Its proof is fairly straightforward, so we omit it; see Exercise 33.14.

Theorem 33.4. *Suppose squares are available in* x *colors, and the empty tile can be tiled in exactly one way. The sum of the weights of such tilings of a* $1 \times n$ *board is* f_{n+1}, *where* $n \geq 0$. ■

This theorem has charming dividends, involving different strategies. To begin with, the theorem plays a pivotal role in establishing combinatorially the summation formula $x \sum_{k=1}^{n} f_k = f_{n+1} + f_n - 1$. We will now prove its equivalent form

$$(x-1)f_n + x \sum_{k=1}^{n-1} f_k = f_{n+1} - 1.$$

Example 33.9. *Establish the summation formula* $(x-1)f_n + x\sum_{k=1}^{n-1} f_k = f_{n+1} - 1.$

Proof. Consider a $1 \times n$ board. By Theorem 33.4, the sum of the weights of its tilings is f_{n+1}. Suppose one of the colors is white. One of the tilings consists of all white squares; so the sum of the weights of tilings that contain at least one nonwhite tile is $f_{n+1} - 1$.

Consider the last tile that is *not* a white square. Suppose it begins at cell k.

Case 1. Suppose $k = n$. Then the last tile is a square and its color has $x - 1$ choices: $\underbrace{\text{subtiling}}_{\text{length } n-1}\square$. The sum of the weights of subtilings of length $n - 1$ is f_n; so the sum of the weights of such tilings is $(x-1)f_n$.

Case 2. Suppose $1 \le k \le n - 1$. Then the tile covering cell k can be a nonwhite square, or a domino covering cells k and $k + 1$:

$$\underbrace{\text{subtiling}}_{\text{length } k-1}\underbrace{\text{nonwhite square/domino}}\underbrace{\text{white squares, dominoes}}.$$

There are $(x-1)+1 = x$ choices for the tile. Since the preceding subtilings have weight f_k, the sum of the weights of such tilings is xf_k.

Thus the total sum of the weights of tilings equals $(x-1)f_n + x\sum_{k=1}^{n-1} f_k$. The desired formula now follows by equating the two sums. ■

We can use the same tiling scheme to prove that $x\sum_{k=1}^{n} f_{2k-1} = f_{2n}$, as the following example demonstrates.

Example 33.10. *Prove combinatorially that* $x\sum_{k=1}^{n} f_{2k-1} = f_{2n}$.

Proof. Consider a $1 \times (2n-1)$ board. By Theorem 33.4, the sum of the weights of its tilings is f_{2n}.

Consider an arbitrary tiling of the board. Since its length is odd, every tiling must contain an odd number of squares. Then the last square must occupy an odd-numbered cell, say, $2k - 1$. This square partitions the tiling into three disjoint subtilings: $\underbrace{\text{subtiling A}}_{\text{length } 2k-2}\square\underbrace{\text{subtiling B}}_{\text{dominoes}}$. The subtiling B consists of dominoes. The sum of the weights of subtilings A is f_{2k-1}. Since squares come in x colors, the sum of the weights of tilings with the last square at cell $2k - 1$ is xf_{2k-1}. Thus the sum of the weights of all tilings equals $x\sum_{k=1}^{n} f_{2k-1}$.

Equating the two sums yields the given formula. ■

As a specific example, consider the tilings of a 1×5 board. Table 33.3 shows the possible values of k, location of the last square, and the sum of the weights of tilings with the last square at cell $2k-1$. The sum of the weights of tilings equals $x^5 + 4x^3 + 3x = f_6$, as desired.

TABLE 33.3.

k	Location of Last Square	Sum of the Weights of Tilings
1	1	x
2	3	$x^3 + x$
3	5	$x^5 + 3x^3 + x$

The same technique can be used to prove that $x \sum_{k=1}^{n} f_{2k} = f_{2n+1} - 1$; see Exercise 33.15.

The next example also deals with a summation formula. The technique applied involves a pair of linear boards.

Example 33.11. *Prove combinatorially that* $x \sum_{k=1}^{n} f_k^2 = f_n f_{n+1}$.

Proof. Consider two linear boards, one of length $n-1$ and the other of length n. The second board is placed below the other in a staggered fashion; see the configuration in Figure 33.24. The sum of the weights of tilings of this configuration is $f_n f_{n+1}$.

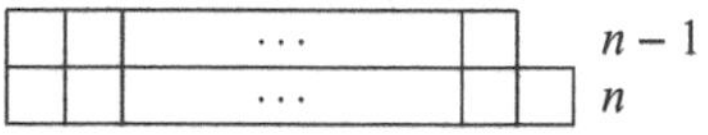

Figure 33.24.

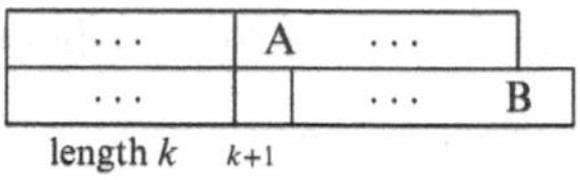

Figure 33.25.

To compute this sum in a different way, suppose the configuration is breakable. Let k be the largest label such that it is breakable at cell k; see Figure 33.25. If no such k exists, we define $k = 0$. So $0 \le k \le n-1$. A square must occupy cell $k+1$ of the lower board, and dominoes must occupy the subtilings A and B; this guarantees the nonexistence of any additional breakable points (to the right of cell k). Consequently, subtilings A and B can be tiled in exactly one way.

The sum of the weights of pairs of boards of length k is f_{k+1}^2. Since there are x colors for a square, this implies that the sum of the weights of such configurations is xf_{k+1}^2. Thus the sum of the weights of tilings of the configuration is $\sum_{k=0}^{n-1} xf_{k+1}^2$.

The given formula now follows by equating the two sums. ■

We now turn to tilings of circular boards. They can be used to construct combinatorial models for Pell–Lucas polynomials and hence Lucas polynomials.

33.7 A MODEL FOR PELL–LUCAS POLYNOMIALS: CIRCULAR BOARDS

Consider a circular board with n cells, labeled 1 through n in the counterclockwise direction; see Figure 33.26.

Figure 33.26.

We would like to tile the board with (curved) squares and (curved) dominoes. Such a tiling is an *n-bracelet*. We assign each square a weight $2x$ and each domino one, with one exception: The weight of the domino is 2 when $n = 2$. The weight of the empty tiling is 2.

Figure 33.27 shows the n-bracelets and the sums of their weights, where $0 \leq n \leq 3$. The weights follow an interesting pattern. Based on it, we conjecture that the sum of the weights of tilings of an n-bracelet is q_n. The next theorem confirms this observation; see Exercise 33.16.

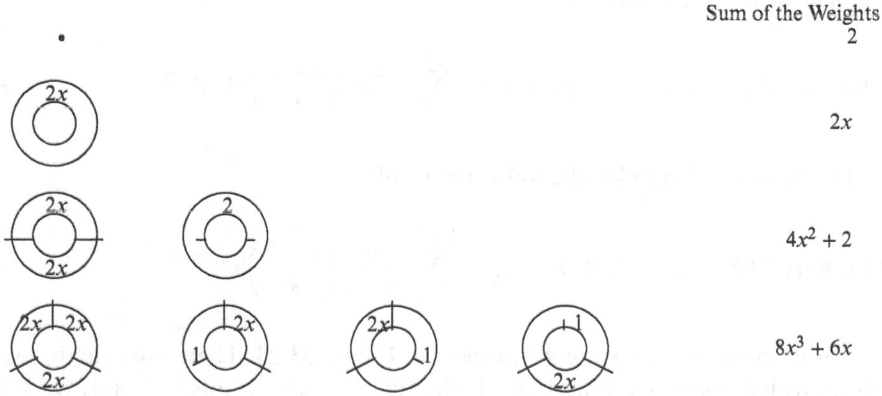

Figure 33.27. n-bracelets, where $0 \leq n \leq 3$.

Theorem 33.5. *The sum of the weights of tilings of a circular board with n cells is q_n, where the weight of a square is $2x$ and each domino is 1, with one exception: The weight of the domino is 2 when $n = 2$. The weight of the empty tiling is 2.* ■

This theorem has an interesting byproduct. It gives a combinatorial model for Lucas polynomials, as the following corollary shows.

Corollary 33.4. *The sum of the weights of tilings of a circular board with n cells is l_n, where the weight of a square is x and each domino is 1, with one exception: The weight of the domino is 2 when $n = 2$. The weight of the empty tiling is 2.* ■

For example, Figure 33.28 shows the tilings of a circular board with four cells. The sum of the weights of the seven tilings is $x^4 + 4x^2 + 2 = l_4$.

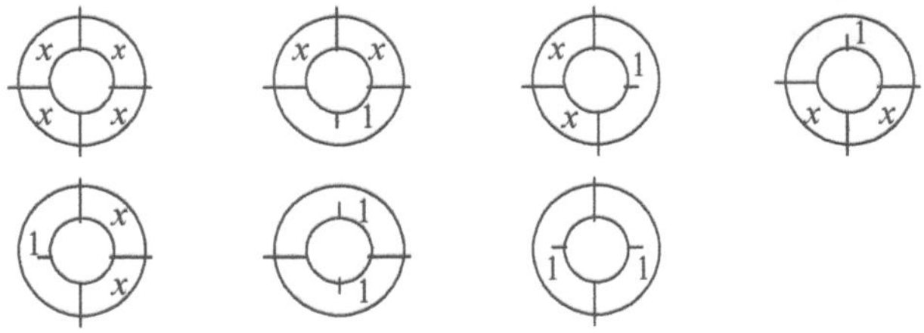

Figure 33.28.

The next result follows from Corollary 33.4.

Corollary 33.5. *The number of distinct n-bracelets is L_n, where $n \geq 0$.* ■

The combinatorial model in Theorem 33.5 can be used to re-confirm the explicit formula for q_n in Section 31.6. The proof is rather straightforward, so we omit it; see Exercise 33.17.

Theorem 33.6. *Let $n \geq 0$. Then* $q_n = \displaystyle\sum_{k=0}^{\lfloor n/2 \rfloor} \frac{n}{n-k}\binom{n-k}{k}(2x)^{n-2k}$. ■

In particular, this yields the following result.

Corollary 33.6. *Let $n \geq 0$. Then* $l_n = \displaystyle\sum_{k=0}^{\lfloor n/2 \rfloor} \frac{n}{n-k}\binom{n-k}{k}x^{n-2k}$. ■

For example, consider the 4-bracelets in Figure 33.28. There are exactly two tilings with a domino in cells 4 and 1; the sum of their weights is $x^2 + 1$. On the other hand, there are five bracelets without a domino in cells 4 and 1; the sum of their weights is $x^4 + 3x^2 + 1$. Thus the sum of the weights of all 4-bracelets equals $x^4 + 4x^2 + 2 = l_4 = \displaystyle\sum_{k=0}^{\lfloor 4/2 \rfloor} \frac{4}{4-k}\binom{4-k}{k}x^{4-2k}$.

Using the circular tiling scheme in Corollary 33.4, we now establish the addition formula for Lucas polynomials. First, recall that the weight of a square is x and that of domino is 1 in the case of circular tilings. But the weight of a domino is 2 when $n = 2$; and that of the empty tiling is 2.

Every bracelet is breakable along the left edge of the tile covering cell 1. For clarity and brevity, we then say that the bracelet has a *breakpoint* along that edge.

We are now ready to present the proof of the addition formula.

Example 33.12. *Prove combinatorially the addition formula* $l_{m+n} = f_{m+1}l_n + f_m l_{n-1}$.

Proof. Consider a circular board with $m + n$ cells. The sum of the weights of its tilings is l_{m+n}.

There are two distinct ways we can construct all $(m + n)$-bracelets; that is, we can partition them into two classes and then compute the sum of the weights of the tilings in each class separately.

Case 1. Consider an arbitrary $(m + n)$-bracelet A and an n-bracelet B. Suppose bracelet A has a subtiling of length m immediately following its breakpoint. Insert this subtiling at the breakpoint of B. This creates an $(m + n)$-bracelet. The sum of the weights of such bracelets is $f_{m+1}l_n$.

Case 2. Suppose bracelet A has *no* subtiling of length m immediately following its breakpoint. Then it must have a subtiling of length $m - 1$ immediately following the breakpont. Insert this subtiling and a domino after the breakpoint of an $(n - 1)$-bracelet. This procedure also creates an $(m + n)$-bracelet. The sum of the weights of such bracelets is $f_m l_{n-1}$.

Thus the sum of the weights of all $(m + n)$-bracelets is $f_{m+1}l_n + f_m l_{n-1}$. The two sums together yield the addition formula. ■

We now illustrate this constructive algorithm for the case $m = 2$ and $n = 3$. Figure 33.29 shows the eleven 5-bracelets. The sum of their weights is $x^5 + 5x^3 + 5x = l_5$. Figure 33.30 shows the four 3-bracelets. The sum of their weights is $x^3 + 3x = l_3$.

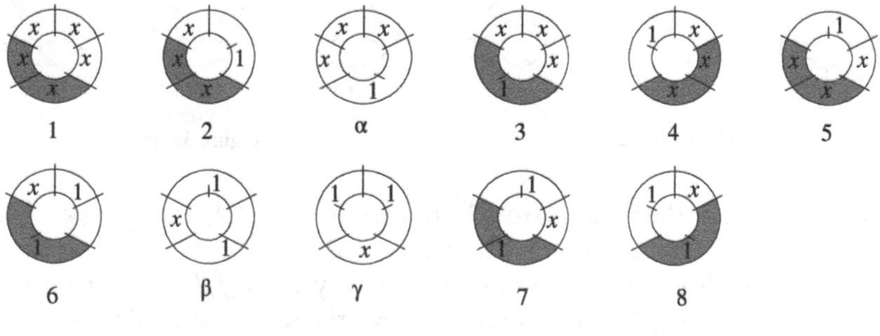

Figure 33.29. 5-bracelets.

Figure 33.30. 3-bracelets.

There are eight 5-bracelets that have subtilings of length 2 following their breakpoints; they are shaded in the bracelets numbered 1 through 8. But only two of them are distinct; one consists of two squares and the other consists of a single domino. Inserting them at the breakpoints of the 3-bracelets yields eight bracelets; see Figure 33.31. The sum of their weights equals $x^5 + 4x^3 + 3x = (x^2 + 1)(x^3 + 3x) = f_3 l_3$.

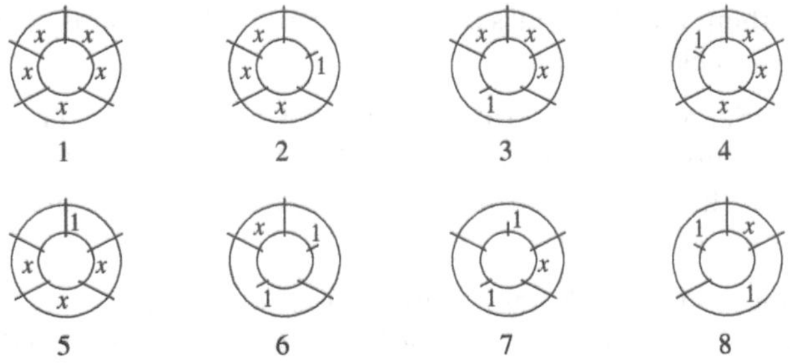

Figure 33.31.

There are three 5-bracelets (see Figure 33.29) that have subtilings of length 1 immediately after their breakpoints; they are labeled α, β, and γ. There are three 2-bracelets (see Figure 33.32); the middle one is in-phase and the last one out-of-phase. Inserting the subtilings at their breakpoints yields three 5-bracelets; see Figure 33.33. The sum of their weights equals $x^3 + 2x = x(x^2 + 2) = f_2 l_2$.

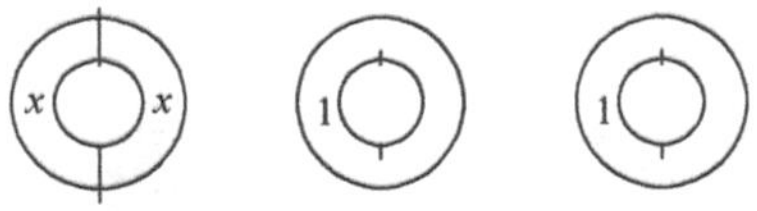

Figure 33.32.

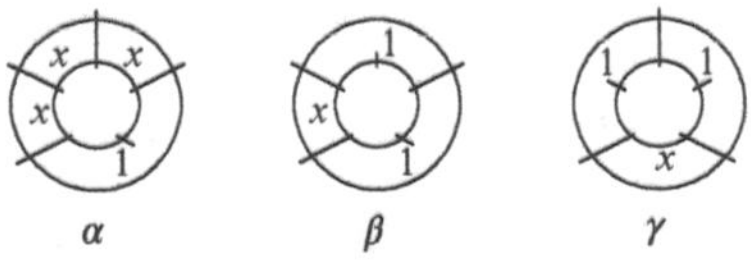

Figure 33.33.

Thus the sum of the weights from both cases equals $f_3 l_3 + f_2 l_2 = (x^5 + 4x^3 + 2x) + (x^3 + 2x) = x^5 + 5x^3 + 5x = l_5$, as expected.

Next we establish combinatorially the identity $f_{2n} = f_n l_n$. The proof is constructive, and uses the fact that two sets have the same cardinality if and only if there is a bijection between them. The proof involves both linear and circular boards. It is short and quite pretty.

We use the Fibonacci model in Theorem 33.1 and the Lucas model in Corollary 33.4. The weight of a square is x, that of a domino is 1, and that of the empty tiling is 1. For a Lucas tiling, the weight of a domino is 2 in the case of a 2-bracelet; and the weight of the empty tiling is 2.

Example 33.13. *Prove combinatorially the identity $f_{2n} = f_n l_n$.*

Proof. Let X denote the set of tilings of a $1 \times (2n - 1)$ board, and Y the set of pairs (A, B) of tilings A and B, where A is a tiling of a $1 \times (n - 1)$ board and B

that of a circular board with n cells. Clearly, the sum of the weights of tilings in X equals f_{2n} and that in Y equals $f_n l_n$.

We now construct a bijection between the sets X and Y. For each element T in X, we have to identify its unique matching element (A, B) in Y. We then must make sure that the matching works in the reverse order as well.

Case 1. Suppose a domino D occupies cells n and $n+1$ of tiling T of the $1 \times (2n-1)$ board; see Figure 33.34. Then tiling A consists of the subtiling a in cells 1 through $n-1$ of T; and B is the n-bracelet formed by the remaining tiles in T, with D occupying cells n and 1; see Figure 33.35.

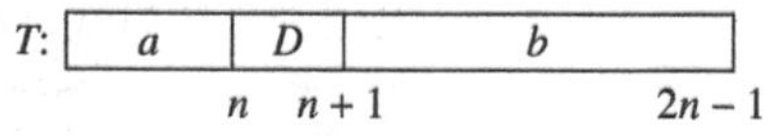

Figure 33.34.

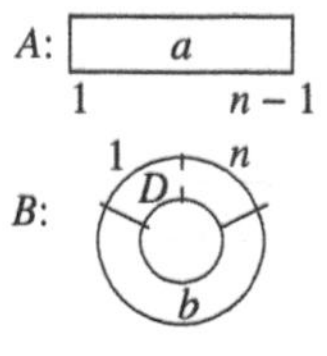

Figure 33.35.

Case 2. Suppose a domino does *not* occupy cells n and $n+1$ in T; see Figure 33.36. Then A consists of the subtiling in cells $n+1$ through $2n-1$ of T; and B is the bracelet formed by the subtiling a in cells 1 through n of T; see Figure 33.37.

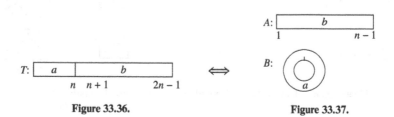

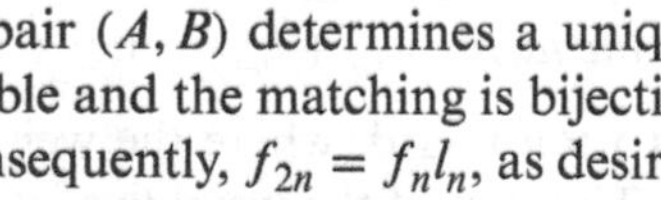

Figure 33.36.

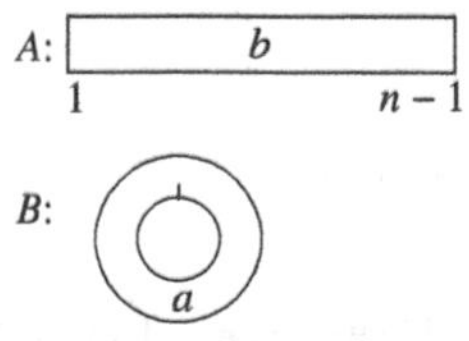

Figure 33.37.

In each case, tiling T is paired with a unique element (A, B) in Y. Conversely, every pair (A, B) determines a unique element T in X. Thus the algorithm is reversible and the matching is bijective.

Consequently, $f_{2n} = f_n l_n$, as desired. ■

We now demonstrate this combinatorial technique for the case $n = 3$. Figure 33.38 shows the tilings of a 1×5 board, the weight of each tiling, and the corresponding pairs (A, B) in Y. Clearly, the sum of the weights of the tilings T equals $x^5 + 4x^3 + 3x = f_6 = (x^2+1)(x^3+3x) = f_3 l_3$, and also equals the sum of the weights of the pairs of tilings (A, B).

In the next example, we use a similar strategy to establish the identity $l_{n+1} + l_{n-1} = (x^2+4)f_n$. [However, in the interest of convenience, we establish an equivalent form: $l_n + l_{n+2} = (x^2+4)f_{n+1}$.] We accomplish this using weighted tilings of

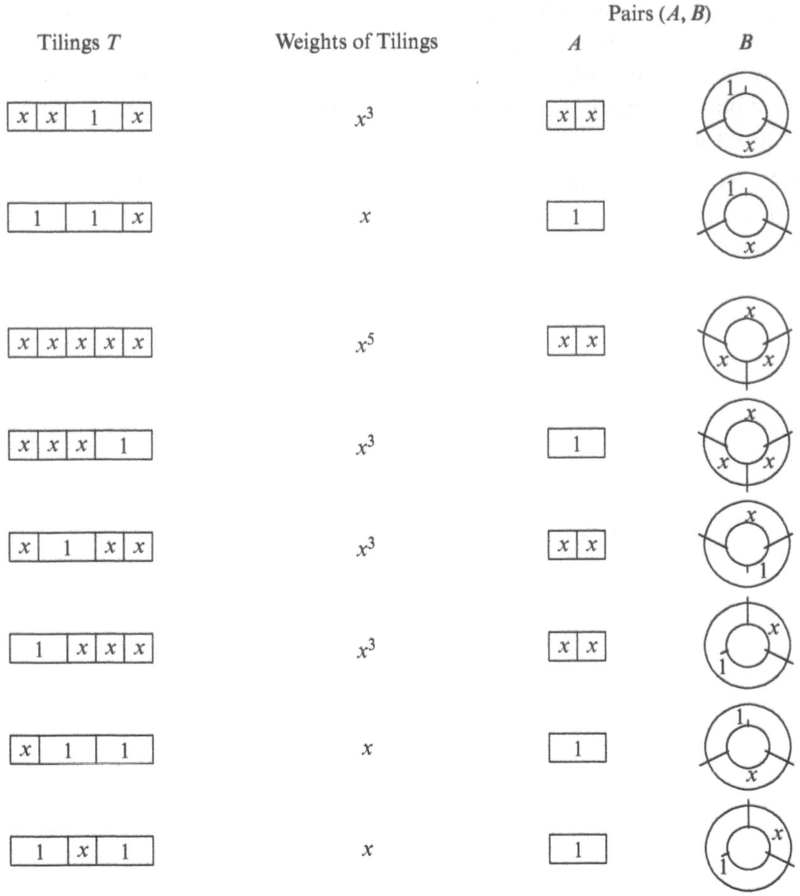

Figure 33.38. Tilings of a 1×5 board and the corresponding pairs (A, B).

linear and circular boards, and a *one-to-five correspondence* between two suitable sets of tilings.

Example 33.14. *Establish combinatorially the identity* $l_n + l_{n+2} = (x^2 + 4)f_{n+1}$.

Proof. Consider the weighted tilings of a $1 \times n$ board, where the weight of a square is x and the weight of a domino is 1. The sum of the weights of its tilings is f_{n+1}.

Let T be an arbitrary tiling of the board. We now devise an algorithm to construct unique n- or $(n + 2)$- bracelets in five distinct ways:

Step 1. Form an in-phase n-bracelet A by joining the two ends of T; see Figure 33.39.

Step 2. Place squares in cells 1 and 2 of a circular board of size $n + 2$, followed by tiling T; see Figure 33.40. This produces an in-phase $(n + 2)$-bracelet B.

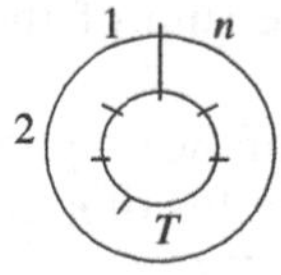

n-bracelet A

Figure 33.39.

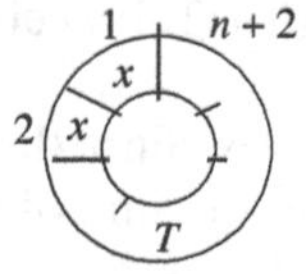

$(n+2)$-bracelet B

Figure 33.40.

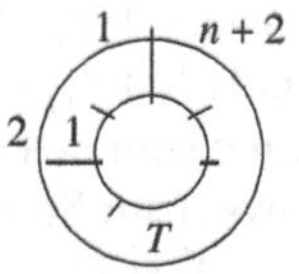

$(n+2)$-bracelet C

Figure 33.41.

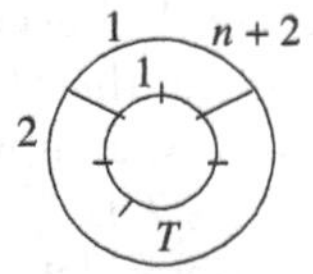

$(n+2)$-bracelet D

Figure 33.42.

Step 3. Place a domino in cells 1 and 2 of a circular board of size $n+2$, followed by T; see Figure 33.41. This also creates another in-phase $(n+2)$-bracelet C.

Step 4. Place a domino in cells $n+2$ and 1 of a circular board of size $n+2$. Then place T in cells 2 through $n+1$. This generates an out-of-phase $(n+2)$-bracelet D; see Figure 33.42.

At this point we have constructed four different n- or $(n+2)$-bracelets from a single linear tiling T. But we are *not* quite done. A quick examination of the bracelets in Figures 33.39–33.42 shows that *not* all n- or $(n+2)$-bracelets can be created by Steps 1–4; some are out-of-phase n-bracelets, and in-phase $(n+2)$-bracelets with a square in cell 1 followed immediately by a domino. They depend on the last tile of T being a domino or a square. This takes us to the next step, which has two parts.

Step 5A. Suppose tiling T ends in a domino. It partitions T into a subtiling a and the domino; see Figure 33.43. Now place the domino in cells n and 1 of a circular board of size n, and subtiling a in its remaining cells. This creates an out-of-phase n-bracelet E_1; see Figure 33.44.

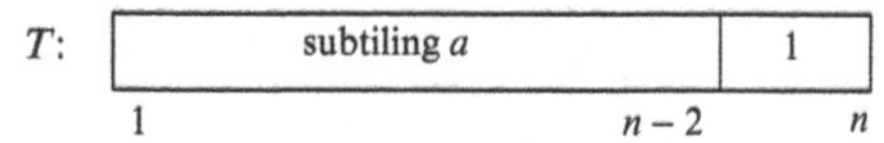

Figure 33.43.

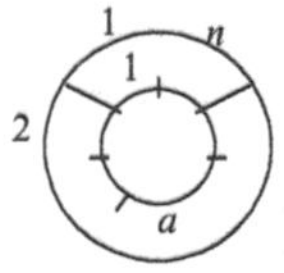

$(n+2)$-bracelet E_1

Figure 33.44.

Step 5B. Suppose tiling T ends in a square. It partitions T into a subtiling b and the square; see Figure 33.45. Now place the square in cell 1 of a circular board of size $n+2$, followed by a domino and subtiling b. This results in an in-phase $(n+2)$-bracelet E_2; see Figure 33.46.

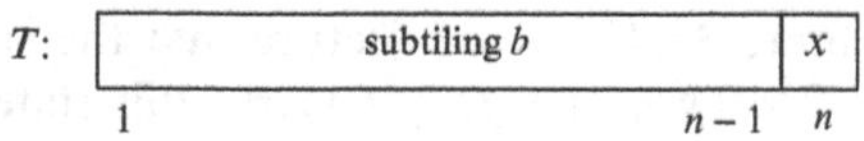

Figure 33.45.

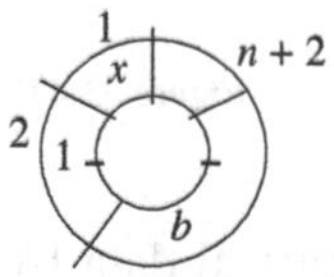

$(n+2)$-bracelet E_2

Figure 33.46.

Steps 1–5 account for every type of n- or $(n+2)$-bracelet. The sum of the weights of their tilings equals $l_n + l_{n+2}$.

We now compute the same sum in a different way. Since the sum of the weights of tilings T is f_{n+1}, it follows from Steps 1–5 that the sum of the weights of n- or $(n+2)$-bracelets equals $f_{n+1} + x^2 f_{n+1} + f_{n+1} + f_{n+1} + f_{n+1} = (x^2+4)f_{n+1}$.

Equating the two sums yields the desired identity. ■

Next we present a new combinatorial model for Fibonacci polynomials.

33.8 A DOMINO MODEL FOR FIBONACCI POLYNOMIALS

It is well known that a $2 \times n$ board can be tiled in F_{n+1} ways [106, 287]. This tiling problem creates an opportunity to construct a new combinatorial model for f_n.

This time, dominoes can be placed horizontally or vertically. Depending on how a domino is placed, we must adjust its weight. So we define the weight of a vertical domino to be x and that of a horizontal domino to be 1. We define the weight of the empty tiling to be 1. As usual, the weight of a tiling is the product of the weights of all dominoes in the tiling.

Figure 33.47 shows the possible such tilings of a $2 \times n$ board and the sum of their weights, where $0 \le n \le 4$.

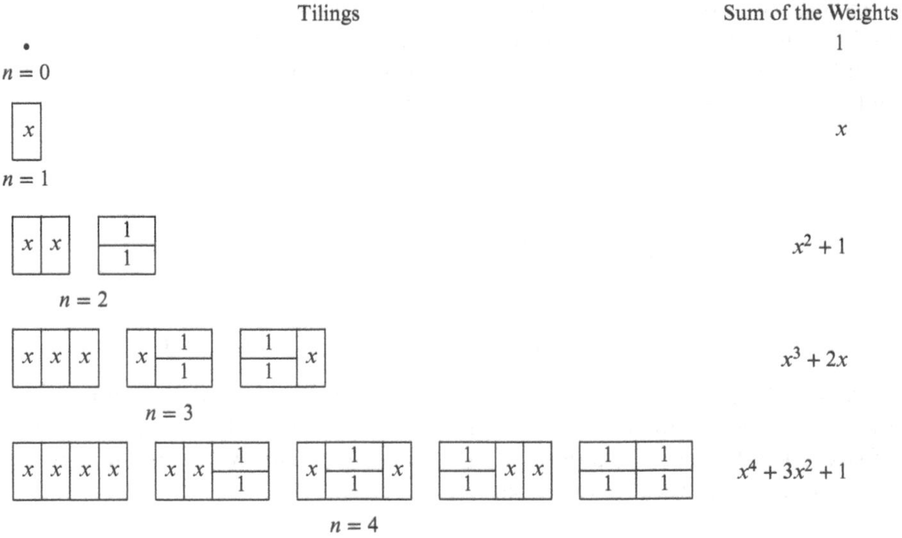

Figure 33.47. Domino tilings of a $2 \times n$ board.

Using the empirical data from Figure 33.47, we conjecture that the sum of the weights of all domino tilings of a $2 \times n$ board is f_{n+1}. Clearly, this statement is true when $0 \le n \le 4$. Now, assume it is true for $n-1$ and $n-2$, where $n \ge 2$. Consider an arbitrary $2 \times n$ tiling T.

Suppose T begins with a vertical domino: $T = \boxed{x}\,\underbrace{\text{subtiling}}_{\text{length } n-1}$. By the hypothesis, the sum of the weights of such tilings is xf_n.

On the other hand, suppose tiling T begins with a horizontal domino. Then we need to place one horizontal domino to cover the space above it; so horizontal dominoes appear in pairs: $T = \boxed{\begin{array}{c}1\\ \hline 1\end{array}}\,\underbrace{\text{subtiling}}_{\text{length } n-2}$. The sum of the weights of such tilings is $1 \cdot 1 \cdot f_{n-1} = f_{n-1}$.

Thus the sum of the weights of all tilings of a $2 \times n$ board is $xf_n + f_{n-1} = f_{n+1}$. So, by induction, the conjecture works for every $n \geq 0$.

This gives the following theorem.

Theorem 33.7. *The sum of the weights of tilings of a* $2 \times n$ *board is* f_{n+1}, *where* $w(\text{vertical domino}) = x$, $w(\text{horizontal domino}) = 1$, *and* $n \geq 0$. ■

With this fact at our fingertips, we can re-establish the explicit formula for f_{n+1} in Example 33.1. We omit the proof in the interest of brevity; see Exercise 33.19.

Theorem 33.8. *Let* k *denote the number of pairs of horizontal dominoes in the tilings of a* $2 \times n$ *board, where* $n \geq 0$. *Then*

$$f_{n+1} = \sum_{k=0}^{\lfloor n/2 \rfloor} \binom{n-k}{k} x^{n-2k}.$$

■

For example, Figure 33.48 shows the domino tilings of a 2×5 board. The sum of their weights is $x^5 + 4x^3 + 3x = f_6$.

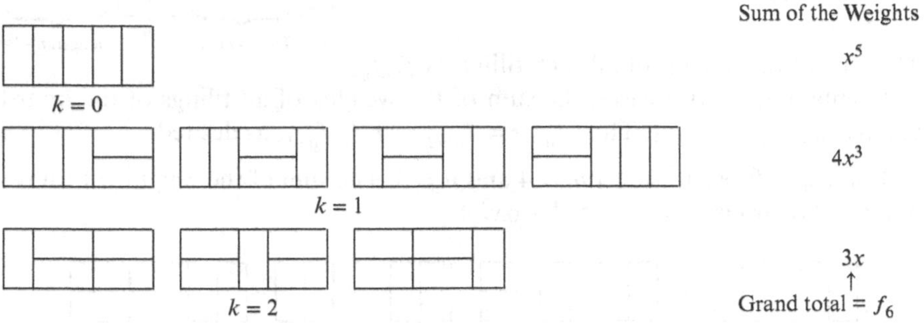

Figure 33.48. Domino tilings of a 2×5 board.

One of them contains $k = 0$ pairs of horizontal dominoes; the resulting weight is x^5. Exactly four of them contain $k = 1$ horizontal pair each; they contribute a total of $4x^3$ to the sum. Exactly three of them contain $k = 2$ horizontal pairs each; they contribute a total of $3x$ to the sum. Thus the grand total is $x^5 + 4x^3 + 3x = f_6$, as expected.

In the next example, we invoke the concept of breakability to establish the Fibonacci addition formula. To this end, first we clarify the concept of breakability in the context of domino tiling.

Breakability

An $2 \times n$ domino tiling is *unbreakable* at cell k if a pair of horizontal dominoes occupies cells $k-1$ and k; otherwise, it *breakable* at cell k. For example, consider the 2×12 tiling in Figure 33.49. It is breakable at cells 2, 3, 5, 7, 8, 9, 11, and 12; and unbreakable at cells 1, 4, 6, and 10.

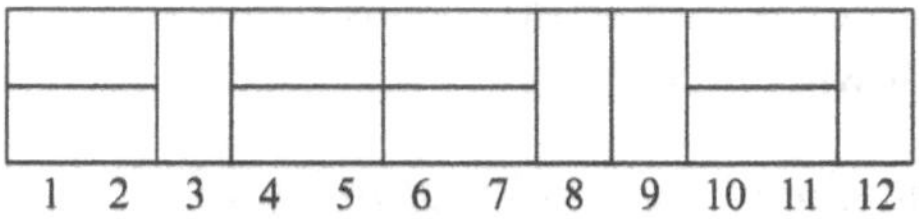

Figure 33.49. A 2×12 tiling.

Example 33.15. *Using the domino tiling, establish the addition formula* $f_{m+n} = f_{m+1}f_n + f_m f_{n-1}$.

Proof. Consider a $2 \times (m+n-1)$ board. By Theorem 33.7, the sum of the weights of tilings of the board is f_{m+n}.

We now compute the same sum in a different way. Let T be an arbitrary tiling of the board.

Case 1. Suppose tiling T is breakable at cell m: $T = \underbrace{\text{subtiling}}_{\text{length } m}\ \underbrace{\text{subtiling}}_{\text{length } n-1}$. The sum of the weights of such tilings is $f_{m+1}f_n$.

Case 2. Suppose tiling T is *not* breakable at cell m: $T = \underbrace{\text{subtiling}}_{\text{length } m-1}$ 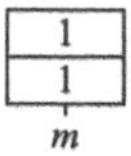 $\underbrace{\text{subtiling}}_{\text{length } n-2}$.

The sum of the weights of all such tilings is $f_m f_{n-1}$.

Combining the two cases, the sum of the weights of all tilings of the board equals $f_{m+1}f_n + f_m f_{n-1}$. Thus $f_{m+n} = f_{m+1}f_n + f_m f_{n-1}$, as desired. ■

For a specific example, let $m = 4$ and $n = 3$. The sum of the weights of tilings of a 6×2 board is $f_7 = x^6 + 5x^4 + 6x^2 + 1$.

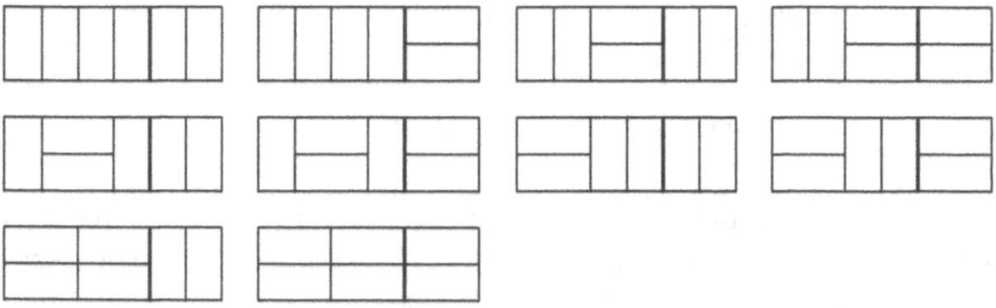

Figure 33.50. 2×6 tilings breakable at cell 4.

Exactly ten of the tilings are breakable at cell 4; see Figure 33.50. The sum of their weights is $x^6 + 4x^4 + 4x^2 + 1$. But three of the tilings of the board are *not* breakable at cell 4; see Figure 33.51. The sum of their weights is $x^4 + 2x^2$.

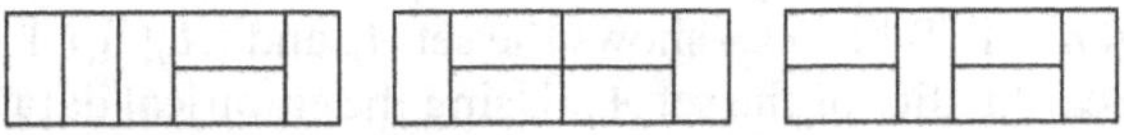

Figure 33.51. 2×6 tilings unbreakable at cell 4.

Thus the sum of the weights of all tilings of the board is

$$\begin{aligned} f_7 &= x^6 + 5x^4 + 6x^2 + 1 \\ &= (x^6 + 4x^4 + 4x^2 + 1) + (x^4 + 2x^2) \\ &= (x^4 + 3x^2 + 1)(x^2 + 1) + (x^3 + 2x)x \\ &= f_5 f_3 + f_4 f_2, \end{aligned}$$

as expected.

The addition formula implies that $f_{2n} = f_n l_n$ and $f_{n+1}^2 + f_n^2 = f_{2n+1}$. It is worthwhile to establish both independently; see Exercises 33.20 and 33.21.

Next we introduce the concept of a median vertical domino.

Median Domino

Consider a $2 \times n$ domino tiling, where n is odd. It must contain an odd number of vertical dominoes. So there must exist a special vertical domino M with an equal number of vertical dominoes on either side. This unique domino M is called the *median (vertical) domino*.

For example, let $n = 3$. Figure 33.52 shows the 2×3 domino tilings. The up arrows point to the median domino in each case.

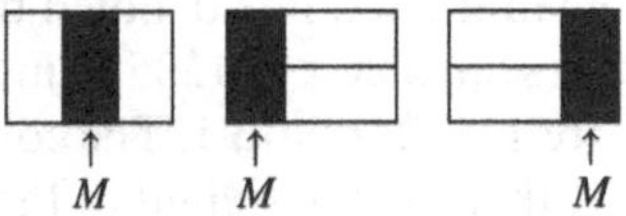

Figure 33.52. 2×3 tilings.

Using the median concept and Theorem 33.7, we can re-confirm Example 33.3. Again, we omit the proof; see Exercise 33.22.

Theorem 33.9.

$$f_{2n+2} = \sum_{\substack{i,j \geq 0 \\ i+j \leq n}} \binom{n-i}{j}\binom{n-j}{i} x^{2n-2i-2j+1}.$$

Next we present a new combinatorial model for Fibonacci polynomials. ■

33.9 ANOTHER MODEL FOR FIBONACCI POLYNOMIALS

In 1991, R. Euler studied an interesting set A_n [150]. It consists of increasing sequences $\{a_n\}$ of positive integers such that $a_1 = 1$, $a_n = n$, and $a_{i+1} - a_i = 1$ or 2, where $1 \le i \le n-1$. Table 33.4 shows the set A_n and $|A_n|$ for $1 \le n \le 5$, where $|A|$ denotes the cardinality of the set A_n. Using the empirical data from the table, we can predict that $|A_n| = F_n$. We can establish this conjecture using recursion [419].

TABLE 33.4. Set A_n and $|A_n|$, Where $1 \le n \le 5$

n	A_n	$\lvert A_n \rvert$
1	1	1
2	12	1
3	123, 13	2
4	1234, 134, 124	3
5	12345, 1345, 1245, 1235, 135	5

↑ F_n

Weight of A_n

This beautiful occurrence of Fibonacci numbers in an unexpected place paves the way for a new combinatorial interpretation of Fibonacci polynomials f_n. To this end, consider an arbitrary element z in A_n. We introduce the concept of the *weight* of an adjacent pair $a_{i+1}a_i$ in z. The *weight* of the pair is x if $a_{i+1} - a_i = 1$; otherwise, its weight is 1. The *weight* of the sequence z is the product of the weights of the adjacent pairs in it, denoted by $w(z)$. The *weight* of A_n is the sum of the weights of all elements in the set; it is denoted by $w(A_n)$.

For example, consider the sequence $z = 123567$ in A_7. The (absolute) differences of adjacent elements are 1, 1, 2, 1, and 1. The corresponding weights are x, x, 1, x, and x; so $w(z) = x^4$. Likewise, the weight of the sequence 12357 is x^2 and that of 1357 is 1. Clearly, $w(1234567) = x^6$.

Table 33.5 shows both A_n and $w(A_n)$, where $1 \le n \le 5$. It exhibits a remarkable property: $w(A_n) = f_n$ in each case. We now confirm this stunning observation.

Clearly, $w(A_1) = f_1$ and $w(A_2) = f_2$. Suppose $w(A_k) = f_k$, where $1 \le k < n$.

Now consider an arbitrary element $S \in A_n$: $S = 1a_2 \cdots bn$. Suppose $b = n-1$. Then $S = \underbrace{1 \cdots (n-1)}n$, so, by definition, the sum of the weights of such sequences in A_n is xf_{n-1}.

On the other hand, suppose $b = n-2$: $S = 1a_2 \cdots (n-2)n$. The sum of the weights of such sequences $\underbrace{1 \cdots (n-2)}n$ in A_n is f_{n-2}.

TABLE 33.5. Sets A_n and Their Weights, Where $1 \le n \le 5$

n	A_n	$w(A_n)$
1	1	1
2	12	x
3	123, 13	$x^2 + 1$
4	1234, 134, 124	$x^3 + 2x$
5	12345, 1345, 1245, 1235, 135	$x^4 + 3x^2 + 1$

↑ f_n

Thus, by the addition principle, $w(A_n) = xf_{n-1} + f_{n-2} = f_n$. This, coupled with the initial conditions, confirms our observation. Thus we have the following result.

Theorem 33.10. *The sum of the weights of all sequences in A_n is f_n.* ■

Clearly, when $x = 1$, $w(A_n) = F_n = |A_n|$, as expected.

By virtue of the identity $f_{n+1} + f_{n-1} = l_n$, we can give a new interpretation to Lucas polynomials: $l_n = w(A_{n+1}) + w(A_{n-1})$.

With Theorem 33.10 at our disposal, we can confirm a number of interesting properties of Fibonacci polynomials. In the interest of brevity, we pursue just three of them.

Example 33.16. *Establish the identity* $f_{n+1} = \sum_{k=0}^{\lfloor n/2 \rfloor} \binom{n-k}{k} x^{n-k}$.

Proof. Consider the set A_{n+1}. By Theorem 33.10, $w(A_{n+1}) = f_{n+1}$.

We now re-compute $w(A_{n+1})$ using the number of differences of 2s. Consider an arbitrary element S in A_{n+1}. Suppose it contains k differences of 2s; they account for $2k$ possible differences of 1s. So S contains a total of $n - 2k$ differences of 1s. This implies that there are $(n - 2k) + k = n - k$ differences of 1s and 2s. Since there are $\binom{n-k}{k}$ sequences with k differences of 2s, the sum of the weights of such sequences equals $\binom{n-k}{k} x^{n-k}$, where $0 \le 2k \le n$. Consequently, $w(A_{n+1})$ equals the cumulative sum of all such weights; that is, $w(A_{n+1}) = \sum_{k=0}^{\lfloor n/2 \rfloor} \binom{n-k}{k} x^{n-k}$.

Equating the two sums yields the desired result. ■

We now illustrate this identity with a numeric example. Let $n = 4$, so $0 \le k \le 2$. There are $\binom{4}{0} = 1$ sequences with zero differences of 2s, namely, 12345; its weight is x^4. There are $\binom{3}{1} = 3$ sequences with one difference of 2s each: 1345, 1245, and 1235; the sum of their weights is $3x^2$. There is $\binom{2}{2} = 1$ sequence with two differences of 2s: 135; its weight is 1. So $w(A_5) = x^4 + 3x^2 + 1 = f_5$, as expected.

Breakability

A sequence in A_n is *unbreakable* at index i if $a_{i+1} - a_i = 2$; otherwise, it is *breakable* at i. For example, the sequence 135|68|9 is *not* breakable at location 1, 2 or 4; but is breakable at 3 and 5.

Using the concept of breakability, we can establish the addition formula $f_{m+n} = f_{m+1}f_n + f_m f_{n-1}$.

Example 33.17. *Establish the addition formula $f_{m+n} = f_{m+1}f_n + f_m f_{n-1}$.*

Proof. Consider the set of sequences A_{m+n}. By Theorem 33.10, $w(A_{m+n}) = f_{m+n}$.

Now consider an arbitrary sequence S in A_{m+n}. Suppose it is breakable at location $m+1$: $S = \underbrace{1 \cdots a_{m+1}}\, a_{m+2} \cdots a_{m+n}$, where $a_{m+2} - a_{m+1} = 1$. The sum of the weights of such sequences is $f_{m+1}f_n$.

On the other hand, suppose S is *not* breakable at location $m+1$. Clearly, the sum of the weights of the subsequences $1 \cdots a_m$ is f_m. Now consider the remaining subsequence $a_{m+1}a_{m+2} \cdots a_{m+n}$. Since $a_{m+2} - (a_{m+1} + 1)$, $a_{m+3} - (a_{m+1} + 1), \ldots, a_{m+n} - (a_{m+1} + 1)$ is an increasing sequence of length $n-1$, it follows that the sum of the weights of subsequences $a_{m+1}a_{m+2} \cdots a_{m+n}$ is f_{n-1}. So the sum of the weights of sequences S that are not breakable at index $m+1$ is $f_m f_{n-1}$.

Thus, by the addition principle, $w(A_{m+n}) = f_{m+1}f_n + f_m f_{n-1}$. Equating the two sums yields the given identity. ■

We now illustrate this argument with $m = 1$ and $n = 4$. There are eight elements in A_6; see Table 33.6.

TABLE 33.6.

12|3456	12|346	12|356	12456	13|456
1246	13|46	1356		

Exactly five of them are breakable at location 2: 12|3456, 12|346, 12|356, 13|456, and 13|46. Their weights are x^5, x^3, x^3, x^3, and x, respectively; and their sum is $x^5 + 3x^3 + x = f_2 f_5$.

Three of the sequences in A_6 are not breakable at location 2: 12456, 1246, and 1356. The sum of their weights is $x^3 + 2x = f_1 f_4$.

Thus $w(A_6) = f_2 f_5 + f_1 f_4 = (x^5 + 3x^3 + x) + (x^3 + 2x) = x^5 + 4x^3 + 3x = f_6$, as expected.

Example 33.18. *Establish the identity* $f_{2n} = \sum_{k=1}^{n} \binom{n}{k} f_k x^k$.

Proof. Consider the set A_{2n}. By Theorem 33.18, $w(A_{2n}) = f_{2n}$.

Since each monomial in f_{2n} has an odd exponent, it follows that each sequence in A_{2n} contributes an odd number of differences of 1s. The maximum number of such differences is $2n - 1$. The remaining differences are 2s, and the maximum number of such differences is $n - 1$. So every sequence must contribute at least $(n - 1) + 1 = n$ differences.

Now consider the sequence of differences from an arbitrary element $x \in A_{2n}$. Suppose there are k 1s in the first n differences, so the corresponding subsequence A yields $n - k$ differences of 2s. Replacing the $n - k$ 2s with 11s yields $k + (2n - 2k) = 2n - k$ differences of 1s. Now replace each 2 with 11 in the remaining subsequence B of differences; B contains $(2n - 1) - (2n - k) = k - 1$ ones; consequently, the resulting subsequence of x contains k elements: $x = \underbrace{k \text{ ones}}_{n \text{ differences}} \underbrace{k \text{ elements}}$. So $w(x) = x^k f_k$.

Since the k differences can be placed among the n differences in $\binom{n}{k}$ different ways, it follows that the sum of the weights of all sequences in A_{2n} with exactly k 1s in the first n differences equals $\binom{n}{k} f_k x^k$.

Since $0 \le k \le n$, it follows that $w(A_{2n}) = \sum_{k=0}^{n} \binom{n}{k} f_k x^k$. Since $f_0 = 0$, equating the two sums yields the desired result. ■

For example, consider the elements of A_6 in Table 33.6, where $n = 3$. Table 33.7 shows the corresponding sequences of differences.

TABLE 33.7.

111|11	111|2	112|1	121|1	211|1
122|	212|	221|		

Table 33.8 shows the possible values of k, the corresponding differences with k 1s in the first three differences, and the sum of their weights. It follows from the table that $w(A_6) = x^5 + 4x^3 + 3x = \sum_{k=1}^{n} \binom{n}{k} f_k x^k$.

TABLE 33.8.

k	Differences With k 1s in the First Three Differences	Sum of the Weights
1	122, 212, 221	$3x = \binom{3}{1} x f_1$
2	1121, 1211, 2111	$3x^3 = \binom{3}{2} x^2 f_2$
3	11111, 1112	$x^5 + x^3 = \binom{3}{3} x^3 f_3$

An Interesting Observation

We now make an interesting observation about the entries in Table 33.8. Suppose we add a ① at the beginning of each entry in the table. The resulting sequences are sequences of partial sums of the corresponding elements in Table 33.6. For example, ①1121 is the sequence of partial sums of the element 12356 in Table 33.6.

Bijection Between A_{n+1} and B_n

Finally, there is a bijection between the set of sequences A_{n+1} and the set of tilings B_n with square tiles and dominoes of a $1 \times n$ board, where $w(\text{square}) = x$ and $w(\text{domino}) = 1$.

Step 1. Let x be an arbitrary sequence in A_{n+1}. Let $x_i x_{i+1}$ be an adjacent pair in x. Represent it by a square tile with weight x if $x_{i+1} - x_i = 1$, and by a domino with weight 1 if $x_{i+1} - x_i = 2$. This yields a unique element y in B_n.

Step 2. On the other hand, let y be an arbitrary tiling in B_n. Replace each square with 1 and a domino with 2. This generates a sequence $b_1 b_2 b_3 \cdots$ of 1s and 2s. Now to recover the corresponding sequence x in A_{n+1}, generate a sequence $x = x_1 x_2 x_3 \cdots$ such that $x_{i+1} = x_i + b_i$, where $x_1 = 1$ and $i \geq 1$. Clearly, the sequence x is unique.

This step can be re-stated as follows: Let $x_1 = 1$ and $i \geq 1$. If cell i contains a square tile, then $x_{i+1} = x_i + 1$; otherwise, $x_{i+1} = x_i + 2$.

Thus the algorithm is reversible, establishing the desired bijection.

For example, consider the eight sequences of differences of the elements of A_6; see Tables 33.6 and 33.7. Replacing 1 with a square with weight x, and a domino with weight 1, generates the tilings of a 1×5 board; see Figure 33.53.

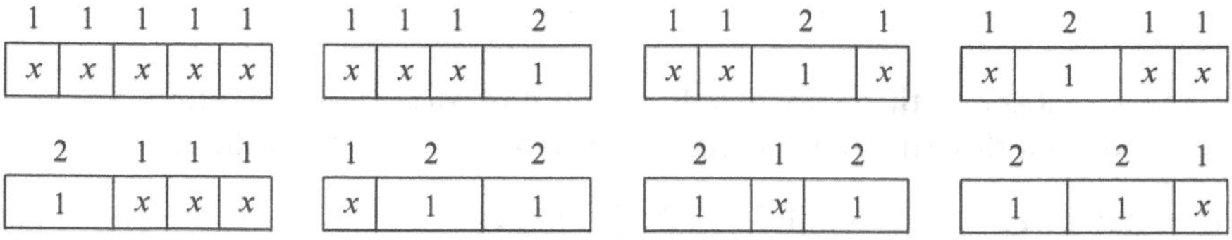

Figure 33.53. Sequences of differences of elements of A_6 and tilings in B_5.

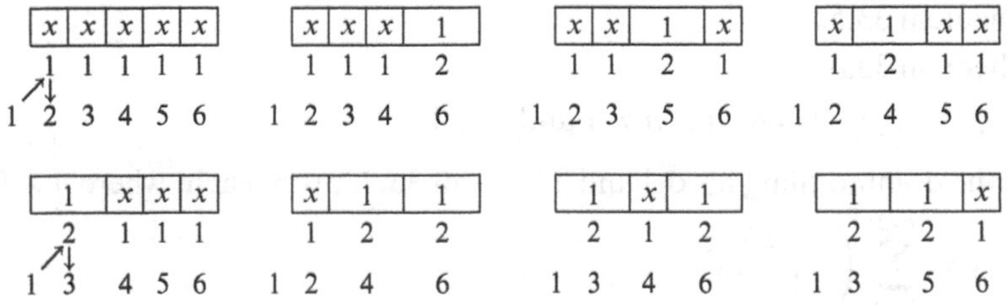

Figure 33.54. 1×5 tilings and the corresponding elements of A_6.

On the other hand, Figure 33.54 shows the 1×5 tilings, differences assigned to the tiles, and the corresponding sequences in A_6 they generate.

EXERCISES 33

Prove each.

1. $x \sum_{k=1}^{n} f_{4k-2} = f_{2n}^2$.
2. $x \sum_{k=1}^{n} f_{4k-3} = f_{2n-1} f_{2n}$.
3. $\sum_{k=1}^{n} F_{4k-1} = F_{2n} F_{2n+1}$.
4. $\sum_{k=1}^{n} L_{4k-2} = F_{4n}$.

Prove each combinatorially.

5. $f_{n+1}^2 + f_n^2 = f_{2n+1}$.
6. $f_{2n} = f_n l_n$.
7. $f_m = f_{n+1} f_{m-n} + f_n f_{m-n-1}$.
8. Prove that $f_{m+n} > x f_m f_n$, where $m > n > 1$ and $x \geq 1$.
9. Prove that $f_{nm} > x f_m^n$, where $m > n > 1$ and $x \geq 1$.

Prove each combinatorially.

10. Let $m \geq 1$ and $n \geq 0$. Then $f_m | f_{mn}$.
11. Theorem 33.2.
12. $q_{m+n} = q_{m+1} p_n + q_m p_{n-1}$.
13. Theorem 33.3.
14. Theorem 33.4.
15. $x \sum_{k=1}^{n} f_{2k} = f_{2n+1} - 1$.

16. Theorem 33.5.
17. Theorem 33.6.
18. $l_{m+n} > xf_m l_n$, where $m > n > 1$ and $x \geq 1$.

Using the domino tiling model and Theorem 33.7, prove each, where $n \geq 0$.

19. $f_{n+1} = \sum_{k=0}^{\lfloor n/2 \rfloor} \binom{n-k}{k} x^{n-2k}$.
20. $f_{2n} = f_n l_n$.
21. $f_{2n+1} = f_{n+1}^2 + f_n^2$.
22. $f_{2n+2} = \sum_{\substack{i,j \geq 0 \\ i+j \leq n}} \binom{n-i}{j}\binom{n-j}{i} x^{2n-2i-2j+1}$.

34

GRAPH-THEORETIC MODELS II

The greater our knowledge increases,
the greater our ignorance unfolds.
–John F. Kennedy (1917–1963)

In the previous chapter, we employed combinatorics to explore the beauty of both Fibonacci and Lucas polynomials, and to establish some of their elegant properties. This time, we explore them using graph-theoretic tools.

34.1 *Q*-MATRIX AND CONNECTED GRAPH

In Chapter 32, we used the *Q-matrix*

$$Q(x) = (q_{ij})_{2\times 2} = \begin{bmatrix} x & 1 \\ 1 & 0 \end{bmatrix}$$

to extract some Fibonacci delights. Interestingly, it can be translated into a connected digraph G with two vertices v_1 and v_2, and three directed edges. The directed edge from v_i to v_j is denoted by v_i–v_j, i–j, or by the "word" ij when there is *no* confusion. We define the *weight* w_{ij} of directed edge v_i–v_j to be q_{ij}, where $1 \le i \le j \le 2$; see Figure 34.1. Since a weight is assigned to each edge, G is a *weighted graph* and $Q(x)$ is its *weighted adjacency matrix*. Again, in the interest of brevity, we let Q_n denote $Q_n(x)$, when there is *no* confusion.

Fibonacci and Lucas Numbers with Applications, Volume Two. Thomas Koshy.

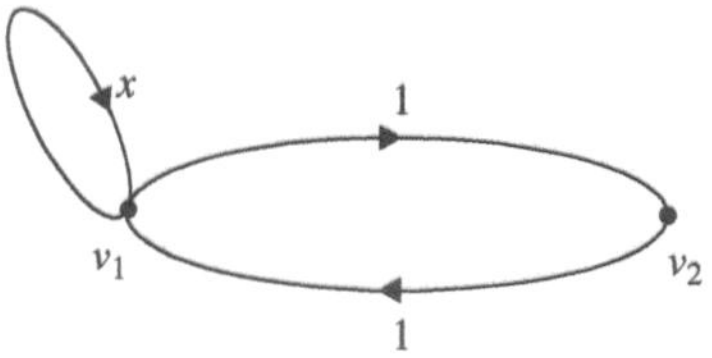

Figure 34.1. Weighted digraph.

Next we make a few more graph-theoretic definitions for clarity.

34.2 WEIGHTED PATHS

A *path* from vertex v_i to vertex v_j in a connected graph is a sequence $v_i - e_i - v_{i+1} - \cdots - v_{j-1} - e_{j-1} - v_j$ of vertices v_k and edges e_k, where edge e_k is incident with vertices v_k and v_{k+1}. The path is *closed* if its endpoints are the same; otherwise, it is *open*. The *length* ℓ of a path is the number of edges in the path; that is, it takes ℓ steps to reach one endpoint of the path from the other. The *weight* of a path is the product of the weights of the edges along the path. (Note that this definition is *different* from that of a path in graph theory, where it is known as a *walk*.) For example, the weight of the path $v_1v_1v_1v_2 = 1112$ is $x \cdot x \cdot 1 = x^2$.

Recall that the Q-matrix has the beautiful property that

$$Q^n = \begin{bmatrix} f_{n+1} & f_n \\ f_n & f_{n-1} \end{bmatrix},$$

where $n \geq 1$. Consequently, we can give a nice graph-theoretic interpretation of the recurrence $f_{n+1} = xf_n + f_{n-1}$:

$$\begin{pmatrix}\text{sum of the weights of}\\ \text{closed paths of length}\\ n \text{ from } v_1 \text{ to } v_1\end{pmatrix} = x\begin{pmatrix}\text{sum of the weights of}\\ \text{paths of length } n\\ \text{from } v_1 \text{ to } v_2\end{pmatrix} + \begin{pmatrix}\text{sum of the weights of}\\ \text{closed paths of length}\\ n \text{ from } v_2 \text{ to } v_2\end{pmatrix}.$$

The Lucas polynomial l_n also can be interpreted using this model. The sum of the weights of closed paths of length n originating at v_1 is f_{n+1}, and that of closed paths of the same length originating at v_2 is f_{n-1}; so the sum of the weights of closed paths of length n is $f_{n+1} + f_{n-1} = l_n$.

For example, consider the closed paths of length 4:

Paths originating at v_1: 11111 11121 11211 12111 12121

Sum of their weights: $x^4 + 3x^2 + 1 = f_5$

Paths originating at v_2: 21112 21212

Sum of their weights: $x^2 + 1 = f_3$

Cumulative sum: $x^4 + 4x^2 + 2 = l_4$.

Since $l_n = xf_n + 2f_{n-1}$, l_n can be interpreted slightly differently also:

$$l_n = x\begin{pmatrix}\text{sum of the weights of paths}\\ \text{of length } n \text{ from } v_1 \text{ to } v_2\end{pmatrix} + 2\begin{pmatrix}\text{sum of the weights of closed}\\ \text{paths of length } n \text{ from } v_2 \text{ to } v_2\end{pmatrix}.$$

For example, let $n = 4$. There are three paths of length 4 from v_1 to v_2: 11112, 11212, and 12112; the sum of their weights is $x^3 + 2x = f_4$. There are two paths of length 4 from v_2 to itself: 21112 and 21212; the sum of their weights is $x^2 + 1 = f_3$. Then $xf_4 + 2f_3 = x(x^3 + 2x) + 2(x^2 + 1) = x^4 + 4x^2 + 2 = l_4$, as expected.

Interesting Special Cases

Clearly, the graph-theoretic model provides one for Fibonacci and Lucas numbers by letting $x = 1$; it provides one for Pell and Pell–Lucas polynomials by replacing x with $2x$, and hence for Pell and Pell–Lucas numbers with $x = 1$. But we need to choose the initial conditions appropriately when the length of the path is zero.

34.3 *Q*-MATRIX REVISITED

The weighted adjacency matrix of a weighted graph can be employed to compute the sum of the weights of paths of a given length n between any two vertices, as the next theorem shows. The proof follows by induction [276].

Theorem 34.1. *Let A be the weighted adjacency matrix of a connected digraph with vertices $v_1, v_2, \ldots, v_k$, and n a positive integer. Then the ij-th entry of the matrix A^n records the sum of the weights of paths of length n from v_i to v_j.* ■

The next result follows by this theorem.

Corollary 34.1. *The ij-th entry of Q^n gives the sum of the weights of paths of length n from v_i to v_j, where $1 \le i \le j \le 2$.* ■

For example, we have $Q^4 = \begin{bmatrix} f_5 & f_4 \\ f_4 & f_3 \end{bmatrix}$. So the sum of the weights of paths of length 4 from v_1 to itself is f_5; the sum for such paths from v_1 to v_2 is f_4, and also from v_2 to v_1; and the sum for such paths from v_2 to itself is f_3; see Table 34.1.

The next result follows from this corollary.

Corollary 34.2. *The ij-th entry of $Q^n(1)$ records the number of paths of length n from v_i to v_j, where $1 \le i \le j \le 2$.* ■

Since $f_{n+1} + f_{n-1} = l_n$, the following result also follows from Corollary 34.1.

Corollary 34.3. *The sum of the weights of all closed paths of length n is l_n.* ■

TABLE 34.1. Paths of Length 4

	Paths from v_1 to v_1	Weight	Paths from v_1 to v_2	Weight
	11111	x^4	11112	x^3
	11121	x^2	11212	x
	11211	x^2	12112	x
	12111	x^2		
	12121	1		
Sum of the Weights		x^4+3x^2+1		x^3+2x
	Paths from v_2 to v_1	Weight	Paths from v_2 to v_2	Weight
	21111	x^3	21112	x^2
	21121	x	21212	1
	21211	x		
Sum of the Weights		x^3+2x		x^2+1

Interesting Observations

Notice that the *eigenvalues* of the matrix Q are α and β, and hence those of Q^n are α^n and β^n. Since the sum of the weights of all closed paths of length n is l_n, it follows that this sum is indeed the sum of the eigenvalues of Q^n. Since $l_n = f_{n+1} + f_{n-1}$, the sum also equals the *trace* of Q^n.

34.4 BYPRODUCTS OF THE MODEL

To showcase the beauty of this approach, we now confirm a few elegant properties of Fibonacci and Lucas polynomials. The essence of our technique lies in computing the sum of the weights in two different ways, and then equating the two sums.

Example 34.1. *Prove that* $f_{2n} = f_n l_n$.

Proof. Consider the sum of the weights of paths of length $2n$ from v_1 to v_2. By Corollary 34.1, the sum is f_{2n}.

We now count it in a different way. Such a path can land at v_1 or v_2 after n steps. Suppose it stops at v_1 after n steps: $v_1 \underbrace{-\cdots-}_{n\text{ steps}} v_1 \underbrace{-\cdots-}_{n\text{ steps}} v_2$. The sum of the weights of paths from v_1 to itself is f_{n+1}, and that from v_1 to v_2 is f_n. So, by the multiplication principle, the sum of the weights of paths from v_1 to v_2 that pass through v_1 after n steps is $f_{n+1}f_n$.

On the other hand, suppose the path lands at v_2 after n steps: $v_1 \underbrace{-\cdots-}_{n\text{ steps}} v_2 \underbrace{-\cdots-}_{n\text{ steps}} v_2$. The sum of the weights of paths from v_1 to v_2 is

f_n, and that from v_2 to itself is f_{n-1}. So, again by the multiplication principle, the sum of the weights of paths from v_1 to v_2 that pass through v_2 after n steps is $f_n f_{n-1}$.

Thus, by the addition principle, the sum of the weights of paths of length $2n$ from v_1 to v_2 is $f_{n+1}f_n + f_n f_{n-1} = f_n(f_{n+1} + f_{n-1}) = f_n l_n$.

Equating the two sums, we get the desired result. ■

For example, there are exactly $F_6 = 8$ paths of length 6 from v_1 to v_2:

111**1**112	121**1**112
111**1**212	121**2**112
111**2**112	112**1**212
112**1**112	121**1**212.

The sum of their weights is $x^5 + 4x^3 + 3x = f_6$.

Six of them land at v_1 after three steps (see the **1**s in boldface); and two at v_2 after three steps (see the **2**s in boldface).

$$\begin{aligned}\text{Sum of their weights} &= (x^5 + 3x^3 + 2x) + (x^3 + x)\\ &= (x^3 + 2x)(x^2 + 1) + (x^2 + 1)x\\ &= (x^2 + 1)(x^3 + 3x)\\ &= f_3 l_3.\end{aligned}$$

Next we establish the Fibonacci addition formula.

Example 34.2. *Prove that* $f_{m+n} = f_{m+1}f_n + f_m f_{n-1}$.

Proof. We compute in two different ways the sum of the weights of paths of length $m + n$ from v_1 to v_2. By Corollary 34.1, the sum of the weights of such paths is f_{m+n}.

Such a path can take us to v_1 or v_2 after m steps. Suppose it lands at v_1 after m steps: $v_1 \underbrace{-\cdots-}_{m\text{ steps}} v_1 \underbrace{-\cdots-}_{n\text{ steps}} v_2$. The sum of the weights of paths from v_1 to itself after m steps is f_{m+1} and that from v_1 to v_2 after n steps is f_n. Consequently, the sum of the weights of paths of length $m + n$ from v_1 to v_2 that land at v_1 after m steps is $f_{m+1}f_n$.

On the other hand, suppose the path takes us to v_2 after m steps: $v_1 \underbrace{-\cdots-}_{m\text{ steps}} v_2 \underbrace{-\cdots-}_{n\text{ steps}} v_2$. The sum of the weights of paths of length $m + n$ from v_1 to v_2 that land at v_2 after m steps is $f_m f_{n-1}$.

Combining the two cases, the sum of the weights of all such paths of length $m + n$ is $f_{m+1}f_n + f_m f_{n-1}$.

The addition formula follows by equating the two sums. ■

We can employ the same technique to establish independently that $f_{n+1}^2 + f_n^2 = f_{2n+1}$ and $l_{m+n} = f_{m+1}l_n + f_m l_{n-1}$; see Exercises 34.11 and 34.12.

Example 34.3. *Prove the Lucas formula* $f_{n+1} = \sum_{k=0}^{\lfloor n/2 \rfloor} \binom{n-k}{k} x^{n-2k}$.

Proof. This time, we focus on the sum of the weights of closed paths of length n originating at v_1. By Corollary 34.1, the sum is f_{n+1}.

Suppose such a path contains k closed paths 121 of two edges, where $k \geq 0$; call them d-edges ("d" for "double") for convenience. The k d-edges account for $2k$ edges, so there are $n - 2k$ edges remaining in the path. Consequently, the total number of *elements* (edges or d-edges) is $(n - 2k) + k = n - k$. The $n - 2k$ edges contribute x^{n-2k} and the k d-edges 1^k to the weight of the path; so the weight of such a path is $x^{n-2k} \cdot 1^k = x^{n-2k}$.

The k d-edges can be selected from the $n - k$ elements in $\binom{n-k}{k}$ ways, where $0 \leq k \leq 2n$. So the sum of the weights of all closed paths originating at v_1 equals $\sum_{k=0}^{\lfloor n/2 \rfloor} \binom{n-k}{k} x^{n-2k}$.

Equating the two sums yields the desired formula. ■

For example, let $n = 5$. It follows from Table 34.2 that the sum of the weights of all closed paths originating at v_1 is $x^5 + 4x^3 + 3x = f_6$. (The d-edges are boldfaced or parenthesized in the table.)

TABLE 34.2. Closed Paths of Length 5 from v_1 to v_1

Number of d-edges k	Closed Paths of Length 5 with k d-edges	Sum of the Weights of Such Paths
0	111111	x^5
1	111**121** 11**121**1 1**121**11 **121**111	$4x^3$
2	1**12**(121) **121**(121) **12**(**121**)1	$3x$

Similarly, there is one closed path of length 4 starting at v_1 with no d-edges: 11111; three with one d-edge: 11**121**, 1**121**1, and **121**11; and one with two d-edges: **12(121)**. The sum of their weights is $x^4 + 3x^2 + 1 = f_5$.

The next identity expresses f_{2n} in terms of the first n Fibonacci polynomials.

Example 34.4. *Prove the identity* $f_{2n} = \sum_{k=1}^{n} \binom{n}{k} f_k x^k$.

Proof. Consider the closed paths of length $2n - 1$ originating at v_1. The sum of the weights of such paths is f_{2n}.

We now compute this sum in a different way. Since $2n - 1$ is odd, each such path P must contain an odd number of edges (loops) 11. The remaining edges must be d-edges. Since there can be a maximum of $n - 1$ d-edges, every path must contain at least $(n - 1) + 1 = n$ elements.

Suppose there are k loops among the first n elements of path P. The corresponding subpath A contains $n - k$ d-edges; its length is $k + 2(n - k) = 2n - k$. The remaining subpath B is of length $(2n - 1) - (2n - k) = k - 1$; so path P is of the form $\underbrace{\text{subpath } A}_{\text{length } 2n-k}\ \underbrace{\text{subpath } B}_{\text{length } k-1}$.

The k loops in subpath A can be placed among the n elements in $\binom{n}{k}$ distinct ways. The sum of the weights of subpaths B is f_k. Consequently, the sum of the weights of such paths P is $\binom{n}{k} f_k x^k$, where $1 \le k \le n$.

Thus the sum of the weights of all closed paths of length $2n - 1$ is $\sum_{k=1}^{n} \binom{n}{k} f_k x^k$.

The given identity now follows by equating the two sums. ■

We now illustrate this combinatorial technique with $n = 3$. There are $8 = F_6$ closed paths of length 5 originating at v_1; see Table 34.3, where the first three edges of each path are boldfaced for convenience.

TABLE 34.3.

k	Closed Paths	Sum of the Weights
1	**112**121 **121**211 **121**121	$3x$
2	**111**211 **112**111 **121**111	$3x^3$
3	**111**111 **111**121	$x^5 + x^3$

It follows from the table that the cumulative sum of the weights of all closed paths is $x^5 + 4x^3 + 3x = f_6$, as expected.

Next we confirm the identity $f_{2n+2} = \sum_{i,j \ge 0} \binom{n-i}{j}\binom{n-j}{i} x^{2n-2i-2j+1}$ using the graph-theoretic model. We accomplish this job using closed paths of length $2n + 1$ from v_1 to v_1, and d-edges. But first we make an important observation. Since every d-edge is of length 2, every such closed path must contain an odd number of loops 11. So there must be a special loop M with an equal number of loops on either side. For convenience, we call M the *median loop*.

We are now ready for the proof; it is basically the same as that in Example 34.3, but the language is graph-theoretic and the proof still refreshing.

Example 34.5. *Confirm the identity* $f_{2n+2} = \sum_{i,j\geq 0} \binom{n-i}{j}\binom{n-j}{i} x^{2n-2i-2j+1}$.

Proof. Consider the closed paths of length $2n+1$ from v_1 to v_1. The sum of the weights of such paths is f_{2n+2}.

Now consider such an arbitrary path P. Let M denote the median loop in it. Suppose there are i d-edges to the left of M and j d-edges to its right: $\underbrace{i \text{ d-edges}}_{\text{to left}}\, 1\, 1\, \underbrace{j \text{ d-edges}}_{\text{to right}}$. Then P contains $2n+1-2i-2j$ loops. So there are $n-i-j$ loops on either side of M. Consequently, there are $(n-i-j)+i = n-j$ edges to the left of M, of which i are d-edges. The i d-edges can be placed among the $n-j$ edges in $\binom{n-j}{i}$ different ways. The weight of this subpath is $\binom{n-j}{i} x^{n-i-j}$.

Similarly, the weight of the subpath to the right of M is $\binom{n-i}{j} x^{n-i-j}$. So the weight of path P equals

$$\binom{n-j}{i} x^{n-i-j} \cdot x \cdot \binom{n-i}{j} x^{n-i-j} = \binom{n-i}{j}\binom{n-j}{i} x^{2n-2i-2j+1},$$

where $0 \leq i+j \leq n$.

Thus the sum of the weights of all closed paths P is

$$\sum_{i,j\geq 0} \binom{n-i}{j}\binom{n-j}{i} x^{2n-2i-2j+1}.$$

Equating the two sums yields the given identity. ■

For example, Table 34.4 gives the closed paths of length 5 from v_1 to itself, where we have identified the loops in boldface. The up arrows indicate the median loops, and the numbers below their locations.

TABLE 34.4.

11111	**111**12	**11**12**1**	**1**12**11**	12**111**	**1**1212	12**1**12	1212**1**
↑	↑	↑	↑	↑	↑	↑	↑
3	2	2	4	4	1	3	5

Table 34.5 shows the possible locations m of the median loops, corresponding value(s) of i and j, and the weights of corresponding path(s). It follows from the table that the sum of the weights of all closed paths of length 5 from v_1 to v_1 is

$$x^5 + 4x^3 + 3x = f_6 = \sum_{0\leq i+j\leq 2} \binom{2-i}{j}\binom{2-j}{i} x^{5-2i-2j}.$$

In the next example, we prove Cassini's formula using the graph-theoretic model and induction.

TABLE 34.5.

m	i	j	Sum(s) of the Weight(s) of Path(s)
1	0	2	x
2	0	1	$2x^3$
3	0	0	x^5
	1	1	x
4	1	0	$2x^3$
5	2	0	x

↑
sum = f_6

Example 34.6. *Prove that* $f_{n+1}f_{n-1} - f_n^2 = (-1)^n$.

Proof. Clearly, the formula works for $n = 1$ and $n = 2$. Suppose it is true for an arbitrary integer $n \geq 2$.

Form two lists A and B of pairs of closed paths from v_1 to v_1. List A consists of such pairs (v, w) of paths of length $n-1$ and $n+1$, respectively. List B consists of pairs (x, y) of paths of the same length n:

List A	List B
$v:\ 1v_2 \cdots v_{n-1}1$	$x:\ 1x_2x_3 \cdots x_n1$
$w:\ 1w_2w_3w_4 \cdots w_{n+1}1$	$y:\ 1y_2y_3 \cdots y_n1$

Clearly, the sum of the weights of the pairs (v, w) is f_nf_{n+2} and that of the pairs (x, y) is f_{n+1}^2.

We now establish a bijection between two suitable subsets of A and B.

Case 1. Suppose $w_2 = 1$. Then moving $w_1 = 1$ to the beginning of v produces a pair (x, y) of paths of length n each:

$$11v_2 \cdots v_{n-1}1$$
$$1w_3w_4 \cdots w_{n+1}1.$$

Case 2. On the other hand, suppose $w_2 = 2$. Then shifting w_1 to v does *not* generate a pair (x, y) in B. So such a pair (v, w) in A does *not* have a matching pair (x, y) in B.

We now count those nonmatchable pairs in A. When $w_2 = 1, w_3 = 1$. So $w = 121w_4 \cdots w_{n+1}1$. The sum of the weights of such paths is f_n. So the sum of the weights of such pairs (v, w) is f_n^2; *no* such pairs have matching elements (x, y) in B. Consequently, the sum of the weights of the pairs in A that have matching elements in B equals $f_nf_{n+2} - f_n^2$.

Let us now reverse the order. Shift $x_1 = 1$ from x to the beginning of y.

Case 1. Suppose $x_2 = 1$. Then $x = \underbrace{1x_3 \cdots x_n1}_{\text{length } n-1}$ and $y = \underbrace{11y_2 \cdots y_n1}_{\text{length } n+1}$. The corresponding pair (x, y) is a valid element in A.

Case 2. Suppose $x_2 = 2$. Then $x_3 = 1$; and $x = \underbrace{21\cdots x_n1}_{\text{length } n-1}$ and $y = \underbrace{1y_2\cdots y_n1}_{\text{length } n}$.

The corresponding pair (x, y) does *not* generate a matching element in A. The sum of the weights of such unmatchable pairs equals $f_{n-1}f_{n+1}$; this equals $f_n^2 + (-1)^n$, by the inductive hypothesis.

Consequently, the sum of the weights of the pairs (x, y) that have matchable counterparts in A equals $f_{n+1}^2 - [f_n^2 + (-1)^n]$.

Since the matching between the two sets of matchable pairs is bijective, the sums of their weights must be equal; that is, $f_nf_{n+2} - f_n^2 = f_{n+1}^2 - [f_n^2 + (-1)^n]$. This implies that $f_nf_{n+2} - f_{n+1}^2 = (-1)^{n+1}$. So Cassini's formula works for $n + 1$ also. Thus, by induction, it works for every $n \geq 1$. ■

We now illustrate the essence of the proof for the case $n = 3$. Table 34.6 lists the pairs (v, w) of paths $v = v_1v_2$ and $w = w_1w_2w_3w_4$ from v_1 to v_1. Six of them are numbered 1 through 6 for convenience; the sum of the weights of these pairs is $x^2(x^4 + 2x^2) + 1 \cdot (x^4 + 2x^2) = (x^2 + 1)(x^4 + 2x^2) = f_3f_5 - f_3^2$. The others are labeled a through d; the sum of the weights of these four pairs equals $x^2(x^2 + 1) + 1 \cdot (x^2 + 1) = (x^2 + 1)^2$. The grand total is $(x^2 + 1)(x^4 + 3x^2 + 1) = f_3f_5$.

TABLE 34.6.

111	111	111	111	111	121	121	121	121	121
11111	11121	11211	12111	12121	11111	11121	11211	12111	12121
1	2	3	*a*	*b*	4	5	6	*c*	*d*

Table 34.7 shows the pairs (x, y) of closed paths $x = x_1x_2x_3$ and $y = y_1y_2y_3$ from v_1 to itself. Again, six of them are labeled 1 through 6; the sum of their weights is $(x^3 + x)(x^3 + 2x) = f_4^2 - [f_3^2 + (-1)^3]$. The sum of the weights of the remaining three, labeled α, β, and γ, is $x(x^3 + 2x)$. Their grand total is $(x^3 + 2x)^2 = f_4^2$.

TABLE 34.7.

1111	1111	1111	1121	1121	1121	1211	1211	1211
1111	1121	1211	1111	1121	1211	1111	1121	1211
1	2	3	4	5	6	α	β	γ

The matchable pairs from both lists are numbered 1 through 6, and the others by letters. Both sets have the same sum of weights: $(x^2 + 1)(x^4 + 2x^2) = (x^3 + x)(x^3 + 2x)$, as expected.

Next we prove the identity $l_{n+1} + l_{n-1} = (x^2 + 4)f_n$. We do this by establishing a *one-to-five correspondence* from the set of paths of length n from v_1 to v_2 to that of closed paths of length $n + 1$ or $n - 1$.

Example 34.7. *Establish the identity* $l_{n+1} + l_{n-1} = (x^2+4)f_n$.

Proof. By Corollary 34.3, the sum of the weights of closed paths of length $n+1$ is l_{n+1}, and that of length $n-1$ is l_{n-1}. So the sum of the weights of paths of length $n+1$ or $n-1$ is $l_{n+1} + l_{n-1}$.

By Corollary 34.1, the sum of the weights of paths of length n from v_1 to v_2 is f_n. Let $w = w_1w_2\cdots w_nw_{n+1}$ be such a path. Clearly, $w_1 = 1$ and $w_{n+1} = 2$. In addition, every v_2 must be preceded by v_1; so $w_n = 1$. Thus $w = \underbrace{1w_2\cdots w_{n-1}12}_{\text{length } n}$.

We now devise an algorithm in five steps to establish the aforementioned correspondence:

Step 1. Append a 1 at the end of w. This generates a closed path of length $n+1$ from v_1 to v_1: $\underbrace{1w_2\cdots w_{n-1}121}_{\text{length } n+1}$. The sum of the weights of such paths equals f_n.

Step 2. Delete $w_{n+1} = 2$. This results in a closed path of length $n-1$ from v_1 to itself: $\underbrace{1w_2\cdots w_{n-1}1}_{\text{length } n-1}$. The sum of the weights of such paths also equals f_n.

Step 3. Place a 2 at the beginning of w. This creates a closed path of length $n+1$ from v_2 to v_2: $\underbrace{21w_2\cdots w_{n-1}12}_{\text{length } n+1}$. The sum of the weights of such paths is again f_n.

Step 4. Replace $w_{n+1} = 2$ with 11. This operation produces a closed path of length $n+1$ from v_1 to itself: $\underbrace{1w_2\cdots w_{n-1}111}_{\text{length } n+1}$. The sum of the weights of such paths is $x^2 f_n$.

These four steps do *not* account for all closed paths of length $n+1$ or $n-1$, namely, the ones that begin with $w_1w_2 = 11$ or 12. This takes us to Step 5, which has therefore two parts.

Step 5A. Suppose $w_2 = 1$. Then delete w_1 and insert 11 at the end of w. This generates a closed path of length $n+1$ from v_1 to itself: $\underbrace{w_2\cdots w_{n-1}1211}_{\text{length } n+1}$. The sum of the weights of such paths is xf_{n-1}.

Step 5B. Suppose $w_2 = 2$. Then delete w_1. This gives a closed path of length $n-1$ from v_2 to itself: $\underbrace{w_2\cdots w_{n+1}}_{\text{length } n-1}$. Such paths contribute f_{n-2} to the sum of the weights.

The sum of the weights of closed paths generated by Step 5 equals $xf_{n-1} + f_{n-2} = f_n$.

Steps 1–5 do *not* produce duplicate paths. So the sum of the weights of closed paths they create is $(x^2+4)f_n$; some are of length $n+1$ and the rest of length $n-1$. Thus the two sums of weights must be equal; that is, $l_{n+1} + l_{n-1} = (x^2+4)f_n$, as desired. ■

We now illustrate the five steps of the algorithm for the case $n = 4$. There are 11 closed paths of length 5 or 4, and four of length 3, a total of 15 closed paths of length 5, 4, or 3:

Length 5:	111111	111121	111211	112111	121111	112121	121121	121211
Length 4:	211112	211212	212112					
Length 3:	1111	1121	1211					
	2112							

The sum of the weights of paths of length 5 is $l_5 = x^5 + 5x^3 + 5x$, and that of paths of length 3 is $l_3 = x^3 + 3x$. So the sum for paths of length 5 or 3 is $x^5 + 6x^3 + 8x$.

Now consider paths of length 4 from v_1 to v_2. There are three such paths $w = w_1w_2w_3w_4w_5$: 11112, 11212, 12112. The sum of their weights is $f_4 = x^3 + 2x$.

1) Appending a 1 at the end of w yields three closed paths of length 5 from v_1 to v_1: 111121, 112121, 121121. The sum of their weights is f_4.
2) Deleting $w_5 = 2$ produces three closed paths of length 3 from v_1 to v_1: 1111, 1121, 1211. The sum of their weights is also f_4.
3) Placing a 2 at the beginning of w gives three closed paths of length 5 from v_2 to v_2: 211112, 211212, 212112. The sum of their weights is once again f_4.
4) Replacing w_5 with 11 generates three more closed paths of length 5 from v_1 to v_1: 111111, 112111, 121111. The sum of their weights is $x^2 f_4$.

5a) When $w_2 = 1$, delete w_1 and append 11 at the end. This yields two closed paths of length 5: 111211, 121211. The sum of their weights equals $x^3 + x$.

5b) When $w_2 = 2$, we delete w_1. This creates exactly one closed path of length 3: 2112. Its weight is x.
So the sum of the weights of paths generated by Step 5 equals $(x^3 + x) + x = f_4$.

Thus, by Steps 1–5, the sum of the weights of all closed paths of length 5 or 3 equals $(x^2 + 4)f_4 = x^5 + 6x^3 + 8x = l_5 + l_3$, as expected.

Interestingly, there is a bijection between the set of closed paths of length n from v_1 to v_1, and that of Fibonacci tilings of a $1 \times n$ board. We let the d-edges do the work for us.

34.5 A BIJECTION ALGORITHM

In the Fibonacci tilings of a $1 \times n$ board with squares and dominoes, we assign a weight x to each square and 1 to each domino [285]. Now consider a closed path of length n from v_1 to itself. Replacing the edge 11 with a square and a d-edge with a domino results in a tiling of the board. Clearly, this process is reversible. Consequently, this algorithm establishes the desired bijection.

To illustrate the bijection algorithm, consider the eight closed paths of length 5 starting at v_1 in Table 34.2. The algorithm produces the corresponding Fibonacci

TABLE 34.8. Closed Paths from v_1 to v_1 and the Corresponding Fibonacci Tilings

Closed Paths	Fibonacci Tilings	Weights of Tilings
111111	x x x x x	x^5
111121	x x x 1	x^3
111211	x x 1 x	x^3
112111	x 1 x x	x^3
121111	1 x x x	x^3
112121	x 1 1	x
121121	1 x 1	x
121211	1 1 x	x

tilings of a 1×5 board with squares and dominoes; they are showcased in Table 34.8. As expected, the sum of the weights of the paths or tilings is $x^5 + 4x^3 + 3x = f_6$.

34.6 FIBONACCI AND LUCAS SUMS

We can elicit the power of matrices and the graph-theoretic model to interpret combinatorially the following summation formulas:

1) $x \sum_{k=1}^{n} f_k = f_{n+1} + f_n - 1$

2) $x \sum_{k=1}^{n} l_k = l_{n+1} + l_n + x - 2$

3) $x \sum_{k=1}^{n} f_{2k-1} = f_{2n}$

4) $x \sum_{k=1}^{n} l_{2k-1} = l_{2n} - 2$

5) $x \sum_{k=1}^{n} f_{2k} = f_{2n+1}$

6) $x \sum_{k=1}^{n} l_{2k} = l_{2n+1} - x.$

In the interest of brevity, we interpret formula 1) and leave the others for Fibonacci enthusiasts.

To this end, first we make an important useful observation about the Q-matrix:

$$x \sum_{k=1}^{n} Q^k = \begin{bmatrix} x \sum_{k=1}^{n} f_{k+1} & x \sum_{k=1}^{n} f_k \\ x \sum_{k=1}^{n} f_k & x \sum_{k=1}^{n} f_{k-1} \end{bmatrix}. \tag{34.1}$$

We need the following result also [276].

Theorem 34.2. *Let A be the weighted adjacency matrix of a weighted graph with vertices $v_1, v_2, \ldots, v_k$, and n a positive integer. Then the ij-th entry of the matrix $A + A^2 + \cdots + A^n$ gives the sum of the weights of the paths of length $\leq n$ from vertex v_i to v_j.* ■

With these two results at our finger tips, we are ready for the interpretation.

1) It follows from equation (34.1) that $x \sum_{k=1}^{n} f_k$ represents x times the sum of the weights of paths of length $\leq n$ from v_1 to v_2. It equals $f_{n+1} + f_n - 1$.

For example, consider the paths of length ≤ 5 from v_1 to v_2; see Table 34.9. We then have

$$x \sum_{k=1}^{5} f_k = x^5 + x^4 + 4x^3 + 3x^2 + 3x$$

$$= (x^5 + 4x^3 + 3x) + (x^4 + 3x^2 + 1) - 1$$

$$= f_6 + f_5 - 1.$$

TABLE 34.9. Paths of Length ≤ 5 from v_1 to v_2

n	Paths of Length n from v_1 to v_2	Sum of the Weights
1	12	1
2	112	x
3	1112 1212	$x^2 + 1$
4	11112 11212 12112	$x^3 + 2x$
5	111112 111212 112112 121112 121212	$x^4 + 3x^2 + 1$
		↑ f_n

2) Clearly, x times the sum of the weights of closed paths of length $\leq n$ is given by

$$x\left(\sum_{k=1}^{n} f_{k+1} + \sum_{k=1}^{n} f_{k-1}\right) = x\sum_{k=1}^{n}\left(f_{k+1} + \sum_{k=1}^{n} f_{k-1}\right)$$

$$= x\sum_{k=1}^{n} l_k$$

$$= l_{n+1} + l_n - x - 2.$$

Thus $x \sum_{k=1}^{n} l_k$ represents the sum of the weights of closed paths of length at most n; it equals $l_{n+1} + l_n - x - 2$.

3) Since $l_{2k} = f_{2k+1} + f_{2k-1}$, we have

$$\begin{aligned} x\sum_{k=1}^{n} l_{2k} &= x\sum_{k=1}^{n} f_{2k+1} + x\sum_{k=1}^{n} f_{2k-1} \\ &= x\begin{pmatrix}\text{sum } S_1 \text{ of the weights} \\ \text{of closed paths of length} \\ n \text{ from } v_1 \text{ to } v_1\end{pmatrix} + x\begin{pmatrix}\text{sum } S_2 \text{ of the weights} \\ \text{of closed paths of length} \\ n \text{ from } v_2 \text{ to } v_2\end{pmatrix} \\ &= (f_{2n+2} - x) + f_{2n} \\ &= l_{2n+1} - x. \end{aligned}$$

TABLE 34.10. Paths of Even Length ≤ 6 from v_1 to v_1

n	Paths of Length $2n$ from v_1 to v_1	Sum S_1 of the Weights
1	111 121	$x^2 + 1$
2	11111 11121 12111 11211 12121	$x^4 + 3x^2 + 1$
3	1111111 1111121 1111211 1112111 1121111 1211111 1112121 1121121 1211121 1211211 1121121 1212111 1212121	$x^6 + 5x^4 + 6x^2 + 1$

↑ f_{2n+1}

For example, let $n = 3$. Consider the weights of closed paths of even length ≤ 6 from v_1 to v_1, and from v_2 to v_2; see Tables 34.10 and 34.11.

TABLE 34.11. Paths of Even Length ≤ 6 from v_2 to v_2

n	Paths of Length $2n$ from v_2 to v_2	Sum S_2 of the Weights
1	212	1
2	21112 21212	$x^2 + 1$
3	2111112 2111212 2112112 2121112 2121212	$x^4 + 3x^2 + 1$

↑ f_{2n-1}

Then

$$\begin{aligned} xS_1 + xS_2 &= x(x^6 + 6x^4 + 10x^3 + 3) + x(x^4 + 4x^3 + 3) \\ &= x^7 + 7x^5 + 14x^3 + 6x \\ &= l_7 - x. \end{aligned}$$

Next we present a new graph-theoretic model for Fibonacci polynomials. We do this by introducing the concept of a Fibonacci walk, and then by assigning a weight to each walk. This approach introduces a new way of studying Fibonacci polynomials and deriving well-known Fibonacci identities. In addition, it shows the occurrence of Fibonacci polynomials and numbers in unexpected contexts.

34.7 FIBONACCI WALKS

Consider a horizontal path starting at the origin and made up of n steps in the easterly direction (E). An E-*step* (or 1-*step*) is a unit step. A D-*step* (or 2-*step*) is made up of two E-steps [280, 363]. An E-step is assigned a *weight* x, and a D-step is assigned a weight 1. A *Fibonacci walk* is a horizontal path made up of E- or D-steps with the assigned weights.

The *weight* of a Fibonacci walk is the product of the weights of the steps in it. The weight of a walk of length 0 is defined as 1.

Figure 34.2 shows the Fibonacci walks of length n and the sum of their weights, where a thick dot indicates the origin and $0 \le n \le 5$.

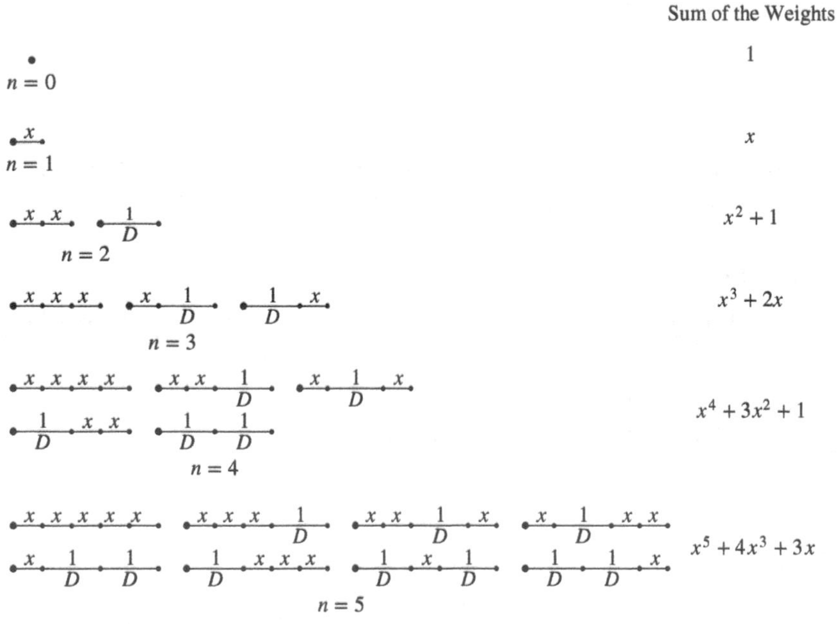

Figure 34.2. Fibonacci walks.

Let $a_n = a_n(x)$ denote the sum of the weights of Fibonacci walks of length n. Using the data from Figure 34.2, we conjecture that $a_n = f_{n+1}$. We now confirm this observation.

Clearly, $a_0 = 1 = f_1$ and $a_1 = x = f_2$. Now consider an arbitrary Fibonacci walk of length n, where $n \geq 2$. Suppose it begins with an E-step: $\underbrace{x}_{E}\ \underbrace{\text{subwalk}}_{\text{length } n-1}$.
By definition, the sum of the weights of such walks equals xa_{n-1}.

On the other hand, suppose the walk begins with a D-step: $\underbrace{1}_{D}\ \underbrace{\text{subwalk}}_{\text{length } n-2}$.
The sum of the weights of such walks equals $1 \cdot a_{n-2} = a_{n-2}$.

These two cases are mutually exclusive; so, by the addition principle, $a_n = xa_{n-1} + a_{n-2}$. This recurrence, coupled with the initial conditions, implies that $a_n = f_{n+1}$, as desired.

We now illustrate the beauty and power of this combinatorial approach by establishing three charming properties of Fibonacci polynomials. Our technique hinges on the well-known *Fubini's principle.*

First, we prove the Lucas-like formula $f_{n+1} = \sum_{k=0}^{\lfloor n/2 \rfloor} \binom{n-k}{k} x^{n-2k}$. Consider an arbitrary Fibonacci walk of length n. The sum of the weights of such walks is f_{n+1}.

We now compute this sum in a different way. Suppose the walk contains k D-steps. Then it contains $n - 2k$ E-steps, so the walk contains a total of $(n - 2k) + k = n - k$ steps, where $2k \leq n$. The k D-steps can be placed among the $n - k$ steps in $\binom{n-k}{k}$ ways. Since each contains $n - 2k$ E-steps, the sum of the weights of such walks is $\binom{n-k}{k} x^{n-2k}$. So the sum of the weights of all walks of length n equals $\sum_{k=0}^{\lfloor n/2 \rfloor} \binom{n-k}{k} x^{n-2k}$. Equating the two sums yields the desired result.

Consequently, there are $f_{n+1}(1) = F_{n+1}$ Fibonacci walks of length n.

For example, $\sum_{k=0}^{2} \binom{5-k}{k} x^{5-2k} = x^5 + 4x^3 + 3x = f_6$, and $F_6 = 8$ Fibonacci walks of length 6, as expected; see Figure 34.2.

Median E-step

Suppose the length of a walk is odd. Then it must contain an odd number of E-steps. So it has a special E-step M such that there is an equal number of E-steps on either side; we call it the *median step.*

The up arrows in Figure 34.3 indicate the median steps in the Fibonacci walks of length 3.

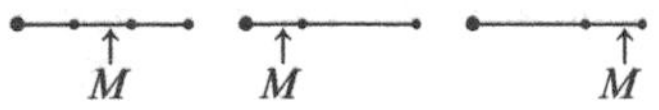

Figure 34.3. Fibonacci walks of length 3.

Now consider an arbitrary Fibonacci walk of length $2n+1$. The sum of the weights of all such walks is f_{2n+2}.

We now compute the sum in a different way. Let M denote the median step of the walk. Suppose there are i D-steps to its left and j D-steps to its right. Then the walk contains a total of $(2n+1)-(2i+2j)=2n-2i-2j+1$ E-steps; so there are $n-i-j$ E-steps on either side of M. Consequently, there are $(n-i-j)+i=n-j$ steps to the left of M, and $(n-i-j)+j=n-i$ steps to its right: $\underbrace{n-i-j\ E\text{–steps}}_{n-j\text{ steps}}\ M\ \underbrace{n-i-j\ E\text{–steps}}_{n-i\text{ steps}}$.

The $n-i-j$ E-steps to the left of M can be selected from the $n-j$ steps in $\binom{n-j}{n-i-j}=\binom{n-j}{i}$ different ways; so the sum of the weights of the corresponding subwalks is $\binom{n-j}{i}x^{n-i-j}$. Likewise, the sum of the weights of the subwalks to the right of M is $\binom{n-i}{j}x^{n-i-j}$. Consequently, the weight of the walk is $\binom{n-i}{j}\binom{n-j}{i}x^{2n-2i-2j+1}$. So the sum of the weights of all Fibonacci walks of length $2n+1$ equals $\sum\limits_{\substack{i,j\geq 0\\ i+j\leq n}}\binom{n-i}{j}\binom{n-j}{i}x^{2n-2i-2j+1}$.

Equating the two sums, we get

$$f_{2n+2}=\sum_{\substack{i,j\geq 0\\ i+j\leq n}}\binom{n-i}{j}\binom{n-j}{i}x^{2n-2i-2j+1}.$$

For example, $f_6=\sum\limits_{\substack{i,j\geq 0\\ i+j\leq 2}}\binom{2-i}{j}\binom{2-j}{i}x^{5-2i-2j}=x^5+4x^3+3x$, as expected.

Breakability

Next we establish the addition formula $f_{m+n}=f_{m+1}f_n+f_mf_{n-1}$. To this end, we introduce the concept of breakability. A walk is *unbreakable* at step k if a D-step occupies steps k and $k+1$; otherwise, it is *breakable* at step k.

For example, the walk in Figure 34.4*a* is breakable at step 2, but unbreakable at step 3; and the one in Figure 34.4*b* is breakable at steps 1 and 3, but unbreakable at step 2.

(*a*) (*b*)

Figure 34.4.

Now consider a Fibonacci walk of length $m+n-1$. The sum of the weights of such walks is f_{m+n}.

Consider an arbitrary walk of length $m+n-1$. Suppose it is breakable at step m: $\underbrace{\text{subwalk}}_{\text{length } m}\ \underbrace{\text{subwalk}}_{\text{length } n-1}$. By the multiplication principle, the sum of the weights of such walks is $f_{m+1}f_n$.

On the other hand, suppose the walk is *not* breakable at m: $\underbrace{\text{subwalk}}_{\text{length } m-1}\ D\ \underbrace{\text{subwalk}}_{\text{length } n-2}$. Again, by the multiplication principle, the sum of the weights of such walks is $f_m \cdot 1 \cdot f_{n-1} = f_m \cdot f_{n-1}$.

Thus, by the addition principle, the sum of the weights of all Fibonacci walks of length $m+n-1$ is $f_{m+1}f_n + f_m f_{n-1}$. Combining the two sums, we get the desired identity.

Since $f_{n+1} + f_{n-1} = l_n$, it follows by the addition formula that $f_{2n} = f_n l_n$.

A Fibonacci Bijection

We now add that there is a bijection between the set of Fibonacci walks of length n, and the set of Fibonacci tilings of a $1 \times n$ board with 1×1 tiles and 1×2 tiles (dominoes), where weight(square tile) = 1 = weight(domino). To see this, consider a Fibonacci walk of length n. Replacing each E-step with a square tile and each D-step with a domino then yields a Fibonacci tiling of length n. This procedure is clearly reversible, establishing the desired bijection.

Pell and Pell–Lucas Extensions

Finally, the concepts and techniques presented above can be extended in an obvious way to Pell polynomials $p_n(x) = f_n(2x)$ and Pell–Lucas polynomials $q_n(x) = l_n(2x)$. Consequently, they can be applied to Pell and Pell–Lucas numbers as well.

EXERCISES 34

List each.

1. Closed paths of length 5 originating at v_1.
2. Paths of length 5 from v_1 to v_2.
3. Paths of length 5 from v_2 to v_1.
4. Closed paths of length 5 originating at v_2.

5–8. Compute the sum of the weights of the paths in Exercises 34.1–34.4.

Prove each.

9. The eigenvalues of the matrix Q are $\alpha = \alpha(x)$ and $\beta = \beta(x)$.

10. The eigenvalues of the matrix Q^n are α^n and β^n.

Using the graph-theoretic model, establish each.

11. $f_{n+1}^2 + f_n^2 = f_{2n+1}$.
12. $l_{m+n} = f_{m+1}l_n + f_m l_{n-1}$.
13. Illustrate the identity $\sum_{k=1}^{n} F_k^2 = F_n F_{n+1}$ using the closed paths of length 6 from v_1 to itself.

Find the Fibonacci tiling corresponding to each closed walk originating at v_1.

14. 1111121
15. 1121121
16. 11211121

Find the closed path originating at v_1, which corresponds to each Fibonacci tiling.

17.

x	1	x	x

18.

x	x	1	x	1

19.

x	1	x	x	1	x

Using the graph-theoretic model, interpret each summation formula.

20. $x\sum_{k=1}^{n} f_{2k-1} = f_{2n}$.
21. $x\sum_{k=1}^{n} l_{2k-1} = l_{2n} - 2$.
22. $x\sum_{k=1}^{n} f_{2k} = f_{2n+1} - 1$.

Compute x times each sum.

23. The sum of the weights of paths of length ≤ 6 from v_1 to v_2.
24. The sum of the weights of paths of length ≤ 6.
25. The sum of the weights of paths of length ≤ 7 from v_2 to itself.
26. The sum of the weights of paths of length ≤ 7.
27. The sum of the weights of paths of length ≤ 8 from v_1 to v_2.
28. Using the graph-theoretic model, prove that $f_{n+3} + f_{n-1} = (x^2 + 2)f_{n+1}$.

35

GIBONACCI POLYNOMIALS

The essence of mathematics lies in its freedom.
–Georg Cantor (1845–1918)

In Chapter 31, we introduced an extended family of Fibonacci polynomials $g_n(x)$, where $g_n(x)$ satisfies the second-order recurrence $g_n(x) = xg_{n-1}(x) + g_{n-2}(x)$. We now extend this family even further and extract a number of delightful properties, including the polynomial extensions of the charming identities $F_{n+1}^3 + F_n^3 - F_{n-1}^3 = F_{3n}$ and $L_{n+1}^3 + L_n^3 - L_{n-1}^3 = 5L_{3n}$ [287, 334, 356].

35.1 GIBONACCI POLYNOMIALS

Let $g_1 = a = a(x)$ and $g_2 = b = b(x)$, where a and b are arbitrary polynomials with integral coefficients. Then $g_0 = b - ax$. We call the resulting members g_n *generalized Fibonacci polynomials*, or *gibonacci polynomials* for short. Clearly, when $a = 1$ and $b = x$, $g_n(x) = f_n(x)$; and when $a = x$ and $b = x^2 + 2$, $g_n(x) = l_n(x)$. Recall that $p_n(x) = f_n(2x)$ and $q_n(x) = l_n(x)$. Again, in the interest of brevity, we *omit* the arguments in polynomial functions.

The first six gibonacci polynomials are

$$\begin{array}{ll} g_1 = a & g_2 = b \\ g_3 = a + bx & g_4 = ax + b(x^2 + 1) \\ g_5 = a(x^2 + 1) + b(x^3 + 2x) & g_6 = a(x^3 + 2x) + b(x^4 + 3x^2 + 1). \end{array}$$

Fibonacci and Lucas Numbers with Applications, Volume Two. Thomas Koshy.

The number $G_n = g_n(1)$ is the nth *gibonacci number*[†]. The first six gibonacci numbers are $a, b, a + b, a + 2b, 2a + 3b$, and $3a + 5b$.

Using the gibonacci recurrence, we now find recurrences for even-numbered and odd-numbered gibonacci polynomials.

Recurrences for $\{g_{2n}\}$ and $\{g_{2n-1}\}$

By the gibonacci recurrence, we have

$$\begin{aligned} g_{2n} &= xg_{2n-1} + g_{2n-2} \\ &= x(xg_{2n-2} + g_{2n-3}) + g_{2n-2} \\ &= (x^2 + 1)g_{2n-2} + (g_{2n-2} - g_{2n-4}) \\ &= (x^2 + 2)g_{2n-4} - g_{2n-4}. \end{aligned} \tag{35.1}$$

In particular, let $n = 4$ and $g_n = l_n$. Then

$$\begin{aligned} (x^2 + 2)l_6 - l_4 &= (x^2 + 2)(x^6 + 6x^4 + 8x^2 + 2) - (x^4 + 4x^2 + 2) \\ &= x^8 + 8x^6 + 20x^4 + 16x^2 + 2 \\ &= l_8. \end{aligned}$$

Similarly, we can show that g_{2n-1} satisfies the fourth-order recurrence

$$g_{2n-1} = (x^2 + 2)g_{2n-3} - g_{2n-5}; \tag{35.2}$$

see Exercise 35.3.

For example,

$$\begin{aligned} (x^2 + 2)f_5 - f_3 &= (x^2 + 2)(x^4 + 3x^2 + 1) - (x^2 + 1) \\ &= x^4 + 5x^4 + 6x^2 + 1 \\ &= f_7. \end{aligned}$$

Next we find an addition formula for gibonacci polynomials.

Gibonacci Addition Formula

First, we make an interesting observation from the short list of gibonacci polynomials: $g_n = af_{n-2} + bf_{n-1}$, where $n \geq 2$. This can be confirmed using the strong version of PMI; see Exercise 35.4. Accordingly, we have the following result.

Theorem 35.1. *Let g_n denote the nth gibonacci polynomial. Then $g_n = af_{n-2} + bf_{n-1}$, where $n \geq 0$.* ■

[†]A.T. Benjamin and J.J. Quinn introduced the term *gibonacci* in [29].

For example,

$$\begin{aligned} g_6 &= af_4 + bf_5 \\ &= a(x^3 + 2x) + b(x^4 + 3x^2 + 1) \\ &= \begin{cases} x^5 + 4x^3 + 3x = f_6 & \text{if } a = 1 \text{ and } b = x \\ x^6 + 6x^4 + 9x^2 + 2 = l_6 & \text{if } a = x \text{ and } b = x^2 + 2. \end{cases} \end{aligned}$$

The next theorem gives the addition formula. It also follows by the strong version of PMI; see Exercise 35.5.

Theorem 35.2. *Let g_n denote the nth gibonacci polynomial. Then $g_{n+m} = f_{m+1}g_n + f_m g_{n-1}$, where $m, n \geq 0$.* ■

We illustrate this with an example. Using the gibonacci recurrence repeatedly, we have

$$\begin{aligned} g_{n+4} &= f_2 g_{n+3} + f_1 g_{n+2} \\ &= f_2(xg_{n+2} + g_{n+1}) + f_1 g_{n+2} = (xf_2 + f_1)g_{n+2} + f_2 g_{n+1} \\ &= f_3(xg_{n+1} + g_n) + f_2 g_{n+1} = (xf_3 + f_2)g_{n+1} + f_3 g_n \\ &= f_4(xg_n + g_{n-1}) + f_3 g_n = (xf_4 + f_3)g_n + f_4 g_{n-1} \\ &= f_5 g_n + f_4 g_{n-1} \\ &= (x^4 + 3x^2 + 1)g_n + (x^3 + 2x)g_{n-1}. \end{aligned}$$

The traditional addition formulas for Fibonacci, Lucas, Pell, and Pell–Lucas polynomials follow from this theorem:

$$\begin{aligned} f_{n+m} &= f_{m+1} f_n + f_m f_{n-1}; \\ l_{n+m} &= f_{m+1} l_n + f_m l_{n-1}; \\ p_{n+m} &= f_{m+1} p_n + f_m p_{n-1}; \\ q_{n+m} &= f_{m+1} q_n + f_m q_{n-1}. \end{aligned}$$

Using the Q-matrix in Chapter 32, we can rewrite the addition formula in Theorem 35.2 in terms of a matrix equation:

$$\begin{aligned} \begin{bmatrix} g_{n+m} \\ g_{n+m-1} \end{bmatrix} &= \begin{bmatrix} f_{m+1} & f_m \\ f_m & f_{m-1} \end{bmatrix} \begin{bmatrix} g_n \\ g_{n-1} \end{bmatrix} \\ &= Q^m \begin{bmatrix} g_n \\ g_{n-1} \end{bmatrix}. \end{aligned}$$

The next example presents a delightful application of Theorem 35.2. We could establish it using PMI, but here use a different technique, one used by S. Clary and P.D. Hemenway in their study of Fibonacci sums [104].

Example 35.1. *Prove that* $x \sum_{i=1}^{n} f_i g_{3i} = f_n f_{n+1} g_{2n+1}$.

Proof. Let A_n denote the LHS of the given formula and B_n its RHS. Then, by the gibonacci recurrence and Theorem 35.2, we have

$$\begin{aligned} B_n - B_{n-1} &= f_n \left(f_{n+1}g_{2n+1} - f_{n-1}g_{2n-1}\right) \\ &= f_n \left[f_{n+1}g_{2n+1} - (f_{n+1} - xf_n)g_{2n-1}\right] \\ &= f_n \left[f_{n+1}(g_{2n+1} - g_{2n-1}) + f_n(xg_{2n-1})\right] \\ &= xf_n(f_{n+1}g_{2n} + f_n g_{2n-1}) \\ &= xf_n g_{3n} \\ &= A_n - A_{n-1}. \end{aligned}$$

By iteration, this implies

$$\begin{aligned} A_n - B_n &= A_{n-1} - B_{n-1} \\ &= A_0 - B_0 \\ &= 0 - f_0 f_1 g_1 \\ &= 0. \end{aligned}$$

Thus $A_n = B_n$, as desired. ∎

In particular, we have

$$x \sum_{i=1}^{n} f_i f_{3i} = f_n f_{n+1} f_{2n+1}; \tag{35.3}$$

$$x \sum_{i=1}^{n} f_i l_{3i} = f_n f_{n+1} l_{2n+1};$$

$$2x \sum_{i=1}^{n} p_i p_{3i} = p_n p_{n+1} p_{2n+1};$$

$$2x \sum_{i=1}^{n} q_i q_{3i} = p_n p_{n+1} q_{2n+1}.$$

It follows from formula (35.3) that

$$\sum_{i=1}^{n} F_i F_{3i} = F_n F_{n+1} F_{2n+1}.$$

K.-Gunter Recke of Göttingen, Germany, studied this special case in 1968 [400], and H.V. Krishna of Manipal Engineering College, Karnataka, India in 1972 [304].

For example, $\sum_{i=1}^{n} F_iF_{3i} = F_1F_3 + F_2F_6 + F_3F_9 + F_4F_{12} + F_5F_{15} = 3{,}560 = F_5F_6F_{11}$.

Binet-like Formula for g_n

Using Theorem 35.1, we can derive an explicit formula for g_n, as the next theorem shows.

Theorem 35.3. *Let $c = c(x) = a + (ax - b)\beta$ and $d = d(x) = a + (ax - b)\alpha$. Then*

$$g_n = \frac{c\alpha^n - d\beta^n}{\alpha - \beta}.$$

Proof. Since $\alpha^2 = \alpha x + 1$, $\beta^2 = \beta x + 1$, $\alpha\beta = -1$, and $\Delta = \alpha - \beta$, by Theorem 35.1, we have

$$\begin{aligned}
\Delta \cdot g_n &= a(\alpha^{n-2} - \beta^{n-2}) + b(\alpha^{n-1} - \beta^{n-1})\\
&= \alpha^n\left(\frac{a}{\alpha^2} + \frac{b}{\alpha}\right) - \beta^n\left(\frac{a}{\beta^2} + \frac{b}{\beta}\right)\\
&= \alpha^n(a\beta^2 - b\beta) - \beta^n(a\alpha^2 - b\alpha)\\
&= \alpha^n[a(\beta x + 1) - b\beta] - \beta^n[a(\alpha x + 1) - b\alpha]\\
&= c\alpha^n - d\beta^n\\
g_n &= \frac{c\alpha^n - d\beta^n}{\alpha - \beta},
\end{aligned}$$

as claimed. ■

The product cd has an interesting dividend:

$$\begin{aligned}
cd &= [a + (ax - b)\beta][a + (ax - b)\alpha]\\
&= a^2 + a(ax - b)\alpha + a(ax - b)\beta - (ax - b)^2\\
&= a^2 + a^2x(\alpha + \beta) - ab(\alpha + \beta) - (ax - b)^2\\
&= a^2 + abx - b^2.
\end{aligned}$$

Let $\mu = \mu(x) = a^2 + abx - b^2$. This expression is called the *characteristic* of the gibonacci family.

To digress a bit, let $x = 1$ and $\mu = \pm 1$. Then $a^2 + ab - b^2 = \pm 1$. Letting $r = -a$ and $s = b$, this yields two popular diophantine equations: $r^2 - rs - s^2 = \pm 1$. The positive solutions of $r^2 - rs - s^2 = 1$ are (F_{2n+1}, F_{2n}), and those of $r^2 - rs - s^2 = -1$ are (F_{2n}, F_{2n-1}), where $n \geq 1$. These two equations play a central role in solving the diophantine equations $x^2 + xy + x - y^2 = 0$ and $x^2 - xy - x - y^2 = 0$; their positive solutions are $(F_{2n}^2, F_{2n}F_{2n+1})$ and $(F_{2n+1}^2, F_{2n}F_{2n+1})$, respectively [57, 247, 248].

Taguiri-like Identity

In 1901, the Italian mathematician A. Taguiri discovered a beautiful formula for the difference of two second-order Fibonacci products [130]:

$$F_{n+h}F_{n+k} - F_nF_{n+h+k} = (-1)^n F_h F_k.$$

D. Everman *et al.* re-discovered this identity in 1960 [155, 361]. This is a generalization of Catalan's identity.

Interestingly, we can extend this property to gibonacci polynomials:

$$g_{n+h}g_{n+k} - g_n g_{n+h+k} = \mu(-1)^n f_h f_k. \tag{35.4}$$

Its proof follows smoothly and elegantly by Theorem 35.3:

$$\begin{aligned}\Delta^2(\text{LHS}) &= (c\alpha^{n+h} - d\beta^{n+h})(c\alpha^{n+k} - d\beta^{n+k}) - (c\alpha^n - d\beta^n)(c\alpha^{n+h+k} - d\beta^{n+h+k})\\ &= cd(-1)^n(\alpha^{h+k} + \beta^{h+k} - \alpha^h\beta^k - \alpha^k\beta^h)\\ &= \mu(-1)^n(\alpha^h - \beta^h)(\alpha^k - \beta^k)\\ \text{LHS} &= \mu(-1)^n f_h f_k\\ &= \text{RHS}.\end{aligned}$$

It follows from (35.4) that

$$\begin{aligned}f_{n+h}f_{n+k} - f_n f_{n+h+k} &= (-1)^n f_h f_k;\\ l_{n+h}l_{n+k} - l_n l_{n+h+k} &= (-1)^{n+1}(x^2+4) f_h f_k;\\ G_{n+h}G_{n+k} - G_n G_{n+h+k} &= \mu(-1)^n F_h F_k,\end{aligned}$$

where $\mu = a^2 + ab - b^2$.

For example, let $n = 4, h = 1$, and $k = 3$. Then

$$\begin{aligned}\text{LHS} &= l_5 l_7 - l_4 l_8\\ &= (x^5 + 5x^3 + 5x)(x^7 + 7x^5 + 14x^3 + 7x)\\ &\quad -(x^4 + 4x^2 + 2)(x^8 + 8x^6 + 20x^4 + 16x^2 + 2)\\ &= -(x^2+4)(x^2+1)\\ &= (-1)^5(x^2+4) f_1 f_3\\ &= \text{RHS}.\end{aligned}$$

Next we study an interesting application of the identity $g_{n+k}g_{n-k} - g_n^2 = (-1)^{n+k+1}\mu f_k^2$. Since

$$\begin{aligned}\frac{g_{2k}}{g_{2k+2}} - \frac{g_{2k-2}}{g_{2k}} &= \frac{g_{2k}^2 - g_{2k+2}g_{2k-2}}{g_{2k}g_{2k+2}}\\ &= \frac{\mu x^2}{g_{2k}g_{2k+2}},\end{aligned}$$

it follows that

$$\sum_{k=1}^{n} \frac{\mu x^2}{g_{2k}g_{2k+2}} = \frac{g_{2n}}{g_{2n+2}} - \frac{g_0}{g_2}. \tag{35.5}$$

In particular, this yields

$$\sum_{k=1}^{n} \frac{x^2}{f_{2k}f_{2k+2}} = \frac{f_{2n}}{f_{2n+2}} \tag{35.6}$$

$$\sum_{k=1}^{n} \frac{x^2(x^2+4)}{l_{2k}l_{2k+2}} = \frac{2}{x^2+2} - \frac{l_{2n}}{l_{2n+2}}. \tag{35.7}$$

It follows from formulas (35.6) and (35.7) that

$$\sum_{k=1}^{n} \frac{4x^2}{p_{2k}p_{2k+2}} = \frac{p_{2n}}{p_{2n+2}};$$

$$\sum_{k=1}^{n} \frac{16x^2(x^2+1)}{q_{2k}q_{2k+2}} = \frac{1}{2x^2+1} - \frac{q_{2n}}{q_{2n+2}};$$

$$\sum_{k=1}^{n} \frac{1}{F_{2k}F_{2k+2}} = \frac{F_{2n}}{F_{2n+2}}; \tag{35.8}$$

$$\sum_{k=1}^{\infty} \frac{x^2}{f_{2k}f_{2k+2}} = \frac{1}{\alpha^2};$$

$$\sum_{k=1}^{\infty} \frac{x^2(x^2+4)}{l_{2k}l_{2k+2}} = \frac{2}{x^2+2} - \frac{1}{\alpha^2};$$

$$\sum_{k=1}^{\infty} \frac{4x^2}{p_{2k}p_{2k+2}} = \frac{1}{\gamma^2};$$

$$\sum_{k=1}^{\infty} \frac{16x^2(x^2+1)}{q_{2k}q_{2k+2}} = \frac{1}{2x^2+1} - \frac{1}{\gamma^2}.$$

see Exercises 35.101–35.104. Khan discovered formula (35.8) in 2013 [266].

Interestingly, the Catalan-like identity can be used to develop a similar one for $\sum_{k=1}^{n} \frac{\mu x^2}{g_{2k-1}g_{2k+1}}$:

$$\sum_{k=1}^{n} \frac{\mu x^2}{g_{2k-1}g_{2k+1}} = \frac{\mu x + g_0 g_1}{g_1 g_2} - \frac{g_{2n}g_{2n+1} + \mu x}{g_{2n+1}g_{2n+2}}; \tag{35.9}$$

see Exercise 35.105.

It follows from formula (35.9) that

$$\sum_{k=1}^{n} \frac{x}{f_{2k-1}f_{2k+1}} = \frac{f_{2n}}{f_{2n+1}}; \tag{35.10}$$

see Exercise 35.106. This implies

$$\sum_{k=1}^{n} \frac{1}{F_{2k-1}F_{2k+1}} = \frac{F_{2n}}{F_{2n+1}}; \tag{35.11}$$

$$\sum_{k=1}^{n} \frac{2x}{p_{2k-1}p_{2k+1}} = \frac{p_{2n}}{p_{2n+1}};$$

$$\sum_{k=1}^{\infty} \frac{x}{f_{2k-1}f_{2k+1}} = \frac{1}{\alpha};$$

$$\sum_{k=1}^{\infty} \frac{2x}{p_{2k-1}p_{2k+1}} = \frac{1}{\gamma};$$

see Exercises 35.107–35.109. R.J. Clarke of Stourbridge, United Kingdom, discovered formula (35.11) in 2015 [103].

The characteristic of the gibonacci family occurs in its Cassini-like formula as well, as the next theorem shows. It follows by identity (35.4); see Exercise 35.9.

Theorem 35.4. $g_{n+1}g_{n-1} - g_n^2 = \mu(-1)^n$. ■

For example, we have

$$\begin{aligned} g_5g_3 - g_4 &= [a(x^2+1) + b(x^3+2x)](a+bx) - [ax + b(x^2+1)]^2 \\ &= \mu. \end{aligned}$$

When $a = 1$ and $b = x$, $\mu = 1$; and when $a = x$ and $b = x^2 + 2$, $\mu = x^2 + x^2(x^2+2) - (x^2+2)^2 = -(x^2+4)$. Consequently, the Cassini-like formulas for f_n and l_n follow from Theorem 35.4; so do the Cassini-like formulas for Pell and Pell–Lucas polynomials.

We can also establish Theorem 35.4 using Theorem 35.1 and the Cassini-like formula for f_n; see Exercise 35.8.

We can generalize the Cassini-like formula as $g_{n+k}g_{n-k} - g_n^2 = (-1)^{n+k+1}\mu f_k^2$; see Exercise 35.15. Since $(g_{n+k}g_{n-k} - g_n^2)^2 = \mu^2 f_k^4$, it then follows that

$$4g_{n+k}g_n^2g_{n-k} + \mu^2 f_k^4 = (g_{n+k}g_{n-k} + g_n^2)^2. \tag{35.12}$$

Consequently, $4g_{n+k}g_n^2g_{n-k} + \mu^2 f_k^4$ is a square.

It follows from identity (35.12) that

$$4f_{n+k}f_n^2f_{n-k} + f_k^4 = (f_{n+k}f_{n-k} + f_n^2)^2; \tag{35.13}$$

$$4l_{n+k}l_n^2l_{n-k} + (x^2+4)^2f_k^4 = (l_{n+k}l_{n-k} + l_n^2)^2;$$

$$4p_{n+k}p_n^2p_{n-k} + p_k^4 = (p_{n+k}p_{n-k} + p_n^2)^2;$$

$$4q_{n+k}q_n^2q_{n-k} + 16(x^2+1)^2p_k^4 = (q_{n+k}q_{n-k} + q_n^2)^2.$$

For example,

$$\begin{aligned}4l_4l_2^2l_0 + (x^2+4)^2f_2^4 &= 4(x^4+4x^2+2)(x^2+2)^2\cdot 2 + (x^2+4)^2x^4\\ &= 9x^8 + 72x^6 + 192x^4 + 192x^2 + 64\\ &= (3x^4+12x^2+8)^2\\ &= [2(x^4+4x^2+2) + (x^2+2)^2]^2\\ &= (l_4l_0 + l_2)^2.\end{aligned}$$

In particular, identity (35.13) implies that $4F_{n+k}F_n^2F_{n-k} + F_k^4 = (F_{n+k}F_{n-k} + F_n^2)^2$. For example, $4F_8F_5^2F_2 + F_3^4 = 2{,}116 = (F_8F_2 + F_5^2)^2$.

The next example, studied in 1964 by Hoggatt, is a beautiful application of Theorem 35.4 with $x = 1$. It deals with the *linear transformation* $w = \dfrac{pz+q}{rz+s}$ and its *fixed points*†, where $w = w(z)$ and $ps - qr \neq 0$.

Example 35.2. *Let p, q, r, and s be any positive integers such that* $ps - qr = 1$. *Prove that the roots of the equation* $t^2 - t - 1 = 0$ *are the fixed points of the linear transformation* $w = \dfrac{pz+q}{rz+s}$ *if and only if* $p = F_{2n+1}$, $q = r = F_{2n}$, *and* $s = F_{2n-1}$, *where* $n \neq 0$.

Proof. Suppose the solutions α and β of the equation $t^2 - t - 1 = 0$ are the fixed points of the linear transformation. Then $p\alpha + q = \alpha(r\alpha + s)$ and $p\beta + q = \beta(r\beta + s)$; that is,

$$r\alpha^2 - (p-s)\alpha - q = 0$$

$$r\beta^2 - (p-s)\beta - q = 0.$$

Since $\alpha + \beta = 1$, subtracting one equation from the other yields $r + s = p$. Adding them yields $p + 2q = 3r + s$; so $q = r$.

Conversely, let $q = r$ and $p = q + s$. Then $w = \dfrac{(q+s)z+q}{qz+s}$. Its fixed points are given by $(q+s)z + q = z(qz^2 + s)$; that is, $z^2 - z - 1 = 0$. So the fixed points of the transformation are α and β.

†A point z is a *fixed point* of the linear transformation if $w(z) = z$.

It remains to show that $p = F_{2n+1}, q = r = F_{2n}$, and $s = F_{2n-1}$. Since $p = q + s$, it follows that s, q (or r), and p satisfy the gibonacci recurrence. For convenience, assume that $s < q < p$. Then $s = G_{k-1}$, $q = G_k$, and $p = G_{k+1}$ for some nonzero integer k.

By Theorem 35.4, $G_{k+1}G_{k-1} - G_k^2 = (-1)^k\mu$, where $\mu = p^2 + pq - q^2$. Since $ps - qr = 1$, $G_{k+1}G_{k-1} - G_k^2 = 1$. So $(-1)^k\mu = 1$. Consequently, $\mu = 1$ and $k = 2n$ for some nonzero integer n. Since $\mu = 1$, $G_k = F_k = F_{2n}$.

Then $p = F_{2n+1}$, $q = r = F_{2n}$, and $s = F_{2n-1}$, as desired. ■

This example has a wonderful byproduct. Since $q = r$ and $r + s = p$, it follows from the equation $ps - qr = 1$ that $r^2 - s^2 - rs + 1 = 0$. By the example, (F_{2n}, F_{2n-1}) is a solution of this quadratic equation: $F_{2n}^2 - F_{2n-1}^2 - F_{2n}F_{2n-1} + 1 = 0$. For instance, $F_6^2 - F_5^2 - F_6F_5 + 1 = 0$.

More generally, $f_{2n}^2 - f_{2n-1}^2 - xf_{2n}f_{2n-1} + 1 = 0$. A similar result exists for Lucas polynomials: $l_{2n}^2 - l_{2n-1}^2 - xl_{2n}l_{2n-1} - (x^2 + 4) = 0$; see Exercises 35.45 and 35.46. The latter equation implies that (L_{2n}, L_{2n-1}) is a solution of the equation $r^2 - s^2 - rs - 5 = 0$.

Next we study a few additional applications of Theorem 35.4.

Additional Applications

To begin with, notice that

$$\begin{aligned}
x^2g_n^2 + xg_ng_{n-1} + g_{n-1}^2 &= x^2g_n^2 + (xg_n + g_{n-1})g_{n-1} \\
&= x^2g_n^2 + g_{n+1}g_{n-1} \\
&= x^2g_n^2 + [g_n^2 + (-1)^n\mu] \\
&= (x^2 + 1)g_n^2 + (-1)^n\mu.
\end{aligned}$$

Then the identities†

$$\begin{aligned}
(x + y)^4 + x^4 + y^4 &= 2(x^2 + xy + y^2)^2; \\
(x + y)^5 - x^5 - y^5 &= 5xy(x + y)(x^2 + xy + y^2); \\
(x + y)^7 - x^7 - y^7 &= 7xy(x + y)(x^2 + xy + y^2)^2
\end{aligned}$$

yield the following gibonacci ones:

$$g_{n+1}^4 + x^4g_n^4 + g_{n-1}^4 = 2\left[(x^2 + 1)g_n^2 + (-1)^n\mu\right]^2;$$

---

†The outstanding French mathematician Augustine Louis Cauchy (1789–1857) proved that if p is a prime > 3, then $(x + y)^p - x^p - y^p = pxy(x + y)(x^2 + xy + y^2)f_p(x, y)$, where $f_p(x, y)$ is a polynomial with integral coefficients. In particular, if $p \equiv 1 \pmod 6$, then $(x + y)^p - x^p - y^p = pxy(x + y)(x^2 + xy + y^2)^2g_p(x, y)$, where $g_p(x, y)$ is a polynomial with integral coefficients [142].

$$g_{n+1}^5 - x^5 g_n^5 - g_{n-1}^5 = 5xg_{n+1}g_ng_{n-1}\left[(x^2+1)g_n^2 + (-1)^n\mu\right];$$
$$g_{n+1}^7 - x^7 g_n^7 - g_{n-1}^7 = 7xg_{n+1}g_ng_{n-1}\left[(x^2+1)g_n^2 + (-1)^n\mu\right]^2.$$

In particular, we have

$$f_{n+1}^4 + x^4 f_n^4 + f_{n-1}^4 = 2\left[(x^2+1)f_n^2 + (-1)^n\right]^2;$$
$$l_{n+1}^4 + x^4 l_n^4 + l_{n-1}^4 = 2\left[(x^2+1)l_n^2 - (-1)^n(x^2+4)\right]^2;$$
$$f_{n+1}^5 - x^5 f_n^5 - f_{n-1}^5 = 5xf_{n+1}f_nf_{n-1}\left[(x^2+1)f_n^2 + (-1)^n\right];$$
$$l_{n+1}^5 - x^5 l_n^5 - l_{n-1}^5 = 5xl_{n+1}l_nl_{n-1}\left[(x^2+1)l_n^2 - (-1)^n(x^2+4)\right];$$
$$f_{n+1}^7 - x^7 f_n^7 - f_{n-1}^7 = 7xf_{n+1}f_nf_{n-1}\left[(x^2+1)f_n^2 + (-1)^n\right]^2;$$
$$l_{n+1}^7 - x^7 l_n^7 - l_{n-1}^7 = 7xl_{n+1}l_nl_{n-1}\left[(x^2+1)l_n^2 - (-1)^n(x^2+4)\right]^2.$$

Since $x^2 + xy + y^2 = (x+y)^2 - xy$, the above identities can be rewritten slightly differently. For instance, $g_{n+1}^4 + x^4 g_n^4 + g_{n-1}^4 = 2\left(g_{n+1}^2 - xg_ng_{n-1}\right)^2$.

A Recurrence for Gibonacci Squares

We can use Theorem 35.4, coupled with the gibonacci recurrence, to develop a second-order recurrence for gibonacci squares:

$$\begin{aligned}
g_{n+2}^2 &= (xg_{n+1} + g_n)^2\\
&= x^2 g_{n+1}^2 + 2xg_{n+1}g_n + g_n^2\\
g_{n+2}^2 - (x^2+2)g_{n+1}^2 &= g_n^2 - 2g_{n+1}(g_{n+1} - xg_n)\\
&= g_n^2 - 2g_{n+1}g_{n-1}\\
&= g_n^2 - 2\left[g_n^2 + (-1)^n\mu\right]
\end{aligned}$$

$$g_{n+2}^2 - (x^2+2)g_{n+1}^2 + g_n^2 = 2(-1)^{n+1}\mu. \tag{35.14}$$

This implies that

$$g_{n+2}^2 - (x^2+2)g_{n+1}^2 + g_n^2 = \begin{cases} 2(-1)^{n+1} & \text{if } g_n = f_n\\ 2\Delta^2(-1)^n & \text{if } g_n = l_n. \end{cases}$$

For example,

$$f_5^2 - (x^2+2)f_4^2 + f_3^2 = (x^4+3x^2+1)^2 - (x^2+2)(x^3+2x)^2 + (x^2+1)^2 = 2(-1)^3;$$
$$\begin{aligned}
l_4^2 - (x^2+2)l_3^2 + l_2^2 &= (x^4+4x^2+2)^2 - (x^2+2)(x^3+3x)^2 + (x^2+2)^2\\
&= 2(-1)^2(x^2+4).
\end{aligned}$$

In the next example, we study the infinite sum of the reciprocals of gibonacci polynomials. It is also an application of Theorem 35.4, and a telescoping sum.

Example 35.3. *Let x be a positive integer. Prove that*

$$\sum_{n=1}^{\infty} \frac{x}{g_n} = \frac{x+1}{a} + \frac{1}{b} - \sum_{n=1}^{\infty} \frac{(-1)^n \mu x}{g_n g_{n+1} g_{n+2}}.$$

Proof.

$$\begin{aligned}
\sum_{n=2}^{\infty} \frac{x}{g_n} - \sum_{n=2}^{\infty} \frac{x g_n}{g_{n-1} g_{n+1}} &= \sum_{n=2}^{\infty} \frac{(g_{n-1} g_{n+1} - g_n^2) x}{g_{n-1} g_n g_{n+1}} \\
&= \sum_{n=2}^{\infty} \frac{(-1)^n \mu\, x}{g_{n-1} g_n g_{n+1}} \\
\sum_{n=2}^{\infty} \frac{x g_n}{g_{n-1} g_{n+1}} &= \sum_{n=2}^{\infty} \frac{g_{n+1} - g_{n-1}}{g_{n-1} g_{n+1}} \\
&= \sum_{n=2}^{\infty} \left(\frac{1}{g_{n-1}} - \frac{1}{g_{n+1}} \right) \\
&= \frac{1}{g_1} + \frac{1}{g_2} \\
&= \frac{1}{a} + \frac{1}{b}.
\end{aligned}$$

Therefore,

$$\begin{aligned}
\sum_{n=2}^{\infty} \frac{x}{g_n} &= \frac{1}{a} + \frac{1}{b} + \sum_{n=2}^{\infty} \frac{(-1)^n \mu x}{g_{n-1} g_n g_{n+1}} \\
\sum_{n=1}^{\infty} \frac{x}{g_n} &= \frac{x+1}{a} + \frac{1}{b} - \sum_{n=2}^{\infty} \frac{(-1)^n \mu x}{g_{n-1} g_n g_{n+1}},
\end{aligned}$$

as desired. ■

R.L. Graham of then Bell Telephone Laboratories, Murray Hill, New Jersey, studied this example in 1963 with $g_n = F_n$ [193].

Next we establish a generalization of Theorem 35.4. Its proof exemplifies the power of both recursion and matrices. To this end, we need the following well-known result from the theory of matrices [274].

Theorem 35.5. *Let k be an arbitrary integer. Let A, B, and C be three identical square matrices, except that the ith rows (or columns) of A and B are different;*

and the ith row (or column) of C is the sum of k times the ith row (or column) of A and that of B. Then $|C| = k|A| + |B|$. ■

We are now ready for the generalization [285].

Theorem 35.6. *Let m, n, and k be arbitrary integers. Then*

$$g_{m+k}g_{n-k} - g_m g_n = (-1)^{n-k+1}\mu f_k f_{m-n+k}. \tag{35.15}$$

Proof. The proof involves two related 2×2 matrices $\{A_k\}$ and $\{B_k\}$, as Theorem 35.5 implies. We define them recursively.

First, we let

$$A_0 = \begin{bmatrix} g_n & g_n \\ g_{n+1} & g_{n+1} \end{bmatrix} \quad \text{and} \quad A_1 = \begin{bmatrix} g_n & g_{n+1} \\ g_{n+1} & g_{n+2} \end{bmatrix}.$$

Clearly, $|A_0| = 0$ and $|A_1| = (-1)^{n+1}\mu$. Using A_{k-2} and A_{k-1}, we now construct A_k recursively: multiply column 2 of A_{k-1} by x and then add column 2 of A_{k-2} to the resulting matrix. This gives A_k.

For example,

$$A_2 = \begin{bmatrix} g_n & xg_{n+1} + g_n \\ g_{n+1} & xg_{n+2} + g_{n+1} \end{bmatrix} = \begin{bmatrix} g_n & g_{n+2} \\ g_{n+1} & g_{n+3} \end{bmatrix}.$$

More generally, it follows by PMI that

$$A_k = \begin{bmatrix} g_n & g_{n+k} \\ g_{n+1} & g_{n+k+1} \end{bmatrix}.$$

By Theorem 35.5, $|A_k| = x|A_{k-1}| + |A_{k-2}|$, where $|A_1| = (-1)^{n+1}\mu$ and $|A_2| = x|A_1| + |A_0| = (-1)^{n+1}\mu x$. Thus A_k satisfies the same recurrence as g_k, with $a = (-1)^{n+1}\mu$ and $b = (-1)^{n+1}\mu x$. Consequently, by Theorem 35.1, $|A_k| = (-1)^{n+1}\mu f_{k-2} + (-1)^{n+1}\mu x f_{k-1} = (-1)^{n+1}\mu f_k$.

Next we introduce the matrix family $\{B_k\}$. We begin with $B_0 = \begin{bmatrix} g_n & g_n \\ g_{n-k} & g_{n-k} \end{bmatrix}$.

We now construct B_1 from A_k^T, the *transpose* of A_k; this gives $\begin{bmatrix} g_{n-k} & g_{n-k+1} \\ g_n & g_{n+1} \end{bmatrix}$.

Now swap its rows, and then its columns; the resulting matrix is $B_1 = \begin{bmatrix} g_{n+1} & g_n \\ g_{n-k+1} & g_{n-k} \end{bmatrix}$. Clearly, $|B_0| = 0$ and $|B_1| = (-1)^{n-k}\mu f_k$.

As before, we now construct B_r from B_{r-1} and B_{r-2}: multiply column 1 of B_{r-1} by x, and then add column 1 of B_{r-2} to the resulting matrix.

For example,

$$B_2 = \begin{bmatrix} xg_{n+1} + g_n & g_n \\ xg_{n-k+1} + g_{n-k} & g_{n-k} \end{bmatrix} = \begin{bmatrix} g_{n+2} & g_n \\ g_{n-k+2} & g_{n-k} \end{bmatrix}.$$

Again, it follows by PMI that

$$B_r = \begin{bmatrix} g_{n+r} & g_n \\ g_{n-k+r} & g_{n-k} \end{bmatrix}.$$

By invoking Theorem 35.5, we get $|B_r| = x|B_{r-1}| + |B_{r-2}|$, where $|B_1| = (-1)^{n-k}\mu f_k$ and $|B_2| = x|B_1| + |B_0| = (-1)^{n-k}\mu x f_k + 0 = (-1)^{n-k}\mu x f_k$. Thus $|B_r|$ satisfies the same recurrence as g_r, with $a = (-1)^{n-k}\mu$ and $b = (-1)^{n-k}\mu x$.

Consequently, by Theorem 35.1, $|B_r| = (-1)^{n-k}\mu f_k \cdot f_{r-2} + (-1)^{n-k}\mu x f_k \cdot f_{r-1} = (-1)^{n-k}\mu f_k f_r$. That is, $g_{n+r}g_{n-r} - g_{n-k+r}g_n = (-1)^{n-k}\mu f_k f_r$. Letting $m = n - k + r$, this yields the desired identity. ■

Identity (35.15) has a number of interesting byproducts. First, notice that $\mu = 1$ when $g_s = f_s$; and $\mu = -(x^2 + 4)$ when $g_s = l_s$. We then have

1) $f_{m+k}f_{n-k} - f_m f_n = (-1)^{n-k+1} f_k f_{m-n+k}$.
   This implies the *d'Ocagne's identity* $F_{m+k}F_{n-k} - F_m F_n = (-1)^{n-k+1}F_k F_{m-n+k}$. It also implies that $p_{m+k}p_{n-k} - p_m p_n = (-1)^{n-k+1}p_k p_{m-n+k}$ and $P_{m+k}P_{n-k} - P_m P_n = (-1)^{n-k+1}P_k P_{m-n+k}$.
2) $l_{m+k}l_{n-k} - l_m l_n = (-1)^{n-k}(x^2 + 4) f_k f_{m-n+k}$.
3) $g_{n+k}g_{n-k} - g_n^2 = (-1)^{n-k+1}\mu f_k^2$.
   In particular, $f_{n+k}f_{n-k} - f_n^2 = (-1)^{n-k+1}f_k^2$, $p_{n+k}p_{n-k} - p_n^2 = (-1)^{n-k+1}p_k^2$, $l_{n+k}l_{n-k} - l_n^2 = (-1)^{n-k}(x^2+4)f_k^2$, and $q_{n+k}q_{n-k} - q_n^2 = 4(-1)^{n-k}(x^2+1)f_k^2$. The well-known *Catalan's identity* $F_{n+k}F_{n-k} - F_n^2 = (-1)^{n-k+1}F_k^2$ also follows from the gibonacci identity.

Identity (35.4) is an equivalent form of identity (35.15).

The next example, studied in 2004 by Br. J. Mahon of Australia, is a fine application of Catalan's identity [341]. We establish the formula in the example using PMI [148].

Example 35.4. *Prove that* $\displaystyle\sum_{k=2}^{n} \frac{F_{2k-2}F_{2k}}{(F_{2k}^2 - 1)(F_{2k+2}^2 - 1)} = -\frac{3}{8} + \frac{F_{2n}F_{2n+2}}{F_{2n+2}^2 - 1}.$

Proof. (Notice that the formula does *not* work when $k = 1$.) When $k = 2$, LHS $= \dfrac{3}{8 \cdot 63}$ = RHS; so the formula holds when $k = 2$.

Assume it works for an arbitrary integer $n \geq 2$. Then

$$\begin{aligned}
\sum_{k=2}^{n+1} \frac{F_{2k-2}F_{2k}}{(F_{2k}^2-1)(F_{2k+2}^2-1)} &= \left(-\frac{3}{8}+\frac{F_{2n}F_{2n+2}}{F_{2n+2}^2-1}\right)+\frac{F_{2n}F_{2n+2}}{(F_{2n+2}^2-1)(F_{2n+4}^2-1)}\\
&= -\frac{3}{8}+\frac{F_{2n}F_{2n+2}F_{2n+4}^2}{(F_{2n+2}^2-1)(F_{2n+4}^2-1)}\\
&= -\frac{3}{8}+\frac{(F_{2n+2}F_{2n+4})(F_{2n}F_{2n+4})}{(F_{2n+2}^2-1)(F_{2n+4}^2-1)}\\
&= -\frac{3}{8}+\frac{(F_{2n+2}F_{2n+4})(F_{2n+2}^2-1)}{(F_{2n+2}^2-1)(F_{2n+4}^2-1)}\\
&= -\frac{3}{8}+\frac{F_{2n+2}F_{2n+4}}{F_{2n+4}^2-1}.
\end{aligned}$$

So the formula works for $n+1$ also.

Thus, by PMI, the formula works for all $n \geq 2$. ■

In particular,

$$\begin{aligned}
\sum_{k=2}^{5} \frac{F_{2k-2}F_{2k}}{(F_{2k}^2-1)(F_{2k+2}^2-1)} &= \frac{1\cdot 3}{(3^2-1)(8^2-1)}+\frac{3\cdot 8}{(8^2-1)(21^2-1)}\\
&\quad +\frac{8\cdot 21}{(21^2-1)(55^2-1)}+\frac{21\cdot 55}{(55^2-1)(144^2-1)}\\
&= -\frac{3}{8}+\frac{144}{377}\\
&= -\frac{3}{8}+\frac{F_{10}F_{12}}{F_{12}^2-1}.
\end{aligned}$$

It would be interesting to discover whether or not this example has a polynomial version. If it does, does the polynomial version have a Lucas counterpart? A gibonacci extension? [?]

35.2 DIFFERENCES OF GIBONACCI PRODUCTS

We now make an interesting observation. Take a good look at the following formulas:

$$\begin{aligned}
g_{n+h}g_{n+k}-g_n g_{n+h+k} &= \mu(-1)^n f_h f_k;\\
g_{m+k}g_{n-k}-g_m g_n &= (-1)^{n-k+1}\mu f_k f_{m-n+k};\\
g_{n+k}g_{n-k}-g_n^2 &= (-1)^{n-k+1}\mu f_k^2.
\end{aligned}$$

Do you see any pattern? The LHS of each is the difference of the products of two gibonacci polynomials. So we are tempted to investigate differences of products of three gibonacci polynomials [292]. The next theorem gives one such charming formula.

Theorem 35.7. *Let $n \geq 0$. Then*

$$g_{n+1}g_{n+2}g_{n+6} - g_{n+3}^3 = \mu(-1)^n(x^3 g_{n+2} - g_{n+1}). \tag{35.16}$$

Proof. By the gibonacci recurrence, we have

$$\begin{aligned} g_{n+6} &= (x^4 + 3x^2 + 1)g_{n+2} + (x^3 + 2x)g_{n+1}; \\ g_{n+1}g_{n+2}g_{n+6} &= (x^4 + 3x^2 + 1)g_{n+2}^2 g_{n+1} + (x^3 + 2x)g_{n+2}g_{n+1}^2; \\ g_{n+3}^3 &= x^3 g_{n+2}^3 + 3x^2 g_{n+2}^2 g_{n+1} + 3x g_{n+2} g_{n+1}^2 + g_{n+1}^3; \end{aligned}$$

see Exercises 35.16 and 35.17.

Then, by Theorem 35.4 and some basic algebra, we have

$$\begin{aligned} \text{LHS} &= g_{n+1}g_{n+2}g_{n+6} - g_{n+3}^3 \\ &= (x^4 + 1)g_{n+2}^2 g_{n+1} + (x^3 - x)g_{n+2}g_{n+1}^2 - x^3 g_{n+2}^3 - g_{n+1}^3 \\ &= x^3 g_{n+2}^2(x g_{n+1} - g_{n+2}) + g_{n+2}g_{n+1}(g_{n+2} - x g_{n+1}) + x^3 g_{n+2} g_{n+1}^2 - g_{n+1}^3 \\ &= -x^3 g_{n+2}^2 g_n + g_{n+2}g_{n+1}g_n + x^3 g_{n+1}^2(x g_{n+1} + g_n) - g_{n+1}^3 \\ &= -x^3 g_{n+2}\left[g_{n+1}^2 + \mu(-1)^{n+1}\right] + g_{n+1}\left[g_{n+1}^2 + \mu(-1)^{n+1}\right] + x^3 g_{n+2}g_{n+1}^2 - g_{n+1}^3 \\ &= \mu(-1)^n(x^3 g_{n+2} - g_{n+1}), \end{aligned}$$

as desired. ■

It follows by Theorem 35.7 that

$$\begin{aligned} f_{n+1}f_{n+2}f_{n+6} - f_{n+3}^3 &= (-1)^n(x^3 f_{n+2} - f_{n+1}); \quad (35.17) \\ l_{n+1}l_{n+2}l_{n+6} - l_{n+3}^3 &= (x^2 + 4)(-1)^{n+1}(x^3 l_{n+2} - l_{n+1}); \\ p_{n+1}p_{n+2}p_{n+6} - p_{n+3}^3 &= (-1)^n(8x^3 p_{n+2} - p_{n+1}); \\ q_{n+1}q_{n+2}q_{n+6} - q_{n+3}^3 &= 4(x^2 + 1)(-1)^{n+1}(8x^3 q_{n+2} - q_{n+1}). \end{aligned}$$

For example,

$$\begin{aligned} f_3 f_4 f_8 - f_5^3 &= (x^2 + 1)(x^3 + 2x)(x^7 + 6x^5 + 10x^3 + 4x) - (x^4 + 3x^2 + 1)^3 \\ &= x^6 + 2x^4 - x^2 - 1 \\ &= (x^3 + 2x) - (x^2 + 1) \\ &= (-1)^2(x^3 f_4 - f_3). \end{aligned}$$

Likewise, $l_3 l_4 l_8 - l_5^3 = -(x^9 - 8x^7 - 17x^5 - x^3 + 12x) = (-1)^3(x^2 + 4)(x^3 l_4 - l_3)$.

It follows from identity (35.17) that

$$F_{n+1}F_{n+2}F_{n+6} - F_{n+3}^3 = (-1)^n F_n.$$

R.S. Melham of the University of Technology, Sydney, Australia, discovered this elegant formula in 2003 [355].

Ray S. Melham was born in Sydney, Australia, and received his BS in Mathematics from the University of New South Wales (1973), and his Ph.D. in Mathematics from the University of Technology, Sydney, Australia (1995). His dissertation, on sequences generated by linear homogeneous recurrences, was under the supervision of A.G. Shannon. Melham has taught at the School of Mathematical Sciences, University of Technology. He has published numerous articles on number theory, and continues his research in the theory of linear homogeneous recurrences.

Similarly, we have

$$L_{n+1}L_{n+2}L_{n+6} - L_{n+3}^3 = 5(-1)^{n+1}L_n;$$
$$P_{n+1}P_{n+2}P_{n+6} - P_{n+3}^3 = (-1)^n(8P_{n+2} - P_{n+1});$$
$$Q_{n+1}Q_{n+2}Q_{n+6} - Q_{n+3}^3 = 2(-1)^{n+1}(8Q_{n+2} - Q_{n+1}).$$

For example, $Q_5Q_6Q_{10} - Q_7^3 = 41 \cdot 99 \cdot 3363 - 239^2 = -1502 = -2(8 \cdot 99 - 41) = 2(-1)^5(8Q_6 - Q_5)$.

Theorem 35.7 has an additional byproduct. It follows from identity (35.16) that $G_{n+1}G_{n+2}G_{n+6} - G_{n+3}^3 = (-1)^n\mu(1)G_n$, so $(G_{n+1}G_{n+2}G_{n+6} - G_{n+3}^3)^2 = \mu^2(1)G_n^2$. This implies

$$4G_{n+1}G_{n+2}G_{n+3}^3G_{n+6} + \mu^2(1)G_n^2 = (G_{n+1}G_{n+2}G_{n+6} + G_{n+3}^3)^2.$$

In particular,

$$4F_{n+1}F_{n+2}F_{n+3}^3F_{n+6} + F_n^2 = (F_{n+1}F_{n+2}F_{n+6} + F_{n+3}^3)^2;$$
$$4L_{n+1}L_{n+2}L_{n+3}^3L_{n+6} + 25L_n^2 = (L_{n+1}L_{n+2}L_{n+6} + L_{n+3}^3)^2.$$

For example,

$$\begin{aligned}4L_4L_5L_6^3L_9+25L_3^2 &= 4\cdot 7\cdot 11\cdot 18^3\cdot 76+25\cdot 4^2\\ &= 11{,}684^2\\ &= (7\cdot 11\cdot 76+18^3)^2\\ &= (L_4L_5L_9+L_6^3)^2.\end{aligned}$$

The next theorem gives a companion formula for the difference of products of three gibonacci polynomials.

Theorem 35.8. *Let $n \geq 0$. Then*

$$g_ng_{n+4}g_{n+5}-g_{n+3}^3 = \mu(-1)^{n+1}(x^3g_{n+4}+g_{n+5}). \tag{35.18}$$

Proof. By the gibonacci recurrence, we have $g_n = (x^2+1)g_{n+4}-(x^3+2x)g_{n+3}$. Then

$$g_ng_{n+4}g_{n+5} = (x^2+1)g_{n+4}^2g_{n+5}-(x^3+2x)g_{n+3}g_{n+4}g_{n+5}.$$

We also have

$$\begin{aligned}g_{n+3}^3 &= (g_{n+5}-xg_{n+4})^3\\ &= g_{n+5}^3-3xg_{n+4}g_{n+5}^2+3x^2g_{n+4}^2g_{n+5}-x^3g_{n+4}^3\\ &= (g_{n+5}-xg_{n+4})(g_{n+5}-2xg_{n+4})g_{n+5}+x^2g_{n+4}^2g_{n+5}-x^3g_{n+4}^3\\ &= g_{n+3}(g_{n+5}-2xg_{n+4})g_{n+5}+x^2g_{n+4}^2g_{n+5}-x^3g_{n+4}^3.\end{aligned}$$

Therefore,

$$\begin{aligned}g_ng_{n+4}g_{n+5}-g_{n+3}^3 &= g_{n+4}^2g_{n+5}-x^3g_{n+3}g_{n+4}g_{n+5}-g_{n+3}g_{n+5}^2+x^3g_{n+4}^3\\ &= (g_{n+4}^2-g_{n+3}g_{n+5})(x^3g_{n+4}+g_{n+5})\\ &= (-1)^{n+1}\mu(x^3g_{n+4}+g_{n+5}),\end{aligned}$$

as claimed. ∎

It follows by Theorem 35.8 that

$$\begin{aligned}f_nf_{n+4}f_{n+5}-f_{n+3}^3 &= (-1)^{n+1}(x^3f_{n+4}+f_{n+5});\\ l_nf_{n+4}l_{n+5}-l_{n+3}^3 &= (x^2+4)(-1)^n(x^3l_{n+4}+l_{n+5});\\ p_np_{n+4}p_{n+5}-p_{n+3}^3 &= (-1)^{n+1}(8x^3p_{n+4}+p_{n+5});\\ q_nq_{n+4}q_{n+5}-q_{n+3}^3 &= 4(x^2+1)(-1)^n(8x^3q_{n+4}+q_{n+5}).\end{aligned}$$

For example,

$$
\begin{aligned}
l_2 l_6 l_7 &= (x^2+2)(x^6+6x^4+9x^2+2)(x^7+7x^5+14x^3+7x)\\
l_5^3 &= x^{15}+15x^{13}+90x^{11}+275x^9+450x^7+375x^5+125x^3\\
l_2 l_6 l_7 - l_5^3 &= x^{11}+11x^9+44x^7+80x^5+71x^3+28x\\
&= (x^2+4)\left[x^3(x^6+6x^4+9x^2+2)+(x^7+7x^5+14x^3+7x)\right]\\
&= \mu(-1)^3(x^3 l_6 + l_7),
\end{aligned}
$$

as expected.

The above identities imply that

$$
\begin{aligned}
F_nF_{n+4}F_{n+5} - F_{n+3}^3 &= (-1)^{n+1}F_{n+6}; \qquad (35.19)\\
L_nL_{n+4}L_{n+5} - L_{n+3}^3 &= 5(-1)^nL_{n+6};\\
P_nP_{n+4}P_{n+5} - P_{n+3}^3 &= (-1)^{n+1}(8P_{n+4}+P_{n+5});\\
Q_nQ_{n+4}Q_{n+5} - Q_{n+3}^3 &= 2(-1)^n(8Q_{n+4}+Q_{n+5}).
\end{aligned}
$$

For example, $F_5F_9F_{10} - F_8^3 = 5\cdot 34\cdot 55 - 21^3 = 89 = (-1)^6F_{11}$, and $P_6P_{10}P_{11} - P_9^3 = 70\cdot 2378\cdot 5741 - 985^3 = -24{,}765 = (-1)^7(8\cdot 2378 + 5741) = (-1)^7(8P_{10}+P_{11})$.

S. Fairgrieve and H.W. Gould discovered the delightful identity (35.19) in 2005 [157].

Theorem 35.8 also has an additional consequence. It follows from identity (35.18) that $G_nG_{n+4}G_{n+5} - G_{n+3}^3 = (-1)^{n+1}\mu(1)G_{n+6}$; so $(G_nG_{n+4}G_{n+5} - G_{n+3}^3)^2 = \mu^2(1)G_{n+6}^2$. Consequently,

$$
4G_nG_{n+3}^3G_{n+4}G_{n+5} + \mu^2(1)G_{n+6}^2 = (G_nG_{n+4}G_{n+5} + G_{n+3}^3)^2.
$$

In particular, this implies

$$
\begin{aligned}
4F_nF_{n+3}^3F_{n+4}F_{n+5} + F_{n+6}^2 &= (F_nF_{n+4}F_{n+5} + F_{n+3}^3)^2;\\
4L_nL_{n+3}^3L_{n+4}L_{n+5} + 25L_{n+6}^2 &= (L_nL_{n+4}L_{n+5} + L_{n+3}^3)^2.
\end{aligned}
$$

For example,

$$
\begin{aligned}
4L_4L_5L_6^3L_9 + 25L_3^2 &= 4\cdot 7\cdot 11\cdot 18^3\cdot 76 + 25\cdot 4^2\\
&= 11{,}684^2\\
&= (7\cdot 11\cdot 76 + 18^3)^2\\
&= (L_4L_5L_9 + L_6^3)^2.
\end{aligned}
$$

The next theorem presents another difference of products of gibonacci polynomials.

Theorem 35.9. *Let $n \geq 0$. Then*

$$g_n g_{n+3}^2 - g_{n+2}^3 = \mu(-1)^{n+1}(x^2 g_{n+2} - g_n). \tag{35.20}$$

Proof. By the gibonacci recurrence, we have

$$\begin{aligned} g_n g_{n+3}^2 &= g_n(x g_{n+2} + g_{n+1})^2 \\ &= x^2 g_n g_{n+2}^2 + 2x g_n g_{n+1} g_{n+2} + g_n g_{n+1}^2. \end{aligned}$$

But

$$\begin{aligned} 2x g_n g_{n+1} g_{n+2} &= (g_{n+2} - x g_{n+1})(g_{n+2} - g_n) g_{n+2} + g_n(g_{n+2} - g_n) g_{n+2} \\ &= g_{n+2}^3 - x g_{n+1} g_{n+2}(g_{n+2} - g_n) - g_n^2 g_{n+2} \\ &= g_{n+2}^3 - x^2 g_{n+1}^2 g_{n+2} - g_n^2 g_{n+2}. \end{aligned}$$

Therefore,

$$\begin{aligned} g_n g_{n+3}^2 - g_{n+2}^3 &= x^2 g_n g_{n+2}^2 - x^2 g_{n+1}^2 g_{n+2} - g_n^2 g_{n+2} + g_n g_{n+1}^2 \\ &= (g_n g_{n+2} - g_{n+1}^2)(x^2 g_{n+2} - g_n) \\ &= (-1)^{n+1} \mu(x^2 g_{n+2} - g_n), \end{aligned}$$

as desired. ∎

As can be predicted, this theorem also has interesting ramifications:

$$\begin{aligned} f_n f_{n+3}^2 - f_{n+2}^3 &= (-1)^{n+1}(x^2 f_{n+2} - f_n); \\ l_n l_{n+3}^2 - l_{n+2}^3 &= (x^2 + 4)(-1)^n(x^2 l_{n+2} - l_n); \\ p_n p_{n+3}^2 - p_{n+2}^3 &= (-1)^{n+1}(4x^2 p_{n+2} - p_n); \\ q_n q_{n+3}^2 - q_{n+2}^3 &= 4(x^2 + 1)(-1)^n(4x^2 q_{n+2} - q_n). \end{aligned} \tag{35.21}$$

For example,

$$\begin{aligned} l_2 l_5^2 - l_4^3 &= (x^2 + 2)(x^5 + 5x^3 + 5x)^2 - (x^4 + 4x^2 + 2)^3 \\ &= x^8 + 8x^6 + 17x^4 + 2x^2 - 8 \\ &= (x^2 + 4)\left[x^2(x^4 + 4x^2 + 2) - (x^2 + 2)\right] \\ &= (x^2 + 4)(x^2 l_4 - l_2). \end{aligned}$$

The above polynomial identities have additional Fibonacci, Lucas, Pell, and Pell–Lucas consequences. For example, it follows from identity (35.21) that

$$F_n F_{n+3}^2 - F_{n+2}^3 = (-1)^{n+1} F_{n+1}.$$

Fairgrieve and Gould found this charming identity also in 2005 [157].

It also follows from identity (35.20) that $G_n G_{n+3}^2 - G_{n+2}^3 = (-1)^{n+1}\mu(1)G_{n+1}$. As before, this yields

$$4G_n G_{n+2}^3 G_{n+3}^2 + \mu^2(1)G_{n+1}^2 = (G_n G_{n+3}^2 + G_{n+2}^3)^2.$$

This implies

$$4F_n F_{n+2}^3 F_{n+3}^2 + F_{n+1}^2 = (F_n F_{n+3}^2 + F_{n+2}^3)^2;$$
$$4L_n L_{n+2}^3 L_{n+3}^2 + 25L_{n+1}^2 = (L_n L_{n+3}^2 + L_{n+2}^3)^2.$$

For example,

$$\begin{aligned}4F_5F_7^3F_8^2 + F_6^2 &= 4 \cdot 5 \cdot 13^3 \cdot 21^2 + 8^2\\ &= 4{,}402^2\\ &= (5 \cdot 21^2 + 13^3)^2\\ &= (F_5F_8^2 + F_7^3)^2.\end{aligned}$$

Fairgrieve and Gould also discovered that $F_n^2 F_{n+3} - F_{n+1}^3 = (-1)^{n+1}F_{n+2}$. The following theorem generalizes this identity to the gibonacci family. Its proof is also short and neat.

Theorem 35.10. *Let $n \ge 0$. Then*

$$g_n^2 g_{n+3} - g_{n+1}^3 = \mu(-1)^{n+1}(g_{n+3} - x^2 g_{n+1}). \tag{35.22}$$

Proof. By the gibonacci recurrence, we have

$$\begin{aligned}g_n^2 g_{n+3} - g_{n+1}^3 &= (g_{n+2} - xg_{n+1})^2 g_{n+3} - g_{n+1}(g_{n+3} - xg_{n+2})^2\\ &= g_{n+2}^2 g_{n+3} + x^2 g_{n+1}^2 g_{n+3} - g_{n+1} g_{n+3}^2 - x^2 g_{n+1} g_{n+2}^2\\ &= (g_{n+1}g_{n+3} - g_{n+2}^2)(x^2 g_{n+1} - g_{n+3})\\ &= (-1)^{n+1}\mu(g_{n+3} - x^2 g_{n+1}).\end{aligned}$$

■

It follows from identity (35.22) that

$$f_n^2 f_{n+3} - f_{n+1}^3 = (-1)^{n+1}(f_{n+3} - x^2 f_{n+1});$$
$$l_n^2 l_{n+3} - l_{n+1}^3 = (x^2+4)(-1)^n(l_{n+3} - x^2 l_{n+1});$$
$$p_n^2 p_{n+3} - p_{n+1}^3 = (-1)^{n+1}(p_{n+3} - 4x^2 p_{n+1});$$
$$q_n^2 q_{n+3} - q_{n+1}^3 = 4(x^2+1)(-1)^n(q_{n+3} - 4x^2 q_{n+1}).$$

For example,

$$\begin{aligned}
l_3^2 l_6 - l_4^3 &= (x^3 + 3x)^2(x^6 + 6x^4 + 9x^2 + 2) - (x^4 + 4x^2 + 2)^3 \\
&= -(2x^6 + 15x^4 + 30x^2 + 8) \\
&= -(x^2 + 4)\left[(x^6 + 6x^4 + 9x^2 + 2) - x^2(x^4 + 4x^2 + 2)\right] \\
&= (x^2 + 4)(-1)^3(l_6 - x^2 l_4),
\end{aligned}$$

as expected.

Theorem 35.10 has another interesting consequence. It also follows from identity (35.22) that $G_n^2 G_{n+3} - G_{n+1}^3 = (-1)^{n+1}\mu(1)G_{n+2}$. Again, as before, this yields

$$4G_n^2 G_{n+1}^3 G_{n+3} + \mu^2(1)G_{n+2}^2 = (G_n^2 G_{n+3} + G_{n+1}^3)^2.$$

Consequently,

$$\begin{aligned}
4F_n^2 F_{n+1}^3 F_{n+3} + F_{n+2}^2 &= (F_n^2 F_{n+3} + F_{n+1}^3)^2; \\
4L_n^2 L_{n+1}^3 L_{n+3} + 25L_{n+2}^2 &= (L_n^2 L_{n+3} + L_{n+1}^3)^2.
\end{aligned}$$

For example,

$$\begin{aligned}
4L_5^2 L_6^3 L_8 + 25L_7^2 &= 4 \cdot 11^2 \cdot 18^3 \cdot 47 + 25 \cdot 29^2 \\
&= 11{,}519^2 \\
&= (11^2 \cdot 47 + 18^3)^2 \\
&= (L_5^2 L_8 + L_6^3)^2.
\end{aligned}$$

The next theorem highlights an interesting difference of two products of four gibonacci polynomials. It is a straightforward application of the Catalan-like formula for gibonacci polynomials.

Theorem 35.11. *Let $n \geq 0$. Then*

$$g_{n+2}g_{n+1}g_{n-1}g_{n-2} - g_n^4 = \mu[(1 - x^2)(-1)^n g_n^2 - \mu x^2]. \tag{35.23}$$

Proof.

$$\begin{aligned}
\text{LHS} &= (g_{n+2}g_{n-2})(g_{n+1}g_{n-1}) - g_n^4 \\
&= [g_n^2 - \mu(-1)^n x^2][g_n^2 + \mu(-1)^n] - g_n^4 \\
&= [\mu(-1)^n - \mu(-1)^n x^2]g_n^2 - \mu^2 x^2 \\
&= \mu(1 - x^2)(-1)^n g_n^2 - \mu^2 x^2.
\end{aligned}$$

■

In particular, Theorem 35.11 implies that

$$\begin{aligned}
f_{n+2}f_{n+1}f_{n-1}f_{n-2} - f_n^4 &= (1 - x^2)(-1)^n f_n^2 - x^2; \\
l_{n+2}l_{n+1}l_{n-1}l_{n-2} - l_n^4 &= (x^2 + 4)[(x^2 - 1)(-1)^n l_n^2 - (x^2 + 4)x^2];
\end{aligned}$$

$$p_{n+2}p_{n+1}p_{n-1}p_{n-2} - p_n^4 = (1 - 4x^2)(-1)^n p_n^2 - 4x^2;$$
$$q_{n+2}q_{n+1}q_{n-1}q_{n-2} - q_n^4 = 4(x^2 + 1)[(4x^2 - 1)(-1)^n q_n^2 - 16(x^2 + 1)x^2].$$

For example,

$$\begin{aligned} l_5 l_4 l_3 l_2 - l_3^4 &= (x^5 + 5x^3 + 5x)(x^4 + 4x^2 + 2)(x^2 + 2)x - (x^3 + 3x)^4 \\ &= -(x^{10} + 9x^8 + 24x^6 + 11x^4 - 20x^2) \\ &= (x^2 + 4)[(1 - x^2)(x^3 + 3x)^2 - (x^2 + 4)^2 x^2] \\ &= (x^2 + 4)[(1 - x^2)l_3^2 - (x^2 + 4)x^2]. \end{aligned}$$

The above identities imply that

$$F_{n+2}F_{n+1}F_{n-1}F_{n-2} - F_n^4 = -1; \tag{35.24}$$
$$L_{n+2}L_{n+1}L_{n-1}L_{n-2} - L_n^4 = -25;$$
$$P_{n+2}P_{n+1}P_{n-1}P_{n-2} - P_n^4 = 3(-1)^{n+1}P_n^2 - 4; \tag{35.25}$$
$$Q_{n+2}Q_{n+1}Q_{n-1}Q_{n-2} - Q_n^4 = 2[3(-1)^n Q_n^2 - 8]. \tag{35.26}$$

As an example, $Q_6Q_5Q_3Q_2 - Q_4^4 = 99 \cdot 41 \cdot 7 \cdot 3 - 17^4 = 1718 = 2(3 \cdot 17^2 - 8) = 2[3(-1)^4 Q_4^2 - 8]$.

The neat identity (35.24) is the *Gelin–Cesàro identity*, stated by E. Gelin, but proved by E. Cesàro (1859–1906) [130, 157].

It follows from identity (35.23) that $G_n^4 - G_{n+2}G_{n+1}G_{n-1}G_{n-2} = \mu^2(1)$. So $(G_{n+2}G_{n+1}G_{n-1}G_{n-2} - G_n^4)^2 = \mu^4(1)$. This implies

$$4G_{n+2}G_{n+1}G_n^4G_{n-1}G_{n-2} + \mu^4(1) = (G_{n+2}G_{n+1}G_{n-1}G_{n-2} + G_n^4)^2. \tag{35.27}$$

In particular,

$$4F_{n+2}F_{n+1}F_n^4F_{n-1}F_{n-2} + 1 = (F_{n+2}F_{n+1}F_{n-1}F_{n-2} + F_n^4)^2;$$
$$4L_{n+2}L_{n+1}L_n^4L_{n-1}L_{n-2} + 625 = (L_{n+2}L_{n+1}L_{n-1}L_{n-2} + L_n^4)^2.$$

For example,

$$\begin{aligned} 4F_9F_8F_7^4F_6F_5 + 1 &= 4 \cdot 34 \cdot 21 \cdot 13^4 \cdot 8 \cdot 5 + 1 \\ &= 3{,}262{,}808{,}641 \\ &= (34 \cdot 21 \cdot 8 \cdot 5 + 13^4)^2 \\ &= (F_9F_8F_6F_5 + F_7^4)^2. \end{aligned}$$

Likewise, $4L_8L_7L_6^4L_5L_4 + 625 = 44{,}069{,}345{,}329 = (L_8L_7L_5L_4 + L_6^4)^2$.

It follows from identities (35.25) and (35.26) that

$$4P_{n+2}P_{n+1}P_n^4P_{n-1}P_{n-2} + [4 + 3(-1)^n P_n^2]^2 = (P_{n+2}P_{n+1}P_{n-1}P_{n-2} + P_n^4)^2;$$
$$4Q_{n+2}Q_{n+1}Q_n^4Q_{n-1}Q_{n-2} + 4[8 - 3(-1)^n Q_n^2]^2 = (Q_{n+2}Q_{n+1}Q_{n-1}Q_{n-2} + Q_n^4)^2.$$

For example,

$$\begin{aligned}4P_7P_6P_5^4P_4P_3 + (4 - 3P_5^2)^2 &= 4 \cdot 169 \cdot 70 \cdot 29^4 \cdot 12 \cdot 5 + (4 - 3 \cdot 29^2)^2\\ &= 2{,}008{,}118{,}560{,}561\\ &= (169 \cdot 70 \cdot 12 \cdot 5 + 29^4)^2\\ &= (P_7P_6P_4P_3 + P_5^4)^2.\end{aligned}$$

Interestingly, we can use the Catalan identity $F_{n+k}F_{n-k} - F_n^2 = (-1)^{n+k+1}F_k^2$ to re-confirm the Gelin–Cesàro identity, as J. Morgado of the University of Porto, Portugal, did in 1980 [361]. To this end, first we rewrite the Catalan identity as

$$F_{n+k}F_{n-k} + (-1)^{n+k}F_k^2 = F_n^2.$$

This implies

$$F_{n+k+1}F_{n-k-1} - (-1)^{n+k}F_{k+1}^2 = F_n^2.$$

Adding these two results and then subtracting one from the other, we get

$$\begin{aligned}F_{n+k}F_{n-k} + F_{n+k+1}F_{n-k-1} - (-1)^{n+k}F_{k+2}F_{k-1} &= 2F_n^2;\\ F_{n+k}F_{n-k} - F_{n+k+1}F_{n-k-1} &= (-1)^{n+k+1}F_{2k+1}.\end{aligned}$$

In particular, we have

$$F_{n+1}F_{n-1} + F_{n+2}F_{n-2} = 2F_n^2; \tag{35.28}$$
$$F_{n+1}F_{n-1} - F_{n+2}F_{n-2} = 2(-1)^n. \tag{35.29}$$

S.E. Ganis found formula (35.29) in 1959 [179].

The Gelin–Cesàro identity follows from these two identities by squaring each and subtracting one from the other.

As a byproduct, it also follows from these two identities that

$$F_{n+1}^2F_{n-1}^2 + F_{n+2}^2F_{n-2}^2 = 2(F_n^4 + 1).$$

For example,

$$\begin{aligned}F_8^2F_6^2 + F_9^2F_5^2 &= 21^2 \cdot 8^2 + 34^2 \cdot 5^2\\ &= 57{,}124 = 2(13^4 + 1)\\ &= 2(F_7^4 + 1).\end{aligned}$$

Using the Catalan-like identity $L_{n+k}L_{n-k} - L_n^2 = 5(-1)^{n+k}F_k^2$, we can establish the Gelin–Cesàro-like identity for the Lucas family:

$$L_n^4 - L_{n+2}L_{n+1}L_{n-1}L_{n-2} = 5^2;$$

see Exercise 35.50.

The Catalan-like identity $g_{n+k}g_{n-k} - g_n^2 = (-1)^{n+k+1}\mu f_k^2$ has additional consequences.

Additional Catalan Byproducts

It follows by the identity that

$$g_{n+2k}g_n - g_{n+k}^2 = (-1)^{n+1}\mu f_k^2; \tag{35.30}$$

$$(g_{n+2k}g_n - g_{n+k}^2)^2 = \mu^2 f_k^4;$$

$$4g_{n+2k}g_{n+k}^2 g_n + \mu^2 f_k^4 = (g_{n+2k}g_n + g_{n+k}^2)^2. \tag{35.31}$$

Consequently, $4g_{n+2k}g_{n+k}^2 g_n + \mu^2 f_k^4$ is a square, where $n, k \geq 0$.

In particular, $4g_{n+2}g_{n+1}^2 g_n + \mu^2$ is a square. This implies $4f_{n+2}f_{n+1}^2 f_n + 1$, $4l_{n+2}l_{n+1}^2 l_n + (x^2+4)^2$, $4p_{n+2}p_{n+1}^2 p_n + 1$, and $q_{n+2}q_{n+1}^2 q_n + 4(x^2+1)^2$ are squares; and so are $4F_{n+2}F_{n+1}^2 F_n + 1$ [361], $4L_{n+2}L_{n+1}^2 L_n + 25$, $4P_{n+2}P_{n+1}^2 P_n + 1$, and $Q_{n+2}Q_{n+1}^2 Q_n + 1$.

For example,

$$\begin{aligned}
4l_4 l_3^2 l_2 + (x^2+4)^2 &= 4(x^4+4x^2+2)(x^3+3x)^2(x^2+2) + (x^2+4)^2 \\
&= 4x^{10} + 48x^{10} + 220x^8 + 472x^6 + 457x^4 + 152x^2 + 16 \\
&= (2x^6 + 12x^4 + 19x^2 + 4)^2 \\
&= [(x^4+4x^2+2)(x^2+2) + (x^3+3x)^2]^2 \\
&= (l_4 l_2 + l_3^2)^2; \\
4(Q_7 Q_6^2 Q_5 + 1) &= 4(239 \cdot 99^2 \cdot 41 + 1) \\
&= 384{,}160{,}000 = 19{,}600^2 \\
&= (239 \cdot 41 + 99^2) \\
&= (Q_7 Q_5 + Q_6^2)^2.
\end{aligned}$$

It follows from identity (35.31) that $4g_{n+4}g_{n+2}^2 g_n + \mu^2 x^4$ is a square. Consequently, $4f_{n+4}f_{n+2}^2 f_n + x^4$, $4l_{n+4}l_{n+2}^2 l_n + (x^2+4)^2 x^4$, $4p_{n+4}p_{n+2}^2 p_n + 4x^4$, and $q_{n+4}q_{n+2}^2 q_n + 64(x^2+1)^2 x^4$ are squares. Therefore, $4F_{n+4}F_{n+2}^2 F_n + 1$ [361], $4L_{n+4}L_{n+2}^2 L_n + 25$, $P_{n+4}P_{n+2}^2 P_n + 1$, and $Q_{n+4}Q_{n+2}^2 Q_n + 16$ are also squares.

For example,

$$\begin{aligned}4l_5l_3^2l_1 + (x^2+4)^2x^4 &= 4(x^5+5x^3+5x)(x^3+3x)^2x + (x^2+4)^2x^4\\ &= 4x^{12}+44x^{10}+177x^8+308x^6+196x^4\\ &= (2x^6+11x^4+14x^2)^2\\ &= [(x^5+5x^3+5x)x + (x^3+3x)^2]^2\\ &= (l_5l_1 + l_3^2)^2;\\ 4(Q_7Q_5^2Q_3+16) &= 4(239\cdot 41^2\cdot 7+16)\\ &= 11{,}249{,}316 = (239\cdot 7 + 41^2)^2\\ &= (Q_7Q_3+Q_5^2)^2.\end{aligned}$$

Next we study yet another Catalan byproduct. To this end, we require the following results:

$$x\sum_{i=1}^{k} g_{n+2i} = g_{n+2k+1} - g_{n+1} \tag{35.32}$$

$$x\sum_{i=1}^{k} g_i^2 = g_{k+1}g_k - g_1g_0; \tag{35.33}$$

see Exercises 35.61 and 35.62; see Chapter 36 also.

It follows from formula (35.33) that

$$\begin{aligned}x\sum_{i=1}^{k} g_{n+i}^2 &= x\sum_{i=1}^{n+k} g_i^2 - x\sum_{i=1}^{n} g_i^2\\ &= (g_{n+k+1}g_{n+k} - g_1g_0) - (g_{n+1}g_n - g_1g_0)\\ &= g_{n+k+1}g_{n+k} - g_{n+1}g_n.\end{aligned} \tag{35.34}$$

It also follows from formula (35.33) that

$$x\sum_{i=1}^{k} f_i^2 = f_{k+1}f_k;$$

see Exercise 35.63.

By the Catalan-like identity, we have

$$g_{n+2i}g_n + (-1)^n\mu f_i^2 = g_{n+i}^2.$$

This yields

$$g_nx\sum_{i=1}^{k} g_{n+2i} + (-1)^n\mu x\sum_{i=1}^{k} f_i^2 = x\sum_{i=1}^{k} g_{n+i}^2$$

$$g_n(g_{n+2k+1} - g_{n+1}) + (-1)^n \mu f_{k+1} f_k = g_{n+k+1} g_{n+k} - g_{n+1} g_n$$

$$g_{n+2k+1} g_n + (-1)^n \mu f_{k+1} f_k = g_{n+k+1} g_{n+k} \tag{35.35}$$

$$(g_{n+2k+1} g_n - g_{n+k+1} g_{n+k})^2 = \mu^2 f_{k+1}^2 f_k^2$$

$$(g_{n+2k+1} g_n + g_{n+k+1} g_{n+k})^2 = 4 g_{n+2k+1} g_{n+k+1} g_{n+k} g_n + \mu^2 f_{k+1}^2 f_k^2. \tag{35.36}$$

Thus $4g_{n+2k+1}g_{n+k+1}g_{n+k}g_n + \mu^2 f_{k+1}^2 f_k^2$ is a square.

In particular,

$$4 g_{n+3} g_{n+2} g_{n+1} g_n + \mu^2 x^2 = (g_{n+3} g_n + g_{n+2} g_{n+1})^2. \tag{35.37}$$

This implies that $4f_{n+3}f_{n+2}f_{n+1}f_n + x^2$, $4l_{n+3}l_{n+2}l_{n+1}l_n + (x^2+4)^2x^2$, $p_{n+3}p_{n+2}p_{n+1}p_n + x^2$, and $q_{n+3}q_{n+2}q_{n+1}q_n + 16(x^2+1)^2x^2$ are squares. Consequently, $4F_{n+3}F_{n+2}F_{n+1}F_n + 1$ [361], $4L_{n+3}L_{n+2}L_{n+1}L_n + 25$, $P_{n+3}P_{n+2}P_{n+1}P_n + 1$ and $Q_{n+3}Q_{n+2}Q_{n+1}Q_n + 4$ are also squares.

For example,

$$\begin{aligned}
4f_5f_4f_3f_2 + x^2 &= 4(x^4+3x^2+1)(x^3+2x)(x^2+1)x + x^2\\
&= 4x^{10} + 24x^8 + 48x^6 + 36x^4 + 9x^2\\
&= (2x^5+6x^3+3x)^2\\
&= [(x^4+3x^2+1)x + (x^3+2x)(x^2+1)]^2\\
&= (f_5f_2 + f_4f_3)^2;\\
4L_6L_5L_4L_3 + 25 &= 4\cdot 18\cdot 11\cdot 7\cdot 4 + 25\\
&= 22{,}201 = (18\cdot 4 + 11\cdot 7)^2\\
&= (L_6L_3 + L_5L_4)^2.
\end{aligned}$$

It also follows from identity (35.36) that

$$4 g_{n+5} g_{n+3} g_{n+2} g_n + \mu^2 (x^2+1) x^2 = (g_{n+5} g_n + g_{n+3} g_{n+2})^2. \tag{35.38}$$

This implies that $4f_{n+5}f_{n+3}f_{n+2}f_n + (x^2+1)^2x^2$, $4l_{n+5}l_{n+3}l_{n+2}l_n + (x^2+4)^2(x^2+1)^2x^2$, $p_{n+5}p_{n+3}p_{n+2}p_n + (4x^2+1)x^2$, and $q_{n+5}q_{n+3}q_{n+2}q_n + 16(x^2+1)^2(4x^2+1)^2x^2$ are squares. Consequently, $F_{n+5}F_{n+3}F_{n+2}F_n + 1$ [361], $L_{n+5}L_{n+3}L_{n+2}L_n + 25$, $P_{n+5}P_{n+3}P_{n+2}P_n + 25$ and $Q_{n+5}Q_{n+3}Q_{n+2}Q_n + 100$ are also squares.

For example,

$$\begin{aligned}
&4l_6l_4l_3l_1 + (x^2+4)^2(x^2+1)^2x^2\\
&\quad = 4(x^6+6x^4+9x^2+2)(x^4+4x^2+2)(x^3+3x)x + (x^2+4)^2(x^2+1)^2x^2
\end{aligned}$$

$$\begin{aligned}
&= 4x^{14} + 52x^{12} + 261x^{10} + 630x^8 + 737x^6 + 368x^4 + 64x^2 \\
&= (2x^7 + 13x^5 + 23x^3 + 8x)^2 \\
&= [(x^6 + 6x^4 + 9x^2 + 2)x + (x^4 + 4x^2 + 2)(x^3 + 3x)]^2 \\
&= (l_6 l_1 + l_4 l_3)^2;
\end{aligned}$$

$$\begin{aligned}
4(Q_7Q_5Q_4Q_2 + 100) &= 4(239 \cdot 41 \cdot 17 \cdot 3 + 100) \\
&= 1{,}999{,}396 = (239 \cdot 3 + 41 \cdot 17)^2 \\
&= (Q_7Q_2 + Q_5Q_4)^2.
\end{aligned}$$

Earlier we established the Taguiri-like identity (35.4) using Theorem 35.3. We now re-confirm it using PMI, and identities (35.30) and (35.35):

$$g_{2n+h}g_n + (-1)^n \mu f_h^2 = g_{n+h}^2;$$
$$g_{2n+h+1}g_n + (-1)^n \mu f_{h+1} f_h = g_{n+h+1}g_{n+h}.$$

These two identities yield

$$(xg_{n+2h+1} + g_{n+2h})g_n + (-1)^n \mu f_h(xf_{h+1} + f_h) = g_{n+h}(xg_{n+h+1} + g_{n+h})$$
$$g_{n+2h+2}g_n + (-1)^n \mu f_{h+2} f_h = g_{n+h+2}g_{n+h}. \tag{35.39}$$

Combining equations (35.35) and (35.39) likewise, we get

$$g_{n+2h+3}g_n + (-1)^n \mu f_{h+3} f_h = g_{n+h+3}g_{n+h}. \tag{35.40}$$

More generally, consider the formula

$$g_{n+h+(h+i)}g_n + (-1)^n \mu f_{h+i} f_h = g_{n+h+i}g_{n+h}. \tag{35.41}$$

It follows by equations (35.30), (35.35), (35.39), and (35.40) that the formula works for $0 \le i \le 3$.

Suppose it is true for all nonnegative integers $< i$, where $i \ge 3$. Then

$$g_{n+h+(h+i-2)}g_n + (-1)^n \mu f_{h+i-2} f_h = g_{n+h+i-2}g_{n+h};$$
$$g_{n+h+(h+i-1)}g_n + (-1)^n \mu f_{h+i-1} f_h = g_{n+h+i-1}g_{n+h}.$$

These two equations yield

$$g_{n+h+(h+i)}g_n + (-1)^n \mu f_{h+i} f_h = g_{n+h+i}g_{n+h}.$$

Thus, by PMI, formula (35.41) works for all $i \ge 0$.

By letting $k = h + i$ in equation (35.41), the Taguiri-like identity (35.4) now follows.

The Taguiri-like identity also has interesting consequences.

Byproducts of the Taguiri-like Identity

To begin with, letting $n = b - 1$, $h = 1$, and $k = a - b + 1$, the identity yields

$$g_{a+1}g_{b-1} - g_a g_b = (-1)^b \mu f_{a-b+1}. \tag{35.42}$$

In particular, this yields

$$\begin{aligned} f_{a+1}f_{b-1} - f_a f_b &= (-1)^b f_{a-b+1}; \\ l_{a+1}l_{b-1} - l_a l_b &= (-1)^{b+1}(x^2+4) f_{a-b+1}; \\ p_{a+1}p_{b-1} - p_a p_b &= (-1)^b p_{a-b+1}; \\ q_{a+1}q_{b-1} - q_a q_b &= 4(-1)^{b+1}(x^2+1) p_{a-b+1}. \end{aligned} \tag{35.43}$$

It follows from identity (35.43) that $F_{a+1}F_{b-1} - F_a F_b = (-1)^b F_{a-b+1}$. M. d'Ocagne published this identity in 1885 [130, 361].

When $a = b$, identity (35.42) gives the Cassini-like formula for the gibonacci family:

$$g_{a+1}g_{a-1} - g_a^2 = (-1)^a \mu;$$

see Theorem 35.4.

Letting $b = a - 1$, identity (35.42) gives

$$\begin{aligned} g_a g_{a-1} &= g_{a+1}g_{a-2} + (-1)^a \mu x \\ &= (xg_a + g_{a-1})(g_a - xga - 1) + (-1)^a \mu x \\ xg_a g_{a-1} &= (g_a^2 - g_{a-1}^2) + (-1)^a \mu. \end{aligned} \tag{35.44}$$

It follows from identity (35.44) that

$$\begin{aligned} xf_a f_{a-1} &= f_a^2 - f_{a-1}^2 + (-1)^a; \\ xl_a l_{a-1} &= l_a^2 - l_{a-1}^2 - (-1)^a (x^2+4); \\ 2xp_a p_{a-1} &= p_a^2 - p_{a-1}^2 + (-1)^a; \\ 2xq_a q_{a-1} &= q_a^2 - q_{a-1}^2 - 4(-1)^a (x^2+1). \end{aligned}$$

For example,

$$\begin{aligned} xf_5 f_4 &= x(x^4 + 3x^2 + 1)(x^3 + 2x) \\ &= x^8 + 5x^6 + 7x^4 + 2x^2 \\ &= (x^4 + 3x^2 + 1)^2 - (x^3 + 2x)^2 - 1 \\ &= f_5^2 - f_4^2 - 1; \end{aligned}$$

$$\begin{aligned} xl_5l_4 &= x(x^5+5x^3+5x)(x^4+4x^2+2) \\ &= x^{10}+9x^8+27x^6+30x^4+10x^2 \\ &= (x^5+5x^3+5x)^2-(x^4+4x^2+2)^2+(x^2+4) \\ &= l_5^2-l_4^2+(x^2+4). \end{aligned}$$

The above identities imply that

$$\begin{aligned} F_aF_{a-1} &= F_a^2-F_{a-1}^2+(-1)^a; \\ L_aL_{a-1} &= L_a^2-L_{a-1}^2-5(-1)^a; \\ 2P_aP_{a-1} &= P_a^2-P_{a-1}^2+(-1)^a; \\ 2Q_aQ_{a-1} &= Q_a^2-Q_{a-1}^2-2(-1)^a. \end{aligned} \tag{35.45}$$

For example, $L_7L_6 = 29 \cdot 18 = 29^2 - 18^2 + 5 = L_7^2 - L_6^2 + 5$, and $2Q_7Q_6 = 2 \cdot 239 \cdot 99 = 239^2 - 99^2 + 2 = Q_7^2 - Q_6^2 + 2$.

d'Ocagne discovered identity (35.45) also in 1885 [130, 361].

The Taguiri-like identity (35.4) has yet another interesting byproduct. It follows by the identity that

$$(g_{n+h+k}g_n - g_{n+h}g_{n+k})^2 = \mu^2 f_h^2 f_k^2.$$

Consequently,

$$4g_{n+h+k}g_{n+h}g_{n+k}g_n + \mu^2 f_h^2 f_k^2 = (g_{n+h+k}g_n + g_{n+h}g_{n+k})^2. \tag{35.46}$$

Clearly, this generalizes the gibonacci identities (35.31) and (35.36).

It follows from identity (35.46) that $4g_{n+h+k}g_{n+h}g_{n+k}g_n + \mu^2 f_h^2 f_k^2$ is always a square. In particular, for instance,

$$4l_{n+h+k}l_{n+h}l_{n+k}l_n + (x^2+4)^2 f_h^2 f_k^2 = (l_{n+h+k}l_n + l_{n+h}l_{n+k})^2.$$

Next we study the gibonacci extensions of two well-known identities involving the third powers of three consecutive Fibonacci and Lucas numbers.

35.3 GENERALIZED LUCAS AND GINSBURG IDENTITIES

In 1876, Lucas established a charming identity involving the cubes of three consecutive Fibonacci numbers: $F_{n+1}^3 + F_n^3 - F_{n-1}^3 = F_{3n}$ [287, 334, 356]. In 1986, C.T. Long of Washington State University in Washington, discovered its Lucas counterpart: $L_{n+1}^3 + L_n^3 - L_{n-1}^3 = 5L_{3n}$ [334].

Interestingly, in 1953 Ginsburg noted that Lucas' identity is the only identity involving the cubes of Fibonacci numbers mentioned in Dickson's classic work *History of the Theory of Numbers* [37, 130, 356]. Ginsburg then developed an equally delightful identity involving the cubes of three Fibonacci numbers, separated by two spaces: $F_{n+2}^3 - 3F_n^3 + F_{n-2}^3 = 3F_{3n}$ [184, 356].

Gibonacci Extensions

We now extend the two identities to Fibonacci and Lucas polynomials. In both cases, their proofs involve some messy but demanding algebra, so we omit some details for the sake of brevity. We capitalize on a powerful technique touched upon by Melham in 1999 [282, 353].

We begin our pursuits with a lemma. Its proof is elementary, so we omit it; see Exercise 35.74.

Lemma 35.1. *Let $r_n = g_n^3 - g_{n+1}g_ng_{n-1}$. Then r_n satisfies the recurrence $r_n = -xr_{n-1} + r_{n-2}$.*

We are now ready to establish a generalization of Lucas' identity. To this end, we need the following identities, where $\Delta = \sqrt{x^2+4}$:

$$\begin{aligned} f_{-n} &= (-1)^{n-1}f_n & f_{2n} &= f_nl_n \\ f_{n+1}+f_{n-1} &= l_n & l_{n+1}+l_{n-1} &= \Delta^2 f_n \\ f_{n+1}^2+f_n^2 &= f_{2n+1} & l_{n+1}^2+l_n^2 &= \Delta^2 f_{2n+1}. \end{aligned}$$

Theorem 35.12.

$$g_{n+1}^3 + xg_n^3 - g_{n-1}^3 = \begin{cases} xf_{3n} & \text{if } g_k = f_k \\ x(x^2+4)l_{3n} & \text{if } g_k = l_k. \end{cases}$$

Proof. By Lemma 35.1, we have

$$\begin{aligned} g_{n+1}^3 - g_{n+2}g_{n+1}g_n &= -x(g_n^3 - g_{n+1}g_ng_{n-1}) + (g_{n-1}^3 - g_ng_{n-1}g_{n-2}) \\ g_{n+1}^3 + xg_n^3 - g_{n-1}^3 &= g_{n+2}g_{n+1}g_n + xg_{n+1}g_ng_{n-1} - g_ng_{n-1}g_{n-2} \\ &= (xg_{n+1}+g_n)g_{n+1}g_n + xg_{n+1}g_ng_{n-1} - g_ng_{n-1}(g_n - xg_{n-1}) \\ &= xg_{n+1}^2g_n + g_n^2(g_{n+1}-g_{n-1}) + xg_ng_{n-1}(g_{n+1}+g_{n-1}) \\ &= xg_{n+1}^2g_n + xg_n^3 + xg_ng_{n-1}(g_{n+1}+g_{n-1}). \end{aligned} \tag{35.47}$$

Suppose $g_k = l_k$. Then (35.47) yields

$$\begin{aligned} g_{n+1}^3 + xg_n^3 - g_{n-1}^3 &= xl_{n+1}^2l_n + xl_n^3 + xl_nl_{n-1}\Delta^2 f_n \\ &= xl_n(l_{n+1}^2 + l_n^2) + x\Delta^2 l_{n-1}f_{2n} \\ &= xl_n\Delta^2 f_{2n+1} + x\Delta^2 l_{n-1}f_{2n} \end{aligned}$$

$$= x\Delta^2(l_n f_{2n+1} + l_{n-1} f_{2n})$$
$$= x(x^2+4)l_{3n},$$

as desired. The other case can be handled similarly; see Exercise 35.75. ∎

For example,

$$\begin{aligned}
f_5^3 + xf_4^3 - f_3^3 &= (x^4+3x^2+1)^3 + x(x^3+2x)^3 - (x^2+1)^3\\
&= x^{12} + 10x^{10} + 36x^8 + 56x^6 + 35x^4 + 6x^2\\
&= xf_{12};\\
l_4^3 + xl_3^3 - l_2^3 &= (x^4+4x^2+2)^3 + x(x^3+3x)^3 - (x^2+2)^3\\
&= x^{12} + 13x^{10} + 63x^8 + 138x^6 + 129x^4 + 36x^2\\
&= x(x^2+4)(x^9+9x^7+27x^5+30x^3+9x)\\
&= x(x^2+4)l_9.
\end{aligned}$$

We note that this theorem can also be established using the addition formula for g_n.

Clearly, both Lucas' and Long's identities follow from this theorem. It also follows from the theorem that

$$\begin{aligned}
p_{n+1}^3 + 2xp_n^3 - p_{n-1}^3 &= 2xp_{3n};\\
q_{n+1}^3 + 2xq_n^3 - q_{n-1}^3 &= 8x(x^2+1)q_{3n};\\
P_{n+1}^3 + 2P_n^3 - P_{n-1}^3 &= 2P_{3n};\\
Q_{n+1}^3 + 2Q_n^3 - Q_{n-1}^3 &= 4Q_{3n}.
\end{aligned}$$

Next we generalize Ginsburg's identity. Although it follows by Theorem 35.12, we provide an independent proof. This approach also illustrates the power of Melham's technique. To this end, we need the next two lemmas. The proof of Lemma 35.2 is also straightforward, so we omit that too; see Exercise 35.76.

Lemma 35.2. *Let $s_n = g_{n+2}^3 - g_{n+4}g_{n+2}g_n$. Then $s_n = (x^2+2)s_{n-2} - s_{n-4}$.* ∎

Lemma 35.3.

$$g_{n+3}^2 - xg_{n+3}g_{n+2} - g_{n-1}g_{n-3} = \begin{cases} x(x^2+2)f_{2n} & \text{if } g_k = f_k\\ x(x^2+2)(x^2+4)f_{2n} & \text{if } g_k = l_k. \end{cases}$$

Proof. We have

$$\begin{aligned}
&g_{n+3}^2 - xg_{n+3}g_{n+2} - g_{n-1}g_{n-3}\\
&\quad = g_{n+3}(g_{n+3} - xg_{n+2}) - g_{n-1}g_{n-3}
\end{aligned}$$

$$= g_{n+3}g_{n+1} - g_{n-1}g_{n-3}$$
$$= g_{n+1}[(x^2+1)g_{n+1} + xg_n] - (g_{n+1} - xg_n)[(x^2+1)g_{n+1} - (x^3+2x)g_n]$$
$$= 2(x^3+2x)g_{n+1}g_n - x(x^3+2x)g_n^2$$
$$= (x^3+2x)g_n[g_{n+1} + (g_{n+1} - xg_n)]$$
$$= (x^3+2x)g_n(g_{n+1} + g_{n-1}). \tag{35.48}$$

Since $f_{n+1} + f_{n-1} = l_n$ and $l_{n+1} + l_{n-1} = (x^2+4)f_n$, the desired results follow from (35.48), as claimed. ■

For example,

$$\begin{aligned} l_5^2 - xl_5l_4 - l_1l_{-1} &= (x^5+5x^3+5x)^2 - x(x^5+5x^3+5x)(x^4+4x^2+2) + x^2 \\ &= x^8 + 8x^6 + 20x^4 + 16x^2 \\ &= x(x^2+2)(x^2+4)(x^3+2x) \\ &= x(x^2+2)(x^2+4)f_4. \end{aligned}$$

It follows by Lemma 35.3 that

$$x(x^2+2)\Delta^2 f_{2n-2} - l_{n+2}^2 + xl_{n+2}l_{n+1} + l_{n-2}l_{n-4} = 0.$$

We employ this result in the proof of Theorem 35.13.

We are now ready to present the next generalization. In addition to Lemmas 35.2 and 35.3, we need the following three identities:

$$\begin{aligned} f_{n+2} - f_{n-2} &= xl_n; \\ l_{n+2} - l_{n-2} &= x\Delta^2 f_n; \\ l_{n+1}f_{2n} + l_nf_{2n-1} &= l_{3n}. \end{aligned}$$

Theorem 35.13.

$$g_{n+2}^3 - (x^2+2)g_n^3 + g_{n-2}^3 = \begin{cases} x^2(x^2+2)f_{3n} & \text{if } g_k = f_k \\ x^2(x^2+2)(x^2+4)l_{3n} & \text{if } g_k = l_k. \end{cases}$$

Proof. Using Lemma 35.2, we have

$$g_{n+2}^3 - g_{n+4}g_{n+2}g_n = (x^2+2)(g_n^3 - g_{n+2}g_ng_{n-2}) - (g_{n-2}^3 - g_ng_{n-2}g_{n-4})$$
$$g_{n+2}^3 - (x^2+2)g_n^3 + g_{n-2}^3$$
$$= g_{n+4}g_{n+2}g_n - (x^2+2)g_{n+2}g_ng_{n-2} + g_ng_{n-2}g_{n-4}$$
$$= g_{n+2}g_n[(x^2+1)g_{n+2} + xg_{n+1}] - (x^2+2)g_{n+2}g_ng_{n-2} + g_ng_{n-2}g_{n-4}$$
$$= (x^2+2)g_{n+2}g_n(g_{n+2} - g_{n-2}) - g_{n+2}^2g_n + xg_{n+2}g_{n+1}g_n + g_ng_{n-2}g_{n-4}. \tag{35.49}$$

Suppose $g_k = l_k$. By Lemma 35.3, we then have

$$\begin{aligned}
\text{LHS} &= (x^2+2)l_{n+2}l_n \cdot x\Delta^2 f_n - l_{n+2}^2 l_n + xl_{n+2}l_{n+1}l_n + l_n l_{n-2}l_{n-4}\\
&= x(x^2+2)\Delta^2 l_{n+2}f_{2n} - l_{n+2}^2 l_n + xl_{n+2}l_{n+1}l_n + l_n l_{n-2}l_{n-4}\\
&= x(x^2+2)\Delta^2 f_{2n}(xl_{n+1}+l_n) - l_{n+2}^2 l_n + xl_{n+2}l_{n+1}l_n + l_n l_{n-2}l_{n-4}\\
&= x^2(x^2+2)\Delta^2 l_{n+1}f_{2n} + x(x^2+2)\Delta^2 f_{2n}l_n\\
&\quad - l_{n+2}^2 l_n + xl_{n+2}l_{n+1}l_n + l_n l_{n-2}l_{n-4}\\
&= x^2(x^2+2)\Delta^2 l_{n+1}f_{2n} + x(x^2+2)\Delta^2 l_n(xf_{2n-1}+f_{2n-2})\\
&\quad - l_{n+2}^2 l_n + xl_{n+2}l_{n+1}l_n + l_n l_{n-2}l_{n-4}\\
&= x^2(x^2+2)\Delta^2(l_{n+1}f_{2n} + l_n f_{2n-1})\\
&\quad + l_n[x(x^2+2)\Delta^2 f_{2n-2} - l_{n+2}^2 + xl_{n+2}l_{n+1} + l_{n-2}l_{n-4}]\\
&= x^2(x^2+2)\Delta^2 l_{3n} + l_n \cdot 0\\
&= x^2(x^2+2)\Delta^2 l_{3n},
\end{aligned}$$

as claimed. The other case follows similarly; see Exercise 35.77. ■

For example,

$$\begin{aligned}
f_5^3 - (x^2+2)f_3^3 + f_1^3 &= (x^4+3x^2+1)^3 - (x^2+2)(x^2+1)^3 + 1\\
&= x^{12} + 9x^{10} + 29x^8 + 40x^6 + 21x^4 + 2x^2\\
&= x^2(x^2+2)(x^8 + 7x^6 + 15x^4 + 10x^2 + 1)\\
&= x^2(x^2+2)f_9;\\
l_5^3 - (x^2+2)l_3^3 + l_1^3 &= (x^5+5x^3+5x)^3 - (x^2+2)(x^3+3x)^3 + x^3\\
&= x^{15} + 15x^{13} + 89x^{11} + 264x^9 + 405x^7 + 294x^5 + 72x^3\\
&= x^2(x^2+2)(x^2+4)(x^9 + 9x^7 + 27x^5 + 30x^3 + 9x)\\
&= x^2(x^2+2)(x^2+4)l_9.
\end{aligned}$$

Obviously, Theorem 35.13 also has Fibonacci, Lucas, Pell, and Pell–Lucas implications:

$$\begin{aligned}
F_{n+2}^3 - 3F_n^3 + F_{n-2}^3 &= 3F_{3n};\\
L_{n+2}^3 - 3L_n^3 + L_{n-2}^3 &= 15L_{3n};\\
p_{n+2}^3 - 2(2x^2+1)p_n^3 + p_{n-2}^3 &= 8x^2(2x^2+1)p_{3n};\\
q_{n+2}^3 - 2(2x^2+1)q_n^3 + q_{n-2}^3 &= 32x^2(x^2+1)(2x^2+1)q_{3n};
\end{aligned}$$

$$P_{n+2}^3 - 6P_n^3 + P_{n-2}^3 = 24P_{3n};$$
$$Q_{n+2}^3 - 6Q_n^3 + Q_{n-2}^3 = 48Q_{3n}.$$

Theorems 35.12 and 35.13 yield the following result.

Corollary 35.1.

$$g_{n+2}^3 - (x^3 + 2x)g_{n+1}^3 - (x^4 + 3x^2 + 2)g_n^3 + (x^3 + 2x)g_{n-1}^3 + g_{n-2}^3 = 0.$$ ■

For example,

$$\begin{aligned} f_5^3 + (x^3 + 2x)f_2^3 + f_1^3 &= (x^4 + 3x^2 + 1)^3 + (x^3 + 2x)x^3 + 1 \\ &= x^{12} + 9x^{10} + 30x^8 + 46x^6 + 32x^4 + 9x^2 + 2 \\ &= (x^3 + 2x)(x^3 + 2x)^3 + (x^4 + 3x^2 + 2)(x^2 + 1)^3 \\ &= (x^3 + 2x)f_4^3 + (x^4 + 3x^2 + 2)f_3^3. \end{aligned}$$

Interestingly, we can generalize Theorems 35.12 and 35.13 into a single relationship linking g_{n+k}^3, g_n^3, g_{n-k}^3, and g_{3n}, where $g_r = f_r$ or l_r. To this end, we need the following lemma.

Lemma 35.4.

$$(x^2 + 1)g_n^3 + 3g_{n-1}g_ng_{n+1} = \begin{cases} f_{3n} & \text{if } g_k = f_k \\ (x^2 + 4)l_{3n} & \text{if } g_k = l_k. \end{cases}$$

Proof. Suppose $g_k = l_k$. Using the identities $l_{a+b} = f_{a+1}l_b + f_al_{b-1}$, $f_nl_n = f_{2n}$, $l_{n+1} + l_{n-1} = \Delta^2 f_n$, and $l_n^2 + l_{n+1}^2 = \Delta^2 f_{2n+1}$, we have

$$\begin{aligned} \Delta^2 l_{3n} &= \Delta^2(f_{2n+1}l_n + f_{2n}l_{n-1}) \\ &= l_n(l_n^2 + l_{n+1}^2) + \Delta^2 f_nl_nl_{n-1} \\ &= l_n^3 + l_n(xl_n + l_{n-1})^2 + l_nl_{n-1}(l_{n+1} + l_{n-1}) \\ &= (x^2 + 1)l_n^3 + 2l_{n-1}l_n(xl_n + l_{n-1}) + l_{n-1}l_nl_{n+1} \\ &= (x^2 + 1)l_n^3 + 3l_{n-1}l_nl_{n+1}. \end{aligned}$$

The other half follows similarly; see Exercise 35.78. ■

For example,

$$\begin{aligned} (x^2 + 1)l_3^3 + 3l_2l_3l_4 &= (x^2 + 1)(x^3 + 3x)^3 + 3(x^2 + 2)(x^3 + 3x)(x^4 + 4x^2 + 2) \\ &= x^{11} + 13x^9 + 63x^7 + 138x^5 + 129x^3 + 36x \end{aligned}$$

$$= (x^2+4)(x^9+9x^7+27x^5+30x^3+9x)$$
$$= (x^2+4)l_9.$$

We are now ready for the generalization.

Theorem 35.14.

$$g_{n+k}^3 - (-1)^k l_k g_n^3 + (-1)^k g_{n-k}^3 = \begin{cases} f_k f_{2k} f_{3n} & \text{if } g_r = f_r \\ (x^2+4) f_k f_{2k} l_{3n} & \text{if } g_r = l_r. \end{cases}$$

Proof. Suppose $g_r = l_r$. The corresponding proof requires Lemma 35.4, and the identities $f_{k+1} + f_{k-1} = l_k$, $f_{k+1}f_{k-1} - f_k^2 = (-1)^k$, and $l_{a-b} = (-1)^b(f_{b-1}l_a - f_b l_{a-1})$. We then have

$$\begin{aligned}
&l_{n+k}^3 + (-1)^k l_{n-k}^3 \\
&= (f_{k+1}l_n + f_k l_{n-1})^3 + (f_{k-1}l_n - f_k l_{n-1})^3 \\
&= l_n^3(f_{k+1}^3 + f_{k-1}^3) + 3f_k l_{n-1} l_n^2 (f_{k+1}^2 - f_{k-1}^2) + 3f_k^2 l_{n-1}^2 l_n (f_{k+1} + f_{k-1}) \\
&= l_n^3(f_{k+1} + f_{k-1})(f_{k+1}^2 - f_{k+1}f_{k-1} + f_{k-1}^2) + 3x f_k^2 l_k l_{n-1} l_n^2 + 3f_k^2 l_k l_{n-1}^2 l_n \\
&= l_n^3 l_k \left[(f_{k+1} - f_{k-1})^2 + f_{k+1}f_{k-1}\right] + 3f_k^2 l_k l_{n-1} l_n (x l_n + l_{n-1}) \\
&= l_k l_n^3 \left\{x^2 f_k^2 + \left[f_k^2 + (-1)^k\right]\right\} + 3f_k^2 l_k l_{n-1} l_n l_{n+1} \\
&= f_k^2 l_k \left[(x^2+1) l_n^3 + 3 l_{n-1} l_n l_{n+1}\right] + (-1)^k l_k l_n^3 \\
&= \Delta^2 l_{3n} f_k^2 l_k + (-1)^k l_k l_n^3.
\end{aligned}$$

This yields the desired cubic identity for Lucas polynomials.

The corresponding identity for Fibonacci polynomials follows similarly; see Exercise 35.79. ∎

For example,

$$\begin{aligned}
&f_7^3 - l_2 f_5^3 + f_3^3 \\
&= (x^6+5x^4+6x^2+1)^3 - (x^2+2)(x^4+3x^2+1)^3 + (x^2+1)^3 \\
&= x^{18} + 15x^{16} + 92x^{14} + 297x^{12} + 540x^{10} + 546x^8 + 280x^6 + 57x^4 + 2x^2 \\
&= x(x^3+2x)(x^{14} + 13x^{12} + 66x^{10} + 165x^8 + 210x^6 + 126x^4 + 28x^2 + 1) \\
&= f_2 f_5 f_{15}.
\end{aligned}$$

Clearly, Theorems 35.12 and 35.13 follow from Theorem 35.14. So do the following identities:

$$p_{n+k}^3 - (-1)^k q_k p_n^3 + (-1)^k p_{n-k}^3 = p_k p_{2k} p_{3n};$$
$$q_{n+k}^3 - (-1)^k q_k q_n^3 + (-1)^k q_{n-k}^3 = 4(x^2+1) p_k p_{2k} k q_{3n};$$

$$P_{n+k}^3 - 2(-1)^k Q_k P_n^3 + (-1)^k P_{n-k}^3 = 2P_k P_{2k} P_{3n};$$
$$Q_{n+k}^3 - 2(-1)^k Q_k Q_n^3 + (-1)^k Q_{n-k}^3 = 2P_k P_{2k} Q_{3n}.$$

For example, let $n = 5$ and $k = 3$. Then

$$\begin{aligned} Q_8^3 + 2Q_3 Q_5^3 - Q_2^3 &= 577^3 + 2 \cdot 7 \cdot 41^3 - 3^3 \\ &= 193{,}064{,}900 \\ &= 2 \cdot 5 \cdot 70 \cdot 275{,}807 \\ &= 2P_3 P_6 Q_{15}. \end{aligned}$$

Next we establish a close link between gibonacci polynomials and geometry.

35.4 GIBONACCI AND GEOMETRY

Let a and b be real numbers. Then $[a^2 + b^2 + (a+b)^2]^2 = 2[a^4 + b^4 + (a+b)^4]$. This is *Candido's identity*, named after the Italian mathematician Giacomo Candido (1871–1941).

This identity has a gibonacci–geometric interpretation. To see this, consider the square $PSTU$ in Figure 35.1, where $PQ = x^2 g_n^2$, $QR = g_{n-1}^2$, and $RS = g_{n+1}^2$. Then

$$\begin{aligned} \text{Area } PSTU &= (g_{n+1}^2 + x^2 g_n^2 + g_{n-1}^2)^2 \\ &= 2(g_{n+1}^4 + x^4 g_n^4 + g_{n-1}^4) \\ &= 2(\text{sum of the shaded areas}); \end{aligned}$$

that is,

$$(g_{n+1}^2 + x^2 g_n^2 + g_{n-1}^2)^2 = 2(g_{n+1}^4 + x^4 g_n^4 + g_{n-1}^4). \tag{35.50}$$

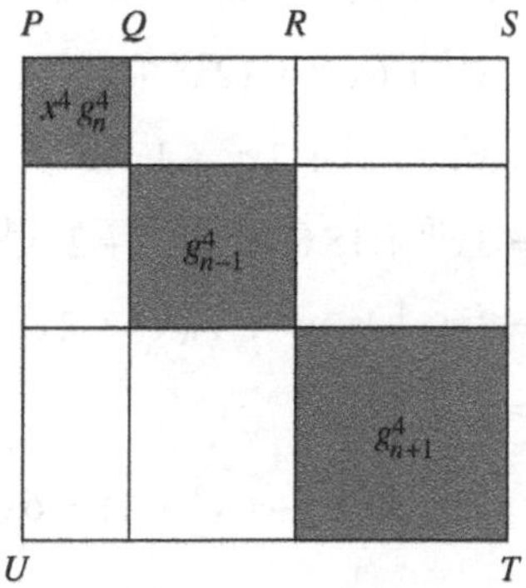

Figure 35.1.

In particular, identity (35.50) implies

$$(f_{n+1}^2 + x^2 f_n^2 + f_{n-1}^2)^2 = 2(f_{n+1}^4 + x^4 f_n^4 + f_{n-1}^4)$$
$$(l_{n+1}^2 + x^2 l_n^2 + l_{n-1}^2)^2 = 2(l_{n+1}^4 + x^4 l_n^4 + l_{n-1}^4)$$
$$(p_{n+1}^2 + 4x^2 p_n^2 + p_{n-1}^2)^2 = 2(p_{n+1}^4 + 16x^4 p_n^4 + p_{n-1}^4)$$
$$(q_{n+1}^2 + 4x^2 q_n^2 + q_{n-1}^2)^2 = 2(q_{n+1}^4 + 16x^4 q_n^4 + q_{n-1}^4).$$

A Cubic Identity

The algebraic identity $(u+v)^3 - u^3 - v^3 = 3uv(u+v)$ also has gibonacci implications. To see this, we let $u = xg_n$ and $v = g_{n-1}$. Then it yields the identity

$$g_{n+1}^3 - x^3 g_n^3 - g_{n-1}^3 = 3xg_{n+1}g_ng_{n-1}. \tag{35.51}$$

In particular, we then have

$$f_{n+1}^3 - x^3 f_n^3 - f_{n-1}^3 = 3xf_{n+1}f_nf_{n-1};$$
$$l_{n+1}^3 - x^3 l_n^3 - l_{n-1}^3 = 3xl_{n+1}l_nl_{n-1};$$
$$p_{n+1}^3 - 8x^3 p_n^3 - p_{n-1}^3 = 6xp_{n+1}p_np_{n-1};$$
$$q_{n+1}^3 - 8x^3 q_n^3 - q_{n-1}^3 = 6xq_{n+1}q_nq_{n-1};$$
$$F_{n+1}^3 - F_n^3 - F_{n-1}^3 = 3F_{n+1}F_nF_{n-1}; \tag{35.52}$$
$$L_{n+1}^3 - L_n^3 - L_{n-1}^3 = 3L_{n+1}L_nL_{n-1}; \tag{35.53}$$
$$P_{n+1}^3 - 8P_n^3 - P_{n-1}^3 = 6P_{n+1}P_nP_{n-1};$$
$$Q_{n+1}^3 - 8Q_n^3 - Q_{n-1}^3 = 6Q_{n+1}Q_nQ_{n-1}.$$

For example,

$$f_5^3 = x^{12} + 9x^{10} + 30x^8 + 45x^6 + 30x^4 + 9x^2 + 1$$
$$x^3 f_4^3 = x^{12} + 6x^{10} + 12x^8 + 8x^6$$
$$f_3^3 = x^6 + 3x^4 + 3x^2 + 1$$
$$\begin{aligned} f_5^3 - x^3 f_4^3 - f_3^3 &= 3x^{10} + 18x^8 + 36x^6 + 27x^4 + 6x^2 \\ &= 3x(x^4 + 3x^2 + 1)(x^3 + 2x)(x^2 + 1) \\ &= 3xf_5f_4f_3. \end{aligned}$$

Likewise, $p_4^3 - 8x^3 p_3^3 - p_2^3 = 48x^3(8x^4 + 6x^2 + 1) = 6x(8x^3 + 4x)(4x^2 + 1)(2x) = 6xp_4p_3p_2$, as expected.

Identity (35.51) has an interesting geometric consequence; see Figure 35.2.

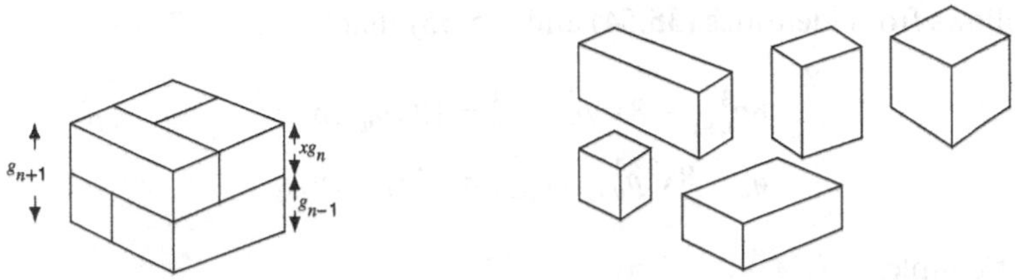

Figure 35.2. $g_{n+1}^3 - x^3g_n^3 - g_{n-1}^3 = 3xg_{n-1}g_ng_{n+1}$.

The identity $(u+v)^3 - u^3 - v^3 = 3uv(u+v)$, coupled with the identities $xf_n + l_n = 2f_{n+1}$ and $xf_n + 2f_{n-1} = l_n$, has additional consequences:

$$8f_{n+1}^3 - x^3f_n^3 - l_n^3 = 6xf_{n+1}f_{2n}; \tag{35.54}$$

$$l_n^3 - x^3f_n^3 - 8f_{n-1}^3 = 6xf_{n-1}f_{2n}. \tag{35.55}$$

For example,

$$\begin{aligned} 8f_4^3 - x^3f_3^3 - l_3^3 &= 8(x^3+2x)^3 - x^3(x^2+1)^3 - (x^3+3x)^3 \\ &= 6x^9 + 36x^7 + 66x^5 + 36x^3 \\ &= 6x(x^3+2x)(x^5+4x^3+3x) \\ &= 6xf_4f_6. \end{aligned}$$

As we can predict, identities (35.54) and (35.55) also can be interpreted geometrically; see Figures 35.3 and 35.4.

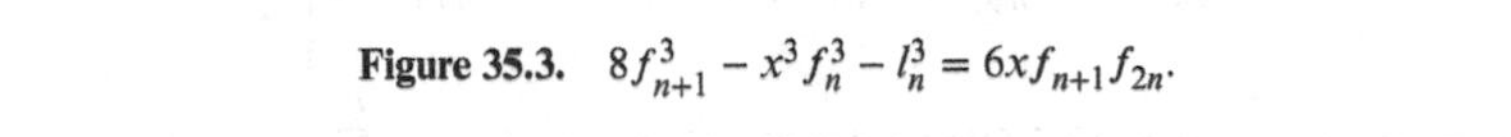

Figure 35.3. $8f_{n+1}^3 - x^3f_n^3 - l_n^3 = 6xf_{n+1}f_{2n}$.

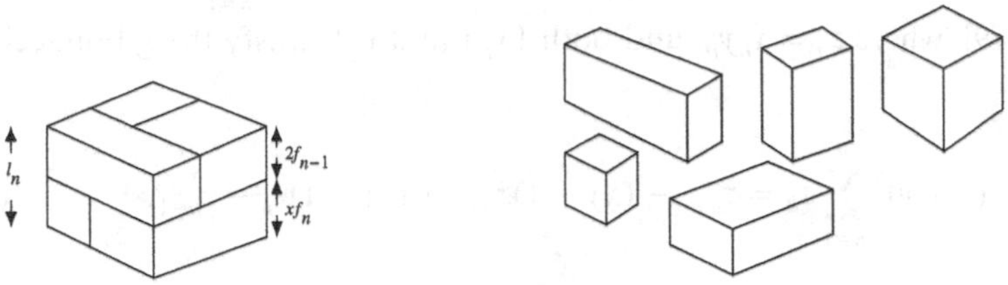

Figure 35.4. $l_n^3 - x^3f_n^3 - 8f_{n-1}^3 = 6xf_{n-1}f_{2n}$.

It follows from identities (35.54) and (35.55) that

$$8p_{n+1}^3 - 8x^3p_n^3 - q_n^3 = 12xp_{n+1}p_{2n};$$
$$q_n^3 - 8x^3p_n^3 - 8p_{n-1}^3 = 12xp_{n-1}p_{2n}.$$

For example,

$$\begin{aligned} q_3^3 - 8x^3p_3^3 - 8p_2^3 &= (8x^3+6x)^3 - 8x^3(4x^2+1)^3 - 8(2x)^3 \\ &= 768x^7 + 768x^5 + 144x^3 \\ &= 12x(2x)(32x^5+32x^3+6x) \\ &= 12xp_2p_6. \end{aligned}$$

Next we use two second-order recurrences to develop a fourth-order recurrence for $g_n(x)g_n(y)$. This has interesting implications, as we will see shortly.

35.5 ADDITIONAL RECURRENCES

Let $x_{n+2} = px_{n+1} + qx_n, y_{n+2} = ry_{n+1} + sy_n$, and $z_n = x_ny_n$, where p, q, r, and s are arbitrary and $n \geq 0$. Then we can show that [285]

$$z_{n+4} = prz_{n+3} + (p^2s + qr^2 + 2qs)z_{n+2} + pqrsz_{n+1} - q^2s^2z_n; \tag{35.56}$$

see Exercise 35.91.

Letting $p = x, r = y$, and $q = 1 = s$, we get a recurrence for $z_n = x_ny_n$:

$$z_{n+4} = xyz_{n+3} + (x^2 + y^2 + 2)z_{n+2} + xyz_{n+1} - z_n. \tag{35.57}$$

In particular, this gives a recurrence for $z_n = x_n^2$:

$$z_{n+4} = x^2z_{n+3} + 2(x^2+1)z_{n+2} + x^2z_{n+1} - z_n. \tag{35.58}$$

We can use recurrence (35.58) to develop a formula for $\sum_{k=1}^{n} z_k$, as Swamy did in 1967 [479], where $z_n = x_ny_n$, and both $\{x_n\}$ and $y_n\}$ satisfy the gibonacci recurrence:

$$(x+y)^2 \sum_{k=1}^{n} z_k = z_{n+2} - (xy-1)z_{n+1} + (xy-1)z_n - z_{n-1}. \tag{35.59}$$

This follows by PMI; see Exercise 35.94.

For example,

$$\begin{aligned}
&z_4 - (xy-1)z_3 + (xy-1)z_2 - z_1 \\
&\quad = (x^3+2x)(y^3+2y) - (xy-1)(x^2+1)(y^2+1) + (xy-1)xy - 1 \\
&\quad = x^3y + 2x^2y^2 + xy^3 + x^2 + 2xy + y^2 \\
&\quad = (x+y)^2(xy+1) \\
&\quad = (x+y)^2(x+y)^2 \sum_{k=1}^{2} z_k.
\end{aligned}$$

Letting $y = x$, formula (35.59) yields

$$4x^2 \sum_{k=1}^{n} z_k = z_{n+2} - (x^2-1)z_{n+1} + (x^2-1)z_n - z_{n-1},$$

where $z_k = f_k^2$.

It follows from recurrence (35.58) that $z_n = f_n^2$ satisfies it, where $z_0 = 0$, $z_1 = 1$, $z_2 = x^2$, and $z_3 = (x^2+1)^2$.

Suppose we let $p = x, r = 2x$, and $q = 1 = s$ in the recurrences for x_n and y_n. Let the initial conditions be such that $x_n = f_n$ and $y_n = p_n$. Then $z_n = f_n p_n$ satisfies the recurrence

$$z_{n+4} = 2x^2 z_{n+3} + (5x^2+2)z_{n+2} + 2x^2 z_{n+1} - z_n, \qquad (35.60)$$

where $z_0 = 0$, $z_1 = 1$, $z_2 = 2x^2$, and $z_3 = (x^2+1)(4x^2+1)$. Then

$$\begin{aligned}
z_4 &= 2x^2(x^2+1)(4x^2+1) + 2x^2(5x^2+2) + 2x^2 \cdot 1 - 0 \\
&= 8x^6 + 20x^4 + 8x^2 \\
&= (x^3+2x)(8x^3+4x) \\
&= f_4 p_4.
\end{aligned}$$

The same recurrence (35.60) works for the hybrids $f_n q_n$, $l_n p_n$, and $l_n q_n$. In particular, it yields a recurrence for $A_n = F_n P_n$, $B_n = L_n P_n$, $C_n = F_n Q_n$, and $D_n = L_n Q_n$:

$$z_{n+4} = 2z_{n+3} + 7z_{n+2} + 2z_{n+1} - z_n. \qquad (35.61)$$

With $A_0 = 0$, $A_1 = 1$, $A_2 = 2$, and $A_3 = 10$, recurrence (35.61) defines the hybrid sequence

$$\{A_n\}_{n\geq 0} = 0, 1, 2, 10, 36, 145, \ldots;$$

see the interesting pattern:

$$\begin{array}{rcl} 0 & = & 0 \cdot 0 \\ 1 & = & 1 \cdot 1 \\ 2 & = & 1 \cdot 2 \\ 10 & = & 2 \cdot 5 \\ 36 & = & 3 \cdot 12 \\ 145 & = & 5 \cdot 29 \\ 560 & = & 8 \cdot 70. \\ & & \uparrow \quad \uparrow \\ & & F_n \; P_n \end{array}$$

In 1998, F.J. Faase of the University of Twente, Netherlands, proved that A_n gives the number of *domino tilings* of the product set $W_4 \times P_{n-1}$, where W_k denotes the wheel graph with $k+1$ vertices and P_k the path graph with k vertices [158, 285].

Similarly,

$$\{B_n\}_{n\geq 0} = 0, 1, 6, 20, 84, 319, \ldots$$
$$\{C_n\}_{n\geq 0} = 0, 1, 3, 14, 51, 205, \ldots$$
$$\{D_n\}_{n\geq 0} = 2, 1, 9, 28, 119, 451, \ldots.$$

Recurrence (35.61) has an additional application: we can use it to develop a formula for the sum $\sum_{k=1}^{n} F_k P_k$. Swamy gave a neat solution to this problem, originally proposed in 1965 by D.G. Mead of Santa Clara University in California [351, 475].

Example 35.5. *Find a formula for* $\sum_{k=1}^{n} F_k P_k$.

Solution. Let $A_k = F_k P_k$. It then follows by recurrence (35.61) that

$$\sum_{k=1}^{n} A_{k+4} = \sum_{k=1}^{n} (2A_{k+3} + 7A_{k+2} + 2A_{k+1} - A_k).$$

After some basic algebra, this simplifies to

$$A_{n+4} = 9\sum_{k=1}^{n} A_k - 10A_1 - 8A_2 - A_3 + A_4 + A_{n+1} + 8A_{n+2} + A_{n+3}.$$

That is,

$$\begin{aligned} 9\sum_{k=1}^{n} A_k &= A_{n+4} - A_{n+3} - 8A_{n+2} - 10A_{n+1} + (10A_1 + 8A_2 + A_3 - A_4) \\ &= A_{n+4} - A_{n+3} - 8A_{n+2} - 10A_{n+1}. \end{aligned}$$

Using recurrence (35.61) twice, this yields

$$\begin{aligned}9\sum_{k=1}^{n} A_k &= A_{n+2} - A_{n+1} + A_n - A_{n-1}\\ &= F_{n+2}P_{n+2} - F_{n+1}P_{n+1} + F_nP_n - F_{n-1}P_{n-1}\\ &= (F_{n+1} + F_n)(2P_{n+1} + P_n) - F_{n+1}P_{n+1} + F_nP_n - (F_{n+1} - F_n)(P_{n+1} - 2P_n)\\ &= 3(F_{n+1}P_n + F_nP_{n+1})\end{aligned}$$

$$\sum_{k=1}^{n} F_kP_k = (F_{n+1}P_n + F_nP_{n+1})/3.$$ ■

For example, $\sum_{k=1}^{n} F_kP_k = 1 + 2 + 10 + 36 + 145 = 194 = (F_6P_5 + F_5P_6)/3$, as expected.

Similarly, we can show that

$$\sum_{k=1}^{n} F_kP_k = (L_{n+1}P_n + L_nP_{n+1} - 2)/3;$$

$$\sum_{k=1}^{n} F_kQ_k = (F_{n+1}Q_n + F_nQ_{n+1} - 1)/3;$$

$$\sum_{k=1}^{n} L_kQ_k = (L_{n+1}Q_n + L_nQ_{n+1} - 3)/3;$$

see Exercises 35.95–35.97.

Suppose we let $q = 1 = s$ in the recurrences of x_n and y_n. We will now find a recurrence for $w_n = x_n + y_n$.

Example 35.6. *Suppose* $x_{n+2} = px_{n+1} + x_n$, $y_{n+2} = ry_{n+1} + y_n$, *and* $w_n = x_n + y_n$, *where* $p = p(x)$, $r = r(y)$, *and* $n \geq 0$. *Prove that* $w_{n+4} - (p + r)w_{n+3} + (pr - 2)w_{n+2} + (p + r)w_{n+1} + w_n = 0$.

Proof. Since $w_{n+1} = x_{n+1} + y_{n+1}$, we have

$$\begin{aligned}w_{n+2} &= x_{n+2} + y_{n+2}\\ &= px_{n+1} + ry_{n+1} + w_n; \qquad (35.62)\\ w_{n+3} &= px_{n+2} + ry_{n+2} + w_{n+1}\\ (p + r)w_{n+3} &= p(p + r)x_{n+2} + r(p + r)y_{n+2} + (p + r)w_{n+1}\\ &= prw_{n+2} + p^2x_{n+2} + r^2y_{n+2} + (p + r)w_{n+1}; \qquad (35.63)\end{aligned}$$

$$\begin{aligned} w_{n+4} &= px_{n+3} + ry_{n+3} + w_{n+2} \\ &= p(px_{n+2} + x_{n+1}) + r(ry_{n+2} + y_{n+1}) + w_{n+2} \\ &= p^2x_{n+2} + r^2y_{n+2} + px_{n+1} + ry_{n+1} + w_{n+2}. \end{aligned} \tag{35.64}$$

Then, by equations (35.62), (35.63), and (35.64), we have

$$\begin{aligned} w_{n+4} - (p+r)w_{n+3} &= (px_{n+1} + ry_{n+1}) + w_{n+2} - prw_{n+2} - (p+r)w_{n+1} \\ &= (w_{n+2} - w_n) + w_{n+2} - prw_{n+2} - (p+r)w_{n+1} \\ &= (2 - pr)w_{n+2} - (p+r)w_{n+1} - w_n. \end{aligned}$$

This yields the desired result. ■

In particular, let $p = x$ and $r = y$. Then the recurrence yields

$$w_{n+4} - (x+y)w_{n+3} + (xy-2)w_{n+2} + (x+y)w_{n+1} + w_n = 0.$$

Swamy discovered this recurrence in 1966 [477].

Letting $y = x$, Swamy's recurrence yields

$$w_{n+4} - 2xw_{n+3} + (x^2-2)w_{n+2} + 2xw_{n+1} + w_n = 0.$$

35.6 PYTHAGOREAN TRIPLES

Finally, gibonacci polynomials can be used to construct Pythagorean triples. To see this, we let $u = xg_{n+2}$ and $v = g_{n+1}$, where x is a positive integer. Then

$$\begin{aligned} u^2 - v^2 &= (u+v)(u-v) \\ &= (xg_{n+2} + g_{n+1})(xg_{n+2} - g_{n+1}) \\ &= g_{n+3}(xg_{n+2} - g_{n+1}); \\ 2uv &= 2xg_{n+2}g_{n+1}; \\ u^2 + v^2 &= x^2g_{n+2}^2 + g_{n+1}^2. \end{aligned}$$

Since $(u^2 - v^2)^2 + (2uv)^2 = (u^2 + v^2)^2$, it follows that

$$\left[g_{n+3}(xg_{n+2} - g_{n+1})\right]^2 + \left(2xg_{n+2}g_{n+1}\right)^2 = \left(x^2g_{n+2}^2 + g_{n+1}^2\right)^2. \tag{35.65}$$

In particular, let $x = 1$ and $g_n = F_n$. Then identity (35.65) yields

$$(F_{n+3}F_n)^2 + (2F_{n+2}F_{n+1})^2 = F_{2n+3}^2.$$

Similarly, $(L_{n+3}L_n)^2 + (2L_{n+2}L_{n+1})^2 = 25F_{2n+3}^2$.

For example, let $n = 5$. Then

$$\begin{aligned}\text{LHS} &= (L_8L_5)^2 + (2L_7L_6)^2\\ &= (47\cdot 11)^2 + (2\cdot 29\cdot 18)^2\\ &= 1{,}357{,}225 = 25\cdot 233^2\\ &= 25F_{13}^2.\end{aligned}$$

EXERCISES 35

1. Solve the quadratic equation $g_{n-1}u^2 - xg_nu - g_{n+1} = 0$, where $g_0 \neq 0$.

Prove each, where g_n denotes the nth gibonacci polynomial.

2. Let r and s be the distinct roots of the quadratic equation $x^2 = px + q$. Let $u_n = (r^n - s^n)/(r - s)$ and $v_n = r^n + s^n$. Then $v_n = u_{n+1} + qu_{n-1}$ [218].
3. $g_{2n-1} = (x^2 + 2)g_{2n-3} - g_{2n-5}$, where $n \geq 3$.
4. $g_n = af_{n-2} + bf_{n-1}$.
5. Theorem 35.2.

Evaluate each, where x is a positive real number.

6. $c(x) + d(x)$.
7. $c(x) - d(x)$.

Prove each.

8. Theorem 35.4 using Theorem 35.1.
9. Theorem 35.4 using identity (35.4).
10. $g_n^2 = (x^2 + 2)g_{n-1}^2 - g_{n-2}^2 - 2\mu(-1)^n$.

Find a second-order recurrence for each.

11. f_n^2.
12. p_n^2.
13. q_n^2.

Prove each.

14. $g_{n+a}g_{n+b} - g_ng_{n+a+b} = (-1)^n\mu f_af_b$.
15. $g_{n+k}g_{n-k} - g_n^2 = (-1)^{n+k+1}\mu f_k^2$.
16. $g_{n+6} = (x^4 + 3x^2 + 1)g_{n+2} + (x^3 + 2x)g_{n+1}$.
17. $g_{n+3}^3 = x^3g_{n+2}^3 + 3x^2g_{n+2}^2g_{n+1} + 3xg_{n+2}g_{n+1}^2 + g_{n+1}^3$.
18. $g_n = (x^2 + 1)g_{n+4} - (x^3 + 2x)g_{n+3}$.

19. $g_n^2 g_{n+3} - g_{n+1}^3 = \mu(-1)^{n+1}(g_{n+3} - x^2 g_{n+1})$.
20. $f_{n-2}f_{n+1} - f_{n-1}f_n = (-1)^{n+1}x$.
21. $f_{n-3}f_{n+1} - f_{n-1}^2 = (-1)^n x^2$.
22. $g_{n-1}g_{n+2} - g_n g_{n+1} = (-1)^n \mu x$.
23. $g_{-n} = (-1)^{n+1}(af_{n+2} - bf_{n+1})$.
24. $g_{n+1} + g_{n-1} = al_{n-2} + bl_{n-1}$.
25. $g_{n+2} + g_{n-2} = (x^2 + 2)g_n$.
26. $g_{n+2} - g_{n-2} = (al_{n-2} + bl_{n-1})x$.
27. $g_{m+n} = g_m f_{n+1} + g_{m-1}f_n$.
28. $g_{m-n} = (-1)^n(g_m f_{n-1} - g_{m-1}f_n)$.
29. $g_{m+n} + g_{m-n} = \begin{cases} (g_{m+1} + g_{m-1})f_n & \text{if } n \text{ is odd} \\ g_m l_n & \text{otherwise.} \end{cases}$
30. $g_{m+n} - g_{m-n} = \begin{cases} g_m l_n & \text{if } n \text{ is odd} \\ (g_{m+1} + g_{m-1})f_n & \text{otherwise.} \end{cases}$
31. $g_{m+n}^2 - g_{m-n}^2 = g_m(g_{m+1} + g_{m-1})f_{2n}$.
32. $g_{n+1}^4 + x^4 g_n^4 + g_{n-1}^4 = 2(xg_{n+1}g_n + g_{n-1}^2)^2$.
33. $g_{n+1}^4 + x^4 g_n^4 + g_{n-1}^4 = 2[g_{n+1}^2 + g_{n-1}^2 - g_n^2 - \mu(-1)^n]^2$.
34. Using Exercises 35.32 and 35.33, deduce the Cassini-like formula for gibonacci polynomials. Assume x is a positive integer.
35. $g_n^2 + g_{n+1}^2 = a^2 f_{2n-3} + b^2 f_{2n-1} + 2ab f_{2n-2}$.

Prove each, where $H_n = G_n^2 G_{n+3}^2 + 4G_{n+1}^2 G_{n+2}^2$, where G_n denotes the nth gibonacci number [269].

36. $H_n = (G_{n+1}^2 + G_{n+2}^2)^2$.
37. Let $G_n = F_n$. Then $H_n = F_{2n+3}^2$.
38. Let $G_n = L_n$. Then $H_n = 25F_{2n+3}^2$.
39. $g_n^2(g_{n+2} + xg_{n+1})^2 + 4x^2 g_{n+2}^2 g_{n+1}^2 = (g_{n+2}^2 + x^2 g_{n+1}^2)^2$.
40. $F_n^2 F_{n+3}^2 + 4F_{n+2}^2 F_{n+1}^2 = F_{2n+3}^2$.
41. $L_n^2 L_{n+3}^2 + 4L_{n+2}^2 L_{n+1}^2 = 25F_{2n+3}^2$.
42. $x\sum_{k=0}^{n} g_k = g_{n+1} + g_n - ax^2 + (a + b)(x - 1)$.

43. $x\sum_{k=0}^{n} g_{2k+1} = g_{2n+2} + ax - b.$

44. $x\sum_{k=0}^{n} g_{2k} = g_{2n+1} - ax^2 + bx - a.$

45. $f_{2n}^2 - f_{2n-1}^2 - xf_{2n}f_{2n-1} + 1 = 0.$

46. $l_{2n}^2 - l_{2n-1}^2 - xl_{2n}l_{2n-1} - (x^2+4) = 0$

47. Let p, q, r, and s denote four consecutive gibonacci numbers. Then $(rs - pq)^2 = (ps)^2 + (2rq)^2$.

48. $(F_nF_{n+3})^2 + (2F_{n+1}F_{n+2})^2 = F_{2n+3}.$

49. $(L_nL_{n+3})^2 + (2L_{n+1}L_{n+2})^2 = 25F_{2n+3}^2.$

50. $L_n^4 - L_{n+2}L_{n+1}L_{n-1}L_{n-2} = 25$, using the Catalan-like identity for Lucas numbers.

51. $L_{n+1}L_{n-1} + L_{n+2}L_{n-2} = 2L_n^2.$

52. $L_{n+1}L_{n-1} - L_{n+2}L_{n-2} = 10(-1)^{n+1}.$

53. $L_{n+1}^2L_{n-1}^2 + L_{n+2}^2L_{n-2}^2 = 2(L_n^4 + 25).$

54. $4g_{n+2}g_{n+1}^2g_n + \mu^2 = (g_{n+2}g_n + g_{n+1}^2)^2.$

55. $4l_{n+2}l_{n+1}^2l_n + (x^2+4)^2 = (l_{n+2}l_n + l_{n+1}^2)^2.$

56. $4(Q_{n+2}Q_{n+1}^2Q_n + 1) = (Q_{n+2}Q_n + Q_{n+1}^2)^2.$

57. $4g_{n+4}g_{n+2}^2g_n + \mu^2x^4 = (g_{n+4}g_n + g_{n+2}^2)^2.$

58. $4f_{n+4}f_{n+2}^2f_n + x^4 = (f_{n+4}f_n + f_{n+2}^2)^2.$

59. $4l_{n+4}l_{n+2}^2l_n + (x^2+4)^2x^4 = (l_{n+4}l_n + l_{n+2}^2)^2.$

60. $4(Q_{n+4}Q_{n+2}^2Q_n + 16) = (Q_{n+4}Q_n + Q_{n+2}^2)^2.$

61. $x\sum_{i=1}^{k} g_{n+2i} = g_{n+2k+1} - g_{n+1}.$

62. $x\sum_{i=1}^{n} g_i^2 = g_{n+1}g_n - g_1g_0.$

63. $x\sum_{i=1}^{n} f_i^2 = f_{n+1}f_n.$

64. $4f_{n+3}f_{n+2}f_{n+1}f_n + x^2 = (f_{n+3}f_n + f_{n+2}f_{n+1})^2.$

65. $4l_{n+3}l_{n+2}l_{n+1}l_n + (x^2+4)^2x^2 = (l_{n+3}l_n + l_{n+2}l_{n+1})^2.$

66. $4F_{n+3}F_{n+2}F_{n+1}F_n + 1 = (F_{n+3}F_n + F_{n+2}F_{n+1})^2.$

67. $4L_{n+3}L_{n+2}L_{n+1}L_n + 25 = (L_{n+3}L_n + L_{n+2}L_{n+1})^2$.
68. $4f_{n+5}f_{n+3}f_{n+2}f_n + (x^2+1)^2x^2 = (f_{n+5}f_n + f_{n+3}f_{n+2})^2$.
69. $4l_{n+5}l_{n+3}l_{n+2}l_n + (x^2+4)^2(x^2+1)^2x^2 = (l_{n+5}l_n + l_{n+3}l_{n+2})^2$.
70. $4(F_{n+5}F_{n+3}F_{n+2}F_n + 1) = (F_{n+5}F_n + F_{n+3}F_{n+2})^2$.
71. $4(L_{n+5}L_{n+3}L_{n+2}L_n + 25) = (L_{n+5}L_n + L_{n+3}L_{n+2})^2$.
72. $4(P_{n+5}P_{n+3}P_{n+2}P_n + 100) = (P_{n+5}P_n + P_{n+3}P_{n+2})^2$.
73. $4(Q_{n+5}Q_{n+3}Q_{n+2}Q_n + 100) = (Q_{n+5}Q_n + Q_{n+3}Q_{n+2})^2$.
74. Let $r_n = g_n^3 - g_{n+1}g_ng_{n-1}$. Then $r_n = -xr_{n-1} + r_{n-2}$.
75. $f_{n+1}^3 + xf_n^3 - f_{n-1}^3 = xf_{3n}$.
76. Let $s_n = g_{n+2}^3 - g_{n+4}g_{n+2}g_n$. Then $s_n = (x^2+2)s_{n-2} - s_{n-4}$.
77. $f_{n+2}^3 - (x^2+2)f_n^3 + f_{n-2}^3 = x^2(x^2+2)f_{3n}$.
78. $(x^2+1)f_n^3 + 3f_{n-1}f_nf_{n+1} = f_{3n}$.
79. $f_{n+k}^3 - (-1)^kl_kf_n^3 + (-1)^kf_{n-k}^3 = f_kf_{2k}f_{3n}$.
80. Find a nonhomogeneous recurrence for $\{g_n^2\}$.

Find a homogeneous recurrence for each.

81. $\{g_{2n}\}$.
82. $\{g_{2n+1}\}$.
83. $\{g_n^2\}$.
84. $\{F_n^2\}$.
85. $\{p_n^2\}$.
86. $\{L_n^2\}$.
87. $\{Q_n^2\}$.

Find a generating function for each.

88. $\{g_{2n}\}$.
89. $\{F_n^2\}$.
90. $\{g_{2n+1}\}$.
91. Let $x_{n+2} = px_{n+1} + qx_n$, $y_{n+2} = ry_{n+1} + sy_n$, and $z_n = x_ny_n$, where p, q, r, and s are arbitrary, and $n \geq 0$. Show that $z_{n+4} = prz_{n+3} + (p^2 + qr^2 + 2qs)z_{n+2} + pqrsz_{n+1} - q^2s^2z_n$.
92. Prove that $z_n = g_n^2$ satisfies the recurrence $z_{n+4} = x^2z_{n+3} + 2(x^2+1)z_{n+2} + x^2z_{n+1} - z_n$.

93. Find a fourth-order recurrence satisfied by p_n^2.

94. Let $x_{n+2} = xx_{n+1} + x_n$, $y_{n+2} = yy_{n+1} + y_n$, and $z_n = x_n y_n$, where $n \geq 0$. Show that $(x+y)^2 \sum_{k=1}^{n} z_k = z_{n+2} - (xy-1)z_{n+1} + (xy-1)z_n - z_{n-1}$ [479].

Prove each, where $x \geq 1$.

95. $\sum_{k=1}^{n} L_k P_k = (L_{n+1}P_n + L_n P_{n+1} - 2)/3.$

96. $\sum_{k=1}^{n} F_k Q_k = (F_{n+1}Q_n + F_n Q_{n+1} - 1)/3.$

97. $\sum_{k=1}^{n} L_k Q_k = (L_{n+1}Q_n + L_n Q_{n+1} - 3)/3.$

98. $\sum_{k=1}^{n} \frac{5}{L_{2k}L_{2k+2}} = \frac{2}{3} - \frac{L_{2n}}{L_{2n+2}}.$

99. $\sum_{k=1}^{n} \frac{4}{P_{2k}P_{2k+2}} = \frac{P_{2n}}{P_{2n+2}}.$

100. $\sum_{k=1}^{n} \frac{8}{Q_{2k}Q_{2k+2}} = \frac{1}{3} - \frac{Q_{2n}}{Q_{2n+2}}.$

101. $\sum_{k=1}^{\infty} \frac{x^2}{f_{2k}f_{2k+2}} = \frac{1}{\alpha^2}.$

102. $\sum_{k=1}^{\infty} \frac{x^2(x^2+4)}{l_{2k}l_{2k+2}} = \frac{2}{x^2+2} - \frac{1}{\alpha^2}.$

103. $\sum_{k=1}^{\infty} \frac{4x^2}{p_{2k}p_{2k+2}} = \frac{1}{\gamma^2}.$

104. $\sum_{k=1}^{\infty} \frac{16x^2(x^2+1)}{q_{2k}q_{2k+2}} = \frac{1}{2x^2+1} - \frac{1}{\gamma^2}.$

105. $\sum_{k=1}^{n} \frac{\mu x^2}{g_{2k-1}g_{2k+1}} = \frac{\mu x + g_0 g_1}{g_1 g_2} - \frac{g_{2n}g_{2n+1} + \mu x}{g_{2n+1}g_{2n+2}}.$

106. $\sum_{k=1}^{n} \frac{x}{f_{2k-1}f_{2k+1}} = \frac{f_{2n}}{f_{2n+1}}.$

107. $\sum_{k=1}^{n} \frac{2x}{p_{2k-1}p_{2k+1}} = \frac{p_{2n}}{p_{2n+1}}.$

108. $\sum_{k=1}^{\infty} \frac{x}{f_{2k-1}f_{2k+1}} = \frac{1}{\alpha}$.

109. $\sum_{k=1}^{\infty} \frac{2x}{p_{2k-1}p_{2k+1}} = \frac{1}{\gamma}$.

110. $\frac{g_n^3}{g_{n+1}(xg_{n+1} - g_n)} + \frac{x^4 g_{n+1}^3}{g_n(g_n - xg_{n+1})} + \frac{g_{n+2}^3}{g_{n+1}(g_{n+2} - xg_{n+1})} = 2xg_{n+2}$.

36

GIBONACCI SUMS

The Good Lord made all the integers;
the rest is man's doing.
–Leopold Kronecker (1823–1891)

In this chapter, we study some gibonacci sums using gibonacci recurrence, Clary and Hemenway's method [104], and the *telescoping sum* $\sum_{k=1}^{n}(a_k - a_{k-1}) = a_n - a_0$. We then deduce the corresponding formulas for Fibonacci and Lucas numbers, Pell and Pell–Lucas polynomials and numbers. We also develop additional summation formulas using exponential generating functions. Again, for brevity, clarity, and convenience, we *omit* the argument from the functional notation.

36.1 GIBONACCI SUMS

We begin our pursuit with the sum of the first n gibonacci polynomials.

Example 36.1. *Prove that* $x\sum_{k=1}^{n} g_k = g_{n+1} + g_n - g_1 - g_0$.

Proof. We derive the formula using two telescoping sums:

$$\begin{aligned} xg_k &= g_{k+1} - g_{k-1} \\ &= (g_{k+1} - g_k) + (g_k - g_{k-1}) \end{aligned}$$

Fibonacci and Lucas Numbers with Applications, Volume Two. Thomas Koshy.

$$\begin{aligned}x\sum_{k=1}^{n} g_k &= \sum_{k=1}^{n}\left[(g_{k+1}-g_k)+(g_k-g_{k-1})\right]\\ &= (g_{n+1}-g_1)+(g_n-g_0)\\ &= g_{n+1}+g_n-g_1-g_0,\end{aligned}$$

as claimed. ∎

Alternately, we can employ the *Clary–Hemenway method* to establish the formula. To this end, let A_n denote the LHS of the sum and $B_n = g_{n+1} + g_n$. Then $B_n - B_{n-1} = g_{n+1} - g_{n-1} = xg_n = A_n - A_{n-1}$. So $A_n - B_n = A_{n-1} - B_{n-1}$. This implies $A_n - B_n = A_0 - B_0 = 0 - (g_1 + g_0)$; so $A_n = B_n - g_1 - g_0$, as desired.

In particular,

$$x\sum_{k=1}^{n} f_k = f_{n+1} + f_n - 1; \tag{36.1}$$

$$x\sum_{k=0}^{n} l_k = l_{n+1} + l_n + x - 2; \tag{36.2}$$

see Exercise 36.3.

For example,

$$\begin{aligned}x\sum_{k=1}^{5} f_k &= x[1 + x + (x^2+1) + (x^3+2x) + (x^4+3x^2+1)]\\ &= (x^5+4x^3+3x) + (x^4+3x^2+1) - 1\\ &= f_6 + f_5 - 1.\end{aligned}$$

Formulas (36.1) and (36.2) yield the following dividends:

$$2x\sum_{k=1}^{n} p_k = p_{n+1} + p_n - 1; \qquad 2x\sum_{k=0}^{n} q_k = q_{n+1} + q_n - 2x - 2;$$

$$\sum_{k=1}^{n} F_k = F_{n+2} - 1; \qquad \sum_{k=0}^{n} L_k = L_{n+2} - 1;$$

$$2\sum_{k=1}^{n} P_k = Q_{n+1} - 1; \qquad \sum_{k=0}^{n} Q_k = P_{n+1}.$$

We can use the summation formula in Example 36.1 to develop a formula for $2\sum_{k=1}^{n} G_{3k}$, where G_i denotes the ith gibonacci number:

$$\begin{aligned}2\sum_{k=1}^{n} G_{3k} &= \sum_{k=1}^{n}[G_{3k} + (G_{3k-1} + G_{3k-2})]\\ &= \sum_{k=1}^{3n} G_k\\ &= G_{3n+1} + G_{3n} - G_1 - G_0.\end{aligned}$$

In particular, $2\sum_{k=1}^{n} F_{3k} = F_{3n+2} - 1$ and $2\sum_{k=1}^{n} L_{3k} = L_{3n+2} - 3$. A.R. Gugheri of Sirjan, Iran, studied this Fibonacci sum in 2010 [203].

Summation formula (36.1) can be generalized, as the following example shows.

Example 36.2. *Prove that*

$$\sum_{i=0}^{n} f_{ki+j} = \frac{f_{nk+k+j} - (-1)^k f_{nk+j} - f_j - (-1)^j f_{k-j}}{l_k - (-1)^k - 1}. \tag{36.3}$$

Proof. Since $\alpha^k\beta^j - \alpha^j\beta^k = (\alpha\beta)^j(\alpha^{k-j} - \beta^{k-j}) = (-1)^j\Delta f_{k-j}$, we have

$$\begin{aligned}
\Delta\sum_{i=0}^{n} f_{ki+j} &= \sum_{i=0}^{n}\left(\alpha^{ki+j} - \beta^{ki+j}\right)\\
&= \alpha^j\sum_{i=0}^{n}\alpha^{ki} - \beta^j\sum_{i=0}^{n}\beta^{ki}\\
&= \alpha^j\left(\frac{\alpha^{nk+k} - 1}{\alpha^k - 1}\right) - \beta^j\left(\frac{\beta^{nk+k} - 1}{\beta^k - 1}\right)\\
&= \frac{(\alpha^{nk+k+j} - \alpha^j)(\beta^k - 1) - (\beta^{nk+k+j} - \beta^j)(\alpha^k - 1)}{(-1)^k - (\alpha^k + \beta^k) + 1}\\
\sum_{i=0}^{n} f_{ki+j} &= \frac{-f_{nk+k+j} + (-1)^k f_{nk+j} + f_j + (\alpha^k\beta^j - \alpha^j\beta^k)/\Delta}{(-1)^k - l_k + 1}.
\end{aligned}$$

This yields the desired formula. ■

In particular, formula (36.3) yields the following results:

$$\begin{aligned}
x\sum_{i=1}^{n} f_i &= f_{n+1} + f_n - 1;\\
x\sum_{i=1}^{n} f_{2i} &= f_{2n+1} - 1;\\
x\sum_{i=1}^{n} f_{2i-1} &= f_{2n};\\
(x^3 + 3x)\sum_{i=1}^{n} f_{3i} &= f_{3n+3} + f_{3n} - x^2 - 1.
\end{aligned} \tag{36.4}$$

It follows from formula (36.4) that

$$2\sum_{i=1}^{n} F_{3i} = F_{3n+2} - 1.$$

We can show similarly that

$$\sum_{i=1}^{n} l_{ki+j} = \frac{l_{nk+k+j} - (-1)^k l_{nk+j} - l_j - (-1)^j l_{k-j}}{l_k - (-1)^k - 1}; \tag{36.5}$$

see Exercise 36.4.

Obviously, formulas (36.3) and (36.5) have Pell and Pell–Lucas counterparts, respectively:

$$\sum_{i=1}^{n} p_{ki+j} = \frac{p_{nk+k+j} - (-1)^k p_{nk+j} - p_j - (-1)^j p_{k-j}}{q_k - (-1)^k - 1};$$

$$\sum_{i=1}^{n} q_{ki+j} = \frac{q_{nk+k+j} - (-1)^k q_{nk+j} - q_j - (-1)^j q_{k-j}}{q_k - (-1)^k - 1}.$$

Next we evaluate the sum $x\sum_{k=0}^{n} g_{2k+1}$.

Example 36.3. *Prove that* $x\sum_{k=0}^{n} g_{2k+1} = g_{2n+2} - g_0$.

Proof. Using the gibonacci recurrence, we have

$$\begin{aligned} x\sum_{k=0}^{n} g_{2k+1} &= \sum_{k=0}^{n} (g_{2k+2} - g_{2k}) \\ &= g_{2n+2} - g_0, \end{aligned}$$

as desired. ■

This example also has interesting dividends:

$$x\sum_{k=0}^{n} f_{2k+1} = f_{2n+2}; \qquad x\sum_{k=0}^{n} l_{2k+1} = l_{2n+2} - 2;$$

$$2x\sum_{k=0}^{n} p_{2k+1} = p_{2n+2}; \qquad 2x\sum_{k=0}^{n} q_{2k+1} = q_{2n+2} - 2;$$

$$\sum_{k=0}^{n} F_{2k+1} = F_{2n+2}; \qquad \sum_{k=0}^{n} L_{2k+1} = L_{2n+2} - 2;$$

$$2\sum_{k=0}^{n} P_{2k+1} = P_{2n+2}; \qquad \sum_{k=0}^{n} Q_{2k+1} = Q_{2n+2} - 1.$$

For example,

$$x\sum_{k=0}^{3} l_{2k+1} = x[x + (x^3 + 3x) + (x^5 + 5x^3 + 5x) + (x^7 + 7x^5 + 14x^3 + 7x)]$$

$$= x^8 + 8x^6 + 20x^4 + 16x^2$$

$$= (x^8 + 8x^6 + 20x^4 + 16x^2 + 2) - 2$$

$$= l_8 - 2;$$

$$2x\sum_{k=0}^{2} p_{2k+1} = 2x[1 + (4x^2 + 1) + (16x^4 + 12x^2 + 1)]$$

$$= 32x^5 + 32x^3 + 6x$$

$$= p_6.$$

Using the same technique as in Example 36.3, we can prove the following: $x\sum_{k=0}^{n} g_{2k} = g_{2n+1} - g_{-1}$; see Exercise 36.2. This implies

$$x\sum_{k=0}^{n} f_{2k} = f_{2n+1} - 1; \qquad x\sum_{k=0}^{n} l_{2k} = l_{2n+1} + x;$$

$$2x\sum_{k=0}^{n} p_{2k} = p_{2n+1} - 1; \qquad 2x\sum_{k=0}^{n} q_{2k} = q_{2n+1} + 2x;$$

$$\sum_{k=0}^{n} F_{2k} = F_{2n+1} - 1; \qquad \sum_{k=0}^{n} L_{2k} = L_{2n+1} + 1;$$

$$2\sum_{k=0}^{n} P_{2k} = P_{2n+1} - 1; \qquad 2\sum_{k=0}^{n} Q_{2k} = Q_{2n+1} + 2.$$

Example 36.3 has an interesting interpretation: The nth row sum of the triangular array A in Table 36.1 is given by $x\sum_{k=0}^{n} g_{2k+1} = g_{2n+2} - g_0$. Likewise, the nth row sum of the triangular array B in Table 36.2 is given by $x\sum_{k=0}^{n} g_{2k} = g_{2n+1} - g_{-1}$.

TABLE 36.1. Array *A*

g_1				
g_3	g_1			
g_5	g_3	g_1		
g_7	g_5	g_3	g_1	
g_9	g_7	g_5	g_3	g_1

TABLE 36.2. Array *B*

g_0				
g_2	g_0			
g_4	g_2	g_0		
g_6	g_4	g_2	g_0	
g_8	g_6	g_4	g_2	g_0

Next we study two interesting patterns:

$$\begin{aligned} xl_1 &= x^2 &&= l_1^2\\ x(l_1+l_3+l_5) &= (x^3+3x)^2 &&= l_3^2\\ x(l_1+l_3+l_5+l_7+l_9) &= (x^5+5x^3+5x)^2 &&= l_5^2\\ &\vdots \end{aligned}$$

and

$$\begin{aligned} x(l_1+l_3) &= (x^3+3x)^2 &&= (x^2+4)f_2^2\\ x(l_1+l_3+l_5+l_7) &= (x^2+4)(x^3+2x)^2 &&= (x^2+4)f_4^2\\ x(l_1+l_3+l_5+l_7+l_9+l_{11}) &= (x^2+4)(x^4+4x^2+3)^2 &&= (x^2+4)f_6^2\\ &\vdots \end{aligned}$$

Using these data, we can easily make a conjecture:

$$x\sum_{k=1}^{n} l_{2k-1} = \begin{cases} l_n^2 & \text{if } n \text{ is odd}\\ (x^2+4)f_n^2 & \text{otherwise.}\end{cases} \tag{36.6}$$

We can establish its validity without much difficulty; see Exercise 36.33.

Formula (36.6) has interesting consequences:

$$2x\sum_{k=1}^{n} q_{2k-1} = \begin{cases} q_n^2 & \text{if } n \text{ is odd}\\ 4(x^2+1)p_n^2 & \text{otherwise;}\end{cases}$$

$$\sum_{k=1}^{n} L_{2k-1} = \begin{cases} L_n^2 & \text{if } n \text{ is odd}\\ 5F_n^2 & \text{otherwise;}\end{cases}$$

$$\sum_{k=1}^{n} Q_{2k-1} = \begin{cases} Q_n^2 & \text{if } n \text{ is odd}\\ 2P_n^2 & \text{otherwise.}\end{cases}$$

We can employ a telescoping sum or the Clary–Hemenway method to evaluate the sum $x\sum_{k=0}^{n} g_k^2$ (see Exercise 36.36):

$$x\sum_{k=0}^{n} g_k^2 = g_n g_{n+1} - g_0 g_{-1}. \tag{36.7}$$

Consequently, the nth row sum of the array in Table 36.3 is given by $x\sum_{k=0}^{n} g_k^2 = g_n g_{n+1} - g_0 g_{-1}$.

TABLE 36.3.

g_0^2				
g_1^2	g_0^2			
g_2^2	g_1^2	g_0^2		
g_3^2	g_2^2	g_1^2	g_0^2	
g_4^2	g_3^2	g_2^2	g_1^2	g_0^2

Formula (36.7) also has Fibonacci and Pell implications:

$$x \sum_{k=0}^{n} f_k^2 = f_n f_{n+1}; \qquad x \sum_{k=0}^{n} l_k^2 = l_n l_{n+1} + 2x;$$

$$2x \sum_{k=0}^{n} p_k^2 = p_n p_{n+1}; \qquad 2x \sum_{k=0}^{n} q_k^2 = q_n q_{n+1} + 4x;$$

$$\sum_{k=0}^{n} F_k^2 = F_n F_{n+1}; \qquad \sum_{k=0}^{n} L_k^2 = L_n L_{n+1} + 2;$$

$$2 \sum_{k=0}^{n} P_k^2 = P_n P_{n+1}; \qquad 2 \sum_{k=0}^{n} Q_k^2 = Q_n Q_{n+1} + 1.$$

For example, we have

$$\begin{aligned} x \sum_{k=0}^{5} f_k^2 &= x\left(f_1^2 + f_2^2 + f_3^2 + f_4^2 + f_5^2\right) \\ &= x\left[1 + x^2 + (x^2+1)^2 + (x^3+2x)^2 + (x^4+3x^2+1)^2\right] \\ &= x^9 + 7x^7 + 16x^5 + 13x^3 + 3x \\ &= (x^4+3x^2+1)(x^5+4x^3+3x) \\ &= f_5 f_6; \end{aligned}$$

$$\begin{aligned} 2x \sum_{k=0}^{4} p_k^2 &= 2x(64x^6 + 80x^4 + 28x^2 + 2x) \\ &= (8x^3+4x)(16x^4+12x^2+1) \\ &= p_4 p_5. \end{aligned}$$

Interestingly, we can obtain formula (36.7) in the reverse order:

$$\begin{aligned} g_{n+1}g_n &= xg_n^2 + g_n g_{n-1} \\ &= x(g_n^2 + g_{n-1}^2) + g_{n-1}g_{n-2} \end{aligned}$$

$$= x(g_n^2 + g_{n-1}^2 + g_{n-2}^2) + g_{n-2}g_{n-3}$$
$$\vdots$$
$$= x(g_n^2 + g_{n-1}^2 + \cdots + g_0^2) + g_0 g_{-1}.$$

This yields the formula.

Consequently, an $f_n \times f_{n+1}$ rectangle can be tiled using n $f_k \times f_k$ tiles, where x is a positive integer and $1 \le k \le n$.

The Catalan-like identity $g_{k+r}g_{k-r} - g_k^2 = \mu(-1)^{k+r+1} f_r^2$, coupled with identity (36.7), can be used to show that

$$x\sum_{k=0}^{n} g_{k+2}g_{k-2} = \begin{cases} g_n g_{n+1} - g_0 g_{-1} & \text{if } n \text{ is odd} \\ g_n g_{n+1} - g_0 g_{-1} - \mu x^3 & \text{otherwise;} \end{cases} \tag{36.8}$$

$$x\sum_{k=0}^{n} g_{k+3}g_{k-3} = \begin{cases} g_n g_{n+1} - g_0 g_{-1} & \text{if } n \text{ is odd} \\ g_n g_{n+1} - g_0 g_{-1} - \mu (x^2+1)^2 & \text{otherwise;} \end{cases} \tag{36.9}$$

see Exercises 36.37 and 36.38.

It follows from summation formulas (36.8) and (36.9) that

$$x\sum_{k=0}^{n} f_{k+2}f_{k-2} = \begin{cases} f_n f_{n+1} & \text{if } n \text{ is odd} \\ f_n f_{n+1} - x^3 & \text{otherwise;} \end{cases}$$

$$x\sum_{k=0}^{n} l_{k+2}l_{k-2} = \begin{cases} l_n l_{n+1} + 2x & \text{if } n \text{ is odd} \\ l_n l_{n+1} + x^2(x^2+4) + 2x & \text{otherwise;} \end{cases}$$

$$x\sum_{k=0}^{n} f_{k+3}f_{k-3} = \begin{cases} f_n f_{n+1} & \text{if } n \text{ is odd} \\ f_n f_{n+1} + (x^2+1)^2 & \text{otherwise;} \end{cases}$$

$$x\sum_{k=0}^{n} l_{k+3}l_{k-3} = \begin{cases} l_n l_{n+1} + 2x & \text{if } n \text{ is odd} \\ l_n l_{n+1} - (x^2+4)(x^2+1)^2 + 2x & \text{otherwise.} \end{cases}$$

Next we develop a formula for the sum $x\sum_{k=1}^{n} f_k f_{k+1}$.

Example 36.4. *Evaluate the sum* $x\sum_{k=1}^{n} f_k f_{k+1}$.

Solution. Using the Fibonacci recurrence, the Cassini-like identity, and identity (36.7), we have

$$x^2\sum_{k=1}^{n} f_k f_{k+1} = x\sum_{k=1}^{n} f_{k+1}(f_{k+1} - f_{k-1})$$

$$
\begin{aligned}
&= x\sum_{k=1}^{n} f_{k+1}^2 - x\sum_{k=1}^{n} f_{k+1}f_{k-1}\\
&= x\left(\sum_{k=1}^{n+1} f_k^2 - 1\right) - x\sum_{k=1}^{n}\left[f_k^2 + (-1)^k\right]\\
&= f_{n+1}f_{n+2} - x - f_nf_{n+1} - x\sum_{k=1}^{n}(-1)^k\\
&= f_{n+1}(f_{n+2} - f_n) - x - x\sum_{k=1}^{n}(-1)^k\\
&= xf_{n+1}^2 - x - x\sum_{k=1}^{n}(-1)^k
\end{aligned}
$$

$$
\begin{aligned}
x\sum_{k=1}^{n} f_kf_{k+1} &= \begin{cases} f_{n+1}^2 & \text{if } n \text{ is odd}\\ f_{n+1}^2 - 1 & \text{otherwise}\end{cases}\\
&= f_{n+1}^2 - \frac{1+(-1)^n}{2}. \qquad (36.10)
\end{aligned}
$$

■

For example, let $n = 4$. Then

$$
\begin{aligned}
x\sum_{k=1}^{4} f_kf_{k+1} &= x\left[(f_1 + f_3)f_2 + (f_3 + f_5)f_4\right]\\
&= x\left[(x^2 + 2)x + (x^4 + 4x^2 + 2)(x^3 + 2x)\right]\\
&= x^8 + 6x^6 + 11x^4 + 6x^2\\
&= (x^4 + 3x^2 + 1)^2 - 1\\
&= f_5^2 - 1.
\end{aligned}
$$

Similarly, we can show that

$$
\begin{aligned}
x\sum_{k=0}^{n} l_kl_{k+1} &= \begin{cases} l_{n+1}^2 - 4 & \text{if } n \text{ is odd}\\ l_{n+1}^2 + x^2 & \text{otherwise;}\end{cases}\\
&= \begin{cases} l_{2n+2} - 2 & \text{if } n \text{ is odd}\\ l_{2n+2} + x^2 - 2 & \text{otherwise;}\end{cases} \qquad (36.11)
\end{aligned}
$$

see Exercises 36.43 and 36.44.

For example,

$$x\sum_{k=0}^{3} l_k l_{k+1} = x\left[2x + x(x^2+2) + (x^2+2)(x^3+3x) + (x^3+3x)(x^4+4x^2+2)\right]$$
$$= x^8 + 8x^6 + 20x^4 + 16x^2$$
$$= (x^8 + 8x^6 + 20x^4 + 16x^2 + 2) - 2$$
$$= l_8 - 2;$$

$$x\sum_{k=0}^{4} l_k l_{k+1} = x(x^9 + 10x^7 + 35x^5 + 50x^3 + 26x)$$
$$= (x^{10} + 10x^8 + 35x^6 + 50x^4 + 25x^2 + 2) + x^2 - 2$$
$$= l_{10} + x^2 - 2.$$

It follows from identities (36.10) and (36.11) that

$$2x\sum_{k=0}^{n} p_k p_{k+1} = \begin{cases} p_{n+1}^2 & \text{if } n \text{ is odd} \\ p_{n+1}^2 - 1 & \text{otherwise;} \end{cases}$$

$$2x\sum_{k=0}^{n} q_k q_{k+1} = \begin{cases} q_{2n+2} - 2 & \text{if } n \text{ is odd} \\ q_{2n+2} + 4x^2 - 2 & \text{otherwise;} \end{cases}$$

$$\sum_{k=0}^{n} F_k F_{k+1} = \begin{cases} F_{n+1}^2 & \text{if } n \text{ is odd} \\ F_{n+1}^2 - 1 & \text{otherwise;} \end{cases} \qquad (36.12)$$

$$\sum_{k=0}^{n} L_k L_{k+1} = \begin{cases} L_{2n+2} - 2 & \text{if } n \text{ is odd} \\ L_{2n+2} - 1 & \text{otherwise;} \end{cases}$$

$$2\sum_{k=0}^{n} P_k P_{k+1} = \begin{cases} P_{n+1}^2 & \text{if } n \text{ is odd} \\ P_{n+1}^2 - 1 & \text{otherwise;} \end{cases}$$

$$4\sum_{k=0}^{n} Q_k Q_{k+1} = \begin{cases} Q_{2n+2} - 1 & \text{if } n \text{ is odd} \\ Q_{2n+2} + 1 & \text{otherwise.} \end{cases}$$

Khan developed formula (36.12) in 2010 [265].

Next we develop a formula for the sum $x(x^2+4)^2 \sum_{k=0}^{n} f_k^2 f_{k+1}^2$.

Example 36.5. *Find a formula for the sum* $S = x(x^2+4)^2 \sum_{k=0}^{n} f_k^2 f_{k+1}^2$.

Solution. We have $\Delta^2 f_k f_{k+1} = l_{2k+1} - (-1)^k x$; so $\Delta^4 f_k^2 f_{k+1}^2 = l_{2k+1}^2 - 2(-1)^k x l_{2k+1} + x^2$. Using Exercises 36.14 and 36.16, we then have

$$\begin{aligned} S &= x\sum_{k=0}^{n} l_{2k+1}^2 - 2x^2 \sum_{k=0}^{n} (-1)^k l_{2k+1} + (n+1)x^3 \\ &= [f_{4n+4} - 2(n+1)x] - 2x^2 \cdot (-1)^n f_{2n+2} + (n+1)x^3 \\ &= f_{4n+4} - 2(-1)^n x^2 f_{2n+2} + (n+1)(x^2-2)x. \end{aligned} \tag{36.13}$$

■

It follows from formula (36.13) that

$$25\sum_{k=0}^{n} F_k^2 F_{k+1}^2 = F_{4n+4} - 2(-1)^n F_{2n+2} - (n+1);$$

$$32x(x^2+1)^2 \sum_{k=0}^{n} p_k^2 p_{k+1}^2 = p_{4n+4} - 8(-1)^n x^2 p_{2n+2} + 4(n+1)(2x^2-1)x;$$

$$800\sum_{k=0}^{n} P_k^2 P_{k+1}^2 = P_{4n+4} - 8(-1)^n P_{2n+2} + 4(n+1).$$

Similarly, we can establish that

$$x\sum_{k=0}^{n} l_k^2 l_{k+1}^2 = f_{4n+4} + 2(-1)^n x^2 f_{2n+2} + (n+1)(x^2-2)x. \tag{36.14}$$

In particular, this yields

$$\sum_{k=0}^{n} L_k^2 L_{k+1}^2 = F_{4n+4} + 2(-1)^n F_{2n+2} - (n+1);$$

$$2x\sum_{k=0}^{n} q_k^2 q_{k+1}^2 = p_{4n+4} + 8(-1)^n x^2 p_{2n+2} + 4(n+1)(2x^2-1)x;$$

$$8\sum_{k=0}^{n} Q_k^2 Q_{k+1}^2 = P_{4n+4} + 8(-1)^n P_{2n+2} + 4(n+1).$$

S. Edwards of Southern Polytechnic State University in Georgia, discovered formula (36.14) in 2014 [149, 332].

36.2 WEIGHTED SUMS

Formula (36.1) can be used to compute the weighted sum $x^2 \sum_{k=1}^{n} kg_k$, as the next example shows. But, first, we make a useful observation. To this end, let $S_i = x \sum_{k=1}^{i} g_k$. Then

$$\begin{aligned} x\sum_{k=1}^{n} kg_k &= \sum_{i=1}^{n}\left(\sum_{k=i}^{n} xg_k\right) \\ &= \sum_{i=1}^{n}(S_n - S_{i-1}) \\ &= nS_n - \sum_{i=1}^{n} S_{i-1}. \end{aligned} \tag{36.15}$$

With this tool at hand, we can evaluate the aforementioned weighted sum with $g_k = f_k$.

Example 36.6. *Evaluate the weighted sum* $x^2\sum_{k=1}^{n} kf_k$.

Solution. By formula (36.1), $S_i = f_{i+1} + f_i - 1$. Notice that $S_0 = f_1 + f_0 - 1 = 0$. Then, by equation (36.15), we have

$$\begin{aligned} x\sum_{k=1}^{n} kf_k &= nS_n - \sum_{i=1}^{n} S_{i-1} \\ x^2\sum_{k=1}^{n} kf_k &= nxS_n - x\sum_{i=1}^{n}(f_{i+1} + f_i - 1) \\ &= nx(f_{n+1} + f_n - 1) - [(f_n + f_{n-1} - 1) + (f_{n+1} + f_n - 1) - nx] \\ &= (nx - 1)f_{n+1} + (nx - 2)f_n - f_{n-1} + 2. \end{aligned} \tag{36.16}$$

■

For example,

$$\begin{aligned} x^2\sum_{k=1}^{4} kf_k &= x^2[1 + 2x + 3(x^2 + 1) + 4(x^3 + 2x)] \\ &= 4x^5 + 3x^4 + 10x^3 + 4x^2 \\ &= (4x - 1)(x^4 + 3x^2 + 1) + (4x - 2)(x^3 + 2x) - (x^2 + 1) + 2 \\ &= (4x - 1)f_5 + (4x - 2)f_4 - f_3 + 2. \end{aligned}$$

It follows from equation (36.16) that

$$4x^2 \sum_{k=1}^{n} kp_k = (2nx-1)p_{n+1} + (2nx-2)p_n - p_{n-1} + 2;$$

$$\sum_{k=1}^{n} kF_k = (n-1)F_{n+1} + (n-2)F_n - F_{n-1} + 2; \tag{36.17}$$

$$4\sum_{k=1}^{n} kP_k = (2n-1)P_{n+1} + (2n-2)P_n - P_{n-1} + 2; \tag{36.18}$$

see Exercises 36.48 and 36.49.

We can similarly establish that

$$x^2 \sum_{k=1}^{n} kl_k = (nx-1)l_{n+1} + (nx-2)l_n - l_{n-1} + 4; \tag{36.19}$$

$$4x^2 \sum_{k=1}^{n} kq_k = (2nx-1)q_{n+1} + (2nx-2)q_n - q_{n-1} + 4;$$

$$\sum_{k=1}^{n} kL_k = (n-1)L_{n+1} + (n-2)L_n - L_{n-1} + 4;$$

$$4\sum_{k=1}^{n} kQ_k = (2n-1)Q_{n+1} + (2n-2)Q_n - Q_{n-1} + 2;$$

see Exercises 36.50–36.55.

We can extend Example 36.6 to weighted sums where the weights form an arbitrary arithmetic sequence. To see this, let a be the first term of an arithmetic sequence with common ratio d; the kth term of the sequence is $a + (k-1)d$. Then, by Examples 36.1 and 36.6, we have

$$\begin{aligned} x^2 \sum_{k=1}^{n} [a + (k-1)d] f_k &= (a-d)x^2 \sum_{k=1}^{n} f_k + dx^2 \sum_{k=1}^{n} kf_k \\ &= (a-d)x(f_{n+1} + f_n - 1) \\ &\quad + d[(nx-1)f_{n+1} + (nx-2)f_n - f_{n-1} + 2] \\ &= [ax + (n-1)dx - d]f_{n+1} + [ax + (n-1)dx - 2d]f_n \\ &\quad - df_{n-1} - (a-d)x + 2d. \end{aligned}$$

Similarly,

$$x^2\sum_{k=1}^{n}[a+(k-1)d]l_k = [ax+(n-1)dx-d]l_{n+1}+[ax+(n-1)dx-2d]l_n$$
$$-\,dl_{n-1}-(a-d)(x-2)x+4d;$$

see Exercise 36.58.

Next we investigate sums with odd weights.

Example 36.7. *Evaluate the sum* $x^2\sum_{k=1}^{n}(2k+1)g_{2k+1}$.

Solution. Let $B_n = x\sum_{k=0}^{n} g_{2k+1}$. By Example 36.3, $B_n = g_{2n+2}-g_0$. We then have

$$\begin{aligned}
x\sum_{k=0}^{n}(2k+1)g_{2k+1} &= x\sum_{k=0}^{n} g_{2k+1} + 2x\sum_{k=1}^{n} g_{2k+1} + \cdots + 2x\sum_{k=n}^{n} g_{2k+1}\\
&= B_n + 2(B_n-B_0)+2(B_n-B_1)+\cdots+2(B_n-B_{n-1})\\
&= (2n+1)B_n - 2\sum_{k=0}^{n-1} B_k\\
x^2\sum_{k=0}^{n}(2k+1)g_{2k+1} &= (2n+1)x(g_{2n+2}-g_0) - 2x\sum_{k=0}^{n-1}(g_{2k+2}-g_0)\\
&= (2n+1)x(g_{2n+2}-g_0) - 2(g_{2n+1}-g_1)+2nxg_0\\
&= (2n+1)xg_{2n+2} - 2g_{2n+1}+2g_1-g_0x. \qquad (36.20)
\end{aligned}$$

■

It follows from formula (36.20) that

$$x^2\sum_{k=0}^{n}(2k+1)f_{2k+1} = (2n+1)xf_{2n+2}-2f_{2n+1}+2;$$

$$2x^2\sum_{k=0}^{n}(2k+1)p_{2k+1} = (2n+1)xp_{2n+2}-p_{2n+1}+1;$$

$$\sum_{k=0}^{n}(2k+1)F_{2k+1} = (2n+1)F_{2n+2}-2F_{2n+1}+2;$$

$$2\sum_{k=0}^{n}(2k+1)P_{2k+1} = (2n+1)P_{2n+2}-P_{2n+1}+1.$$

For example, we have

$$\begin{aligned} x^2 \sum_{k=0}^{3}(2k+1)f_{2k+1} &= x^2[1 + 3(x^2+1) + 5(x^4+3x^2+1) + 7(x^6+5x^4+6x^2+1)] \\ &= 7x^8 + 40x^6 + 60x^4 + 16x^2 \\ &= 7x(x^7 + 6x^5 + 10x^3 + 4x) - 2(x^6 + 5x^4 + 6x^2 + 1) + 2 \\ &= 7f_8 - 2f_7 + 2. \end{aligned}$$

We can show similarly that

$$x^2 \sum_{k=0}^{n}(2k+1)l_{2k+1} = (2n+1)xl_{2n+2} - 2l_{2n+1};$$

$$2x^2 \sum_{k=0}^{n}(2k+1)q_{2k+1} = (2n+1)xq_{2n+2} - q_{2n+1};$$

$$\sum_{k=0}^{n}(2k+1)L_{2k+1} = (2n+1)L_{2n+2} - L_{2n+1};$$

$$2\sum_{k=0}^{n}(2k+1)Q_{2k+1} = (2n+1)Q_{2n+2} - 2Q_{2n+1};$$

see Exercises 36.59–36.62.

We can use exponential generating functions to develop a host of identities involving Fibonacci and Lucas polynomials. So we begin with a brief introduction to them.

36.3 EXPONENTIAL GENERATING FUNCTIONS

Using the Taylor expansion of $e^t = \sum_{n=0}^{\infty} \frac{t^n}{n!}$, we have

$$e^{\alpha t} = \sum_{n=0}^{\infty} \frac{\alpha^n t^n}{n!} \quad \text{and} \quad e^{\beta t} = \sum_{n=0}^{\infty} \frac{\beta^n t^n}{n!}.$$

Then

$$\frac{e^{\alpha t} - e^{\beta t}}{\alpha - \beta} = \sum_{n=0}^{\infty} f_n \frac{t^n}{n!} \quad \text{and} \quad e^{\alpha t} + e^{\beta t} = \sum_{n=0}^{\infty} l_n \frac{t^n}{n!}.$$

Thus $\frac{e^{\alpha t} - e^{\beta t}}{\alpha - \beta}$ and $e^{\alpha t} + e^{\beta t}$ generate f_n and l_n, respectively.

More generally, $\dfrac{e^{\alpha^k t} - e^{\beta^k t}}{\alpha - \beta}$ and $e^{\alpha^k t} + e^{\beta^k t}$ generate f_{nk} and l_{nk}, respectively.

Let $a(t) = \sum_{n=0}^{\infty} a_n \dfrac{t^n}{n!}$ and $b(t) = \sum_{n=0}^{\infty} b_n \dfrac{t^n}{n!}$. Then

$$a(t)b(t) = \sum_{n=0}^{\infty} \left[\sum_{k=0}^{n} \binom{n}{k} a_k b_{n-k} \right] \frac{t^n}{n!}; \tag{36.21}$$

$$a(t)b(-t) = \sum_{n=0}^{\infty} \left[\sum_{k=0}^{n} (-1)^{n-k} \binom{n}{k} a_k b_{n-k} \right] \frac{t^n}{n!}. \tag{36.22}$$

Using the generating functions for f_n and l_n, and these two products, we can generate several identities for the Fibonacci and Pell families.

1) Let $a(t) = \dfrac{e^{\alpha x t} - e^{\beta x t}}{\alpha - \beta}$ and $b(t) = e^t$. Then

$$\frac{e^{(\alpha x+1)t} - e^{(\beta x+1)t}}{\alpha - \beta} = \sum_{n=0}^{\infty} \left[\sum_{k=0}^{n} \binom{n}{k} f_k x^k \right] \frac{t^n}{n!}$$

$$\frac{e^{\alpha^2 t} - e^{\beta^2 t}}{\alpha - \beta} = \sum_{n=0}^{\infty} \left[\sum_{k=0}^{n} \binom{n}{k} f_k x^k \right] \frac{t^n}{n!}$$

$$\sum_{n=0}^{\infty} f_{2n} \frac{t^n}{n!} = \sum_{n=0}^{\infty} \left[\sum_{k=0}^{n} \binom{n}{k} f_k x^k \right] \frac{t^n}{n!}.$$

Equating the coefficients of $\dfrac{t^n}{n!}$ on both sides yields the combinatorial identity

$$\sum_{k=0}^{n} \binom{n}{k} f_k x^k = f_{2n}. \tag{36.23}$$

For example, $\sum_{k=0}^{3} \binom{3}{k} f_k x^k = x^5 + 4x^3 + 3x = f_6$.

2) Choosing $a(t) = b(t) = \dfrac{e^{\alpha x t} - e^{\beta x t}}{\alpha - \beta}$, we can show similarly that

$$(x^2 + 4) \sum_{k=0}^{n} \binom{n}{k} f_k f_{n-k} = 2^n l_n - 2x^n; \tag{36.24}$$

see Exercise 36.102.

For example,

$$(x^2+4)\sum_{k=0}^{3}\binom{3}{k}f_k f_{3-k} = (x^2+4)(3f_1f_2+3f_2f_1)$$
$$= 6(x^2+4)x$$
$$= 2^3 l_3 - 2x^3.$$

3) Choosing $a(t) = \dfrac{e^{\alpha xt} - e^{\beta xt}}{\alpha - \beta}$ and $b(t) = e^{\alpha xt} + e^{\beta xt}$, we get

$$\frac{e^{2\alpha xt} - e^{2\beta xt}}{\alpha - \beta} = \sum_{n=0}^{\infty}\left[x^n \sum_{k=0}^{n}\binom{n}{k} f_k l_{n-k}\right]\frac{t^n}{n!}$$
$$\sum_{n=0}^{\infty} 2^n x^n f_n \frac{t^n}{n!} = \sum_{n=0}^{\infty}\left[x^n \sum_{k=0}^{n}\binom{n}{k} f_k l_{n-k}\right]\frac{t^n}{n!}.$$

This implies

$$\sum_{k=0}^{n}\binom{n}{k} f_k l_{n-k} = 2^n f_n. \tag{36.25}$$

4) Likewise, we can show that

$$\sum_{k=0}^{n}\binom{n}{k} l_k l_{n-k} = 2^n l_n + 2x^n; \tag{36.26}$$

see Exercise 36.103.

For example,

$$\sum_{k=0}^{3}\binom{3}{k} l_k l_{3-k} = l_0 l_3 + 3l_1 l_2 + 3l_2 l_1 + l_3 l_0$$
$$= 2\cdot 2(x^3+3x) + 2\cdot 3x(x^2+2)$$
$$= 10x^3 + 24x$$
$$= 2^3 l_3 + 2x^3.$$

Exercises 36.104–36.108 give additional identities involving Fibonacci and Lucas polynomials, and hence Pell and Pell–Lucas polynomials.

In the next example, we find an explicit formula for the convolution $\sum_{\substack{0\le i,j,k\le n\\ i+j+k=n}} \dfrac{f_i f_j f_k}{i!j!k!}$ involving three factors. Although the sum looks a bit intimidating, we can accomplish the task using the power of generating functions.

Example 36.8. *Find an explicit formula for the sum* $S_n = \sum_{\substack{0\le i,j,k\le n\\ i+j+k=n}} \frac{f_i f_j f_k}{i!j!k!}$.

Solution. Let $g(t) = \sum_{n\ge 0} \frac{f_i f_j f_k}{i!j!k!} t^n$. Then

$$
\begin{aligned}
g(t) &= \left(\sum_{i\ge 0} \frac{f_i}{i!} t^i\right)\left(\sum_{j\ge 0} \frac{f_j}{j!} t^j\right)\left(\sum_{k\ge 0} \frac{f_k}{k!} t^k\right)\\
&= \left(\sum_{s\ge 0} \frac{\alpha^s - \beta^s}{\alpha - \beta} \cdot \frac{t^s}{s!}\right)^3\\
&= \left(\frac{e^{\alpha t} - e^{\beta t}}{\alpha - \beta}\right)^3\\
(\alpha - \beta)^3 g(t) &= e^{3\alpha t} - e^{3\beta t} - 3\left[e^{(2\alpha+\beta)t} - e^{(\alpha+2\beta)t}\right]\\
&= \left(e^{3\alpha t} - e^{3\beta t}\right) - 3\left[e^{(\alpha+x)t} - e^{(\beta+x)t}\right].
\end{aligned}
$$

Equating the coefficients of t^n on both sides, we then get

$$
\begin{aligned}
S_n &= \frac{(3\alpha)^n - (3\beta)^n}{\Delta^3 n!} - \frac{3}{\Delta^3 n!}\left[(\alpha + x)^n - (\beta + x)^n\right]\\
&= \frac{3^n f_n}{\Delta^2 n!} - \frac{3}{\Delta^3 n!}\sum_{k=1}^{n}\binom{n}{k} x^{n-k}(\alpha^k - \beta^k)\\
&= \frac{3^n f_n}{(x^2+4)n!} - \frac{3}{(x^2+4)n!}\sum_{k=1}^{n}\binom{n}{k} x^{n-k} f_k. \qquad (36.27)
\end{aligned}
$$

■

For example, let $n = 4$. Then $\sum_{k=1}^{4}\binom{4}{k} x^{4-k} f_k = 15x^3 + 6x$. So

$$
\begin{aligned}
\sum_{\substack{0\le i,j,k\le 4\\ i+j+k=4}} \frac{f_i f_j f_k}{i!j!k!} &= \frac{3^4(x^3+2x)}{(x^2+4)4!} - \frac{3(15x^3+6x)}{(x^2+4)4!}\\
&= \frac{3x^3 + 12x}{2(x^2+4)}.
\end{aligned}
$$

It follows from formula (36.27) that

$$
\sum_{\substack{0\le i,j,k\le n\\ i+j+k=n}} \frac{F_i F_j F_k}{i!j!k!} = \frac{3^n F_n - 3F_{2n}}{5n!}; \qquad (36.28)
$$

$$\sum_{\substack{0\le i,j,k\le n\\ i+j+k=n}} \frac{p_ip_jp_k}{i!j!k!} = \frac{3^np_n}{4(x^2+1)n!} - \frac{3}{4(x^2+1)n!}\sum_{k=1}^{n}\binom{n}{k}(2x)^{n-k}p_k;$$

$$\sum_{\substack{0\le i,j,k\le n\\ i+j+k=n}} \frac{P_iP_jP_k}{i!j!k!} = \frac{3^nP_n}{8n!} - \frac{3}{8n!}\sum_{k=1}^{n}\binom{n}{k}2^{n-k}P_k.$$

For example, $\displaystyle\sum_{\substack{0\le i,j,k\le 4\\ i+j+k=4}} \frac{F_iF_jF_k}{i!j!k!} = \frac{3^4F_4-3F_8}{5\cdot 4!} = \frac{3}{2}.$

Ohtsuka found formula (36.28) in 2014 [312, 371].

Interestingly, formula (36.27) has a Lucas counterpart:

$$\sum_{\substack{0\le i,j,k\le n\\ i+j+k=n}} \frac{l_il_jl_k}{i!j!k!} = \frac{3^nl_n}{n!} + \frac{3}{n!}\sum_{k=0}^{n}\binom{n}{k}x^{n-k}l_k; \qquad (36.29)$$

see Exercise 36.126.

For example,

$$\sum_{\substack{0\le i,j,k\le 3\\ i+j+k=3}} \frac{l_il_jl_k}{i!j!k!} = 9x^3+18x = \frac{3^3}{3!}(x^3+3x) + \frac{3}{3!}(9x^3+9x)$$

$$= \frac{3^3l_3}{3!} + \frac{3}{3!}\sum_{k=0}^{3}\binom{3}{k}x^{3-k}l_k.$$

It follows from formula (36.29) that

$$\sum_{\substack{0\le i,j,k\le n\\ i+j+k=n}} \frac{L_iL_jL_k}{i!j!k!} = \frac{3^nL_n}{n!} + \frac{3L_{2n}}{n!};$$

$$\sum_{\substack{0\le i,j,k\le n\\ i+j+k=n}} \frac{q_iq_jq_k}{i!j!k!} = \frac{3^nq_n}{n!} + \frac{3}{n!}\sum_{k=0}^{n}\binom{n}{k}(2x)^{n-k}q_k;$$

$$4\sum_{\substack{0\le i,j,k\le n\\ i+j+k=n}} \frac{Q_iQ_jQ_k}{i!j!k!} = \frac{3^nQ_n}{n!} - \frac{3}{n!}\sum_{k=0}^{n}\binom{n}{k}2^{n-k}Q_k.$$

For example, $\displaystyle\sum_{\substack{0\le i,j,k\le 3\\ i+j+k=3}} \frac{L_iL_jL_k}{i!j!k!} = 27 = \frac{3^3L_3}{3!} + \frac{3L_6}{3!}.$

Next we use the differential operator $\frac{d}{dt}$ to generalize some of the above identities.

The Differential Operator $\dfrac{d}{dt}$

Suppose $a(t) = \sum_{n=0}^{\infty} a_n \frac{t^n}{n!}$. Then $\frac{d^r}{dt^r}a(t) = \sum_{n=0}^{\infty} a_{n+r}\frac{t^n}{n!}$, where $r \geq 0$.

In particular, $a(t) = \frac{d^r}{dt^r}\left(\frac{e^{\alpha xt} - e^{\beta xt}}{\alpha - \beta}\right)$ and $b(t) = e^t$. Then

$$\begin{aligned} a(t)b(t) &= \frac{x^r\left[\alpha^r e^{(\alpha x+1)t} - \beta^r e^{(\beta x+1)t}\right]}{\alpha - \beta} \\ &= \frac{x^r\left(\alpha^r e^{\alpha^2 t} - \beta^r e^{\beta^2 t}\right)}{\alpha - \beta} \\ &= x^r \sum_{n=0}^{\infty} f_{2n+r}\frac{t^n}{n!}. \end{aligned} \tag{36.30}$$

Since $\frac{e^{\alpha xt} - e^{\beta xt}}{\alpha - \beta} = \sum_{n=0}^{\infty} f_n x^n \frac{t^n}{n!}$, we also have $a(t) = \sum_{n=0}^{\infty} f_{n+r}x^{n+r}\frac{t^n}{n!}$. So

$$a(t)b(t) = \sum_{n=0}^{\infty}\left[\sum_{k=0}^{n}\binom{n}{k}x^{k+r}f_{k+r}\right]\frac{t^n}{n!}. \tag{36.31}$$

It now follows from equations (36.30) and (36.31) that

$$\sum_{k=0}^{n}\binom{n}{k}f_{k+r}x^k = f_{2n+r}. \tag{36.32}$$

For example,

$$\begin{aligned} \sum_{k=0}^{3}\binom{3}{k}f_{k+4}x^k &= f_4 + 3xf_5 + 3x^2f_6 + x^3f_7 \\ &= x^9 + 8x^7 + 21x^5 + 20x^3 + 5x \\ &= f_{10}. \end{aligned}$$

We can show similarly that

$$\sum_{k=0}^{n}\binom{n}{k}l_{k+r}x^k = l_{2n+r}; \tag{36.33}$$

see Exercise 36.109.

Using the differential operator, and choosing $a(t)$ and $b(t)$ appropriately, we can easily establish additional identities:

$$(x^2+4)\sum_{k=0}^{n}\binom{n}{k}f_{k+r}f_{n-k+r} = 2^n l_{n+2r} - 2(-1)^r x^n; \tag{36.34}$$

$$\sum_{k=0}^{n}\binom{n}{k}l_{k+r}l_{n-k+r} = 2^n l_{n+2r} + 2(-1)^r x^n; \tag{36.35}$$

$$\sum_{k=0}^{n}\binom{n}{k}f_{k+r}l_{n-k+r} = 2^n f_{n+2r}; \tag{36.36}$$

$$\sum_{k=0}^{n}(-1)^{n-k}\binom{n}{k}f_{2k+2r} = f_{n+2r}x^n; \tag{36.37}$$

$$\sum_{k=0}^{n}(-1)^{n-k}\binom{n}{k}l_{2k+2r} = l_{n+2r}x^n; \tag{36.38}$$

see Exercises 36.110–36.114.

36.4 INFINITE GIBONACCI SUMS

Next we study some infinite gibonacci sums.

Example 36.9. *Evaluate the infinite sum* $\sum_{n=2}^{\infty}\frac{x}{g_{n-1}g_{n+1}}$, *where* $x \ge 1$.

Solution.

$$\begin{aligned}\sum_{n=2}^{m}\frac{x}{g_{n-1}g_{n+1}} &= \sum_{n=2}^{m}\left(\frac{1}{g_{n-1}g_n} - \frac{1}{g_ng_{n+1}}\right)\\ &= \frac{1}{g_1g_2} - \frac{1}{g_mg_{m+1}}\end{aligned}$$

$$\begin{aligned}\sum_{n=2}^{\infty}\frac{x}{g_{n-1}g_{n+1}} &= \frac{1}{ab} - \lim_{m\to\infty}\frac{1}{g_mg_{m+1}}\\ &= \frac{1}{ab} - 0\\ &= \frac{1}{ab}.\end{aligned}$$

■

In particular, we have

$$\sum_{n=2}^{\infty} \frac{x}{f_{n-1}f_{n+1}} = \frac{1}{x};$$

$$\sum_{n=2}^{\infty} \frac{x}{l_{n-1}l_{n+1}} = \frac{1}{x(x^2+2)}.$$

Consequently,

$$\sum_{n=2}^{\infty} \frac{1}{F_{n-1}F_{n+1}} = 1.$$

R.L. Graham of then Bell Telephone Laboratories, Murray Hill, New Jersey, studied this sum in 1963 [193].

Similarly, we can show that

$$\sum_{n=2}^{\infty} \frac{xg_n}{g_{n-1}g_{n+1}} = \frac{a+b}{ab};$$

see Exercise 36.115. Consequently,

$$\sum_{n=2}^{\infty} \frac{xf_n}{f_{n-1}f_{n+1}} = \frac{x+1}{x};$$

$$\sum_{n=2}^{\infty} \frac{xl_n}{l_{n-1}l_{n+1}} = \frac{x^2+x+2}{x(x^2+2)}.$$

In 1963, Graham also studied the special case $\sum_{n=2}^{\infty} \frac{F_n}{F_{n-1}F_{n+1}} = 2$ [194].

In this example, the two factors in the denominator on the LHS are two spaces apart. In the next example, we let them be four spaces apart.

Example 36.10. *Evaluate the infinite sum* $\sum_{n=1}^{\infty} \frac{x(x^2+2)}{g_n g_{n+4}}$, *where* $x \geq 1$.

Solution. Since $g_{n+4} = (x^2+2)g_{n+2} - g_n$, by Example 36.9, we have

$$\sum_{n=1}^{\infty} \frac{x}{g_n g_{n+2}} - x\sum_{n=1}^{\infty} \frac{x^2+2}{g_n g_{n+4}} = x\sum_{n=1}^{\infty} \frac{g_{n+4} - (x^2+2)g_{n+2}}{g_n g_{n+2} g_{n+4}}$$

$$= -\sum_{n=1}^{\infty} \frac{x}{g_{n+2}g_{n+4}}$$

$$= \frac{x}{g_1 g_3} + \frac{x}{g_2 g_4} - \sum_{n=1}^{\infty} \frac{x}{g_n g_{n+2}}$$

$$\sum_{n=1}^{\infty} \frac{x(x^2+2)}{g_n g_{n+4}} = 2\sum_{n=1}^{\infty} \frac{x}{g_n g_{n+2}} - \frac{x}{g_1 g_3} - \frac{x}{g_2 g_4}$$

$$= \frac{2}{g_1 g_2} - \frac{x}{g_1 g_3} - \frac{x}{g_2 g_4}$$

$$= \frac{2g_3 g_4 - x g_2 g_4 - x g_1 g_3}{g_1 g_2 g_3 g_4}.$$ ■

In particular, this implies

$$\sum_{n=1}^{\infty} \frac{x(x^2+2)}{f_n f_{n+4}} = \frac{x^4+3x^2+3}{(x^2+1)(x^3+2x)}$$

$$\sum_{n=1}^{\infty} \frac{x(x^2+2)}{l_n l_{n+4}} = \frac{x^7+7x^5+15x^3+8x}{x(x^2+2)(x^3+3x)(x^4+4x^2+2)}.$$

Using the same scheme, we can evaluate the sum $\sum_{n=1}^{\infty} \frac{x(x^2+1)(x^2+3)}{g_n g_{n+6}}$; see Exercises 36.117 and 36.118.

The next example also deals with infinite sums, and is beautiful in its own right.

Example 36.11. *Prove that*

$$\sum_{n=1}^{\infty} \frac{1}{g_n g_{n+2}^2 g_{n+3}} + \sum_{n=1}^{\infty} \frac{1}{g_n g_{n+1}^2 g_{n+3}} = \frac{1}{x g_1 g_2^2 g_3},$$

where $x \geq 1$.

Proof. Let $r = g_n$, $s = g_{n+1}$, $t = g_{n+2}$, and $u = g_{n+3}$; so $r + sx = t$ and $s + tx = u$. Then

$$\sum_{n=1}^{m} \frac{1}{g_n g_{n+2}^2 g_{n+3}} + \sum_{n=1}^{m} \frac{1}{g_n g_{n+1}^2 g_{n+3}}$$

$$= \sum_{n=1}^{m} \left(\frac{1}{rt^2 u} + \frac{1}{rs^2 u} \right)$$

$$= \sum_{n=1}^{m} \left(\frac{s}{rst^2 u} + \frac{t}{rs^2 tu} \right)$$

$$= \frac{1}{x} \sum_{n=1}^{m} \left(\frac{t-r}{rst^2 u} + \frac{u-s}{rs^2 tu} \right)$$

$$= \frac{1}{x}\sum_{n=1}^{m}\left(\frac{1}{rstu} - \frac{1}{st^2u} + \frac{1}{rs^2t} - \frac{1}{rstu}\right)$$

$$= \frac{1}{x}\sum_{n=1}^{m}\left(\frac{1}{rs^2t} - \frac{1}{st^2u}\right)$$

$$= \frac{1}{x}\left[\left(\frac{1}{g_1g_2^2g_3} - \frac{1}{g_2g_3^2g_4}\right) + \left(\frac{1}{g_2g_3^2g_4} - \frac{1}{g_3g_4^2g_5}\right) + \cdots\right]$$

$$= \frac{1}{x}\left(\frac{1}{g_1g_2^2g_3} - \frac{1}{g_{m+1}g_{m+2}^2g_{m+3}}\right)$$

$$\text{Given Sum} = \frac{1}{x}\lim_{m\to\infty}\left(\frac{1}{g_1g_2^2g_3} - \frac{1}{g_{m+1}g_{m+2}^2g_{m+3}}\right)$$

$$= \frac{1}{xg_1g_2^2g_3},$$

as desired. ∎

In particular,

$$\sum_{n=1}^{\infty}\frac{1}{g_ng_{n+2}^2g_{n+3}} + \sum_{n=1}^{\infty}\frac{1}{g_ng_{n+1}^2g_{n+3}} = \begin{cases} \dfrac{1}{x^3(x^2+1)} & \text{if } g_n = f_n \\ \dfrac{1}{x^2(x^2+2)^2(x^3+3x)} & \text{if } g_n = l_n. \end{cases}$$

L. Carlitz, in 1963, studied this infinite sum when $G_n = F_n$ [16, 79].

In the next example, we study an infinite sum with interesting ramifications [402, 423]. It takes advantage of the property $\sum_{n=1}^{\infty}\frac{t^n}{n} = \ln\left(\frac{1}{1-t}\right)$, with $|t| < 1$.

Example 36.12. *Let k, r, and s be any complex numbers such that $|r|, |s| < |k|$. Suppose $S_n = Ar^n + Bs^n$, where A and B are arbitrary complex numbers. Evaluate the infinite sum $\sum_{n=1}^{\infty}\frac{2n+1}{n(n+1)|k|^n}S_n$.*

Solution. Let $t < 1$. We then have

$$\sum_{n=1}^{\infty}\frac{2n+1}{n(n+1)}t^n = \sum_{n=1}^{\infty}\left(\frac{1}{n} + \frac{1}{n+1}\right)t^n$$

$$= \ln\left(\frac{1}{1-t}\right) + \frac{1}{t}\ln\left(\frac{1}{1-t}\right) - 1$$

$$= -\left(1 + \frac{1}{t}\right)\ln(1-t) - 1.$$

Consequently,

$$\sum_{n=1}^{\infty} \frac{2n+1}{n(n+1)|k|^n} S_n = \sum_{n=1}^{\infty} \frac{2n+1}{n(n+1)} \left[A(r/k)^n + B(s/k)^n\right]$$
$$= A\sum_{n=1}^{\infty} \frac{2n+1}{n(n+1)}(r/k)^n + B\sum_{n=1}^{\infty} \frac{2n+1}{n(n+1)}(s/k)^n$$
$$= -A\left(1+\frac{k}{r}\right)\ln\left(1-\frac{r}{k}\right) - B\left(1+\frac{k}{s}\right)\ln\left(1-\frac{s}{k}\right) - A - B. \tag{36.39}$$

■

Formula (36.39) has several interesting consequences.

1) Let $r = \alpha$, $s = \beta$, $A = \dfrac{1}{\alpha - \beta} = -B$, and $k = 2$. Then $S_n = F_n$, $1 + \dfrac{k}{r} = \alpha - \beta$, $1 - \dfrac{r}{k} = \dfrac{\beta^2}{2}$, $1 + \dfrac{k}{s} = \beta - \alpha$, and $1 - \dfrac{s}{k} = \dfrac{\alpha^2}{2}$. It then follows from formula (36.39) that

$$\sum_{n=1}^{\infty} \frac{2n+1}{n(n+1)2^n} F_n = -\left(\ln\frac{\alpha^2}{2} + \ln\frac{\beta^2}{2}\right)$$
$$= -\ln\frac{\alpha^2\beta^2}{4}$$
$$= \ln 4$$
$$\approx 1.38629436112.$$

H.-J. Seiffert of Berlin, Germany found this formula in 1994 [402, 423].

2) Let $r = \alpha$, $s = \beta$, $A = 1 = B$, and $k = 2$. Then $S_n = L_n$, $1 + \dfrac{k}{r} = \alpha - \beta$, $1 - \dfrac{r}{k} = \dfrac{\beta^2}{2}$, $1 + \dfrac{k}{s} = \beta - \alpha$, and $1 - \dfrac{s}{k} = \dfrac{\alpha^2}{2}$. Formula (36.39) then yields

$$\sum_{n=1}^{\infty} \frac{2n+1}{n(n+1)2^n} L_n = (\alpha-\beta)\left(\ln\frac{\alpha^2}{2} - \ln\frac{\beta^2}{2}\right) - 2$$
$$= (\alpha-\beta)\ln\frac{\alpha^2}{\beta^2} - 2$$
$$= \sqrt{5}\ln\frac{7+3\sqrt{5}}{2} - 2$$
$$\approx 2.30408940964.$$

Similarly, we can show that

$$\sum_{n=1}^{\infty} \frac{2n+1}{n(n+1)4^n} F_n = 2\ln\frac{16}{11} + \frac{1}{\sqrt{5}}\ln\frac{27-7\sqrt{5}}{22}$$
$$\approx 0.453312327005;$$
$$\sum_{n=1}^{\infty} \frac{2n+1}{n(n+1)4^n} L_n = 2\sqrt{5}\ln\frac{16}{11} + \ln\frac{27-7\sqrt{5}}{22} - 2$$
$$\approx 0.586052269341;$$

see Exercises 36.119 and 36.120.

The next example presents two interesting infinite sums involving the central binomial coefficient $\binom{2n}{n}$, originally studied by Seiffert in 1991 [418]. The proofs are a straightforward application of a series expansion of $(\arcsin x)^2$.

Example 36.13. *Prove that*

$$\sum_{n=1}^{\infty} \frac{G_{2n}}{\binom{2n}{n} n^2} = \begin{cases} \dfrac{\pi^2}{5} & \text{if } G_k = L_k \\ \dfrac{4\sqrt{5}\pi^2}{125} & \text{if } G_k = F_k. \end{cases}$$

Proof. It is well known that [260]

$$\sum_{n=1}^{\infty} \frac{(2u)^{2n}}{\binom{2n}{n} n^2} = 2(\arcsin u)^2$$

and it converges when $|u| \le 1$. Since $|\alpha/2| < 1$ and $|\beta/2| < 1$, this implies

$$\sum_{n=1}^{\infty} \frac{\alpha^{2n}}{\binom{2n}{n} n^2} = 2(\arcsin \alpha/2)^2 \quad \text{and} \quad \sum_{n=1}^{\infty} \frac{\beta^{2n}}{\binom{2n}{n} n^2} = 2(\arcsin \beta/2)^2.$$

We have $\sin \pi/10 = -\beta/2$ and $\sin 3\pi/10 = \alpha/2$ [287]; so $\arcsin \alpha/2 = 3\pi/10$ and $\arcsin \beta/2 = -\pi/10$. Consequently,

$$\sum_{n=1}^{\infty} \frac{L_{2n}}{\binom{2n}{n} n^2} = 2\left[(\arcsin \alpha/2)^2 + (\arcsin \beta/2)^2\right]$$
$$= 2\left[(3\pi/10)^2 + (-\pi/10)^2\right]$$
$$= \frac{\pi^2}{5};$$

$$\sum_{n=1}^{\infty} \frac{F_{2n}}{\binom{2n}{n} n^2} = \frac{2}{\sqrt{5}} \left[(\arcsin \alpha/2)^2 - (\arcsin \beta/2)^2\right]$$
$$= \frac{2}{\sqrt{5}} \left[(3\pi/10)^2 - (-\pi/10)^2\right]$$
$$= \frac{4\sqrt{5}\pi^2}{125},$$

as desired. ■

In the following example, we evaluate the sum of two convergent series involving f_n and l_n.

Example 36.14. *Let $a = a(x) \geq 2$ and $b = b(x) \geq 2$, where $x \geq 1$. Assume that both $a(x)$ and $b(x)$ have integral coefficients. Prove that*

$$\sum_{n=1}^{\infty} \frac{a^n f_n - b^n l_n}{(ab)^n} = \frac{abx^2 - (ab^2 + ab - a - 2b)x + a^2b - 2b^2 - b + 2}{(a^2 - ax - 1)(b^2 - bx - 1)}.$$

Proof. Since $\alpha < a$, $\alpha < b$, $|\beta| < a$, and $\beta < b$, it follows that the series $\sum_{n=1}^{\infty} \frac{\alpha^n}{a^n}, \sum_{n=1}^{\infty} \frac{\alpha^n}{b^n}, \sum_{n=1}^{\infty} \frac{\beta^n}{a^n}$, and $\sum_{n=1}^{\infty} \frac{\beta^n}{b^n}$ are convergent. Clearly,

$$\sum_{n=1}^{\infty} \frac{\alpha^n}{a^n} = \frac{\alpha}{a-\alpha}, \quad \sum_{n=1}^{\infty} \frac{\beta^n}{a^n} = \frac{\beta}{a-\beta}, \quad \sum_{n=1}^{\infty} \frac{\alpha^n}{b^n} = \frac{\alpha}{b-\alpha}, \quad \sum_{n=1}^{\infty} \frac{\beta^n}{b^n} = \frac{\beta}{b-\beta}.$$

Recalling that $\Delta = \sqrt{x^2+4}$, we then have

$$\sum_{n=1}^{\infty} \frac{a^n f_n}{(ab)^n} = \frac{1}{\Delta} \sum_{n=1}^{\infty} \frac{\alpha^n - \beta^n}{b^n}$$
$$= \frac{1}{\Delta} \left(\frac{\alpha}{b-\alpha} - \frac{\beta}{b-\beta} \right)$$
$$= \frac{b}{b^2 - bx - 1}.$$

Similarly,

$$\sum_{n=1}^{\infty} \frac{b^n l_n}{(ab)^n} = \frac{\alpha}{a-\alpha} + \frac{\beta}{a-\beta}$$
$$= \frac{ax+2}{a^2 - ax - 1}.$$

Thus

$$\sum_{n=1}^{\infty} \frac{a^n f_n - b^n l_n}{(ab)^n} = \frac{b}{b^2 - bx - 1} - \frac{ax+2}{a^2 - ax - 1}$$

$$= \frac{abx^2 - (ab^2 + ab - a - 2b)x + a^2b - 2b^2 - b + 2}{(a^2 - ax - 1)(b^2 - bx - 1)},$$

as desired. ■

In particular, let $a = 3$ and $b = 2$. Then

$$\sum_{n=1}^{\infty} \frac{3^n F_n - 2^n L_n}{6^n} = 1.$$

Joseph J. Kostal of the University of Illinois at Chicago studied this special case in 1992 [301].

EXERCISES 36

Prove each.

1. $x\sum_{k=0}^{n} g_{2k} = g_{2n+1} - g_{-1}$, using telescoping sums.
2. $x\sum_{k=0}^{n} g_{2k} = g_{2n+1} - g_{-1}$, using the Clary–Hemenway method.
3. $x\sum_{k=0}^{n} l_k = l_{n+1} + l_n + x - 2$.
4. $\sum_{i=1}^{n} l_{ki+j} = \dfrac{l_{nk+k+j} - (-1)^k l_{nk+j} - l_j - (-1)^j l_{k-j}}{l_k - (-1)^k - 1}$.
5. $x\sum_{k=0}^{n} g_{2k+1} = g_{2n+2} - g_0$, using the Clary–Hemenway method.
6. $x\sum_{k=0}^{n} f_{2k+1} = f_{2n+2}$.
7. $x\sum_{k=0}^{n} l_{2k+1} = l_{2n+2} - 2$.
8. $x\sum_{k=0}^{n} l_{2k} = l_{2n+1} + x$.

9. $x\sum_{k=1}^{n} f_{4k-1} = f_{2n}f_{2n+1}$.

10. $x\sum_{k=0}^{n} f_{4k+1} = l_{2n+1}l_{2n+2} + 2x$.

11. $(2x^2-1)\sum_{k=0}^{n} x^k f_k = (f_{n+1} + xf_n)x^{n+1} - x$.

12. $(2x^2-1)\sum_{k=0}^{n} x^k l_k = (l_{n+1} + xl_n)x^{n+1} + x^2 - 2$.

13. $x\sum_{k=0}^{n} l_{2k+1} = l_{2n+2} - 2$.

14. $\sum_{k=0}^{n} (-1)^k l_{2k+1} = (-1)^n f_{2n+2}$.

15. $x\sum_{k=0}^{n} l_{4k+2} = f_{4n+4}$.

16. $x\sum_{k=0}^{n} l_{2k+1}^2 = f_{4n+4} - 2(n+1)x$.

Let R_n and D_n denote the nth row sum and the nth diagonal sum of the triangular array A in Table 36.4, where $n \geq 0$. (W.G. Brady studied this array for $x = 1$ [44].)

Then prove each.

17. $xR_n = f_{2n+2}$.

18. $x(x^2+4)D_n = l_{2n+3} - x$.

19. Multiply column j of array A with $j+1$, where $0 \leq j \leq n$. Let S_n denote the nth row sum of the resulting array. Then prove $x^2 S_n = f_{2n+3} - 1$.

20. Verify the formula in Exercise 36.19 for $n = 3$.

TABLE 36.4. Array *A*

f_1				
f_3	f_1			
f_5	f_3	f_1		
f_7	f_5	f_3	f_1	
f_9	f_7	f_5	f_3	f_1

TABLE 36.5. Array *B*

l_1				
l_3	l_1			
l_5	l_3	l_1		
l_7	l_5	l_3	l_1	
l_9	l_7	l_5	l_3	l_1

Let R_n and D_n denote the nth row sum and the nth diagonal sum of the triangular array B in Table 36.5, where $n \geq 0$. Then prove each.

21. $xR_n = l_{2n+2} - 2$.
22. $xD_n = f_{2n+3} - 1$.
23. Multiply column j of array B with $j+1$, where $0 \leq j \leq n$. Let S_n denote the nth row sum of the resulting array. Then prove $x^2 S_n = l_{2n+3} - (2n+3)x$.
24. Illustrate the formula in Exercise 36.23 for $n = 3$.

Let R_n and D_n denote the nth row sum and the nth diagonal sum of the triangular array C in Table 36.6, where $n \geq 0$. (Brady studied this array when $x = 1$ [44].)

Then prove each.

25. $xR_n = f_{2n+1} - 1$.
26. $x(x^2+4)D_n = l_{2n+3} - (x^2+2)$.
27. Multiply column j of array C with $j+1$, where $0 \leq j \leq n$. Let S_n denote the nth row sum of the resulting array. Then prove $x^2 S_n = f_{2n+2} - (n+1)x$.
28. Verify the formula in Exercise 36.27 for $n = 4$.

TABLE 36.6. Array C

f_0				
f_2	f_0			
f_4	f_2	f_0		
f_6	f_4	f_2	f_0	
f_8	f_6	f_4	f_2	f_0

TABLE 36.7. Array D

l_0				
l_2	l_0			
l_4	l_2	l_0		
l_6	l_4	l_2	l_0	
l_8	l_6	l_4	l_2	l_0

Let R_n and D_n denote the nth row sum and the nth diagonal sum of the triangular array D in Table 36.7, where $n \geq 0$. Then prove each.

29. $xR_n = l_{2n+1} + x$.
30. $$xD_n = \begin{cases} f_{2n+2} & \text{if } n \text{ is odd} \\ f_{2n+2} + x & \text{otherwise.} \end{cases}$$
31. Multiply column j of array D with $j+1$, where $0 \leq j \leq n$. Let S_n denote the nth row sum of the resulting array. Then prove $x^2 S_n = l_{2n+2} + (n+1)x^2 - 2$.
32. Illustrate the formula in Exercise 36.31 for $n = 4$.

Prove each.

33. $$x\sum_{k=1}^{n} l_{2k-1} = \begin{cases} l_n^2 & \text{if } n \text{ is odd} \\ (x^2+4)f_n^2 & \text{otherwise.} \end{cases}$$

34. $x\sum_{k=0}^{n} f_{2k} = \begin{cases} f_{n+1}l_n & \text{if } n \text{ is odd} \\ f_n l_{n+1} & \text{otherwise.} \end{cases}$

35. $x\sum_{k=0}^{n} l_{2k} = l_{2n+1} + x.$

36. $x\sum_{k=0}^{n} g_k^2 = g_n g_{n+1} - g_0 g_{-1}.$

37. $x\sum_{k=0}^{n} g_{k+2}g_{k-2} = \begin{cases} g_n g_{n+1} - g_0 g_{-1} & \text{if } n \text{ is odd} \\ g_n g_{n+1} - g_0 g_{-1} - \mu x^3 & \text{otherwise.} \end{cases}$

38. $x\sum_{k=0}^{n} g_{k+3}g_{k-3} = \begin{cases} g_n g_{n+1} - g_0 g_{-1} & \text{if } n \text{ is odd} \\ g_n g_{n+1} - g_0 g_{-1} + \mu(x^2+1)^2 & \text{otherwise.} \end{cases}$

39. $x(x^2+4)\sum_{k=0}^{n} f_k^2 = l_{2n+1} + \begin{cases} x & \text{if } n \text{ is odd} \\ -x & \text{otherwise.} \end{cases}$

40. $x\sum_{k=0}^{n} l_k^2 = l_{2n+1} + \begin{cases} x & \text{if } n \text{ is odd} \\ 3x & \text{otherwise.} \end{cases}$

41. $x^2\sum_{k=1}^{2n-1} (2n-k)f_k^2 = f_{2n}^2.$

42. $x^2\sum_{k=1}^{2n-1} (2n-k)l_k^2 = l_{2n}^2 - 4(nx^2+1).$

43. $x\sum_{k=0}^{n} l_k l_{k+1} = \begin{cases} l_{n+1}^2 - 4 & \text{if } n \text{ is odd} \\ l_{n+1}^2 + x^2 & \text{otherwise.} \end{cases}$

44. $x\sum_{k=0}^{n} l_k l_{k+1} = \begin{cases} l_{2n+2} - 2 & \text{if } n \text{ is odd} \\ l_{2n+2} + x^2 - 2 & \text{otherwise.} \end{cases}$

45. $x(x^2+4)\sum_{k=1}^{n} f_k f_{k+2} = \begin{cases} f_{n+1}f_{n+2} & \text{if } n \text{ is odd} \\ f_{n+1}f_{n+2} - x & \text{otherwise.} \end{cases}$

46. $x(x^2+4)\sum_{k=1}^{n} f_k f_{k+2} = \begin{cases} l_{2n+3} - x & \text{if } n \text{ is odd} \\ l_{2n+3} + x - 1 & \text{otherwise.} \end{cases}$

47. $x\sum_{k=1}^{n} l_k l_{k+2} = \begin{cases} l_{2n+3} + x^3 - x^2 + 6x & \text{if } n \text{ is odd} \\ l_{2n+3} - x^2 & \text{otherwise.} \end{cases}$

48. $\sum_{k=1}^{n} kF_k = nF_{n+2} - F_{n+3} + 2.$

49. $2\sum_{k=1}^{n} kP_k = nQ_{n+1} - P_{n+1} + 1.$

50. $x^2\sum_{k=1}^{n} kl_k = (nx-1)l_{n+1} + (nx-2)l_n - l_{n-1} + 4.$

51. $4x^2\sum_{k=1}^{n} kq_k = (2nx-1)q_{n+1} + (2nx-2)q_n - q_{n-1} + 4.$

52. $\sum_{k=1}^{n} kL_k = (n-1)L_{n+1} + (n-2)L_n - L_{n-1} + 4.$

53. $\sum_{k=1}^{n} kL_k = nL_{n+2} - L_{n+3} + 4.$

54. $4\sum_{k=1}^{n} kQ_k = (2n-1)Q_{n+1} + (2n-2)Q_n - Q_{n-1} + 2.$

55. $2\sum_{k=1}^{n} kQ_k = 2nP_{n+1} - Q_{n+1} + 1.$

56. $x^2\sum_{k=1}^{n}(n-k+1)f_k = (x+1)f_{n+1} + (x+2)f_n + f_{n-1} - (n+1)x - 2.$

57. $x^2\sum_{k=1}^{n}(n-k+1)l_k = (x+1)l_{n+1} + (x+2)l_n + l_{n-1} + (n+1)(x^2-2x) - 4.$

58. $$x^2\sum_{k=1}^{n}[a+(k-1)d]l_k = [ax+(n-1)dx-d]l_{n+1} + [ax+(n-1)dx-2d]l_n - dl_{n-1} + (a-d)(x-2)x + 4d.$$

59. $x^2\sum_{k=0}^{n}(2k+1)l_{2k+1} = (2n+1)xl_{2n+2} - 2l_{2n+1}.$

60. $2x^2\sum_{k=0}^{n}(2k+1)q_{2k+1} = (2n+1)xq_{2n+2} - q_{2n+1}.$

61. $\sum_{k=0}^{n}(2k+1)L_{2k+1} = (2n+1)L_{2n+2} - 2L_{2n+1}.$

62. $2\sum_{k=0}^{n}(2k+1)Q_{2k+1} = (2n+1)Q_{2n+2} - 2Q_{2n+1}.$

63. $x^2 \sum_{k=0}^{n} kg_{2k+1} = nxg_{2n+2} - g_{2n+1} + g_1.$

64. $x^2 \sum_{k=0}^{n} kf_{2k+1} = nxf_{2n+2} - f_{2n+1} + 1.$

65. $x^2 \sum_{k=0}^{n} kl_{2k+1} = nxl_{2n+2} - l_{2n+1} + x.$

66. $4x^2 \sum_{k=0}^{n} kp_{2k+1} = 2nxp_{2n+2} - p_{2n+1} + 1.$

67. $4x^2 \sum_{k=0}^{n} kq_{2k+1} = 2nxq_{2n+2} - q_{2n+1} + 2x.$

68. $x^2 \sum_{k=0}^{n} kg_{2k} = nxg_{2n+1} - g_{2n} + xg_1 - (x^2 - 1)g_0 - xg_{-1}.$

69. $x^2 \sum_{k=0}^{n} kf_{2k} = nxf_{2n+1} - f_{2n}.$

70. $x^2 \sum_{k=0}^{n} kl_{2k} = nxl_{2n+1} - l_{2n} + 2.$

71. $4x^2 \sum_{k=0}^{n} kp_{2k} = 2nxp_{2n+1} - p_{2n}.$

72. $4x^2 \sum_{k=0}^{n} kq_{2k} = 2nxq_{2n+1} - q_{2n} + 2.$

73. $f_{a+2b} - f_a = \begin{cases} f_{a+b}l_b & \text{if } b \text{ is odd} \\ f_b l_{a+b} & \text{otherwise.} \end{cases}$

74. $l_{a+2b} - l_a = \begin{cases} l_a l_{a+b} & \text{if } b \text{ is odd} \\ (x^2+4) f_a f_{a+b} & \text{otherwise.} \end{cases}$

75. $x \sum_{k=1}^{n} f_k f_{k+c} = \begin{cases} f_n f_{n+c+1} - f_c & \text{if } n \text{ is odd} \\ f_n l_{n+c+1} & \text{otherwise.} \end{cases}$

76. $x \sum_{k=1}^{n} l_k l_{k+c} = \begin{cases} (x^2+4) f_n f_{n+c+1} - 2l_{c+1} - xl_c & \text{if } n \text{ is odd} \\ (x^2+4) f_n f_{n+c+1} & \text{otherwise.} \end{cases}$

77. $xf_{2n} = f_{n+1}^2 - f_{n-1}^2.$

78. $x\sum_{k=0}^{n} f_k^2 f_{2k} = f_n^2 f_{n+1}^2.$

79. $2x\sum_{k=0}^{n} p_k^2 p_{2k} = p_n^2 p_{n+1}^2.$

80. $\sum_{k=0}^{n} F_k^2 F_{2k} = F_n^2 F_{n+1}^2$ (Gugheri, [205]).

81. $2\sum_{k=0}^{n} P_k^2 P_{2k} = P_n^2 P_{n+1}^2.$

82. $x(x^2+4)\sum_{k=0}^{n} l_k^2 f_{2k} = l_n^2 l_{n+1}^2 - 4x^2.$

83. $8x(x^2+1)\sum_{k=0}^{n} q_k^2 p_{2k} = q_n^2 q_{n+1}^2 - 16x^2.$

84. $4\sum_{k=0}^{n} Q_k^2 P_{2k} = Q_n^2 Q_{n+1}^2 - 1.$

85. $\sum_{k=0}^{n} \binom{n}{k} f_k x^k = f_{2n}.$

86. $\sum_{k=0}^{n} \binom{n}{k} (-1)^{k+1} f_k x^{n-k} = f_n.$

87. $\sum_{k=0}^{n} \binom{n}{k} (-1)^{k} l_k x^{n-k} = l_n.$

88. $x^4 \sum_{i=0}^{n}\sum_{j=0}^{i}\sum_{k=0}^{j}\sum_{l=0}^{k} f_l^2 = f_{n+2}^2 - [2n^2+8n+7+(-1)^n]\frac{x^2}{8} - \frac{1-(-1)^n}{2}.$

89. $\sum_{i=1}^{n} x(x^2+1)^{n-i} f_{2i+k-3} + (x^2+1)^n f_k = f_{2n+k}.$

90. $x\sum_{k=1}^{n} f_{4k} = f_{2n} f_{2n+2}.$

91. $x(x^2+4)\sum_{k=1}^{n} f_{2k-1}^2 = f_{4n} + 2nx.$

92. $x(x^2+4)\sum_{k=1}^{n} f_{2k-2} f_{2k} = f_{4n} - nx(x^2+2).$

93. $3x^3(x^2+4)^2 \sum_{i=1}^{n-1}\sum_{j=1}^{i}\sum_{k=1}^{j} f_{2k-1}^2 = 3f_{4n} - 3nx(x^2+2) + n(n^2-1)x^3(x^2+4).$

94. $x(x^2+4)\sum_{k=1}^{n} f_{2k-2}f_{2k} = f_{4n} - nx(x^2+2).$

95. $\sum_{k=0}^{n}\binom{n}{k} f_{4mk} = f_{2mn}l_{2m}^n.$

96. $\sum_{k=0}^{n}\binom{n}{k} l_{4mk} = l_{2mn}l_{2m}^n.$

97. $\sum_{k=0}^{n}\binom{n}{k}^2 L_k = \sum_{k=0}^{n}(-1)^{n-k}\binom{n}{k}\binom{n+k}{k}L_{n-k}$ (Carlitz, [80]).

98. $\sum_{k=0}^{n}\binom{n}{k}^2 F_k = \sum_{k=0}^{n}(-1)^{n-k+1}\binom{n}{k}\binom{n+k}{k}F_{n-k}$ (Carlitz, [80]).

99. $\sum_{k=0}^{n}\binom{n}{k}^2 L_{2k} = \sum_{k=0}^{n}\binom{n}{k}\binom{n+k}{k}L_{n-k}$ (Carlitz, [81]).

100. $\sum_{k=0}^{n}\binom{n}{k}^2 F_{2k} = \sum_{k=0}^{n}\binom{n}{k}\binom{n+k}{k}F_{n-k}$ (Carlitz, [81]).

101. $8\sum_{k=0}^{n} F_{3k+1}F_{3k+2}F_{6k+3} = F_{3n+3}^4$ (Zeitlin, [516]).

102. $(x^2+4)\sum_{k=0}^{n}\binom{n}{k} f_k f_{n-k} = 2^n l_n - 2x^n.$

103. $\sum_{k=0}^{n}\binom{n}{k} l_k l_{n-k} = 2^n l_n + 2x^n.$

104. $\sum_{k=0}^{n}(-1)^{k+1}\binom{n}{k} f_k x^{n-k} = f_n.$

105. $\sum_{k=0}^{n}(-1)^{n-k}\binom{n}{k} f_{2k} = f_n x^n.$

106. $\sum_{k=0}^{n}\binom{n}{k} l_k x^k = l_{2n}.$

107. $\sum_{k=0}^{n}(-1)^k\binom{n}{k} l_k x^{n-k} = l_n.$

108. $\sum_{k=0}^{n}(-1)^{n-k}\binom{n}{k}l_{2k} = l_n x^n$.

109. $\sum_{k=0}^{n}\binom{n}{k}l_{k+r}x^k = l_{2n+r}$.

110. $(x^2+4)\sum_{k=0}^{n}\binom{n}{k}f_{k+r}f_{n-k+r} = 2^n l_{n+2r} - 2(-1)^r x^n$.

111. $\sum_{k=0}^{n}\binom{n}{k}l_{k+r}l_{n-k+r} = 2^n l_{n+2r} + 2(-1)^r x^n$.

112. $\sum_{k=0}^{n}\binom{n}{k}f_{k+r}l_{n-k+r} = 2^n f_{n+2r}$.

113. $\sum_{k=0}^{n}(-1)^{n-k}\binom{n}{k}f_{2k+2r} = f_{n+2r}x^n$.

114. $\sum_{k=0}^{n}(-1)^{n-k}\binom{n}{k}l_{2k+2r} = l_{n+2r}x^n$.

115. $\sum_{n=2}^{\infty}\frac{xg_n}{g_{n-1}g_{n+1}} = \frac{a+b}{ab}$.

116. $g_{n+4} = (x^2+2)g_{n+2} - g_n$.

117. $g_{n+6} = (x^2+1)(x^2+3)g_{n+2} - (x^2+2)g_n$.

118. $\sum_{n=1}^{\infty}\frac{x(x^2+1)(x^2+3)}{g_n g_{n+6}} = \frac{3}{g_1 g_2} - \frac{x}{g_1 g_3} - \frac{x}{g_2 g_4} - \frac{x(x^2+2)}{g_1 g_5} - \frac{x(x^2+2)}{g_2 g_6}$.

119. $\sum_{n=1}^{\infty}\frac{2n+1}{n(n+1)4^n}F_n = 2\ln\frac{16}{11} + \frac{1}{\sqrt{5}}\ln\frac{27-7\sqrt{5}}{22}$.

120. $\sum_{n=1}^{\infty}\frac{2n+1}{n(n+1)4^n}L_n = 2\sqrt{5}\ln\frac{16}{11} + \ln\frac{27-7\sqrt{5}}{22} - 2$.

121. Show that $\sum_{n=1}^{\infty}\frac{x}{f_{2n}} = 1 + \sum_{n=1}^{\infty}\frac{1}{f_{2n+1}f_{2n}f_{2n-1}}$ (André-Jeannin, [8]).

Evaluate each sum.

122. $\sum_{n=0}^{\infty}\frac{n2^n}{5^n}F_n$.

123. $\sum_{n=0}^{\infty}\frac{n2^n}{5^n}L_n$.

124. $\sum_{n=0}^{\infty} \frac{n2^n}{5^n} P_n.$

125. $\sum_{n=0}^{\infty} \frac{n2^n}{5^n} Q_n$ (Euler, [151]).

126. Prove that $\sum_{\substack{0 \le i,j,k \le n \\ i+j+k=n}} \frac{l_i l_j l_k}{i!j!k!} = \frac{3^n l_n}{n!} + \frac{3}{n!} \sum_{k=0}^{n} \binom{n}{k} x^{n-k} l_k.$

37

ADDITIONAL GIBONACCI DELIGHTS

Imagination is more important than knowledge.
–Albert Einstein (1879–1955)

In this chapter, we return to some well-known and elegant Fibonacci and Lucas identities, and generalize them. We introduce a new family of positive integers corresponding to ordinary binomial coefficients, and use them for deriving new identities. We also study a new class of integers and a new class of polynomial functions. Again, in the interest of brevity and convenience, we denote $g_n(x)$ by g_n.

37.1 SOME FUNDAMENTAL IDENTITIES REVISITED

Recall from Chapter 31 that $f_{n+1}^2 + f_n^2 = f_{2n+1}$. Clearly, this follows from the addition formula: $f_{2n+1} = f_{n+(n+1)} = f_{n+1}f_{n+1} + f_nf_n = f_{n+1}^2 + f_n^2$. It also follows by the Cassini-like formula, coupled with the addition formula:

$$\begin{aligned} f_{n+2}f_n - f_{n+1}^2 &= -(f_{n+1}f_{n-1} - f_n^2) \\ f_{n+1}^2 + f_n^2 &= f_{n+2}f_n + f_{n+1}f_{n-1} \\ &= f_{2n+1}. \end{aligned}$$

Fibonacci and Lucas Numbers with Applications, Volume Two. Thomas Koshy.

This identity is equivalent to $f_n^2 + (-1)^{n+k+1} f_k^2 = f_{n-k} f_{n+k}$; see Exercise 37.2. This modified form implies that

$$F_n^2 + (-1)^{n+k+1} F_k^2 = F_{n-k} F_{n+k}; \tag{37.1}$$
$$p_n^2 + (-1)^{n+k+1} p_k^2 = p_{n-k} p_{n+k};$$
$$P_n^2 + (-1)^{n+k+1} P_k^2 = P_{n-k} P_{n+k}.$$

Identity (37.1) occurs in [221] without proof.

Using the addition formulas for f_{a+b} and f_{a-b}, we now generalize the property $f_{n+1}^2 + f_n^2 = f_{2n+1}$.

Example 37.1. *Prove that*

$$f_{n+k+1}^2 + f_{n-k}^2 = f_{2n+1} f_{2k+1}. \tag{37.2}$$

Proof. We have

$$\begin{aligned} f_{n+k+1} &= f_{n+(k+1)} = f_{n+1} f_{k+1} + f_n f_k \\ f_{n-k} &= f_{(n+1)-(k+1)} = (-1)^{k+1}(f_{n+1} f_k - f_n f_{k+1}) \\ f_{n+k+1}^2 + f_{n-k}^2 &= (f_{n+1} f_{k+1} + f_n f_k)^2 + (f_{n+1} f_k - f_n f_{k+1})^2 \\ &= f_{n+1}^2 (f_{k+1}^2 + f_k^2) + f_n^2 (f_{k+1}^2 + f_k^2) \\ &= (f_{n+1}^2 + f_n^2)(f_{k+1}^2 + f_k^2) \\ &= f_{2n+1} f_{2k+1}, \end{aligned}$$

as claimed. ∎

This can also be established using the Binet-like formula; see Exercise 37.4.

It follows from identity (37.2) that

$$F_{n+k+1}^2 + F_{n-k}^2 = F_{2n+1} F_{2k+1}; \tag{37.3}$$
$$p_{n+k+1}^2 + p_{n-k}^2 = p_{2n+1} p_{2k+1};$$
$$P_{n+k+1}^2 + P_{n-k}^2 = P_{2n+1} P_{2k+1}.$$

Melham of the University of Technology, Sydney, Australia, discovered (37.3) in 1998 [354].

Identity (37.2) has a Lucas counterpart. Using the addition formulas for l_{a+b} and l_{a-b}, we will now establish it.

Example 37.2. *Prove that*

$$l_{n+k+1}^2 + l_{n-k}^2 = (x^2 + 4) f_{2n+1} f_{2k+1}. \tag{37.4}$$

Proof. We have

$$\begin{aligned}
l_{n+k+1} &= l_{n+(k+1)} = f_{n+1}l_{k+1} + f_n l_k;\\
l_{n-k} &= l_{(n+1)-(k+1)} = (-1)^k(f_{n+1}l_k - f_n l_{k+1});\\
l_{n+k+1}^2 + l_{n-k}^2 &= (f_{n+1}l_{k+1} + f_n l_k)^2 + (f_{n+1}l_k - f_n l_{k+1})^2\\
&= f_{n+1}^2(l_{k+1}^2 + l_k^2) + f_n^2(l_{k+1}^2 + l_k^2)\\
&= (f_{n+1}^2 + f_n^2)(l_{k+1}^2 + l_k^2)\\
&= f_{2n+1} \cdot (x^2+4)f_{2k+1}\\
&= (x^2+4)f_{2n+1}f_{2k+1},
\end{aligned}$$

as desired. ∎

For example,

$$\begin{aligned}
l_5^2 + l_0^2 &= (x^5 + 5x^3 + 5x)^2 + 4\\
&= x^{10} + 10x^8 + 35x^6 + 50x^4 + 25x^2 + 4\\
&= (x^2+4)(x^8 + 6x^6 + 11x^4 + 6x^2 + 1)\\
&= (x^2+4)(x^4 + 3x^2 + 1)^2\\
&= (x^2+4)f_5^2.
\end{aligned}$$

Identity (37.4) also can be established using the Binet-like formula; see Exercise 37.5.

It implies that

$$\begin{aligned}
L_{n+k+1}^2 + L_{n-k}^2 &= 5F_{2n+1}F_{2k+1}; \qquad (37.5)\\
q_{n+k+1}^2 + q_{n-k}^2 &= 4(x^2+1)p_{2n+1}p_{2k+1};\\
Q_{n+k+1}^2 + Q_{n-k}^2 &= 2P_{2n+1}P_{2k+1}.
\end{aligned}$$

Melham discovered identity (37.5) also in 1999 [354].

In the next example, we establish a charming identity involving the squares of five consecutive Fibonacci polynomials. The identities $f_{n+1}^2 + f_n^2 = f_{2n+1}$ and $f_{n+1}^2 + f_{n-1}^2 = xf_{2n}$ play a pivotal role in the proof.

Example 37.3. *Establish the identity*

$$f_n^2 - f_{n+1}^2 - 4f_{n+2}^2 - f_{n+3}^2 + f_{n+4}^2 = (x^2-1)l_{2n+4}. \qquad (37.6)$$

Proof.

$$\begin{aligned}
\text{LHS} &= -\left(f_{n+2}^2 - f_n^2\right) + \left(f_{n+4}^2 - f_{n+2}^2\right) - \left(f_{n+2}^2 + f_{n+1}^2\right) - \left(f_{n+3}^2 + f_{n+2}^2\right) \\
&= -xf_{2n+2} + xf_{2n+6} - f_{2n+3} - f_{2n+5} \\
&= -xf_{2n+2} + (x^2 f_{2n+5} + xf_{2n+4}) - f_{2n+3} - f_{2n+5} \\
&= -xf_{2n+2} + (x^2 - 1)f_{2n+5} + (x^2 f_{2n+3} + xf_{2n+2}) - f_{2n+3} \\
&= (x^2 - 1)(f_{2n+3} + f_{2n+5}) \\
&= (x^2 - 1)l_{2n+4},
\end{aligned}$$

as desired. ■

For example,

$$\begin{aligned}
f_2^2 - f_3^2 - 4f_4^2 - f_5^2 + f_6^2 &= x^2 - (x^2+1)^2 - 4(x^3+2x)^2 - (x^4+3x^2+1)^2 \\
&\quad + (x^5+4x^3+3x)^2 \\
&= x^{10} + 7x^8 + 12x^6 - 4x^4 - 14x^2 - 2 \\
&= (x^2-1)(x^8+8x^6+20x^4+16x^2+2) \\
&= (x^2-1)l_8.
\end{aligned}$$

It follows from identity (37.6) that

$$\begin{aligned}
F_n^2 + F_{n+4}^2 &= F_{n+1}^2 + 4F_{n+2}^2 + F_{n+3}^2; \qquad (37.7)\\
p_n^2 + p_{n+4}^2 &= p_{n+1}^2 + 4p_{n+2}^2 + p_{n+3}^2 + (4x^2-1)q_{2n+4}; \\
P_n^2 + P_{n+4}^2 &= P_{n+1}^2 + 4P_{n+2}^2 + P_{n+3}^2 + 6Q_{2n+4}.
\end{aligned}$$

Swamy developed identity (37.7) in 1966 [476].

Identity (37.6) has a predictable Lucas companion:

$$l_n^2 - l_{n+1}^2 - 4l_{n+2}^2 - l_{n+3}^2 + l_{n+4}^2 = (x^2-1)(x^2+4)l_{2n+4}. \qquad (37.8)$$

Its proof follows along the same lines; so we omit it in the interest of brevity; see Exercise 37.6.

It follows from identity (37.8) that

$$\begin{aligned}
L_n^2 + L_{n+4}^2 &= L_{n+1}^2 + 4L_{n+2}^2 + L_{n+3}^2; \qquad (37.9)\\
q_n^2 + q_{n+4}^2 &= q_{n+1}^2 + 4q_{n+2}^2 + q_{n+3}^2 + 4(4x^2-1)q_{2n+4}; \\
Q_n^2 + Q_{n+4}^2 &= Q_{n+1}^2 + 4Q_{n+2}^2 + Q_{n+3}^2 + 4(4x^2-1)Q_{2n+4}.
\end{aligned}$$

For example, $Q_5^2 + Q_9^2 = 41^2 + 1393^2 = 1{,}942{,}130 = 99^2 + 4 \cdot 239^2 + 577^2 + 12 \cdot 114243 = Q_6^2 + 4Q_7^2 + Q_8^2$.

It follows from identities (37.7) and (37.9) that the squares of both Fibonacci and Lucas numbers satisfy the fourth-order recurrence

$$y_{n+4} - y_{n+3} - 4y_{n+2} - y_{n+1} + y_n = 0. \tag{37.10}$$

The next example has a close resemblance to Example 37.3. It also involves the squares of four consecutive Fibonacci polynomials.

Example 37.4. *Prove that*

$$x^2 f_{n+3}^2 + 2(x^2+1)f_{n+2}^2 + x^2 f_{n+1}^2 = f_{n+4}^2 + f_n^2. \tag{37.11}$$

Proof. We have

$$\begin{aligned}
\text{LHS} - \text{RHS} &= x^2(f_{n+3}^2 + f_{n+2}^2) + x^2(f_{n+2}^2 + f_{n+1}^2) + (f_{n+2}^2 - f_n^2) - (f_{n+4}^2 - f_{n+2}^2)\\
&= x^2 f_{2n+5}^2 + x^2 f_{2n+3}^2 + x f_{2n+2} - x f_{2n+6}\\
&= x(x f_{2n+5} - f_{2n+6}) + x(x f_{2n+3} + f_{2n+2})\\
&= -x f_{2n+4} + x f_{2n+4}\\
&= 0.
\end{aligned}$$

This gives the desired result. ■

For example,

$$\begin{aligned}
&x^2 f_6^2 + (2x^2+2) f_5^2 + x^2 f_4^2\\
&= x^2(x^5+4x^3+3x)^2 + (2x^2+2)(x^4+3x^2+1)^2 + x^2(x^3+2x)^2\\
&= x^{12} + 10x^{10} + 37x^8 + 62x^6 + 46x^4 + 12x^2 + 1\\
&= (x^6+5x^4+6x^2+1)^2 + (x^2+1)^2\\
&= f_7^2 + f_3^2.
\end{aligned}$$

Identity (37.11) implies that

$$\begin{aligned}
4x^2 p_{n+3}^2 + 2(4x^2+1)p_{n+2}^2 + 4x^2 p_{n+1}^2 &= p_{n+4}^2 + p_n^2;\\
4P_{n+3}^2 + 10P_{n+2}^2 + 4P_{n+1}^2 &= P_{n+4}^2 + P_n^2.
\end{aligned}$$

The polynomial identity (37.11) also has a close Lucas relative:

$$x^2 l_{n+3}^2 + 2(x^2+1)l_{n+2}^2 + x^2 l_{n+1}^2 = l_{n+4}^2 + l_n^2. \tag{37.12}$$

We omit its proof in the interest of brevity; see Exercise 37.7.

It follows from identity (37.12) that

$$4x^2q_{n+3}^2 + 2(4x^2+1)q_{n+2}^2 + 4x^2q_{n+1}^2 = q_{n+4}^2 + q_n^2;$$

$$4Q_{n+3}^2 + 10Q_{n+2}^2 + 4Q_{n+1}^2 = Q_{n+4}^2 + Q_n^2.$$

For example,

$$\begin{aligned} 4Q_8^2 + 10Q_7^2 + 4Q_6^2 &= 4 \cdot 577^2 + 10 \cdot 239^2 + 4 \cdot 99^2 \\ &= 1{,}942{,}130 = 1393^2 + 41^2 \\ &= Q_9^2 + Q_5^2. \end{aligned}$$

It follows from identities (37.11) and (37.12) that the squares of both Fibonacci and Lucas polynomials satisfy the fourth-order recurrence

$$z_{n+4} - x^2z_{n+3} - (2x^2+2)z_{n+2} - x^2z_{n+1} + z_n = 0. \tag{37.13}$$

Clearly, recurrence (37.10) is a special case of this.

37.2 LUCAS AND GINSBURG IDENTITIES REVISITED

In Chapter 35, we studied the polynomial extension of the Lucas property

$$f_{n+1}^3 + xf_n^3 - f_{n-1}^3 = xf_{3n}. \tag{37.14}$$

We now re-confirm this using the addition formula for Fibonacci polynomials, and leave the proof using the Binet-like formula as an exercise; see Exercise 37.8. The first approach is a lot faster, prettier, and efficient; and reveals once again the beauty and power of the addition formula.

Example 37.5. *Prove identity (37.14).*

Proof. By the Fibonacci addition formula, we have

$$\begin{aligned} f_{3n} &= f_{(n-1)+(2n+1)} \\ &= f_nf_{2n+1} + f_{n-1}f_{2n} \\ &= f_nf_{n+(n+1)} + f_{n-1}f_{(n-1)+(n+1)} \\ &= f_n(f_{n+1}^2 + f_n^2) + f_{n-1}(f_nf_{n+1} + f_{n-1}f_n) \\ &= f_nf_{n+1}^2 + f_n^3 + f_nf_{n-1}(f_{n+1} + f_{n-1}) \end{aligned}$$

$$
\begin{aligned}
&= f_n f_{n+1}^2 + f_n^3 + \frac{1}{x} f_{n-1}(f_{n+1} - f_{n-1})(f_{n+1} + f_{n-1}) \\
&= f_n f_{n+1}^2 + f_n^3 + \frac{1}{x} f_{n-1} f_{n+1}^2 - \frac{1}{x} f_{n-1}^3 \\
x f_{3n} &= f_{n+1}^2 (x f_n + f_{n-1}) + x f_n^3 - f_{n-1}^3 \\
&= f_{n+1}^3 + x f_n^3 - f_{n-1}^3,
\end{aligned}
$$

as desired. ■

We can similarly establish its Lucas counterpart

$$l_{n+1}^3 + x l_n^3 - l_{n-1}^3 = x(x^2 + 4) l_{3n}, \tag{37.15}$$

using the addition formula $(x^2 + 4) f_{a+b+1} = l_a l_b + l_{a+1} l_{b+1}$; see Exercises 37.9–37.12.

A Generalization of Identity (37.14)

Next we generalize the Fibonacci extension (37.14) of the Lucas identity. But before we state and establish the generalization, we need to lay some groundwork in the form of three lemmas and a theorem. We begin with the first lemma.

Lemma 37.1. $g_{3n+6} = (x^3 + 3x) g_{3n+3} + g_{3n}$.

Proof. By the gibonacci addition formula, we have

$$
\begin{aligned}
g_{3n+6} &= g_{(3n+3)+3} \\
&= f_4 g_{3n+3} + f_3 g_{3n+2} \\
&= (x^3 + 2x) g_{3n+3} + (x^2 + 1) g_{3n+2} \\
&= (x^3 + 2x) g_{3n+3} + x(x g_{3n+2}) + g_{3n+2} \\
&= (x^3 + 2x) g_{3n+3} + x(g_{3n+3} - g_{3n+1}) + g_{3n+2} \\
&= (x^3 + 3x) g_{3n+3} + g_{3n},
\end{aligned}
$$

as claimed. ■

We can also establish this lemma using a repeated application of the polynomial recurrence; see Exercise 37.26.

The next result is a repeated application of Lemma 37.1.

Lemma 37.2.

$$(x^3 + 2x) g_{3n+9} + (x^4 + 3x^2 + 2) g_{3n+6} - (x^3 + 2x) g_{3n+3} - g_{3n} = g_{3n+12}.$$

Proof. By Lemma 37.1, we have

$$\begin{aligned}\text{LHS} &= (x^3+2x)g_{3n+9} + (x^4+3x^2+2)g_{3n+6} - (x^3+2x)g_{3n+3}\\ &\quad - [g_{3n+6} - (x^3+3x)g_{3n+3}]\\ &= (x^3+2x)g_{3n+9} + (x^4+3x^2+1)g_{3n+6} + xg_{3n+3})\\ &= (x^3+2x)g_{3n+9} + (x^4+3x^2+1)g_{3n+6} + [xg_{3n+9} - (x^4+3x^2)g_{3n+6}]\\ &= (x^3+3x)g_{3n+9} + g_{3n+6}\\ &= (x^3+3x)g_{3n+12}. \end{aligned}$$ ■

Later we will encounter the coefficients on the LHS of this identity in a different context.

Homogeneous FL-Identities

Next we study a powerful proof technique developed by L.A.G. Dresel of the University of Reading, Berkshire, England [134]. Before presenting it, we need to lay some groundwork.

Let $x = 1$. Suppose we let $X = \alpha^n$ and $Y = \beta^n$, so $XY = (-1)^n$. Using these two substitutions, we can transform any Fibonacci–Lucas (FL) identity into an algebraic identity in X and Y.

For example, the FL-identity $2L_{2n} = L_n^2 + 5F_n^2$ transforms into $2(X^2+Y^2) = (X+Y)^2 + (X-Y)^2$ and $L_n^2 - 5F_n^2 = 4(-1)^n$ into $(X+Y)^2 - (X-Y)^2 = 4(XY)^n$. The converse also works. For example, the algebraic identity $X^3+Y^3 = (X+Y)^3 - 3XY(X+Y)$ confirms the FL-identity $L_{3n} = L_n^3 - 3(-1)^n L_n$.

An FL-identity (or FL-expression) is *homogeneous* if the corresponding XY-transform is homogeneous. For example, the FL-identities $2L_{2n} = L_n^2 + 5F_n^2$ and $L_n^2 - 5F_n^2 = 4(-1)^n$ are each homogeneous of degree 2.

Since $F_{jn+k} = \dfrac{\alpha^k X^j - \beta^k Y^j}{\sqrt{5}}$ and $L_{jn+k} = \alpha^k X^j + \beta^k Y^j$, both F_{jn+k} and L_{jn+k} can be transformed into homogeneous expressions of degree j in the variable n. Since $(-1)^{jn} = (XY)^j$, $(-1)^{jn}$ transforms into a homogeneous expression of degree $2j$ in n.

For example, the degree of L_{3n} and L_n^3 each is 3, and that of $(-1)^n L_n$ is $2+1=3$. So $L_{3n} = L_n^3 - 3(-1)^n L_n$ is a homogeneous identity of degree 3 in n. Likewise, $F_{3n} = 5F_n^3 + 3(-1)^n F_n$ is also homogeneous of degree 3 in n. Both F_{n+1} and F_{n-1} are each of degree 1 in n, and $(-1)^n$ of degree 2. So $F_{n+1}F_{n-1} - F_n^2 = (-1)^n$ is homogeneous of degree 2 in n.

Now consider an arbitrary homogeneous FL-identity of degree d in the variable n. Its XY-transform is of the form

$$\sum_{i=0}^{d} a_i X^{d-i} Y^i = \sum_{i=0}^{d} b_i X^{d-i} Y^i,$$

where a_i and b_i are independent of n. This yields a polynomial equation of degree d:

$$\sum_{i=0}^{d} a_i Z^i = \sum_{i=0}^{d} b_i Z^i, \tag{37.16}$$

where $Z = X^{-1}Y^i = (\alpha^{-1}\beta)^n$. This is the *polynomial equation* of the FL-identity.

Since $\alpha^{-1}\beta \neq 0$ or ± 1, it follows that distinct values of n yield distinct values of Z. Consequently, the polynomial equation (37.16) is satisfied by infinitely many values of Z and hence is an algebraic identity. This leads us to the following theorem by Dresel. Its proof is a simple, direct application of the *fundamental theorem of algebra.*

Theorem 37.1 (Dresel, 1993 [134]). *Suppose an FL-equation is homogeneous of degree d in a variable n and is satisfied by* $d+1$ *distinct values of n. Then it is an identity true for all values of n.*

Proof. The FL-equation can be transformed into a polynomial equation of degree d of the form (37.16). Since it is satisfied by $d+1$ different values of n, the polynomial form is satisfied by $d+1$ different values of Z. Since the polynomials are of degree d, it follows by the fundamental theorem of algebra that they are identical. Consequently, the FL-equation is an identity for all n. ■

The following corollary is an immediate consequence of *Dresel's theorem.*

Corollary 37.1 (Dresel, 1993 [134]). *Any homogeneous FL-identity is true for all n, positive, negative, or zero.* ■

The next example illustrates the beauty and power of Dresel's theorem.

Example 37.6. *Prove that*

$$2(F_n^2 + F_{n+1}^2 + F_{n+2}^2 + F_{n+3}^2)^2 = 3(F_n^4 + F_{n+1}^4 + F_{n+2}^4 + F_{n+3}^4). \tag{37.17}$$

Proof. The given FL-equation is homogeneous of degree 4 in n. So, by Dresel's theorem, it suffices to confirm its validity for five distinct values of n. When $n = 0$, LHS = 72 = RHS; when $n = 1$, LHS = 450 = RHS; when $n = 2$, LHS = 3042 = RHS; and when $n = -1$ or -2, LHS = 18 = RHS. Thus, the given FL-equation is true for all n. ■

Similarly, we can establish the identities

$$\begin{aligned}&(F_n^2 + F_{n+1}^2 + F_{n+2}^2 + F_{n+3}^2 + F_{n+4}^2)^2\\ &\quad = F_n^4 + 7F_{n+1}^4 + 25F_{n+2}^4 + 7F_{n+3}^4 + F_{n+4}^4;\end{aligned} \tag{37.18}$$

$$\begin{aligned}&3(F_n^2 + F_{n+1}^2 + F_{n+2}^2 + F_{n+3}^2 + F_{n+4}^2 + F_{n+5}^2)^2\\ &\quad = 32(F_{n+1}^4 + 4F_{n+2}^4 + 4F_{n+3}^4 + F_{n+4}^4);\end{aligned} \tag{37.19}$$

see Exercises 37.21 and 37.22. Melham discovered identities (37.17)–(37.19) [357].

The next lemma is an application of Dresel's theorem.

Lemma 37.3. *Let* $0 \le n \le 3$ *and* $k \ge 0$. *Then*

$$F_{3k+1}F^3_{n+k+1} + F_{3k+2}F^3_{n+k} - F^3_{n-2k-1} = F_{3k+1}F_{3k+2}F_{3n}.$$

Proof. Suppose $n = 3$. Then the equation becomes

$$F_{3k+1}F^3_{k+4} + F_{3k+2}F^3_{k+3} - F^3_{2k-2} - F_9 f_{3k+1}F_{3k+2} = 0. \tag{37.20}$$

This is a homogeneous equation of degree 6 in k. So, by Dresel's theorem, to establish this identity for all k, it suffices to confirm it for seven distinct values of k; say, $0 \le k \le 6$, for convenience.

When $k = 0$ and $k = 1$, LHS = 0 = RHS. Similarly, the identity works for the remaining values of k. So it is true for all values of k.

The remaining three values of n can be handled similarly. Thus the lemma is true for all $0 \le n \le 3$ and $k \ge 0$. ■

We are now ready to state and establish a generalization of Lucas' identity in the following theorem [354].

Theorem 37.2 (Melham, 1999 [354]).

$$F_{3k+1}F^3_{n+k+1} + F_{3k+2}F^3_{n+k} - F^3_{n-2k-1} = F_{3k+1}F_{3k+2}F_{3n}. \tag{37.21}$$

Proof. We establish this using PMI. By Lemma 37.3, the identity works when $n = 0$, 1, 2, and 3. Suppose it is true for $n = m$, $m + 1$, $m + 2$, and $m + 3$. Let A, B, C, and D denote the corresponding LHSs, respectively. Then $A = F_{3k+1}F_{3k+2}F_{3m}$, $B = F_{3k+1}F_{3k+2}F_{3(m+1)}$, $C = F_{3k+1}F_{3k+2}F_{3(m+2)}$, and $D = F_{3k+1}F_{3k+2}F_{3(m+3)}$. Using Lemma 37.2 and the identity $F^3_{n+4} = 3F^3_{n+3} + 6F^3_{n+2} - 3F^3_{n+1} - F^3_n$, we then have

$$3D + 6C - 3B - A = F_{3k+1}F_{3k+2}[3F_{3(m+3)} + 6F_{3(m+2)} - 3F_{(3(m+2)} - F_{3m}].$$

That is,

$$F_{3k+1}F^3_{(m+4)+k+1} + F_{3k+2}F^3_{(m+4)+k} - F^3_{(m+4)-2k-1} = F_{3k+1}F_{3k+2}F_{3(m+4)}.$$

So identity (37.21) is true when $n = m + 4$.

Thus, by PMI, it is true for all $n \ge 0$. ■

In 1999, Melham also developed its Lucas counterpart, as the next theorem shows [354]. Its proof uses Dresel's powerful proof technique of 1993. In the interest of brevity, we only outline the proof.

Theorem 37.3 (Melham, 1999 [354]).

$$F_{3k+1}L_{n+k+1}^3 + F_{3k+2}L_{n+k}^3 - L_{n-2k-1}^3 = 5F_{3k+1}F_{3k+2}L_{3n}. \tag{37.22}$$

Proof. Identity (37.22) is a homogeneous equation of degree 3 in the variable n. So it suffices to establish it for four distinct values of n, say, $0 \le n \le 3$.

When $n = 3$, for example, the identity becomes

$$F_{3k+1}L_{k+4}^3 + F_{3k+2}L_{k+3}^3 - L_{2k-2}^3 = 380F_{3k+1}F_{3k+2}.$$

This is a homogeneous equation of degree 6 in the variable k. To establish its validity, we need only confirm it for seven distinct values of k, say, $0 \le k \le 6$. These cases can be confirmed.

The same argument works for $0 \le n < 3$. This completes the proof. ∎

In the interest of completeness, we present two additional identities, but omit their proofs for the sake of brevity:

$$(-1)^{k+1}F_kF_{n+k}^3 - F_kF_{n-k}^3 + F_{2k}F_n^3 = (-1)^{k+1}F_k^2F_{2k}F_{3n}; \tag{37.23}$$

$$F_mF_{n+k}^3 + (-1)^{k+m+1}F_kF_{n+m}^3 + (-1)^{k+m}F_{k-m}F_n^3 = F_{k-m}F_kF_mF_{3n+k+m}. \tag{37.24}$$

Both are generalizations of the Lucas identity, and the Ginsburg identity $F_{n+2}^3 - 3F_n^3 + F_{n-2}^3 = 3F_{3n}$. Dresel discovered identity (37.23) in 1993, and Melham found identity (37.24) in 2003 [134, 356]. Identity (37.23) follows from (37.24) when $k + m = 0$.

[?] Do identities (37.23) and (37.24) have Lucas counterparts? Do they have their own polynomial versions? These questions remain unresolved.

Next we develop a recurrence for g_n^2. (We will develop another one later.) This approach also contributes additional results.

Recurrence for Gibonacci Squares

Using the familiar Fibonacci polynomial recurrence, we have

$$\begin{aligned}
g_{n+3}^2 &= (xg_{n+2} + g_{n+1})^2 \\
&= x^2g_{n+2}^2 + g_{n+1}^2 + 2xg_{n+2}g_{n+1} \\
&= x^2g_{n+2}^2 + g_{n+1}^2 + g_{n+2}(g_{n+2} - g_n) + xg_{n+2}g_{n+1} \\
&= (x^2+1)g_{n+2}^2 + g_{n+1}^2 - g_{n+2}g_n + xg_{n+1}(xg_{n+1} + g_n) \\
&= (x^2+1)g_{n+2}^2 + (x^2+1)g_{n+1}^2 - g_n(g_{n+2} - xg_{n+1}) \\
&= (x^2+1)g_{n+2}^2 + (x^2+1)g_{n+1}^2 - g_n^2,
\end{aligned} \tag{37.25}$$

as desired.

In particular, this yields

$$f_{n+3}^2 = (x^2+1)f_{n+2}^2 + (x^2+1)f_{n+1}^2 - f_n^2; \tag{37.26}$$

$$l_{n+3}^2 = (x^2+1)l_{n+2}^2 + (x^2+1)l_{n+1}^2 - l_n^2; \tag{37.27}$$

$$p_{n+3}^2 = (4x^2+1)p_{n+2}^2 + (4x^2+1)p_{n+1}^2 - p_n^2;$$

$$q_{n+3}^2 = (4x^2+1)q_{n+2}^2 + (4x^2+1)q_{n+1}^2 - q_n^2;$$

$$F_{n+3}^2 = 2F_{n+2}^2 + 2F_{n+1}^2 - F_n^2; \tag{37.28}$$

$$L_{n+3}^2 = 2L_{n+2}^2 + 2L_{n+1}^2 - L_n^2;$$

$$P_{n+3}^2 = 5P_{n+2}^2 + 5P_{n+1}^2 - P_n^2;$$

$$Q_{n+3}^2 = 5Q_{n+2}^2 + 5Q_{n+1}^2 - Q_n^2.$$

For example,

$$\begin{aligned}
&(x^2+1)f_5^2 + (x^2+1)f_4^2 - f_3^2 \\
&= (x^2+1)(x^4+3x^2+1)^2 + (x^2+1)(x^3+2x)^2 - (x^2+1)^2 \\
&= x^{10} + 8x^8 + 22x^6 + 24x^4 + 9x^2 \\
&= (x^5+4x^3+3x)^2 \\
&= f_6^2.
\end{aligned}$$

H.W. Gould of the University of West Virginia, Morgantown, West Virginia, discovered identity (37.28) in 1963 [189]. It has a charming visual representation; see Figure 6.2 [287].

It follows from recurrence (37.25) that, knowing the initial values of g_0^2, g_1^2, and g_2^2, we can define g_{n+3}^2 recursively, where $n \ge 0$. In particular, for instance, let $a_n = f_n^2$. Then

$$a_0 = 0, \quad a_1 = 1, \quad a_2 = x^2$$

$$a_{n+3} = (x^2+1)a_{n+2} + (x^2+1)a_{n+1} - a_n,$$

where $n \ge 0$.

Pay special attention to the coefficients in recurrence (37.25); they appear a number of times in the following discussions, but in different contexts.

An Alternate Version of Identity (37.25)

Identity (37.25) can be rewritten in a slightly modified, but equivalent, form. To see this, we have

$$xg_{n+3}^2 = g_{n+3}(g_{n+4} - g_{n+2}) = g_{n+3}g_{n+4} - g_{n+2}g_{n+3}$$

$$x(x^2+1)g_{n+2}^2 = (x^2+1)g_{n+2}(g_{n+3} - g_{n+1}) = (x^2+1)(g_{n+2}g_{n+3} - g_{n+1}g_{n+2})$$

$$x(x^2+1)g_{n+1}^2 = (x^2+1)(g_{n+1}g_{n+2} - g_n g_{n+1})$$
$$xg_n^2 = xg_n(g_{n+2} - xg_{n+1}) = xg_n g_{n+2} - x^2 g_n g_{n+1}.$$

Then

$$x[g_{n+3}^2 - (x^2+1)g_{n+2}^2 - (x^2+1)g_{n+1}^2 + g_n^2]$$
$$= g_{n+3}g_{n+4} - (x^2+1)g_{n+2}g_{n+3} - g_{n+2}g_{n+3} + xg_n g_{n+2} + g_n g_{n+1}$$
$$0 = g_{n+3}g_{n+4} - (x^2+1)g_{n+2}g_{n+3} - g_{n+2}(g_{n+3} - xg_n) + g_n g_{n+1}$$
$$0 = g_{n+3}g_{n+4} - (x^2+1)g_{n+2}g_{n+3} - g_{n+2}\cdot(x^2+1)g_{n+1} + g_n g_{n+1}.$$

Thus

$$g_{n+3}g_{n+4} - (x^2+1)g_{n+2}g_{n+3} - (x^2+1)g_{n+1}g_{n+2} + g_n g_{n+1} = 0.$$

In particular, this implies that

$$f_{n+3}f_{n+4} - (x^2+1)f_{n+2}f_{n+3} - (x^2+1)f_{n+1}f_{n+2} + f_n f_{n+1} = 0$$
$$l_{n+3}l_{n+4} - (x^2+1)l_{n+2}l_{n+3} - (x^2+1)l_{n+1}l_{n+2} + l_n l_{n+1} = 0.$$

It also follows from the characteristic equation of recurrence (37.25) that

$$f_{n+3}l_{n+4} - (x^2+1)f_{n+2}l_{n+2} - (x^2+1)f_{n+1}l_{n+1} + f_n l_n = 0$$
$$f_{n+3}l_{m+4} - (x^2+1)f_{n+2}l_{m+2} - (x^2+1)f_{n+1}l_{m+1} + f_n l_m = 0.$$

Recurrence (37.25) can be employed to develop one for $l_n^2 \pm f_n^2$.

Recurrence for $l_n^2 \pm f_n^2$

Let $h_n = h_n(x)$ satisfy the Fibonacci polynomial recurrence, as g_n does. Then $h_{n+3}^2 = (x^2+1)h_{n+2}^2 + (x^2+1)h_{n+1}^2 - h_n^2$. So

$$g_{n+3}^2 + h_{n+3}^2 = (x^2+1)(g_{n+2}^2 + h_{n+2}^2) + (x^2+1)(g_{n+1}^2 + h_{n+1}^2) - (g_n^2 + h_n^2) = 0.$$

Thus $g_n^2 + h_n^2$ satisfies recurrence (37.25), and so does $g_n^2 - h_n^2$.

In particular, let $g_n = l_n$, $h_n = f_n$, $a_n = l_n^2 + f_n^2$ and $b_n = l_n^2 - f_n^2$. Then

$$a_0 = 4, \quad a_1 = x^2+1, \quad a_2 = x^4 + 5x^2 + 4$$
$$a_{n+3} = (x^2+1)a_{n+2} + (x^2+1)a_{n+1} - a_n;$$

b_n also satisfies the same recurrence, where $b_0 = 4$, $b_1 = x^2 - 1$, and $b_2 = x^4 + 3x^2 + 4$.

For example,

$$\begin{aligned} a_3 &= (x^2+1)a_2 + (x^2+1)a_1 - a_0 \\ &= (x^2+1)(x^4+5x^2+4) + (x^2+1)(x^2+1) - 4 \\ &= x^6 + 7x^4 + 11x^2 + 1 \\ &= (x^3+3x)^2 + (x^2+1)^2 \\ &= l_3^2 + f_3^2. \end{aligned}$$

Likewise, $a_4 = x^8 + 9x^6 + 24x^4 + 20x^2 + 4 = (x^4 + 4x^2 + 2)^2 + (x^3 + 2x)^2 = l_4^2 + f_4^2$.

As special cases, both $a_n = a_n(1)$ and $b_n = b_n(1)$ satisfy the recurrence $z_{n+3} = 2z_{n+2} + 2z_{n+1} - z_n$, where $a_0 = 4$, $a_1 = 2$, $a_2 = 10$; and $b_0 = 4$, $b_1 = 0$, and $b_2 = 8$.

Identity (37.28) Revisited

Hoggatt and Bicknell discovered identity (37.28) in 1964 in a very different context [223, 287]. To see this, consider the matrix P, studied by Bicknell and Hoggatt [41], and T.A. Brennan of Lockheed Missiles and Space Company, Sunnyvale, California:

$$P = \begin{bmatrix} 0 & 0 & 1 \\ 0 & 1 & 2 \\ 1 & 1 & 2 \end{bmatrix}.$$

Then, by PMI,

$$P^n = \begin{bmatrix} F_{n-1}^2 & F_{n-1}F_n & F_n^2 \\ 2F_{n-1}F_n & F_{n+1}^2 - F_{n-1}F_n & 2F_nF_{n+1} \\ F_n^2 & F_nF_{n+1} & F_{n+1}^2 \end{bmatrix}.$$

Interestingly, matrix P satisfies the characteristic equation $x^3 - 2x^2 - 2x - 1 = 0$ of recurrence (37.28): $P^3 - 2P^2 - 2P + I = 0$; see Exercise 37.28. Then $P^{n+3} - 2P^{n+2} - 2P^{n+1} + P^n = 0$. Consequently, the corresponding elements of P^{n+3}, P^{n+2}, P^{n+1}, and P^n must satisfy the same recurrence. Equating the elements in position (1,3) of this matrix equation yields recurrence (37.28).

Likewise, by equating the elements in position (2,3), we get

$$2F_{n+3}F_{n+4} - 2\cdot 2F_{n+2}F_{n+3} - 2\cdot 2F_{n+1}F_{n+2} + 2F_nF_{n+1} = 0.$$

That is,

$$F_{n+3}F_{n+4} - 2F_{n+2}F_{n+3} - 2F_{n+1}F_{n+2} + F_nF_{n+1} = 0,$$

as found earlier.

37.3 FIBONOMIAL COEFFICIENTS

The characteristic equation $x^3 - 2x^2 - 2x + 1 = 0$ of recurrence (37.28) is in fact a special case of a far more general result, established by Brennan. To see this, we first introduce the concept of a Fibonomial coefficient (binomial coefficient for Fibonacci numbers) [45, 279].

Generalized binomial coefficients were originally studied by G. Fontené in 1915, and then independently by M. Ward in 1936 [191, 353], where the upper and lower numbers are arbitrary. In 1949, D. Jarden investigated the special case when the upper and lower numbers are Fibonacci numbers [353].

The *n*th *fibonomial coefficient* $\begin{bmatrix} n \\ r \end{bmatrix}$ is defined by

$$\begin{bmatrix} n \\ r \end{bmatrix} = \begin{cases} 0 & \text{if } r < 0 \text{ or } n < r \\ 1 & \text{if } r = 0 \text{ or } n \\ \dfrac{F_n^*}{F_r^* F_{n-r}^*} = \dfrac{F_n F_{n-1} \cdots F_{n-r+1}}{F_r F_{r-1} \cdots F_2 F_1} & \text{otherwise,} \end{cases}$$

where $F_k^* = F_k F_{k-1} \cdots F_2 F_1$ and $F_0^* = 1$ [45, 213, 258, 279, 331, 492]. R.F. Torretto and J.A. Fuchs of the University of Santa Clara, California, introduced the bracketed bi-level notation for fibonomial coefficients in 1964 [492]. In 1970, D. Lind established that every fibonomial coefficient is an integer [331].

For example, $\begin{bmatrix} 7 \\ 3 \end{bmatrix} = \dfrac{F_7^*}{F_3^* F_4^*} = \dfrac{F_7 F_6 F_5}{F_3 F_2 F_1} = 260 = \begin{bmatrix} 7 \\ 4 \end{bmatrix}$.

Clearly, $\begin{bmatrix} n \\ 0 \end{bmatrix} = 1 = \begin{bmatrix} n \\ n \end{bmatrix}$, as in the case of ordinary binomial coefficients.

Fibonomial coefficients satisfy two Pascal-like recurrences:

$$\begin{bmatrix} n \\ r \end{bmatrix} = \begin{bmatrix} n-1 \\ r \end{bmatrix} F_{r+1} + \begin{bmatrix} n-1 \\ r-1 \end{bmatrix} F_{n-r-1}$$

and

$$\begin{bmatrix} n \\ r \end{bmatrix} = \begin{bmatrix} n-1 \\ r \end{bmatrix} F_{r-1} + \begin{bmatrix} n-1 \\ r-1 \end{bmatrix} F_{n-r+1};$$

these can be confirmed algebraically. These recurrences, when coupled with the property that $\begin{bmatrix} n \\ r \end{bmatrix} = \begin{bmatrix} n \\ n-r \end{bmatrix}$, can be used to construct the *fibonomial triangle* in Table 37.1. Figure 37.1 shows its fractal-like binary version.

TABLE 37.1. Fibonomial Triangle

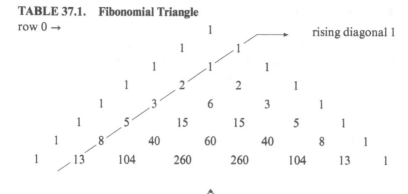

row 0 →								1							
							1		1						
						1		1		1					
					1		2		2		1				
				1		3		6		3		1			
			1		5		15		15		5		1		
		1		8		40		60		40		8		1	
	1		13		104		260		260		104		13		1

rising diagonal 1

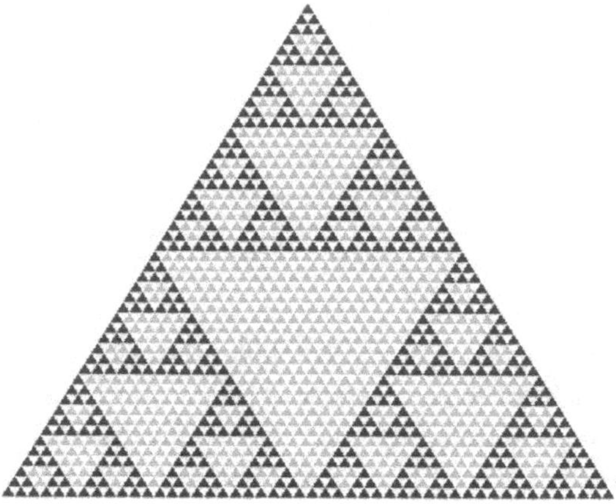

Figure 37.1. A binary version of the fibonomial triangle.

Brennan's Equation

In 1964, Brennan established that

$$\sum_{r=0}^{n+1} (-1)^{r(r+1)/2} \begin{bmatrix} n \\ r \end{bmatrix} x^{n-r+1} = 0$$

is the characteristic equation of the product of n Fibonacci recurrences $y_{n+2} = y_{n+1} + y_n$ [46]. When $n = 2$, it yields

$$\sum_{r=0}^{3} (-1)^{r(r+1)/2} \begin{bmatrix} 3 \\ r \end{bmatrix} x^{3-r} = 0$$

$$\begin{bmatrix} 3 \\ 0 \end{bmatrix} x^3 - \begin{bmatrix} 3 \\ 1 \end{bmatrix} x^2 - \begin{bmatrix} 3 \\ 2 \end{bmatrix} x + \begin{bmatrix} 3 \\ 3 \end{bmatrix} = 0$$

$$x^3 - 2x^2 - 2x + 1 = 0.$$

This is the characteristic equation of recurrence (37.28) we encountered earlier; so we can recover the latter from the former.

An interesting observation: The entries in row 3 of the fibonomial triangle are the absolute values of the coefficients in the characteristic equation. The sign of each coefficient is given by $(-1)^{r(r+1)} = (-1)^{t_r}$, where $t_r = r(r+1)/2$ is the rth triangular number. Each power of x is of the form x^{3-r}, where $0 \le r \le 3$.

Using this observation, row 4 of the array yields

$$x^4 + (-1)^1 \cdot 3x^3 + (-1)^3 \cdot 6x^2 + (-1)^6 \cdot 3x + (-1)^{10} \cdot 1 = 0$$
$$x^4 - 3x^3 - 6x^2 + 3x + 1 = 0.$$

(This can be confirmed using Brennan's equation.)

Correspondingly, we have the following fourth-order recurrences:

$$F_{n+4}^3 - 3F_{n+3}^3 - 6F_{n+2}^3 + 3F_{n+1}^3 + F_n^3 = 0;$$
$$L_{n+4}^3 - 3L_{n+3}^3 - 6L_{n+2}^3 + 3L_{n+1}^3 + L_n^3 = 0.$$

More generally,

$$g_{n+4}^3 = (x^3 + 2x)g_{n+3}^3 + (x^4 + 3x^2 + 2)g_{n+2}^3 - (x^3 + 2x)g_{n+1}^3 - g_n^3. \qquad (37.29)$$

Its proof involves some complicated algebra, so we omit it; see Exercise 37.31. These coefficients appear several times in later discussions. For example,

$$f_5^3 = (x^4 + 3x^2 + 1)^3 = x^{12} + 9x^{10} + 30x^8 + 45x^6 + 30x^4 + 9x^2 + 1$$
$$f_4^3 = (x^3 + 2x)^3 = x^9 + 6x^7 + 12x^5 + 8x^3$$
$$f_3^3 = (x^2 + 1)^3 = x^6 + 3x^4 + 3x^2 + 1$$
$$f_2^3 = x^3$$
$$(x^3 + 2x)f_5^3 + (x^4 + 3x^2 + 1)f_4^3 - (x^3 + 2x)f_3^3 - f_2^3$$
$$= x^{15} + 12x^{13} + 57x^{11} + 136x^9 + 171x^7 + 108x^5 + 27x^3$$
$$= (x^5 + 4x^3 + 3x)^3$$
$$= f_5^3.$$

It follows from recurrence (37.29) that

$$f_{n+4}^3 = (x^3 + 2x)f_{n+3}^3 + (x^4 + 3x^2 + 2)f_{n+2}^3 - (x^3 + 2x)f_{n+1}^3 - f_n^3;$$
$$l_{n+4}^3 = (x^3 + 2x)l_{n+3}^3 + (x^4 + 3x^2 + 2)l_{n+2}^3 - (x^3 + 2x)l_{n+1}^3 - l_n^3;$$
$$p_{n+4}^3 = 4(2x^3 + x)p_{n+3}^3 + 2(8x^4 + 6x^2 + 1)p_{n+2}^3 - 4(2x^3 + x)p_{n+1}^3 - p_n^3;$$
$$q_{n+4}^3 = 4(2x^3 + x)q_{n+3}^3 + 2(8x^4 + 6x^2 + 1)q_{n+2}^3 - 4(2x^3 + x)q_{n+1}^3 - q_n^3;$$
$$P_{n+4}^3 = 12P_{n+3}^3 + 30P_{n+2}^3 - 12P_{n+1}^3 - P_n^3;$$
$$Q_{n+4}^3 = 12Q_{n+3}^3 + 30Q_{n+2}^3 - 12Q_{n+1}^3 - Q_n^3.$$

A Byproduct of Recurrence (37.29)

We can employ identity (37.29) to develop a recurrence for $l_n^3 \pm f_n^3$. To see this, suppose $h_n = h_n(x)$ satisfies the Fibonacci polynomial recurrence. Then h_n also satisfies recurrence (37.29), and so does $g_n^3 \pm h_n^3$.

In particular, let $a_n = l_n^3 + f_n^3$. Then a_n can be defined recursively:

$$a_0 = 8, \quad a_1 = x^3 + 1, \quad a_2 = x^6 + 6x^4 + x^3 + 12x^2 + 8$$
$$a_3 = x^9 + 9x^7 + x^6 + 27x^5 + 3x^4 + 27x^3 + 3x^2 + 1$$
$$a_{n+4} = (x^3 + 2x)a_{n+3} + (x^4 + 3x^2 + 2)a_{n+2} - (x^3 + 2x)a_{n+1} - a_n,$$

where $n \geq 0$.

For example,

$$\begin{aligned} a_4 &= (x^3 + 2x)a_3 + (x^4 + 3x^2 + 2)a_2 - (x^3 + 2x)a_1 - a_0 \\ &= x^{12} + 12x^{10} + x^9 + 54x^8 + 6x^7 + 112x^6 + 12x^5 + 108x^4 + 8x^3 + 48x^2 + 8 \\ &= (x^4 + 3x^2 + 2)^3. \end{aligned}$$

On the other hand, let $b_n = l_n^3 - f_n^3$. Then

$$b_0 = 8, \quad b_1 = x^3 - 1, \quad b_2 = x^6 + 6x^4 - x^3 + 12x^2 + 8$$
$$b_3 = x^9 + 9x^7 - x^6 + 27x^5 - 3x^4 + 27x^3 - 3x^2 - 1$$
$$b_{n+4} = (x^3 + 2x)b_{n+3} + (x^4 + 3x^2 + 2)b_{n+2} - (x^3 + 2x)b_{n+1} - b_n,$$

where $n \geq 0$.

Next we generalize fibonomial coefficients, and use them to extract additional properties.

37.4 GIBONOMIAL COEFFICIENTS

Let x be a positive integer. Then f_n is also a positive integer. Let $f_n^* = f_n f_{n-1} \cdots f_2 f_1$, where $f_0^* = 1$. The *gibonomial coefficient* $\left[\left[{n \atop r}\right]\right]$ is defined by

$$\left[\left[{n \atop r}\right]\right] = \begin{cases} 0 & \text{if } r < 0 \text{ or } n < r \\ 1 & \text{if } r = 0 \text{ or } n \\ \dfrac{f_n^*}{f_r^* f_{n-r}^*} = \dfrac{f_n f_{n-1} \cdots f_{n-r+1}}{f_r f_{r-1} \cdots f_2 f_1} & \text{otherwise.} \end{cases}$$

Clearly, $\left[\left[{n \atop r}\right]\right] = \left[\left[{n \atop n-r}\right]\right]$; see Exercise 37.32.

For example, $\left[\!\!\left[\begin{matrix}3\\2\end{matrix}\right]\!\!\right] = \dfrac{f_3^*}{f_2^* f_1^*} = x^2 + 1 = \left[\!\!\left[\begin{matrix}3\\1\end{matrix}\right]\!\!\right]$. Similarly, $\left[\!\!\left[\begin{matrix}4\\2\end{matrix}\right]\!\!\right] = x^4 + 3x^2 + 2$.

Gibonomial Recurrences

It follows by the definition that

$$\left[\!\!\left[\begin{matrix}n\\r\end{matrix}\right]\!\!\right] = \left[\!\!\left[\begin{matrix}n-1\\r\end{matrix}\right]\!\!\right] f_{r+1} + \left[\!\!\left[\begin{matrix}n-1\\r-1\end{matrix}\right]\!\!\right] f_{n-r-1} \tag{37.30}$$

$$= \left[\!\!\left[\begin{matrix}n-1\\r\end{matrix}\right]\!\!\right] f_{r-1} + \left[\!\!\left[\begin{matrix}n-1\\r-1\end{matrix}\right]\!\!\right] f_{n-r+1}; \tag{37.31}$$

see Exercises 37.33 and 37.34; see Exercise 37.35 for an alternate recurrence involving Lucas polynomials.

For example,

$$\left[\!\!\left[\begin{matrix}3\\2\end{matrix}\right]\!\!\right] = \left[\!\!\left[\begin{matrix}2\\2\end{matrix}\right]\!\!\right] f_3 + \left[\!\!\left[\begin{matrix}2\\1\end{matrix}\right]\!\!\right] f_0 = x^2 + 1;$$

$$\begin{aligned}\left[\!\!\left[\begin{matrix}4\\2\end{matrix}\right]\!\!\right] &= \left[\!\!\left[\begin{matrix}3\\2\end{matrix}\right]\!\!\right] f_1 + \left[\!\!\left[\begin{matrix}3\\1\end{matrix}\right]\!\!\right] f_3 = (x^2+1) + (x^2+1)^2\\ &= x^4 + 3x^2 + 2.\end{aligned}$$

We can use the gibonomial coefficients to construct the *gibonomial triangle* in Table 37.2.

TABLE 37.2. Gibonomial Triangle

				1				
			1		1			
		1		x		1		
	1		x^2+1		x^2+1		1	
1		x^3+2x		x^4+3x^2+2		x^3+2x		1

Interesting Observations

1) We have

$$\begin{aligned}\left[\!\!\left[\begin{matrix}n\\1\end{matrix}\right]\!\!\right] &= \left[\!\!\left[\begin{matrix}n-1\\1\end{matrix}\right]\!\!\right] f_0 + \left[\!\!\left[\begin{matrix}n-1\\0\end{matrix}\right]\!\!\right] f_n\\ &= f_n = \left[\!\!\left[\begin{matrix}n\\n-1\end{matrix}\right]\!\!\right].\end{aligned}$$

So both the rising diagonal 1 and the falling diagonal 1 consist of f_n, where $n \geq 1$.

2) $\left[\left[\begin{matrix}n\\2\end{matrix}\right]\right] = \dfrac{f_n^*}{f_2^* f_{n-2}^*} = \dfrac{1}{x} f_n f_{n-1}$; so every element on diagonal 2 is the product of two consecutive elements on diagonal 1, divided by x. Since x is a factor of f_n or f_{n-1}, it follows that $\dfrac{f_n^*}{f_2^* f_{n-2}^*}$ is a polynomial with integral coefficients.

For example,

$$\begin{aligned}\left[\left[\begin{matrix}5\\2\end{matrix}\right]\right] &= x^6 + 5x^4 + 7x^2 + 2\\ &= \frac{1}{x}(x^4 + 3x^2 + 1)(x^3 + 2x) = \frac{1}{x} f_5 f_4\\ &= \left[\left[\begin{matrix}5\\1\end{matrix}\right]\right] \cdot \left[\left[\begin{matrix}4\\1\end{matrix}\right]\right].\end{aligned}$$

3) It follows from both recurrences (37.30) and (37.31) that every gibonomial coefficient is an integer-valued polynomial.

4) Since $f_{k+1} + f_{k-1} = l_k$, it also follows from recurrences (37.30) and (37.31) that

$$2\left[\left[\begin{matrix}n\\r\end{matrix}\right]\right] = \left[\left[\begin{matrix}n-1\\r\end{matrix}\right]\right] l_r + \left[\left[\begin{matrix}n-1\\r-1\end{matrix}\right]\right] l_{n-r}. \tag{37.32}$$

Consequently, $2f_n = f_{n-r} l_r + f_r l_{n-r}$, as we learned in Chapter 32.

It follows from equation (37.32) that

$$2\left[\begin{matrix}n\\r\end{matrix}\right] = \left[\begin{matrix}n-1\\r\end{matrix}\right] L_r + \left[\begin{matrix}n-1\\r-1\end{matrix}\right] L_{n-r}.$$

Brennan discovered this formula in 1963 [45].

Central Gibonomial Coefficients

The *central gibonomial coefficients* $\left[\left[\begin{matrix}2n\\n\end{matrix}\right]\right]$ contain a hidden treasure:

$$\begin{aligned}\left[\left[\begin{matrix}2n\\n\end{matrix}\right]\right] &= \left[\left[\begin{matrix}2n-1\\n\end{matrix}\right]\right] f_{n+1} + \left[\left[\begin{matrix}2n-1\\n-1\end{matrix}\right]\right] f_{n-1}\\ &= \left[\left[\begin{matrix}2n-1\\n\end{matrix}\right]\right] f_{n+1} + \left[\left[\begin{matrix}2n-1\\r-1\end{matrix}\right]\right] f_{n-1}\\ &= \left[\left[\begin{matrix}2n-1\\n\end{matrix}\right]\right] (f_{n+1} + f_{n-1})\\ &= \left[\left[\begin{matrix}2n-1\\n\end{matrix}\right]\right] l_n.\end{aligned} \tag{37.33}$$

Thus the central gibonomial coefficient $\left[\!\left[\begin{matrix}2n\\ n\end{matrix}\right]\!\right]$ is the product of its northeast (or northwest) neighbor in the gibonomial array and l_n.

For example,

$$\begin{aligned}\left[\!\left[\begin{matrix}6\\ 3\end{matrix}\right]\!\right] &= x^9 + 8x^7 + 22x^5 + 23x^3 + 6x\\ &= (x^6 + 5x^4 + 7x^2 + 2)(x^3 + 2x)\\ &= \left[\!\left[\begin{matrix}5\\ 3\end{matrix}\right]\!\right] l_3.\end{aligned}$$

It follows from property (37.33) that $f_{2n} = f_n l_n$; see Exercise 37.36.

Star of David Property

Gibonomial coefficients satisfy the *Star of David property*:

$$\left[\!\left[\begin{matrix}n-1\\ r-1\end{matrix}\right]\!\right]\left[\!\left[\begin{matrix}n\\ r+1\end{matrix}\right]\!\right]\left[\!\left[\begin{matrix}n+1\\ r\end{matrix}\right]\!\right] = \left[\!\left[\begin{matrix}n-1\\ r\end{matrix}\right]\!\right]\left[\!\left[\begin{matrix}n+1\\ r+1\end{matrix}\right]\!\right]\left[\!\left[\begin{matrix}n\\ r-1\end{matrix}\right]\!\right]; \qquad (37.34)$$

see Figure 37.2.

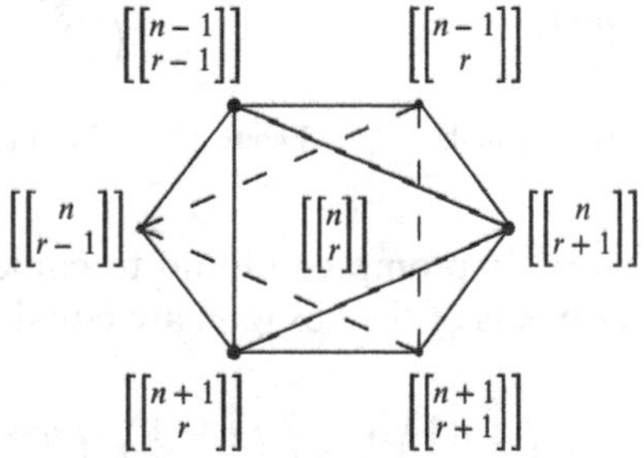

Figure 37.2. Star of David.

This property can also be established algebraically:

$$\begin{aligned}\text{LHS} &= \frac{f^*_{n-1}}{f^*_{r-1}f^*_{n-r}} \cdot \frac{f^*_n}{f^*_{r+1}f^*_{n-r-1}} \cdot \frac{f^*_{n+1}}{f^*_r f^*_{n-r+1}}\\ &= \frac{f^*_{n-1}}{f^*_r f^*_{n-r-1}} \cdot \frac{f^*_{n+1}}{f^*_{r+1}f^*_{n-r}} \cdot \frac{f^*_n}{f^*_{r-1}f^*_{n-r+1}}\\ &= \left[\!\left[\begin{matrix}n-1\\ r\end{matrix}\right]\!\right]\left[\!\left[\begin{matrix}n+1\\ r+1\end{matrix}\right]\!\right]\left[\!\left[\begin{matrix}n\\ r-1\end{matrix}\right]\!\right]\\ &= \text{RHS}.\end{aligned}$$

For example,

$$\begin{aligned}
&\left[\!\left[{4 \atop 2}\right]\!\right]\left[\!\left[{5 \atop 4}\right]\!\right]\left[\!\left[{6 \atop 3}\right]\!\right] \\
&= (x^4 + 3x^2 + 2) \cdot (x^4 + 3x^2 + 1) \cdot (x^6 + 5x^4 + 7x^2 + 2)(x^3 + 3x) \\
&= (x^{11} + 9x^9 + 30x^7 + 45x^5 + 29x^3 + 6x)(x^6 + 5x^4 + 7x^2 + 2) \\
&= (x^3 + 2x) \cdot (x^6 + 6x^4 + 10x^2 + 3)(x^2 + 1) \cdot (x^6 + 5x^4 + 7x^2 + 2) \\
&= \left[\!\left[{4 \atop 3}\right]\!\right]\left[\!\left[{6 \atop 4}\right]\!\right]\left[\!\left[{5 \atop 2}\right]\!\right].
\end{aligned}$$

Hoggatt and Hansell discovered the binomial version of the Star of David property in 1971 [227]; see Figures 37.3 and 37.4.

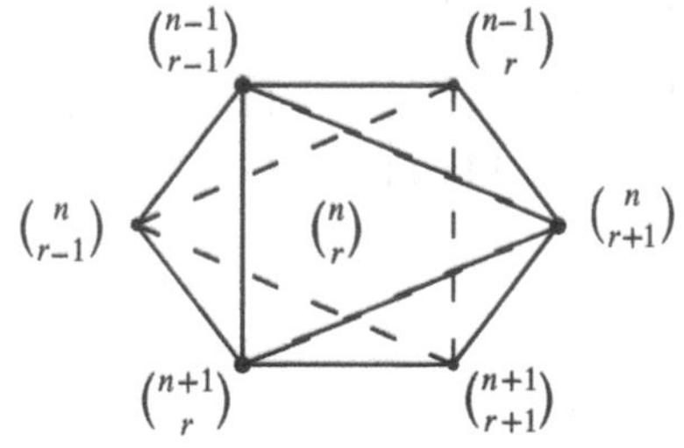

Figure 37.3. Products of triples equal.

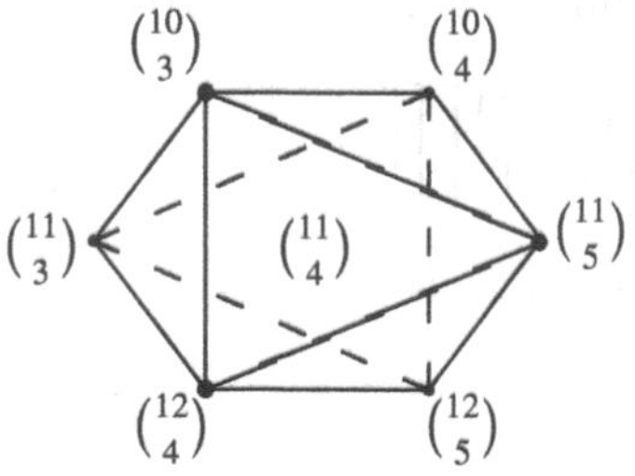

Figure 37.4. Products of triples = 27,442,800.

The Star of David property prompted Gould to conjecture that the gcds of the triples of alternate elements of the hexagon are equal [192]:

$$\left(\binom{n-1}{r-1}, \binom{n}{r+1}, \binom{n+1}{r}\right) = \left(\binom{n-1}{r}, \binom{n+1}{r+1}, \binom{n}{r-1}\right),$$

where (a, b, c) denotes the gcd of the positive integers a, b, and c. A.P. Hillman of the University of New Mexico, Albuquerque, and Hoggatt confirmed it in 1972; they also established Gould's conjecture that this property holds for fibonomial coefficients also [210]. Consequently, the lcms of the triples of both binomial and fibonomial coefficients are equal.

For example, let $a = \binom{10}{3}$, $b = \binom{11}{5}$, $c = \binom{12}{4}$, $p = \binom{10}{4}$, $q = \binom{12}{5}$, and $r = \binom{11}{3}$. Then $(a, b, c) = 3 = (p, q, r)$ and $[a, b, c] = 27{,}720 = [p, q, r]$, where $[a, b, c]$ denotes the lcm of the positive integers a, b, and c.

Interestingly, the gcd property works for gibonacci coefficients as well; R. Flórez established it in 2017 [166].

We now return to Brennan's equation.

Brennan's Equation Revisited

Following the spirit of Brennan's equation, the characteristic equation of the product of n Fibonacci polynomial recurrences $y_{n+2} = xy_{n+1} + y_n$ is given by

$$\sum_{r=0}^{n+1} (-1)^{t_r} \left[\left[\begin{matrix} n \\ r \end{matrix} \right]\right] z^{n-r+1} = 0.$$

When $n = 2$, this yields

$$\sum_{r=0}^{n+1} (-1)^{t_r} \left[\left[\begin{matrix} n \\ r \end{matrix} \right]\right] = 0$$

$$\left[\left[\begin{matrix} 3 \\ 0 \end{matrix} \right]\right] z^3 - \left[\left[\begin{matrix} 3 \\ 1 \end{matrix} \right]\right] z^2 - \left[\left[\begin{matrix} 3 \\ 2 \end{matrix} \right]\right] z + \left[\left[\begin{matrix} 3 \\ 3 \end{matrix} \right]\right] = 0$$

$$z^3 - (x^2 + 1)z^2 - (x^2 + 1)z + 1 = 0.$$

Likewise, $n = 3$ and $n = 4$ yield, respectively,

$$z^4 - (x^3 + 2x)z^3 - (x^4 + 3x^2 + 2)z^2 + (x^3 + 2x)z + 1 = 0$$

$$\begin{aligned} z^5 &- (x^4 + 3x^2 + 1)z^4 - (x^6 + 5x^4 + 7x^2 + 2)z^3 \\ &+ (x^6 + 5x^4 + 7x^2 + 2)z^2 + (x^4 + 3x^2 + 1)z - 1 = 0. \end{aligned}$$

These three equations imply that

$$g_{n+3}^2 = (x^2 + 1)g_{n+2}^2 + (x^2 + 1)g_{n+1}^2 - g_n^2; \tag{37.35}$$

$$g_{n+4}^3 = (x^3 + 2x)g_{n+3}^3 + (x^4 + 3x^2 + 2)g_{n+2}^3 - (x^3 + 2x)g_{n+1}^3 - g_n^3; \tag{37.36}$$

$$\begin{aligned} g_{n+5}^4 &= (x^4 + 3x^2 + 1)g_{n+4}^4 + (x^6 + 5x^4 + 7x^2 + 2)g_{n+3}^4 \\ &\quad - (x^6 + 5x^4 + 7x^2 + 2)g_{n+2}^4 - (x^4 + 3x^2 + 1)g_{n+1}^4 + g_n^4. \end{aligned} \tag{37.37}$$

It follows from equation (37.37) that

$$\begin{aligned} f_{n+5}^4 &= (x^4 + 3x^2 + 1)f_{n+4}^4 + (x^6 + 5x^4 + 7x^2 + 2)f_{n+3}^4 \\ &\quad - (x^6 + 5x^4 + 7x^2 + 2)f_{n+2}^4 - (x^4 + 3x^2 + 1)f_{n+1}^4 + f_n^4; \\ l_{n+5}^4 &= (x^4 + 3x^2 + 1)l_{n+4}^4 + (x^6 + 5x^4 + 7x^2 + 2)l_{n+3}^4 \\ &\quad - (x^6 + 5x^4 + 7x^2 + 2)l_{n+2}^4 - (x^4 + 3x^2 + 1)l_{n+1}^4 + l_n^4; \\ p_{n+5}^4 &= (16x^4 + 12x^2 + 1)p_{n+4}^4 + (64x^6 + 80x^4 + 28x^2 + 2)p_{n+3}^4 \\ &\quad - (64x^6 + 80x^4 + 28x^2 + 2)p_{n+2}^4 - (16x^4 + 12x^2 + 1)p_{n+1}^4 + p_n^4; \\ q_{n+5}^4 &= (16x^4 + 12x^2 + 1)q_{n+4}^4 + (64x^6 + 80x^4 + 28x^2 + 2)q_{n+3}^4 \\ &\quad - (64x^6 + 80x^4 + 28x^2 + 2)q_{n+2}^4 - (16x^4 + 12x^2 + 1)q_{n+1}^4 + q_n^4. \end{aligned}$$

Consequently,

$$F_{n+5}^4 = 5F_{n+4}^4 + 15F_{n+3}^4 - 15F_{n+2}^4 - 5F_{n+1}^4 + F_n^4;$$
$$L_{n+5}^4 = 5L_{n+4}^4 + 15L_{n+3}^4 - 15L_{n+2}^4 - 5L_{n+1}^4 + L_n^4;$$
$$P_{n+5}^4 = 29P_{n+4}^4 + 174P_{n+3}^4 - 174P_{n+2}^4 - 29P_{n+1}^4 + P_n^4;$$
$$Q_{n+5}^4 = 29Q_{n+4}^4 + 174Q_{n+3}^4 - 174Q_{n+2}^4 - 29Q_{n+1}^4 + Q_n^4.$$

For example,

$$\begin{aligned} 5L_7^4 + 15L_6^4 - 15L_5^4 - 5L_4^4 + L_3^4 &= 5 \cdot 47^4 + 15 \cdot 29^4 - 15 \cdot 18^4 - 5 \cdot 11^4 + 7^4 \\ &= 76^4 \\ &= L_8^4. \end{aligned}$$

Similar results follow from equations (37.35) and (37.36).

Next we find a generating function for gibonomial coefficients.

Generating Function for Gibonomial Coefficients

Using the *gaussian binomial coefficients* (or *q-binomial coefficients*)

$$\left\{ {m \atop r} \right\}_q = \frac{1-q^m}{1-q} \cdot \frac{1-q^{m-1}}{1-q^2} \cdots \frac{1-q^{m-r+1}}{1-q^r},$$

we have

$$\prod_{r=0}^{m-1} (1 - q^r z) = \sum_{r=0}^{m} (-1)^r \left\{ {m \atop r} \right\}_q q^{r(r-1)/2} z^r \tag{37.38}$$

$$\prod_{r=0}^{m-1} \frac{1}{1 - q^r z} = \sum_{r=0}^{\infty} \left\{ {m+r-1 \atop r} \right\}_q z^r, \tag{37.39}$$

where q is a dummy variable [77, 84, 279].

Letting $q = \beta/\alpha$,

$$\begin{aligned} \left\{ {m \atop r} \right\}_{\beta/\alpha} &= \frac{(\alpha^m - \beta^m)(\alpha^{m-1} - \beta^{m-1}) \cdots (\alpha^{m-r+1} - \beta^{m-r+1})}{(\alpha - \beta)(\alpha^2 - \beta^2) \cdots (\alpha^r - \beta^r)} \cdot \alpha^{-r(m-r)} \\ &= \frac{f_m f_{m-1} \cdots f_{m-r+1} \cdot \Delta^r}{f_1 f_2 \cdots f_r \cdot \Delta^r} \cdot \alpha^{-r(m-r)} \\ &= \frac{f_m^*}{f_r^* f_{m-r}^*} \alpha^{-r(m-r)} \\ &= \left[\left[{m \atop r} \right]\right] \alpha^{-r(m-r)}. \end{aligned} \tag{37.40}$$

Likewise,

$$\left\{ {m+r-1 \atop r} \right\}_{\beta/\alpha} = \left[\left[{m+r-1 \atop r} \right]\right] \alpha^{-r(m-1)}. \tag{37.41}$$

Since

$$\begin{aligned} (-1)^r \left(\frac{\beta}{\alpha}\right)^{r(r-1)/2} &= (-1)^r(-\alpha^{-2})^{r(r-1)/2} \\ &= (-1)^{r(r+1)/2}\alpha^{-r(r-1)}, \end{aligned}$$

replacing z with $\alpha^{m-1}z$ and letting $q = \beta/\alpha$, identities (37.38) and (37.40) then yield

$$\begin{aligned} \prod_{r=0}^{m-1}(1-\beta^r\alpha^{m-r-1}z) &= \sum_{r=0}^{m}(-1)^{r(r+1)/2}\alpha^{-r(r-1)}\left[\left[{m \atop r} \right]\right]\alpha^{-r(m-r)} \cdot (\alpha^{m-1}z)^r \\ &= \sum_{r=0}^{m}(-1)^{r(r+1)/2}\left[\left[{m \atop r} \right]\right] z^r. \end{aligned}$$

Identities (37.39) and (37.41) then imply that

$$\begin{aligned} \frac{1}{\sum_{r=0}^{m}(-1)^{r(r+1)/2}\left[\left[{m \atop r} \right]\right] z^r} &= \sum_{r=0}^{\infty}\left[\left[{m+r-1 \atop r} \right]\right]\alpha^{-r(m-1)} \cdot \left[\alpha^{m-1}z\right]^r \\ &= \sum_{r=0}^{\infty}\left[\left[{m+r-1 \atop r} \right]\right] z^r \\ &= \sum_{r=0}^{\infty}\left[\left[{m+r-1 \atop m-1} \right]\right] z^r \end{aligned}$$

$$\frac{z^{m-1}}{\sum_{r=0}^{m}(-1)^{r(r+1)/2}\left[\left[{m \atop r} \right]\right] z^r} = \sum_{n=0}^{\infty}\left[\left[{n \atop m-1} \right]\right] z^n, \tag{37.42}$$

where $m \geq 1$. This is the desired generating function.

When $m = 2$ and $m = 3$, equation (37.42) gives

$$\begin{aligned} \frac{z}{1-xz-z^2} &= \sum_{n=0}^{\infty} f_n z^n \\ &= \sum_{n=0}^{\infty}\left[\left[{n \atop 1} \right]\right] z^n; \end{aligned}$$

$$\begin{aligned}\frac{z^2}{1-(x^2+1)z-(x^2+1)z^2+z^3} &= z^2+(x^2+1)z^3+(x^4+3x^2+2)z^4+(x^6+5x^4+7x^2+2)z^5+\cdots\\ &= \sum_{n=0}^{\infty}\left[\left[\begin{matrix}n\\2\end{matrix}\right]\right]z^n,\end{aligned}$$

respectively.

In particular, equation (37.42) gives a generating function for fibonomial coefficients [84, 212]:

$$\frac{z^{m-1}}{\sum_{r=0}^{m}(-1)^{r(r+1)/2}\left[\begin{matrix}m\\r\end{matrix}\right]z^r}=\sum_{n=0}^{\infty}\left[\begin{matrix}n\\m-1\end{matrix}\right]z^n,$$

where $m \geq 1$.

Next we show how gibonomial coefficients can be used to develop additional identities.

Addition Formula

In 1964, Torretto and Fuchs developed an interesting additive formula involving the sum of products of $m+1$ terms of sequences satisfying general second-order recurrence [492]. The identity

$$\sum_{r=0}^{m}(-1)^{r(r+3)/2}\left[\begin{matrix}m\\r\end{matrix}\right]F_{n+m-r}^{m+1}=F_m^{*}F_{(m+1)(n+m/2)}$$

is a special case of their formula (5).

This has an analogous result for f_n:

$$\sum_{r=0}^{m}(-1)^{r(r+3)/2}\left[\left[\begin{matrix}m\\r\end{matrix}\right]\right]f_{n+m-r}^{m+1}=f_m^{*}f_{(m+1)(n+m/2)}. \tag{37.43}$$

When $m=1$, this yields the familiar identity $f_{n+1}^2+f_n^2=f_{2n+1}$; and when $m=2$, it yields

$$\begin{aligned}\left[\left[\begin{matrix}2\\0\end{matrix}\right]\right]f_{n+2}^3+\left[\left[\begin{matrix}2\\1\end{matrix}\right]\right]f_{n+1}^3-\left[\left[\begin{matrix}2\\2\end{matrix}\right]\right]f_{n+1}^3&=f_1f_2f_{3(n+1)}\\ f_{n+2}^3+xf_{n+1}^3-f_n^3&=xf_{3n+3},\end{aligned}$$

which is the same as identity (37.14) that we studied earlier.

Letting $m = 3$ and $m = 4$, we get

$$f_{n+3}^4 + (x^2+1)f_{n+2}^4 - (x^2+1)f_{n+1}^4 - f_n^4 = x(x^2+1)f_{4n+6}; \tag{37.44}$$

$$f_{n+4}^5 + (x^3+2x)f_{n+3}^5 - (x^4+3x^2+2)f_{n+2}^5 - (x^3+2x)f_{n+1}^5 + f_n^5$$
$$= x(x^2+1)(x^3+2x)f_{5n+10}. \tag{37.45}$$

It follows from identities (37.44) and (37.45) that

$$F_{n+3}^4 + 2F_{n+2}^4 - 2F_{n+1}^4 - F_n^4 = 2F_{4n+6}; \tag{37.46}$$

$$p_{n+3}^4 + (4x^2+1)p_{n+2}^4 - (4x^2+1)p_{n+1}^4 - p_n^4 = 2x(4x^2+1)p_{4n+6};$$

$$P_{n+3}^4 + 5P_{n+2}^4 - 5P_{n+1}^4 - P_n^4 = 10P_{4n+6};$$

$$F_{n+4}^5 + 3F_{n+3}^5 - 6F_{n+2}^5 - 3F_{n+1}^5 + F_n^5 = 6F_{5n+10}; \tag{37.47}$$

$$p_{n+4}^5 + 4x(2x^2+1)p_{n+3}^5 - 2(8x^4+6x^2+1)p_{n+2}^5 - 4x(2x^2+1)p_{n+1}^5 + p_n^5$$
$$= 8x^2(2x^2+1)(4x^2+1)p_{5n+10};$$

$$P_{n+4}^5 + 12P_{n+3}^5 - 30P_{n+2}^5 - 12P_{n+1}^5 + P_n^5 = 120P_{5n+10}.$$

Identities (37.46) and (37.47) appear in [353].

Letting $m = 5$, identity (37.43) yields

$$\sum_{r=0}^{5} (-1)^{r(r+3)/2} \left[\left[\begin{matrix}5\\r\end{matrix}\right]\right] f_{n+5-r}^6 = f_1 f_2 f_3 f_4 f_5 f_{6n+15};$$

$$\left[\left[\begin{matrix}5\\0\end{matrix}\right]\right] f_{n+5}^6 + \left[\left[\begin{matrix}5\\1\end{matrix}\right]\right] f_{n+4}^6 - \left[\left[\begin{matrix}5\\2\end{matrix}\right]\right] f_{n+3}^6$$
$$- \left[\left[\begin{matrix}5\\3\end{matrix}\right]\right] f_{n+2}^6 + \left[\left[\begin{matrix}5\\4\end{matrix}\right]\right] f_{n+1}^6 - \left[\left[\begin{matrix}5\\5\end{matrix}\right]\right] f_n^6 = f_1 f_2 f_3 f_4 f_5 f_{6n+15};$$

that is,

$$f_{n+5}^6 + (x^4+3x^2+1)f_{n+4}^6 - (x^6+5x^4+7x^2+2)f_{n+3}^6$$
$$- (x^6+5x^4+7x^2+2)f_{n+2}^6 + (x^4+3x^2+1)f_{n+1}^6 - f_n^6$$
$$= x(x^2+1)(x^3+2x)(x^4+3x^2+1)f_{6n+15}.$$

In particular, we have

$$F_{n+5}^6 + 5F_{n+4}^6 - 15F_{n+3}^6 - 15F_{n+2}^6 + 5F_{n+1}^6 - F_n^6 = 30F_{6n+15};$$

$$\begin{aligned}&p_{n+5}^6 + (16x^4 + 12x^2 + 1)p_{n+4}^6 - 2(32x^6 + 40x^4 + 14x^2 + 1)p_{n+3}^6\\&- 2(32x^6 + 40x^4 + 14x^2 + 1)p_{n+2}^6 + (16x^4 + 12x^2 + 1)p_{n+1}^6 - p_n^6\\&= x(x^2+1)(x^3+2x)(x^4+3x^2+1)p_{6n+15};\end{aligned}$$

$$P_{n+5}^6 + 29P_{n+4}^6 - 174P_{n+3}^6 - 174P_{n+2}^6 + 29P_{n+1}^6 - P_n^6 = 30P_{6n+15}.$$

The corresponding identities for the special case $x = 1$ appear in [353].

Next we study some additional identities.

37.5 ADDITIONAL IDENTITIES

The following example is a delightful application of the identities $f_{2n} = f_n l_n$ and $xf_n + l_n = 2f_{2n+1}$, and the expansions of $(a-b)^3$ and $(a-b)^5$.

Example 37.7. *Prove that*

$$x^5 f_n^5 + l_n^5 = 2f_{n+1}(16f_{n+1}^4 - 20xf_{n+1}^2 f_{2n} + 5x^2 f_{2n}^2). \tag{37.48}$$

Proof.

$$\begin{aligned}x^5f_n^5 + l_n^5 &= x^5f_n^5 + (2f_{n+1} - xf_n)^5\\&= 32f_{n+1}^5 - 80xf_{n+1}^4 f_n + 80x^2 f_{n+1}^3 f_n^2 - 40x^3 f_{n+1}^2 f_n^3 + 10x^4 f_{n+1} f_n^4\\&= 32f_{n+1}^5 + 10xf_nf_{n+1}(x^3f_n^3 - 4x^2f_n^2f_{n+1} + 8xf_{n+1}^2 - 8f_{n+1}^3)\\&= 32f_{n+1}^5 + 10xf_nf_{n+1}[(x^3f_n^3 - 6x^2f_n^2f_{n+1} + 12xf_{n+1}^2 - 8f_{n+1}^3)\\&\quad + (2x^2f_n^2f_{n+1} - 4f_nf_{n+1}^2)]\\&= 32f_{n+1}^5 + 10xf_nf_{n+1}[(xf_n - 2f_{n+1})^3 + 2xf_{n+1}(xf_n - 2f_{n+1})]\\&= 32f_{n+1}^5 + 10xf_nf_{n+1}(xf_n - 2f_{n+1})(x^2f_n^2 - 4xf_nf_{n+1} + 4f_{n+1}^2 + 2xf_{n+1})\\&= 32f_{n+1}^5 + 10xf_nf_{n+1}[-l_n(4f_{n+1}^2 + x^2f_n^2 - 2xf_nf_{n+1})]\\&= 32f_{n+1}^5 - 40xf_nl_nf_{n+1}^3 - 10xf_nl_nf_{n+1}[xf_n(xf_n - 2f_{n+1})]\\&= 32f_{n+1}^5 - 40xf_nl_nf_{n+1}^3 - 10xf_nl_nf_{n+1}(-xf_nl_n)\\&= 32f_{n+1}^5 - 40xf_{2n}f_{n+1}^3 - 10x^2f_{2n}^2f_{n+1}\\&= 2f_{n+1}(16f_{n+1}^4 - 20xf_{n+1}^2f_{2n} + 5x^2f_{2n}^2),\end{aligned}$$

as desired. ■

It follows from identity (37.48) that

$$F_n^5 + L_n^5 = 2F_{n+1}(16F_{n+1}^4 - 20F_{n+1}^2 F_{2n} + 5F_{2n}^2); \tag{37.49}$$
$$32x^5 p_n^5 + q_n^5 = 8p_{n+1}(4p_{n+1}^4 - 10xp_{n+1}^2 p_{2n} + 5x^2 p_{2n}^2);$$
$$4(P_n^5 + Q_n^5) = P_{n+1}(4P_{n+1}^4 - 10P_{n+1}^2 P_{2n} + 5P_{2n}^2).$$

For example,

$$\begin{aligned} P_5(4P_5^4 - 10P_5^2 P_8 + 5P_8^2) &= 29(4 \cdot 29^4 - 10 \cdot 29^2 \cdot 408 + 5 \cdot 408^2) \\ &= 6{,}674{,}756 = 4(12^5 + 17^5) \\ &= 4(P_4^5 + Q_4^5), \end{aligned}$$

as expected.

J.L. Díaz-Barrero of the Polytechnic University of Catalunya, Barcelona, Spain, developed identity (37.49) in 2006 [127].

Next we pursue a different approach to extract new identities. This approach also has interesting ramifications. To this end, suppose $h_n = h_n(y)$ also satisfies the same recurrence as g_n. In Section 35.5, we found that $z_n = g_n h_n$ satisfies the recurrence

$$z_{n+4} = xyz_{n+3} + (x^2 + y^2 + 2)z_{n+2} + xyz_{n+1} - z_n. \tag{37.50}$$

Recurrence (37.50) has several interesting consequences:

1) Suppose $x = y$. Then recurrence (37.50) gives a fourth-order recurrence for gibonacci squares:

$$g_{n+4}^2 = x^2 g_{n+3}^2 + 2(x^2 + 1)g_{n+2}^2 + x^2 g_{n+1}^2 - g_n^2. \tag{37.51}$$

In particular, we have

$$\begin{aligned} f_{n+4}^2 &= x^2 f_{n+3}^2 + 2(x^2 + 1)f_{n+2}^2 + x^2 f_{n+1}^2 - f_n^2; \\ l_{n+4}^2 &= x^2 l_{n+3}^2 + 2(x^2 + 1)l_{n+2}^2 + x^2 l_{n+1}^2 - l_n^2; \\ p_{n+4}^2 &= 4x^2 p_{n+3}^2 + 2(4x^2 + 1)p_{n+2}^2 + 4x^2 p_{n+1}^2 - p_n^2; \\ q_{n+4}^2 &= 4x^2 q_{n+3}^2 + 2(4x^2 + 1)q_{n+2}^2 + 4x^2 q_{n+1}^2 - q_n^2; \\ F_{n+4}^2 &= F_{n+3}^2 + 4F_{n+2}^2 + F_{n+1}^2 - F_n^2; \\ L_{n+4}^2 &= L_{n+3}^2 + 4L_{n+2}^2 + L_{n+1}^2 - L_n^2; \\ P_{n+4}^2 &= 4P_{n+3}^2 + 10P_{n+2}^2 + 4P_{n+1}^2 - P_n^2; \\ Q_{n+4}^2 &= 4Q_{n+3}^2 + 10Q_{n+2}^2 + 4Q_{n+1}^2 - Q_n^2. \end{aligned}$$

For example,

$$\begin{aligned}4Q_9^2 + 10Q_8^2 + 4Q_7^2 - Q_6^2 &= 4 \cdot 1393^2 + 10 \cdot 577^2 + 4 \cdot 239^2 - 99^2 \\ &= 11{,}309{,}769 = 3363^2 \\ &= Q_{10}^2.\end{aligned}$$

2) Letting $g_n = f_n$ and $h_n = l_n$, (37.50) gives a recurrence for $f_n(x)l_n(y)$. In particular, when $y = x$, it becomes a recurrence for $f_n l_n = f_{2n}$:

$$f_{2n+4} = x^2 f_{2n+3} + 2(x^2+1)f_{2n+2} + x^2 f_{2n+1} - f_{2n}.$$

This implies

$$\begin{aligned}F_{2n+4} &= F_{2n+3} + 4F_{2n+2} + F_{2n+1} - F_{2n};\\ p_{2n+4} &= 4x^2 p_{2n+3} + 2(4x^2+1)p_{2n+2} + 4x^2 p_{2n+1} - p_{2n};\\ P_{2n+4} &= 4P_{2n+3} + 10P_{2n+2} + 4P_{2n+1} - P_{2n}.\end{aligned}$$

3) Let $g_n = f_n$ and $h_n(2x) = p_n$. Then (37.50) gives a recurrence for the hybrid $a_n = f_n p_n$:

$$a_{n+4} = 2x^2 a_{n+3} + (5x^2+2)a_{n+2} - a_n + 2x^2.$$

For example,

$a_1 = 1$ $\quad a_2 = 2x^2$
$a_3 = 4x^4 + 5x^2 + 1$ $\quad a_4 = 8x^6 + 20x^4 + 8x^2$
$a_5 = 16x^8 + 60x^6 + 53x^4 + 15x^2 + 1$ $\quad a_6 = 32x^{10} + 160x^8 + 230x^6 + 120x^4 + 18x^2.$

In particular, let $A_n = a_n(1)$. Then $A_n = F_n P_n$ satisfies the recurrence

$$A_{n+4} = 2A_{n+3} + 7A_{n+2} + 2A_{n+1} - A_n,$$

where $A_0 = 0, A_1 = 1, A_2 = 2, A_3 = 10$, and $n \geq 0$; we found this recurrence in Chapter 35.

We can find similarly recurrences for $f_n q_n$, $l_n p_n$, and $l_n q_n$; see Exercises 37.39–37.42.

Recurrence (37.51) Revisited

Interestingly, identity (37.51) can be rewritten in a slightly modified, but equivalent, form:

$$g_{n+4}^2 = (x^2+2)g_{n+3}^2 - (x^2+2)g_{n+1}^2 + g_n^2; \tag{37.52}$$

see Exercise 37.47. Consequently,

$$\begin{aligned}
f_{n+4}^2 &= (x^2+2)f_{n+3}^2 - (x^2+2)f_{n+1}^2 + f_n^2;\\
l_{n+4}^2 &= (x^2+2)l_{n+3}^2 - (x^2+2)l_{n+1}^2 + l_n^2;\\
p_{n+4}^2 &= 2(2x^2+1)p_{n+3}^2 - 2(2x^2+1)p_{n+1}^2 + p_n^2;\\
q_{n+4}^2 &= 2(2x^2+2)q_{n+3}^2 - 2(2x^2+1)q_{n+1}^2 + q_n^2;\\
F_{n+4}^2 &= 3F_{n+3}^2 - 3F_{n+1}^2 + F_n^2;\\
L_{n+4}^2 &= 3L_{n+3}^2 - 3L_{n+1}^2 + L_n^2;\\
P_{n+4}^2 &= 6P_{n+3}^2 - 6P_{n+1}^2 + P_n^2;\\
Q_{n+4}^2 &= 6Q_{n+3}^2 - 6Q_{n+1}^2 + Q_n^2.
\end{aligned}$$

For example,

$$\begin{aligned}
6P_8^2 - 6P_6^2 + P_5^2 &= 6\cdot 408^2 - 6\cdot 70^2 + 29^2\\
&= 970{,}225\\
&= P_9^2;
\end{aligned}$$

and similarly, $6Q_8^2 - 6Q_6^2 + Q_5^2 = 1{,}940{,}449 = Q_9^2$.

It also follows from identity (37.52) that $g_{n+4}^2 \equiv g_n^2 \pmod{x^2+2}$. The sequence $\{g_n^2 \pmod{x^2+2}\}$ is periodic with period 4; see, for example, Table 37.3.

TABLE 37.3.

n	0	1	2	3	4	5	6	7	8
$g_n^2 \pmod{x^2+2}$	0	1	−2	1	0	1	−2	1	0

The next example is an interesting application of the following identities:

$$l_{2n} = l_n^2 - 2(-1)^n; \tag{37.53}$$

$$l_{2n} = (x^2+4)f_n^2 + 2(-1)^n; \tag{37.54}$$

$$\begin{aligned}
(x^2+4)f_{n+2}f_{n-1} &= l_{2n+1} + (-1)^n l_3;\\
l_{n+1}l_n &= l_{2n+1} + (-1)^n x;\\
l_{n+2}l_{n-1} &= l_{2n+1} - (-1)^n l_3;\\
(x^2+4)f_{n+1}f_n &= l_{2n+1} - (-1)^n x.
\end{aligned}$$

It illustrates the power of the identities when applied in tandem.

Example 37.8. *Prove that*

$$(x^2+4)f_{n-1}l_nl_{n+1}f_{n+2}+(x^2+4)l_{n-1}f_nf_{n+1}l_{n+2}=2[(x^2+4)f_{2n+1}^2+xl_3-4]. \tag{37.55}$$

Proof. It follows by identities (37.53) and (37.54) that $l_{4n+2}=l_{2n+1}^2+2=(x^2+4)f_{2n+1}^2+2$; so $l_{2n+1}^2=(x^2+4)f_{2n+1}^2-4$. Using the above identities, we then have

$$\begin{aligned}
\text{LHS} &= [(x^2+4)f_{n-1}f_{n+2}](l_nl_{n+1})+[(x^2+4)f_nf_{n+1}](l_{n-1}l_{n+2})\\
&= [l_{2n+1}+(-1)^nl_3][l_{2n+1}+(-1)^nx]+[l_{2n+1}-(-1)^nx][l_{2n+1}-(-1)^nl_3]\\
&= 2(l_{2n+1}^2+xl_3)\\
&= 2[(x^2+4)f_{2n+1}^2+xl_3-4]\\
&= \text{RHS},
\end{aligned}$$

as claimed. ■

It follows from identity (37.55) that

$$(x^2+1)p_{n-1}q_nq_{n+1}p_{n+2}+(x^2+1)q_{n-1}p_np_{n+1}q_{n+2}=2(x^2+1)p_{2n+1}^2+xq_3-2;$$

$$F_{n-1}L_nL_{n+1}F_{n+2}+L_{n-1}F_nF_{n+1}L_{n+2}=2F_{2n+1}^2; \tag{37.56}$$

$$2(P_{n-1}Q_nQ_{n+1}P_{n+2}+Q_{n-1}P_nP_{n+1}Q_{n+2})=P_{2n+1}^2+3. \tag{37.57}$$

For example,

$$\begin{aligned}
2(P_4Q_5Q_6P_7+Q_4P_5P_6Q_7) &= 2(12\cdot 41\cdot 99\cdot 169+17\cdot 29\cdot 70\cdot 239)\\
&= 32{,}959{,}084\\
&= 5741^2+3\\
&= P_{11}^2+3.
\end{aligned}$$

P.S. Bruckman of Sointula, Canada, a prolific problem proposer and solver, developed identity (37.56) in 2005 [72]. It follows from identity (37.57) that every odd-numbered Pell number is odd.

37.6 STRAZDINS' IDENTITY

The next identity follows by a repeated application of the Fibonacci polynomial recurrence; I. Strazdins of Riga Technical University, Latvia, discovered it in 2000 [473]:

$$f_{n+1}^2-4xf_nf_{n-1}=x^2f_{n-2}^2+(x^2-1)f_{n-1}(xf_n-f_{n-3}).$$

This identity can be extended in an obvious way to the gibonacci family:

$$g_{n+1}^2 - 4xg_ng_{n-1} = x^2g_{n-2}^2 + (x^2-1)g_{n-1}(xg_n - g_{n-3}). \tag{37.58}$$

We omit its proof in the interest of brevity; see Exercise 37.48.

For example,

$$\begin{aligned} x^2f_2^2 + (x^2-1)f_3(xf_4 - f_1) &= x^2 \cdot x^2 + (x^2-1)(x^2+1)[x(x^3+2x) - 1] \\ &= x^8 + 2x^6 - x^4 - 2x^2 + 1 \\ &= (x^4+3x^2+1)^2 - 4x(x^3+2x)(x^2+1) \\ &= f_5^2 - 4xf_4f_3. \end{aligned}$$

Likewise, $x^2l_3^2 + (x^2-1)l_4(xl_5 - l_2) = x^{12} + 8x^{10} + 18x^8 + 4x^6 - 15x^4 - 4x^2 + 4 = l_6^2 - 4xl_5l_4$.

As we can predict, identity (37.58) has additional byproducts:

$$\begin{aligned} l_{n+1}^2 - 4xl_nl_{n-1} &= x^2l_{n-2}^2 + (x^2-1)l_{n-1}(xl_n - l_{n-3}); \\ p_{n+1}^2 - 8xp_np_{n-1} &= 4x^2p_{n-2}^2 + (4x^2-1)p_{n-1}(2xp_n - p_{n-3}); \\ q_{n+1}^2 - 8xq_nq_{n-1} &= 4x^2q_{n-2}^2 + (4x^2-1)q_{n-1}(2xq_n - q_{n-3}); \\ F_{n+1}^2 - 4F_nF_{n-1} &= F_{n-2}^2; \\ L_{n+1}^2 - 4L_nL_{n-1} &= L_{n-2}^2; \\ P_{n+1}^2 - 8P_nP_{n-1} &= 4P_{n-2}^2 + 3P_{n-1}(2P_n - P_{n-3}); \\ Q_{n+1}^2 - 8Q_nQ_{n-1} &= 4Q_{n-2}^2 + 3Q_{n-1}(2Q_n - Q_{n-3}). \end{aligned}$$

Since $P_{n+1}P_{n-1} - P_n^2 = (-1)^n$ and $Q_{n+1}Q_{n-1} - Q_n^2 = 2(-1)^{n-1}$, the last two identities can be simplified:

$$\begin{aligned} P_{n+1}^2 - 14P_nP_{n-1} &= P_{n-2}^2 - 3(-1)^n; \\ Q_{n+1}^2 - 14Q_nQ_{n-1} &= Q_{n-2}^2 + 6(-1)^n. \end{aligned}$$

For example,

$$\begin{aligned} P_6^2 - 14P_5P_4 &= 70^2 - 14 \cdot 29 \cdot 12 \\ &= 28 = P_3^2 - 3(-1)^5; \\ Q_6^2 - 14Q_5Q_4 &= 99^2 - 14 \cdot 41 \cdot 17 \\ &= 43 = Q_3^2 + 6(-1)^5. \end{aligned}$$

EXERCISES 37

Prove each.

1. $l_n^2 - x^2 f_n^2 = 4f_{n+1}f_{n-1}$.
2. $f_n^2 + (-1)^{n+k+1} f_k^2 = f_{n-k}f_{n+k}$.
3. $l_n^2 + (-1)^{n+k} l_k^2 = l_{n-k}l_{n+k} + 4(-1)^n$.
4. Identity (37.2), using the Binet-like formula.
5. Identity (37.4), using the Binet-like formula.
6. $l_n^2 - l_{n+1}^2 - 4l_{n+2}^2 - l_{n+3}^2 + l_{n+4}^2 = (x^2-1)(x^2+4)l_{2n+4}$.
7. $x^2 l_{n+3}^2 + (2x^2+2)l_{n+2}^2 + x^2 l_{n+1}^2 = l_{n+4}^2 + l_n^2$.
8. Identity (37.14), using the Binet-like formula.
9. $(x^2+4)f_{a+b+1} = l_a l_b + l_{a+1}l_{b+1}$.
10. $l_{a+b} = l_a f_{b-1} + l_{a+1} f_b$.
11. $f_{n+1}^2 - f_{n-1}^2 = x f_{2n}$.
12. Identity (37.15).
13. Redo identity (37.14) using the Binet-like formula.
14. Redo identity (37.15) using the Binet-like formula.
15. $F_{n+k}F_{n-k} - F_n^2 = (-1)^{n+k+1}F_k^2$.
16. $L_{n+k}L_{n-k} - L_n^2 = 5(-1)^{n+k}F_k^2$.
17. $l_{4n} = l_n^4 - 4(-1)^n l_n^2 + 2$.
18. $L_{5n} = L_n\left[L_n^4 - 5(-1)^n L_n^2 + 5\right]$.
19. $L_{n+1}^2 + L_{n-1}^2 = 3L_{2n} - 4(-1)^n$.
20. $F_{n-k-2}F_{n-k-1}F_{n-k+1}F_{n-k+2} = F_{n-k}^4 - 1$.
21. $\left(\sum_{k=0}^{4} F_{n+k}^2\right)^2 = F_n^4 + 7F_{n+1}^4 + 25F_{n+2}^4 + 7F_{n+3}^4 + F_{n+4}^4$ (Melham, [357]).
22. $3\left(\sum_{k=0}^{5} F_{n+k}^2\right)^2 = 32(F_{n+1}^4 + 4F_{n+2}^4 + 4F_{n+3}^4 + F_{n+4}^4)$ (Melham, [357]).
23. $g_{n+1}^3 - x^3 g_n^3 - g_{n-1}^3 = 3x g_{n+1}g_n g_{n-1}$, using recursion.
24. $2g_{n-1}^3 + x^3 g_n^3 + 6g_{n-1}g_{n+1}^2 = \begin{cases} l_n^3 & \text{if } g_k = f_k \\ (x^2+4)^3 f_n^3 & \text{if } g_k = l_k. \end{cases}$
25. $g_{n+2}^2 - (x^2-1)g_{n+1}^3 + (x-1)g_n^3 - g_{n-1}^3 - 3x g_{n+2}g_{n+1}g_n =$

$$\begin{cases} x f_{3n} & \text{if } g_k = f_k \\ x(x^2+4)l_{3n} & \text{if } g_k = l_k. \end{cases}$$

26. Lemma 37.1, using recursion.

27. $g_{n+1}^2 - 4xg_ng_{n-1} = x^2g_{n-2}^2 + (x^2-1)g_{n-1}(xg_n - g_{n-3})$.

28. Let $P = \begin{bmatrix} 0 & 0 & 1 \\ 0 & 1 & 2 \\ 1 & 1 & 2 \end{bmatrix}$. Then $P^3 - 2P^2 - 2P + I = 0$.

29. Let $Q = \begin{bmatrix} x & 1 \\ 1 & 0 \end{bmatrix}$. Then $Q^2 - xQ - I = 0$.

30. $Q^{n+2} - xQ^{n+1} - Q^n = 0$.

31. Identity (37.29).

32. $\left[\left[{n \atop r}\right]\right] = \left[\left[{n \atop n-r}\right]\right]$.

33. $\left[\left[{n \atop r}\right]\right] = \left[\left[{n-1 \atop r}\right]\right] f_{r+1} + \left[\left[{n-1 \atop r-1}\right]\right] f_{n-r-1}$.

34. $\left[\left[{n \atop r}\right]\right] = \left[\left[{n-1 \atop r}\right]\right] f_{r-1} + \left[\left[{n-1 \atop r-1}\right]\right] f_{n-r+1}$.

35. $2\left[\left[{n \atop r}\right]\right] = \left[\left[{n-1 \atop r}\right]\right] l_r + \left[\left[{n-1 \atop r-1}\right]\right] l_{n-r}$.

36. $f_{2n} = f_nl_n$, using identity (37.33).

37. $g_{n+1}^5 - x^5g_n^5 - g_{n-1}^5 = 5xg_{n+1}g_ng_{n-1}(x^2g_n^2 + xg_ng_{n-1} + g_{n-1}^2)$.

38. $g_{n+1}^7 - x^7g_n^7 - g_{n-1}^7 = 7xg_{n+1}g_ng_{n-1}(x^2g_n^2 + xg_ng_{n-1} + g_{n-1}^2)$.

Define each recursively, where $b_n = f_nq_n$, $c_n = l_np_n$, $d_n = l_nq_n$, $e_n = p_nq_n$, $B_n = F_nQ_n$, $C_n = L_nP_n$, $D_n = L_nQ_n$, and $E_n = P_nQ_n$.

39. b_n.

40. c_n.

41. d_n.

42. e_n.

43. B_n.

44. C_n.

45. D_n.

46. E_n.

47. Identity (37.52).

48. Identity (37.58).

49. $f_{3n} = f_{2n}l_n - (-1)^nf_n$.

50. $f_{3n} = [l_{2n} + (-1)^n]f_n$.
51. $l_{3n} = [l_{2n} - (-1)^n]l_n$.
52. $f_{n+3k}l_{3k} - f_{n+6k} = (-1)^k f_n$.
53. $l_{n+3k}l_{3k} - l_{n+6k} = (-1)^k l_n$.
54. $f_{5n} = [l_{4n} + (-1)^n l_{2n} + 1]f_n$.
55. $l_{5n} = [l_{4n} + -(-1)^n l_{2n} + 1]l_n$.
56. $f_{6n} = (l_{2n}^2 - 1)f_{2n}$.
57. $l_{6n} = (l_{4n} - 1)l_{2n}$.
58. $x(x^2+4)f_{3n} = f_{n+1}l_{n+1}^2 + xf_nl_n^2 - f_{n-1}l_{n-1}^2$.
59. $xl_{3n} = l_{n+1}f_{n+1}^2 + xl_nf_n^2 - l_{n-1}f_{n-1}^2$.
60. $2[(x^2+1)f_n^2 + (-1)^n]^2 = f_{n+1}^4 + x^4f_n^4 + f_{n-1}^4$.
61. $2[(x^2+1)l_n^2 - (-1)^n(x^2+4)]^2 = l_{n+1}^4 + x^4l_n^4 + l_{n-1}^4$.
62. Let $\{u_n\}$ and $\{v_n\}$ be two sequences satisfying the same second-order recurrence $y_{n+2} = gy_{n+1} + hy_n$. Find a third-order recurrence satisfied by $w_n = u_nv_n$ (Alexanderson, [2]).
63. Find a fifth-order recurrence satisfied by $\{F_n^4\}$ (Melham, [357]).

38

FIBONACCI AND LUCAS POLYNOMIALS III

The mathematician does not study pure mathematics because it is useful; he studies it because he delights in it and he delights in it because it is useful.
–Henri Poincaré (1854–1912)

In this chapter, we investigate additional properties of the extended Fibonacci family, originally studied by Seiffert, a prolific problem proposer and solver. He discovered a number of interesting properties of the family. The following property, found in 2001, gives an explicit formula for f_{2n+1}, and bears a slight resemblance to the Lucas-like formula (31.27) [437, 440]. The proof requires a recurrence for f_{2n+5}, so we first derive one for g_{2n+5} in the following lemma. Once again, we *omit* the argument from the functional notation whenever the omission causes *no* ambiguity.

Lemma 38.1. *Let $n \geq 0$. Then $g_{2n+5} = (x^2+2)g_{2n+3} - g_{2n+1}$.*

Proof. By the gibonacci recurrence, we have

$$\begin{aligned} g_{2n+5} &= xg_{2n+4} + g_{2n+3} \\ &= x(xg_{2n+3} + g_{2n+2}) + g_{2n+3} \end{aligned}$$

Fibonacci and Lucas Numbers with Applications, Volume Two. Thomas Koshy.

$$= (x^2+1)g_{2n+3} + (g_{2n+3} - g_{2n+1})$$
$$= (x^2+2)g_{2n+3} - g_{2n+1}. \qquad \blacksquare$$

For example,

$$\begin{aligned}(x^2+2)f_5 - f_3 &= (x^2+2)(x^4+3x^2+1) - (x^2+1)\\ &= x^6+5x^4+6x^2+1\\ &= f_7.\end{aligned}$$

Likewise, $(x^2+2)l_5 - l_3 = x^7 + 7x^5 + 14x^3 + 7x = l_7$.

We can show similarly that $g_{2n+4} = (x^2+2)g_{2n+2} - g_{2n}$; see Exercise 38.1.

38.1 SEIFFERT'S FORMULAS

We are now ready to present Seiffert's formula for f_{2n+1} [437, 440].

Theorem 38.1 (Seiffert, 2001 [437]). *Let x be any complex number and $n \geq 0$. Then*

$$f_{2n+1} = \sum_{k=0}^{n} (-1)^{\lceil k/2 \rceil} \binom{n - \lceil k/2 \rceil}{\lfloor k/2 \rfloor} (x^2+2)^{n-k}.$$

Proof. For clarity and convenience, we first rewrite the formula in terms of the floor function:

$$f_{2n+1} = \sum_{k=0}^{n} (-1)^{\lfloor (k+1)/2 \rfloor} \binom{n - \lfloor (k+1)/2 \rfloor}{\lfloor k/2 \rfloor} (x^2+2)^{n-k}.$$

Let S_{2n+1} denote the sum on the RHS. Then $S_1 = 1 = f_1$ and $S_3 = (x^2+2) - 1 = x^2+1 = f_3$.

We now show that S_{2n+1} satisfies the same recurrence as f_{2n+1}:

$$\begin{aligned}(x^2+2)S_{2n+3} - S_{2n+1} &= \sum_{k=0}^{n+1} (-1)^{\lfloor (k+1)/2 \rfloor} \binom{n - \lfloor (k+1)/2 \rfloor + 1}{\lfloor k/2 \rfloor} (x^2+2)^{n-k+2}\\ &\quad - \sum_{k=0}^{n} (-1)^{\lfloor (k+1)/2 \rfloor} \binom{n - \lfloor (k+1)/2 \rfloor}{\lfloor k/2 \rfloor} (x^2+2)^{n-k}\\ &= \sum_{k=0}^{n+2} (-1)^{\lfloor (k+1)/2 \rfloor} \binom{n - \lfloor (k+1)/2 \rfloor + 1}{\lfloor k/2 \rfloor} (x^2+2)^{n-k+2}\\ &\quad + \sum_{r=2}^{n+2} (-1)^{\lfloor (r+1)/2 \rfloor} \binom{n - \lfloor (r+1)/2 \rfloor + 1}{\lfloor r/2 \rfloor - 1} (x^2+2)^{n-r+2}\end{aligned}$$

$$= \sum_{k=0}^{n+2} (-1)^{\lfloor (k+1)/2 \rfloor} \binom{n - \lfloor (k+1)/2 \rfloor + 1}{\lfloor k/2 \rfloor} (x^2 + 2)^{n-k+2}$$
$$+ \sum_{r=0}^{n+2} (-1)^{\lfloor (r+1)/2 \rfloor} \binom{n - \lfloor (r+1)/2 \rfloor + 1}{\lfloor r/2 \rfloor - 1} (x^2 + 2)^{n-r+2}$$
$$= \sum_{k=0}^{n+2} (-1)^{\lfloor (k+1)/2 \rfloor} K (x^2 + 2)^{n-k+2},$$

where, by Pascal's identity,

$$K = \binom{n - \lfloor (k+1)/2 \rfloor + 1}{\lfloor k/2 \rfloor} + \binom{n - \lfloor (k+1)/2 \rfloor + 1}{\lfloor k/2 \rfloor - 1} - \binom{n - \lfloor (k+1)/2 \rfloor + 2}{\lfloor k/2 \rfloor}.$$

Thus

$$(x^2 + 2)S_{2n+3} - S_{2n+1} = \sum_{k=0}^{n+2} (-1)^{\lfloor (k+1)/2 \rfloor} \binom{n - \lfloor (k+1)/2 \rfloor + 2}{\lfloor k/2 \rfloor} (x^2 + 2)^{n-k+2}$$
$$= S_{2n+5}.$$

Consequently, S_{2n+1} satisfies the same recursive definition as f_{2n+1}; so $S_{2n+1} = f_{2n+1}$, as desired. ■

As an example, we have

$$f_5 = \sum_{k=0}^{2} (-1)^{\lceil k/2 \rceil} \binom{2 - \lceil k/2 \rceil}{\lfloor k/2 \rfloor} (x^2 + 2)^{2-k}$$
$$= (x^2 + 2)^2 - (x^2 + 2) - 1$$
$$= x^4 + 3x^2 + 1.$$

The next corollary follows trivially from Seiffert's formula.

Corollary 38.1. *Let $n \geq 0$. Then*

$$p_{2n+1} = \sum_{k=0}^{n} (-1)^{\lceil k/2 \rceil} \binom{n - \lceil k/2 \rceil}{\lfloor k/2 \rfloor} 2^{n-k} (2x^2 + 1)^{n-k};$$

$$F_{2n+1} = \sum_{k=0}^{n} (-1)^{\lceil k/2 \rceil} \binom{n - \lceil k/2 \rceil}{\lfloor k/2 \rfloor} 2^{n-k} 3^{n-k};$$

$$P_{2n+1} = \sum_{k=0}^{n} (-1)^{\lceil k/2 \rceil} \binom{n - \lceil k/2 \rceil}{\lfloor k/2 \rfloor} 6^{n-k}.$$

■

For example,

$$\begin{aligned} P_7 &= \sum_{k=0}^{3}(-1)^{\lceil k/2 \rceil}\binom{3-\lceil k/2 \rceil}{\lfloor k/2 \rfloor}6^{n-k} \\ &= 6^3 - 6^2 - 2\cdot 6 + 1 = 169. \end{aligned}$$

Seiffert discovered the next result in 1994 [424]. The featured proof, a bit long, is based on the one given in 1996 by N. Jensen of Kiel, Germany. It involves a lot of clever but basic algebra, and eight steps [257].

Theorem 38.2 (Seiffert, 1994 [424]). *Let x and y be arbitrary complex numbers, $z = \sqrt{x^2 + y^2 + 4}$, and $n \ge 0$. Then*

$$\sum_{k=0}^{\lfloor n/2 \rfloor}\binom{n}{k} f_{n-2k}(x) f_{n-2k}(y) = z^{n-1} f_n(xy/z). \tag{38.1}$$

Proof. [First, we make an observation. Since x and y are complex numbers, z can be zero; and when $z = 0$, the RHS of identity (38.1) becomes undefined. Fortunately, this singularity is removable. Using PMI, we can show that there is a function $h_n : \mathbb{C}^3 \to \mathbb{C}$ such that $z^{n-1} f_n(xy/z) = h_n(x, y, z^2)$, where $\mathbb{C}$ denotes the set of complex numbers. Clearly, $h_1(x, y, z^2) = 1$ and $h_2(x, y, z^2) = x$. Then

$$\begin{aligned} h_{n+2}(x, y, z^2) &= z^{n+1} f_{n+2}(xy/z) \\ &= z^{n+1}\left[\frac{xy}{z} f_{n+1}(xy/z) + f_n(xy/z)\right] \\ &= xy h_{n+1}(x, y, z^2) + z^2 h_n(x, y, z^2). \end{aligned}$$

So, by PMI, h_n is defined for all complex numbers, x, y, and z.]

We can now establish identity (38.1) in eight small steps.

1) First we prove that

$$\sum_{k=0}^{\lfloor (n-1)/2 \rfloor}\binom{n}{k}\lambda^k(1+\lambda^{n-2k}) = \begin{cases} (1+\lambda)^n & \text{if } n \text{ is odd} \\ (1+\lambda)^n - \binom{n}{n/2}\lambda^{n/2} & \text{otherwise,} \end{cases}$$

where λ is a real number.

Case 1. Suppose n is odd. Then

$$\sum_{k=(n+1)/2}^{n}\binom{n}{k}\lambda^k = \sum_{k=(n+1)/2}^{n}\binom{n}{n-k}\lambda^k = \sum_{k=0}^{(n-1)/2}\binom{n}{k}\lambda^{n-k}.$$

By the binomial theorem, we then have

$$\begin{aligned}(1+\lambda)^n &= \sum_{k=0}^{(n-1)/2}\binom{n}{k}\lambda^k + \sum_{k=(n+1)/2}^{n}\binom{n}{k}\lambda^k\\ &= \sum_{k=0}^{(n-1)/2}\binom{n}{k}\lambda^k + \sum_{k=0}^{(n-1)/2}\binom{n}{k}\lambda^{n-k}\\ &= \sum_{k=0}^{(n-1)/2}\binom{n}{k}\lambda^k(1+\lambda^{n-k}).\end{aligned}$$

Case 2. Suppose n is even. As in Case 1, we have

$$\sum_{k=s+1}^{n}\binom{n}{k}\lambda^k = \sum_{k=0}^{n/2}\binom{n}{k}\lambda^{n-k}.$$

Then

$$\begin{aligned}(1+\lambda)^n &= \sum_{k=0}^{n/2-1}\binom{n}{k}\lambda^k + \sum_{k=n/2}^{n}\binom{n}{k}\lambda^k\\ &= \sum_{k=0}^{n/2-1}\binom{n}{k}\lambda^k + \sum_{k=0}^{n/2}\binom{n}{k}\lambda^{n-k}\\ &= \sum_{k=0}^{\lfloor n/2-1\rfloor}\binom{n}{k}\lambda^k(1+\lambda^{n-2k}) + \binom{n}{n/2}\lambda^{n/2}.\end{aligned}$$

This yields the desired result when n is even.

2) Let a, b, c, and d be any real numbers such that $ab = cd$. Let $b \neq 0$. Letting $\lambda = a/b$, it follows from Step 1 that

$$\sum_{k=0}^{\lfloor (n-1)/2\rfloor}\binom{n}{k}(ab)^k(a^{n-2k}+b^{n-2k}) = \begin{cases}(a+b)^n & \text{if } n \text{ is odd}\\ (a+b)^n - \binom{n}{n/2}(ab)^{n/2} & \text{otherwise.}\end{cases}$$

This formula works when $b = 0$ also.

Replacing a with c, and b with d, we obtain a similar equation. Subtracting the resulting equation from the one above and using the fact that $ab = cd$, we get

$$\sum_{k=0}^{\lfloor (n-1)/2\rfloor}\binom{n}{k}(ab)^k(a^{n-2k}+b^{n-2k}-c^{n-2k}-d^{n-2k}) = (a+b)^n-(c+d)^n.$$

Suppose n is odd. Then $\lfloor (n-1)/2 \rfloor = \lfloor n/2 \rfloor$. On the other hand, let n be even. Then $\lfloor (n-1)/2 \rfloor = n/2 - 1$. But when $k = n/2$, $a^{n-2k} + b^{n-2k} - c^{n-2k} - d^{n-2k} = 0$. So when n is even also, we can keep the same upper limit for the sum. Thus, when $n \geq 0$,

$$\sum_{k=0}^{\lfloor n/2 \rfloor} \binom{n}{k} (ab)^k (a^{n-2k} + b^{n-2k} - c^{n-2k} - d^{n-2k}) = (a+b)^n - (c+d)^n.$$

3) Let x be a real number and $\Delta = \Delta(x) = \sqrt{x^2+4}$. Then, by Binet's formula,

$$f_n(x) = \frac{\alpha^n(x) - \beta^n(x)}{\alpha(x) - \beta(x)} = \frac{(x+\Delta)^n - (x-\Delta)^n}{\Delta}.$$

4) Let $a = a(x,y) = [x + \Delta(x)][y + \Delta(y)]$, $b = b(x,y) = [x - \Delta(x)][y - \Delta(y)]$, $c = c(x,y) = [x + \Delta(x)][y - \Delta(y)]$, and $d = d(x,y) = [x - \Delta(x)][y + \Delta(y)]$, where $ab = [x^2 - \Delta^2(x)][y^2 - \Delta^2(y)] = 16 = cd$.

5) We have $\Delta(x)\Delta(y) = \sqrt{(x^2+4)(y^2+4)+16} = z\Delta(xy/z)$.

6) We now prove that

$$a^{n-2k} + b^{n-2k} - c^{n-2k} - d^{n-2k} = 4^{n-2k}\Delta(x)\Delta(y) f_{n-2k}(x) f_{n-2k}(y).$$

$$\begin{aligned}
\text{LHS} &= \left\{[x+\Delta(x)]^{n-2k} - [x-\Delta(x)]^{n-2k}\right\}[y+\Delta(y)]^{n-2k} \\
&\quad - \left\{[x+\Delta(x)]^{n-2k} - [x-\Delta(x)]^{n-2k}\right\}[y-\Delta(y)]^{n-2k} \\
&= \left\{[x+\Delta(x)]^{n-2k} - [x-\Delta(x)]^{n-2k}\right\}\left\{[y+\Delta(y)]^{n-2k} - [y-\Delta(y)]^{n-2k}\right\} \\
&= 2^{n-2k}\Delta(x) f_{n-2k}(x) \cdot 2^{n-2k}\Delta(y) f_{n-2k}(y) \\
&= \text{RHS}.
\end{aligned}$$

7) By Steps 4 and 5, we have

$$\begin{aligned}
a + b &= [x+\Delta(x)][y+\Delta(y)] + [x-\Delta(x)][y-\Delta(y)] \\
&= 2[xy + \Delta(x)\Delta(y)] \\
&= 2z[xy/z + \Delta(xy/z)] \\
&= 4z\alpha(xy/z).
\end{aligned}$$

Similarly, $c + d = 4z\beta(xy/z)$.

8) Combining Steps 3–7, we get

$$\sum_{k=0}^{\lfloor n/2 \rfloor} \binom{n}{k} 16^k \cdot 4^{n-2k} \cdot z\Delta(xy/z) \cdot f_{n-2k}(x) f_{n-2k}(y) = 4^n z^n [\alpha^n(xy/z) - \beta^n(xy/z)]$$

$$\sum_{k=0}^{\lfloor n/2 \rfloor} \binom{n}{k} f_{n-2k}(x) f_{n-2k}(y) = z^{n-1} \cdot \frac{\alpha^n(xy/z) - \beta^n(xy/z)}{\Delta(xy/z)}$$
$$= z^{n-1} f_n(xy/z).$$

Since the LHS is a polynomial in x and y, and the RHS is one in x, y, and z^2, the given result is true for all complex variables x and y. ■

For a specific example, let $n = 5$. Then

$$\begin{aligned}\sum_{k=0}^{2} \binom{5}{k} f_{5-2k}(x) f_{5-2k}(y) &= f_5(x)f_5(y) + 5f_3(x)f_3(y) + 10f_1(x)f_1(y)\\ &= (x^4 + 3x^2 + 1)(y^4 + 3y^2 + 1) + 5(x^2 + 1)(y^2 + 1) + 10\\ &= (xy)^4 + 3z^2(xy)^2 + z^4\\ &= z^4[(xy/z)^4 + 3(xy/z)^2 + 1]\\ &= z^4 f_5(xy/z).\end{aligned}$$

Interesting Byproducts

Seiffert's formula (38.1) has some delightful byproducts. To see them, first we let $x = 3i$ and $y = 2$, where $i = \sqrt{-1}$. Then $\Delta(x) = \alpha(x) - \beta(x) = i(\alpha^2 - \beta^2) = \sqrt{5}i$; $\alpha(x) = \dfrac{x + \Delta(x)}{2} = (3i + \sqrt{5})/2 = i\alpha^2$; and $\beta(x) = \dfrac{x - \Delta(x)}{2} = (3i - \sqrt{5})/2 = i\beta^2$. Consequently,

$$f_j(3i) = \frac{\alpha^j(3i) - \beta^j(3i)}{\alpha(3i) - \beta(3i)} = \frac{(i\alpha^2)^j - (i\beta^2)^j}{\sqrt{5}i} = i^{j-1} F_{2j}.$$

We also have $z = \sqrt{(3i)^2 + 4 + 4} = i$, $\dfrac{xy}{z} = 6$, $\alpha(xy/z) = \alpha(6) = (6 + \sqrt{36 + 4})/2 = 3 + \sqrt{10}$, and $\beta(xy/z) = \beta(6) = (6 - \sqrt{36 + 4})/2 = 3 - \sqrt{10}$.

Since $f_j(2) = P_j$, formula (38.1) yields a hybrid formula:

$$\sum_{k=0}^{\lfloor n/2 \rfloor} i^{n-2k-1} \binom{n}{k} F_{2n-4k} P_{n-2k} = i^{n-1} f_n(6)$$
$$\sum_{k=0}^{\lfloor n/2 \rfloor} (-1)^k \binom{n}{k} F_{2n-4k} P_{n-2k} = f_n(6). \tag{38.2}$$

For example,

$$\sum_{k=0}^{2} (-1)^k \binom{5}{k} F_{10-4k} P_{5-2k} = \binom{5}{0} F_{10} P_5 - \binom{5}{1} F_6 P_3 + \binom{5}{2} F_2 P_1$$

$$= 1 \cdot 55 \cdot 29 - 5 \cdot 8 \cdot 5 + 10 \cdot 1 \cdot 1 = 1405$$
$$= 6^4 + 3 \cdot 6^2 + 1 = f_5(6).$$

Suppose we let $x = 1 = y$. Then $z = \sqrt{6}$ and formula (38.1) gives

$$\sum_{k=0}^{\lfloor n/2 \rfloor} \binom{n}{k} F_{n-2k}^2 = 6^{(n-1)/2} f_n(\sqrt{6}/6). \tag{38.3}$$

For example,

$$\sum_{k=0}^{\lfloor 5/2 \rfloor} \binom{5}{k} F_{10-4k}^2 = \binom{5}{0} F_5^2 + \binom{5}{1} F_3^2 + \binom{5}{2} F_1^2$$
$$= 1 \cdot 5^2 + 5 \cdot 2^2 + 10 \cdot 1^2 = 55$$
$$= 36\left[\left(\sqrt{6}/6\right)^4 + 3\left(\sqrt{6}/6\right)^2 + 1\right]$$
$$= 6^{(5-1)/2} f_5(\sqrt{6}/6).$$

Since $f_n(2u) = p_n(u)$, it follows from formula (38.1) that

$$\sum_{k=0}^{\lfloor n/2 \rfloor} \binom{n}{k} p_{n-2k}(x) p_{n-2k}(y) = z^{n-1} p_n(4xy/z), \tag{38.4}$$

where $z = 2\sqrt{x^2 + y^2 + 1}$. In particular, this implies

$$\sum_{k=0}^{\lfloor n/2 \rfloor} \binom{n}{k} p_{n-2k}^2 = \left(2\sqrt{2x^2+1}\right)^{n-1} p_n\left(\frac{2x^2}{\sqrt{2x^2+1}}\right); \tag{38.5}$$

$$\sum_{k=0}^{\lfloor n/2 \rfloor} \binom{n}{k} P_{n-2k}^2 = \left(2\sqrt{3}\right)^{n-1} p_n\left(2\sqrt{3}/3\right). \tag{38.6}$$

For example, let $n = 5$. Then

$$\sum_{k=0}^{2} \binom{n}{k} P_{5-2k}^2 = \binom{5}{0} P_5^2 + \binom{5}{1} P_3^2 + \binom{5}{2} P_1^2$$
$$= 1 \cdot 29^2 + 5 \cdot 5^2 + 10 \cdot 1^2 = 976$$
$$= \left(2\sqrt{3}/3\right)^{(5-1)/2} p_5\left(2\sqrt{3}/3\right).$$

Seiffert's formula (38.1) has several additional consequences, as the next corollary shows.

Corollary 38.2. *Let $n \geq 0$. Then*

$$5\sum_{k=0}^{\lfloor n/2\rfloor}\binom{n}{k}F_{n-2k}^2 = 3^n - (-2)^n; \tag{38.7}$$

$$\sum_{k=0}^{n}\binom{2n+1}{n-k}F_{2k+1} = 5^n; \tag{38.8}$$

$$\sum_{k=0}^{n}\binom{2n}{n-k}F_{2k}F_{4k} = 5^{n-1}(4^n - 1); \tag{38.9}$$

$$\sum_{k=0}^{n}\binom{2n+1}{n-k}F_{2k+1}L_{4k+2} = 5^n(2^{2n+1} + 1); \tag{38.10}$$

$$\sum_{k=0}^{\lfloor n/2\rfloor}(-1)^k\binom{n}{k}F_{2n-4k}P_{n-2k} = f_n(6); \tag{38.11}$$

$$\sum_{\substack{k=0\\(5,n-2k)=1}}^{\lfloor n/2\rfloor}(-1)^{\lfloor n-2k+2\rfloor}\binom{n}{k} = F_n. \tag{38.12}$$

Proof. [We confirm identities (38.9), (38.10), and (38.12) and leave the others as exercises; see Exercises 38.2–38.4.]
To Prove Identity (38.9):
Let $x = 1$, $y = \sqrt{5}$ and $n = 2m$. Then $\Delta(y) = 3$, $\alpha(y) = \alpha^2$, and $\beta(y) = -\beta^2$. So $f_{2j}(y) = (\alpha^{4j} - \beta^{4j})/3 = \dfrac{\sqrt{5}}{3}f_{4j}$. By formula (38.1), we then have

$$\begin{aligned}\text{LHS} &= \sum_{k=0}^{m}\binom{2m}{k}f_{2m-2k}(x)f_{2m-2k}(y)\\ &= \sum_{k=0}^{m}\binom{2m}{m-k}f_{2k}(x)f_{2k}(y)\\ &= \frac{\sqrt{5}}{3}\sum_{k=0}^{m}\binom{2m}{m-k}F_{2k}F_{4k}.\end{aligned}$$

Since $z = \sqrt{10}$, $\Delta(xy/z) = 3/\sqrt{2}$, $\alpha(xy/z) = \sqrt{2}$, and $\beta(xy/z) = -1/\sqrt{2}$. So

$$\text{RHS} = \left(\sqrt{10}\right)^{2m-1} f_{2m}\left(1/\sqrt{2}\right)$$

$$= \frac{10^m}{\sqrt{10}} \cdot \frac{2^m - (1/2)^m}{3/\sqrt{2}}$$
$$= \frac{5^m(4^m - 1)}{3\sqrt{5}}.$$

Thus

$$\frac{\sqrt{5}}{3}\sum_{k=0}^{m}\binom{2m}{m-k}F_{2k}F_{4k} = \frac{5^m(4^m-1)}{3\sqrt{5}}.$$

This yields identity (38.9), as claimed.

To Prove Identity (38.10):

Let $x = 1$, $y = \sqrt{5}$ and $n = 2m + 1$. Since $f_{2j+1}(y) = (\alpha^{4j+2} + \beta^{4j+2})/3 = \frac{1}{3}l_{4j+2}$, we have

$$\text{LHS} = \frac{1}{3}\sum_{k=0}^{m}\binom{2m+1}{k}F_{2m-2k+1}L_{4m-4k+2}$$
$$= \frac{1}{3}\sum_{k=0}^{m}\binom{2m+1}{m-k}F_{2k+1}L_{4k+2};$$
$$\text{RHS} = z^{2m}f_{2m+1}(xy/z)$$
$$= 10^m\sqrt{2}\cdot\frac{(\sqrt{2})^{2m+1} - (-1/\sqrt{2})^{2m+1}}{3}$$
$$= \frac{5^m(2^{2m+1}+1)}{3}.$$

Thus

$$\sum_{k=0}^{m}\binom{2m+1}{m-k}F_{2k+1}L_{4k+2} = 5^m(2^{2m+1}+1).$$

To Prove Identity (38.12):

Let $x = \alpha i$ and $y = \beta i$. Using recursion, we can compute the values of $f_j(\alpha i)$, as in Table 38.1. It follows from the table that the sequence $\{f_j(\alpha i)\}_{j\geq 0}$ is periodic with period 20, and $f_j(\alpha i) = 0$ if and only if $5|j$. Similarly, sequence $\{f_j(\beta i)\}_{j\geq 0}$ enjoys the same properties.

TABLE 38.1.

j	0	1	2	3	4	5	6	7	8	9
$f_j(\alpha i)$	0	1	αi	$-\alpha$	$-i$	0	$-i$	α	αi	-1
j	10	11	12	13	14	15	16	17	18	19
$f_j(\alpha i)$	0	-1	$-\alpha i$	α	i	0	i	$-\alpha$	$-\alpha i$	1

Thus

$$f_j(\alpha i)f_j(\beta i) = \begin{cases} 0 & \text{if } j \equiv 0 \pmod 5 \\ 1 & \text{if } j \equiv 1,2,8,9 \pmod{10} \\ -1 & \text{if } j \equiv 3,4,6,7 \pmod{10}. \end{cases}$$

So $f_j(\alpha i)f_j(\beta i) = (-1)^{\lfloor (j+2)/5 \rfloor}$ if $(5,j) = 1$. Consequently,

$$\sum_{k=0}^{\lfloor n/2 \rfloor} \binom{n}{k} f_{n-2k}(\alpha i) f_{n-2k}(\beta i) = \sum_{\substack{k=0 \\ (5,n-2k)=1}}^{\lfloor n/2 \rfloor} \binom{n}{k}.$$

Since $z = 1, xy/z = 1$. So $z^{n-1} f_n(xy/z) = f_n(1) = F_n$. Thus

$$\sum_{\substack{k=0 \\ (5,n-2k)=1}}^{\lfloor n/2 \rfloor} (-1)^{\lfloor (n-2k+2)/5 \rfloor} \binom{n}{k} = F_n,$$

as desired. ■

For example, let $n = 6$. Then

$$\sum_{\substack{k=0 \\ (5,2k-1)=1}}^{3} (-1)^{\lfloor (8-2k)/5 \rfloor} \binom{6}{k} = -\binom{6}{0} - \binom{6}{1} + \binom{6}{2} = 8 = F_6.$$

Seiffert's Identity and the Pell Family

Identity (38.1) has byproducts to the Pell family also:

$$\sum_{k=0}^{\lfloor n/2 \rfloor} \binom{n}{k} p_{n-2k}(x) p_{n-2k}(y) = z^{n-1} p_n(4xy/z); \tag{38.13}$$

$$\sum_{k=0}^{\lfloor n/2 \rfloor} \binom{n}{k} P_{n-2k}^2 = 2^{n-3}[3^n - (-1)^n], \tag{38.14}$$

where $z = 2\sqrt{x^2 + y^2 + 1}$; see Exercises 38.5 and 38.6.

For example,

$$\sum_{k=0}^{3} \binom{6}{k} P_{6-2k}^2 = P_6^2 + 6P_4^2 + 15P_2^2 = 5824 = 2^2(3^6 - 1);$$

$$\sum_{k=0}^{3} \binom{7}{k} P_{7-2k}^2 = P_7^2 + 7P_5^2 + 21P_3^2 + 35P_1^2 = 35{,}008 = 2^4(3^7 + 1).$$

The next theorem, developed by Seiffert in 1996, presents another Fibonacci polynomial identity [426, 429]. Its derivation, again a bit long, hinges on the Binet-like formulas for f_n and l_n, the Lucas-like formula for f_n, and a bit of differential and integral calculus. We accomplish this task in eight small steps.

Theorem 38.3 (Seiffert, 1996 [426]). *Let x and y be arbitrary complex numbers, and n a positive integer. Then*

$$f_n(x)f_n(y) = n\sum_{k=0}^{n-1} \frac{1}{k+1}\binom{n+k}{2k+1}(x+y)^k f_{k+1}\left(\frac{xy-4}{x+y}\right). \tag{38.15}$$

Proof. 1) By the Lucas-like formula for f_n, we have

$$\begin{aligned} f_{2n} &= \sum_{k=0}^{n-1} \binom{2n-k-1}{k} x^{2n-k-1} \\ &= \sum_{k=0}^{n-1} \binom{n+k}{2k+1} x^{2k+1}. \end{aligned} \tag{38.16}$$

2) By equation (31.22), $l'_{2n} = 2nf_{2n}$.

3) Integrating both sides of this equation with respect to t from 0 to x and using equation (38.16), we get

$$\begin{aligned} \int_0^x l'_{2n}(t)dt &= 2n\int_0^x f_{2n}(t)dt \\ l_{2n}(x) - l_{2n}(0) &= 2n\sum_{k=0}^{n-1}\binom{n+k}{2k+1}\int_0^x t^{2k+1}dt \\ l_{2n}(x) - 2 &= n\sum_{k=0}^{n-1}\frac{1}{k+1}\binom{n+k}{2k+1}x^{2k+2}. \end{aligned} \tag{38.17}$$

4) Since both sides of identity (38.15) are analytic functions of x and y, it suffices to establish it for real variables x and y, where $x \geq y > 0$. We now let $2u = \Delta(x)\Delta(y) + xy - 4$ and $2v = \Delta(x)\Delta(y) - xy + 4$. Clearly, $u > 0$ and $v > 4$.

Using equation (38.17), we then have

$$\frac{l_{2n}(\sqrt{u}) - l_{2n}(\sqrt{v}i)}{u+v} = n\sum_{k=0}^{n-1}\frac{1}{k+1}\binom{n+k}{2k+1}A_{k+1}, \tag{38.18}$$

where $A_j = \dfrac{u^j - (-v)^j}{u+v}$, $j \geq 0$, and $i = \sqrt{-1}$.

5) Since $u - v = xy - 4$ and $uv = \frac{1}{4}[\Delta^2(x)\Delta^2(y) - (xy-4)^2] = (x+y)^2$, we have

$$\begin{aligned}
&(xy-4)A_{j-1} + (x+y)^2 A_{j-2} \\
&= \frac{1}{u+v}\left\{(u-v)\left[u^{j-1} - \left(\sqrt{v}i\right)^{2j-2}\right] + uv\left[u^{j-2} - \left(\sqrt{v}i\right)^{2j-4}\right]\right\} \\
&= \frac{1}{u+v}[u^j - (-v)^j] \\
&= A_j.
\end{aligned}$$

Thus A_j satisfies the second-order recurrence $A_j = (xy-4)A_{j-1} + (x+y)^2 A_{j-2}$, where $A_0 = 0$, $A_1 = 1$, and $j \geq 2$. Solving this recurrence (see Exercise 38.7), we get

$$A_j = (x+y)^{j-1} f_j\left(\frac{xy-4}{x+y}\right), \tag{38.19}$$

where $j \geq 0$.

6) Next we compute $\alpha^2(z)$ and $\beta^2(z)$, where $z = \sqrt{u}$ and $z = \sqrt{v}i$. Since $2\alpha(\sqrt{u}) = \sqrt{u} + \Delta(\sqrt{u})$, it follows that $4\alpha^2(\sqrt{u}) = 2u + 4 + 2\sqrt{u(u+4)}$. But

$$\begin{aligned}
4\alpha(x)\alpha(y) &= [x + \Delta(x)][y + \Delta(y)] \\
&= 2u + 4 + x\sqrt{y^2+4} + y\sqrt{x^2+4}.
\end{aligned}$$

Since

$$\begin{aligned}
4u(u+4) &= 2x^2y^2 + 4x^2 + 4y^2 + 2xy\sqrt{(x^2+4)(y^2+4)} \\
&= x^2(y^2+4) + y^2(x^2+4) + 2xy\sqrt{(x^2+4)(y^2+4)} \\
&= \left(x\sqrt{y^2+4} + y\sqrt{x^2+4}\right)^2,
\end{aligned}$$

it follows that $x\sqrt{y^2+4} + y\sqrt{x^2+4} = 2\sqrt{u(u+4)}$. Consequently, $4\alpha(x)\alpha(y) = 2u + 4 + 2\sqrt{u(u+4)}$. Thus

$$\alpha^2(\sqrt{u}) = \frac{1}{4}\left[2u + 4 + 2\sqrt{u(u+4)}\right] = \alpha(x)\alpha(y).$$

Similarly, we have

$$\begin{aligned}
\beta^2(\sqrt{u}) &= \frac{1}{4}\left[2u + 4 - 2\sqrt{u(u+4)}\right] = \beta(x)\beta(y); \\
\alpha^2(\sqrt{v}i) &= \frac{1}{4}\left[2v - 4 + 2\sqrt{v(v-4)}\right] = \alpha(x)\beta(y); \\
\beta^2(\sqrt{v}i) &= \frac{1}{4}\left[2v - 4 - 2\sqrt{v(v-4)}\right] = \beta(x)\alpha(y).
\end{aligned}$$

7) Using the Binet-like formulas, we then have

$$\begin{aligned}
f_n(x)f_n(y) &= \frac{\alpha^n(x) - \beta^n(x)}{\Delta(x)} \cdot \frac{\alpha^n(y) - \beta^n(y)}{\Delta(y)} \\
&= \frac{[\alpha^{2n}(\sqrt{u}) + \beta^{2n}(\sqrt{u})] - [\alpha^{2n}(\sqrt{vi}) + \beta^{2n}(\sqrt{vi})]}{\Delta(x)\Delta(y)} \\
&= \frac{l_{2n}(\sqrt{u}) - l_{2n}(\sqrt{vi})}{u+v}. \qquad (38.20)
\end{aligned}$$

8) Combining equations (38.18), (38.19), and (38.20), we get the desired result. ■

For example, let $n = 4$ and $f_j = f_j\left(\frac{xy-4}{x+y}\right)$. Then

$$\begin{aligned}
&4\sum_{k=0}^{3} \frac{1}{k+1}\binom{4+k}{2k+1}(x+y)^k f_{k+1} \\
&= 4\left[f_1 + 5(x+y)f_2 + 2(x+y)^2 f_3 + \frac{1}{4}(x+y)^3 f_4\right] \\
&= 16 + 20(x+y)\left(\frac{xy-4}{x+y}\right) + 8(x+y)^2\left[\left(\frac{xy-4}{x+y}\right)^2 + 1\right] \\
&\quad + (x+y)^3\left[\left(\frac{xy-4}{x+y}\right)^3 + 2\left(\frac{xy-4}{x+y}\right)\right] \\
&= (x^3+2x)(y^3+2y) \\
&= f_4(x)f_4(y),
\end{aligned}$$

as expected.

As we might predict, this theorem has an array of interesting special cases. They are summarized in the following corollary. We will confirm a few of them and leave the others as routine exercises; see Exercises 38.13 and 38.14. To this end, we need the following facts: $2f_n(4) = F_{3n}$, $f_n(3i) = i^{n-1}F_{2n}$, $3f_{2n}(\sqrt{5}) = \sqrt{5}F_{4n}$, $3f_{2n-1}(\sqrt{5}) = L_{4n-2}$, and $5^{(n-1)/2}f_n(4/\sqrt{5}) = [5^n - (-1)^n]/6$; see Exercises 38.8–38.12.

Corollary 38.3 (Seiffert, 1996 [426]).

$$f_n(x)f_n(x+1) = n\sum_{k=0}^{n-1} \frac{(-1)^{n-k+1}}{k+1}\binom{n+k}{2k+1} f_{k+1}(x^2+x+4); \qquad (38.21)$$

$$f_n(x)f_n(4/x) = n\sum_{k=0}^{\lfloor (n-1)/2 \rfloor} \frac{1}{2k+1}\binom{n+2k}{4k+1}\left(\frac{x^2+4}{x}\right)^{2k}, \ x \neq 0; \qquad (38.22)$$

$$f_n^2 = n\sum_{k=0}^{n-1}\frac{(-1)^{n-k+1}}{k+1}\binom{n+k}{2k+1}(x^2+4)^k; \tag{38.23}$$

$$f_n^2 = n\sum_{k=0}^{n-1}\frac{1}{k+1}\binom{n+k}{2k+1}\frac{x^{2k+2}-(-4)^{k+1}}{x^2+4}; \tag{38.24}$$

$$f_{2n-1} = (2n-1)\sum_{k=0}^{2n-2}\frac{(-1)^k}{k+1}\binom{2n+k-1}{2k+1}f_{k+1}(4/x)x^k. \tag{38.25}$$

Proof.

To Prove Identity (38.22):

Let $y = 4/x$, where $x \neq 0$. Since $f_j(0) = \dfrac{1-(-1)^j}{2}$, identity (38.15) then yields

$$f_n(x)f_n(4/x) = n\sum_{k=0}^{\lfloor (n-1)/2\rfloor}\frac{1}{k+1}\binom{n+k}{2k+1}\left(\frac{x^2+4}{x}\right)^k\left[\frac{1-(-1)^k}{2}\right],$$

where k is even. Thus

$$f_n(x)f_n(4/x) = n\sum_{k=0}^{\lfloor (n-1)/2\rfloor}\frac{1}{2k+1}\binom{n+2k}{4k+1}\left(\frac{x^2+4}{x}\right)^{2k}.$$

To Prove Identity (38.23):

Let $y = -x$. Then $u = 0$, $v = x^2+1$, $u - v = -(x^2+4)$, and $uv = 0$. Then $A_j = -(x^2+4)A_{j-1}$, where $A_1 = 1$ and $j \geq 2$. Solving this recurrence yields $A_j = (-1)^{j-1}(x^2+4)^{j-1}$. Thus

$$f_n(x)f_n(-x) = n\sum_{k=0}^{n-1}\frac{1}{k+1}\binom{n+k}{2k+1}(-1)^k(x^2+4)^k$$

$$f_n^2 = n\sum_{k=0}^{n-1}\frac{(-1)^{n-k+1}}{k+1}\binom{n+k}{2k+1}(x^2+4)^k,$$

as desired.

To Prove Identity (38.25):

When we replace n with $2n-1$, identity (38.15) becomes

$$f_{2n-1}(x)f_{2n-1}(y) = (2n-1)\sum_{k=0}^{2n-2}\frac{1}{k+1}\binom{2n+k-1}{2k+1}(x+y)^k f_{k+1}\left(\frac{xy-4}{x+y}\right).$$

Now let $y = 0$. Since $f_j(0) = \dfrac{1-(-1)^j}{2}$, this yields

$$\begin{aligned} f_{2n-1} &= (2n-1)\sum_{k=0}^{2n-2} \frac{1}{k+1}\binom{2n+k-1}{2k+1} x^k f_{k+1}(-4/x) \\ &= (2n-1)\sum_{k=0}^{2n-2} \frac{(-1)^k}{k+1}\binom{2n+k-1}{2k+1} f_{k+1}(4/x)x^k. \end{aligned}$$

■

Next we pursue a few special cases of the summation formula.

Pell Byproducts

Since $f_n(2x) = p_n(x)$, identity (38.15) gives the Pell formula

$$p_n(x)p_n(y) = n\sum_{k=0}^{n-1} \frac{2^k}{k+1}\binom{n+k}{2k+1}(x+y)^k f_{k+1}\left(\frac{2xy-2}{x+y}\right). \tag{38.26}$$

Since $p_n(1/2) = F_n$, letting $x = 1/2$ and $y = 1$, this yields a Fibonacci–Pell hybrid formula:

$$\begin{aligned} F_n P_n &= n\sum_{k=0}^{n-1} \frac{2^k}{k+1}\binom{n+k}{2k+1}(3/2)^k f_{k+1}(-2/3) \\ &= n\sum_{k=0}^{n-1} \frac{(-3)^k}{k+1}\binom{n+k}{2k+1} f_{k+1}(2/3). \end{aligned} \tag{38.27}$$

For example, let $n = 4$. Then

$$\begin{aligned} 4\sum_{k=0}^{3} \frac{(-3)^k}{k+1}\binom{4+k}{2k+1} f_{k+1}(2/3) &= 4\left[4f_1(2/3) - 15f_2(2/3) + 18f_3(2/3) - \frac{27}{4}f_4(2/3)\right] \\ &= 4\left(4\cdot 1 - 15\cdot\frac{2}{3} + 18\cdot\frac{13}{9} - \frac{27}{4}\cdot\frac{44}{27}\right) = 3\cdot 12 \\ &= F_4 P_4, \end{aligned}$$

as expected.

Identities (38.21) through (38.25) provide additional byproducts, as the next corollary reveals. Again, in the interest of brevity, we omit their proofs; see Exercises 38.15–38.27.

Corollary 38.4 (Seiffert, 1996 [426]).

$$F_nP_n = n\sum_{k=0}^{n-1}\frac{(-1)^{n-k+1}}{k+1}\binom{n+k}{2k+1}f_{k+1}(6); \tag{38.28}$$

$$F_nF_{3n} = 2n\sum_{k=0}^{\lfloor(n-1)/2\rfloor}\frac{1}{k+1}\binom{n+2k}{2k+1}25^k; \tag{38.29}$$

$$P_n^2 = n\sum_{k=0}^{\lfloor(n-1)/2\rfloor}\frac{1}{2k+1}\binom{n+2k}{4k+1}16^k; \tag{38.30}$$

$$F_{4n} = \frac{36n}{25^n-1}\sum_{k=0}^{n-1}\frac{1}{2k+1}\binom{2n+2k}{4k+1}81^k5^{n-k-1}; \tag{38.31}$$

$$L_{4n-2} = \frac{18(2n-1)}{5^{2n-1}+1}\sum_{k=0}^{n-1}\frac{1}{2k+1}\binom{2n+2k-1}{4k+1}81^k5^{n-k-1}; \tag{38.32}$$

$$F_n^2 = n\sum_{k=0}^{n-1}\frac{(-1)^{n-k+1}}{k+1}\binom{n+k}{2k+1}5^k; \tag{38.33}$$

$$F_{2n}^2 = n\sum_{k=0}^{n-1}\frac{1}{k+1}\binom{n+k}{2k+1}5^k; \tag{38.34}$$

$$P_n^2 = n\sum_{k=0}^{n-1}\frac{(-1)^{n-k+1}}{k+1}\binom{n+k}{2k+1}8^k; \tag{38.35}$$

$$F_{2n}^2 = \frac{n}{5}\sum_{k=0}^{n-1}\frac{(-1)^{n-k+1}}{k+1}\binom{n+k}{2k+1}\left(9^{k+1}-4^{k+1}\right); \tag{38.36}$$

$$F_{2n-1} = \frac{2n-1}{2}\sum_{k=0}^{2n-2}\frac{(-1)^k}{k+1}\binom{2n+k-1}{2k+1}F_{3k+3}; \tag{38.37}$$

$$F_{6n-3} = (4n-2)\sum_{k=0}^{2n-2}\frac{(-1)^k}{k+1}\binom{2n+k-1}{2k+1}4^kF_{k+1}; \tag{38.38}$$

$$L_{4n-2} = \frac{2n-1}{2}\sum_{k=0}^{2n-2}\frac{(-1)^k}{k+1}\binom{2n+k-1}{2k+1}\left[5^{k+1}-(-1)^{k+1}\right]; \tag{38.39}$$

$$P_{2n-1} = (2n-1)\sum_{k=0}^{2n-2}\frac{(-1)^k}{k+1}\binom{2n+k-1}{2k+1}2^kP_{k+1}. \tag{38.40}$$

■

For example,

$$5\sum_{k=0}^{2}\frac{1}{2k+1}\binom{7+2k}{4k+1}16^k = 5\left[\frac{1}{1}\binom{5}{1}\cdot 1+\frac{1}{3}\binom{7}{5}\cdot 16+\frac{1}{5}\binom{9}{9}\cdot 1\right]$$
$$= 841 = 29^2$$
$$= P_5^2;$$

$$5\sum_{k=0}^{4}\frac{(-1)^{6-k}}{k+1}\binom{5+k}{2k+1}f_{k+1}(6) = 5\left[\frac{1}{1}\binom{5}{1}f_1(6)-\frac{1}{2}\binom{6}{3}f_2(6)+\frac{1}{3}\binom{7}{5}f_3(6)\right]$$
$$-5\left[\frac{1}{4}\binom{8}{7}f_4(6)-\frac{1}{5}\binom{9}{9}f_5(6)\right]$$
$$= 5(5-60+259-456+281) = 5\cdot 29$$
$$= F_5P_5;$$

$$\frac{7}{2}\sum_{k=0}^{6}\frac{(-1)^k}{k+1}\binom{7+k}{2k+1}F_{3k+3} = 13$$
$$= F_7.$$

Corollary 38.4 can be used to extract further implications, as the next corollary shows. Again, we omit their proofs; see Exercises 38.28–38.31.

Corollary 38.5.

$$2\sin^2 n\pi/3 = 3n\sum_{k=0}^{2n-1}\frac{(-1)^k}{k+1}\binom{2n+k}{2k+1}3^k; \tag{38.41}$$

$$4\cos^2(2n-1)\pi/6 = 3(2n-1)\sum_{k=0}^{2n-2}\frac{(-1)^k}{k+1}\binom{2n+k-1}{2k+1}3^k; \tag{38.42}$$

$$2\sin^2 n\pi/3 = n\sum_{k=0}^{2n-1}\frac{(-1)^{k+1}}{k+1}\binom{2n+k}{2k+1}\left(4^{k+1}-1\right); \tag{38.43}$$

$$4\cos^2(2n-1)\pi/6 = (2n-1)\sum_{k=0}^{2n-2}\frac{(-1)^k}{k+1}\binom{2n+k-1}{2k+1}\left(4^{k+1}-1\right). \tag{38.44}$$

■

For example,

$$6\sum_{k=0}^{3}\frac{(-1)^k}{k+1}\binom{4+k}{2k+1}3^k = \frac{3}{2} = 2\sin^2 4\pi/3;$$

$$9\sum_{k=0}^{2}\frac{(-1)^k}{k+1}\binom{3+k}{2k+1}3^k = 0 = 4\cos^2 3\pi/6;$$

$$3\sum_{k=0}^{5}\frac{(-1)^{k+1}}{k+1}\binom{6+k}{2k+1}(4^{k+1}-1) = 0 = 2\sin^2 3\pi/3.$$

The next theorem presents a related Fibonacci polynomial identity, developed by Seiffert in 1996 [428, 432]. It is a delightful application of identity (38.1). This identity also has several interesting consequences, as we will see shortly.

Theorem 38.4 (Seiffert, 1996 [428]). *Let x and y be arbitrary complex numbers, and $n \geq 1$. Then*

$$\sum_{k=0}^{n}\binom{2n}{n-k}f_k(x)f_k(y) = (x-y)^{n-1}f_n\left(\frac{xy+4}{x-y}\right). \tag{38.45}$$

Proof. Let $\Delta = \Delta(x)$. Then $\Delta[(x^2+2)i] = x\Delta\, i$, $\alpha[(x^2+2)i] = \dfrac{(x^2+2)+x\Delta}{2}i$, and $\beta[(x^2+2)i] = \dfrac{(x^2+2)-x\Delta}{2}i$. By the Binet-like formula, we then have

$$\begin{aligned}
f_{2k}(x) &= \frac{\alpha^{2k}(x)-\beta^{2k}(x)}{\alpha-\beta}\\
&= \frac{(x+\Delta)^{2k}-(x-\Delta)^{2k}}{\Delta}\\
&= \frac{1}{\Delta}\left[(x^2+2x\Delta+\Delta^2)^k-(x^2-2x\Delta+\Delta^2)^k\right]\\
&= \frac{2^k}{\Delta}\left\{\left[(x^2+2)+x\sqrt{x^2+4}\right]^k-\left[(x^2+2)-x\sqrt{x^2+4}\right]^k\right\}\\
&= \frac{1}{\Delta\, i^k}\left\{\alpha^k[(x^2+2)i]-\beta^k[(x^2+2)i]\right\}\\
&= \frac{x\Delta\, i}{\Delta\, i^k}\cdot\frac{\alpha^k[(x^2+2)i]-\beta^k[(x^2+2)i]}{\Delta[(x^2+2)i]}\\
&= \frac{x}{i^{k-1}}f_k[(x^2+2)i].
\end{aligned} \tag{38.46}$$

Replacing n with $2n$ and k with $n-k$ in identity (38.1), we get

$$\sum_{k=0}^{n}\binom{2n}{n-k}f_{2k}(x)f_{2k}(y) = z^{n-1}f_{2n}(xy/z).$$

Using identity (38.46), this becomes

$$xy\sum_{k=0}^{n}(-1)^{k-1}\binom{2n}{n-k}f_k[(x^2+2)i]f_k[(y^2+2)i] = i^{1-n}xyz^{2n-2}f_n\left\{[((xy/z)^2+2)i]\right\}. \tag{38.47}$$

Replacing x with $i\sqrt{2+xi}$ and y with $i\sqrt{2-yi}$, $(x^2+2)i$ becomes x and $(y^2+2)i$ becomes y; and z becomes $\sqrt{(y-x)i}$. Then

$$(xy/z)^2 = \frac{(2+xi)(2-yi)}{(y-x)i}$$

$$\left[(xy/z)^2+2\right]i = \frac{xy+4}{y-x}$$

$$f_n\left[(xy/z)^2+2\right]i = (-1)^{n+1}f_n\left(\frac{xy+4}{x-y}\right).$$

Thus equation (38.47) yields

$$\sum_{k=0}^{n}(-1)^{k-1}\binom{2n}{n-k}f_k(x)f_k(y) = i^{n-1}(y-x)^{n-1}i^{n+1}f_n\left(\frac{xy+4}{x-y}\right)$$

$$\sum_{k=0}^{n}\binom{2n}{n-k}f_k(x)f_k(y) = (x-y)^{n-1}f_n\left(\frac{xy+4}{x-y}\right),$$

as desired. ■

For example, let $n=3$. Then

$$\begin{aligned}\sum_{k=0}^{3}\binom{6}{3-k}f_k(x)f_k(y) &= 15f_1(x)f_1(y)+6f_2(x)f_2(y)+f_3(x)f_3(y)\\ &= 15+6xy+(x^2+1)(y^2+1)\\ &= (xy+4)^2+(x-y)^2\\ &= (x-y)^2\left[\left(\frac{xy+4}{x-y}\right)^2+1\right]\\ &= (x-y)^2f_3\left(\frac{xy+4}{x-y}\right).\end{aligned}$$

Seiffert's identity (38.45) also has interesting dividends, as the next corollary shows. As before, we confirm a few and leave the rest as routine exercises; see Exercises 38.32–38.36.

Corollary 38.6.

$$\sum_{\substack{k=0\\5\nmid(2n-k-1)}}^{2n-1} (-1)^{\lfloor(2n-k+1)/5\rfloor}\binom{4n-2}{k} = 5^{n-1}L_{2n-1}; \tag{38.48}$$

$$\sum_{\substack{k=0\\5\nmid(2n-k)}}^{2n} (-1)^{\lfloor(2n-k+2)/5\rfloor}\binom{4n}{k} = 5^{n}F_{2n}; \tag{38.49}$$

$$\sum_{k=0}^{n}\binom{2n}{n-k}F_{3k}P_k = 2^n f_n(6); \tag{38.50}$$

$$\sum_{k=0}^{n}\binom{2n}{n-k}f_k(x)f_k(x+1) = f_n(x^2+x+4); \tag{38.51}$$

$$\sum_{k=0}^{n}(-1)^{k-1}\binom{2n}{n-k}f_k(x)f_k(4/x) = \frac{1-(-1)^n}{2}\left(\frac{x^2+4}{x}\right)^{n-1},\ x\neq 0; \tag{38.52}$$

$$\sum_{k=0}^{n}\binom{2n}{n-k}f_k^2 = (x^2+4)^{n-1}; \tag{38.53}$$

$$\sum_{k=0}^{n}(-1)^{k-1}\binom{2n}{n-k}f_k^2 = \frac{4^n-(-x^2)^n}{4+x^2}; \tag{38.54}$$

$$\sum_{k=0}^{\lfloor(n-1)/2\rfloor}\binom{2n}{n-2k-1}f_{2k+1} = x^{n-1}f_n(4/x). \tag{38.55}$$

Proof.

To Prove Identity (38.48):

If we let $x=\alpha i$ and $y=\beta i$, it follows from the proof of identity (38.12) that

$$f_k(\alpha i)f_k(\beta i) = \begin{cases}(-1)^{\lfloor(k+2)/5\rfloor} & \text{if } 5\nmid k\\ 0 & \text{otherwise.}\end{cases}$$

Consequently, by identity (38.45), we have

$$\sum_{\substack{k=0\\5\nmid k}}^{n}(-1)^{\lfloor(k+2)/5\rfloor}\binom{2n}{n-k} = (\sqrt{5}i)^{n-1}f_n(-\sqrt{5}i)$$

$$\sum_{\substack{k=0\\5\nmid k}}^{2n-1}(-1)^{\lfloor(k+2)/5\rfloor}\binom{4n-2}{2n-k-1} = (-5)^{2n-1}f_n(-\sqrt{5}i).$$

Since $\alpha(-\sqrt{5}i) = \beta i$ and $\beta(-\sqrt{5}i) = \alpha i$, it follows by the Binet-like formulas that $f_{2n-1}(-\sqrt{5}i) = (-1)^{n-1}L_{2n-1}$. Thus we have

$$\sum_{\substack{k=0\\5\nmid k}}^{2n-1} (-1)^{\lfloor (k+2)/5\rfloor}\binom{4n-2}{2n-k-1} = 5^{n-1}L_{2n-1}$$

$$\sum_{\substack{k=0\\5\nmid(2n-k-1)}}^{2n-1} (-1)^{\lfloor (2n-k+1)/5\rfloor}\binom{4n-2}{k} = 5^{n-1}L_{2n-1}.$$

To Prove Identity (38.52):
Let $y = -4/x$. Then $f_k(y) = f_k(-4/x) = (-1)^{k-1}f_k(4/x)$; and $f_n(0) = [\alpha^n(0) - \beta^n(0)]/2 = [1-(-1)^n]/2$. Thus, by identity (38.46), we have

$$\sum_{k=0}^{n}(-1)^{k-1}\binom{2n}{n-k}f_k(x)f_k(4/x) = \frac{1-(-1)^n}{2}\left(\frac{x^2+4}{x}\right)^{n-1},$$

where $x \neq 0$.

To Prove Identity (38.54):
Let $y = -x$. Then $\dfrac{xy}{x-y} = \dfrac{4-x^2}{2x}$, $\alpha\left(\dfrac{4-x^2}{2x}\right) = 2/x$, $\beta\left(\dfrac{4-x^2}{2x}\right) = -x/2$, and $\Delta\left(\dfrac{4-x^2}{2x}\right) = \dfrac{4+x^2}{2x}$. So

$$\begin{aligned} f_n\left(\frac{4-x^2}{2x}\right) &= \frac{(2/x)^n - (-x/2)^n}{(4+x^2)/2x} \\ &= \frac{4^n-(-x^2)^n}{(2x)^n}\cdot\frac{2x}{4+x^2} \\ (2x)^{n-1}f_n\left(\frac{4-x^2}{2x}\right) &= \frac{4^n-(-x^2)^n}{4+x^2}. \end{aligned}$$

By identity (38.46), we then have

$$\sum_{k=0}^{n}\binom{2n}{n-k}f_k(x)f_k(-x) = (2x)^{n-1}f_n\left(\frac{4-x^2}{2x}\right)$$

$$\sum_{k=0}^{n}(-1)^{k-1}\binom{2n}{n-k}f_k^2 = \frac{4^n-(-x^2)^n}{4+x^2}.$$ ■

Additional Byproducts

Identities (38.51) and (38.55) have their own byproducts, as the next corollary shows.

Corollary 38.7.

$$\sum_{k=0}^{n} \binom{2n}{n-k} F_k P_k = f_n(6); \tag{38.56}$$

$$\sum_{k=0}^{\lfloor (n-1)/2 \rfloor} \binom{2n}{2n-2k-1} F_{2k+1} = f_n(4); \tag{38.57}$$

$$\sum_{k=0}^{n} (-1)^{k-1} \binom{2n}{n-k} P_k^2 = [1-(-1)^n]2^{2n-3}; \tag{38.58}$$

$$\sum_{k=0}^{n} \binom{2n}{n-k} F_k^2 = 5^{n-1}; \tag{38.59}$$

$$\sum_{k=0}^{n} \binom{2n}{n-k} P_k^2 = 8^{n-1}; \tag{38.60}$$

$$\sum_{k=0}^{\lfloor (n-1)/2 \rfloor} \binom{2n}{2n-2k-1} = 4^{n-1}; \tag{38.61}$$

$$\sum_{k=0}^{n} (-1)^{k-1} \binom{2n}{n-k} F_k^2 = [4^n - (-1)^n]/5; \tag{38.62}$$

$$\sum_{k=0}^{n} (-1)^{k-1} \binom{2n}{n-k} P_k^2 = [4^n - (-4)^n]/8; \tag{38.63}$$

$$\sum_{k=0}^{\lfloor (n-1)/2 \rfloor} \binom{2n}{2n-2k-1} P_{2k+1} = 2^{n-1} P_n; \tag{38.64}$$

$$\sum_{k=0}^{\lfloor (n-1)/2 \rfloor} \binom{2n}{2n-2k-1} p_{2k+1} = (2x)^{n-1} f_n(2/x); \tag{38.65}$$

$$\sum_{k=0}^{\lfloor (n-1)/2 \rfloor} \binom{2n}{2n-2k-1} p_{2k+1}(2) = 4^{n-1} F_n. \tag{38.66}$$

The next theorem gives two binomial formulas: one for odd-numbered Fibonacci numbers, the other for even-numbered Lucas numbers [436, 448]. We can establish them using formula (38.53). The proof is a bit long, so we break it up into eight small steps.

Theorem 38.5 (Seiffert, 2001 [436]). *Let $n \geq 1$. Then*

$$\sum_{\substack{k=0 \\ 5\nmid(2n-k-1)}}^{2n-1} (-1)^k \binom{4n-2}{k} = 5^{n-1} F_{2n-1}; \tag{38.67}$$

$$\sum_{\substack{k=0 \\ 5\nmid(2n-k)}}^{2n} (-1)^{k+1} \binom{4n-2}{k} = 5^{n-1} L_{2n}. \tag{38.68}$$

Proof.
1) Let $A_n = i^{n-1} f_n(i\alpha)$, where $i = \sqrt{-1}$. Then, by the Fibonacci recurrence, we have

$$\begin{aligned} A_{n+1} &= i^n f_{n+1}(i\alpha) \\ &= i^n[(i\alpha)f_n(i\alpha) + f_{n-1}(i\alpha)] \\ &= i^{n+1}\alpha f_n(i\alpha) + i^n f_{n-1}(i\alpha) \\ &= -\alpha A_n - A_{n-1}. \end{aligned}$$

2) Since $A_0 = 0$ and $A_1 = 1$, we can employ this recurrence to compute several values of A_n, as Table 38.2 shows. Clearly, the sequence $\{A_n\}$ is periodic with period 5.

TABLE 38.2.

n	0	1	2	3	4	5	6	7	8	9
A_n	0	1	$-\alpha$	α	-1	0	1	$-\alpha$	α	-1

It follows by PMI that

$$A_n = \begin{cases} 0 & \text{if } n \equiv 0 \pmod 5 \\ 1 & \text{if } n \equiv 1 \pmod 5 \\ -\alpha & \text{if } n \equiv 2 \pmod 5 \\ \alpha & \text{if } n \equiv 3 \pmod 5 \\ -1 & \text{otherwise.} \end{cases} \tag{38.69}$$

3) Letting $x = i\alpha$ in formula (38.53), we get

$$\sum_{k=0}^{n} \binom{2n}{n-k} f_k^2(i\alpha) = (4 - \alpha^2)^{n-1}.$$

Since $4 - \alpha^2 = 4 - \dfrac{3+\sqrt{5}}{2} = -\sqrt{5}\beta$ and $f_k^2(i\alpha) = (-1)^{k-1}A_k^2$, this yields

$$\sum_{k=0}^{n} \binom{2n}{n-k} A_k^2 = (-\sqrt{5}\beta)^{n-1}.$$

4) Using formula (38.69), we can rewrite this as

$$\sum_{k=0}^{n}(-1)^{k-1}\binom{2n}{n-k}c_k + \alpha^2 \sum_{k=0}^{n}(-1)^{k-1}\binom{2n}{n-k}d_k = (-\sqrt{5}\beta)^{n-1},$$

where $c_k = \begin{cases} 1 & k = 1,4 \pmod 5 \\ 0 & \text{otherwise,} \end{cases}$ and $d_k = \begin{cases} 1 & k = 2,3 \pmod 5 \\ 0 & \text{otherwise.} \end{cases}$

5) Let $S_n = \sum_{k=0}^{n}(-1)^{k-1}\binom{2n}{n-k}c_k$ and $T_n = \sum_{k=0}^{n}(-1)^{k-1}\binom{2n}{n-k}d_k$. Then $S_n + \alpha^2 T_n = (-\sqrt{5}\beta)^{n-1}$. But $\alpha^2 = 2\alpha + 1 = (3+\sqrt{5})/2$ and $\beta^{n-1} = (L_{n-1} - \sqrt{5}F_{n-1})/2$. So $2S_n + (3+\sqrt{5})T_n = (-\sqrt{5})^{n-1}(L_{n-1} - \sqrt{5}F_{n-1})$.

6) Equating the rational and irrational parts from both sides, we get

$$2S_n + 3T_n = \begin{cases} 5^{(n-1)/2}L_{n-1} & \text{if } n \text{ is odd} \\ 5^{n/2}F_{n-1} & \text{otherwise,} \end{cases}$$

$$T_n = \begin{cases} -5^{(n-1)/2}F_{n-1} & \text{if } n \text{ is odd} \\ -5^{n/2}L_{n-1} & \text{otherwise.} \end{cases}$$

7) Then

$$\begin{aligned} 2(S_n + T_n) &= (2S_n + 3T_n) - T_n \\ &= \begin{cases} 5^{(n-1)/2}(L_{n-1} + F_{n-1}) & \text{if } n \text{ is odd} \\ 5^{(n-2)/2}(5F_{n-1} + L_{n-1}) & \text{otherwise.} \end{cases} \end{aligned}$$

Since $L_{n-1} + F_{n-1} = 2F_n$ and $5F_{n-1} + L_{n-1} = 2L_n$ (see Exercises 38.48 and 38.49), this yields

$$S_n + T_n = \begin{cases} 5^{(n-1)/2}F_n & \text{if } n \text{ is odd} \\ 5^{(n-2)/2}L_n & \text{otherwise.} \end{cases}$$

That is,

$$\sum_{k=0}^{n}(-1)^{k-1}\binom{2n}{n-k}(c_k + d_k) = \begin{cases} 5^{(n-1)/2}F_n & \text{if } n \text{ is odd} \\ 5^{(n-2)/2}L_n & \text{otherwise.} \end{cases}$$

But $c_k + d_k = 1$ if and only if $k \not\equiv 0 \pmod 5$. Thus

$$\sum_{k=0}^{n}(-1)^{k-1}\binom{2n}{n-k} = \begin{cases} 5^{(n-1)/2}F_n & \text{if } n \text{ is odd} \\ 5^{(n-2)/2}L_n & \text{otherwise,} \end{cases}$$

where $k \not\equiv 0 \pmod 5$.

8) Replacing n with $2n-1$, this yields

$$\sum_{k=0}^{2n-1}(-1)^{k-1}\binom{4n-2}{2n-k-1} = 5^{n-1}F_{2n-1}$$

$$\sum_{\substack{k=0 \\ 5\nmid(2n-k-1)}}^{2n-1}(-1)^{k}\binom{4n-2}{k} = 5^{n-1}F_{2n-1},$$

as claimed.

Formula (38.68) follows similarly by replacing n with $2n$. ■

For example, let $n = 5$. Then

$$\begin{aligned}\sum_{\substack{k=0 \\ 5\nmid(9-k)}}^{9}(-1)^{k}\binom{18}{k} &= \binom{18}{0} - \binom{18}{1} + \binom{18}{2} - \binom{18}{3} - \binom{18}{5} + \binom{18}{6} \\ &\quad - \binom{18}{7} + \binom{18}{8} \\ &= 21{,}250 = 5^4 \cdot 34 \\ &= 5^4 F_9.\end{aligned}$$

Similarly, $\displaystyle\sum_{\substack{k=0 \\ 5\nmid(8-k)}}^{8}(-1)^{k}\binom{16}{k} = 5875 = 5^3 \cdot 47 = 5^3 L_8.$

38.2 ADDITIONAL FORMULAS

The next two results provide additional explicit formulas for f_n. Theorem 38.6 employs its generating function, as well as Jacobi polynomials, named after the German mathematician Carl Gustav Jacob Jacobi (1804–1851). Since Jacobi polynomials are beyond the scope of our discourse, we omit its proof [427, 430] in the interest of brevity, but illustrate it with an example.

Theorem 38.6 (Seiffert, 1996 [427]). *Let x be a real variable and $i = \sqrt{-1}$. Then*

$$f_{n+1} = \sum_{k=0}^{n}\binom{n+k+1}{2k+1} i^{n-k}(x-2i)^k.$$

■

For example,

$$\begin{aligned}\sum_{k=0}^{3}\binom{4+k}{2k+1}i^{3-k}(x-2i)^k &= -4i-10(x-2i)+6i(x-2i)^2+(x-2i)^3\\ &= x^3+2x\\ &= f_4.\end{aligned}$$

Theorem 38.6 has an interesting consequence [430], as the next corollary shows.

Corollary 38.8. *Let x be a positive real number, $\Delta = \Delta(x) = \sqrt{x^2+4}$, and $\theta_k = (n-k)\pi/2 - k\arccos(x/\Delta)$. Then*

$$f_{n+1} = \sum_{k=0}^{n}\binom{n+k+1}{2k+1}\Delta^k\cos\theta_k.$$

Proof. Since $i = e^{\pi i/2}$ and $x - 2i = \Delta e^{-i\arccos(x/\Delta)}$, it follows by Theorem 38.6 that

$$\begin{aligned}f_{n+1} &= \sum_{k=0}^{n}\binom{n+k+1}{2k+1}e^{(n-k)\pi i/2}\cdot\left[\Delta\, e^{-i\arccos(x/\Delta)}\right]^k\\ &= \sum_{k=0}^{n}\binom{n+k+1}{2k+1}\Delta^k e^{[(n-k)\pi i/2 - k\arccos(x/\Delta)]i}\\ &= \sum_{k=0}^{n}\binom{n+k+1}{2k+1}\Delta^k\cos\theta_k,\end{aligned}$$

as claimed. ■

For example,

$$\begin{aligned}f_4 &= \sum_{k=0}^{3}\binom{4+k}{2k+1}\Delta^k\cos\theta_k\\ &= 4\cos\theta_0 + 10\Delta\cos\theta_1 + 6\Delta^2\cos\theta_2 + \Delta^3\cos\theta_3\\ &= 4\cdot 0 + 10\Delta\left(-\frac{x}{\Delta}\right) + 6\Delta^2\left(\frac{4x}{\Delta^2}\right) + \Delta^3\left(\frac{4x^3}{\Delta^3} - \frac{3x}{\Delta}\right)\\ &= x^3+2x.\end{aligned}$$

This corollary has an interesting byproduct to the Pell family, as the following corollary shows.

Corollary 38.9. *Let $n \geq 1$. Then*

$$P_n = \sum_{\substack{k=0 \\ 3k \not\equiv 2n \pmod 4}}^{n-1} (-1)^{\lfloor (3k-2n+3)/4 \rfloor} 2^{\lfloor 3k/2 \rfloor} \binom{n+k}{2k+1}.$$

Proof. Since $\cos \pi/4 = \sqrt{2}/2$, $\theta_k = (n-k)\pi/2 - k \arccos(\sqrt{2}/2) = (n-k)\pi/2 - k\pi/4 = (2n-3k)\pi/4$. Since $f_n(2) = P_n$, it follows by Corollary 38.8 that

$$\begin{aligned} P_{n+1} &= \sum_{k=0}^{n} \binom{n+k+1}{2k+1} 2^{3k/2} \cos(3k-2n)\pi/4 \\ &= 2^n \sum_{k=0}^{n} \binom{n+k+1}{2k+1} A_{3k-2n}, \end{aligned} \tag{38.70}$$

where $A_j = 2^{j/2} \cos j\pi/4$.

Using the addition formula for the cosine function, it follows that $A_{4r} = (-1)^r 2^{2r} = A_{4r+1}$, $A_{4r+2} = 0$, and $A_{4r+3} = (-1)^{r+1} 2^{2r+1}$, where r is any integer. Combining these equations, we get an explicit formula for A_j:

$$A_j = \begin{cases} (-1)^{\lfloor (j+1)/4 \rfloor} 2^{\lfloor j/2 \rfloor} & \text{if } j \not\equiv 2 \pmod 4 \\ 0 & \text{otherwise.} \end{cases}$$

Consequently,

$$\begin{aligned} A_{3k-2n} &= \begin{cases} (-1)^{\lfloor (3k-2n+1)/4 \rfloor} 2^{\lfloor (3k-2n)/2 \rfloor} & \text{if } 3k-2n \not\equiv 2 \pmod 4 \\ 0 & \text{otherwise} \end{cases} \\ &= \begin{cases} (-1)^{\lfloor (3k-2n+1)/4 \rfloor} 2^{\lfloor 3k/2 \rfloor - n} & \text{if } 3k-2n \not\equiv 2 \pmod 4 \\ 0 & \text{otherwise.} \end{cases} \end{aligned}$$

Substituting for A_{3k-2n} in equation (38.70), and then replacing n with $n-1$, we get the desired result. ■

For example,

$$\begin{aligned} P_5 &= \sum_{\substack{k=0 \\ 3k \not\equiv 2 \pmod 4}}^{4} (-1)^{\lfloor (3k-7)/4 \rfloor} 2^{\lfloor 3k/2 \rfloor} \binom{5+k}{2k+1} \\ &= 5 - 40 + 128 - 64 = 29. \end{aligned}$$

The next result also exemplifies the power of generating functions in extracting Fibonacci properties [431, 435].

Theorem 38.7 (Seiffert, 1998 [431]). *Define the sequence* $\{c_k\}_{k\geq 1}$ *by*

$$c_k = \begin{cases} 1 & \text{if } k \equiv 2 \pmod 5 \\ -1 & \text{if } k \equiv 3 \pmod 5 \\ 0 & \text{otherwise.} \end{cases}$$

Let $n \geq 1$. *Then*

$$\frac{1}{n}\sum_{k=1}^{n} k\binom{2n}{n-k}c_k = F_{2n-2}; \tag{38.71}$$

$$\frac{1}{2n-1}\sum_{k=1}^{2n-1}(-1)^k k\binom{4n-2}{2n-k-1}c_k = 5^{n-1}F_{2n-2}; \tag{38.72}$$

$$\frac{1}{2n}\sum_{k=1}^{2n}(-1)^k k\binom{4n}{2n-k}c_k = 5^{n-1}L_{2n-1}. \tag{38.73}$$

Proof. Recall from Section 31.5 that

$$\sum_{k=1}^{\infty} f_k(x)z^k = \frac{z}{1-xz-z^2}.$$

(The series converges when $|z| < \min\{\alpha(x), \beta(x)\}$.) Thus

$$\sum_{k=1}^{\infty} f_k(\alpha i)(zi)^k = \frac{zi}{1+\alpha z+z^2}$$

$$\sum_{k=1}^{\infty} f_k(\beta i)(zi)^k = \frac{zi}{1+\beta z+z^2}$$

$$\sum_{k=1}^{\infty}[f_k(\beta i) - f_k(\alpha i)](zi)^k = \frac{zi}{1+\beta z+z^2} - \frac{zi}{1+\alpha z+z^2}$$

$$\sum_{k=1}^{\infty}\frac{i^{k-1}}{\sqrt{5}}[f_k(\beta i) - f_k(\alpha i)]z^k = \frac{z^2}{1+z+z^2+z^3+z^4}.$$

Clearly, the sequence $\{c_k\}$ is generated by

$$\begin{aligned}\sum_{k=1}^{\infty} c_k z^k &= \sum_{j=0}^{\infty}\left(z^{5j+2} - z^{5j+3}\right) \\ &= \frac{(1-z)z^2}{1-z^5} \\ &= \frac{z^2}{1+z+z^2+z^3+z^4}\end{aligned}$$

and the series converges when $|z| < 1$. Thus

$$\sum_{k=1}^{\infty} \frac{i^{k-1}}{\sqrt{5}}[f_k(\beta i) - f_k(\alpha i)]z^k = \sum_{k=1}^{\infty} c_k z^k.$$

Equating the coefficients of z^k from both sides, we get

$$c_k = \frac{i^{k-1}}{\sqrt{5}}[f_k(\beta i) - f_k(\alpha i)].$$

It follows by PMI that $f_{2k}(2i) = ki^{k-1}$, where $k \geq 1$. Letting $y = 2i$, $\dfrac{xy+4}{x-y} = \dfrac{2(2+xi)}{x-2i} = 2i$. So, by equation (38.45), we have

$$\sum_{k=1}^{n} \binom{2n}{n-k} f_k(x) \cdot ki^{k-1} = (x-2i)^{n-1} \cdot ni^{n-1}$$

$$\sum_{k=1}^{n} k\binom{2n}{n-k} i^{k-1} f_k(x) = n(2+xi)^{n-1} \tag{38.74}$$

$$\begin{aligned}\sum_{k=1}^{n} k\binom{2n}{n-k} c_k &= \sum_{k=1}^{n} \frac{k}{\sqrt{5}}\binom{2n}{n-k}[f_k(\beta i) - f_k(\alpha i)]\\ &= \frac{n}{\sqrt{5}}[(2-\beta)^{n-1} - (2-\alpha)^{n-1}]\\ &= n(\alpha^{2n-2} - \beta^{2n-2})/\sqrt{5}\\ &= nF_{2n-2}.\end{aligned}$$

This gives identity (38.71).

To Prove Identities (38.72) and (38.73):
Since $f_k(-x) = (-1)^{k-1} f_k(x)$, by identity (38.74) we also have

$$\begin{aligned}\sum_{k=1}^{n}(-1)^k k\binom{2n}{n-k} c_k &= \sum_{k=1}^{n} k\binom{2n}{n-k}\frac{i^{k-1}}{\sqrt{5}}[f_k(-\alpha i) - f_k(-\beta i)]\\ &= \frac{n}{\sqrt{5}}[(2+\alpha)^{n-1} - (2+\beta)^{n-1}].\end{aligned}$$

But $2 + \alpha = \sqrt{5}\alpha$ and $2 + \beta = -\sqrt{5}\beta$. Then replacing n with $2n - 1$, we get identity (38.72); and replacing n with $2n$, identity (38.73). ■

For example, we have

$$\frac{1}{5}\sum_{k=1}^{5}(-1)^k k\binom{10}{5-k}c_k = \frac{1}{5}\left[2\binom{10}{3}c_2 - 3\binom{10}{2}c_3\right]$$
$$= 75 = 5^2F_4;$$
$$\frac{1}{6}\sum_{k=1}^{6}(-1)^k k\binom{12}{6-k}c_k = \frac{1}{6}\left[2\binom{12}{3}c_2 - 3\binom{12}{2}c_3\right]$$
$$= 275 = 5^2L_5.$$

Byproducts of Identity (38.74)

Identity (38.74) has three interesting consequences:

For the first, when $x = 2i$, identity (38.74) yields

$$\sum_{k=1}^{n} k\binom{2n}{n-k}i^{k-1}ki^{k-1} = n(2 + i \cdot 2i)^{n-1}$$
$$\sum_{k=1}^{n}(-1)^k\binom{2n}{n-k}k^2 = 0. \tag{38.75}$$

As an example, we have

$$\sum_{k=1}^{n}(-1)^k\binom{2n}{n-k}k^2 = -\binom{10}{4}1^2 + \binom{10}{3}2^2 - \binom{10}{2}3^2 + \binom{10}{1}4^2 - \binom{10}{0}5^2$$
$$= -210 + 480 - 405 + 160 - 25 = 0.$$

To extract the next two identities, we let $x = 2$. By *De Moivre's theorem*, identity (38.74) then yields

$$\sum_{k=1}^{n} k\binom{2n}{n-k}i^{k-1}P_k = n2^{n-1}(1+i)^{n-1}$$
$$= n2^{3(n-1)/2}[\cos(n-1)\pi/4 + i\sin(n-1)\pi/4].$$

Equating the real and imaginary parts from both sides, we get

$$\sum_{\substack{k=1 \\ k\equiv 1 \pmod 2}}^{n} k\binom{2n}{n-k}i^{k-1}P_k = n2^{3(n-1)/2}\cos(n-1)\pi/4 \tag{38.76}$$
$$\sum_{\substack{k=1 \\ k\equiv 0 \pmod 2}}^{n} k\binom{2n}{n-k}i^{k-2}P_k = n2^{3(n-1)/2}\sin(n-1)\pi/4. \tag{38.77}$$

For example, we have

$$\sum_{\substack{k=1\\k\equiv 1 \pmod 2}}^{6} k\binom{12}{6-k} i^{k-1}P_k = 1\binom{12}{5}P_1 - 3\binom{12}{3}P_3 + 5\binom{12}{1}P_5$$
$$= 792 - 3\cdot 220\cdot 5 + 5\cdot 12\cdot 29 = -768$$
$$= 6\cdot 2^7\sqrt{2}\cos 5\pi/4.$$

Similarly, $\displaystyle\sum_{\substack{k=1\\k\equiv 0 \pmod 2}}^{7} k\binom{14}{7-k} i^{k-2}P_k = -3584 = 7\cdot 2^9 \sin 3\pi/4.$

The next theorem gives two summation formulas, one involving f_{3k} and the other l_{3k} [438, 441]. The proof uses the Binet-like formulas, Lucas' formula, and a bit of messy algebra. It is somewhat long, so we break it up into seven steps.

Theorem 38.8 (Seiffert, 2002 [438]). *Let x be any complex number and $n \geq 0$. Then*

$$\sum_{k=0}^{\lfloor n/2\rfloor} \binom{n-k}{k} x^k f_{3k} = \frac{xf_{2n+1} - f_{2n} + (-x)^{n+2}f_n + (-x)^{n+1}f_{n-1}}{2x^2-1}; \quad (38.78)$$

$$\sum_{k=0}^{\lfloor n/2\rfloor} \binom{n-k}{k} x^k l_{3k} = \frac{xl_{2n+1} - l_{2n} + (-x)^{n+2}l_n + (-x)^{n+1}l_{n-1}}{2x^2-1}. \quad (38.79)$$

Proof. When each side is multiplied by $2x^2-1$, both sides become a polynomial in x. So it suffices to let x be a positive real number.

1) Recall that $2\alpha(x) = x + \Delta(x)$, $2\beta(x) = x - \Delta(x)$, $\alpha(x)\beta(x) = -1$. Let $y = \sqrt{\alpha(x)/x} - \sqrt{-x\beta(x)}$. Then

$$\Delta(y) = \sqrt{y^2+4}$$
$$= \sqrt{\frac{\alpha(x)}{x} - x\beta(x) + 2}$$
$$= \sqrt{\alpha(x)/x} + \sqrt{-x\beta(x)}$$
$$\alpha(y) = \frac{y+\Delta(y)}{2} = \sqrt{\alpha(x)/x}$$
$$\beta(y) = \frac{y-\Delta(y)}{2} = -\sqrt{-x\beta(x)}.$$

2) But

$$x\alpha(x)\Delta^2(y) = x\alpha(x)\left[\frac{\alpha(x)}{x} - x\beta(x) + 2\right]$$

$$
\begin{aligned}
&= \alpha^2(x) + x^2 + 2x\alpha(x) \\
&= [\alpha(x) + x]^2 \\
\Delta(y) &= \frac{\alpha(x) + x}{\sqrt{x\alpha(x)}}.
\end{aligned}
$$

Thus $\Delta(y) = \sqrt{\alpha(x)/x} + \sqrt{-x\beta(x)} = \dfrac{\alpha(x) + x}{\sqrt{x\alpha(x)}}$. Then $[\alpha(x) + x][\beta(x) + x] = \alpha(x)\beta(x) + x[\alpha(x) + \beta(x)] + x^2 = -1 + x^2 + x^2 = 2x^2 - 1$.

3) Using the Binet-like formula, we now have

$$
\begin{aligned}
f_{n+1}(y) &= \frac{\sqrt{x\alpha(x)}}{\alpha(x) + x}\left\{\left[\alpha(x)/x\right]^{(n+1)/2} - (-1)^{n+1}\left[-x\beta(x)\right]^{(n+1)/2}\right\} \\
&= \frac{\beta(x) + x}{2x^2 - 1}\left[\frac{\alpha^{(n+2)/2}(x)}{x^{n/2}} - (-1)^{n+1}x^{(n+2)/2}\alpha^{-n/2}(x)\right] \\
x^{n/2}\alpha^{3n/2}(x)f_{n+1}(y) &= \frac{\beta(x) + x}{2x^2 - 1}\left[\alpha^{2n+1}(x) - (-x)^{n+1}\alpha^n(x)\right] \\
&= \frac{x\alpha^{2n+1}(x) - \alpha^{2n}(x) + (-x)^{n+2}\alpha^n(x) + (-x)^{n+1}\alpha^{n-1}(x)}{2x^2 - 1}.
\end{aligned}
\tag{38.80}
$$

4) Since $2\beta(x) = x - \Delta(x)$, $8\beta^3(x) = 4[x^3 - (x^2 + 1)\Delta(x) + 3x]$. Since $\alpha(x) + \beta(x) = x$, this yields

$$
\begin{aligned}
\beta^3(x) &= \frac{x^3 - (x^2 + 1)\Delta(x) + 3x}{2} \\
&= (x^2 + 1)\beta(x) + x \\
&= x^2\beta(x) + [\beta(x) + x] \\
&= x^2\beta(x) - \alpha(x) + 2x \\
\frac{-\beta^3(x)}{x} &= \frac{\alpha(x)}{x} - x\beta(x) - 2 \\
&= y^2 \\
y &= \sqrt{-\beta^3(x)/x}.
\end{aligned}
$$

5) Then, by the Lucas-like formula, we have

$$
f_{n+1}(y) = \sum_{k=0}^{\lfloor n/2 \rfloor} \binom{n-k}{k} y^{n-2k}
$$

$$\begin{aligned}
x^{n/2}\alpha^{3n/2}(x)f_{n+1}(y) &= \sum_{k=0}^{\lfloor n/2\rfloor}\binom{n-k}{k}x^{n/2}\alpha^{3n/2}(x)\left[-\beta^3(x)/x\right]^{(n-2k)/2}\\
&= \sum_{k=0}^{\lfloor n/2\rfloor}\binom{n-k}{k}x^{n/2-(n-2k)/2}[\alpha(x)\beta(x)]^{3n/2}(-1)^{(n-2k)/2}\beta^{-3k}(x)\\
&= \sum_{k=0}^{\lfloor n/2\rfloor}\binom{n-k}{k}x^k(-1)^{[3n+(n-2k)+6k]/2}\alpha^{3k}(x)\\
&= \sum_{k=0}^{\lfloor n/2\rfloor}\binom{n-k}{k}x^k\alpha^{3k}(x). \qquad (38.81)
\end{aligned}$$

6) Combining equations (38.80) and (38.81), we get

$$\sum_{k=0}^{\lfloor n/2\rfloor}\binom{n-k}{k}x^k\alpha^{3k}(x) = \frac{x\alpha^{2n+1}(x)-\alpha^{2n}(x)+(-x)^{n+2}\alpha^n(x)+(-x)^{n+1}\alpha^{n-1}(x)}{2x^2-1}. \qquad (38.82)$$

7) By the Binet-like formulas, $2\alpha^j(x) = l_j(x) + \Delta\, f_j(x)$. Substituting this in equation (38.82), and then equating the rational and irrational parts, we get the desired results, as claimed. ■

We now illustrate both identities with $n = 2$:

$$\begin{aligned}
\sum_{k=0}^{1}\binom{2-k}{k}x^k f_{3k} &= f_0 + xf_3\\
&= x^3 + x\\
&= (2x^2-1)(x^3+3x)\\
&= x(x^4+3x^2+1) - (x^3+2x) + x^4\cdot x - x^3\\
&= xf_5 - f_4 + x^4 f_2 - x^3 f_1;
\end{aligned}$$

$$\begin{aligned}
\sum_{k=0}^{1}\binom{2-k}{k}x^k l_{3k} &= l_0 + xl_3\\
&= x^4 + 3x^2 + 2\\
&= \frac{x(x^5+5x^3+5x) - (x^4+4x^2+2) + x^4(x^2+2) - x^3\cdot x}{2x^2-1}\\
&= \frac{xl_5 - l_4 + x^4 l_2 - x^3 l_1}{2x^2-1}.
\end{aligned}$$

The next corollary reveals some implications of Theorem 38.8 to the extended Fibonacci family.

Corollary 38.10. *Let x be any real number and $n \geq 0$. Then*

$$\sum_{k=0}^{\lfloor n/2 \rfloor} \binom{n-k}{k} (2x)^k p_{3k} = \frac{2xp_{2n+1} - p_{2n} + (-2x)^{n+2} p_n + (-2x)^{n+1} p_{n-1}}{8x^2 - 1};$$

$$\sum_{k=0}^{\lfloor n/2 \rfloor} \binom{n-k}{k} (2x)^k q_{3k} = \frac{2xq_{2n+1} - q_{2n} + (-2x)^{n+2} q_n + (-2x)^{n+1} q_{n-1}}{8x^2 - 1};$$

$$\sum_{k=0}^{\lfloor n/2 \rfloor} \binom{n-k}{k} F_{3k} = F_{2n-1} + (-1)^n F_{n-2};$$

$$\sum_{k=0}^{\lfloor n/2 \rfloor} \binom{n-k}{k} L_{3k} = L_{2n-1} + (-1)^n L_{n-2};$$

$$\sum_{k=0}^{\lfloor n/2 \rfloor} \binom{n-k}{k} 2^k P_{3k} = \frac{1}{7}[2P_{2n+1} - P_{2n} + (-2)^{n+2} P_n + (-2)^{n+1} P_{n-1}];$$

$$\sum_{k=0}^{\lfloor n/2 \rfloor} \binom{n-k}{k} 2^k Q_{3k} = \frac{1}{7}[2Q_{2n+1} - Q_{2n} + (-2)^{n+2} Q_n + (-2)^{n+1} Q_{n-1}]. \qquad \blacksquare$$

The next theorem also exemplifies the close link between Fibonacci polynomials and trigonometry [445, 449]. Its proof uses the Binet-like formula for f_k, the binomial theorem, the imaginary number i, and Euler's formula.

Theorem 38.9 (Seiffert, 2004 [445]). *Let x be any real number and $n \geq 0$. Then*

$$\sum_{k=0}^{2n} (-1)^{\lfloor k/2 \rfloor} \binom{2n}{k} f_k = \sqrt{2}(-\Delta)^n f_n \cos(ny + \pi/4); \qquad (38.83)$$

$$\sum_{k=0}^{2n} (-1)^{\lceil k/2 \rceil} \binom{2n}{k} f_k = \sqrt{2}(-\Delta)^n f_n \sin(ny + \pi/4), \qquad (38.84)$$

where $\Delta = \sqrt{x^2+4}$ *and* $y = \arccos \frac{x}{\Delta}$.

Proof. Let $\alpha = \alpha(x)$, $\beta = \beta(x)$, and $S_n(x) = \sum_{j=0}^{2n} (-1)^j \binom{2n}{j} f_j(x)$. By the Binet-like formula, we then have

$$\begin{aligned} S_n(x) &= \frac{1}{\Delta} \sum_{j=0}^{2n} (-1)^j \binom{2n}{j} \left[(-i\alpha)^j - (-i\beta)^j\right] \\ &= \frac{1}{\Delta} \left[(1 - i\alpha)^{2n} - (1 - i\beta)^{2n}\right]. \end{aligned}$$

Since $1 - \alpha^2 = -x\alpha$, $(1 - i\alpha)^2 = 1 - 2i\alpha - \alpha^2 = -\alpha(x + 2i)$. Similarly, $(1 - i\beta)^2 = -\beta(x + 2i)$. Consequently,

$$\begin{aligned} S_n(x) &= \frac{1}{\Delta}\left\{[-\alpha(x+2i)]^n - [-\beta(x+2i)]^n\right\} \\ &= \frac{1}{\Delta}[(-1)^n(x+2i)^n(\alpha^n - \beta^n)] \\ &= (-1)^n(x+2i)^n f_n. \end{aligned}$$

Since $\cos y = \dfrac{x}{\Delta}$, $\sin y = \dfrac{2}{\Delta}$. Then, by Euler's formula, $(x + 2i)^n = \Delta^n e^{iny}$. So $S_n(x) = (-1)^n f_n e^{iny}$.

Equating the real and imaginary parts, we get

$$\sum_{k=0}^{n}(-1)^k\binom{2n}{2k} f_{2k} = (-1)^n\Delta^n f_n \cos(ny); \tag{38.85}$$

$$\sum_{k=0}^{n-1}(-1)^{k+1}\binom{2n}{2k+1} f_{2k+1} = (-1)^n\Delta^n f_n \sin(ny). \tag{38.86}$$

Subtracting equation (38.86) and (38.85) yields formula (38.83):

$$\begin{aligned} \sum_{k=0}^{2n}(-1)^{\lfloor k/2\rfloor}\binom{2n}{k} f_k &= (-1)^n\Delta^n f_n[\cos(ny) - \sin(ny)] \\ &= \sqrt{2}(-\Delta)^n f_n \cos(ny + \pi/4). \end{aligned}$$

Likewise, adding equation (38.86) and (38.85) yields formula (38.84). ■

For example, let $n = 2$. Then

$$\begin{aligned} \sum_{k=0}^{4}(-1)^{\lfloor k/2\rfloor}\binom{4}{k} f_k &= f_0 + 4f_1 - 6f_2 - 4f_3 + f_4 \\ &= 0 + 4 - 6x - 4(x^2+1) + (x^3 + 2x) \\ &= x^3 - 4x^2 - 4x. \end{aligned}$$

Since $\cos y = \dfrac{x}{\Delta}$, $\sin y = \dfrac{2}{\Delta}$, $\cos 2y = 2\cos^2 y - 1 = \dfrac{x^2 - \Delta}{\Delta^2}$, $\sin 2y = 2\sin y \cos y = \dfrac{4x}{\Delta}$, and $\cos(2y + \pi/4) = \cos 2y \cos\pi/4 - \sin 2y \sin\pi/4 = \dfrac{x^2 - 4x - 4}{\sqrt{2}\Delta^2}$. Then

$$\text{RHS} = \sqrt{2}\Delta^2 f_2 \cdot \frac{x^2 - 4x - 4}{\sqrt{2}\Delta^2} = x^3 - 4x^2 - 4x = \text{LHS}.$$

Similarly,

$$\sum_{k=0}^{4}(-1)^{\lceil k/2\rceil}\binom{4}{k}f_k = x^3+4x^2-4x = \sqrt{2}\Delta^2 f_2\cdot\frac{x^2+4x-4}{\sqrt{2}\Delta^2}.$$

Theorem 38.9 also has interesting byproducts, as the following corollary shows.

Corollary 38.11. *Let x be any real number and* $n \geq 0$. *Then*

$$\sum_{k=0}^{2n}(-1)^{\lfloor k/2\rfloor}\binom{2n}{k}p_k = \sqrt{2}(-2D)^n p_n(x)\cos(ny+\pi/4); \tag{38.87}$$

$$\sum_{k=0}^{2n}(-1)^{\lceil k/2\rceil}\binom{2n}{k}p_k = \sqrt{2}(-2D)^n p_n(x)\sin(ny+\pi/4); \tag{38.88}$$

$$\sum_{k=0}^{2n}(-1)^{\lfloor k/2\rfloor}\binom{2n}{k}F_k = \sqrt{2\cdot 5^n}(-1)^n F_n\cos(nz+\pi/4);$$

$$\sum_{k=0}^{2n}(-1)^{\lceil k/2\rceil}\binom{2n}{k}F_k = \sqrt{2\cdot 5^n}(-1)^n F_n\sin(ny+\pi/4);$$

$$\sum_{k=0}^{2n}(-1)^{\lfloor k/2\rfloor}\binom{2n}{k}P_k = 2^{(3n+1)/2}(-1)^n P_n\cos(nz+\pi/4);$$

$$\sum_{k=0}^{2n}(-1)^{\lceil k/2\rceil}\binom{2n}{k}P_k = 2^{(3n+1)/2}(-1)^n P_n\sin(ny+\pi/4).$$

Here $D=\sqrt{x^2+1}$, $y=\arccos\frac{x}{D}$, *and* $z=\arccos\frac{1}{\sqrt{5}}$. ∎

In particular, we have

$$\sum_{k=0}^{4}(-1)^{\lfloor k/2\rfloor}\binom{4}{k}p_k = 8x^3-16x^2-8x = \sqrt{2}(-2D)^2(2x)\cdot\frac{4x^2-8x-4}{4\sqrt{2}D^2};$$

$$\sum_{k=0}^{4}(-1)^{\lceil k/2\rceil}\binom{4}{k}F_k = 1 = \sqrt{2\cdot 5^2}(-1)^2F_2\cdot\frac{1}{5\sqrt{2}};$$

$$\sum_{k=0}^{8}(-1)^{\lfloor k/2\rfloor}\binom{8}{k}P_k = -768 = 2^{13/2}(-1)^4\cos 5\pi/4.$$

Suppose we let $x = 1$. Then $y = \pi/4$. Then formulas (38.87) and (38.88) yield

$$\sum_{k=0}^{2n}(-1)^{\lfloor k/2\rfloor}\binom{2n}{k}P_k = 2^{(3n+1)/2}(-1)^n P_n \cos(n+1)\pi/4;$$

$$\sum_{k=0}^{2n}(-1)^{\lceil k/2\rceil}\binom{2n}{k}P_k = 2^{(3n+1)/2}(-1)^n P_n \sin(n+1)\pi/4.$$

For example,

$$\sum_{k=0}^{6}(-1)^{\lfloor k/2\rfloor}\binom{6}{k}P_k = 6\cdot 1 - 15\cdot 2 - 20\cdot 5 + 15\cdot 12 + 6\cdot 29 - 1\cdot 70$$
$$= 160 = 2^5(-1)^3\cdot 5\cos\pi.$$

Likewise, $\sum_{k=0}^{6}(-1)^{\lceil k/2\rceil}\binom{6}{k}P_k = 0 = 2^5(-1)^3\cdot 5\sin\pi.$

About a year after the discovery of Theorem 38.9, Seiffert developed an equally spectacular formula for $f_{n+1}(z) + if_n(z)$, where z is a complex variable; see Theorem 38.10 [446, 452]. Its proof uses the Binet-like formula and the property that $\frac{1}{y}f_{2j}(y) = i^{1-j}f_j(i(y^2+2))$. Using Euler's formula, Seiffert then deduced from it two interesting formulas for P_n.

Theorem 38.10 (Seiffert, 2005 [446]). *Let x be a complex number and $n \geq 0$. Then*

$$f_{n+1}(z) + if_n(z) = \frac{1}{4^n}\sum_{k=0}^{n}\binom{2n+1}{2k+1}(z-2i)^k(z+2i)^{n-k}; \tag{38.89}$$

$$P_n = 2^{-\lfloor n/2\rfloor}\sum_{\substack{0\le k\le n-1\\ n-2k\not\equiv 3 \pmod 4}}(-1)^{\lfloor (n-2k)/4\rfloor}\binom{2n-1}{2k+1}; \tag{38.90}$$

$$P_n = 2^{-\lfloor (n+1)/2\rfloor}\sum_{\substack{0\le k\le n\\ n\not\equiv 2k \pmod 4}}(-1)^{\lfloor (n-2k-1)/4\rfloor}\binom{2n+1}{2k+1}. \tag{38.91}$$

Proof.

1) Recall that $f_{2n+1}(z) = \dfrac{\alpha^{2n+1} - \beta^{2n+1}}{\Delta}$, where $\alpha = \alpha(z)$, $\beta = \beta(z)$, and $\Delta = \Delta(z) = \sqrt{z^2+4}$.

By the binomial theorem, we have

$$\begin{aligned}(x+y)^{2n+1}-(x-y)^{2n+1} &= \sum_{r=0}^{2n+1}\binom{2n+1}{r}x^{2n+1-r}[y^r-(-y)^r]\\ &= 2\sum_{k=0}^{n}\binom{2n+1}{2k+1}x^{2n+1-(2k+1)}y^{2k+1}\\ &= 2y\sum_{k=0}^{n}\binom{2n+1}{2k+1}x^{2n-2k}y^{2k}.\end{aligned}$$

With $x = z$ and $y = \Delta$, this implies

$$\begin{aligned}f_{2n+1}(z) &= \frac{2\Delta}{2^{2n+1}\Delta}\sum_{k=0}^{n}\binom{2n+1}{2k+1}(z^2+4)^k z^{2n-2k}\\ &= 4^{-n}\sum_{k=0}^{n}\binom{2n+1}{2k+1}(z^2+4)^k z^{2n-2k}. \qquad (38.92)\end{aligned}$$

So

$$\begin{aligned}f_{2n+1}(\sqrt{iz-2}) &= 4^{-n}\sum_{k=0}^{n}\binom{2n+1}{2k+1}(iz+2)^k(iz-2)^{n-k}\\ &= 4^{-n}\sum_{k=0}^{n}\binom{2n+1}{2k+1}i^k(z-2i)^k\cdot i^{n-k}(z+2i)^{n-k}\\ &= 4^{-n}i^n\sum_{k=0}^{n}\binom{2n+1}{2k+1}(z-2i)^k(z+2i)^{n-k}\\ (-i)^n f_{2n+1}(\sqrt{iz-2}) &= 4^{-n}\sum_{k=0}^{n}\binom{2n+1}{2k+1}(z-2i)^k(z+2i)^{n-k}. \qquad (38.93)\end{aligned}$$

We can now evaluate the LHS in a different way. First, letting $y = \sqrt{iz-2}$ in the equation $\frac{1}{y}f_{2j}(y) = i^{1-j}f_j(i(y^2+2))$, we get

$$\begin{aligned}\frac{1}{\sqrt{iz-2}}f_{2j}(\sqrt{iz-2}) &= i^{1-j}f_j(-z)\\ &= i^{1-j}\cdot(-1)^{j-1}f_j(z)\\ &= i^{j-1}f_j(z). \qquad (38.94)\end{aligned}$$

Since

$$\frac{1}{\sqrt{iz-2}}\left[f_{2n+2}(\sqrt{iz-2}) - f_{2n}(\sqrt{iz-2})\right] = f_{2n+1}(\sqrt{iz-2}),$$

by the Fibonacci recurrence, equation (38.94) then yields

$$\begin{aligned}(-i)^n f_{2n+1}(\sqrt{iz-2}) &= \frac{(-i)^n}{\sqrt{iz-2}}\left[f_{2n+2}(\sqrt{iz-2}) - f_{2n}(\sqrt{iz-2})\right]\\ &= (-i)^n[i^n f_{n+1}(z) - i^{n-1} f_n(z)]\\ &= f_{n+1}(z) + i f_n(z). \qquad (38.95)\end{aligned}$$

The desired result now follows by equations (38.93) and (38.95).

We now deduce formulas (38.90) and (38.91) from (38.89).

2) Letting $z = 2$ in formula (38.89), we get

$$\begin{aligned}P_{n+1} + iP_n &= \frac{1}{4^n}\sum_{k=0}^{n}\binom{2n+1}{2k+1}(2-2i)^k(2+2i)^{n-k}\\ &= \frac{1}{2^n}\sum_{k=0}^{n}\binom{2n+1}{2k+1}(1-i)^k(1+i)^{n-k}.\end{aligned}$$

Since $1 - i = \sqrt{2}e^{-\pi/4}$ and $1 + i = \sqrt{2}e^{\pi/4}$ by Euler's formula, this can be rewritten as

$$\begin{aligned}P_{n+1} + iP_n &= \frac{1}{2^n}\sum_{k=0}^{n}\binom{2n+1}{2k+1}2^{k/2}e^{-ik\pi/4}\cdot 2^{(n-k)/2}e^{-i(n-k)\pi/4}\\ &= 2^{-n/2}\sum_{k=0}^{n}\binom{2n+1}{2k+1}e^{i(n-2k)\pi/4}.\end{aligned}$$

Equating the real and imaginary parts, we get

$$P_{n+1} = 2^{-n/2}\sum_{k=0}^{n}\binom{2n+1}{2k+1}A_{n-2k}, \qquad (38.96)$$

$$P_n = 2^{-n/2}\sum_{k=0}^{n}\binom{2n+1}{2k+1}B_{n-2k}, \qquad (38.97)$$

where $A_j = \cos j\pi/4$, $B_j = \sin i\pi/4$, and j is an integer. Now

$$A_j = \begin{cases}(-1)^{\lfloor (j+1)/4\rfloor}2^{\lfloor j/2\rfloor - j/2} & \text{if } j \not\equiv 2 \pmod 4\\ 0 & \text{otherwise}\end{cases}$$

and

$$B_j = \begin{cases} (-1)^{\lfloor (j-1)/4 \rfloor} 2^{\lfloor j/2 \rfloor - j/2} & \text{if } j \not\equiv 0 \pmod 4 \\ 0 & \text{otherwise.} \end{cases}$$

So

$$A_{n-2k-1} = \begin{cases} (-1)^{\lfloor (n-2k)/4 \rfloor} 2^{\lfloor (n-2k-1)/2 \rfloor - (n-2k-1)/2} & \text{if } n-2k \not\equiv 3 \pmod 4 \\ 0 & \text{otherwise.} \end{cases}$$

Consequently,

$$\begin{aligned} 2^{-n/2} A_{n-2k-1} &= \begin{cases} (-1)^{\lfloor (n-2k)/4 \rfloor} 2^{-\lfloor (n+1)/2 \rfloor + 1} & \text{if } n-2k \not\equiv 3 \pmod 4 \\ 0 & \text{otherwise} \end{cases} \\ &= \begin{cases} (-1)^{\lfloor (n-2k)/4 \rfloor} 2^{-\lfloor n/2 \rfloor} & \text{if } n-2k \not\equiv 3 \pmod 4 \\ 0 & \text{otherwise.} \end{cases} \end{aligned}$$

Replacing n with $n-1$, equation (38.96) yields formula (38.90).

3) Since

$$\begin{aligned} &2^{-n/2} B_{n-2k-1} \\ &= \begin{cases} (-1)^{\lfloor (n-2k-1)/4 \rfloor} 2^{\lfloor (n-2k)/2 \rfloor - (n-2k)/2} \cdot 2^{-n/2} & \text{if } n-2k \not\equiv 3 \pmod 4 \\ 0 & \text{otherwise} \end{cases} \\ &= \begin{cases} (-1)^{\lfloor (n-2k-1)/4 \rfloor} 2^{-\lfloor (n+1)/2 \rfloor} & \text{if } n \not\equiv 2k \pmod 4 \\ 0 & \text{otherwise,} \end{cases} \end{aligned}$$

formula (38.91) follows from (38.97), as desired. ■

For example, let $n = 5$. Then, by formula (38.90), we have

$$\begin{aligned} P_5 &= 2^{\lfloor -5/2 \rfloor} \sum_{\substack{0 \le k \le 4 \\ 2k \not\equiv 2 \pmod 4}} (-1)^{\lfloor (5-2k)/4 \rfloor} \binom{9}{2k+1} \\ &= \frac{1}{4}\left[-\binom{9}{1} + \binom{9}{5} - \binom{9}{9}\right] \\ &= 29, \text{ as expected.} \end{aligned}$$

Similarly, using formula (38.91),

$$P_4 = 2^{-2} \sum_{\substack{0 \le k \le 4 \\ 2k \not\equiv 0 \pmod 4}} (-1)^{\lfloor (3-2k)/4 \rfloor} \binom{9}{2k+1}$$

$$= \frac{1}{4}\left[\binom{9}{3} - \binom{9}{7}\right]$$
$$= 12, \text{ again as expected.}$$

A Formula for p_{2n+1}

Letting $z = 2x$ in formula (38.92), we can obtain a formula for $p_{2n+1}(x)$, and hence for P_{2n+1}:

$$p_{2n+1} = \sum_{k=0}^{n} \binom{2n+1}{2k+1}(x^2+1)^k x^{2n-2k} \tag{38.98}$$

$$P_{2n+1} = \sum_{k=0}^{n} \binom{2n+1}{2k+1} 2^k. \tag{38.99}$$

By virtue of formula (38.99), we can compute P_{2n+1} using the odd-numbered binomial coefficients in row $2n+1$ of Pascal's triangle with weights 2^k, where $0 \le k \le n^{\dagger}$.

For example, $P_7 = \sum_{k=0}^{3} \binom{7}{2k+1} 2^k = \binom{7}{1}2^0 + \binom{7}{3}2^1 + \binom{7}{5}2^2 + \binom{7}{7}2^3 = 169.$

Formula (38.99), coupled with the identity $p_{n+1} + p_{n-1} = q_n$, yields a formula for q_{2n}:

$$q_{2n} = \sum_{k=0}^{n} \left[\binom{2n+1}{2k+1} + \binom{2n-1}{2k+1}\right](x^2+1)^{k+1} x^{2n-2k-2}. \tag{38.100}$$

This implies

$$Q_{2n} = \sum_{k=0}^{n} \left[\binom{2n+1}{2k+1} + \binom{2n-1}{2k+1}\right] 2^k.$$

A Formula for p_{2n}

Recall from Section 32.8 that

$$f_{2n}(z) = \frac{2}{4^n} \sum_{k=0}^{n-1} \binom{2n}{2k+1}(z^2+4)^k z^{2n-2k-1}.$$

This yields

$$p_{2n} = \sum_{k=0}^{n-1} \binom{2n}{2k+1}(x^2+1)^k x^{2n-2k-1} \tag{38.101}$$

$$P_{2n} = \sum_{k=0}^{n-1} \binom{2n}{2k+1} 2^k. \tag{38.102}$$

---

†For an alternate method, see [285].

The next theorem is a natural continuation of Theorem 38.10 [451, 455]. Here also the proof uses the Binet-like formulas, the Lucas-like formula for $f_{2n}(y)$, and the techniques from the proof of Theorem 38.10. For convenience, we divide it into seven small steps.

Theorem 38.11 (Seiffert, 2006 [451]). *Let x be a nonzero complex number and n a positive integer. Then*

$$\sum_{k=0}^{2n-1} \binom{4n-k-1}{k} 2^{4n-2k-1} x^k f_k = x^{2n-1} l_{2n-1}(x) f_{2n}(4/x); \tag{38.103}$$

$$\sum_{k=0}^{2n-1} \binom{4n-k-1}{k} 2^{4n-2k-1} x^k l_k = x^{2n-1}(x^2+4) f_{2n-1}(x) f_{2n}(4/x); \tag{38.104}$$

$$\sum_{k=0}^{2n} \binom{4n-k-1}{k} 2^{4n-2k+2} x^k f_k = x^{2n+1} f_{2n}(x) l_{2n+1}(4/x); \tag{38.105}$$

$$\sum_{k=0}^{2n-1} \binom{4n-k-1}{k} 2^{4n-2k+2} x^k l_k = x^{2n+1} l_{2n}(x) l_{2n+1}(4/x). \tag{38.106}$$

Proof. Recall that $2\alpha(x) = x + \Delta(x)$, $2\beta(x) = x - \Delta(x)$, $\alpha(x) + \beta(x) = x$, and $\alpha(x)\beta(x) = -1$.

1) Clearly, it suffices to establish the identities when x is a positive rational number such that $\Delta(x)$ is irrational. So we let $y = 2\sqrt{-\beta(x)/x}$.

2) Then

$$\begin{aligned}
\Delta(y) &= \sqrt{y^2+4} = 2\sqrt{\frac{-\beta(x)}{x} + 1} \\
&= 2\sqrt{\frac{x-\beta(x)}{x}} = 2\sqrt{\alpha(x)/x}; \\
\alpha(y) &= \frac{y+\Delta(y)}{2} \\
&= \frac{1}{\sqrt{x}}\left[\sqrt{-\beta(x)} + \sqrt{\alpha(x)}\right]; \\
\beta(y) &= \frac{y-\Delta(y)}{2} \\
&= \frac{1}{\sqrt{x}}\left[\sqrt{-\beta(x)} - \sqrt{\alpha(x)}\right].
\end{aligned}$$

3) Since $\Delta(4/x) = \frac{2}{x}\sqrt{x^2+4} = \frac{2\Delta(x)}{x}$, $\alpha(4/x) = \frac{2+\Delta(x)}{x}$. Similarly, $\beta(4/x) = \frac{2-\Delta(x)}{x}$. Then

$$\begin{aligned}\left[\sqrt{-\beta(x)}+\sqrt{\alpha(x)}\right]^2 &= \alpha(x)-\beta(x)+2\sqrt{-\alpha(x)\beta(x)}\\ &= 2+\Delta(x)\\ &= x\alpha(4/x);\\ \left[\sqrt{-\beta(x)}-\sqrt{\alpha(x)}\right]^2 &= \alpha(x)-\beta(x)-2\sqrt{-\alpha(x)\beta(x)}\\ &= \Delta(x)-2\\ &= -x\beta(4/x).\end{aligned}$$

4) By the Binet-like formula, we then have

$$\begin{aligned}f_{2n}(y) &= \frac{\alpha^{2n}(y)-\beta^{2n}(y)}{\Delta(y)}\\ &= \frac{\sqrt{x}}{2x^n\sqrt{\alpha(x)}}\left\{\left[\sqrt{-\beta(x)}+\sqrt{\alpha(x)}\right]^{2n}-\left[\sqrt{-\beta(x)}-\sqrt{\alpha(x)}\right]^{2n}\right\}\\ &= \frac{\sqrt{x}}{2x^n\sqrt{\alpha(x)}}\left\{[x\alpha(4/x)]^n-[-x\beta(4/x)]^n\right\}\\ &= \frac{\sqrt{x}}{2\sqrt{\alpha(x)}}\left\{[\alpha(4/x)]^n-[-\beta(4/x)]^n\right\}. \qquad (38.107)\end{aligned}$$

5) Since

$$\left[\sqrt{-\beta(x)}/x\right]^{2n-2k-1} = \left[\frac{1}{x\alpha(x)}\right]^{2n-2k-1} = \frac{[x\alpha(x)]^k\sqrt{x\alpha(x)}}{[x\alpha(x)]^n},$$

by the Lucas-like formula, we have

$$\begin{aligned}f_{2n}(y) &= \sum_{k=0}^{n-1}\binom{2n-k-1}{k}y^{2n-2k-1}\\ &= \frac{\sqrt{x\alpha(x)}}{[x\alpha(x)]^n}\sum_{k=0}^{n-1}\binom{2n-k-1}{k}2^{2n-2k-1}x^k\alpha^k(x). \qquad (38.108)\end{aligned}$$

6) Combining identities (38.107) and (38.108), we get

$$\sum_{k=0}^{n-1} \binom{2n-k-1}{k} 2^{2n-2k-1} x^k \alpha^k(x) = \frac{1}{2} x^n \alpha^{n-1}(x) \left\{ [\alpha(4/x)]^n - [-\beta(4/x)]^n \right\}.$$

7) Now replace n with $2n$, and $2\alpha^k(x)$ with $l_k(x) + \sqrt{x^2+4} f_k(x)$. Equating the rational and irrational parts of the resulting equation yields identities (38.103) and (38.104). Similarly, replacing n with $2n+1$ yields identities (38.105) and (38.106). ■

This theorem also has implications to the extended Fibonacci family, as the next corollary shows.

Corollary 38.12. *Let x be a nonzero complex number and n a positive integer. Then*

$$\sum_{k=0}^{2n-1} \binom{4n-k-1}{k} 2^{2n-k} x^k p_k = x^{2n-1} q_{2n-1}(x) p_{2n}(2/x);$$

$$\sum_{k=0}^{2n-1} \binom{4n-k-1}{k} 2^{2n-k-2} x^k q_k = x^{2n-1}(x^2+1) p_{2n-1}(x) p_{2n}(2/x);$$

$$\sum_{k=0}^{2n} \binom{4n-k+1}{k} 2^{2n-k+1} x^k p_k = x^{2n+1} p_{2n}(x) q_{2n+1}(2/x);$$

$$\sum_{k=0}^{2n} \binom{4n-k+1}{k} 2^{2n-k+1} x^k q_k = x^{2n+1} q_{2n}(x) q_{2n+1}(2/x);$$

$$\sum_{k=0}^{2n-1} \binom{4n-k-1}{k} 2^{4n-2k-1} F_k = L_{2n-1} f_{2n}(4);$$

$$\sum_{k=0}^{2n-1} \binom{4n-k-1}{k} 2^{4n-2k-1} L_k = 5F_{2n-1} f_{2n}(4);$$

$$\sum_{k=0}^{2n} \binom{4n-k+1}{k} 2^{4n-2k+2} F_k = F_{2n} l_{2n+1}(4);$$

$$\sum_{k=0}^{2n} \binom{4n-k+1}{k} 2^{4n-2k+2} L_k = L_{2n} l_{2n+1}(4);$$

$$\sum_{k=0}^{2n-1} \binom{4n-k-1}{k} 2^{2n-k} P_k = Q_{2n-1} P_{2n}(2);$$

$$\sum_{k=0}^{2n-1} \binom{4n-k-1}{k} 2^{2n-k-3} Q_k = P_{2n-1}p_{2n}(2);$$

$$\sum_{k=0}^{2n} \binom{4n-k+1}{k} 2^{2n-k+1} P_k = P_{2n}q_{2n+1}(2);$$

$$\sum_{k=0}^{2n} \binom{4n-k+1}{k} 2^{2n-k+1} Q_k = Q_{2n}q_{2n+1}(2).$$

■

38.3 LEGENDRE POLYNOMIALS

The next theorem establishes a close link between Fibonacci polynomials and the well-known *Legendre polynomials* $P_n(x)$, named after the French mathematician Adrien Marie Legendre (1752–1833). It also establishes a link between Fibonacci polynomials and Legendre polynomials [211, 440]. They are defined by the second-order recurrence $(n+1)P_{n+1}(x) = (2n+1)P_n(x) - nP_{n-1}(x)$, where $P_0(x) = 1$ and $P_1(x) = x$.

Table 38.3 gives the first eight Legendre polynomials. See Exercises 38.53–38.60 for some simple properties of Legendre polynomials.

TABLE 38.3. First Eight Legendre Polynomials

n	$P_n(x)$
1	x
2	$\frac{1}{2}(3x^2-1)$
3	$\frac{1}{2}(5x^3-3x)$
4	$\frac{1}{8}(35x^4-30x^2+3)$
5	$\frac{1}{8}(63x^5-70x^3+15x)$
6	$\frac{1}{16}(231x^6-315x^4+105x^2-5)$
7	$\frac{1}{16}(429x^7-693x^5+315x^3-35x)$
8	$\frac{1}{128}(6435x^8-12012x^6+6930x^4-1260x^2+35)$

We are now ready to present the links.

Theorem 38.12 (Seiffert, 2002 [440]). *Let $\Delta = \Delta(x) = \sqrt{x^2+4}$, where x is a real number. Then*

$$f_{2n} = \frac{x}{\Delta} \sum_{k=0}^{2n-1} P_k(\Delta/2)P_{2n-k-1}(\Delta/2), \quad n \geq 1; \qquad (38.109)$$

$$l_{2n+1} = x \sum_{k=0}^{2n} P_k(\Delta/2)P_{2n-k}(\Delta/2), \quad n \geq 0. \tag{38.110}$$

Proof. Let $\alpha = \alpha(x)$ and $\beta = \beta(x)$. Now consider the power series

$$g(z) = \sum_{r=1}^{\infty}[\alpha^r - (-1)^r\beta^r]z^{r-1}. \tag{38.111}$$

Since $\alpha + \beta = x$, $\alpha\beta = -1$, $\alpha - \beta = \Delta(x) = \Delta$, and $\dfrac{1}{1-t} = \sum_{n=0}^{\infty} t^n$, we can rewrite this as

$$\begin{aligned} g(z) &= \alpha \sum_{r=0}^{\infty}(\alpha z)^r + \beta \sum_{r=0}^{\infty}(-\beta z)^r \\ &= \frac{\alpha}{1-\alpha z} + \frac{\beta}{1+\beta z} \\ &= \frac{x}{1-\Delta z + z^2}. \end{aligned}$$

Using the Binet-like formulas for f_n and l_n, we can rewrite the power series as

$$g(z) = \Delta \sum_{n=1}^{\infty} f_{2n}z^{2n-1} + \sum_{n=0}^{\infty} l_{2n+1}z^{2n}.$$

Thus

$$\frac{x}{1-\Delta z + z^2} = \Delta \sum_{n=1}^{\infty} f_{2n}z^{2n-1} + \sum_{n=0}^{\infty} l_{2n+1}z^{2n}. \tag{38.112}$$

Legendre polynomials are generated by [408]

$$\sum_{r=0}^{\infty} P_r(x)z^r = \frac{1}{\sqrt{1-2xz+z^2}}.$$

Squaring both sides, we get

$$\begin{aligned} \frac{1}{1-2xz+z^2} &= \left[\sum_{r=0}^{\infty} P_r(x)z^r\right]^2 \\ &= \sum_{r=0}^{\infty}\left[\sum_{k=0}^{r} P_k(x)P_{r-k}(x)z^r\right]. \end{aligned}$$

Replacing x with $\Delta/2$, this yields

$$\frac{1}{1-\Delta z+z^2}=\sum_{r=0}^{\infty}\left[\sum_{k=0}^{r}P_k(\Delta/2)P_{r-k}(\Delta/2)z^r\right]$$
$$\frac{x}{1-\Delta z+z^2}=\sum_{r=0}^{\infty}\left[x\sum_{k=0}^{r}P_k(\Delta/2)P_{r-k}(\Delta/2)z^r\right]. \tag{38.113}$$

Formulas (38.109) and (38.110) now follow by equating the corresponding coefficients from equations (38.112) and (38.113). ■

For example,

$$\begin{aligned}
f_6 &= \frac{x}{\Delta}\sum_{k=0}^{5}P_k(\Delta/2)P_{5-k}(\Delta/2)\\
&= \frac{2x}{\Delta}[P_0(\Delta/2)P_5(\Delta/2)+P_1(\Delta/2)P_4(\Delta/2)+P_2(\Delta/2)P_3(\Delta/2)]\\
&= \frac{2x}{\Delta}\cdot\frac{\Delta}{256}[(63x^4+224x^2+128)+(35x^4+160x^2+128)+2(15x^4+64x^2+64)]\\
&= \frac{x}{128}(128x^4+512x^2+384)\\
&= x^5+4x^3+3x;
\end{aligned}$$

$$\begin{aligned}
l_5 &= x\sum_{k=0}^{4}P_k(\Delta/2)P_{4-k}(\Delta/2)\\
&= x[2P_0(\Delta/2)P_4(\Delta/2)+2P_1(\Delta/2)P_3(\Delta/2)+P_2^2(\Delta/2)]\\
&= \frac{x}{64}[(35x^4+160x^2+128)+(20x^4+112x^2+128)+(9x^4+48x^2+64)]\\
&= \frac{x}{64}(64x^4+320x^2+320)\\
&= x^5+5x^3+5x.
\end{aligned}$$

As we can predict, Theorem 38.12 also yields a number of interesting dividends, as the next two corollaries show.

Corollary 38.13. *Let x be any real number. Then*

$$p_{2n}=\frac{x}{\sqrt{x^2+1}}\sum_{k=0}^{2n-1}P_k(\sqrt{x^2+1})P_{2n-k-1}(\sqrt{x^2+1});$$

$$q_{2n+1}=2x\sum_{k=0}^{2n}P_k(\sqrt{x^2+1})P_{2n-k}(\sqrt{x^2+1});$$

$$F_{2n} = \frac{1}{\sqrt{5}} \sum_{k=0}^{2n-1} P_k(\sqrt{5}/2)P_{2n-k-1}(\sqrt{5}/2);$$

$$L_{2n+1} = \sum_{k=0}^{2n} P_k(\sqrt{5}/2)P_{2n-k}(\sqrt{5}/2); \tag{38.114}$$

$$P_{2n} = \frac{1}{\sqrt{2}} \sum_{k=0}^{2n-1} P_k(\sqrt{2})P_{2n-k-1}(\sqrt{2}); \tag{38.115}$$

$$Q_{2n+1} = \sum_{k=0}^{2n} P_k(\sqrt{2})P_{2n-k}(\sqrt{2}).$$

■

For example,

$$\begin{aligned} Q_5 &= \sum_{k=0}^{4} P_k(\sqrt{2})P_{4-k}(\sqrt{2}) \\ &= 2P_0(\sqrt{2})P_4(\sqrt{2}) + 2P_1(\sqrt{2})P_3(\sqrt{2}) + P_2^2(\sqrt{2}) \\ &= \frac{83}{4} + 14 + \frac{25}{4} = 41, \end{aligned}$$

as expected.

H.J. Hindin of Huntington Station, New York, discovered formulas (38.114) and (38.115) independently in 2001 [211].

Theorem 38.12 yields additional dividends if we let x be a complex number, as the following corollary reveals. We omit their proofs in the interest of brevity; see Exercises 38.61–38.68.

Corollary 38.14. *Let x be a complex number. Then*

$$F_{6n} = \frac{4\sqrt{5}}{5} \sum_{k=0}^{2n-1} P_k(\sqrt{5})P_{2n-k-1}(\sqrt{5}), \quad n \geq 1; \tag{38.116}$$

$$F_{4n} = 2\sqrt{2} \sum_{k=0}^{2n-1} P_k(3/2)P_{2n-k-1}(3/2), \quad n \geq 1; \tag{38.117}$$

$$F_{4n} = \frac{3(-i)^{2n-1}}{\sqrt{5}} \sum_{k=0}^{2n-1} P_k(\sqrt{5}i/2)P_{2n-k-1}(\sqrt{5}i/2), \quad n \geq 1; \tag{38.118}$$

$$L_{6n+3} = 4 \sum_{k=0}^{2n} P_k(\sqrt{5})P_{2n-k}(\sqrt{5}), \quad n \geq 0; \tag{38.119}$$

$$F_{4n+2} = \sum_{k=0}^{2n} P_k(3/2)P_{2n-k}(3/2), \quad n \geq 0; \tag{38.120}$$

$$L_{4n+2} = 3(-1)^n \sum_{k=0}^{2n} P_k(\sqrt{5}i/2)P_{2n-k}(\sqrt{5}i/2), \quad n \geq 0; \tag{38.121}$$

$$2\sin 2n\pi/3 = \sum_{k=0}^{2n-1} P_k(\sqrt{3}/2)P_{2n-k-1}(\sqrt{3}/2), \quad n \geq 1; \tag{38.122}$$

$$2\sin(2n+1)\pi/3 = \sum_{k=0}^{2n} P_k(\sqrt{3}/2)P_{2n-k}(\sqrt{3}/2), \quad n \geq 0. \tag{38.123}$$

■

EXERCISES 38

Prove each.

1. $g_{2n+4} = (x^2+2)g_{2n+2} - g_{2n}$.
2. Identity (38.7).
3. Identity (38.8).
4. Identity (38.11).
5. Identity (38.13).
6. Identity (38.14).
7. Solve the recurrence $(xy-4)A_{n-1} + (x+y)^2A_{n-2} = A_n$, where $A_0 = 1$, $A_1 = 1$, and $n \geq 2$.

Prove each.

8. $2f_n(4) = F_{3n}$.
9. $f_n(3i) = i^{n-1}F_{2n}$.
10. $3f_{2n}(\sqrt{5}) = \sqrt{5}F_{4n}$.
11. $3f_{2n-1}(\sqrt{5}) = L_{4n-2}$.
12. $5^{(n-1)/2}f_n(4/\sqrt{5}) = [5^n - (-1)^n]/6$.
13. Identity (38.21).
14. Identity (38.24).

15–27. Identities (38.28)–(38.40).

28–31. Identities (38.41)–(38.44).

32–34. Identities (38.49)–(38.51).

35. Identity (38.53).

36–47. Identities (38.55)–(38.66).

48. $L_{n-1} + F_{n-1} = 2F_n$.
49. $L_{n-1} + 5F_{n-1} = 2L_n$.
50. Using formula (38.90), compute P_4 and P_7.
51. Using formula (38.91), compute P_5 and P_6.
52. Derive formula (38.100).

In Exercises 38.53–38.60, $P_n(x)$ denotes the nth Legendre polynomial. Prove each.

53. $P_n(1) = 1$.
54. $P_n(-1) = (-1)^n$.
55. The degree of $P_n(x)$ is n.
56. The leading coefficient in $P_n(x)$ is $A_n = \dfrac{(2n)!}{2^n(n!)^2}$.
57. $A_n = \dfrac{n+1}{2^n} C_n$, where C_n denotes the nth Catalan number [278].
58. $P'_n(x) = xP'_{n-1}(x) + nP_{n-1}(x)$, where the prime denotes differentiation with respect to x.
59. $P'_n(1) = t_n$, where t_n denotes the nth triangular number $n(n+1)/2$ and $n \geq 1$.
60. $P'_n(-1) = (-1)^{n-1} t_n$, where $n \geq 1$.

61–68. Identities (38.116)– (38.123).

39

GIBONACCI DETERMINANTS

It is not enough to have a good mind.
The main thing is to use it well.
–René Descartes (1596–1650)

In this chapter, we investigate a number of determinants with Fibonacci and Lucas implications. As in previous chapters, we *omit* the arguments in the functional notation when they are *not* needed for clarity.

39.1 A CIRCULANT DETERMINANT

A *circulant* matrix is an $n \times n$ square matrix such that row $i + 1$ is a cyclic shift of row i by one position to the right (or left), and row 1 is a cyclic shift of row n in the same direction, where $1 \leq i \leq n$. We begin our study with such a matrix that has applications to the extended Fibonacci family, as we will see shortly.

Example 39.1. *Let*

$$M = \begin{bmatrix} a & b & b+a \\ a+b & a & b \\ b & a+b & a \end{bmatrix},$$

where $a = a(x)$ and $b = b(x)$ are polynomials with complex coefficients. Prove that $|M| = 2(a^3 + b^3)$, where $|M|$ denotes the determinant of the matrix M.

Fibonacci and Lucas Numbers with Applications, Volume Two. Thomas Koshy.

Proof. Using elementary operations, we get

$$\begin{aligned}
|M| &= (2a+2b)\begin{vmatrix} 1 & b & b+a \\ 1 & a & b \\ 1 & a+b & a \end{vmatrix} \\
&= (2a+2b)\begin{vmatrix} 1 & b & b+a \\ 0 & a-b & -a \\ 0 & a & -b \end{vmatrix} \\
&= (2a+2b)(a^2-ab+b^2) \\
&= 2(a^3+b^3),
\end{aligned}$$

as desired. ∎

Suppose we let $a = xg_{n+1}$ and $b = g_n$. Then

$$\begin{vmatrix} xg_{n+1} & g_n & g_{n+2} \\ g_{n+2} & xg_{n+1} & g_n \\ g_n & g_{n+2} & xg_{n+1} \end{vmatrix} = 2\left(x^3g_{n+1}^3 + g_n^3\right).$$

In particular, we have

$$\begin{vmatrix} xf_{n+1} & f_n & f_{n+2} \\ f_{n+2} & xf_{n+1} & f_n \\ f_n & f_{n+2} & xf_{n+1} \end{vmatrix} = 2\left(x^3f_{n+1}^3 + f_n^3\right);$$

$$\begin{vmatrix} xl_{n+1} & l_n & l_{n+2} \\ l_{n+2} & xl_{n+1} & l_n \\ l_n & l_{n+2} & xl_{n+1} \end{vmatrix} = 2\left(x^3l_{n+1}^3 + l_n^3\right);$$

$$\begin{vmatrix} 2xp_{n+1} & p_n & p_{n+2} \\ p_{n+2} & 2xp_{n+1} & p_n \\ p_n & p_{n+2} & 2xp_{n+1} \end{vmatrix} = 2\left(8x^3p_{n+1}^3 + p_n^3\right);$$

$$\begin{vmatrix} 2xq_{n+1} & q_n & q_{n+2} \\ q_{n+2} & 2xq_{n+1} & q_n \\ q_n & q_{n+2} & 2xq_{n+1} \end{vmatrix} = 2\left(8x^3q_{n+1}^3 + q_n^3\right);$$

$$\begin{vmatrix} 2P_{n+1} & P_n & P_{n+2} \\ P_{n+2} & 2P_{n+1} & P_n \\ P_n & P_{n+2} & 2P_{n+1} \end{vmatrix} = 2\left(8P_{n+1}^3 + P_n^3\right);$$

$$\begin{vmatrix} 2Q_{n+1} & Q_n & Q_{n+2} \\ Q_{n+2} & 2Q_{n+1} & Q_n \\ Q_n & Q_{n+2} & 2Q_{n+1} \end{vmatrix} = 2\left(8Q_{n+1}^3 + Q_n^3\right).$$

For example,

$$\begin{vmatrix} L_7 & L_6 & L_8 \\ L_8 & L_7 & L_6 \\ L_6 & L_8 & L_7 \end{vmatrix} = 60{,}442 = 2\left(L_7^3 + L_6^3\right);$$

$$\begin{vmatrix} 2Q_7 & Q_6 & Q_8 \\ Q_8 & 2Q_7 & Q_6 \\ Q_6 & Q_8 & 2Q_7 \end{vmatrix} = 220{,}371{,}302 = 2\left(8Q_7^3 + Q_6^3\right).$$

J.M. Patel of Ahmedabad, India, studied the Fibonacci and Lucas special cases for $x = 1$ in 2006 [378].

39.2 A HYBRID DETERMINANT

Next we study an interesting hybrid determinant. It is based on the determinant

$$|M| = \begin{vmatrix} a^2+b^2-c^2-d^2 & 2(bc-ad) & 2(ac+bd) \\ 2(ad+bc) & a^2-b^2+c^2-d^2 & 2(cd-ab) \\ 2(bd-ac) & 2(ab+cd) & a^2-b^2-c^2+d^2 \end{vmatrix},$$

where a, b, c, and d are integer polynomial functions of x. C.W. Trigg of San Diego, California, investigated it in 1970 [494].

To evaluate $|M|$, we use the following facts from the theory of determinants, where A and B are square matrices of the same order:

- $|AB| = |A| \cdot |B|$; and
- $|A^{\mathrm{T}}| = |A|$, where A^{T} denotes the *transpose* of A.

Let $k = a^2 + b^2 + c^2 + d^2$. Then

$$\begin{aligned} |M|^2 &= |M| \cdot |M^{\mathrm{T}}| \\ &= |M \cdot M^{\mathrm{T}}| \\ &= \begin{vmatrix} k^2 & 0 & 0 \\ 0 & k^2 & 0 \\ 0 & 0 & k^2 \end{vmatrix} \\ &= k^6. \end{aligned}$$

So $|M| = \pm k^3 = \pm(a^2 + b^2 + c^2 + d^2)^3$.

In particular, let $a = f_n$, $b = l_n$, $c = p_n$, and $d = q_n$. Then

$$\begin{aligned} |M| &= \begin{vmatrix} f_n^2+l_n^2-p_n^2-q_n^2 & 2(l_np_n-f_nq_n) & 2(f_np_n+l_nq_n) \\ 2(f_nq_n+l_np_n) & f_n^2-l_n^2+p_n^2-q_n^2 & 2(p_nq_n-f_nl_n) \\ 2(l_nq_n-f_np_n) & 2(f_nl_n+p_nq_n) & f_n^2-l_n^2-p_n^2+q_n^2 \end{vmatrix} \\ &= \pm(f_n^2+l_n^2+p_n^2+q_n^2)^3. \end{aligned}$$

To determine the correct sign, consider the case $n = 0$. Then

$$|M| = \begin{vmatrix} 0 & 0 & 8 \\ 0 & -8 & 0 \\ 8 & 0 & 0 \end{vmatrix} = 8^3$$

is positive. Consequently, $|M| = (f_n^2 + l_n^2 + p_n^2 + q_n^2)^3$.

C.K. Cook of Sumter, South Carolina, studied this determinant in 2006 for the special case $x = 1$ [111, 113]:

$$\begin{vmatrix} F_n^2 + L_n^2 - P_n^2 - 4Q_n^2 & 2(L_nP_n - 2F_nQ_n) & 2(F_nP_n + 2L_nQ_n) \\ 2(2F_nQ_n + L_nP_n) & F_n^2 - L_n^2 + P_n^2 - 4Q_n^2 & 2(2P_nQ_n - F_nL_n) \\ 2(2L_nQ_n - F_nP_n) & 2(F_nL_n + 2P_nQ_n) & F_n^2 - L_n^2 - P_n^2 + 4Q_n^2 \end{vmatrix}$$
$$= (F_n^2 + L_n^2 + P_n^2 + 4Q_n^2)^3.$$

In particular,

$$\begin{vmatrix} F_5^2 + L_5^2 - P_5^2 - 4Q_5^2 & 2(L_5P_5 - 2F_5Q_5) & 2(F_5P_5 + 2L_5Q_5) \\ 2(2F_5Q_5 + L_5P_5) & F_5^2 - L_5^2 + P_5^2 - 4Q_5^2 & 2(2P_5Q_5 - F_5L_5) \\ 2(2L_5Q_5 - F_5P_5) & 2(F_5L_5 + 2P_5Q_5) & F_5^2 - L_5^2 - P_5^2 + 4Q_5^2 \end{vmatrix}$$
$$= \begin{vmatrix} -7419 & -182 & 2094 \\ 1458 & -5979 & 4646 \\ 1514 & 4866 & 5787 \end{vmatrix}$$
$$= 458{,}492{,}366{,}431$$
$$= (F_5^2 + L_5^2 + P_5^2 + 4Q_5^2)^3.$$

Next, we look at the interesting determinant

$$D = \begin{vmatrix} a & b & c & d \\ b & a & d & c \\ c & d & a & b \\ d & c & b & a \end{vmatrix},$$

where $a = a(x)$, $b = b(x)$, $c = c(x)$, and $d = d(x)$. This determinant was first evaluated in 1866 [381]:

$$D = (a + b + c + d)(a + b - c - d)(a - b + c - d)(a - b - c + d). \tag{39.1}$$

Suppose we let $a = xf_{n+3}, b = f_{n+2}, c = xf_{n+1}$, and $d = f_n$. Then

$$\begin{aligned} a + b + c + d &= x(f_{n+3} + f_{n+1}) + (f_{n+2} + f_n) \\ &= xl_{n+2} + l_{n+1} \\ &= l_{n+3}; \end{aligned}$$

$$\begin{aligned}
a+b-c-d &= x(f_{n+3}-f_{n+1})+(f_{n+2}-f_n)\\
&= x^2 f_{n+2}+xf_{n+1}\\
&= xf_{n+3};\\
a-b+c-d &= x(f_{n+3}+f_{n+1})-(f_{n+2}+f_n)\\
&= xl_{n+2}-l_{n+1};\\
a-b-c+d &= x(f_{n+3}-f_{n+1})-(f_{n+2}-f_n)\\
&= x^2 f_{n+2}-xf_{n+1}.
\end{aligned}$$

Using the addition formula $f_{n+1}l_{n+2}+f_{n+2}l_{n+1}=2f_{2n+3}$, equation (39.1) then yields

$$\begin{aligned}
&\begin{vmatrix} xf_{n+3} & f_{n+2} & xf_{n+1} & f_n\\ f_{n+2} & xf_{n+3} & f_n & xf_{n+1}\\ xf_{n+1} & f_n & xf_{n+3} & f_{n+2}\\ f_n & xf_{n+1} & f_{n+2} & xf_{n+3}\end{vmatrix}\\
&= l_{n+3}\cdot xf_{n+3}\cdot(xl_{n+2}-l_{n+1})\cdot x(xf_{n+2}-f_{n+1})\\
&= x^2 f_{2n+6}\left[x^2 f_{2n+4}-x(f_{n+1}l_{n+2}+f_{n+2}l_{n+1})+f_{2n+2}\right]\\
&= x^2 f_{2n+6}\left(x^2 f_{2n+4}-2xf_{2n+3}+f_{2n+2}\right). \qquad (39.2)
\end{aligned}$$

This implies

$$\begin{vmatrix} F_{n+3} & F_{n+2} & F_{n+1} & F_n\\ F_{n+2} & F_{n+3} & F_n & F_{n+1}\\ F_{n+1} & F_n & F_{n+3} & F_{n+2}\\ F_n & F_{n+1} & F_{n+2} & F_{n+3}\end{vmatrix} = F_{2n+6}\left(F_{2n+4}-2F_{2n+3}+F_{2n+2}\right)$$

$$= F_{2n}F_{2n+6};$$

$$\begin{vmatrix} 2xp_{n+3} & p_{n+2} & 2xp_{n+1} & p_n\\ p_{n+2} & 2xp_{n+3} & p_n & 2xp_{n+1}\\ 2xp_{n+1} & p_n & 2xp_{n+3} & p_{n+2}\\ p_n & 2xp_{n+1} & p_{n+2} & 2xp_{n+3}\end{vmatrix} = 4x^2 p_{2n+6}\left(4x^2 p_{2n+4}-4xp_{2n+3}+p_{2n+2}\right);$$

$$\begin{vmatrix} 2P_{n+3} & P_{n+2} & 2P_{n+1} & P_n\\ P_{n+2} & 2P_{n+3} & P_n & 2P_{n+1}\\ 2P_{n+1} & P_n & 2P_{n+3} & P_{n+2}\\ P_n & 2P_{n+1} & P_{n+2} & 2P_{n+3}\end{vmatrix} = 4P_{2n+6}\left(4P_{2n+3}+5P_{2n+2}\right).$$

In particular, let $n=5$. Then

$$\begin{vmatrix} F_8 & F_7 & F_6 & F_5\\ F_7 & F_8 & F_5 & F_6\\ F_6 & F_5 & F_8 & F_7\\ F_5 & F_6 & F_7 & F_8\end{vmatrix} = 54{,}285 = F_{10}F_{16};$$

$$\begin{vmatrix} 2P_8 & P_7 & 2P_6 & P_5 \\ P_7 & 2P_8 & P_5 & 2P_6 \\ 2P_6 & P_5 & 2P_8 & P_7 \\ P_5 & 2P_6 & P_7 & 2P_8 \end{vmatrix} = 382{,}586{,}783{,}232 = 4P_{16}(4P_{13} + 5P_{12}).$$

Similarly, we can show that

$$\begin{vmatrix} xl_{n+3} & l_{n+2} & xl_{n+1} & l_n \\ l_{n+2} & xl_{n+3} & l_n & xl_{n+1} \\ xl_{n+1} & l_n & xl_{n+3} & l_{n+2} \\ l_n & xl_{n+1} & l_{n+2} & xl_{n+3} \end{vmatrix} = x^2(x^2+4)^2 f_{2n+6}\left(x^2 f_{2n+4} - 2x f_{2n+3} + f_{2n+2}\right);$$

see Exercises 39.1–39.4 and 39.14.

Consequently,

$$\begin{vmatrix} L_{n+3} & L_{n+2} & L_{n+1} & L_n \\ L_{n+2} & L_{n+3} & L_n & L_{n+1} \\ L_{n+1} & L_n & L_{n+3} & L_{n+2} \\ L_n & L_{n+1} & L_{n+2} & L_{n+3} \end{vmatrix} = 25F_{2n}F_{2n+6};$$

$$\begin{vmatrix} 2xq_{n+3} & q_{n+2} & 2xq_{n+1} & q_n \\ q_{n+2} & 2xq_{n+3} & q_n & 2xq_{n+1} \\ 2xq_{n+1} & q_n & 2xq_{n+3} & q_{n+2} \\ q_n & 2xq_{n+1} & q_{n+2} & 2xq_{n+3} \end{vmatrix}$$
$$= 64x^2(x^2+1)^2 p_{2n+6}\left(4x^2 p_{2n+4} - 4x p_{2n+3} + p_{2n+2}\right);$$

$$\begin{vmatrix} 2Q_{n+3} & Q_{n+2} & 2Q_{n+1} & Q_n \\ Q_{n+2} & 2Q_{n+3} & Q_n & 2Q_{n+1} \\ 2Q_{n+1} & Q_n & 2Q_{n+3} & Q_{n+2} \\ Q_n & 2Q_{n+1} & Q_{n+2} & 2Q_{n+3} \end{vmatrix} = 16P_{2n+6}\left(4P_{2n+3} + 5P_{2n+2}\right).$$

The next example also deals with an interesting 5×5 determinant $|M|$, developed by Patel in 2006 [379]. The featured proof is based on the one given in the following year by H. Kwong of the State University of New York, Fredonia, New York [311]; it employs the eigenvalues of the matrix M to evaluate the determinant.

Example 39.2. *Let*

$$M = \begin{bmatrix} -L_{2n} & F_{2n} & L_n^2 & 2F_{2n} & L_n^2 \\ F_{2n} & -3[3F_n^2 + 2(-1)^n] & F_{2n} & 2F_n^2 & F_{2n} \\ L_n^2 & F_{2n} & -L_{2n} & 2F_{2n} & L_n^2 \\ 2F_{2n} & 2F_n^2 & 2F_{2n} & -6F_{n+1}F_{n-1} & 2F_{2n} \\ L_n^2 & F_{2n} & L_n^2 & 2F_{2n} & -L_{2n} \end{bmatrix},$$

where $n \geq 2$. Prove that $|M| = (L_n^2 + L_{2n})^5$.

Proof. The proof uses the following identities:

$$5F_n^2 = L_{2n} - 2(-1)^n \qquad L_n^2 = L_{2n} + 2(-1)^n$$
$$L_n^2 + 5F_n^2 = 2L_{2n} \qquad F_{2n} = F_nL_n$$
$$F_{n+1}F_{n-1} = F_n^2 + (-1)^n \qquad 5F_{n+1}F_{n-1} = L_{2n} + 3(-1)^n.$$

Letting $\lambda = L_n^2 + L_{2n}$, we can then show that

$$10F_n^2 + 6(-1)^n = \lambda \qquad 6F_{n+1}F_{n-1} = \lambda - 4F_n^2$$
$$3[3F_n^2 + 2(-1)^n] = \lambda - F_n^2 \qquad 5F_n^2 + 3L_n^2 = 2\lambda;$$

see Exercises 39.5–39.9.

Now consider the vectors

$$v_1 = \begin{bmatrix} 0 \\ 2 \\ 0 \\ -1 \\ 0 \end{bmatrix}, \quad v_2 = \begin{bmatrix} -F_n \\ L_n \\ 0 \\ 0 \\ 0 \end{bmatrix}, \quad v_3 = \begin{bmatrix} 0 \\ L_n \\ -F_n \\ 0 \\ 0 \end{bmatrix}, \quad v_4 = \begin{bmatrix} 0 \\ L_n \\ 0 \\ 0 \\ -F_n \end{bmatrix}, \quad \text{and } v_5 = \begin{bmatrix} L_n \\ F_n \\ L_n \\ 2F_n \\ L_n \end{bmatrix}.$$

Then

$$Mv_1 = \begin{bmatrix} 0 \\ -2\lambda \\ 0 \\ \lambda \\ 0 \end{bmatrix}.$$

Similarly, $Mv_i = -\lambda v_i$, where $2 \le i \le 4$; and $Mv_5 = \lambda v_5$. So $-\lambda, -\lambda, -\lambda, -\lambda$, and λ are the eigenvalues of the matrix M.

We now claim that the five vectors are linearly independent. To see this, suppose $av_1 + bv_2 + cv_3 + dv_4 + ev_5 = 0$, where a, b, c, d and e are real numbers. They yield the following 5×5 linear system:

$$-bF_n + eL_n = 0 \tag{39.3}$$

$$2a + bL_n + cL_n + dL_n + eF_n = 0 \tag{39.4}$$

$$-cF_n + eL_n = 0 \tag{39.5}$$

$$-a + 2eF_n = 0 \tag{39.6}$$

$$-dF_n + eL_n = 0. \tag{39.7}$$

It follows from equations (39.3), (39.5), and (39.7) that $b = c = d$. By equations (39.4) and (39.6), we get $b = -\dfrac{5a}{6L_n}$. Equation (39.3) then yields $a(3L_n^2 + 5F_n^2) = 0$; this implies $a = 0$ and hence $e = 0$.

Thus $a = b = c = d = e = 0$; so the vectors are linearly independent, as claimed.

Consequently, $|M|$ is the product of the eigenvalues of the matrix [13]; that is, $|M| = \lambda^5 = (L_n^2 + L_{2n})^5$, as desired. (Since $L_n^2 + L_{2n} = 2[L_{2n} + (-1)^n]$, this can be rewritten as $|M| = 32[L_{2n} + (-1)^n]^5$.) ■

In particular, let $n = 3$. Then

$$
\begin{aligned}
|M| &= \begin{bmatrix} -18 & 8 & 16 & 16 & 16 \\ 8 & -30 & 8 & 8 & 8 \\ 16 & 8 & -18 & 16 & 16 \\ 16 & 8 & 16 & -18 & 16 \\ 16 & 8 & 16 & 16 & -18 \end{bmatrix} \\
&= 45{,}435{,}424 \\
&= 32[L_6 + (-1)^3]^5,
\end{aligned}
$$

as expected.

Next we evaluate an $n \times n$ determinant of Fibonacci polynomials.

Example 39.3. *Evaluate the determinant*

$$
A_n = \begin{vmatrix} f_0 & f_1 & f_2 & & f_{n-2} & f_{n-1} \\ f_1 & f_0 & f_1 & & f_{n-3} & f_{n-2} \\ f_2 & f_1 & f_0 & \cdots & f_{n-4} & f_{n-3} \\ & & & \vdots & & \\ f_{n-2} & f_{n-3} & f_{n-4} & \cdots & f_0 & f_1 \\ f_{n-1} & f_{n-2} & f_{n-3} & & f_1 & f_0 \end{vmatrix}.
$$

Solution. Let R_i denote row i. Adding $-xR_2 - R_3$ to R_1, we get

$$
\begin{aligned}
A_n &= \begin{vmatrix} -2x & 0 & 0 & & 0 & 0 \\ f_1 & f_0 & f_1 & & f_{n-3} & f_{n-2} \\ f_2 & f_1 & f_0 & \cdots & f_{n-4} & f_{n-3} \\ & & & \vdots & & \\ f_{n-2} & f_{n-3} & f_{n-4} & \cdots & f_0 & f_1 \\ f_{n-1} & f_{n-2} & f_{n-3} & & f_1 & f_0 \end{vmatrix} \\
&= -2xA_{n-1}.
\end{aligned}
$$

Since $A_2 = -1$, it follows by PMI that $A_n = (-2x)^{n-2}A_2 = -(-2x)^{n-2}$, where $n \geq 2$. ■

Seiffert studied the determinant A_5 in 1993 for the case $x = 1$ [133, 422]. It follows by Example 39.3 that

$$
\begin{vmatrix} p_0 & p_1 & p_2 & & p_{n-2} & p_{n-1} \\ p_1 & p_0 & p_1 & & p_{n-3} & p_{n-2} \\ p_2 & p_1 & p_0 & \cdots & p_{n-4} & p_{n-3} \\ & & & \vdots & & \\ p_{n-2} & p_{n-3} & p_{n-4} & \cdots & p_0 & p_1 \\ p_{n-1} & p_{n-2} & p_{n-3} & & p_1 & p_0 \end{vmatrix} = -(-4x)^{n-2};
$$

$$\begin{vmatrix} F_0 & F_1 & F_2 & & F_{n-2} & F_{n-1} \\ F_1 & F_0 & F_1 & & F_{n-3} & F_{n-2} \\ F_2 & F_1 & F_0 & \cdots & F_{n-4} & F_{n-3} \\ & & & \vdots & & \\ F_{n-2} & F_{n-3} & F_{n-4} & \cdots & F_0 & F_1 \\ F_{n-1} & F_{n-2} & F_{n-3} & & F_1 & F_0 \end{vmatrix} = -(-2)^{n-2};$$

$$\begin{vmatrix} P_0 & P_1 & P_2 & & P_{n-2} & P_{n-1} \\ P_1 & P_0 & P_1 & & P_{n-3} & P_{n-2} \\ P_2 & P_1 & P_0 & \cdots & P_{n-4} & P_{n-3} \\ & & & \vdots & & \\ P_{n-2} & P_{n-3} & P_{n-4} & \cdots & P_0 & P_1 \\ P_{n-1} & P_{n-2} & P_{n-3} & & P_1 & P_0 \end{vmatrix} = -(-4)^{n-2},$$

where $n \geq 2$.

In particular,

$$\begin{vmatrix} P_0 & P_1 & P_2 & P_3 & P_4 & P_5 \\ P_1 & P_0 & P_1 & P_2 & P_3 & P_4 \\ P_2 & P_1 & P_0 & P_1 & P_2 & P_3 \\ P_3 & P_2 & P_1 & P_0 & P_1 & P_2 \\ P_4 & P_3 & P_2 & P_1 & P_0 & P_1 \\ P_5 & P_4 & P_3 & P_2 & P_1 & P_0 \end{vmatrix} = \begin{vmatrix} 0 & 1 & 2 & 5 & 12 & 29 \\ 1 & 0 & 1 & 2 & 5 & 12 \\ 2 & 1 & 0 & 1 & 2 & 5 \\ 5 & 2 & 1 & 0 & 1 & 2 \\ 12 & 5 & 2 & 1 & 0 & 1 \\ 29 & 12 & 5 & 2 & 1 & 0 \end{vmatrix}$$

$$= -(-4)^4.$$

The determinant A_n in Example 39.3 has a Lucas counterpart B_n:

$$B_n = \begin{vmatrix} l_0 & l_1 & l_2 & & l_{n-2} & l_{n-1} \\ l_1 & l_0 & l_1 & & l_{n-3} & l_{n-2} \\ l_2 & l_1 & l_0 & \cdots & l_{n-4} & l_{n-3} \\ & & & \vdots & & \\ l_{n-2} & l_{n-3} & l_{n-4} & \cdots & l_0 & l_1 \\ l_{n-1} & l_{n-2} & l_{n-3} & & l_1 & l_0 \end{vmatrix}.$$

We can show that $B_{2n-1} = 2(-4x)^{n-1}$ and $B_{2n} = (4 - x^2)(-4x)^{n-1}$, where $B_1 = 2, B_2 = 4 - x^2$, and $n \geq 2$; see Exercise 39.19.

In the next example, we study a 5×5 determinant developed by Patel in 2005 [377]. The proof given uses Cassini's formula for Fibonacci numbers and the identity $L_n = F_{n+1} + F_{n-1}$.

Example 39.4. *Let*

$$M = \begin{bmatrix} F_nF_{n+2} + F_{n+1}^2 & F_n^2 & F_{n+1}^2 & F_{n+2}^2 & -(F_nF_{n+2} + F_{n+1}^2) \\ F_nF_{n+2} + F_{n+1}^2 & F_nF_{n+3} & -F_{n+1}L_{n+1} & F_{n-1}F_{n+2} & F_nF_{n+2} + F_{n+1}^2 \\ 0 & 2F_{n+1}F_{n+2} & 2F_nF_{n+2} & -2F_nF_{n+1} & 0 \\ F_nF_{n+2} + F_{n+1}^2 & -F_nF_{n+3} & F_{n+1}L_{n+1} & -F_{n-1}F_{n+2} & F_nF_{n+2} + F_{n+1}^2 \\ -(F_nF_{n+2} + F_{n+1}^2) & F_n^2 & F_{n+1}^2 & F_{n+2}^2 & F_nF_{n+2} + F_{n+1}^2 \end{bmatrix},$$

where $n \geq 2$. Prove that $|M| = -32\left[2F_{n+1}^2 - (-1)^n\right]^5$.

Proof. In the interest of brevity, we let $a = F_n$, $b = F_{n+1}$, and $c = F_{n+2}$; so $b - a = F_{n-1}$, $c = a + b$, and $a^2 + ab + b^2 = a(a+b) + b^2 = b^2 + ac = F_{n+1}^2 + F_nF_{n+2} = 2F_{n+1}^2 - (-1)^n$. Using elementary operations, we then have

$$|M| = \begin{vmatrix} ac+b^2 & a^2 & b^2 & c^2 & -(ac+b^2) \\ ac+b^2 & a(b+c) & -b(a+c) & c(b-a) & ac+b^2 \\ 0 & 2bc & 2ac & -2ab & 0 \\ ac+b^2 & -a(b+c) & b(a+c) & -c(b-a) & ac+b^2 \\ -(ac+b^2) & a^2 & b^2 & c^2 & ac+b^2 \end{vmatrix}$$

$$= (ac+b^2)^2 \begin{vmatrix} 1 & a^2 & b^2 & c^2 & -1 \\ 1 & a(b+c) & -b(a+c) & c(b-a) & 1 \\ 0 & 2bc & 2ac & -2ab & 0 \\ 1 & -a(b+c) & b(a+c) & -c(b-a) & 1 \\ -1 & a^2 & b^2 & c^2 & 1 \end{vmatrix}$$

$$= 2(ac+b^2)^2 \begin{vmatrix} 1 & a^2 & b^2 & c^2 & -1 \\ 1 & a(b+c) & -b(a+c) & c(b-a) & 1 \\ 0 & bc & ac & -ab & 0 \\ 2 & 0 & 0 & 0 & 2 \\ 0 & 2a^2 & 2b^2 & 2c^2 & 0 \end{vmatrix}$$

$$= 8(ac+b^2)^2 \begin{vmatrix} 1 & a^2 & b^2 & c^2 & -2 \\ 1 & a(b+c) & -b(a+c) & c(b-a) & 1 \\ 0 & bc & ac & -ab & 0 \\ 1 & 0 & 0 & 0 & 0 \\ 0 & a^2 & b^2 & c^2 & 0 \end{vmatrix}$$

$$= -8(ac+b^2)^2 \begin{vmatrix} a^2 & b^2 & c^2 & -2 \\ a(b+c) & -b(a+c) & c(b-a) & 0 \\ bc & ac & -ab & 0 \\ a^2 & b^2 & c^2 & 0 \end{vmatrix}$$

$$= 16(ac+b^2)^2 \begin{vmatrix} a(b+c) & -b(a+c) & c(b-a) \\ bc & ac & -ab \\ a^2 & b^2 & c^2 \end{vmatrix}$$

$$= 16(ac+b^2)^2 \begin{vmatrix} a^2+2ab & -b^2-2ab & b^2-a^2 \\ ab+b^2 & a^2+ab & -ab \\ a^2 & b^2 & (a+b)^2 \end{vmatrix}$$

$$= 16(a^2+ab+b^2)^2 \begin{vmatrix} a^2+2ab & -b^2-2ab & b^2-a^2 \\ 1 & 1 & 1 \\ a^2 & b^2 & (a+b)^2 \end{vmatrix}$$

$$= 32(a^2+ab+b^2)^3 \begin{vmatrix} a^2+ab & -ab & b^2+ab \\ 1 & 1 & 1 \\ a^2 & b^2 & (a+b)^2 \end{vmatrix}$$

$$= 32(a^2+ab+b^2)^3 \begin{vmatrix} a^2+ab & -a^2-2ab & b^2-a^2 \\ 1 & 0 & 0 \\ a^2 & b^2-a^2 & b^2+2ab \end{vmatrix}$$
$$= -32(a^2+ab+b^2)^3[(a^2+2ab)(b^2+2ab)+(b^2-a^2)^2]$$
$$= -32(a^2+ab+b^2)^3(a^4+2a^3b+3a^2b^2+2ab^3+b^4)$$
$$= -32(a^2+ab+b^2)^5$$
$$= -32\left[2F_{n+1}^2-(-1)^n\right]^5,$$

as claimed. ■

In the next example, we study an interesting recurrence. It turns out that its solution is a gibonacci determinant.

Example 39.5. *Solve the recurrence $z_{n+2} = g_{n+2}z_{n+1} + z_n$, where $z_1 = c$ and $z_2 = d$.*

Solution. It follows by the recurrence that

$$\begin{aligned} z_3 &= g_3z_2 + z_1 \\ z_4 &= g_4z_3 + z_2 \\ z_5 &= g_5z_4 + z_3 \\ &\vdots \\ z_{n+1} &= g_{n+1}z_n + z_{n-1} \\ z_{n+2} &= g_{n+2}z_{n+1} + z_n. \end{aligned}$$

We can express the values of z_3 through z_{n+2} as determinants:

$$z_3 = \begin{vmatrix} d & c \\ -1 & g_3 \end{vmatrix};$$
$$z_4 = g_4(g_3d+c)+d$$
$$= \begin{vmatrix} d & c & 0 \\ -1 & g_3 & 1 \\ 0 & -1 & g_4 \end{vmatrix}.$$

More generally,

$$z_{n+1} = \begin{vmatrix} d & c & 0 & 0 & & 0 & 0 \\ -1 & g_3 & 1 & 0 & & 0 & 0 \\ 0 & -1 & g_4 & 1 & \cdots & 0 & 0 \\ & & & & \vdots & & \\ 0 & 0 & 0 & 0 & \cdots & g_{n+1} & 1 \\ 0 & 0 & 0 & 0 & & -1 & g_{n+2} \end{vmatrix}.$$

We can confirm this by expanding the determinant with respect to the last row. ■

An Interesting Byproduct

The solutions of the recurrence have a surprising consequence: The ratio $\frac{z_{n+2}}{z_{n+1}}$ of any two adjacent solutions can be expressed as a finite continued fraction:

$$\begin{aligned}
\frac{z_3}{z_2} &= g_3 + \frac{c}{d} \\
\frac{z_4}{z_3} &= g_4 + \frac{1}{z_3/z_2} \\
&= g_4 + \frac{1}{g_3 + \frac{c}{d}} \\
&\vdots \\
\frac{z_{n+2}}{z_{n+1}} &= g_{n+2} + \cfrac{1}{g_{n+1} + \cfrac{1}{g_n + \cfrac{1}{\ddots + \cfrac{1}{g_3 + \frac{c}{d}}}}}.
\end{aligned}$$

In 1973, G. Ledin, Jr., of San Francisco, studied the special case of this example when $g_n = F_n$ [49, 317].

The determinant

$$|A_n| = \begin{vmatrix} a_1 & 1 & 0 & 0 & & 0 & 0 \\ -1 & a_2 & 1 & 0 & & 0 & 0 \\ 0 & -1 & a_2 & 1 & \cdots & 0 & 0 \\ & & & & \vdots & & \\ 0 & 0 & 0 & 0 & \cdots & a_{n-1} & 1 \\ 0 & 0 & 0 & 0 & & -1 & a_n \end{vmatrix}$$

has a very close resemblance to the determinant z_{n+2} above. It was studied in 1964 by C.A. Church, Jr., of Duke University, North Carolina [101]. Expanding it with respect to the last row yields the recurrence $|A_n| = a_n|A_{n-1}| + |A_{n-2}|$, where $|A_1| = a_1$, $|A_2| = a_1a_2 + 1$, and $n \geq 3$.

In particular, let $a_i = x$ for every i. Then $|A_n| = x|A_{n-1}| + |A_{n-2}|$, where $|A_1| = x$, $|A_2| = x^2 + 1$, and $n \geq 3$. Consequently, $A_n = f_{n+1}$, where $n \geq 1$.

For example,

$$\begin{aligned}
|A_3| &= \begin{vmatrix} x & 1 & 0 \\ -1 & x & 1 \\ 0 & -1 & x \end{vmatrix} \\
&= x^3 + 2x = f_4.
\end{aligned}$$

Closely related to $|A_n|$ is the determinant

$$|C_n| = \begin{vmatrix} x & -1 & 0 & 0 & & 0 & 0 \\ -1 & x & -1 & 0 & & 0 & 0 \\ 0 & 1 & x & -1 & \cdots & 0 & 0 \\ & & & & \vdots & & \\ 0 & 0 & 0 & 0 & \cdots & x & -1 \\ 0 & 0 & 0 & 0 & & 1 & x \end{vmatrix}.$$

$|C_n|$ satisfies the same recursive definition as $|A_n|$; so $|C_n| = f_{n+1}$, where $n \geq 1$.

The matrix C_n is a *tridiagonal matrix*, meaning that all its nonzero elements lie on the main diagonal, superdiagonal, or subdiagonal.

39.3 BASIN'S DETERMINANT

The determinant $|C_n|$ is a special case of the determinant

$$|B_n(b,c,a)| = \begin{vmatrix} c & a & 0 & 0 & & 0 & 0 \\ b & c & a & 0 & & 0 & 0 \\ 0 & b & c & a & \cdots & 0 & 0 \\ & & & & \vdots & & \\ 0 & 0 & 0 & 0 & \cdots & c & a \\ 0 & 0 & 0 & 0 & & b & c \end{vmatrix},$$

studied in 1963 by S.L. Basin of Sylvania Electronic Systems, Mountain View, California. The main diagonal consists of c's, the superdiagonal of a's, and the subdiagonal of b's. Expanding it with respect to row 1, we get a recurrence satisfied by $|B_n|$: $|B_n| = c|B_{n-1}| - ab|B_{n-2}|$, where $|B_1| = c$, $|B_2| = c^2 - ab$, and $|B_n| = |B_n(b,c,a)|$.

Solving the recurrence, we get

$$|B_n(b,c,a)| = \frac{u^{n+1} - v^{n+1}}{u - v},$$

where $u = \dfrac{c + \sqrt{c^2 - ab}}{2}$ and $v = \dfrac{c - \sqrt{c^2 - ab}}{2}$; see Exercise 39.23.

In particular, replace b, c, and a with 1, $a + b$, and ab, respectively; so $u = a$ and $v = b$. Thus $|B_n(1, a+b, ab)| = \dfrac{a^{n+1} - b^{n+1}}{a - b}$; that is,

$$\begin{vmatrix} a+b & ab & 0 & 0 & & 0 & 0 \\ 1 & a+b & ab & 0 & & 0 & 0 \\ 0 & 1 & a+b & ab & \cdots & 0 & 0 \\ & & & & \vdots & & \\ 0 & 0 & 0 & 0 & \cdots & a+b & ab \\ 0 & 0 & 0 & 0 & & 1 & a+b \end{vmatrix} = \frac{a^{n+1} - b^{n+1}}{a - b}.$$

Church studied this determinant in 1964 [100].

Suppose we let $a = \alpha(x)$ and $b = \beta(x)$. Then $a + b = x$ and $ab = -1$. So $|B_n(1, x, -1)| = f_{n+1} = |C_n|$.

In 2002, N.D. Cahill (Eastman Kodak Company) *et al.* studied the special case $B_n(i, x, i)$, where $i = \sqrt{-1}$. Clearly, $B_n(1, 2x, -1) = p_{2n+1} = B_n(1, 2x, i)$ [76].

Next we investigate two special cases of the determinant $|B_n(b, c, a)|$.

Two Special Cases

Let $a = -1 = b$ and $c = x^2 + 2$. Then

$$|B_n(-1, x^2+2, -1)| = \begin{vmatrix} x^2+2 & -1 & 0 & 0 & & 0 & 0 \\ -1 & x^2+2 & -1 & 0 & & 0 & 0 \\ 0 & -1 & x^2+2 & -1 & \cdots & 0 & 0 \\ & & & & \vdots & & \\ 0 & 0 & 0 & 0 & \cdots & x^2+2 & -1 \\ 0 & 0 & 0 & 0 & & -1 & x^2+2 \end{vmatrix}.$$

Since $u = \dfrac{x^2 + 2 + x\Delta}{2} = \alpha^2$ and $v = \dfrac{x^2 + 2 - x\Delta}{2} = \beta^2$, it follows that

$$|B_n(-1, x^2+2, -1)| = \frac{\alpha^{2n+2} - \beta^{2n+2}}{\alpha^2 - \beta^2} = \frac{1}{x} f_{2n+2}.$$

This can be confirmed using PMI also; see Exercise 39.24.

For example,

$$\begin{aligned} |B_3(-1, x^2+2, -1)| &= \begin{vmatrix} x^2+2 & -1 & 0 \\ -1 & x^2+2 & -1 \\ 0 & -1 & x^2+2 \end{vmatrix} \\ &= x^6 + 6x^4 + 10x^2 + 4 \\ &= \frac{1}{x} f_8. \end{aligned}$$

The determinant $|B_n(-1, 3, -1)| = F_{2n+2}$ has a delightful application to the study of spanning trees of wheel graphs, as K.R. Rebman of California State University at Haywood found in 1975 [287, 399].

The matrix C_n in the following example is related to $B_n(-1, x^2+2, -1)$.

Example 39.6. *Evaluate the $n \times n$ determinant*

$$|C_n(x)| = \begin{vmatrix} x^2+2 & -1 & 0 & 0 & & 0 & -1 \\ -1 & x^2+2 & -1 & 0 & & 0 & 0 \\ 0 & -1 & x^2+2 & -1 & \cdots & 0 & 0 \\ & & & & \vdots & & \\ 0 & 0 & 0 & 0 & \cdots & x^2+2 & -1 \\ -1 & 0 & 0 & 0 & & -1 & x^2+2 \end{vmatrix},$$

where we define $|C_1(x)| = x^2 = l_2 - 2$, $|C_2(x)| = \begin{vmatrix} x^2+2 & -2 \\ -2 & x^2+2 \end{vmatrix} = x^4 + 4x^2 = l_4 - 2$, *and* $n \geq 3$.

Solution. Expanding $|C_n|$ by row 1, we get

$$\begin{aligned} |C_n| &= (x^2+2)|B_{n-1}| + |D_{n-1}| + (-1)^n|E_{n-1}| \\ &= \frac{x^2+2}{x} f_{2n} + |D_{n-1}| + (-1)^n|E_{n-1}|, \end{aligned}$$

where

$$|D_{n-1}| = \begin{vmatrix} -1 & -1 & 0 & 0 & & 0 & 0 \\ 0 & x^2+2 & -1 & 0 & & 0 & 0 \\ 0 & -1 & x^2+2 & -1 & \cdots & 0 & 0 \\ & & & & \vdots & & \\ 0 & 0 & 0 & 0 & \cdots & x^2+2 & -1 \\ -1 & 0 & 0 & 0 & & -1 & x^2+2 \end{vmatrix};$$

$$|E_{n-1}| = \begin{vmatrix} -1 & x^2+2 & -1 & 0 & & 0 \\ 0 & -1 & x^2+2 & -1 & \cdots & 0 \\ & & & & \vdots & \\ 0 & 0 & 0 & 0 & \cdots & x^2+2 \\ -1 & 0 & 0 & 0 & & -1 \end{vmatrix}.$$

Expanding $|D_{n-1}|$ by column 1, we get

$$\begin{aligned} |D_{n-1}| &= -|B_{n-2}| + (-1)(-1)^{n-2}(-1)^{n-2} \\ &= -\frac{1}{x} f_{2n-2} - 1. \end{aligned}$$

Likewise, we get

$$\begin{aligned} |E_{n-1}| &= (-1)^{n-1} + (-1)^{n-1}|B_{n-2}| \\ &= (-1)^{n-1}\left(\frac{1}{x} f_{2n-2} + 1\right). \end{aligned}$$

Thus

$$\begin{aligned} |C_n| &= \frac{x^2+2}{x} f_{2n} + \left(-\frac{1}{x} f_{2n-2} - 1\right) + (-1)^n \cdot (-1)^{n-1}\left(\frac{1}{x} f_{2n-2} + 1\right) \\ &= \frac{1}{x}[(x^2+2)f_{2n} - f_{2n-2}] - \frac{1}{x} f_{2n-2} - 2 \\ &= \frac{1}{x}(f_{2n+2} - f_{2n-2}) - 2 \\ &= l_{2n} - 2. \end{aligned}$$

■

For example,

$$C_3 = \begin{vmatrix} x^2+2 & -1 & -1 \\ -1 & x^2+2 & -1 \\ -1 & -1 & x^2+2 \end{vmatrix}$$

$$= x^6 + 6x^4 + 9x^2$$

$$= l_6 - 2;$$

$$C_4 = \begin{vmatrix} x^2+2 & -1 & 0 & -1 \\ -1 & x^2+2 & -1 & 0 \\ 0 & -1 & x^2+2 & -1 \\ -1 & 0 & -1 & x^2+2 \end{vmatrix}$$

$$= x^8 + 8x^6 + 10x^4 + 16x^2$$

$$= l_8 - 2.$$

Next we investigate a similar determinant of order n:

$$|K_n(x,i)| = \begin{vmatrix} x & i & 0 & 0 & & 0 & 0 \\ i & x & i & 0 & & 0 & 0 \\ 0 & i & x & i & \cdots & 0 & 0 \\ & & & & \vdots & & \\ 0 & 0 & 0 & 0 & \cdots & x & i \\ 0 & 0 & 0 & 0 & & i & x \end{vmatrix},$$

where $i = \sqrt{-1}$, $|K_0(x,i)| = 1$, $|K_1(x,i)| = x$, and $n \geq 2$.

Expanding this with respect to row n, we get $|K_n| = x|K_{n-1}| + |K_{n-2}|$, where $K_j = K_j(x,i)$. Since $|K_0(x,i)| = f_1$ and $|K_1(x,i)| = f_2$, it follows that $|K_n(x,i)| = f_{n+1}$. Hence $|K_n(2x,i)| = p_{n+1}$, as we studied in Exercises 22.15 and 22.16. Paul F. Byrd of then San Jose State College, California, studied the determinant $|K_n(2x,i)|$ in 1963 [75].

For example,

$$\begin{vmatrix} x & i & 0 & 0 \\ i & x & i & 0 \\ 0 & i & x & i \\ 0 & 0 & i & x \end{vmatrix} = x^4 + 3x^2 + 1 = f_5.$$

The next example illustrates the power of pattern recognition. It shows how inductive reasoning, coupled with PMI, can do wonders.

Example 39.7. *Evaluate the determinant*

$$E_n = \begin{vmatrix} x^2+1 & 1 & 0 & 0 & & 0 & 0 \\ 1 & x^2+1 & 1 & 0 & & 0 & 0 \\ 1 & 1 & x^2+1 & 1 & \cdots & 0 & 0 \\ & & & & \vdots & & \\ 1 & 1 & 1 & 1 & \cdots & x^2+1 & 1 \\ 1 & 1 & 1 & 1 & & 1 & x^2+1 \end{vmatrix},$$

where $E_n = E_n(x)$ *and* $n \geq 1$.

Solution. Clearly, $|E_1| = x^2 + 1 = f_3$, $|E_2| = (x^2+1)^2 - 1 = x^4 + 2x^2 = xf_4$; and

$$\begin{aligned} |E_3| &= \begin{vmatrix} x^2+1 & 1 & 0 \\ 1 & x^2+1 & 1 \\ 1 & 1 & x^2+1 \end{vmatrix} \\ &= \begin{vmatrix} x^2+1 & 1 & 0 \\ 1 & x^2+1 & 1 \\ 0 & -x^2 & x^2 \end{vmatrix} \\ &= (x^2+1)(x^4+2x^2) - x^2 \\ &= x^2 f_5. \end{aligned}$$

Clearly, a pattern slowly emerges: $|E_1| = x^{1-1}f_{1+2}$, $|E_2| = x^{2-1}f_{2+2}$, and $|E_3| = x^{3-1}f_{3+2}$. More generally, we conjecture that $|E_n| = x^{n-1}f_{n+2}$, where $n \geq 1$.

We now establish this using PMI. Assume it works for all positive integers $< n$, where $n \geq 4$. Expanding $|E_n|$ by row 1, we get

$$\begin{aligned} |E_n| &= (x^2+1)|E_{n-1}| - |M_{n-1}| \\ &= (x^2+1)x^{n-2}f_{n+1} - |M_{n-1}|, \end{aligned}$$

where

$$|M_{n-1}| = \begin{vmatrix} 1 & 1 & 0 & 0 & & 0 & 0 \\ 1 & x^2+1 & 1 & 0 & & 0 & 0 \\ 1 & 1 & x^2+1 & 1 & \cdots & 0 & 0 \\ & & & & \vdots & & \\ 1 & 1 & 1 & 1 & \cdots & x^2+1 & 1 \\ 1 & 1 & 1 & 1 & & 1 & x^2+1 \end{vmatrix}$$

$$= \begin{vmatrix} 1 & 0 & 0 & 0 & & 0 & 0 \\ 1 & x^2 & 1 & 0 & & 0 & 0 \\ 1 & 0 & x^2+1 & 1 & \cdots & 0 & 0 \\ & & & & \vdots & & \\ 1 & 0 & 1 & 1 & \cdots & x^2+1 & 1 \\ 1 & 0 & 1 & 1 & & 1 & x^2+1 \end{vmatrix}$$

$$= \begin{vmatrix} 1 & 0 & 0 & 0 & & 0 & 0 \\ 0 & x^2 & 1 & 0 & & 0 & 0 \\ 0 & 0 & x^2+1 & 1 & \cdots & 0 & 0 \\ & & & & \vdots & & \\ 0 & 0 & 1 & 1 & \cdots & x^2+1 & 1 \\ 0 & 0 & 1 & 1 & & 1 & x^2+1 \end{vmatrix}$$

$$= 1 \cdot x^2 |E_{n-3}|$$

$$= x^{n-2} f_{n-1}.$$

Thus

$$\begin{aligned} |E_n| &= (x^2+1)x^{n-2} f_{n+1} - x^{n-2} f_{n-1} \\ &= x^{n-2}[(x^2+1)f_{n+1} - f_{n-1}] \\ &= x^{n-2} \cdot x f_{n+2} \\ &= x^{n-1} f_{n+2}. \end{aligned}$$

Consequently, the result follows by the strong version of PMI. ■

In particular, $|E_n(1)| = F_{n+2}$. Cahill *et al.* studied this special case in 2002 [76]. It also follows that $|E_n(2x)| = (2x)^{n-1} p_{n+2}$ and hence $|E_n(2)| = 2^{n-1} P_{n+2}$.

For example,

$$|E_5(2)| = \begin{vmatrix} 5 & 1 & 0 & 0 & 0 \\ 1 & 5 & 1 & 0 & 0 \\ 1 & 1 & 5 & 1 & 0 \\ 1 & 1 & 1 & 5 & 1 \\ 1 & 1 & 1 & 1 & 5 \end{vmatrix}$$

$$= 2^4 \cdot 169 = 2^4 P_7.$$

The following two examples show two closely related determinants that Cahill *et al.* [76] studied in 2002.

Example 39.8. *Evaluate the $n \times n$ determinants*

$$|G_n| = \begin{vmatrix} 1 & -1 & 0 & 0 & & 0 & 0 \\ 1 & 2 & -1 & 0 & & 0 & 0 \\ 1 & 1 & 2 & -1 & \cdots & 0 & 0 \\ & & & & \vdots & & \\ 1 & 1 & 1 & 1 & \cdots & -1 & 0 \\ 1 & 1 & 1 & 1 & & 2 & -1 \\ 1 & 1 & 1 & 1 & & 1 & 2 \end{vmatrix}$$

and

$$|H_n| = \begin{vmatrix} 2 & -1 & 0 & 0 & & 0 & 0 \\ 1 & 2 & -1 & 0 & & 0 & 0 \\ 1 & 1 & 2 & -1 & \cdots & 0 & 0 \\ & & & & \vdots & & \\ 1 & 1 & 1 & 1 & \cdots & -1 & 0 \\ 1 & 1 & 1 & 1 & & 2 & -1 \\ 1 & 1 & 1 & 1 & & 1 & 2 \end{vmatrix}.$$

Solution. (Notice that $|H_{n-1}|$ is the cofactor of the element in position (1,1) of $|G_n|$; and $|G_n|$ is the cofactor of the element in position (1,2) of $|H_{n+1}|$.)

We have $|G_1| = 1$, $|G_2| = 3$, and $|G_3| = \begin{vmatrix} 1 & -1 & 0 \\ 1 & 2 & -1 \\ 1 & 1 & 2 \end{vmatrix} = 8$. More generally, we can prove that $|G_n| = F_{2n}$, where $n \geq 2$; see Exercise 39.22.

On the other hand, $|H_1| = 2$, $|H_2| = 5$, and $|H_3| = \begin{vmatrix} 2 & -1 & 0 \\ 1 & 2 & -1 \\ 1 & 1 & 2 \end{vmatrix} = 13$. More generally, we can confirm that $|H_n| = F_{2n+1}$, where $n \geq 1$; see Exercise 39.22. ■

39.4 LOWER HESSENBERG MATRICES

Matrices E_n in Example 39.7, and matrices G_n and H_n in Example 39.8 are *lower Hessenberg matrices*. They are named after Karl Hessenberg (1904–1959), a German engineer who investigated them in the computation of eigenvalues and eigenvectors of linear operators for his dissertation in 1942. A *lower Hessenberg matrix A* is a square matrix $A = (a_{ij})_{n\times n}$ such that $a_{ij} = 0$ when $j > i + 1$, and $a_{i,i+1} \neq 0$ for some i.

The following theorem gives a recurrence for computing the determinants of lower Hessenberg matrices [76]. We establish it using PMI.

Theorem 39.1 (Cahill *et al.*, 2002 [76]). *Let A_n denote the lower Hessenberg matrix*

$$A_n = \begin{bmatrix} a_{11} & a_{12} & 0 & 0 & & 0 \\ a_{21} & a_{22} & a_{23} & 0 & & 0 \\ a_{21} & a_{22} & a_{23} & a_{24} & \cdots & 0 \\ & & & & \vdots & 0 \\ a_{n-1,1} & a_{n-1,2} & & & \cdots & a_{n-1,n} \\ a_{n,1} & a_{n,2} & & & & a_{n,n} \end{bmatrix},$$

where $|A_0| = 0$ and $|A_1| = a_{11}$. Then

$$|A_n| = a_{nn}|A_{n-1}| + \sum_{r=1}^{n-1}\left[(-1)^{n-r}a_{nr}\prod_{j=r}^{n-1}a_{j,j+1}|A_{r-1}|\right], \tag{39.8}$$

where $n \geq 2$.

Proof. Since $|A_1| = a_{11}$ and

$$\begin{aligned} |A_2| &= a_{11}a_{22} - a_{21}a_{12} \\ &= a_{22}|A_1| + \sum_{r=1}^{1}\left[(-1)^{2-r}a_{2r}\prod_{j=r}^{1}a_{j,j+1}|A_{r-1}|\right], \end{aligned}$$

the formula works when $n = 1$ and $n = 2$.

Now assume it works when $n = k$, where $k \geq 2$. Expanding $|A_{k+1}|$ by row k, we get

$$\begin{aligned} |A_{k+1}| &= a_{k+1,k+1}|A_k| - a_{k,k+1}a_{k+1,k}|A_{k-1}| - \sum_{r=1}^{k-1}\left[(-1)^{k-r}a_{k+1,r}\prod_{j=r}^{k-1}a_{j,j+1}|A_{r-1}|\right] \\ &= a_{k+1,k+1}|A_k| - a_{k,k+1}a_{k+1,k}|A_{k-1}| + a_{k,k+1}\sum_{r=1}^{k-1}\left[(-1)^{k-r+1}a_{k+1,r}\prod_{j=r}^{k-1}a_{j,j+1}|A_{r-1}|\right] \\ &= a_{k+1,k+1}|A_k| - a_{k,k+1}a_{k+1,k}|A_{k-1}| + \sum_{r=1}^{k-1}\left[(-1)^{k-r+1}a_{k+1,r}\prod_{j=r}^{k}a_{j,j+1}|A_{r-1}|\right] \\ &= a_{k+1,k+1}|A_k| + \sum_{r=1}^{k}\left[(-1)^{k-r+1}a_{k+1,r}\prod_{j=r}^{k}a_{j,j+1}|A_{r-1}|\right]. \end{aligned}$$

So the formula also works when $n = k + 1$.

Thus, by PMI, it works for every $n \geq 1$. ■

We now illustrate formula (39.8) using the matrix G_n in Example 39.8. First notice that $g_{11} = 1$; $g_{ii} = 2$ if $i \geq 2$; $g_{i,i+1} = -1$; and $g_{ij} = 1$ if $i > j$. By formula

(39.8), we then have

$$\begin{aligned}|G_n| &= 2|G_{n-1}| + \sum_{r=1}^{n-1}\left[(-1)^{n-r}g_{nr}\prod_{j=r}^{n-1}g_{j,j+1}|G_{r-1}|\right]\\ &= 2|G_{n-1}| + \sum_{r=1}^{n-1}|G_{r-1}|.\end{aligned}$$

In particular, this yields

$$\begin{aligned}|G_4| &= 2|G_3| + \sum_{r=1}^{3}|G_{r-1}|\\ &= 2\cdot 8 + |G_0| + |G_1| + |G_2|\\ &= 16 + 1 + 1 + 3\\ &= 21 = F_8,\end{aligned}$$

as expected.

The next example is interesting in its own right. C. Libis of the University of Rhode Island, Rhode Island, studied it in 2004 [322]. The featured proof is based on the one published in the following year by J. Seibert of the University of Hradec Králové, Czech Republic; it is a straightforward application of PMI [412].

Example 39.9. *Let $A_n = (a_{ij})_{n\times n}$ be the $n \times n$ matrix, where*

$$a_{ij} = \begin{cases} i+1 & \text{if } i = j\\ \min(i,j) & \text{otherwise.}\end{cases}$$

Find $|A_n|$.

Proof. Using the strong version of PMI, we establish that $|A_n| = F_{2n+1}$. Since $|A_1| = |2| = 2 = F_3$ and $|A_2| = \begin{vmatrix}2 & 1\\ 1 & 3\end{vmatrix} = 5 = F_5$, the result is true when $n = 1$ and $n = 2$.

Now assume the result is true for all positive integers $< n$. Then

$$|A_n| = \begin{vmatrix} 2 & 1 & 1 & & 1 & 1\\ 1 & 3 & 2 & & 2 & 2\\ 1 & 2 & 4 & \cdots & 3 & 3\\ & & & \vdots & & \\ 1 & 2 & 3 & \cdots & n & n-1\\ 1 & 2 & 3 & \cdots & n-1 & n+1\end{vmatrix}.$$

Let R_i denote the ith row of $|A_n|$, and C_j its jth column. Adding $-C_{n-1}$ to C_n, we get

$$|A_n| = \begin{vmatrix} 2 & 1 & 1 & & 1 & 0 \\ 1 & 3 & 2 & & 2 & 0 \\ 1 & 2 & 4 & \cdots & 3 & 0 \\ & & & \vdots & & \\ 1 & 2 & 3 & \cdots & n & -1 \\ 1 & 2 & 3 & \cdots & n-1 & 2 \end{vmatrix}.$$

Now add $-R_{n-1}$ to R_n. This yields

$$|A_n| = \begin{vmatrix} 2 & 1 & 1 & & 1 & 0 \\ 1 & 3 & 2 & & 2 & 0 \\ 1 & 2 & 4 & \cdots & 3 & 0 \\ & & & \vdots & & \\ 1 & 2 & 3 & \cdots & n & -1 \\ 0 & 0 & 0 & \cdots & -1 & 3 \end{vmatrix}.$$

Expanding this with respect to C_n, we get

$$\begin{aligned} |A_n| &= 3(-1)^{2n-2}|A_{n-1}| + (-1)^{2n-1} \begin{vmatrix} 2 & 1 & 1 & & 1 \\ 1 & 3 & 2 & & 2 \\ 1 & 2 & 4 & \cdots & 3 \\ & & & \vdots & \\ 1 & 2 & 3 & \cdots & n-1 \end{vmatrix} \\ &= 3|A_{n-1}| - |A_{n-2}| \\ &= 3F_{2n-1} - F_{2n-3} \\ &= F_{2n+1}. \end{aligned}$$

Thus, by the strong version of PMI, the result is true for every $n \geq 1$. ■

For example,

$$|A_3| = \begin{vmatrix} 2 & 1 & 1 \\ 1 & 3 & 2 \\ 1 & 2 & 4 \end{vmatrix} = 13 = F_7,$$

as expected.

Next we construct a Fibonacci matrix M_n with a delightful property. To this end, recall by the Cassini-like formula that $(f_{k+1}, f_{k+2}) = 1$, where (a, b) denotes the greatest common divisor (gcd) of the integer polynomials $a = a(x)$ and $b = b(x)$. It then follows by recursion that $(f_{k+1}, f_{k+2}, \ldots, f_{k+n}) = 1$.

39.5 DETERMINANT WITH A PRESCRIBED FIRST ROW

Does there exist a matrix $M_n = (a_{ij})_{n \times n}$ such that $a_{1j} = f_{k+j}$ and $|M_n| = (f_{k+1}, f_{k+2}, \ldots, f_{k+n})$, where $n \geq 4$? To our dismay, the answer is yes!

To see this, we choose

$$M_n = \begin{bmatrix} f_{k+1} & f_{k+2} & f_{k+3} & f_{k+4} & & f_{k+n-2} & f_{k+n-1} & f_{k+n} \\ f_{k+2} & f_{k+3} & 0 & 0 & & 0 & 0 & f_{k+n-1} \\ 0 & 0 & 1 & 0 & \cdots & 0 & 0 & f_{k+n-2} \\ & & & & \vdots & & & \\ 0 & 0 & 0 & 0 & \cdots & 1 & 0 & f_{k+3} \\ 0 & 0 & 0 & 0 & & 0 & f_{k+3} & f_{k+2} \\ 0 & 0 & 0 & 0 & & 0 & f_{k+2} & f_{k+1} \end{bmatrix}.$$

Expanding $|M_n|$ by the last row, we get

$$\begin{aligned} |M_n| &= f_{k+1}f_{k+3}(f_{k+1}f_{k+3} - f_{k+2}^2) - f_{k+2}^2(f_{k+1}f_{k+3} - f_{k+2}^2) \\ &= (f_{k+1}f_{k+3} - f_{k+2}^2)^2 \\ &= 1 \\ &= (f_{k+1}, f_{k+2}, \ldots, f_{k+n}), \end{aligned}$$

as desired.

D. Lind of the University of Virginia, Charlottesville, Virginia, studied this problem for the case $x = 1$ in 1969 [330]. Three years earlier, B.R. Toskey of Seattle University, Washington, investigated a generalization of the problem: Find a matrix $M = (a_{ij})_{n \times n}$ such that $|M| = (a_{11}, a_{12}, \ldots, a_{1n})$, where $n \geq 2$ [493]. (A very closely related problem appeared in 1964 from J.F. Ramaley of the University of California, Berkeley, [398], and a solution by R.F. Jackson of the University of Toledo appeared the following year [254].)

Interestingly, C.C. MacDuffee's *The Theory of Matrices* contains a constructive proof of Toskey's generalization to the problem [339]. In 1968, D.C.B. Marsh of the Colorado School of Mines, Golden, Colorado, gave basically the same generalized proof [349].

The following theorem gives the generalization. We establish it using PMI.

Theorem 39.2. *Let $a_{11}, a_{12}, \ldots, a_{1n}$ be positive integers such that $(a_{11}, a_{12}, \ldots, a_{1n}) = d_n$. Then there is a matrix $M_n = (a_{ij})_{n \times n}$ such that $|M_n| = d_n$, where $n \geq 2$.*

Proof. Let $d_2 = (a_{11}, a_{12})$. Then, by the Euclidean algorithm, there are integers a_{21} and a_{22} such that $a_{11}a_{22} - a_{12}a_{21} = d_2$. Let

$$M_2 = \begin{bmatrix} a_{11} & a_{12} \\ a_{21} & a_{22} \end{bmatrix}.$$

Then $|M_2| = d_2$. So the theorem is true when $n = 2$.

Now assume the existence of an integral matrix M_{n-1} with its first row consisting of the elements a_{11}, a_{12}, …, and a_{1n}, and $|M_{n-1}| = d_{n-1} = (a_{11}, a_{12}, \dots, a_{1,n-1})$. Then there are integers a_{nn} and b such that $d_{n-1}a_{nn} - ba_{1n} = d_n = (a_{11}, a_{12}, \dots, a_{1,n-1}, a_{1n})$.

Let

$$M_n = \begin{bmatrix} M_{n-1} & C \\ R & a_{nn} \end{bmatrix},$$

where C is the $(n-1) \times 1$ matrix $[a_{1n}, 0, 0, \dots, 0]^{\mathrm{T}}$ and R is the $1 \times (n-1)$ matrix $\dfrac{b}{d_{n-1}}[a_{11}, a_{12}, \dots, a_{1,n-1}]$.

Expanding $|M_n|$ by the last column, we get

$$\begin{aligned} |M_n| &= a_{nn}|M_{n-1}| + (-1)^{n-1} \cdot (-1)^{n-2} \cdot \frac{b}{d_{n-1}} \cdot a_{1n}|M_{n-1}| \\ &= a_{nn}d_{n-1} - ba_{1n} \\ &= d_n. \end{aligned}$$

So the theorem works for n also.

Thus, by PMI, it is true for $n \geq 2$. ■

EXERCISES 39

Let $a = xl_{n+3}, b = l_{n+2}, c = xl_{n+1}$, and $d = l_n$. Compute each.

1. $a + b + c + d$.
2. $a + b - c - d$.
3. $a - b + c - d$.
4. $a - b - c + d$.

Prove each.

5. $(x^2 + 4)f_{n+1}f_{n-1} = l_{2n} + (x^2 + 2)(-1)^n$.
6. $2(x^2 + 4)f_n^2 + 6(-1)^n = l_n^2 + l_{2n}$.
7. $6F_{n+1}F_{n-1} = L_n^2 + L_{2n} - 4F_n^2$.
8. $3[3F_n^2 + 2(-1)^n] = L_n^2 + L_{2n} - F_n^2$.
9. $3l_n^2 + (x^2 + 4)f_n^2 = 2(l_n^2 + l_{2n})$.
10. Let $D_n = \begin{vmatrix} g_n^2 & g_{n+1}^2 & g_{n+2}^2 \\ g_{n+1}^2 & g_{n+2}^2 & g_{n+3}^2 \\ g_{n+2}^2 & g_{n+3}^2 & g_{n+4}^2 \end{vmatrix}$. Show that $D_n = (-1)^n D_0$.

Evaluate each determinant.

11. $\begin{vmatrix} f_{p+2n} & f_{p+n} & f_p \\ f_{q+2n} & f_{q+n} & f_q \\ f_{r+2n} & f_{r+n} & f_r \end{vmatrix}$.

12. $\begin{vmatrix} f_n & f_{n+k} & f_{n+2k} \\ f_{n+3k} & f_{n+4k} & f_{n+5k} \\ f_{n+6k} & f_{n+7k} & f_{n+8k} \end{vmatrix}$.

13. $\begin{vmatrix} l_n & l_{n+k} & l_{n+2k} \\ l_{n+3k} & l_{n+4k} & l_{n+5k} \\ l_{n+6k} & l_{n+7k} & l_{n+8k} \end{vmatrix}$.

14. $\begin{vmatrix} xl_{n+3} & l_{n+2} & xl_{n+1} & l_n \\ l_{n+2} & xl_{n+3} & l_n & xl_{n+1} \\ xl_{n+1} & l_n & xl_{n+3} & l_{n+2} \\ l_n & xl_{n+1} & l_{n+2} & xl_{n+3} \end{vmatrix}$.

15. $\begin{vmatrix} f_n & f_{n+k} & f_{n+2k} \\ f_{n+3k} & f_{n+4k} & f_{n+5k} \\ f_{n+6k} & f_{n+7k} & f_{n+8k} \end{vmatrix}$.

16. $\begin{vmatrix} f_n^2 & f_{n+1}^2 & f_{n+2}^2 \\ f_{n+1}^2 & f_{n+2}^2 & f_{n+3}^2 \\ f_{n+2}^2 & f_{n+3}^2 & f_{n+4}^2 \end{vmatrix}$.

17. $\begin{vmatrix} l_n^2 & l_{n+1}^2 & l_{n+2}^2 \\ l_{n+1}^2 & l_{n+2}^2 & l_{n+3}^2 \\ l_{n+2}^2 & l_{n+3}^2 & l_{n+4}^2 \end{vmatrix}$.

18. $\begin{vmatrix} g_n + k & xg_{n+1} + k & g_{n+2} + k \\ g_{n+1} + k & xg_{n+2} + k & g_{n+3} + k \\ g_{n+2} + k & xg_{n+3} + k & g_{n+4} + k \end{vmatrix}$.

19. $\begin{vmatrix} l_0 & l_1 & l_2 & & l_{n-2} & l_{n-1} \\ l_1 & l_0 & l_1 & & l_{n-3} & l_{n-2} \\ l_2 & l_1 & l_0 & \cdots & l_{n-4} & l_{n-3} \\ & & & \vdots & & \\ l_{n-2} & l_{n-3} & l_{n-4} & \cdots & l_0 & l_1 \\ l_{n-1} & l_{n-2} & l_{n-3} & & l_1 & l_0 \end{vmatrix}$.

20. $\begin{vmatrix} x^2+1 & ix & 0 & 0 & & 0 & 0 \\ ix & x^2 & ix & 0 & & 0 & 0 \\ 0 & ix & x^2 & ix & \cdots & 0 & 0 \\ & & & & \vdots & & \\ 0 & 0 & 0 & 0 & \cdots & x^2 & ix \\ 0 & 0 & 0 & 0 & & ix & x^2 \end{vmatrix}$.

21. $$\begin{vmatrix} x^2 & ix & 0 & 0 & & 0 & 0 \\ ix & x^2+1 & ix & 0 & & 0 & 0 \\ 0 & ix & x^2 & ix & \cdots & 0 & 0 \\ & & & & \vdots & & \\ 0 & 0 & 0 & 0 & \cdots & x^2 & ix \\ 0 & 0 & 0 & 0 & & ix & x^2 \end{vmatrix}.$$

22. The determinants $|G_n|$ and $|H_n|$ in Example 39.8.

23. Solve the recurrence $B_n - cB_{n-1} - abB_{n-2}$, where $B_1 = c$, $B_2 = c^2 - ab$, and $n \geq 3$.

24. Prove by PMI that $|B_n(-1, x^2+2, -1)| = \frac{1}{x} f_{2n+2}$.

Evaluate each determinant, using Theorem 39.1.

25. $|H_4|$.

26. $|S_4| = \begin{vmatrix} 1 & -1 & 0 & 0 \\ 1 & x^2+1 & -1 & 0 \\ 1 & 1 & x^2+1 & -1 \\ 1 & 1 & 1 & x^2+1 \end{vmatrix}$, where $|S_0| = 1 = |S_1|$.

27. $|T_4| = \begin{vmatrix} x^2+1 & -1 & 0 & 0 \\ 1 & x^2+1 & -1 & 0 \\ 1 & 1 & x^2+1 & -1 \\ 1 & 1 & 1 & x^2+1 \end{vmatrix}$, where $|T_0| = 1$ and $|T_1| = x^2 + 1$.

40

FIBONOMETRY II

> Nature's great book is written in mathematics.
>
> –Galileo Galilei (1564–1642)

Recall from Chapter 26 that there is a close relationship between Fibonacci and Lucas numbers, and trigonometry [287]. In this chapter, we explore such links with Fibonacci and Lucas polynomials. Through this we discover similar links with Pell and Pell–Lucas polynomials, and hence with Pell and Pell–Lucas numbers. Again, in the interest of brevity, we *omit* arguments from the functional notation throughout our discussion, unless needed for clarity. Also, we assume that x is a real number ≥ 1.

40.1 FIBONOMETRIC RESULTS

We begin our exploration with a simple fibonometric identity. Its proof uses the following well-known trigonometric results:

$$\tan^{-1} u + \tan^{-1} v = \tan^{-1} \frac{u+v}{1-uv}, \quad uv < 1 \tag{40.1}$$

$$\tan^{-1} u - \tan^{-1} v = \tan^{-1} \frac{u-v}{1+uv}, \quad uv > -1 \tag{40.2}$$

$$\tan^{-1} u = \cot^{-1} \frac{1}{u}. \tag{40.3}$$

Theorem 40.1.

$$\tan^{-1} f_{2n+2} - \tan^{-1} f_{2n} = \tan^{-1} \frac{x}{f_{2n+1}}.$$

Proof. By formula (40.2) and the Cassini-like formula, we have

$$\begin{aligned}\tan^{-1} f_{2n+2} - \tan^{-1} f_{2n} &= \tan^{-1} \frac{f_{2n+2} - f_{2n}}{1 + f_{2n+2}f_{2n}} \\ &= \tan^{-1} \frac{x f_{2n+1}}{f_{2n+1}^2} \\ &= \tan^{-1} \frac{x}{f_{2n+1}}. \qquad \blacksquare\end{aligned}$$

It follows by Theorem 40.1 that

$$\begin{aligned}\sum_{k=1}^{n} \tan^{-1} \frac{x}{f_{2k-1}} &= \sum_{k=1}^{n} \left(\tan^{-1} f_{2k} - \tan^{-1} f_{2k-2}\right) \\ &= \tan^{-1} f_{2n} - \tan^{-1} f_0 \\ &= \tan^{-1} f_{2n} - 0 \\ &= \tan^{-1} f_{2n}.\end{aligned}$$

Thus we have the following result [343, 358].

Corollary 40.1. $\sum_{k=1}^{n} \tan^{-1} \frac{x}{f_{2k-1}} = \tan^{-1} f_{2n}.$ ■

In particular, we have

$$\sum_{k=1}^{n} \tan^{-1} \frac{1}{F_{2k-1}} = \tan^{-1} F_{2n}; \tag{40.4}$$

$$\sum_{k=1}^{n} \tan^{-1} \frac{x}{p_{2k-1}} = \tan^{-1} p_{2n};$$

$$\sum_{k=1}^{n} \tan^{-1} \frac{2}{P_{2k-1}} = \tan^{-1} P_{2n}.$$

Br. J. Mahon of Australia developed identity (40.4) in 2006 [342].

Interestingly, we can rewrite Theorem 40.1 in a slightly different way:

$$\tan^{-1} \frac{x}{f_{2k+1}} = \tan^{-1} \frac{1}{f_{2k}} - \tan^{-1} \frac{1}{f_{2k+2}}. \tag{40.5}$$

This implies

$$\sum_{k=1}^{n} \tan^{-1} \frac{x}{f_{2k+1}} = \sum_{k=1}^{n} \left(\tan^{-1} \frac{1}{f_{2k}} - \tan^{-1} \frac{1}{f_{2k+2}} \right)$$
$$= \tan^{-1} \frac{1}{f_2} - \tan^{-1} \frac{1}{f_{2n+2}}$$
$$= \tan^{-1} \frac{1}{x} - \tan^{-1} \frac{1}{f_{2n+2}}$$

$$\sum_{k=1}^{\infty} \tan^{-1} \frac{x}{f_{2k+1}} = \tan^{-1} \frac{1}{x}. \tag{40.6}$$

Hoggatt developed this summation formula in 1966 [216].

It follows from formula (40.6) that

$$\sum_{k=1}^{n} \tan^{-1} \frac{1}{F_{2k+1}} = \frac{\pi}{4}; \tag{40.7}$$

$$\sum_{k=1}^{n} \tan^{-1} \frac{2x}{p_{2k+1}} = \tan^{-1} \frac{1}{2x};$$

$$\sum_{k=1}^{n} \tan^{-1} \frac{2}{P_{2k+1}} = \tan^{-1} \frac{1}{2}.$$

D.H. Lehmer (1905–1991) discovered formula (40.7) in 1936 [287].

Identity (40.5) has a Lucas counterpart:

$$\tan^{-1} \frac{1}{l_{2n}} + \tan^{-1} \frac{1}{l_{2n+2}} = \tan^{-1} \frac{(x^2+4)f_{2n+1}}{(x^2+4)f_{2n+1}^2 + x^2 - 1};$$

see Exercise 40.2.

This identity yields the following dividends:

$$\tan^{-1} \frac{1}{L_{2n}} + \tan^{-1} \frac{1}{L_{2n+2}} = \tan^{-1} \frac{1}{F_{2n+1}}; \tag{40.8}$$

$$\tan^{-1} \frac{1}{q_{2n}} + \tan^{-1} \frac{1}{q_{2n+2}} = \tan^{-1} \frac{4(x^2+1)p_{2n+1}}{4(x^2+1)p_{2n+1}^2 + 4x^2 - 1};$$

$$\tan^{-1} \frac{1}{2Q_{2n}} + \tan^{-1} \frac{1}{2Q_{2n+2}} = \tan^{-1} \frac{8P_{2n+1}}{8P_{2n+1}^2 + 3}.$$

Identity (40.5), coupled with the trigonometric identity $\tan^{-1} x + \tan^{-1} 1/x = \pi/2$, yields an interesting dividend:

$$\begin{aligned}\tan^{-1}\frac{1}{f_{2n}} &= \tan^{-1}\frac{x}{f_{2n+1}} + \tan^{-1}\frac{1}{f_{2n+2}}\\ &= \left(\frac{\pi}{2} - \tan^{-1}\frac{f_{2n+1}}{x}\right) + \left(\frac{\pi}{2} - \tan^{-1} f_{2n+2}\right)\end{aligned}$$

$$\tan^{-1}\frac{1}{f_{2n}} + \tan^{-1}\frac{f_{2n+1}}{x} + \tan^{-1} f_{2n+2} = \pi.$$

This yields the following byproducts:

$$\tan^{-1}\frac{1}{F_{2n}} + \tan^{-1} F_{2n+1} + \tan^{-1} F_{2n+2} = \pi; \tag{40.9}$$

$$\tan^{-1}\frac{1}{p_{2n}} + \tan^{-1}\frac{p_{2n+1}}{x} + \tan^{-1} p_{2n+2} = \pi;$$

$$\tan^{-1}\frac{1}{P_{2n}} + \tan^{-1}\frac{P_{2n+1}}{2} + \tan^{-1} P_{2n+2} = \pi.$$

W.W. Horner of Pittsburgh, Pennsylvania, discovered identity (40.9) in 1968 [238].

It follows from identity (40.5) that

$$\cot^{-1}\frac{f_{2n+1}}{x} = \cot^{-1} f_{2n} - \cot^{-1} f_{2n+2}.$$

We can derive this in a different way.

To begin, using the Binet-like formula, we have

$$f_{2n+1}f_{2n+2} - f_{2n}f_{2n+3} = x; \tag{40.10}$$

see Exercise 40.3.

For example, $f_5 f_6 - f_4 f_7 = (x^4 + 3x^2 + 1)(x^5 + 4x^3 + 3x) - (x^3 + 2x)(x^6 + 5x^4 + 6x^2 + 1) = x$.

It follows from identity (40.10) that

$$\begin{aligned}f_{2n} &= \frac{f_{2n+1}f_{2n+2} - x}{xf_{2n+2} + f_{2n+1}}\\ &= \frac{(f_{2n+1}/x)f_{2n+2} - 1}{(f_{2n+1}/x) + f_{2n+2}}.\end{aligned}$$

This implies

$$\cot^{-1} f_{2n} = \cot^{-1}\frac{f_{2n+1}}{x} + \cot^{-1} f_{2n+2}.$$

Consequently,

$$\begin{aligned}\sum_{k=0}^{n} \cot^{-1} \frac{f_{2k+1}}{x} &= \sum_{k=0}^{n} \left(\cot^{-1} f_{2k} - \cot^{-1} f_{2k+2}\right)\\ &= \cot^{-1} 0 - \cot^{-1} f_{2n+2}\\ &= \pi/2 - \cot^{-1} f_{2n+2}.\end{aligned}$$

This yields the summation formula

$$\sum_{k=0}^{\infty} \cot^{-1} \frac{f_{2k+1}}{x} = \frac{\pi}{2}. \tag{40.11}$$

In particular,

$$\sum_{k=0}^{\infty} \cot^{-1} F_{2k+1} = \frac{\pi}{2};$$

$$\sum_{k=0}^{\infty} \cot^{-1} \frac{p_{2k+1}}{2x} = \frac{\pi}{2};$$

$$\sum_{k=0}^{\infty} \cot^{-1} \frac{P_{2k+1}}{2} = \frac{\pi}{2}.$$

Returning to Theorem 40.1, we see it has a companion result for odd-numbered Fibonacci polynomials, as the following theorem shows.

Theorem 40.2.

$$\tan^{-1} \frac{1}{f_{2n-1}} + \tan^{-1} \frac{1}{f_{2n+1}} = \tan^{-1} \frac{l_{2n}}{f_{2n}^2}.$$ ■

The proof is straightforward, so we omit it; see Exercise 40.4.
In particular, we have

$$\tan^{-1} \frac{1}{F_{2n-1}} + \tan^{-1} \frac{1}{F_{2n+1}} = \tan^{-1} \frac{L_{2n}}{F_{2n}^2};$$

$$\tan^{-1} \frac{1}{p_{2n-1}} + \tan^{-1} \frac{1}{p_{2n+1}} = \tan^{-1} \frac{q_{2n}}{p_{2n}^2};$$

$$\tan^{-1} \frac{1}{P_{2n-1}} + \tan^{-1} \frac{1}{P_{2n+1}} = \tan^{-1} \frac{2Q_{2n}}{P_{2n}^2}.$$

It also follows from Theorem 40.2 that

$$\sum_{k=1}^{n}(-1)^{k-1}\tan^{-1}\frac{l_{2k}}{f_{2k}^2}=\sum_{k=1}^{n}(-1)^{k-1}\left(\tan^{-1}\frac{1}{f_{2k-1}}+\tan^{-1}\frac{1}{f_{2k+1}}\right)$$
$$=\tan^{-1}+(-1)^{n-1}\tan^{-1}\frac{1}{f_{2n+1}}$$
$$=\frac{\pi}{4}-(-1)^{n}\tan^{-1}\frac{1}{f_{2n+1}}.$$

Consequently,

$$\sum_{k=1}^{\infty}(-1)^{k-1}\tan^{-1}\frac{l_{2k}}{f_{2k}^2}=\frac{\pi}{4}, \tag{40.12}$$

and hence

$$\sum_{k=1}^{\infty}(-1)^{k-1}\cot^{-1}\frac{f_{2k}^2}{l_{2k}}=\frac{\pi}{4}.$$

This formula appears in a slightly different form in [343]; Melham and Shannon re-discovered it in 1995 [358].

We now explore a few additional trigonometric facts involving gibonacci polynomials. But before we do, recall that $\mu=\mu(x)=a^2+abx-b^2$. When $g_n=f_n$, $\mu=1$; and when $g_n=l_n$, $\mu=-(x^2+4)$. Also, $f_{n+1}+f_{n-1}=l_n$, $l_{n+1}+l_{n-1}=(x^2+4)f_n$, and $f_nl_n=f_{2n}$.

Theorem 40.3. *Let g_n denote the nth gibonacci polynomial. Then*

$$\tan\left(\tan^{-1}\frac{g_n}{g_{n+1}}-\tan^{-1}\frac{g_{n+1}}{g_{n+2}}\right)=\frac{\mu(-1)^{n+1}}{g_{n+1}(g_n+g_{n+2})}.$$

Proof. By Theorem 36.4, we have

$$\text{LHS}=\frac{\dfrac{g_n}{g_{n+1}}-\dfrac{g_{n+1}}{g_{n+2}}}{1+\dfrac{g_ng_{n+1}}{g_{n+1}g_{n+2}}}$$
$$=\frac{g_{n+2}g_n-g_{n+1}^2}{g_{n+1}(g_{n+2}+g_n)}$$
$$=\frac{\mu(-1)^{n+1}}{g_{n+1}(g_n+g_{n+2})}.$$

This gives the desired trigonometric identity. ■

The next corollary follows from this theorem.

Corollary 40.2.

$$\tan\left(\tan^{-1}\frac{f_n}{f_{n+1}} - \tan^{-1}\frac{f_{n+1}}{f_{n+2}}\right) = \frac{(-1)^{n+1}}{f_{2n+2}}; \tag{40.13}$$

$$\tan\left(\tan^{-1}\frac{l_n}{l_{n+1}} - \tan^{-1}\frac{l_{n+1}}{l_{n+2}}\right) = \frac{(-1)^{n}}{f_{2n+2}}. \tag{40.14}$$

The next result follows from this corollary. It also follows by PMI; see Exercise 40.17. ■

Theorem 40.4.

$$\tan^{-1}\frac{f_n}{f_{n+1}} = \sum_{i=1}^{n}(-1)^{i+1}\tan^{-1}\frac{1}{f_{2i}}.$$

■

The next result follows from this theorem; see Exercise 40.19.

Corollary 40.3.

$$\sum_{n=1}^{\infty}(-1)^{n+1}\tan^{-1}\frac{1}{f_{2n}} = \tan^{-1}(-\beta(x)).$$

■

This corollary implies

$$\sum_{n=1}^{\infty}\tan^{-1}\frac{(-1)^n}{F_{2n}} = \tan^{-1}(-\beta);$$

$$\sum_{n=1}^{\infty}\tan^{-1}\frac{(-1)^n}{p_{2n}} = \tan^{-1}(-\delta(x));$$

$$\sum_{n=1}^{\infty}\tan^{-1}\frac{(-1)^n}{P_{2n}} = \tan^{-1}(-\delta),$$

where $\delta(x) = x - \sqrt{x^2+1}$.

Similar results follow from identity (40.14); see Exercises 40.18–40.24.

This has a somewhat related result for odd-numbered Fibonacci polynomials, as the next theorem shows. Its proof hinges on the following result, and is straightforward; see Exercise 40.25.

Lemma 40.1.

$$\tan^{-1}\frac{x}{f_{2n+1}} = \tan^{-1}\frac{1}{f_{2n}} - \tan^{-1}\frac{1}{f_{2n+2}}.$$

■

We are now ready for the theorem.

Theorem 40.5.

$$\sum_{n=1}^{\infty} \tan^{-1} \frac{x}{f_{2n+1}} = \frac{\pi}{4}.$$

Proof. Since $\tan^{-1}$ is a continuous function, by Lemma 40.1, we have

$$\begin{aligned}
\sum_{n=1}^{k} \tan^{-1} \frac{x}{f_{2n+1}} &= \sum_{n=1}^{k} \left(\tan^{-1} \frac{1}{f_{2n}} - \tan^{-1} \frac{1}{f_{2n+2}} \right) \\
&= \frac{\pi}{4} - \tan^{-1} \frac{1}{f_{2k+2}} \\
\sum_{n=1}^{\infty} \tan^{-1} \frac{x}{f_{2n+1}} &= \frac{\pi}{4} - \tan^{-1} 0 \\
&= \frac{\pi}{4}.
\end{aligned}$$

■

The next theorem gives a somewhat similar result involving odd-numbered Lucas polynomials. The proof hinges on the following lemma (see Exercise 40.27), which is the Lucas counterpart of Lemma 40.1.

Lemma 40.2.

$$\tan^{-1} \frac{x l_{2n+1}}{l_{2n+1}^2 + x^2 + 5} = \tan^{-1} \frac{1}{l_{2n}} - \tan^{-1} \frac{1}{l_{2n+2}}.$$

■

We now present the theorem.

Theorem 40.6.

$$\sum_{n=1}^{\infty} \tan^{-1} \frac{x l_{2n+1}}{l_{2n+1}^2 + x^2 + 5} = \tan^{-1} \frac{1}{x^2 + 2}.$$

Proof. Recall that $\tan^{-1}$ is a continuous function. By Lemma 40.2 and Theorem 40.5, we have

$$\begin{aligned}
\sum_{n=1}^{k} \tan^{-1} \frac{x l_{2n+1}}{l_{2n+1}^2 + x^2 + 5} &= \sum_{n=1}^{k} \left(\tan^{-1} \frac{1}{l_{2n}} - \tan^{-1} \frac{1}{l_{2n+2}} \right) \\
&= \tan^{-1} \frac{1}{l_2} - \tan^{-1} \frac{1}{l_{2k+2}} \\
\sum_{n=1}^{\infty} \tan^{-1} \frac{x l_{2n+1}}{l_{2n+1}^2 + x^2 + 5} &= \tan^{-1} \frac{1}{l_2} - 0.
\end{aligned}$$

This gives the desired result. ■

In the next example, we develop series expansions of $\arctan 2x/5$ and $\arctan \dfrac{2\sqrt{x^2+4}}{3}$. We use the following facts to accomplish this:

$$\arctan x + \arctan y = \frac{x+y}{1-xy}$$

$$\arctan x = \sum_{n=0}^{\infty} \frac{(-1)^n}{2n+1} x^{2n+1},$$

where $|x| \le 1$. In addition, we need the fact that, if the series $\sum_{n=0}^{\infty} a_n$ and $\sum_{n=0}^{\infty} b_n$ converge to A and B, respectively, then $\sum_{n=0}^{\infty} (a_n \pm b_n)$ converges to $A \pm B$.

Example 40.1. *Prove that*

$$\arctan \frac{xz}{z^2+1} = \sum_{n=0}^{\infty} \frac{(-1)^n}{2n+1} \cdot \frac{l_{2n+1}}{z^{2n+1}}; \tag{40.15}$$

$$\arctan \frac{z\sqrt{x^2+4}}{z^2-1} = \sum_{n=0}^{\infty} \frac{(-1)^n \sqrt{x^2+4}}{2n+1} \cdot \frac{f_{2n+1}}{z^{2n+1}}. \tag{40.16}$$

Proof.

1) By the Binet-like formula for l_k, we have

$$\begin{aligned}
\text{RHS} &= \sum_{n=0}^{\infty} \frac{(-1)^n}{2n+1} (\alpha/z)^{2n+1} + \sum_{n=0}^{\infty} \frac{(-1)^n}{2n+1} (\beta/z)^{2n+1} \\
&= \arctan(\alpha/z) + \arctan(\beta/z) \\
&= \arctan \frac{(\alpha+\beta)/z}{1-\alpha\beta/z^2} \\
&= \arctan \frac{xz}{z^2+1} \\
&= \text{LHS}.
\end{aligned}$$

2)

$$\begin{aligned}
\text{RHS} &= \sum_{n=0}^{\infty} \frac{(-1)^n}{2n+1} (\alpha/z)^{2n+1} - \sum_{n=0}^{\infty} \frac{(-1)^n}{2n+1} (\beta/z)^{2n+1} \\
&= \arctan(\alpha/z) - \arctan(\beta/z) \\
&= \arctan \frac{(\alpha-\beta)/z}{1+\alpha\beta/z^2} \\
&= \arctan \frac{z\sqrt{x^2+4}}{z^2-1} \\
&= \text{LHS}.
\end{aligned}$$

■

This example has several implications. Letting $z = 2$, formulas (40.15) and (40.16) yield

$$\arctan\frac{2x}{5} = \sum_{n=0}^{\infty}\frac{(-1)^n l_{2n+1}}{(2n+1)2^{2n+1}}$$

$$\arctan\frac{2\sqrt{x^2+4}}{3} = \sum_{n=0}^{\infty}\frac{(-1)^n\sqrt{x^2+4}f_{2n+1}}{(2n+1)2^{2n+1}}$$

$$\arctan\frac{4x}{5} = \sum_{n=0}^{\infty}\frac{(-1)^n q_{2n+1}}{(2n+1)2^{2n+1}}$$

$$\arctan\frac{4\sqrt{x^2+1}}{3} = \sum_{n=0}^{\infty}\frac{(-1)^n\sqrt{x^2+1}p_{2n+1}}{(2n+1)2^{2n}}.$$

40.2 HYPERBOLIC FUNCTIONS

Fibonacci polynomials, coupled with hyperbolic functions and differential calculus, yield interesting dividends. The next example, studied by Seiffert in 1991, is such a confluence [67, 417]. As we might predict, the proof is a bit long and involved. It uses the following hyperbolic facts:

$$2\sinh x = e^x - e^{-x}$$

$$2\cosh x = e^x + e^{-x}$$

$$\cosh^2 x - \sinh^2 x = 1$$

$$\cosh(x \pm y) = \cosh x\cosh y \pm \sinh x\sinh y$$

$$\sinh 2x = 2\sinh x\cosh x$$

$$\cosh 2x = \cosh^2 x + \sinh^2 x.$$

Example 40.2. *Prove that*

$$\sum_{k=1}^{n-1}\frac{1}{x^2+\sin^2 k\pi/2n} = \frac{(2n-1)f_{2n+1}(2x)+(2n+1)f_{2n-1}(2x)}{4x(x^2+1)f_{2n}(2x)} - \frac{1}{2x^2}; \quad (40.17)$$

$$\sum_{k=1}^{n-1}\csc^2 k\pi/2n = 2(n^2-1)/3.$$

Proof. The characteristic equation of the recurrence satisfied by f_{2n} is $z^2 - 2xz - 1 = 0$; so the characteristic roots are $r = x + D$ and $s = x - D$, where $D = \sqrt{x^2+1}$.

Let $x = \sinh u$, where $u \geq 0$. Then $r = e^u$, $s = -e^{-u}$, and $D = \cosh u$. It follows by the Binet-like formula that

$$\begin{aligned} f_{2n}(2x) &= \frac{r^{2n} - s^{2n}}{r - s} \\ &= \frac{e^{2nu} - e^{-2nu}}{e^u + e^u} \\ &= \frac{\sinh 2nu}{\cosh u}. \end{aligned} \tag{40.18}$$

Similarly,

$$f_{2n+1}(2x) = \frac{\cosh(2n+1)u}{\cosh u}. \tag{40.19}$$

Consequently, $f_{2n+1}(2x) + f_{2n-1}(2x) = 2\cosh 2nu$. Also recall that $f_{2n+1}(2x) - f_{2n-1}(2x) = 2xf_{2n}(2x)$.

It follows from equation (40.18) that $f_{2n}(x) = 0$ if and only if $e^{2nu} = e^{-2nu}$; that is, if and only if $2nu = k\pi i$, where $-(n-1) \leq k \leq n-1$. So $f_{2n}(x) = 0$ if and only if $2nu = \pm\pi i$, where $0 \leq k \leq n-1$; that is, if and only if $x = \sinh u = \sinh(\pm k\pi i/2n) = \pm i \sin k\pi/2n$. (We add that Hoggatt and Bicknell studied the roots of $f_n(x)$ in 1973 [224].)

Since the leading term of $f_{2n}(2x)$ is $(2x)^{2n-1}$, this implies

$$\begin{aligned} f_{2n}(2x) &= 2^{2n-1}x\prod_{k=1}^{n-1}(x + i\sin k\pi/2n)(x - i\sin k\pi/2n) \\ &= 2^{2n-1}x\prod_{k=1}^{n-1}(x^2 + \sin^2 k\pi/2n). \end{aligned} \tag{40.20}$$

Taking natural logarithms of both sides of equation (40.20) and differentiating the resulting equation with respect to x, we get

$$\begin{aligned} \frac{2f'_{2n}(2x)}{f_{2n}(2x)} &= \frac{1}{x} + \sum_{k=1}^{n-1}\frac{2x}{x^2 + \sin^2 k\pi/2n} \\ \frac{f'_{2n}(2x)}{xf_{2n}(2x)} - \frac{1}{2x^2} &= \sum_{k=1}^{n-1}\frac{2x}{x^2 + \sin^2 k\pi/2n}, \end{aligned} \tag{40.21}$$

where prime denotes differentiation with respect to x. In the interest of brevity, we denote the LHS of equation (40.21) by $A_n(x)$.

Since $D' = \dfrac{x}{D} = \tanh u$ and $u' = \dfrac{1}{\cosh u} = \dfrac{1}{\sqrt{x^2+1}} = \dfrac{1}{D}$, differentiating equation (40.18) with respect to x gives

$$2f'_{2n}(2x) = \frac{1}{D^2}(2n\cosh 2nu - \tanh u \sinh 2nu) \qquad (40.22)$$

$$= \frac{1}{D^2}\left\{n[f_{2n+1}(2x) + f_{2n-1}(2x)] - \frac{x}{D}\cdot Df_{2n}(2x)\right\}$$

$$= \frac{n}{D^2}\left[f_{2n+1}(2x) + f_{2n-1}(2x)\right] - \frac{x}{D^2}f_{2n}(2x)$$

$$\frac{f'_{2n}(2x)}{xf_{2n}(2x)} = \frac{2n\left[f_{2n+1}(2x) + f_{2n-1}(2x)\right]}{4x(x^2+1)f_{2n}(2x)} - \frac{1}{2(x^2+1)}$$

$$= \frac{(2n-1)f_{2n+1}(2x) + (2n+1)f_{2n-1}(2x)}{4x(x^2+1)f_{2n}(2x)}$$

$$+ \frac{f_{2n+1}(2x) - f_{2n-1}(2x)}{4x(x^2+1)f_{2n}(2x)} - \frac{1}{2(x^2+1)}$$

$$= \frac{(2n-1)f_{2n+1}(2x) + (2n+1)f_{2n-1}(2x)}{4x(x^2+1)f_{2n}(2x)}$$

$$+ \frac{2xf_{2n}(2x)}{4x(x^2+1)f_{2n}(2x)} - \frac{1}{2(x^2+1)}$$

$$\frac{f'_{2n}(2x)}{xf_{2n}(2x)} - \frac{1}{2x^2} = \frac{(2n-1)f_{2n+1}(2x) + (2n+1)f_{2n-1}(2x)}{4x(x^2+1)f_{2n}(2x)} - \frac{1}{2x^2}. \qquad (40.23)$$

Equation (40.17) now follows by equations (40.21) and (40.23).

To establish the second equation, we let $B_n(x)$ denote the RHS of equation (40.23); so $A_n(x) = B_n(x)$.

To begin, we rewrite $B_n(x)$ in a different form. It follows from equation (40.22) that

$$\frac{f'_{2n}(2x)}{xf_{2n}(2x)} = \frac{2f'_{2n}(2x)}{(2x/D)\sinh 2nu}$$

$$= \frac{1}{D^2}\cdot\frac{D}{2x\sinh 2nu}(2n\cosh 2nu - \tanh u \sinh 2nu)$$

$$= \frac{n\coth 2nu}{Dx} - \frac{1}{2(x^2+1)}$$

$$B_n(x) = \frac{n\coth 2nu}{Dx} - \frac{1}{2(x^2+1)} - \frac{1}{2x^2}$$

$$= \frac{n\coth 2nu}{Dx} - \frac{2x^2+1}{2x^2(x^2+1)}.$$

Since $2x^2 + 1 = 2\sinh^2 u + 1 = \sinh^2 u + \cosh^2 u = \cosh 2u$, and $4x^2(x^2+1) = 4x^2D^2 = (2\sinh u\cosh u)^2 = \sinh^2 2u$, this implies

$$A_n(x) = B_n(x) = \frac{n\coth 2nu}{Dx} - \frac{2\cosh 2u}{\sinh^2 2u}.$$

Thus

$$\begin{aligned} A_n(x) &= \frac{2n\coth 2nu}{2Dx} - 2\cosh 2u \cdot \operatorname{csch}^2 2u \\ &= 2n\coth 2nu \cdot \operatorname{csch} 2u - 2\cosh 2u \cdot \operatorname{csch}^2 2u. \qquad (40.24) \end{aligned}$$

Let $C_n(u)$ denote the RHS of equation (40.24). Since $x = \sinh u$, it follows that $\lim_{x\to 0} A_n(x) = A_n(0) = C_n(0) = \lim_{u\to\infty} C_n(u)$, provided that both limits exist. From equation (40.21), $A_n(0)$ exists.

We now show that $\lim_{u\to\infty} C_n(u)$ also exists. To establish this, we use the following hyperbolic expansions in terms of the *big-oh notation* [276]:

$$\cosh z = 1 + \frac{z^2}{2} + O(z^4) \qquad \coth z = \frac{1}{z}\left[1 + \frac{z^2}{3} + O(z^4)\right]$$

$$\operatorname{csch} z = \frac{1}{z}\left[1 - \frac{z^2}{6} + O(z^4)\right] \qquad \operatorname{csch}^2 z = \frac{1}{z^2}\left[1 - \frac{z^2}{3} + O(z^4)\right].$$

By equation (40.24), we then have

$$\begin{aligned} C_n(u) &= \frac{2n}{2nu}\left[1 + \frac{4n^2u^2}{3} + O(u^4)\right] \cdot \frac{1}{2u}\left[1 - \frac{2u^2}{3} + O(u^4)\right] \\ &\quad - 2\left[1 + 2u^2 + O(u^4)\right] \cdot \frac{1}{4u^2}\left[1 - \frac{4u^2}{3} + O(u^4)\right] \\ &= \frac{1}{2u^2}\left[1 + \frac{2}{3}(2n^2-1)u^2 - 1 - \frac{2u^2}{3}\right] + O(u^2) \\ &= \frac{2}{3}(n^2-1) + O(u^2). \end{aligned}$$

So $C_n(0) = \frac{2}{3}(n^2-1)$. Thus $A_n(0) = \sum_{k=1}^{n-1} \csc^2 k\pi/2n = \frac{2}{3}(n^2-1)$, as desired. ■

It follows by equation (40.17) that

$$\sum_{k=1}^{n-1} \frac{1}{x^2/4 + \sin^2 k\pi/2n} = \frac{(2n-1)f_{2n+1} + (2n+1)f_{2n-1}}{4(x/2)(x^2/4+1)f_{2n}} - \frac{1}{2(x/2)^2}$$

$$\sum_{k=1}^{n-1} \frac{2}{x^2 + 4\sin^2 k\pi/2n} = \frac{2nl_{2n} - xf_{2n}}{x(x^2+4)f_{2n}} - \frac{1}{x^2}. \qquad (40.25)$$

Consequently,

$$\sum_{k=1}^{n-1} \frac{2}{1+4\sin^2 k\pi/2n} = \frac{2nL_{2n} - F_{2n}}{5F_{2n}} - 1$$
$$= \frac{2nL_{2n}}{5F_{2n}} - \frac{6}{5}.$$

In particular, let $n = 3$. Then

$$\text{LHS} = \frac{2}{1+4\sin^2 \pi/6} + \frac{2}{1+4\sin^2 \pi/3}$$
$$= \frac{3}{2}$$
$$= \frac{6L_6}{5F_6} - \frac{6}{5}$$
$$= \text{RHS}.$$

Since $f_{2k}(2x) = p_k(x)$, it also follows from equation (40.17) that

$$\sum_{k=1}^{n-1} \frac{1}{x^2 + \sin^2 k\pi/2n} = \frac{(2n-1)p_{2n+1} + (2n+1)p_{2n-1}}{4x(x^2+1)p_{2n}} - \frac{1}{2x^2}$$
$$= \frac{nq_{2n} - xp_{2n}}{2x(x^2+1)p_{2n}} - \frac{1}{2x^2}.$$

This implies

$$\sum_{k=1}^{n-1} \frac{1}{1+\sin^2 k\pi/2n} = \frac{2nQ_{2n} - P_{2n}}{4P_{2n}} - \frac{1}{2}$$
$$= \frac{nQ_{2n}}{2P_{2n}} - \frac{3}{4}.$$

For example,

$$\sum_{k=1}^{2} \frac{1}{1+\sin^2 k\pi/6} = \frac{48}{35}$$
$$= \frac{3\cdot 99}{2\cdot 70} - \frac{3}{4}$$
$$= \frac{3Q_6}{2P_6} - \frac{3}{4}.$$

Using hyperbolic functions and the Binet-like formula for f_n, we now evaluate an interesting infinite integral. Seiffert studied it in 1992 [421].

Example 40.3. *Evaluate the infinite integral* $\int_0^\infty \frac{dx}{(x^2+1)f_{2n+1}(2x)}$.

Solution. Let I_n denote the given integral. Since $\alpha(2x) = x + \sqrt{x^2+1}$, $\beta(2x) = x - \sqrt{x^2+1}$, and $\Delta(2x) = 2\sqrt{x^2+1}$, we have

$$I_n = \int_0^\infty \frac{2dx}{\sqrt{x^2+1}\left[(x+\sqrt{x^2+1})^{2n+1} - (x-\sqrt{x^2+1})^{2n+1}\right]}.$$

Now make the substitution $x = \sinh t$. Since $\cosh^2 z - \sinh^2 z = 1$, $x + \sqrt{x^2+1} = \sinh t + \cosh t$. Since $(\cosh z + \sinh z)^n = \cosh nz + \sinh nz$, $\alpha^{2n+1}(2x) - \beta^{2n+1}(2x) = (\sinh t + \cosh t)^{2n+1} - (\sinh t - \cosh t)^{2n+1} = 2\cosh(2n+1)t$. Using $\frac{d}{dx}(\sinh z) = \cosh z$, we then have

$$I_n = \int_0^\infty \frac{dt}{\cosh(2n+1)t}.$$

Now let $u = (2n+1)t$. Then

$$\begin{aligned} I_n &= \int_0^\infty \operatorname{sech} u \, du \\ &= \frac{1}{2n+1}\left[\arctan(\sinh u)\right]_0^\infty \\ &= \frac{\pi}{4n+2}. \end{aligned}$$

■

It follows by this example that $\int_0^\infty \frac{dx}{(x^2+1)p_{2n+1}(x)} = \frac{\pi}{4n+2}$.

Next we study some interesting results involving the inverse hyperbolic functions $\tanh^{-1}$ and $\coth^{-1}$.

40.3 INVERSE HYPERBOLIC SUMMATION FORMULAS

The inverse tangent formulas (40.1)–(40.3) have their hyperbolic counterparts [1, 358]:

$$\tanh^{-1} u + \tanh^{-1} v = \tanh^{-1} \frac{u+v}{1+uv} \tag{40.26}$$

$$\tanh^{-1} u - \tanh^{-1} v = \tanh^{-1} \frac{u-v}{1-uv} \tag{40.27}$$

$$\tanh^{-1} u = \coth^{-1} \frac{1}{u}. \tag{40.28}$$

We can use them to develop inverse hyperbolic summation formulas.

Theorem 40.7. *Let $n \geq 1$. Then*

$$\tanh^{-1}\frac{1}{f_{2n+2}} + \tanh^{-1}\frac{1}{f_{2n+4}} = \tanh^{-1}\frac{l_{2n+3}}{f_{2n+3}^2} \tag{40.29}$$

$$\tanh^{-1}\frac{1}{f_{2n+1}} - \tanh^{-1}\frac{1}{f_{2n+3}} = \tanh^{-1}\frac{x}{f_{2n+2}^2}. \tag{40.30}$$

Proof. [We will prove identity (40.29) and leave the other as an exercise; see Exercise 40.28.]

Using identity (40.26), we have

$$\begin{aligned}\tanh(\text{LHS}) &= \frac{\dfrac{1}{f_{2n+2}} + \dfrac{1}{f_{2n+4}}}{1 + \dfrac{1}{f_{2n+2}f_{2n+4}}}\\ &= \frac{f_{2n+2} + f_{2n+4}}{f_{2n+2}f_{2n+4} + 1}\\ &= \frac{l_{2n+3}}{f_{2n+3}^2}.\end{aligned}$$

This yields the desired result. ■

It follows from identity (40.29) that

$$\begin{aligned}\sum_{k=1}^{n}(-1)^{k-1}\tanh^{-1}\frac{l_{2n+3}}{f_{2n+3}^2} &= \sum_{k=1}^{n}(-1)^{k-1}\left(\tanh^{-1}\frac{1}{f_{2k+2}} + \tanh^{-1}\frac{1}{f_{2k+4}}\right)\\ &= \tanh^{-1}\frac{1}{f_4} - (-1)^n\tanh^{-1}\frac{1}{f_{2n+4}}\\ \sum_{k=1}^{\infty}(-1)^{k-1}\tanh^{-1}\frac{l_{2n+3}}{f_{2n+3}^2} &= \tanh^{-1}\frac{1}{f_4}\\ &= \tanh^{-1}\frac{1}{x^3+2x}.\end{aligned} \tag{40.31}$$

Similarly, we can show that

$$\sum_{k=1}^{\infty}\tanh^{-1}\frac{x}{f_{2k+2}} = \tanh^{-1}\frac{1}{x^2+1}; \tag{40.32}$$

see Exercise 40.29.

Since $\tanh^{-1} u = \coth^{-1} \frac{1}{u}$, it follows by identities (40.29)–(40.32) that

$$\coth^{-1} f_{2n+2} + \coth^{-1} f_{2n+4} = \coth^{-1} \frac{f_{2n+3}^2}{l_{2n+3}};$$

$$\coth^{-1} f_{2n+1} - \coth^{-1} f_{2n+3} = \coth^{-1} \frac{f_{2n+3}}{x};$$

$$\sum_{k=1}^{\infty} (-1)^{k-1} \coth^{-1} \frac{f_{2n+3}^2}{l_{2n+3}} = \coth^{-1} (x^3 + 2x);$$

$$\sum_{k=1}^{\infty} (-1)^{k-1} \coth^{-1} \frac{f_{2k+2}}{x} = \coth^{-1} (x^2 + 1).$$

Interesting Byproducts

Identities (40.31) and (40.32) have interesting applications to the Fibonacci and Pell families. To this end, we use the formula

$$\tanh^{-1} u = \frac{1}{2} \ln \frac{1+u}{1-u},$$

where $|u| < 1$.

Since $\tanh^{-1}(-u) = -\tanh^{-1} u$, it follows by identity (40.31) that

$$\begin{aligned}\sum_{k=1}^{\infty} \tanh^{-1} \frac{(-1)^{k-1} L_{2k+3}}{F_{2k+3}^2} &= \frac{1}{2} \ln \frac{1+1/3}{1-1/3} \\ &= \ln 2\end{aligned}$$

$$\sum_{k=1}^{\infty} \ln \frac{1 + \dfrac{(-1)^{k-1} L_{2k+3}}{F_{2k+3}^2}}{1 - \dfrac{(-1)^{k-1} L_{2k+3}}{F_{2k+3}^2}} = \ln 2$$

$$\prod_{k=1}^{\infty} \frac{F_{2k+3}^2 - (-1)^k L_{2k+3}}{F_{2k+3}^2 + (-1)^k L_{2k+3}} = 2. \tag{40.33}$$

Similarly, identity (40.32) yields

$$\prod_{k=1}^{\infty} \frac{F_{2k+2} + 1}{F_{2k+2} - 1} = 3; \tag{40.34}$$

see Exercise 40.30.

It follows from identities (40.29)–(40.32) that

$$\tanh^{-1}\frac{1}{p_{2n+2}}+\tanh^{-1}\frac{1}{p_{2n+4}}=\tanh^{-1}\frac{q_{2n+3}}{p^2_{2n+3}};$$

$$\tanh^{-1}\frac{1}{p_{2n+1}}-\tanh^{-1}\frac{1}{p_{2n+3}}=\tanh^{-1}\frac{2x}{p_{2n+2}};$$

$$\sum_{k=1}^{\infty}(-1)^{k-1}\tanh^{-1}\frac{q_{2n+3}}{p^2_{2n+3}}=\tanh^{-1}\frac{1}{8x^3+4x}; \tag{40.35}$$

$$\sum_{k=1}^{\infty}\tanh^{-1}\frac{2x}{p_{2k+2}}=\tanh^{-1}\frac{1}{4x^2+1}. \tag{40.36}$$

Using formulas (40.35) and (40.36), we can evaluate two infinite products, similar to (40.33) and (40.34):

$$\prod_{k=1}^{\infty}\frac{P^2_{2k+3}-2(-1)^kQ_{2k+3}}{P^2_{2k+3}+2(-1)^kQ_{2k+3}}=\frac{13}{11}; \tag{40.37}$$

$$\prod_{k=1}^{\infty}\frac{P^2_{2k+3}+2}{P^2_{2k+3}-2}=\frac{3}{2}; \tag{40.38}$$

see Exercises 40.31 and 40.32.

The next result [358] is an application of the Catalan-like identity $f_{n+k}f_{n-k}-f_n^2=(-1)^{n+k+1}f_k^2$ and the identity $f_{n+k}-f_{n-k}=\begin{cases} f_nl_k & \text{if } k \text{ is odd}\\ f_kl_n & \text{otherwise}\end{cases}$.
In the interest of brevity, we omit its proof; see Exercise 40.33.

Theorem 40.8 (Melham and Shannon, 1995 [358]). *Let $n \geq k \geq 2$. Then*

$$\tanh^{-1}\frac{f_n}{f_{n+k}}-\tanh^{-1}\frac{f_{n-k}}{f_n}=\begin{cases}\tanh^{-1}\dfrac{(-1)^{n-1}f_k^2}{f_n^2l_k} & \text{if } k \text{ is odd}\\[2ex] \tanh^{-1}\dfrac{(-1)^nf_k}{f_{2n}} & \text{otherwise.}\end{cases}$$

■

It follows by this theorem that

$$\sum_{i=1}^{n}\tanh^{-1}\frac{f_k}{f_{2ik}}=\tanh^{-1}\frac{f_{nk}}{f_{(n+1)k}}, \tag{40.39}$$

where k is even, and

$$\sum_{i=1}^{n} \tanh^{-1} \frac{(-1)^{i-1} f_k^2}{f_{ik}^2 l_k} = \tanh^{-1} \frac{f_{nk}}{f_{(n+1)k}}, \tag{40.40}$$

where k is odd; see Exercises 40.34 and 40.35.

Correspondingly, we have

$$\begin{aligned}\sum_{i=1}^{\infty} \tanh^{-1} \frac{f_k}{f_{2ik}} &= \lim_{n\to\infty} \left(\tanh^{-1} \frac{f_{nk}}{f_{(n+1)k}}\right) \\ &= \tanh^{-1} \left(\lim_{n\to\infty} \frac{f_{nk}}{f_{(n+1)k}}\right) \\ &= \tanh^{-1} \alpha^{-k},\end{aligned} \tag{40.41}$$

where k is even and $\alpha = \alpha(x)$; and similarly,

$$\sum_{i=1}^{\infty} \tanh^{-1} \frac{(-1)^{i-1} f_k^2}{f_{ik}^2 l_k} = \tanh^{-1} \alpha^{-k}, \tag{40.42}$$

where k is odd and $\alpha = \alpha(x)$.

Let $k \geq 2$ and be even. It then follows from formula (40.41) that

$$\sum_{i=1}^{\infty} \tanh^{-1} \frac{F_k}{F_{2ik}} = \tanh^{-1} \alpha^{-k}, \tag{40.43}$$

$$\sum_{i=1}^{\infty} \tanh^{-1} \frac{p_k}{p_{2ik}} = \tanh^{-1} \gamma^{-k}, \tag{40.44}$$

where $\gamma = \gamma(x) = x + \sqrt{x^2 + 1}$.

On the other hand, let $k \geq 2$ and be odd. From formula (40.42), we then have

$$\sum_{i=1}^{\infty} \tanh^{-1} \frac{(-1)^{i-1} F_k^2}{F_{ik}^2 L_k} = \tanh^{-1} \alpha^{-k}, \tag{40.45}$$

$$\sum_{i=1}^{\infty} \tanh^{-1} \frac{(-1)^{i-1} p_k^2}{p_{ik}^2 q_k} = \tanh^{-1} \gamma^{-k}. \tag{40.46}$$

An Interesting Byproduct

Formula (40.45) has an interesting consequence. To see this, we let $k = 3$. Then it yields

$$\sum_{i=1}^{\infty} \tanh^{-1} \frac{(-1)^{i-1}}{F_{3i}^2} = \tanh^{-1} \frac{1}{\alpha^3}$$

$$\sum_{i=1}^{\infty} \ln \frac{F_{3i}^2 - (-1)^i}{F_{3i}^2 + (-1)^i} = \ln \alpha$$

$$\prod_{i=1}^{\infty} \frac{F_{3i}^2 - (-1)^i}{F_{3i}^2 + (-1)^i} = \alpha, \text{ the golden ratio.}$$

The next result is the Lucas companion of Theorem 40.8. Its proof hinges on the Catalan-like identity $l_{n+k}l_{n-k} - l_n^2 = (-1)^{n+k}\Delta^2 f_k^2$ and the identity

$$l_{n+k} - l_{n-k} = \begin{cases} l_k l_n & \text{if } k \text{ is odd} \\ \Delta^2 f_k f_n & \text{otherwise.} \end{cases}$$

Again, we omit the proof; see Exercise 40.36.

Theorem 40.9 (Melham and Shannon, 1995 [358]). *Let $n \geq k \geq 2$ and $1 \leq x \leq 2$. Then*

$$\tanh^{-1} \frac{l_{n-k}}{l_n} - \tanh^{-1} \frac{l_n}{l_{n+k}} = \begin{cases} \tanh^{-1} \dfrac{(-1)^{n-1}\Delta^2 f_k^2}{l_n^2 l_k} & \text{if } k \text{ is odd} \\ \tanh^{-1} \dfrac{(-1)^n f_k}{f_{2n}} & \text{otherwise.} \end{cases}$$

■

This theorem implies that

$$\sum_{i=1}^{n} \tanh^{-1} \frac{f_k}{f_{2ik}} = \tanh^{-1} \frac{2}{l_k} - \tanh^{-1} \frac{l_{nk}}{l_{(n+1)k}}, \tag{40.47}$$

where k is even; and

$$\sum_{i=1}^{n} \tanh^{-1} \frac{(-1)^{i-1}\Delta^2 f_k^2}{l_{ik}^2 l_k} = \tanh^{-1} \frac{2}{l_k} - \tanh^{-1} \frac{l_{nk}}{l_{(n+1)k}}, \tag{40.48}$$

where k is odd.

It follows from formulas (40.39) and (40.47) that

$$\tanh^{-1} \frac{f_{nk}}{f_{(n+1)k}} + \tanh^{-1} \frac{l_{nk}}{l_{(n+1)k}} = \tanh^{-1} \frac{2}{l_k},$$

where k is even. Taking limits of both sides as $n \to \infty$, this yields

$$\tanh^{-1} \alpha^{-k} + \tanh^{-1} \alpha^{-k} = \tanh^{-1} \frac{2}{l_k}$$

$$\tanh^{-1} \frac{2}{l_k} = 2 \tanh^{-1} \alpha^{-k}, \tag{40.49}$$

where k is even and $\alpha = \alpha(x)$.

It follows from formula (40.48) that

$$\sum_{i=1}^{\infty} \tanh^{-1} \frac{(-1)^{i-1} \Delta^2 f_k^2}{l_{ik}^2 l_k} = \tanh^{-1} \frac{2}{l_k} - \tanh^{-1} \alpha^{-k}, \tag{40.50}$$

where k is odd and $\alpha = \alpha(x)$.

This summation formula also has intriguing consequences:

$$\sum_{i=1}^{\infty} \tanh^{-1} \frac{5(-1)^{i-1} F_k^2}{L_k L_{ik}^2} = \tanh^{-1} \frac{2}{L_k} - \tanh^{-1} \alpha^{-k}, \tag{40.51}$$

$$\sum_{i=1}^{\infty} \tanh^{-1} \frac{4(x^2+1)(-1)^{i-1} p_k^2}{q_k q_{ik}^2} = \tanh^{-1} \frac{2}{q_k} - \tanh^{-1} \gamma^{-k}, \tag{40.52}$$

where k is odd and $\gamma = \gamma(x) = x + \sqrt{x^2+1}$.

Interestingly, formula (40.51) has a charming byproduct. To see this, we let $k = 3$. Then the formula yields

$$\sum_{i=1}^{\infty} \tanh^{-1} \frac{5(-1)^{i-1}}{L_{3i}^2} = \tanh^{-1} \frac{1}{2} - \tanh^{-1} \alpha^{-3}$$

$$\sum_{i=1}^{\infty} \ln \frac{L_{3i}^2 - 5(-1)^i}{L_{3i}^2 + 5(-1)^i} = \ln 3 - \ln \alpha$$

$$\prod_{i=1}^{\infty} \frac{L_{3i}^2 - 5(-1)^i}{L_{3i}^2 + 5(-1)^i} = \frac{3}{\alpha}$$

$$= 3(\alpha - 1).$$

EXERCISES 40

Prove each.

1. $\displaystyle\sum_{k=1}^{n} \tan^{-1} \frac{2x}{p_{2k+1}} = \tan^{-1} \frac{1}{2x} - \tan^{-1} \frac{1}{p_{2n+2}}.$

2. $\tan^{-1}\frac{1}{l_{2n}} + \tan^{-1}\frac{1}{l_{2n+2}} = \tan^{-1}\frac{(x^2+4)f_{2n+1}}{(x^2+4)f_{2n+1}^2 + x^2 - 1}$.
3. $f_{2n+1}f_{2n+2} - f_{2n}f_{2n+3} = x$.
4. Theorem 40.2.
5. $\sum_{k=1}^{n}\tan^{-1}\frac{1}{L_{2n}} = \tan^{-1}(-\beta)$.
6. $\tan^{-1}\frac{1}{l_{2n}} - \tan^{-1}\frac{1}{l_{2n+2}} = \tan^{-1}\frac{xl_{2n+1}}{l_{2n+1}^2 + x^2 + 5}$.
7. $\tan^{-1}\frac{1}{q_{2n}} - \tan^{-1}\frac{1}{q_{2n+2}} = \tan^{-1}\frac{2xq_{2n+1}}{q_{2n+1}^2 + 4x^2 + 5}$.
8. $\tan^{-1}\frac{1}{L_{2n}} - \tan^{-1}\frac{1}{L_{2n+2}} = \tan^{-1}\frac{L_{2n+1}}{L_{2n+1}^2 + 6}$.
9. $\tan^{-1}\frac{1}{2Q_{2n}} - \tan^{-1}\frac{1}{2Q_{2n+2}} = \tan^{-1}\frac{4Q_{2n+1}}{4Q_{2n+1}^2 + 9}$.
10. $\sum_{k=1}^{\infty}\frac{L_{2n+1}}{L_{2n+1}^2 + 6} = \tan^{-1}\frac{1}{3}$.
11. $\sum_{n=1}^{\infty}\tan^{-1}\frac{xl_{2n+1}}{l_{2n+1}^2 + x^2 + 5} = \tan^{-1}\frac{1}{x^2+2}$.
12. $\tan^{-1}\frac{x}{f_{2n-1}} = \tan^{-1} f_{2n} - \tan^{-1} f_{2n-2}$.
13. $\tan^{-1}\frac{1}{F_{2n+1}} + \tan^{-1} L_{2n} + \tan^{-1} L_{2n+2} = \pi$.
14. $\tan^{-1}\frac{(-1)^{n+1}}{f_{2n+2}} = \tan^{-1}\frac{f_{n+2}}{f_{n+1}} - \tan^{-1}\frac{f_{n+1}}{f_n}$.
15. $\tan^{-1}\frac{(-1)^n}{f_{2n+2}} = \tan^{-1}\frac{l_{n+2}}{l_{n+1}} - \tan^{-1}\frac{l_{n+1}}{l_n}$.
16. $\sum_{k=1}^{n}\tan^{-1}\frac{x}{f_{2k-1}} = \tan^{-1} f_{2n}$.
17. Theorem 40.4.
18. $\sum_{n=1}^{m}(-1)^{n+1}\tan^{-1}\frac{1}{f_{2n}} = \tan^{-1}\frac{2}{x} - \tan^{-1}\frac{l_m}{l_{m+1}}$.
19. $\sum_{n=1}^{\infty}(-1)^{n+1}\tan^{-1}\frac{1}{f_{2n}} = \tan^{-1}(-\beta(x))$.

20. $\sum_{n=1}^{m}(-1)^{n+1}\tan^{-1}\frac{1}{p_{2n}} = \frac{\pi}{4} - \tan^{-1}\frac{q_m}{q_{m+1}}.$

21. $\sum_{n=1}^{m}(-1)^{n+1}\tan^{-1}\frac{1}{F_{2n}} = \tan^{-1} 2 - \tan^{-1}\frac{L_m}{L_{m+1}}.$

22. $\sum_{n=1}^{m}(-1)^{n+1}\tan^{-1}\frac{1}{P_{2n}} = \frac{\pi}{4} - \tan^{-1}\frac{Q_m}{Q_{m+1}}.$

23. $\sum_{n=1}^{\infty}(-1)^{n+1}\tan^{-1}\frac{1}{p_{2n}} = \frac{\pi}{4} - \tan^{-1}(-\delta(x)).$

24. $\sum_{n=1}^{\infty}(-1)^{n+1}\tan^{-1}\frac{1}{P_{2n}} = \frac{\pi}{4} - \tan^{-1}(-\delta).$

25. Lemma 40.1.

26. $l_{2m}l_{2n} = (x^2+4)f_{m+n}^2 + l_{m-n}^2.$

27. Lemma 40.2.

28. $\tanh^{-1}\frac{1}{f_{2n+1}} - \tanh^{-1}\frac{1}{f_{2n+3}} = \tanh^{-1}\frac{x}{f_{2n+2}^2}.$

29. $\sum_{k=1}^{\infty}\tanh^{-1}\frac{x}{f_{2k+2}} = \tanh^{-1}\frac{1}{f_3}.$

30. $\prod_{k=1}^{\infty}\frac{F_{2k+2}+1}{F_{2k+2}-1} = 3.$

31. $\prod_{k=1}^{\infty}\frac{P_{2k+3}^2 - 2(-1)^k Q_{2k+3}}{P_{2k+3}^2 + 2(-1)^k Q_{2k+3}} = \frac{13}{11}.$

32. $\prod_{k=1}^{\infty}\frac{P_{2k+3}^2+2}{P_{2k+3}^2-2} = \frac{3}{2}.$

33. Theorem 40.8.

34. Let $n \geq k \geq 2$ and k be even. Then $\sum_{i=1}^{n}\tanh^{-1}\frac{f_k}{f_{2ik}} = \tanh^{-1}\frac{f_{nk}}{f_{(n+1)k}}.$

35. Let $n \geq k \geq 2$ and k be odd. Then $\sum_{i=1}^{n}\tanh^{-1}\frac{(-1)^{i-1}f_k^2}{f_{ik}^2 l_k} = \tanh^{-1}\frac{f_{nk}}{f_{(n+1)k}}.$

36. Theorem 40.9.

41

CHEBYSHEV POLYNOMIALS

Mathematics is the music of reason.
–Paul Lockhart, *A Mathematician's Lament*

The well-known Chebyshev polynomials are named after the Russian mathematician Pafnuty Lvovich Chebyshev (1821–1894). They have interesting applications to algebraic number theory, approximation theory, ergodic theory, and numerical analysis [403]. Rivlin describes them as "like fine jewel[s] that reveal different characteristics under illumination from varying positions" [403].

The Chebyshev and Fibonacci families are closely related. To see this relationship, we briefly study the Chebyshev family, and then study the connections between the two families. The Chebyshev family consists of two subfamilies $\{T_n(x)\}$ and $\{U_n(x)\}$. They are directly related to l_n and f_n, respectively, as we will see shortly.

Chebyshev polynomials $T_n(x)$ and $U_n(x)$ satisfy the recurrence

$$z_n = 2xz_{n-1} - z_{n-2}, \tag{41.1}$$

where $z_n = z_n(x)$; z_0 and z_1 are arbitrary; and $n \geq 2$. When $z_0 = 1$ and $z_1 = x$, $z_n = T_n(x)$; and when $z_0 = 1$ and $z_1 = 2x$, $z_n = U_n(x)$ [285]. In the interest of brevity and clarity, we *drop* the argument from the functional notation when omitting it causes *no* ambiguity.

To begin, we turn to the *Chebyshev polynomials of the first kind* $T_n(x)$.

Fibonacci and Lucas Numbers with Applications, Volume Two. Thomas Koshy.

41.1 CHEBYSHEV POLYNOMIALS $T_n(x)$

Table 41.1 shows the Chebyshev polynomials T_n, where $0 \leq n \leq 9$.

TABLE 41.1. Chebyshev Polynomials $T_n(x)$

n	$T_n(x)$	n	$T_n(x)$
0	1	5	$16x^5 - 20x^3 + 5x$
1	x	6	$32x^6 - 48x^4 + 18x^2 - 1$
2	$2x^2 - 1$	7	$64x^7 - 112x^5 + 56x^3 - 7x$
3	$4x^3 - 3x$	8	$128x^8 - 256x^6 + 160x^4 - 32x^2 + 1$
4	$8x^4 - 8x^2 + 1$	9	$256x^9 - 576x^7 + 432x^5 - 120x^3 + 9x$

A Diminnie Delight

The next example features a beautiful application of the Chebyshev polynomials $T_n(x)$. To make the exposition short and elegant, we first present a slightly modified version of these polynomials. Consider the polynomials $c_n(x)$, defined by the recurrence $c_n(x) = xc_{n-1}(x) - c_{n-2}(x)$, where $c_0(x) = 2$, $c_1(x) = x$, and $n \geq 2$. Then

$$c_2(x) = x^2 - 2 \qquad c_3(x) = x^3 - 3x$$

$$c_4(x) = x^4 - 4x^2 + 2 \qquad c_5(x) = \boxed{x^5 - 5x^3 + 5x}$$

$$c_6(x) = x^6 - 6x^4 + 9x^2 - 2 \qquad c_7(x) = x^7 - 7x^5 + 14x^3 - 7x$$

$$\vdots$$

It follows by the recurrence that $c_n(x) = 2T_n(x/2) = i^n l_n(-ix)$, where $i = \sqrt{-1}$. For example,

$$2T_5(x/2) = 2[16(x/2)^5 - 20(x/2)^3 + 5(x/2)] = x^5 - 5x^3 + 5x = c_5(x).$$

The polynomials $c_n(x)$ satisfy a spectacular property:

$$c_n\left(y + \frac{1}{y}\right) = y^n + \frac{1}{y^n}; \tag{41.2}$$

this follows by PMI.

We are now ready for the application. In 1993, David Doster of Choate Rosemary Hall, Wallingford, Connecticut, proposed an interesting problem: solve the recurrence $d_{n+1} = 5d_n^3 - 3d_n$, where $d_0 = 1$ [132]. This inspired Charles R. Diminnie[†] to investigate a similar, but harder, problem in 1994 [131]. Achilleas Sinefakopoulos, then a student at the University of Athens, Greece, provided a neat solution a few months later [231, 463].

[†]Private Communication, November, 2016.

Charles R. Diminnie was born in Paterson, New Jersey, in 1944. After graduating from St. Bonaventure University in New York in 1966, he entered Michigan State University for graduate studies. He received his M.S. in 1967, and Ph.D. three years later, for his dissertation in functional analysis, *An Application of the Process of Regularization to the Analysis of Distributions*.

After teaching at Bowling Green State University for a year, Diminnie joined the faculty at St. Bonaventure University as an assistant professor, becoming full professor in 1979. Since his retirement in 1996, he has been an adjunct faculty member at Angelo State University in Texas.

Diminnie has received the Medal for General Excellence in Sciences (1966), University President's Award for Excellence in Teaching (2014), and Texas Technical University Chancellor's Council Award for Teaching (2014). He has written or co-authored a number of articles in functional analysis.

Example 41.1. *Solve the recurrence $d_{n+1} = 5d_n(5d_n^4 - 5d_n^2 + 1)$, where $d_0 = 1$.*

Solution. The first four values of d_n are:

$$d_0 = 1$$
$$d_1 = 5$$
$$d_2 = 75{,}025$$

$$d_3 = 5 \cdot 75025(5 \cdot 75025^4 - 5 \cdot 75025^2 + 1),$$

which is really huge; in fact, $d_3 = 59{,}425{,}114{,}757{,}512{,}643{,}212{,}875{,}125$. (Here is a hint: $d_2 = 75{,}025 = F_{52}$.)

There appears to be some relationship between the recurrence $d_{n+1} = 5d_n(5d_n^4 - 5d_n^2 + 1)$ and $c_5(x) = \boxed{x^5 - 5x^3 + 5x}$. In fact, there is one:

$$\begin{aligned}\sqrt{5}d_{n+1} &= 5\sqrt{5}d_n(5d_n^4 - 5d_n^2 + 1)\\ &= (\sqrt{5}d_n)^5 - 5(\sqrt{5}d_n)^3 + 5(\sqrt{5}d_n)\\ &= c_5(\sqrt{5}d_n).\end{aligned}$$

Consequently, the given recurrence can be rewritten as $\sqrt{5}d_{n+1} = c_5(\sqrt{5}d_n)$, where $d_0 = 1$.

Notice that $d_0 = F_{5^0}$, $d_1 = F_{5^1}$, $d_2 = F_{5^2}$, and $d_3 = F_{5^3}$. Using these initial values, we predict that $d_n = F_{5^n}$.

More generally, we conjecture that the solution of the recurrence $\sqrt{5}d_{n+1} = c_m(\sqrt{5}d_n)$ is $d_n = F_{m^n}$, where m is odd and > 1.

Proof. Clearly, the formula works when $n = 0$. Assume, it works for an arbitrary integer $n \geq 0$. Since m is odd, by Binet's formula, we then have

$$\begin{aligned}
\sqrt{5}d_{n+1} &= c_m(\sqrt{5}d_n) \\
&= c_m(\sqrt{5}F_{m^n}) \\
&= c_m\left(\alpha^{m^n} - \beta^{m^n}\right) \\
&= c_m\left(\alpha^{m^n} + \frac{1}{\alpha^{m^n}}\right) \\
&= \left(\alpha^{m^n}\right)^m + \left(\frac{1}{\alpha^{m^n}}\right)^m \\
&= \alpha^{m^{n+1}} + \frac{1}{\alpha^{m^{n+1}}} \\
&= \alpha^{m^{n+1}} - \beta^{m^{n+1}} \\
d_{n+1} &= F_{m^{n+1}}.
\end{aligned}$$

So the formula works for $n + 1$ also. Thus, by PMI, it works for all $n \geq 0$.

In particular, the solution of the given recurrence is $d_n = F_{5^n}$, as predicted. ∎

We can extend this problem to Fibonacci polynomials [296].

Fibonacci Extensions

Solve the recurrence

$$a_{n+1} = a_n(\Delta^4 a_n^4 - 5\Delta^2 a_n^2 + 5), \tag{41.3}$$

where $a_n = a_n(x)$, $a_0 = 1$, and $n \geq 0$.

Then

$$\begin{aligned}
a_0 &= 1 \\
a_1 &= x^4 + 3x^2 + 1 \\
a_2 &= x^{24} + 23x^{22} + 231x^{20} + 1330x^{18} + 4845x^{16} + 11628x^{14} + 18564x^{12} \\
&\quad + 19448x^{10} + 12870x^8 + 5005x^6 + 1001x^4 + 78x^2 + 1 \\
a_3 &= x^{124} + 123x^{122} + 7381x^{120} + \cdots + 1,
\end{aligned}$$

a polynomial of degree 124 and with 63 terms.

Looking at these four initial values of a_n, it does not appear to be easy to conjecture a formula for a_n. But, here is an interesting observation: $a_0 = f_{5^0}$, $a_1 = f_{5^1}$, and $a_2 = f_{5^2}$. This, coupled with the solution $d_n = F_{5^n}$ of recurrence (41.3), helps us conjecture that $a_n = f_{5^n}$, where $n \geq 0$.

To confirm this formula, we rely on the polynomials $c_n(x)$. To this end, first we establish a close relationship between a_{n+1} and c_5. Using recurrence (41.3), we have

$$\begin{aligned}\Delta a_{n+1} &= \Delta^5 a_n^5 - 5\Delta^3 a_n^3 + 5\Delta a_n\\ &= (\Delta a_n)^5 - 5(\Delta a_n)^3 + 5(\Delta a_n)\\ &= c_5(\Delta a_n).\end{aligned}$$

Consequently, we claim that the solution of the recurrence $\Delta a_{n+1} = c_5(\Delta a_n)$ is $a_n = f_{5^n}$.

More generally, we now confirm that the solution of the recurrence

$$\Delta a_{n+1} = c_m(\Delta a_n) \tag{41.4}$$

is $a_n = f_{k\cdot m^n}$, where $a_0 = f_k$, k and m are odd positive integers, $k \neq m, m \geq 3$ and $n \geq 0$.

Proof. Clearly, the formula is true when $n = 0$. Assume, it is true for an arbitrary integer $n \geq 0$. Since k and m are odd, by the Binet-like formula for $f_{k\cdot m^n}$, we then have

$$\begin{aligned}\Delta a_{n+1} &= c_m(\Delta f_{k\cdot m^n})\\ &= c_m\left(\alpha^{k\cdot m^n} - \beta^{k\cdot m^n}\right)\\ &= c_m\left(\alpha^{k\cdot m^n} + \frac{1}{\alpha^{k\cdot m^n}}\right)\\ &= \alpha^{k\cdot m^{n+1}} + \frac{1}{\alpha^{k\cdot m^{n+1}}}\\ &= \alpha^{k\cdot m^{n+1}} - \beta^{k\cdot m^{n+1}}\\ a_{n+1} &= f_{k\cdot m^{n+1}}.\end{aligned}$$

So the formula works for $n + 1$ also. Thus, by induction, formula (41.4) works for all $n \geq 0$; that is, the solution of recurrence (41.4) is $a_n = f_{k\cdot m^n}$. ■

For example, with $k = 3, m = 5$, and $a_0 = f_3 = x^2 + 1$, we have

$$\begin{aligned}a_1 &= a_0(\Delta^4 a_0^4 - 5\Delta^2 a_0^2 + 5)\\ &= (x^2 + 1)[(x^2 + 4)^2(x^2 + 1)^4 - 5(x^2 + 4)(x^2 + 1)^2 + 5]\end{aligned}$$

$$= x^{14} + 13x^{12} + 66x^{10} + 165x^8 + 210x^6 + 126x^4 + 28x^2 + 1$$
$$= f_{3\cdot 5}.$$

In particular, the solution of recurrence (41.3) is $a_n = f_{5^n}$, where $n \geq 0$, as conjectured. Clearly, the solution to Diminnie's recurrence follows from this.

Suppose we let $m = 3$ in recurrence (41.4). Since $c_3(x) = x^3 - 3x$,

$$\Delta a_{n+1} = c_3(\Delta a_n)$$
$$a_{n+1} = \Delta^2 a_n^3 - 3a_n. \tag{41.5}$$

The solution of this recurrence is $a_n = f_{k\cdot 3^n}$, where $a_0 = f_k$ and $n \geq 0$.

Likewise, when $m = 7$, we get

$$a_{n+1} = \Delta^6 a_n^7 - 7\Delta^4 a_n^5 + 14\Delta^2 a_n^3 - 7a_n; \tag{41.6}$$

its solution is $a_n = f_{k\cdot 7^n}$, where $a_0 = f_k$ and $n \geq 0$.

Obviously, we can continue this procedure for any odd integer ≥ 9.

In particular, let $x = 1 = k$. Then the solutions of the recurrences $a_{n+1} = 5a_n^3 - 3a_n$ and $a_{n+1} = 125a_n^7 - 175a_n^5 + 70a_n^3 - 7a_n$ are $a_n = F_{3^n}$ and $a_n = F_{7^n}$, respectively.

As we can predict, the polynomial extension (41.3) has Pell consequences.

Pell Extensions

Let $b_n = b_n(x) = a_n(2x)$, $b_0 = 1$, and $n \geq 0$. Then recurrences (41.5), (41.3), and (41.6) yield

$$b_{n+1} = b_n(4D^2b_n^2 - 3), \tag{41.7}$$
$$b_{n+1} = b_n(16D^4b_n^4 - 20D^2b_n^2 + 5), \tag{41.8}$$
$$b_{n+1} = b_n(64D^6b_n^6 - 112D^4b_n^4 + 56D^2b_n^2 - 7), \tag{41.9}$$

respectively. The corresponding solutions are $b_n = p_{k\cdot 3^n}$, $b_n = p_{k\cdot 5^n}$, and $b_n = p_{k\cdot 7^n}$, respectively.

When $x = 1 = k$, these yield the solutions $b_n = P_{3^n}$, $b_n = P_{5^n}$, and $b_n = P_{7^n}$, respectively.

For example, $b_1 = 29 = P_{5^1}$; so $b_2 = 29(64 \cdot 29^4 - 40 \cdot 29^2 + 5) = 1{,}311{,}738{,}121 = P_{5^2}$, as expected.

Lucas Counterparts

Recall that the solutions of recurrences (41.3), (41.5), and (41.6) pivoted on the polynomial $c_m(x)$, where m is odd and ≥ 3. Interestingly, focusing on $c_m(x)$ with m even and ≥ 2 yields equally rewarding results.

For example, consider the recurrence

$$a_{n+1} = a_n^4 - 4a_n^2 + 2, \tag{41.10}$$

where $a_n = a_n(x)$, $a_1 = l_{4e}$, e is a positive integer such that $4 \nmid e$, and $n \geq 1$.

Clearly, $a_{n+1} = c_4(a_n)$. Using the Binet-like formula for $l_{e \cdot 4^n}$, property (41.2), and induction, we can show that $a_n = l_{e \cdot 4^n}$.

For example, let $e = 1$. Then $a_1 = l_4 = x^4 + 4x^2 + 2$, and

$$\begin{aligned} a_2 &= a_1^4 - 4a_1^2 + 2 \\ &= (x^4 + 4x^2 + 2)^4 - 4(x^4 + 4x^2 + 2)^2 + 2 \\ &= x^{16} + 16x^{14} + 104x^{12} + 352x^{10} + 660x^8 + 672x^6 + 336x^4 + 64x^2 + 2 \\ &= l_{4^2}. \end{aligned}$$

Similarly, the recurrences

$$a_{n+1} = a_n^2 - 2, \quad a_1 = l_{2e} \quad (2 \nmid e), \tag{41.11}$$

$$a_{n+1} = a_n^6 - 6a_n^4 + 9a_n^2 - 2, \quad a_1 = l_{6e} \quad (6 \nmid e) \tag{41.12}$$

yield the abbreviated recurrences $a_{n+1} = c_2(a_n)$ and $a_{n+1} = c_6(a_n)$, respectively, where $a_n = a_n(x)$. Correspondingly, we have $a_n = l_{e \cdot 2^n}$ and $a_n = l_{e \cdot 6^n}$, respectively.

In particular, let $x = 1 = e$. Then L_{2^n}, L_{4^n}, and L_{6^n} are the solutions of the recurrences (41.11), (41.10), and (41.12), respectively; M. Klamkin (1921–2004) found these solutions [270].

Pell–Lucas Byproducts

Since $l_k(2x) = q_k(x)$, it follows from recurrences (41.11), (41.10), and (41.12) that

$$\begin{aligned} b_{n+1} &= b_n^2 - 2, \quad b_1 = q_{2e} \quad (2 \nmid e), \\ b_{n+1} &= b_n^4 - 4b_n^2 + 2, \quad b_1 = q_{4e} \quad (4 \nmid e), \\ b_{n+1} &= b_n^6 - 6b_4^4 + 9b_n^2 - 2, \quad b_1 = q_{6e} \quad (6 \nmid e), \end{aligned} \tag{41.13}$$

respectively, where $b_n = a_n(2x)$. The corresponding solutions are $b_n = q_{e \cdot 2^n}$, $b_n = q_{e \cdot 4^n}$, and $b_n = q_{e \cdot 6^n}$, respectively.

In particular, let $x = 1 = e$. Then $b_n = 2Q_{2^n}$, $b_n = 2Q_{4^n}$, and $b_n = 2Q_{6^n}$, respectively.

For example, consider recurrence (41.13), where $b_1 = 34 = 2Q_4$. Then $b_2 = 34^4 - 4 \cdot 34^2 + 2 = 1{,}331{,}714 = 2Q_{4^2}$.

Two Charming Recurrences

Next we study two equally delightful recurrences with Fibonacci and Pell implications.

Recurrence *A*

Consider the recurrence

$$x_{n+1} = x_n(\Delta^2 x_n^2 + 3), \tag{41.14}$$

where $x_0 = f_e$, e is a positive even integer, and $n \geq 0$.

Suppose $e = 2$. Then $x_0 = f_2 = x$ and $x_1 = x[(x^2+4)x^2+3] = x^5 + 4x^3 + 3x = f_{2\cdot 3}$.

More generally, we conjecture that $x_n = f_{e\cdot 3^n}$, where $n \geq 0$. It is clearly true when $n = 0$. Assume, it is true for an arbitrary integer $n \geq 0$. Then

$$\begin{aligned}
x_{n+1} &= \Delta^2 f_{e\cdot 3^n}^3 + 3f_{e\cdot 3^n}\\
\Delta x_{n+1} &= \left(\alpha^{e\cdot 3^n} - \beta^{e\cdot 3^n}\right)^3 + 3\left(\alpha^{e\cdot 3^n} - \beta^{e\cdot 3^n}\right)\\
&= \alpha^{e\cdot 3^{n+1}} - \beta^{e\cdot 3^{n+1}} - 3(\alpha\beta)^{e\cdot 3^n}\left(\alpha^{e\cdot 3^n} - \beta^{e\cdot 3^n}\right) + 3\left(\alpha^{e\cdot 3^n} - \beta^{e\cdot 3^n}\right)\\
&= \alpha^{e\cdot 3^{n+1}} - \beta^{e\cdot 3^{n+1}}\\
x_{n+1} &= f_{e\cdot 3^{n+1}}.
\end{aligned}$$

Thus, by induction, the conjecture works for all $n \geq 0$.

For example, let $e = 4$. Then $x_0 = f_4 = x^3 + 2x$. So

$$\begin{aligned}
x_1 &= x_0(\Delta^2 x_0^2 + 3)\\
&= (x^3+2x)[(x^2+4)(x^3+2x)^2+3]\\
&= x^{11} + 10x^9 + 36x^7 + 56x^5 + 35x^3 + 6x\\
&= f_{4\cdot 3^1}.
\end{aligned}$$

In particular, the solution of the recurrence $x_{n+1} = x_n(5x_n^2+3)$ is $x_n = F_{e\cdot 3^n}$, where $x_0 = F_e$ and $n \geq 0$.

Pell Byproducts

It follows from recurrence (41.14) that the solution of the recurrence $x_{n+1} = x_n(4D^2x_n^2+3)$ is $x_n = p_{e\cdot 3^n}$, where $x_0 = p_e$ and $n \geq 0$. In particular, the solution of the recurrence $x_{n+1} = x_n(8x_n^2+3)$ is $x_n = P_{e\cdot 3^n}$.

Next we study a similar recurrence; that too has interesting consequences.

Recurrence *B*

Consider the recurrence

$$z_{n+2} = z_{n+1}(\Delta^2 z_n^2 + 2), \tag{41.15}$$

where $z_1 = f_{2k}, z_2 = f_{4k}, k$ is an odd positive integer, and $n \geq 1$.

When $k = 1$, $z_1 = f_2 = x$ and $z_2 = f_4 = x^3 + 2x$. So

$$\begin{aligned} z_3 &= (x^3 + 2x)[(x^2 + 4)x^2 + 2] \\ &= x^7 + 6x^5 + 10x^3 + 4x \\ &= f_{2^3}. \end{aligned}$$

More generally, it follows by induction that the solution of recurrence (41.15) is $z_n = f_{k \cdot 2^n}$, where $n \geq 1$.

For example, let $k = 3$. Then $z_1 = f_6 = x^5 + 4x^3 + 3x$ and $z_2 = f_{12} = x^{11} + 10x^9 + 36x^7 + 56x^5 + 35x^3 + 6x$. Consequently,

$$\begin{aligned} z_3 &= z_2(\Delta^2 z_1^2 + 2) \\ &= (x^{11} + 10x^9 + 36x^7 + 56x^5 + 35x^3 + 6x)[(x^2 + 4)(x^5 + 4x^3 + 3x)^2 + 2] \\ &= x^{23} + 22x^{21} + 210x^{19} + 1140x^{17} + 3876x^{15} + 8568x^{13} + 12376x^{11} \\ &\quad + 11440x^9 + 6435x^7 + 2002x^5 + 286x^3 + 12x \\ &= f_{3 \cdot 2^3}. \end{aligned}$$

Suppose we let $x = 1$ in recurrence (41.15). Then the solution of the recurrence $z_{n+2} = z_{n+1}(5z_n^2 + 2)$ is $z_n = F_{k \cdot 2^n}$, where $z_1 = F_{2k}, z_2 = F_{4k}$, and $n \geq 0$.

Recurrence (41.14) also has Pell implications.

Pell Consequences

The solution of the recurrence $z_{n+2} = 2z_{n+1}(2D^2 z_n^2 + 1)$ is $z_n = p_{k \cdot 2^n}$, where $z_1 = p_{2k}$, $z_2 = p_{4k}$, and $n \geq 0$. Consequently, the solution of the recurrence $z_{n+2} = 2z_{n+1}(4z_n^2 + 1)$ is $z_n = P_{k \cdot 2^n}$, where $z_1 = P_{2k}, z_2 = P_{4k}$, and $n \geq 0$.

For example, let $k = 5$. Then $z_1 = P_{10} = 2{,}378$ and $z_2 = P_{20} = 15{,}994{,}428$. So $z_3 = 2z_2(4z_1^2 + 1) = 2 \cdot 15{,}994{,}428(4 \cdot 2378^2 + 1) = 723{,}573{,}111{,}879{,}672 = P_{40}$.

Lucas Counterparts

Interestingly, recurrences (41.14) and (41.15) have their own Lucas counterparts:

$$u_{n+1} = u_n(u_n^2 - 3), \tag{41.16}$$

where $u_0 = l_e$ and $n \geq 0$; and

$$v_{n+2} = v_{n+1}(v_n^2 - 2) - 2, \tag{41.17}$$

where $v_1 = l_{2k}, v_2 = l_{4k}$, and $n \geq 1$.

Their solutions are $u_n = l_{e\cdot 3^n}$ and $v_n = l_{k\cdot 2^n}$, respectively. Their proofs follow similarly, so we omit them.

For example, let $e = 4$. Then

$$\begin{aligned} u_1 &= l_4(l_4^2 - 3) \\ &= x^{12} + 12x^{10} + 54x^8 + 112x^6 + 105x^4 + 36x^2 + 2 \\ &= l_{l_{4\cdot 3^1}}; \end{aligned}$$

likewise,

$$\begin{aligned} v_3 &= l_{12}(l_6^2 - 2) - 2 \\ &= x^{24} + 24x^{22} + 252x^{20} + 1520x^{18} + 5814x^{16} + 14688x^{14} \\ &\quad + 24752x^{12} + 27456x^{10} + 19305x^8 + 8008x^6 + 1716x^4 + 144x^2 + 2 \\ &= l_{3\cdot 2^3}. \end{aligned}$$

Pell–Lucas Versions

It follows from recurrences (41.16) and (41.17) that the solutions of the recurrences

$$u_{n+1} = u_n(u_n^2 - 3), \quad u_0 = q_e \tag{41.18}$$

and

$$v_{n+2} = v_{n+1}(v_n^2 - 2) - 2, \quad v_1 = q_{2k} \quad \text{and} \quad v_2 = q_{4k} \tag{41.19}$$

are $u_n = q_{e\cdot 3^n}$ and $v_n = q_{k\cdot 2^n}$, respectively.

In particular, let $x = 1$. Then the solutions of the recurrences (41.16), (41.17), (41.18), and (41.19) are $u_n = L_{e\cdot 3^n}$, $v_n = L_{k\cdot 2^n}$, $u_n = 2Q_{e\cdot 3^n}$, and $v_n = 2Q_{k\cdot 2^n}$, respectively.

For example, when $k = 1$,

$$\begin{aligned} v_5 &= v_4(v_3^2 - 2) - 2 \\ &= 1{,}331{,}714(1154^2 - 2) - 2 \\ &= 2 \cdot 886{,}731{,}088{,}897 \\ &= 2Q_{2^5}. \end{aligned}$$

? We invite Fibonacci enthusiasts to interpret the above recurrences combinatorially.

Binet-like Formula

Using standard techniques, we can develop an explicit formula for T_n. First, we note that the characteristic equation of the Chebyshev recurrence is $t^2 - 2xt + 1 = 0$; so the characteristic roots are $a = a(x) = x + \sqrt{x^2 - 1}$ and $b = b(x) = x - \sqrt{x^2 - 1}$. Consequently, the general solution of the recurrence is $z_n = Aa^n + Bb^n$. Using the initial conditions, $z_0 = 1$ and $z_1 = x$, we get $A = 1/2 = B$; so the Binet-like formula is

$$T_n = \frac{a^n + b^n}{2}, \tag{41.20}$$

where $n \geq 0$.

For example,

$$\begin{aligned} 2T_4 &= \left(x + \sqrt{x^2 - 1}\right)^4 + \left(x - \sqrt{x^2 - 1}\right)^4 \\ &= 2[x^4 + 6x^2(x^2 - 1) + (x^2 - 1)^2] \\ T_4 &= 8x^4 - 8x^2 + 1. \end{aligned}$$

(Note that both $a(x)$ and $b(x)$ occur in the formula for the inverse hyperbolic cosine function *arccosh*: $\operatorname{arccosh} x = \ln(x + \sqrt{x^2 - 1})$. The signs correspond to the two branches of the graph of $y = \operatorname{arccosh} x$ [115].)

We can now see how T_n and l_n are closely linked. First, notice that $l_0 = 2 = 2(-i)^0 T_0(ix/2)$ and $l_1 = x = 2(-i)^1 T_1(ix/2)$.

Now suppose $2(-i)^k T_k(ix/2) = l_k$ for all nonnegative integers $k < n$. Then

$$\begin{aligned} T_n(ix/2) &= 2(ix/2)T_{n-1}(ix/2) - T_{n-2}(ix/2) \\ &= ix\left[\frac{l_{n-1}}{2(-i)^{n-1}}\right] - \frac{l_{n-2}}{2(-i)^{n-2}} \\ 2T_n(ix/2) &= x\frac{l_{n-1}}{(-i)^n} + \frac{l_{n-2}}{(-i)^n} \\ 2(-i)^n T_n(ix/2) &= xl_{n-1} + l_{n-2} \\ &= l_n. \end{aligned} \tag{41.21}$$

Thus, by PMI, $2(-i)^n T_n(ix/2) = l_n$ for every $n \geq 0$.

For example,

$$\begin{aligned} 2(-i)^3 T_3(ix/2) &= 2(-i)^3[4(ix/2)^3 - 3(ix/2)] \\ &= x^3 + 3x \\ &= l_3. \end{aligned}$$

It follows from formula (41.21) that $L_n = 2|T_n(i/2)|$, $q_n = 2(-i)^n T_n(ix)$, and $Q_n = |T_n(i)|$.

For example,

$$\begin{aligned}2T_5(i/2) &= 2[16(i/2)^5 - 20(i/2)^3 + 5(i/2)]\\ &= -11i\\ |2T_5(i/2)| &= 11 = L_5;\\ q_3 &= 2(-i)^3 T_3(ix)\\ &= 2(-i)^3[4(ix)^3 - 3(ix)]\\ &= 8x^3 + 6x.\end{aligned}$$

Identity (41.21) is in fact reversible: $T_n = \frac{i^n}{2} l_n(-2ix)$. For example, $T_5 = \frac{i^5}{2}[(-2ix)^5 + 5(-2ix)^3 + 5(-2ix)] = 16x^5 - 20x^3 + 5x$.

Chebyshev polynomials T_n satisfy a number of interesting properties [285, 403]. By virtue of identity (41.21), they have Lucas and Pell–Lucas consequences.

For example, we can easily establish that

$$2T_m T_n = T_{m+n} + T_{m-n}, \tag{41.22}$$

where $m \geq n$; see Exercise 41.17.

Consequently, $l_{m+n} + (-1)^n l_{m-n} = l_m l_n$, as we learned in Chapter 33.

Suppose $m \geq 1$. It then follows from identity (41.22) that

$$\begin{aligned}2T_{m+1} + T_{m-1} &= 2T_m T_1\\ &= 2xT_m.\end{aligned}$$

As a result, property (41.22) generalizes the Chebyshev recurrence.

Using the binomial theorem, we can develop another explicit formula for T_n.

A Second Explicit Formula for T_n

By the Binet-like formula, we have

$$2T_n = \left(x + \sqrt{x^2-1}\right)^n + \left(x - \sqrt{x^2-1}\right)^n$$

$$T_n = \sum_{k=0}^{\lfloor n/2 \rfloor} \binom{n}{2k} x^{n-2k}(x^2-1)^k. \tag{41.23}$$

For example,

$$T_5 = \sum_{k=0}^{2} \binom{5}{2k} x^{5-2k}(x^2-1)^k$$

$$= x^5 + 10x^3(x^2 - 1) + 5x(x^2 - 1)^2$$
$$= 16x^5 - 20x^3 + 5x,$$

as expected; see Table 41.1.

It follows from formula (41.23) that

$$T_n(ix/2) = \sum_{k=0}^{\lfloor n/2 \rfloor} \binom{n}{2k} (ix/2)^{n-2k}[(ix/2)^2 - 1]^k$$

$$2(-i)^n T_n(ix/2) = \sum_{k=0}^{\lfloor n/2 \rfloor} \binom{n}{2k} (x^2 + 4)^k x^{n-2k}$$

$$l_n = \frac{1}{2^{n-1}} \sum_{k=0}^{\lfloor n/2 \rfloor} \binom{n}{2k} (x^2 + 4)^k x^{n-2k}, \qquad (41.24)$$

as we found in Chapter 31.

In particular, formula (41.24) implies that

$$q_n = 2 \sum_{k=0}^{\lfloor n/2 \rfloor} \binom{n}{2k} (x^2 + 1)^k x^{n-2k}; \qquad (41.25)$$

$$Q_n = \sum_{k=0}^{\lfloor n/2 \rfloor} \binom{n}{2k} 2^k.$$

For example,

$$q_5 = 2 \sum_{k=0}^{2} \binom{5}{2k} (x^2 + 1)^k x^{5-2k}$$
$$= 2[x^5 + 10x^3(x^2 + 1) + 5x(x^2 + 1)^2]$$
$$= 32x^5 + 40x^3 + 10x.$$

Additional Byproducts

Formula (41.23) has two additional consequences.

1) The highest-degree monomial in $(x^2 - 1)^k x^{n-2k}$ is $x^{2k} \cdot x^{n-2k} = x^n$; its coefficient is $\sum_{k=0}^{\lfloor n/2 \rfloor} \binom{n}{2k}$, which equals 2^{n-1} by the binomial theorem. So the leading term in T_n is $2^{n-1}x^n$. This will come in handy shortly. For example, the leading term in T_7 is $2^6 x^7 = 64x^7$; see Table 41.1.

2) It is well known in elementary number theory that $a^{p-1} \equiv 1 \pmod{p}$, where a is a positive integer and p is a prime such that $p \nmid a$. This is *Fermat's little theorem*, named after the French mathematician Pierre de Fermat (1601–1665). It can be extended to any positive integer a: $a^p \equiv a \pmod{p}$; this is true even if $p = 2$ [277].

Interestingly, this generalization has a surprising analog for the Chebyshev polynomial T_p: $T_p \equiv x \pmod{p}$, where x is a positive integer and p is an odd prime. To establish this, we need the property that $\binom{p}{k} \equiv 0 \pmod{p}$, where $0 < k < p$ [277]. It then follows by formula (41.23) that

$$\begin{aligned} T_p &= \sum_{k=0}^{(p-1)/2} \binom{p}{2k} x^{p-2k}(x^2-1)^k \\ &= x^p + \sum_{k=1}^{(p-1)/2} \binom{p}{2k} x^{p-2k}(x^2-1)^k \\ &\equiv x^p + 0 \pmod{p} \\ &\equiv x \pmod{p}. \end{aligned}$$

For example,

$$\begin{aligned} T_{11} &= 1024x^{11} - 2816x^9 + 2816x^7 - 1232x^5 + 220x^3 - 11x \\ &= 1 \cdot x^{11} - 0 \cdot x^9 + 0 \cdot x^7 - 0 \cdot x^5 + 0 \cdot x^3 - 0 \cdot x \pmod{11} \\ &\equiv x^{11} \pmod{11} \\ &\equiv x \pmod{11}. \end{aligned}$$

Next we explore an intimate relationship between T_n and trigonometry.

41.2 $T_n(x)$ AND TRIGONOMETRY

Let θ be any angle, where $0 \le \theta \le \pi$. Consider the expansions of cosines of (integral) multiples of θ in terms of powers of $\cos\theta$; see Table 41.2. Surprisingly, the RHS of each expansion is of the form $T_n(\cos\theta)$; that is, $T_n(x) = \cos n\theta$, where $0 \le n \le 5$, $\theta = \arccos x$, and $-1 \le x \le 1$. We can confirm this observation for all $n \ge 0$ by PMI and De Moivre's theorem [285].

TABLE 41.2.

$\cos 0\theta = 1$	$\cos 3\theta = 4\cos^3\theta - 3\cos\theta$
$\cos 1\theta = \cos\theta$	$\cos 4\theta = 8\cos^4\theta - 8\cos^2\theta + 1$
$\cos 2\theta = 2\cos^2\theta - 1$	$\cos 5\theta = 16\cos^5\theta - 20\cos^3\theta + 5\cos\theta$

The recursive formula $\cos n\theta = 2\cos(n-1)\theta\cos\theta - \cos(n-2)\theta$ reveals the basis of the Chebyshev recurrence $T_n(x) = 2xT_{n-1} - T_{n-2}$. We can use the trigonometric relationship $T_n(\cos\theta) = \cos n\theta$ to extract additional properties of T_n [285]. For example,

$$T_m(T_n) = T_{mn};$$
$$2T_mT_n = T_{m+n} + T_{m-n};$$
$$(T_{m+n} - 1)(T_{m-n} - 1) = (T_m - T_n)^2.$$

We can use the binomial theorem, coupled with De Moivre's theorem and the property that $T_n = \cos n\theta$, to derive another explicit formula for T_n [285]:

$$T_n = \sum_{k=0}^{\lfloor n/2\rfloor}\sum_{j=k}^{\lfloor n/2\rfloor}(-1)^k\binom{n}{2j}\binom{j}{k}x^{n-2k}. \tag{41.26}$$

For example,

$$\begin{aligned}T_5 &= \sum_{k=0}^{2}\sum_{j=k}^{2}(-1)^k\binom{5}{2j}\binom{j}{k}x^{5-2k}\\ &= (1+10+5)x^5 - (10+5\cdot 2)x^3 + 5x\\ &= 16x^5 - 20x^3 + 5x.\end{aligned}$$

It follows from formula (41.26) that

$$l_n = \frac{1}{2^{n-1}}\sum_{k=0}^{\lfloor n/2\rfloor}\sum_{j=k}^{\lfloor n/2\rfloor}\binom{n}{2j}\binom{j}{k}4^kx^{n-2k};$$

$$L_n = \frac{1}{2^{n-1}}\sum_{k=0}^{\lfloor n/2\rfloor}\sum_{j=k}^{\lfloor n/2\rfloor}\binom{n}{2j}\binom{j}{k}4^k;$$

$$q_n = 2\sum_{k=0}^{\lfloor n/2\rfloor}\sum_{j=k}^{\lfloor n/2\rfloor}\binom{n}{2j}\binom{j}{k}x^{n-2k};$$

$$Q_n = \sum_{k=0}^{\lfloor n/2\rfloor}\sum_{j=k}^{\lfloor n/2\rfloor}\binom{n}{2j}\binom{j}{k}.$$

For example,

$$\begin{aligned}16\,l_5 &= \sum_{k=0}^{2}\sum_{j=k}^{2}\binom{n}{2j}\binom{j}{k}4^kx^{5-2k}\\ &= 16(x^5 + 5x^3 + 5x)\\ l_5 &= x^5 + 5x^3 + 5x;\end{aligned}$$

$$
\begin{aligned}
Q_7 &= \sum_{k=0}^{3}\sum_{j=k}^{3}\binom{7}{2j}\binom{j}{k}\\
&= (1+21+35+7)+(21+70+21)+(35+21)+7\\
&= 239,
\end{aligned}
$$

as expected.

Formula (41.26) has a close link with the left-justified Pascal's triangle. To see this, we rewrite the formula as

$$
\begin{aligned}
T_n &= \sum_{j\ge 0}\binom{n}{2j}\binom{j}{0}x^n - \sum_{j\ge 0}\binom{n}{2j}\binom{j}{1}x^{n-2} + \sum_{j\ge 0}\binom{n}{2j}\binom{j}{2}x^{n-4} - \cdots\\
&= \left[1\binom{n}{0} + 1\binom{n}{2} + 1\binom{n}{4} + \cdots\right]x^n - \left[1\binom{n}{2} + 2\binom{n}{4} + 3\binom{n}{6} + \cdots\right]x^{n-2}\\
&\quad + \left[1\binom{n}{4} + 3\binom{n}{6} + 6\binom{n}{8} + \cdots\right]x^{n-4} - \left[1\binom{n}{6} + 4\binom{n}{8} + 10\binom{n}{10} + \cdots\right]x^{n-6} +\\
&\quad \vdots\\
&= \left[\binom{n}{0}\ \binom{n}{2}\ \binom{n}{4}\ \binom{n}{6}\cdots\right]
\begin{bmatrix}
1 & 0 & 0 & 0 & 0 & \\
1 & 1 & 0 & 0 & 0 & \\
1 & 2 & 1 & 0 & 0 & \cdots\\
1 & 3 & 3 & 1 & 0 & \\
1 & 4 & 6 & 4 & 1 & \\
 & & \vdots & & &
\end{bmatrix}
\begin{bmatrix}
x^n\\
-x^{n-2}\\
x^{n-4}\\
-x^{n-6}\\
\vdots
\end{bmatrix}.
\end{aligned}
$$

Letting $x = \cos\theta$, this gives the expansion of $\cos n\theta$ in terms of powers of $\cos\theta$.

41.3 HIDDEN TREASURES IN TABLE 41.1

Table 41.1 contains some hidden treasures [32, 500]. We can highlight them by arranging the nonzero coefficients in a left-justified triangular array A; see Table 41.3.

Let $a_{n,k}$ denote the element on row n and column k of array A; it is the coefficient of x^{n-2k} in T_n. We can define it recursively:

$$
\begin{aligned}
a_{0,0} &= 1\\
a_{1,0} &= 1\\
a_{n-k} &= 2a_{n-1,k} - a_{n-2,k-1},
\end{aligned}
\tag{41.27}
$$

where $n \ge 2$ and $k \ge 1$.

TABLE 41.3. Array A

$n \backslash k$	0	1	2	3
0	1			
1	1			
2	2	−1		
3	4	−3		
4	8	−8	1	
5	16	−20	5	
6	32	−48	18	−1
7	64	−112	56	−7

For example, $a_{6,2} = 2a_{5,2} - a_{4,1} = 2 \cdot 5 - (-8) = \textcircled{18}$; see the arrows in Table 41.3.

Since $T_n(\cos\theta) = \cos n\theta$, it follows that $T_n(1) = 1$. Consequently, every row sum is one; that is, $\sum\limits_{k\geq 0} a_{n,k} = 1$.

Formula (41.26) then implies that

$$a_{n,k} = \sum_{j=k}^{\lfloor n/2 \rfloor} (-1)^k \binom{n}{2j}\binom{j}{k} \quad \text{and} \quad \sum_{k=0}^{\lfloor n/2 \rfloor}\sum_{j=k}^{\lfloor n/2 \rfloor} (-1)^k \binom{n}{2j}\binom{j}{k} = 1.$$

For example, let $n = 7$. Then

$$\sum_{k=0}^{3}\sum_{j=k}^{3}(-1)^k\binom{7}{2j}\binom{j}{k}$$
$$= \sum_{j=0}^{3}\binom{7}{2j}\binom{j}{0} - \sum_{j=1}^{3}\binom{7}{2j}\binom{j}{1} + \sum_{j=2}^{3}\binom{7}{2j}\binom{j}{2} - \sum_{j=3}^{3}\binom{7}{2j}\binom{j}{3}$$
$$= 64 - (21 + 70 + 21) + (35 + 21) - 7 = 1.$$

Here $a_{7,0} = 64, a_{7,1} = -112, a_{7,2} = 56$, and $a_{7,3} = -7$.

In 1974, W.J. Wagner of Los Altos High School, Los Altos, California, conjectured an explicit formula for $a_{n,k}$ [500]:

$$a_{n,k} = (-1)^k \frac{n}{n-k}\binom{n-k}{k} 2^{n-2k-1}, \tag{41.28}$$

where $a_{0,0} = 1$. The conjecture is based on pattern recognition.

Since the leading coefficient in T_n is 2^{n-1}, $a_{n,0} = 2^{n-1} = (-1)^0 \frac{n}{n-0}\binom{n-0}{0}2^{n-0-1}$. So assume $k \geq 1$. Tables 41.4, 41.5, and 41.6 show the

TABLE 41.4.

n	$\lvert a_{n,1}\rvert$
2	$1 = 2^{-1}\cdot 2$
3	$3 = 2^{0}\cdot 3$
4	$8 = 2^{1}\cdot 4$
5	$20 = 2^{2}\cdot 5$
6	$48 = 2^{3}\cdot 6$

TABLE 41.5.

n	$\lvert a_{n,2}\rvert$
4	$1 = 2^{-1}\cdot 2$
5	$5 = 2^{0}\cdot 5$
6	$18 = 2^{1}\cdot 9$
7	$56 = 2^{2}\cdot 14$
8	$160 = 2^{3}\cdot 6$

TABLE 41.6.

n	$\lvert a_{n,3}\rvert$
6	$1 = 2^{-2}\cdot 2$
7	$7 = 2^{0}\cdot 7$
8	$32 = 2^{1}\cdot 16$
9	$120 = 2^{2}\cdot 30$
10	$400 = 2^{3}\cdot 50$

values of $|a_{n,1}|$, $|a_{n,2}|$, and $|a_{n,3}|$. Using these tables, we conjecture that

$$\begin{aligned}
a_{n,1} &= -n\cdot 2^{n-3} &&= -\frac{n}{n-1}\binom{n-1}{1}2^{n-2\cdot 1-1};\\
a_{n,2} &= \frac{n(n-3)}{2}\cdot 2^{n-5} &&= \frac{n}{n-2}\binom{n-2}{2}2^{n-2\cdot 2-1};\\
a_{n,3} &= -\frac{n(n-4)(n-5)}{6}\cdot 2^{n-7} &&= -\frac{n}{n-3}\binom{n-3}{3}2^{n-2\cdot 3-1}.
\end{aligned}$$

Similarly,

$$a_{n,4} = \frac{n(n-5)(n-6)(n-7)}{4!}\cdot 2^{n-9} = \frac{n}{n-4}\binom{n-4}{4}2^{n-2\cdot 4-1}.$$

Clearly, a nice pattern emerges from these data. We confirm the observation by using recursion. Recall that formula (41.28) works when $k = 0$. So assume $k \geq 1$. By Tables 41.4, 41.5, and 41.6, it works for $2 \leq n \leq 10$ also.

Assume that the formula works for all positive integers $< n$. Then, by recurrence (41.28), we have

$$\begin{aligned}
a_{n,k} &= 2a_{n-1,k} - a_{n-2,k-1}\\
&= (-1)^k\frac{n-1}{n-k-1}\binom{n-k-1}{k}2^{n-2k-1} - (-1)^{k-1}\frac{n-2}{n-k-1}\binom{n-k-1}{k-1}2^{n-2k-1}\\
&= (-1)^k\frac{2^{n-2k-1}}{n-k-1}\left[(n-1)\cdot\frac{n-2k}{n-k}\binom{n-k}{k} + (n-2)\cdot\frac{k}{n-k}\binom{n-k}{k}\right]\\
&= (-1)^k\frac{2^{n-2k-1}}{(n-k)(n-k-1)}\binom{n-k}{k}\left[(n-1)(n-2k)+(n-2)k\right]\\
&= (-1)^k\frac{2^{n-2k-1}}{(n-k)(n-k-1)}\binom{n-k}{k}[n(n-k-1)]\\
&= (-1)^k\frac{n}{n-k}\binom{n-k}{k}2^{n-2k-1}.
\end{aligned}$$

So the formula works for n. Thus, by PMI, the formula works for all $n \geq 0$.

Formula (41.28) follows quickly and directly when we let $x = 1$ in the explicit formula

$$2T_n = \sum_{k=0}^{\lfloor n/2 \rfloor} (-1)^k \frac{n}{n-k} \binom{n-k}{k} (2x)^{n-2k}, \tag{41.29}$$

which we will develop in Theorem 43.7.

Note an added byproduct of formula (41.29). Since $T_n(\cos\theta) = \cos n\theta$, the formula can be used to expand $\cos n\theta$ in terms of powers of $\cos\theta$:

$$\begin{aligned}\cos n\theta &= \sum_{k=0}^{\lfloor n/2 \rfloor} (-1)^k \frac{n}{n-k} \binom{n-k}{k} 2^{n-2k-1} \cos^{n-2k}\theta \qquad (41.30)\\ &= \sum_{k=0}^{\lfloor n/2 \rfloor} (-1)^k a_{n,k} \cos^{n-2k}\theta.\end{aligned}$$

For example,

$$\begin{aligned}\cos 7\theta &= \sum_{k=0}^{3} (-1)^k \frac{7}{7-k} \binom{7-k}{k} 2^{6-2k} \cos^{7-2k}\theta\\ &= \frac{7}{7}\binom{7}{0} 2^6 \cos^7\theta - \frac{7}{6}\binom{6}{1} 2^4 \cos^5\theta + \frac{7}{5}\binom{5}{2} 2^2 \cos^3\theta - \frac{7}{4}\binom{4}{3} 2^0 \cos^5\theta\\ &= 64\cos^7\theta - 112\cos^5\theta + 56\cos^3\theta - 7\cos\theta.\end{aligned}$$

Next we bring to light another unexpected occurrence of Fibonacci numbers. Take a look at the sums of numbers d_n along the northeast diagonals in Table 41.3. They appear to be Fibonacci numbers; see Figure 41.1. So we conjecture that $d_n = F_{n+1}$, where $n \geq 0$; that is, $\sum_{k=0}^{\lfloor n/3 \rfloor} a_{n-k,k} = F_{n+1}$.

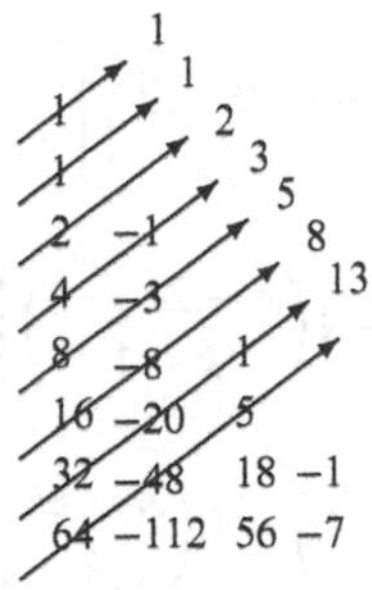

Figure 41.1.

We now establish that $d_n = \sum_{k=0}^{M} a_{n-k,k}$, where $M = \lfloor n/3 \rfloor$ and $n \geq 3$.

It follows by the property

$$\binom{n-k}{k} + \binom{n-k-1}{k} = \frac{n}{n-k}\binom{n-k}{k}$$

that

$$\begin{aligned}\frac{n-k}{n-2k}\binom{n-2k}{k} &= \binom{n-2k}{k} + \binom{n-2k-1}{k-1}\\ &= \binom{n-2k-1}{k} + 2\binom{n-2k-1}{k-1}.\end{aligned}$$

By formula (41.28), we then have

$$\begin{aligned}d_n &= \sum_{k=0}^{M} a_{n-k,k}\\
&= \sum_{k=0}^{M}(-1)^k \frac{n-k}{n-2k}\binom{n-2k}{k}2^{n-3k-1}\\
&= \sum_{k=0}^{M}(-1)^k \left[\binom{n-2k-1}{k} + 2\binom{n-2k-1}{k-1}\right] 2^{n-3k-1}\\
&= \sum_{k=0}^{M}(-1)^k \left[2\binom{n-2k-1}{k} + 4\binom{n-2k-1}{k-1}\right] 2^{n-3k-2}\\
&= \sum_{k=0}^{M}(-1)^k \left\{\binom{n-2k-1}{k} + \left[\binom{n-2k-2}{k-1} + \binom{n-2k-2}{k}\right]\right\} 2^{n-3k-2}\\
&\quad + 4\sum_{k=0}^{M}(-1)^k \binom{n-2k-1}{k-1}2^{n-3k-2}\\
&= \sum_{k=0}^{M}(-1)^k \left[\binom{n-2k-1}{k} + \binom{n-2k-2}{k-1}\right] 2^{n-3k-2}\\
&\quad + \sum_{k=0}^{M}(-1)^k \left[\binom{n-2k-2}{k} + 4\binom{n-2k-1}{k-1}\right] 2^{n-3k-2}\\
&= d_{n-1} + \sum_{k=0}^{M}(-1)^k \left[2\binom{n-2k-2}{k} + 8\binom{n-2k-1}{k-1}\right] 2^{n-3k-3}\end{aligned}$$

$$= d_{n-1} + \sum_{k=0}^{M}(-1)^k \left\{ \binom{n-2k-2}{k} + \left[\binom{n-2k-3}{k-1} + \binom{n-2k-3}{k} \right] \right\} 2^{n-3k-2}$$

$$+ 8 \sum_{k=0}^{M}(-1)^k \binom{n-2k-1}{k-1} 2^{n-3k-3}$$

$$= d_{n-1} + d_{n-2} + \sum_{k=0}^{M}(-1)^k \left[\binom{n-2k-3}{k} + 8\binom{n-2k-1}{k-1} \right] 2^{n-3k-3}$$

$$= d_{n-1} + d_{n-2} + S.$$

Clearly, $\binom{n-0-1}{-1} = 0$. Since $n-3 < n-1 \leq 3M$, $n-2M-3 < M$; so $\binom{n-2M-3}{M} = 0$. The sum of the remaining terms in S is zero; so $S = 0$. Thus d_n satisfies the Fibonacci recurrence. This, coupled with the initial conditions $d_0 = F_1$ and $d_1 = F_2$, implies that $d_n = F_{n+1}$, as desired.

For the sake of curiosity, we now briefly show that $S = 0$ in the above proof in two cases. To begin with, let $n = 9$; so $M = 3$. Then

$$S = \sum_{k=0}^{3}(-1)^k \left[\binom{6-2k}{k} + 8\binom{8-2k}{k-1} \right] 2^{6-3k}$$

$$= (1 + 8 \cdot 0)2^6 - (4 + 8 \cdot 1)2^3 + (1 + 8 \cdot 4)2^0 - (0 + 8 \cdot 1)2^{-3}$$

$$= 64 - 12 \cdot 8 + 33 - 1 = 0.$$

Similarly, when $n = 10$,

$$S = \sum_{k=0}^{3}(-1)^k \left[\binom{7-2k}{k} + 8\binom{9-2k}{k-1} \right] 2^{7-3k}$$

$$= 128 - 208 + 86 - 6 = 0.$$

We can minimize the work required by using generating functions, as Hoggatt did in 1975 [32].

Generating Function for d_n

Let $g_k = g_k(x)$ denote the generating function of the elements in column k of array A, where $k \geq 0$. Since $a_{0,0} = 1$ and $a_{n,0} = 2a_{n-1,0}$, it follows that $g_0 = \dfrac{1-x}{1-2x}$.

Consequently, by PMI,

$$g_k = \frac{1-x}{1-2x}\left(\frac{-1}{1-2x}\right)^k, \tag{41.31}$$

where $k \geq 0$; see Exercise 41.64.

So the generating function of the sum of the elements along the rising diagonals are given by

$$\begin{aligned}
\sum_{k=0}^{\infty} x^{3k} g_k &= \sum_{k=0}^{\infty} \frac{1-x}{1-2x}\left(\frac{-x^3}{1-2x}\right)^k \\
&= \frac{1-x}{1-2x}\sum_{k=0}^{\infty}\left(\frac{-x^3}{1-2x}\right)^k \\
&= \frac{1-x}{1-2x}\cdot\frac{1}{1+\dfrac{x^3}{1-2x}} \\
&= \frac{1-x}{1-2x+x^3} \\
&= \frac{1}{1-x-x^2}.
\end{aligned}$$

This is the generating function for the Fibonacci numbers F_{n+1}, where $n \geq 0$; so the sums of the elements along the rising diagonals are the Fibonacci numbers F_{n+1}, as desired.

Using the generating function (41.31), we can easily confirm that every row sum is one:

$$\begin{aligned}
x^{2k} g_k &= \frac{1-x}{1-2x}\left(\frac{-x^2}{1-2x}\right)^k \\
\sum_{k=0}^{\infty} x^{2k} g_k &= \frac{1-x}{1-2x}\cdot\frac{1}{1+\dfrac{x^2}{1-2x}} \\
&= \frac{1}{1-x} \\
&= \sum_{k=0}^{\infty} x^n.
\end{aligned}$$

Since every coefficient on the RHS is 1, it follows that every row sum is indeed one.

Next, consider the sums w_n along the falling diagonals of array A. Clearly, $w_0 = 1$, $w_1 = 0 = w_2 = w_3$. More generally, we conjecture that $w_n = 0$ when $n \geq 1$.

This can be established fairly quickly. Using the binomial theorem and formula (41.28), we have

$$\begin{aligned} w_n &= \sum_{k=0}^{n} a_{n+k,k} \\ &= \sum_{k=0}^{n} (-1)^k \cdot \frac{n+k}{k} \binom{n}{k} 2^{n-k-1} \\ &= \sum_{k=0}^{n} (-1)^k \left[\binom{n}{k} + \binom{n-1}{k-1} \right] 2^{n-k-1} \\ &= \frac{1}{2} \sum_{k=0}^{n} (-1)^k \binom{n}{k} 2^{n-k} - \frac{1}{2} \sum_{k=0}^{n-1} (-1)^k \binom{n-1}{k} 2^{n-k-1} \\ &= \frac{1}{2}(2-1)^n - \frac{1}{2}(2-1)^{n-1} \\ &= 0, \end{aligned}$$

as desired.

Next we replace the negative numbers in array A with their absolute values. Table 41.7 shows the resulting array B. Clearly, the elements $|a_{n,k}|$ satisfy the recurrence $|a_{n,k}| = 2|a_{n-1,k}| + |a_{n-2,k-1}|$, where $|a_{0,0}| = 1 = |a_{1,0}|$, $n \geq 2$, and $k \geq 2$.

TABLE 41.7. Triangular Array B

n \ k	0	1	2	3	Row sums
0	1				1
1	1				1
2	2	1			3
3	4	3			7
4	8	8 ↘	1		17
5	16	20	5 ↓		41
6	32	48	(18)	1	99
7	64	112	56	7	239
					↑ Q_n

It appears from Table 41.7 that the nth row sum of array B is Q_n. This is so since $Q_n = |T_n(i)|$, which we found earlier. This can be confirmed independently also; see Exercise 41.65. We now establish it using the generating function technique invoked earlier.

Let $g_k^* = g_k^*(x)$ denote the generating function of the elements in column k of array B, where $k \geq 0$. It follows by formula (41.31) that

$$g_k^* = \frac{1-x}{1-2x}\left(\frac{1}{1-2x}\right)^k.$$

Consequently, the row sums are given by the generating function

$$\begin{aligned}\sum_{k=0}^{\infty} x^{2k} g_k^* &= \frac{1-x}{1-2x}\sum_{k=0}^{\infty}\left(\frac{x^2}{1-2x}\right)^k\\ &= \frac{1-x}{1-2x}\cdot\frac{1}{1-\dfrac{x^2}{1-2x}}\\ &= \frac{1-x}{1-2x-x^2}\\ &= \sum_{k=0}^{\infty} Q_n x^n.\end{aligned}$$

Now consider the sums D_n of rising diagonal elements in array B; see Figure 41.2. The sums D_n also manifest an interesting pattern: 1, 1, 2, 5, 11, 24, 53, ... The sequence $\{D_n\}_{n\geq 0}$ can be defined recursively:

$$\begin{aligned}D_0 &= 1 = D_1\\ D_2 &= 2\\ D_n &= 2D_{n-1} + D_{n-3}, \qquad (41.32)\end{aligned}$$

where $n \geq 3$.

We can establish recurrence (41.32) using formula (41.28) and an argument similar to the one for the proof that d_n satisfies the Fibonacci recurrence; see Exercise 41.66.

Using the technique illustrated earlier, we can find a generating function for the sequence $\{D_n\}_{n\geq 0}$.

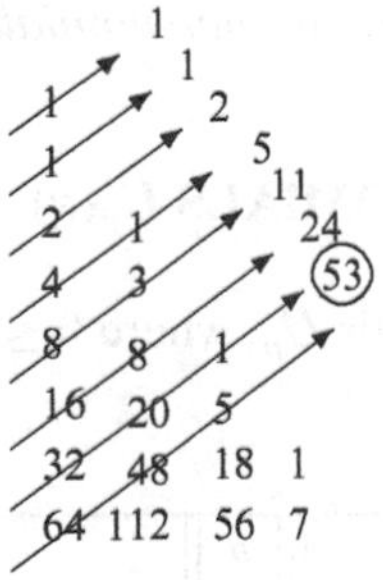

Figure 41.2.

Generating Function for $\{D_n\}_{n\geq 0}$

The sums D_n can be generated by

$$\begin{aligned}\sum_{k=0}^{\infty} x^{3k} g_k^* &= \frac{1-x}{1-2x} \sum_{k=0}^{\infty} \left(\frac{x^3}{1-2x}\right)^k \\ &= \frac{1-x}{1-2x} \cdot \frac{1}{1-\dfrac{x^3}{1-2x}} \\ &= \frac{1-x}{1-2x-x^3}.\end{aligned}$$

Thus

$$\frac{1-x}{1-2x-x^3} = \sum_{k=0}^{\infty} D_n x^n.$$

The falling diagonal sums z_n also display an interesting pattern: 1, 2, 6, 18, 54, 162, 486, ... ; see Figure 41.3. It follows from these data that

$$z_n = \begin{cases} 1 & \text{if } n = 0 \\ 2 \cdot 3^{n-1} & \text{otherwise;} \end{cases}$$

see Exercise 41.67.

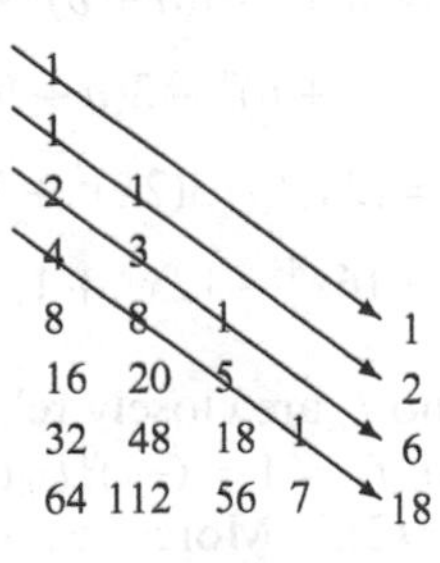

Figure 41.3.

Next we investigate the *Chebyshev polynomials of the second kind* $U_n(x)$.

41.4 CHEBYSHEV POLYNOMIALS $U_n(x)$

Table 41.8 shows the polynomials U_n, where $0 \leq n \leq 9$.

TABLE 41.8.

n	$U_n(x)$	n	$U_n(x)$
0	1	5	$32x^5 - 32x^3 + 6x$
1	$2x$	6	$64x^6 - 80x^4 + 24x^2 - 1$
2	$4x^2 - 1$	7	$128x^7 - 192x^5 + 80x^3 - 8x$
3	$8x^3 - 4x$	8	$256x^8 - 448x^6 + 240x^4 - 40x^2 + 1$
4	$16x^4 - 12x^2 + 1$	9	$512x^9 - 1024x^7 + 672x^5 - 160x^3 + 10x$

Binet-like Formula

Recall that the general solution of the Chebyshev recurrence is $z_n = Aa^n + Bb^n$. The initial conditions $z_0 = 1$ and $z_1 = 2x$ imply that $A = \dfrac{a}{a-b}$ and $B = -\dfrac{b}{a-b}$. Consequently,

$$U_n = \frac{a^{n+1} - b^{n+1}}{a - b}, \tag{41.33}$$

where $n \geq 0$. (Recall that $a + b = 2x$, $a - b = 2\sqrt{x^2 - 1}$ and $ab = 1$.)

For example,

$$\begin{aligned}
(a-b)U_4 &= a^5 - b^5 \\
U_4 &= a^4 + a^3b + a^2b^2 + ab^3 + b^4 \\
&= (a+b)(a^3 + b^3) + 1 \\
&= (a+b)^2[(a+b)^2 - 3ab] + 1 \\
&= (a+b)^4 - 3(a+b)^2 + 1 \\
&= (2x)^4 - 3(2x)^2 + 1 \\
&= 16x^4 - 12x^5 + 1.
\end{aligned}$$

Earlier, we found that T_n and l_n are closely related. The same is true with U_n and f_n. To see this, notice that $f_1 = 1 = (-i)^0 U_0(ix/2)$, $f_2 = x = (-i)^1 U_1(ix/2)$, and $f_3 = x^2 + 1 = (-i)^2 U_2(ix/2)$. More generally, we conjecture that $f_{n+1} = (-i)^n U_n(ix/2)$. This can be confirmed easily; see Exercise 41.28.

For example,

$$\begin{aligned} f_5 &= (-i)^4 U_4(ix/2) \\ &= 16(ix/2)^4 - 12(ix/2)^2 + 1 \\ &= x^4 + 3x^2 + 1. \end{aligned}$$

It now follows that $F_{n+1} = |U_n(i/2)|$, $p_{n+1} = (-i)^n U_n(ix)$, and $P_{n+1} = |U_n(i)|$.

For example, $U_6(ix) = 64(ix)^6 - 80(ix)^4 + 24(ix)^2 - 1 = -(64x^6 + 80x^4 + 24x^2 + 1)$; so $p_7 = (-i)^6 U_6(ix) = 64x^6 + 80x^4 + 24x^2 + 1$. Consequently, $P_7 = 169$.

The identity $f_{n+1} = (-i)^n U_n(ix/2)$ is also reversible: $U_n = i^n f_{n+1}(-2ix)$; see Exercise 41.29.

Since $l_n = f_{n+1} + f_{n-1}$, it follows by the identity $f_{n+1} = i^{-n} U_n(ix/2)$ that $l_n = i^{-n}[U_n(ix/2) - U_{n-2}(ix/2)]$ and hence $q_n = i^{-n}[U_n(ix) - U_{n-2}(ix)]$. For example, $i^{-5}[U_5(ix/2) - U_3(ix/2)] = x^5 + 5x^3 + 5x = l_5$.

Another Explicit Formula for U_n

Using the Binet-like formula for U_n and the binomial theorem, we can derive a second explicit formula for U_n [285]:

$$U_n = \sum_{k=0}^{\lfloor n/2 \rfloor} \binom{n+1}{2k+1} (x^2 - 1)^k x^{n-2k}, \qquad (41.34)$$

where $n \geq 0$; see Exercise 41.38.

For example,

$$\begin{aligned} U_5 &= \sum_{k=0}^{2} \binom{6}{2k+1} (x^2 - 1)^k x^{5-2k} \\ &= 6x^5 + 20(x^2 - 1)x^3 + 6(x^2 - 1)^2 x \\ &= 32x^5 - 32x^3 + 6x. \end{aligned}$$

It follows from formula (41.34) that

$$f_{n+1} = \frac{1}{2^n} \sum_{k=0}^{\lfloor n/2 \rfloor} \binom{n+1}{2k+1} (x^2 + 4)^k x^{n-2k};$$

$$p_{n+1} = \sum_{k=0}^{\lfloor n/2 \rfloor} \binom{n+1}{2k+1} (x^2 + 1)^k x^{n-2k}.$$

It follows from formula (41.34) that the highest power of x in U_n is n, and the leading coefficient is

$$\sum_{k\geq 0}\binom{n+1}{2k+1} = \sum_{i \text{ odd}}\binom{n+1}{i} = 2^n.$$

So the leading term in U_n is $(2x)^n$; see Table 41.8.

Recall that there are a number of properties linking Fibonacci and Lucas polynomials. As we might predict, T_n and U_n also enjoy similar relationships. For example, we have [285]

$$U_n - U_{n-2} = 2T_n; \tag{41.35}$$

$$U_n - xU_{n-1} = T_n; \tag{41.36}$$

see Exercises 41.31 and 41.36.

Next we highlight a charming relationship bridging T_n, U_n, and Pell's equation†.

41.5 PELL'S EQUATION

Consider the Pell's equation $v^2 - dw^2 = 1$, where v and w are integers, and $d > 0$ and nonsquare. Let (v_1, w_1) be its fundamental solution, and (v_n, w_n) an arbitrary solution. Then (v_n, w_n) can be computed recursively [285]:

$$\begin{bmatrix} v_n \\ w_n \end{bmatrix} = \begin{bmatrix} v_1 & dw_1 \\ w_1 & v_1 \end{bmatrix}\begin{bmatrix} v_{n-1} \\ w_{n-1} \end{bmatrix},$$

where $n \geq 2$.

In particular, consider the Pell's equation $v^2 - (x^2-1)w^2 = 1$. Clearly, $(v_0, w_0) = (1, 0)$ and $(v_1, w_1) = (x, 1)$ are solutions of the equation. Consequently, its infinitely many solutions are given by the matrix recurrence

$$\begin{aligned}\begin{bmatrix} v_n \\ w_n \end{bmatrix} &= \begin{bmatrix} x & x^2-1 \\ 1 & x \end{bmatrix}\begin{bmatrix} v_{n-1} \\ w_{n-1} \end{bmatrix} \\ &= 2x\begin{bmatrix} v_{n-1} \\ w_{n-1} \end{bmatrix} - \begin{bmatrix} v_{n-2} \\ w_{n-2} \end{bmatrix},\end{aligned}$$

where $n \geq 2$.

Since $v_n = 2xv_{n-1} - v_{n-2}$, where $v_0 = 1$ and $v_1 = x$, it follows that $v_n = T_n$. Clearly, w_n also satisfies the Chebyshev recurrence, with $w_0 = 0$ and $w_1 = 1$; so $w_n = U_{n-1}$. Thus every solution of the Pell's equation is given by $(v_n, w_n) = (T_n, U_{n-1})$.

---

†For a detailed discussion of Pell's equation, see [285].

This implies

$$T_n^2 - (x^2 - 1)U_{n-1}^2 = 1, \tag{41.37}$$

where $n \geq 1$. This can be confirmed independently using the Binet-like formulas; see Exercise 41.50.

It follows from identity (41.37) that

$$\begin{aligned}\frac{l_n^2}{(-i)^{2n}} + (x^2+4)\frac{f_n^2}{(-i)^{2n-2}} &= 4\\ \frac{l_n^2}{(-1)^n} + (x^2+4)\frac{f_n^2}{(-1)^{n-1}} &= 4\\ l_n^2 - (x^2+4)f_n^2 &= 4(-1)^n.\end{aligned}$$

Consequently, identity (41.37) generalizes the Fibonacci–Lucas identity $l_n^2 - (x^2+4)f_n^2 = 4(-1)^n$ we studied at length in Chapter 31.

Earlier, we found that T_n enjoys a special relationship with trigonometry. Predictably, so does U_n.

41.6 $U_n(x)$ AND TRIGONOMETRY

This time, we investigate the expansions of $\sin n\theta$; see Table 41.9. Again, we look for a pattern. In each case, $\sin(n+1)\theta/\sin\theta$ is the polynomial $U_n(\cos\theta)$.

TABLE 41.9.

$\sin 0\theta = 0$
$\sin 1\theta = \sin\theta$
$\sin 2\theta = 2\cos\theta\sin\theta$
$\sin 3\theta = (4\cos^2\theta - 1)\sin\theta$
$\sin 4\theta = (8\cos^3\theta - 4\cos\theta)\sin\theta$
$\sin 5\theta = (16\cos^4\theta - 12\cos^2\theta + 1)\sin\theta$

More generally, we conjecture that $\dfrac{\sin(n+1)\theta}{\sin\theta} = U_n(x)$, where $x = \cos\theta$. This can be confirmed by the addition formula for the sine function, identity (41.36), and PMI [285]; see Exercise 41.41.

This trigonometric bridge opens new opportunities for exploration. For example, using the *sum identity* $\sin(A+B) + \sin(A-B) = 2\sin A\cos B$, we have $\sin(n+1)\theta + \sin(n-1)\theta = 2\sin n\theta\cos\theta$. This yields the Chebyshev recurrence $U_{n+1} + U_{n-1} = 2xU_n$, where $U_0 = \sin 1\theta/\sin\theta = 1$ and $U_1 = \sin 2\theta/\sin\theta = 2x$.

Next we present a simple, but interesting, application of the polynomials $U_n(x)$.

A Fibonacci Curiosity

In 2015, using recursion it was established that

$$\sum_{\substack{k\geq 0\\ k \text{ even}}} \binom{n-k}{k} = \sum_{\substack{k\geq 0\\ k \text{ odd}}} \binom{n-k}{k}$$

if and only if $n \equiv 2 \pmod 3$ [294]. A few months later, N. Lord of Tonbridge School, Kent, United Kingdom, gave a simpler, alternate proof of this property using the Chebyshev polynomials $U_n(x)$ [338]. To see this, it follows from the trigonometric form $U_n(\cos\theta) = \dfrac{\sin(n+1)\theta}{\sin\theta}$ that

$$U_n(x) = \sum_{k=0}^{\lfloor n/2 \rfloor} (-1)^k \binom{n-k}{k} (2x)^{n-k}.$$

Since $\cos 2\pi/3 = -1/2$, this summation formula yields

$$(-1)^n U_n(-1/2) = \sum_{\substack{k\geq 0\\ k \text{ even}}} \binom{n-k}{k} - \sum_{\substack{k\geq 0\\ k \text{ odd}}} \binom{n-k}{k}.$$

But $U_n(-1/2) = \dfrac{\sin(n+1)2\pi/3}{\sin 2\pi/3}$. So $U_n(-1/2) = 0$ if and only if $n \equiv 2 \pmod 3$. The desired result now follows.

We can establish an additional Chebyshev property, that

$$T_{m+n} - T_{m-n} = 2(x^2-1)U_{m-1}U_{n-1}, \tag{41.38}$$

where $m \geq n$; see Exercise 41.45.

This yields the Fibonacci–Lucas hybrid identity

$$l_{m+n} - (-1)^n l_{m-n} = (x^2+4) f_m f_n; \tag{41.39}$$

see Exercise 41.53.

It follows from identity (41.39) that

$$L_{m+n} - (-1)^n L_{m-n} = 5F_m F_n;$$
$$q_{m+n} - (-1)^n q_{m-n} = 4(x^2+1) p_m p_n;$$
$$Q_{m+n} - (-1)^n Q_{m-n} = 4P_m P_n.$$

For example,

$$\begin{aligned}
q_8 + q_2 &= q_{5+3} - (-1)^3 q_{5-3} \\
&= (256x^8 + 512x^6 + 320x^4 + 64x^2 + 2) + (4x^2 + 2) \\
&= 256x^8 + 512x^6 + 320x^4 + 68x^2 + 4 \\
&= 4(x^2 + 1)(16x^4 + 12x^2 + 1)(4x^2 + 1) \\
&= 4(x^2 + 1)p_5 p_3.
\end{aligned}$$

41.7 ADDITION AND CASSINI-LIKE FORMULAS

Interestingly, Chebyshev polynomials have their own addition and Cassini-like formulas [285]. Their proofs are fairly straightforward; see Exercises 41.46–41.49.

Addition Formulas

$$T_{m+n} = T_m U_n - T_{m-1} U_{n-1}; \tag{41.40}$$

$$U_{m+n} = U_m U_n - U_{m-1} U_{n-1}. \tag{41.41}$$

It follows from identities (41.40) and (41.41) that

$$\begin{aligned}
l_{m+n} &= l_m f_{n+1} + l_{m-1} f_n; \\
f_{m+n} &= f_{m+1} f_n + f_m f_{n-1}; \\
q_{m+n} &= q_m p_{n+1} + q_{m-1} p_n; \\
p_{m+n} &= p_{m+1} p_n + p_m p_{n-1}.
\end{aligned}$$

Cassini-like Formulas

$$T_{n+1} T_{n-1} - T_n^2 = x^2 - 1; \tag{41.42}$$

$$U_{n+1} U_{n-1} - U_n^2 = -1. \tag{41.43}$$

It follows from these two identities that

$$\begin{aligned}
l_{n+1} l_{n-1} - l_n^2 &= (-1)^{n+1}(x^2 + 4); \\
f_{n+1} f_{n-1} - f_n^2 &= (-1)^n; \\
q_{n+1} q_{n-1} - q_n^2 &= 4(-1)^{n+1}(x^2 + 1); \\
p_{n+1} p_{n-1} - p_n^2 &= (-1)^n.
\end{aligned}$$

As we might predict, Table 41.8 also contains hidden treasures. We now study some of them.

41.8 HIDDEN TREASURES IN TABLE 41.8

As before, we arrange the nonzero coefficients in Table 41.8 in a left-justified array D; see Table 41.10.

TABLE 41.10. Triangular Array D

$n \backslash k$	0	1	2	3
0	1			
1	2			
2	4	−1		
3	8	−4		
4	16	−12	1	
5	32	−32	6	
6	64	−80	24	−1
7	128	−192	80	−8

Let $d_{n,k}$ denote the element in row n and column k of array D, where $n, k \geq 0$. Then

$$d_{0,0} = 1$$
$$d_{1,0} = 2$$
$$d_{n,k} = 2d_{n-1,k} - d_{n-2,k-1}, \tag{41.44}$$

where $n \geq 2$ and $k \geq 1$.

For example, $d_{6,2} = 2d_{5,2} - d_{4,1} = 2 \cdot 6 - (-12) = 24$; see the arrows in Table 41.10.

We now list a few additional properties of array D. Their proofs are quite similar to their counterparts in Section 41.3, so we omit them; see Exercises 41.70–41.77.

1) We can compute the elements $d_{n,k}$ explicitly: $d_{n,k} = (-1)^k \binom{n-k}{k} 2^{n-2k}$.
2) Let x_n denote the sum of the elements along the nth rising diagonal. Then

$$x_0 = 1, \quad x_1 = 2$$
$$x_n = x_{n-1} + x_{n-2} + 1, \tag{41.45}$$

where $n \geq 2$. Solving this recurrence, we get $x_n = F_{n+3} - 1$.

For example, $x_6 = 64 - 32 + 1 = 33 = F_8 - 1$; likewise, $x_7 = 54 = F_{10} - 1$.

3) The sum of the elements on the nth falling diagonal is 1, where $n \geq 0$.
4) The generating function of the elements in column k is $h_k = \dfrac{1}{1-2x}\left(\dfrac{-1}{1-2x}\right)^k$, where $k \geq 0$.

Now consider the array $E = (e_{n,k})$, where $e_{n,k} = |d_{n,k}|$; see Figure 41.4.

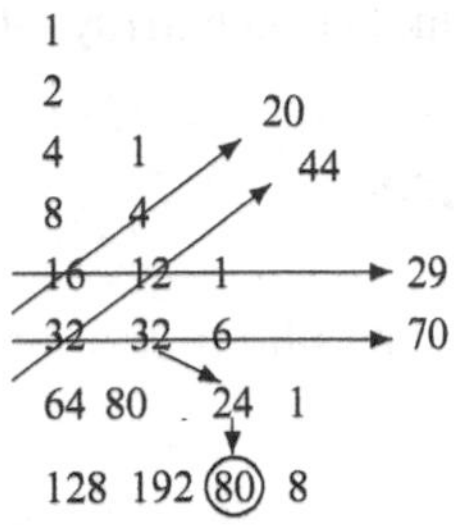

Figure 41.4. Triangular array E.

1) We can define the elements $e_{n,k}$ recursively:

$$\begin{aligned} e_{0,0} &= 1, \quad e_{1,0} = 2 \\ e_{n,k} &= 2e_{n-1,k} + e_{n-2,k-1}, \end{aligned} \tag{41.46}$$

where $n \geq 2$ and $k \geq 0$. For example, (80) $= 2 \cdot 24 + 32$; see Figure 41.4.
2) The nth row sum of array E is the Pell number P_{n+1}, where $n \geq 0$; see Figure 41.4 and Exercise 41.75.

We now give a generating function proof of this fact. The generating function of the elements in column k of the array is $h_k^* = \dfrac{1}{1-2x}\left(\dfrac{1}{1-2x}\right)^k$, where $k \geq 0$; see Exercise 41.77. So

$$x^{2k}h_k^* = \frac{1}{1-2x}\left(\frac{x^2}{1-2x}\right)^k$$

$$\sum_{k=0}^{\infty} x^{2k}h_k^* = \frac{1}{1-2x} \cdot \frac{1}{1-\dfrac{x^2}{1-2x}}$$

$$= \frac{1}{1-2x-x^2}$$

$$= \sum_{k=0}^{\infty} P_{n+1}x^n.$$

This yields the desired result.

3) We can define recursively the nth rising diagonal sum u_n:

$$u_0 = 1, \quad u_1 = 2, \quad u_2 = 4$$
$$u_n = 2u_{n-1} + u_{n-3}, \quad n \geq 3.$$

4) The nth falling diagonal sum is 3^n, where $n \geq 0$; see Exercise 41.76.

Next we highlight a close link between arrays A and D.

41.9 A CHEBYSHEV BRIDGE

Since $h_k = \dfrac{1}{1-x}g_k$, the entries in column k of array D are the partial sums of those in column k of array A.

For example,

$$\begin{aligned} d_{2,1} &= -1 &&= a_{2,1} \\ d_{3,1} &= -4 &&= a_{2,1} + a_{3,1} \\ d_{4,1} &= -12 &&= a_{2,1} + a_{3,1} + a_{4,1} \\ d_{5,1} &= -32 &&= a_{2,1} + a_{3,1} + a_{4,1} + a_{5,1}. \end{aligned}$$

More generally,

$$\sum_{n=0}^{\infty} d_{n+2k,k} x^n = \sum_{n=0}^{\infty} \left(\sum_{j=0}^{n} a_{j+2k,k} \right) x^n.$$

Consequently,

$$d_{n+2k,k} = \sum_{j=0}^{n} a_{j+2k,k}. \tag{41.47}$$

For example,

$$\begin{aligned} d_{7,3} &= \sum_{j=0}^{1} a_{j+6,3} \\ &= a_{6,3} + a_{7,3} \\ &= -1 + (-7) = -8; \\ d_{11,4} &= \sum_{j=0}^{3} a_{j+8,4} \\ &= a_{8,4} + a_{9,4} + a_{10,4} + a_{11,4} \\ &= 1 + 9 + 50 + 220 = 280. \end{aligned}$$

Equation (41.47) implies an interesting combinatorial identity:

$$\binom{n+k}{k}2^n = \sum_{j=0}^{n} \frac{j+2k}{j+k}\binom{j+k}{k}2^{j-1}. \tag{41.48}$$

For example, let $n = 3$ and $k = 5$. Then

$$\begin{aligned}
\text{RHS} &= \sum_{j=0}^{3} \frac{10+j}{5+j}\binom{5+j}{5}2^{j-1} \\
&= \frac{10}{5}\binom{5}{5}2^{-1} + \frac{11}{6}\binom{6}{5}2^{0} + \frac{12}{7}\binom{7}{5}2 + \frac{13}{8}\binom{8}{5}2^2 \\
&= 1 + 11 + 72 + 364 = 448 \\
&= \binom{3+5}{5}2^3 \\
&= \text{LHS}.
\end{aligned}$$

Next we express both T_n and U_n as products of trigonometric expressions. This approach yields interesting dividends.

41.10 T_n AND U_n AS PRODUCTS

Since $T_n(x) = \cos n\theta$, it follows that $T_n = 0$ when $\theta = (2k-1)\pi/2n$, where $1 \le k \le n$; see Exercise 41.52. Since the leading term in T_n is $2^{n-1}x^n$, this implies that

$$T_n = 2^{n-1}\prod_{k=1}^{n}[x - \cos(2k-1)\pi/2n]. \tag{41.49}$$

Using this, we can find a trigonometric product for l_n. To this end, we have

$$\begin{aligned}
T_n(ix)T_n(-ix) &= 2^{2n-2}\prod_{k=1}^{n}[ix - \cos(2k-1)\pi/2n]\prod_{k=1}^{n}[-ix - \cos(2k-1)\pi/2n] \\
&= 2^{2n-2}\prod_{k=1}^{n}[x^2 + \cos^2(2k-1)\pi/2n] \\
\frac{l_n}{2(-i)^n}\cdot\frac{l_n}{2i^n} &= \frac{1}{4}\prod_{k=1}^{n}[x^2 + 2 + 2\cos(2k-1)\pi/n] \\
l_n^2 &= \prod_{k=1}^{n}[x^2 + 2 + 2\cos(2k-1)\pi/n].
\end{aligned} \tag{41.50}$$

Case 1. Let n be odd, say, $n = 2m - 1$. For convenience, we let $B_k = x^2 + 2 + 2\cos(2k-1)\pi/n$. Since $B_k > 0$, we then have

$$\begin{aligned}
l_n^2 &= \prod_{k=1}^{m-1} B_k \cdot x^2 \cdot \prod_{k=m+1}^{2m-1} B_k \\
&= x^2 \prod_{k=1}^{m-1} B_k \cdot \prod_{j=1}^{m-1} B_j \\
l_n &= x \prod_{k=1}^{m-1} B_k \\
&= x \prod_{k=1}^{\lfloor n/2 \rfloor} [x^2 + 2 + 2\cos(2k-1)\pi/n],
\end{aligned}$$

where n is odd.

Case 2. Let n be even, say, $n = 2m$. Then, by equation (41.50),

$$\begin{aligned}
l_n^2 &= \prod_{k=1}^{m} B_k \cdot \prod_{k=m+1}^{2m} B_k \\
&= \prod_{k=1}^{m} B_k \cdot \prod_{j=1}^{m} B_j \\
l_n &= \prod_{k=1}^{\lfloor n/2 \rfloor} B_k,
\end{aligned}$$

where n is even.

Combining the two cases, we have

$$l_n = \begin{cases} x \prod_{k=1}^{\lfloor n/2 \rfloor} [x^2 + 2 + 2\cos(2k-1)\pi/n] & \text{if } n \text{ is odd} \\ \prod_{k=1}^{\lfloor n/2 \rfloor} [x^2 + 2 + 2\cos(2k-1)\pi/n] & \text{otherwise.} \end{cases} \tag{41.51}$$

For example,

$$\begin{aligned}
l_6 &= \prod_{k=1}^{3} [x^2 + 2 + 2\cos(2k-1)\pi/6] \\
&= (x^2 + 2 + 2\cos\pi/6)(x^2 + 2 + 2\cos 3\pi/6)(x^2 + 2 + 2\cos 5\pi/6)
\end{aligned}$$

$$= (x^2 + 2)[(x^2 + 2)^2 - 3]$$
$$= x^6 + 6x^4 + 9x^2 + 2.$$

It follows from formula (41.51) that

$$L_n = \prod_{k=1}^{\lfloor n/2 \rfloor} [3 + 2\cos(2k-1)\pi/n];$$

$$q_n = \begin{cases} 2^{(n+1)/2} x \prod_{k=1}^{\lfloor n/2 \rfloor} [2x^2 + 1 + \cos(2k-1)\pi/n] & \text{if } n \text{ is odd} \\ 2^{n/2} \prod_{k=1}^{\lfloor n/2 \rfloor} [2x^2 + 1 + \cos(2k-1)\pi/n] & \text{otherwise;} \end{cases}$$

$$Q_n = \begin{cases} 2^{(n-1)/2} \prod_{k=1}^{\lfloor n/2 \rfloor} [3 + \cos(2k-1)\pi/n] & \text{if } n \text{ is odd} \\ 2^{(n-2)/2} \prod_{k=1}^{\lfloor n/2 \rfloor} [3 + \cos(2k-1)\pi/n] & \text{otherwise.} \end{cases}$$

For example,

$$Q_6 = 4\prod_{k=1}^{3}[3 + \cos(2k-1)\pi/6]$$
$$= 4(3 + \cos\pi/6)(3 + \cos\pi/2)(3 + \cos 5\pi/6)$$
$$= 4(3 + \sqrt{3}/2) \cdot 3 \cdot (3 - \sqrt{3}/2)$$
$$= 99.$$

To express U_n as a trigonometric product, we apply a similar technique. The zeros of $U_n(x)$ are $x = \cos\dfrac{k\pi}{n+1}$, where $1 \le k \le n$. Since the leading term in U_n is $(2x)^n$, it follows that

$$U_n = 2^n \prod_{k=1}^{n} \left(x - \cos\frac{k\pi}{n+1}\right). \tag{41.52}$$

Consequently,

$$U_{n-1}(ix)U_{n-1}(-ix) = 2^{2n-2}\prod_{k=1}^{n-1}(ix - \cos k\pi/n)\prod_{k=1}^{n-1}(-ix - \cos k\pi/n)$$
$$= 2^{2n-2}\prod_{k=1}^{n-1}(x^2 + \cos^2 k\pi/n)$$

$$= \prod_{k=1}^{n-1} (4x^2 + 2 + 2\cos 2k\pi/n)$$

$$U_{n-1}(ix/2)U_{n-1}(-ix/2) = \prod_{k=1}^{n-1} (x^2 + 2 + 2\cos 2k\pi/n)$$

$$i^{n-1} f_n \cdot (-i)^{n-1} f_n = \prod_{k=1}^{n-1} (x^2 + 2 + 2\cos 2k\pi/n)$$

$$f_n^2 = \prod_{k=1}^{n-1} (x^2 + 2 + 2\cos 2k\pi/n).$$

Case 1. Let n be even, say, $n = 2m$, and $B_k = x^2 + 2 + 2\cos 2k\pi/n$. Then $B_m = x^2 + 2 + 2\cos \pi = x^2$. Since $B_{n-k} = B_k$, we have

$$f_n^2 = \prod_{k=1}^{n-1} B_k$$

$$= \prod_{k=1}^{m-1} B_k \cdot x^2 \cdot \prod_{k=1}^{2m-1} B_k$$

$$= x^2 \prod_{k=1}^{m-1} B_k \cdot \prod_{k=1}^{m-1} B_k.$$

Since $B_k > 0$, this implies

$$f_n = x \prod_{k=1}^{\lfloor (n-1)/2 \rfloor} (x^2 + 2 + 2\cos\, 2k\pi/n),$$

where n is even.

Case 2. Let n be odd, say, $n = 2m - 1$. Then

$$f_n^2 = \prod_{k=1}^{2m-2} B_k$$

$$= \prod_{k=1}^{m-1} B_k \cdot \prod_{k=1}^{2m-2} B_k$$

$$= \prod_{k=1}^{m-1} B_k \cdot \prod_{j=1}^{m-1} B_j$$

$$\begin{aligned} f_n &= \prod_{k=1}^{m-1} B_k \\ &= \prod_{k=1}^{\lfloor (n-1)/2 \rfloor} (x^2 + 2 + 2\cos 2k\pi/n), \end{aligned}$$

where n is odd.

Combining the two cases, we get

$$f_n = \begin{cases} \displaystyle\prod_{k=1}^{\lfloor (n-1)/2 \rfloor} (x^2 + 2 + 2\cos 2k\pi/n) & \text{if } n \text{ is odd} \\ x \displaystyle\prod_{k=1}^{\lfloor (n-1)/2 \rfloor} (x^2 + 2 + 2\cos 2k\pi/n) & \text{otherwise.} \end{cases} \tag{41.53}$$

For example,

$$\begin{aligned} f_6 &= x \prod_{k=1}^{2} (x^2 + 2 + 2\cos k\pi/3) \\ &= x(x^2 + 2 + 2\cos \pi/3)(x^2 + 2 - 2\cos \pi/3) \\ &= x[(x^2 + 2)^2 - 1] \\ &= x^5 + 4x^3 + 3x. \end{aligned}$$

It follows from formula (41.53) that

$$F_n = \prod_{k=1}^{\lfloor (n-1)/2 \rfloor} (3 + 2\cos 2k\pi/n);$$

$$p_n = \begin{cases} 2^{(n-1)/2} \displaystyle\prod_{k=1}^{\lfloor (n-1)/2 \rfloor} (2x^2 + 1 + \cos 2k\pi/n) & \text{if } n \text{ is odd} \\ 2^{n/2} x \displaystyle\prod_{k=1}^{\lfloor (n-1)/2 \rfloor} (2x^2 + 1 + \cos 2k\pi/n) & \text{otherwise;} \end{cases}$$

$$P_n = \begin{cases} 2^{(n-1)/2} \displaystyle\prod_{k=1}^{\lfloor (n-1)/2 \rfloor} (3 + \cos 2k\pi/n) & \text{if } n \text{ is odd} \\ 2^{n/2} \displaystyle\prod_{k=1}^{\lfloor (n-1)/2 \rfloor} (3 + \cos 2k\pi/n) & \text{otherwise.} \end{cases}$$

For example,

$$\begin{aligned}P_6 &= 8\prod_{k=1}^{2}(3 + \cos k\pi/3]\\&= 8(3 + \cos \pi/3)(3 - \cos \pi/3)\\&= 8(9 - \cos^2 \pi/3)\\&= 70,\end{aligned}$$

as expected.

D.A. Lind discovered the trigonometric product for F_n in 1967 [287, 324]; and P.S. Bruckman confirmed it using Chebyshev polynomials in 1992 [63, 285].

Finally, we can derive a generating function for both Chebyshev polynomials, using standard techniques.

41.11 GENERATING FUNCTIONS

Let $g(t)$ be a generating function of the sequence $\{c_n\}_{n\geq 0}$: $g(t) = \sum_{n=0}^{\infty} c_n t^n$, where $c_n = c_n(x)$ is a Chebyshev polynomial. We can then show that

$$g(t) = \frac{c_0 + (c_1 - 2xc_0)t}{1 - 2xt + t^2};$$

see Exercise 41.78.

In particular, this yields

$$\frac{1 - xt}{1 - 2xt + t^2} = \sum_{n=0}^{\infty} T_n t^n$$

$$\frac{1}{1 - 2xt + t^2} = \sum_{n=0}^{\infty} U_n t^n.$$

EXERCISES 41

Note: A function f is *even* if $f(-x) = f(x)$ for every x, and *odd* if $f(-x) = -f(x)$ for every x. In Exercises 41.15 and 41.16, v equals 1 if n is odd, and -1 if n is even. In Exercises 41.54–41.60, the prime denotes differentiation with respect to x.

1. Using formula (41.20), derive the Binet-like formula for l_n.

Prove each.

2. $T_{-n} = T_n$.

3. $T_n(1) = 1$.
4. $T_n(-1) = (-1)^n$.
5. T_{2n} is an even function and T_{2n+1} an odd function.
6. T_n ends in $\begin{cases} 0 & \text{if } n \equiv \pm 1 \pmod 4 \\ -1 & \text{if } n \equiv 0 \pmod 4 \\ 1 & \text{otherwise.} \end{cases}$
7. $T_{n+1} + T_{n-1} = 2xT_n$.
8. $T_{n+2} + T_{n-2} = 2(2x^2 - 1)T_n$.
9. $T_{n+2} - T_{n-2} = 4x^2 T_n$.
10. $T_{n+1}^2 + T_n^2 = xT_{2n+1} + 1$.
11. $T_{2n} = 2T_n^2 - 1$.
12. $T_{3n} = 4T_n^3 - 3T_n$.
13. $T_{4n} = 8T_n^4 - 8T_n^2 + 1$.
14. $T_{5n} = 16T_n^5 - 20T_n^3 + 5T_n$.
15. $T_{2n+1} = 2x(T_{2n} - T_{2n-2} + T_{2n-4} - \cdots + \nu T_2) - \nu T_1$.
16. $T_{2n} = 2x(T_{2n-1} - T_{2n-3} + T_{2n-5} - \cdots + \nu T_1) - \nu T_0$.
17. $T_{m+n} + T_{m-n} = 2T_m T_n$.
18. $T_{2n+1} = 2T_{n+1}T_n - x$.
19. $T_{2n-1} = 2T_{n-1}T_n - x$.
20. $T_{2n+1} + T_{2n-1} = 2x(2T_n^2 - 1) = 2xT_{2n}$.
21. $T_{2n+1} - T_{2n-1} = 2(x^2 - 1)T_n U_{n-1}$.
22. $T_m(T_n) = T_{mn}$.
23. U_{2n} is an even function and U_{2n+1} an odd function.
24. U_n ends in $\begin{cases} 0 & \text{if } n \equiv \pm 1 \pmod 4 \\ 1 & \text{if } n \equiv 0 \pmod 4 \\ -1 & \text{otherwise.} \end{cases}$
25. $U_n(1) = n + 1$.
26. $U_n(-1) = (n + 1)(-1)^n$.
27. $U_{-n} = -U_n$.
28. $f_{n+1} = (-i)^n U_n(ix/2)$.
29. $U_n = i^n f_{n+1}(-2ix)$.
30. $U_{n+1} + U_{n-1} = 2xU_n$.
31. $U_{n+1} - U_{n-1} = 2T_{n+1}$.

32. $U_{n+2} + U_{n-2} = 2(2x^2 - 1)U_n$.

33. $U_{n+2} - U_{n-2} = 4xT_{n+1}$.

34. $U_{2n-1} = 2U_{n-1}T_n$.

35. $(x^2 - 1)U_{n+1}^2 + U_n^2 = xT_{2n+3} - 1$.

36. $U_n - xU_{n-1} = T_n$.

37. $2T_n = (x^2 - 1)U_{n-1}^2 + 2$.

38. $U_n = \sum_{k=0}^{\lfloor n/2 \rfloor} \binom{n+1}{2k+1}(x^2 - 1)^k x^{n-2k}$.

39. $f_{n+1} = \frac{1}{2^n} \sum_{k=0}^{\lfloor n/2 \rfloor} \binom{n+1}{2k+1}(x^2 + 4)^k x^{n-2k}$.

40. $p_{n+1} = \sum_{k=0}^{\lfloor n/2 \rfloor} \binom{n+1}{2k+1}(x^2 + 1)^k x^{n-2k}$.

41. Let $x = \cos\theta$. Then $\frac{\sin(n+1)\theta}{\sin\theta} = U_n(x)$.

42. $T_{n+1} - T_{n-1} = (x^2 - 1)U_{n-1}$.

43. $T_{n+1}^2 - T_{n-1}^2 = x(x^2 - 1)U_{2n-1}$.

44. $U_{n+1}^2 - U_{n-1}^2 = xU_{2n+1}$.

45. $T_{m+n} - T_{m-n} = 2(x^2 - 1)U_{m-1}U_{n-1}$.

46. $T_{m+n} = T_m U_n - T_{m-1}U_{n-1}$.

47. $U_{m+n} = U_m U_n - U_{m-1}U_{n-1}$.

48. $T_{n+1}T_{n-1} - T_n^2 = x^2 - 1$.

49. $U_{n+1}U_{n-1} - U_n^2 = -1$.

50. $T_n^2 - (x^2 - 1)U_{n-1}^2 = 1$.

51. $T_{n+1} = xT_n - (1 - x^2)U_{n-1}$.

52. The zeros of T_n are $x_k = \cos(2k - 1)\pi/2n$, where $1 \leq k \leq n$.

53. $l_{m+n} - (-1)^n l_{m-n} = (x^2 + 4)f_m f_n$.

54. $T_n' = nU_{n-1}$.

55. $(x^2 - 1)U_n' = (n + 1)T_{n+1} - xU_n$.

56. $T_n' = 2nT_{n-1} + \frac{n}{n-2}T_{n-2}'$.

57. $(x^2 - 1)T_n'' = n[(n + 1)T_n - U_n]$.

58. $T_{2n+1}' = 2(2n + 1)(T_{2n} + T_{2n-2} + T_{2n-4} + \cdots + T_2) + 1$.

59. $T_{2n}' = 2(2n)(T_{2n-1} + T_{2n-3} + T_{2n-5} + \cdots + T_1)$.

60. $2T_n = \frac{1}{n+1}T'_{n+1} - \frac{1}{n-1}T'_{n-1}$.
61. Deduce the Cassini-like formula for l_n from identity (41.42).
62. Deduce the Cassini-like formula for f_n from identity (41.43).
63. Define T_{nk} recursively, where $k \geq 0$.

Prove each.

64. Let g_k denote the generating function of the elements in column k of array A in Table 41.3, where $k \geq 0$. Then $g_k = \frac{1-x}{1-2x}\left(\frac{-1}{1-2x}\right)^k$ (Hoggatt, [32]).
65. Let R_n denote the sum of the elements in row n of array B in Table 41.7. Then $R_n = Q_n$, where $n \geq 0$.
66. Let D_n denote the nth diagonal sum of array B. Then $D_n = 2D_{n-1} + D_{n-3}$, where $n \geq 3$.
67. Let z_n denote the nth falling sum of array B. Then $z_n = \begin{cases} 1 & \text{if } n = 0 \\ 2 \cdot 3^{n-1} & \text{otherwise.} \end{cases}$

Using standard techniques, find a generating function for each sequence.

68. $\{D_n\}_{n\geq 0}$.
69. $\{z_n\}_{n\geq 0}$.
70. Prove that the nth row sum of array D is $n + 1$, where $n \geq 0$.
71. Prove that the element $d_{n,k}$ in row n and column k of array D is given by

$$d_{n,k} = (-1)^k \binom{n-k}{k} 2^{n-2k}.$$

72. Solve the recurrence $x_n = x_{n-1} + x_{n-2} + 1$, where $x_0 = 1, x_1 = 2$, and $n \geq 2$.

Prove each.

73. The generating function of the elements in column k of array C is $h_k = \frac{1}{1-2x}\left(\frac{-1}{1-2x}\right)^k$, where $k \geq 0$.
74. The nth falling diagonal sum of array C is one.
75. The n row sum of array D is P_{n+1}, where $n \geq 0$.
76. The nth falling diagonal sum of array D is 3^n, where $n \geq 0$.
77. The generating function of the elements in column k of array D is $h^*_k = \frac{1}{1-2x}\left(\frac{1}{1-2x}\right)^k$, where $k \geq 0$.
78. Let $g(t)$ be a generating function of a Chebyshev polynomial $c_n(x)$, where $c_0(x)$ and $c_1(x)$ are arbitrary. Then $g(t) = \frac{c_0 + (c_1 - 2xc_0)t}{1 - 2xt + t^2}$.

42

CHEBYSHEV TILINGS

> Life is good for only two things, discovering mathematics and teaching mathematics.
> –Siméon Denis Poisson (1781–1840)

In Chapter 33, we constructed combinatorial models for Fibonacci and Lucas polynomials using linear and circular tilings. We accomplished this by assigning suitable weights to 1×1 tiles (squares) and 1×2 tiles (dominoes). We now build on this idea to create similar models for the Chebyshev polynomials T_n and U_n [29, 285].

For the sake of convenience, we begin with combinatorial models for U_n.

42.1 COMBINATORIAL MODELS FOR U_n

Model I: Tiles of One Color

Suppose we would like to tile a $1 \times n$ board with square tiles and dominoes. Assume tiles come in one color. Now we need to pick suitable weights for squares and dominoes. Following the Chebyshev recurrence $z_n = 2xz_{n-1} - z_{n-2}$, we assign the weight $2x$ to each square and -1 to each domino. When convenient, we denote the weight of a tile by w(tile). For example, $w(\text{domino}) = -1$.

As usual, the weight of a tiling of the board is the product of the tiles in it. The weight of the empty tiling is again one.

Fibonacci and Lucas Numbers with Applications, Volume Two. Thomas Koshy.

Figure 42.1 shows the possible tilings of length n, where $0 \le n \le 4$.

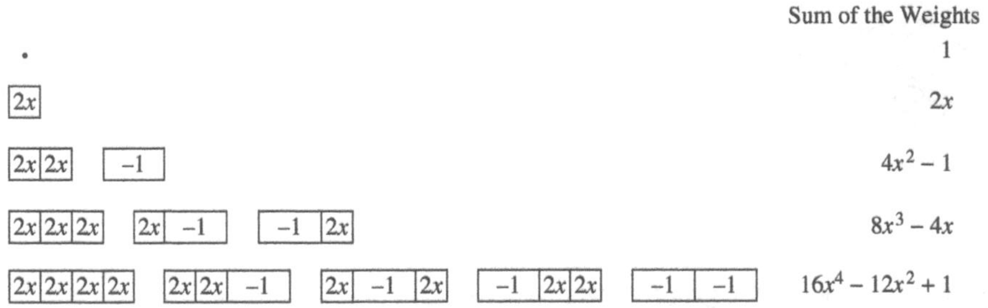

Figure 42.1. Chebyshev models for U_n using Model I.

Using the empirical data from Figure 42.1, we conjecture that the sum of the weights of tilings in this model is U_n, where $n \ge 0$. The next theorem does in fact validate this observation. L.W. Shapiro of Howard University, Washington, D.C., developed this model in 1981 [285, 461].

Theorem 42.1. *The sum of the weights of tilings of a* $1 \times n$ *board is* U_n *in Model I, where* $n \ge 0$.

Proof. It follows by Figure 42.1 that the theorem is true when $n = 0$ and $n = 1$. Assume it works for tilings of length $n - 1$ and $n - 2$, where $n \ge 2$. Consider now an arbitrary tiling T of length n.

Suppose the tiling T ends in a domino: $T = \underbrace{\text{subtiling}}_{\text{length } n-2}\boxed{\;-1\;}$. By definition, the sum of the weights of such tilings equals $-U_{n-2}$.

On the other hand, suppose T ends in a square. $T = \underbrace{\text{subtiling}}_{\text{length } n-1}\boxed{2x}$. The sum of the weights of such tilings is $2xU_{n-1}$.

By the addition principle, the sum of the weights of all tilings of length n equals $2xU_{n-1} - U_{n-2} = U_n$. Thus, by PMI, the theorem works for all tilings of length $n \ge 0$. ■

Using this theorem, we can obtain every tiling of length n recursively from those of length $n - 1$ and $n - 2$ by one of the following two steps, where $n \ge 2$:

Step 1. Append a square to every tiling of length $n - 1$.
Step 2. Append a domino to every tiling of length $n - 2$.

This algorithm is clearly reversible.

We can invoke the techniques used with Fibonacci polynomials to establish several Chebyshev identities. For example, by counting the tilings based on the number of dominoes in them, we can develop an explicit formula for U_n [285]. The proof is fairly straightforward, so we omit it; see Exercise 42.1.

Theorem 42.2. *Let $n \geq 0$. Then* $U_n = \sum_{k=0}^{\lfloor n/2 \rfloor} (-1)^k \binom{n-k}{k} (2x)^{n-2k}$. ■

For example,

$$\begin{aligned} U_5 &= \sum_{k=0}^{2} (-1)^k \binom{5-k}{k} (2x)^{5-2k} \\ &= (2x)^5 - 4(2x)^3 + 3(2x) \\ &= 32x^5 - 32x^3 + 6x. \end{aligned}$$

Geometrically, this means there is exactly one tiling of length 5 with zero dominoes, exactly four tilings with one domino each, and exactly three tilings with two dominoes each; see Figure 42.2. The sum of the weights is $(2x)^5 - 4(2x)^3 + 3(2x) = U_5$, as expected.

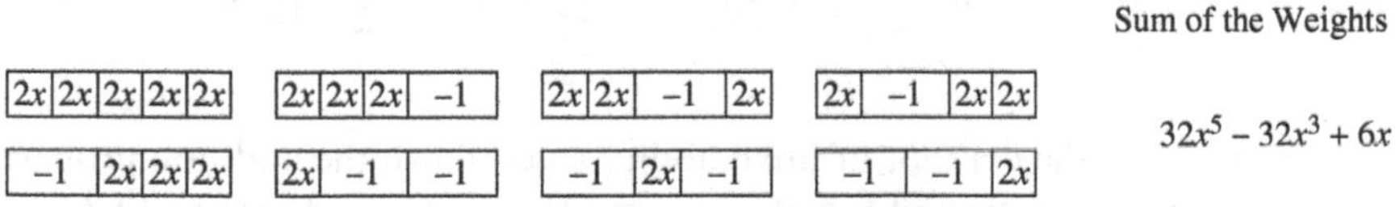

Figure 42.2. Chebyshev tilings of length 5.

Theorem 42.2 has an interesting consequence, as the following corollary shows.

Corollary 42.1. $U_n(ix/2) = i^n f_{n+1}$.

Proof. By Theorem 42.2 and the Lucas-like formula for f_{n+1} in Chapter 31, we have

$$\begin{aligned} U_n(ix/2) &= \sum_{k=0}^{\lfloor n/2 \rfloor} (-i)^{2k} \binom{n-k}{k} (ix)^{n-2k} \\ &= i^n \sum_{k=0}^{\lfloor n/2 \rfloor} \binom{n-k}{k} x^{n-2k} \\ &= i^n f_{n+1}. \end{aligned}$$

■

Corollary 42.1 implies that $U_n(i/2) = i^n F_{n+1}$; so $|U_n(i/2)| = F_{n+1}$. Consequently, we have the next result.

Corollary 42.2. *There are F_{n+1} tilings of length n in Model I.* ■

For example, there are $F_5 = 5$ tilings of length 4, and $F_6 = 8$ tilings of length 5; see Figures 42.1 and 42.2.

The formula for U_n in Theorem 42.2 can be rewritten in a different way, as the next corollary shows.

Corollary 42.3.

$$U_n = \sum_{\substack{k\geq 0\\ k \text{ even}}} \binom{n-k}{k}(2x)^{n-2k} - \sum_{\substack{k\geq 0\\ k \text{ odd}}} \binom{n-k}{k}(2x)^{n-2k}. \qquad \blacksquare$$

In words, U_n equals the sum of the weights of all tilings of length n that have an even number of dominoes, minus the sum of the weights of all tilings of length n that have an odd number of dominoes.

For example,

$$\begin{aligned} U_5 &= [(2x)^5 + 3(2x)] - 4(2x)^3 \\ &= \sum_{\substack{k\geq 0\\ k \text{ even}}} (-1)^k \binom{5-k}{k}(2x)^{5-2k} - \sum_{\substack{k\geq 0\\ k \text{ odd}}} (-1)^k \binom{5-k}{k}(2x)^{5-2k}. \end{aligned}$$

We can invoke the concept of breakability to establish the addition property for U_n [285]. In the interest of brevity, we omit the proof; see Exercise 42.7.

Theorem 42.3. *Let $m, n \geq 1$. Then $U_{m+n} = U_m U_n - U_{m-1}U_{n-1}$.* ■

For example,

$$\begin{aligned} U_8 &= U_5U_3 - U_4U_2 \\ &= (32x^5 - 32x^3 + 6x)(8x^3 - 4x) - (16x^4 - 12x^2 + 1)(4x^2 - 1) \\ &= (256x^8 - 384x^6 + 176x^4 - 24x^2) - (64x^6 - 64x^4 + 16x^2 - 1) \\ &= 256x^8 - 448x^6 + 240x^4 - 40x^2 + 1; \end{aligned}$$

see Table 41.8.

It follows from Theorem 42.3 that

$$\begin{aligned} f_{m+n+1} &= f_{m+1}f_{n+1} - f_m f_n; \\ p_{m+n+1} &= p_{m+1}p_{n+1} - p_m p_n. \end{aligned}$$

Finally, we can apply the concept of the median square in a board of odd length in order to derive the following result. Again, we omit its proof; see Exercise 42.8.

Theorem 42.4. $$U_{2n+1} = \sum_{i,j\geq 0} (-1)^{i+j}\binom{n-i}{j}\binom{n-j}{i}(2x)^{2n-2i-2j+1}. \qquad \blacksquare$$

For example,

$$\begin{aligned} U_7 &= \sum_{i,j\geq 0}(-1)^{i+j}\binom{3-i}{j}\binom{3-j}{i}(2x)^{7-2i-2j} \\ &= (2x)^7 - 6(2x)^5 + 10(2x)^3 - 4(2x) \\ &= 128x^7 - 192x^5 + 80x^3 - 8x; \end{aligned}$$

see Table 42.1.

TABLE 42.1.

i	j	$i+j$	$\binom{3-i}{j}$	$\binom{3-j}{i}$	Corresponding Weight
0	0	0	3	1	$(2x)^7$
0	1	1	3	1	$-(2x)^5$
0	2	2	3	1	$(2x)^3$
0	3	3	1	1	$-2x$
1	0	1	1	3	$-(2x)^5$
1	1	2	2	2	$(2x)^3$
1	2	3	1	1	$-2x$
2	0	2	1	3	$(2x)^3$
2	1	3	1	1	$-2x$
3	0	3	1	1	$-2x$

↑
sum = U_7

It follows from Theorem 42.4 that

$$f_{2n+2} = \sum_{i,j\geq 0}(-1)^{i+j}\binom{n-i}{j}\binom{n-j}{i}x^{2n-2i-2j+1};$$

$$p_{2n+2} = \sum_{i,j\geq 0}(-1)^{i+j}\binom{n-i}{j}\binom{n-j}{i}(2x)^{2n-2i-2j+1}.$$

Next we present a colored tiling model for U_n [285].

Model II: Tiles of Two Colors

Suppose square tiles are available in two colors, black and white. Each square tile has weight x, and each domino has weight -1 as in Model I.

Figure 42.3 shows the possible tilings of length n, and the sum of their weights, where $0 \leq n \leq 3$.

Using the experimental data from Figure 42.3, we conjecture that the sum of the weights of tilings in Model II is U_n, where $n \geq 0$. The next theorem confirms it. The proof is straightforward, so we omit it; see Exercise 42.9.

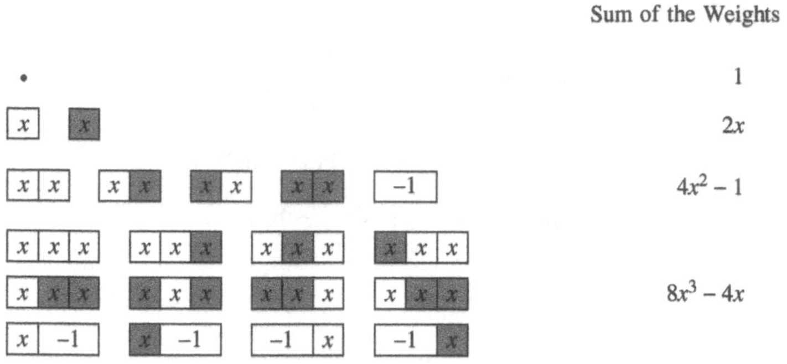

Figure 42.3. Chebyshev models for U_n using Model II.

Theorem 42.5. *The sum of the weights of colored tilings of length n in Model II is U_n, where $n \geq 0$.* ■

Since $P_{n+1} = |U_n(i)|$, the next corollary follows from this theorem.

Corollary 42.4. *There are P_{n+1} different ways of tiling a $1 \times n$ board using Model II.* ■

For example, using Model II, we can tile a 1×3 board in $12 = P_4$ different ways (see Figure 42.3); and a 1×3 board in $12 = P_4$ different ways.

Next we study tiling models for Chebyshev polynomials T_n [285].

42.2 COMBINATORIAL MODELS FOR T_n

Model III: Tiles of One Color

In this tiling model, we assume that tiles come in one color. We assign the weight $2x$ to each square, with one exception: If a tiling begins with a square, its weight is x. The weight of a domino remains the same as before, namely, -1.

Figure 42.4 shows the possible tilings of length n and the sum of their weights, where $0 \leq n \leq 4$.

Using the data from Figure 42.4, we can predict that the sum of the weights of tilings of length n is T_n. The next theorem confirms this observation.

Theorem 42.6. *The sum of the weights of tilings of a $1 \times n$ board using Model III is T_n, where $n \geq 0$.*

Proof. Clearly, the theorem is true when $n = 0$ and $n = 1$. Assume it is true for tilings of length $n - 1$ and $n - 2$, where $n \geq 2$. Let T be an arbitrary tiling of length n.

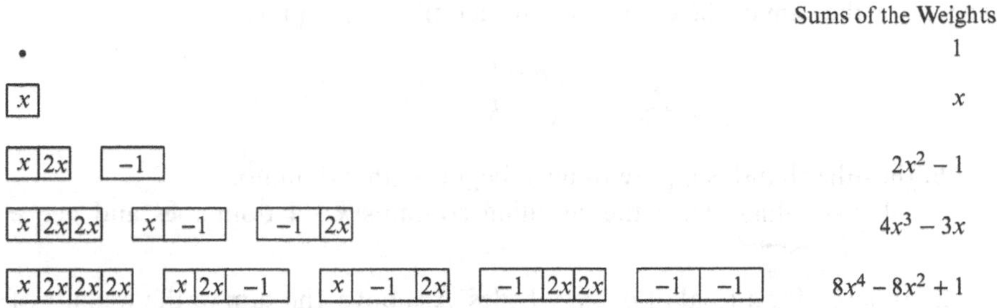

Figure 42.4. Chebyshev models for T_n using Model III.

Suppose tiling T ends in a domino: $T = \underbrace{\text{subtiling}}_{\text{length } n-2}\boxed{\;-1\;}$. By definition, the sum of the weights of such tilings equals $-T_{n-2}$.

On the other hand, suppose T ends in a square: $T = \underbrace{\text{subtiling}}_{\text{length } n-1}\boxed{2x}$. The sum of the weights of such tilings equals $2xT_{n-1}$.

Thus $T_n = 2xT_{n-1} - T_{n-2}$, where $n \geq 2$. This, coupled with the initial conditions, gives us the desired result. ■

Suppose $x = \cos\theta$. Then $T_n = \cos n\theta$. So, by virtue of Theorem 42.6, we can interpret $\cos n\theta$ as the sum of the weights of tilings of a $1 \times n$ board, where $w(\text{domino}) = -1$ and $w(\text{square}) = 2\cos\theta$. There is one exception: If a tiling begins with a square, then its weight is $\cos\theta$, where $n \geq 0$.

Theorem 42.2 has a companion result for T_n, as the next theorem shows. Its proof is charming.

Theorem 42.7. *Let $n \geq 1$. Then* $2T_n = \displaystyle\sum_{k=0}^{\lfloor n/2 \rfloor} (-1)^k \frac{n}{n-k}\binom{n-k}{k}(2x)^{n-2k}$.

Proof. Consider a $1 \times n$ board. By Theorem 42.6, the sum of the weights of its tilings is T_n.

We will now count the sum in a different way. To this end, let T be an arbitrary tiling of the board. Suppose it contains k dominoes, where $k \geq 0$.

Suppose tiling T begins with a square: $T = \boxed{x}\underbrace{\text{subtiling}}_{\text{length } n-1}$. The subtiling contains k dominoes; so it contains $(n-1) - 2k = n - 2k - 1$ squares. Consequently, the subtiling contains a total of $(n - 2k - 1) + k = n - k - 1$ tiles. So the k dominoes can be placed among the $n - k - 1$ tiles in $\binom{n-k-1}{k}$ ways; in other words, there are $\binom{n-k-1}{k}$ such subtilings. The sum of their weights equals

$$\binom{n-k-1}{k} x \cdot (-1)^k (2x)^{n-2k-1} = \frac{(-1)^k}{2}\binom{n-k-1}{k}(2x)^{n-2k}.$$

Therefore, the sum of the weights of all such tilings T equals

$$\frac{1}{2}\sum_{k\geq 0}(-1)^k\binom{n-k-1}{k}(2x)^{n-2k}.$$

On the other hand, suppose tiling T begins with a domino: $T = \boxed{\;-1\;}\underbrace{\text{subtiling}}_{\text{length } n-2}$. Then the subtiling contains $k-1$ dominoes and hence $n-2k$ squares, for a total of $n-k-1$ tiles. As above, the sum of the weights of all such tilings equals

$$\sum_{k\geq 1}(-1)\cdot\binom{n-k-1}{k-1}(-1)^{k-1}(2x)^{n-2k} = \sum_{k\geq 1}(-1)^k\binom{n-k-1}{k-1}(2x)^{n-2k}.$$

Combining the two cases, the sum of the weights of all tilings of the board is given by

$$T_n = \frac{1}{2}\sum_{k\geq 0}(-1)^k\binom{n-k-1}{k}(2x)^{n-2k} + \sum_{k\geq 1}(-1)^k\binom{n-k-1}{k-1}(2x)^{n-2k}$$

$$2T_n = \sum_{k\geq 0}(-1)^k\left[\binom{n-k-1}{k} + 2\binom{n-k-1}{k-1}\right](2x)^{n-2k}$$

$$= \sum_{k\geq 0}(-1)^k\left[\binom{n-k}{k} + \binom{n-k-1}{k-1}\right](2x)^{n-2k}$$

$$= \sum_{k=0}^{\lfloor n/2\rfloor}(-1)^k\frac{n}{n-k}\binom{n-k}{k}(2x)^{n-2k},$$

as desired. ■

For example,

$$2T_5 = \sum_{k=0}^{2}(-1)^k\frac{5}{5-k}\binom{5-k}{k}(2x)^{5-2k}$$

$$= (2x)^5 - 5(2x)^3 + 5(2x)$$

$$T_5 = 16x^5 - 20x^3 + 5x.$$

It follows by Theorem 42.7 that

$$l_n = \sum_{k=0}^{\lfloor n/2\rfloor}\frac{n}{n-k}\binom{n-k}{k}x^{n-2k}; \tag{42.1}$$

$$q_n = \sum_{k=0}^{\lfloor n/2\rfloor}\frac{n}{n-k}\binom{n-k}{k}(2x)^{n-2k}.$$

For example,

$$l_6 = \sum_{k=0}^{3} \frac{6}{6-k}\binom{6-k}{k} x^{6-2k}$$
$$= x^6 + 6x^4 + 9x^2 + 2;$$
$$q_5 = \sum_{k=0}^{2} \frac{5}{5-k}\binom{5-k}{k} (2x)^{5-2k}$$
$$= (2x)^5 + 5(2x)^3 + 5(2x)$$
$$= 32x^5 + 40x^3 + 10x;$$

see Tables 31.1 and 31.6.

Using the concept of breakability, we can establish the addition formula for T_n. We leave its proof as an exercise; see Exercise 42.11.

Theorem 42.8. *Let $m, n \geq 1$. Then $T_{m+n} = T_m U_n - T_{m-1} U_{n-1}$.* ■

Next we present briefly a coloring model for T_n.

Model IV: Tiles of Two Colors

In this model, assume that square tiles come in two colors, say, black and white, and that no tiling begins with a black square. Suppose $w(\text{square}) = x$ and $w(\text{domino}) = -1$.

Figure 42.5 shows such tilings of length n and the sum of their weights, where $0 \leq n \leq 4$. It appears from the figure that the sum of the weights of tilings of a $1 \times n$ board using this model is T_n. The following theorem confirms this observation. The proof is straightforward, so we omit it; see Exercise 42.12.

Theorem 42.9. *The sum of the weights of tilings of length n in Model IV is T_n.* ■

Since $2q_n(x) = i^n T_n(ix)$, $Q_n = |T_n(i)|$. Consequently, it follows by Theorem 42.9 that there are Q_n tilings of length n Model IV.

Corollary 42.5. *The number of colored tilings of length n in Model IV is Q_n, where $n \geq 0$.* ■

For example, there are $7 = Q_3$ tilings of length 3, and $17 = Q_4$ tilings of length 4.

This corollary can be confirmed independently; see Exercise 42.13.

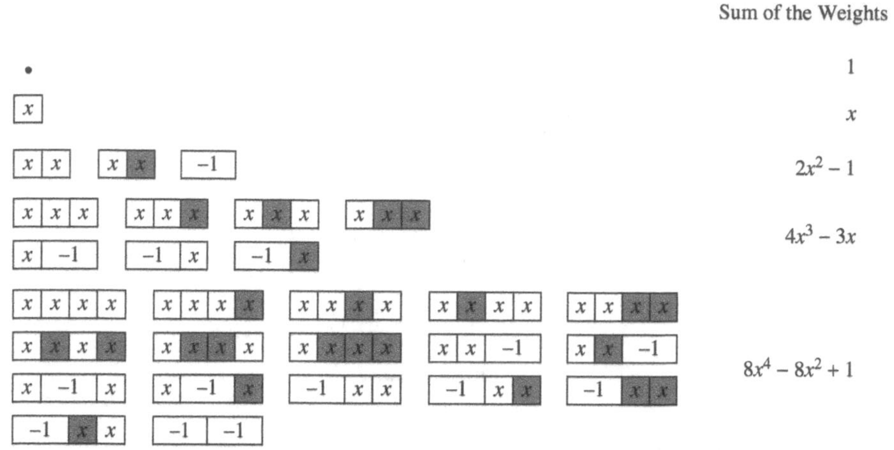

Figure 42.5. Colored tilings of length n using Model IV.

We can use Theorems 42.5 and 42.9 in tandem to extract a hybrid property bridging the two Chebyshev subfamilies, as the next theorem shows [285]; see Exercise 42.14.

Theorem 42.10. $xU_{n-1} + T_n = U_n$, *where* $n \geq 1$. ■

For example, let $n = 4$. The sum of the weights of colored tilings of length 4 is $U_4 = 16x^4 - 12x^2 + 1$; see Figure 42.6. There are exactly $29 = P_5$ such tilings.

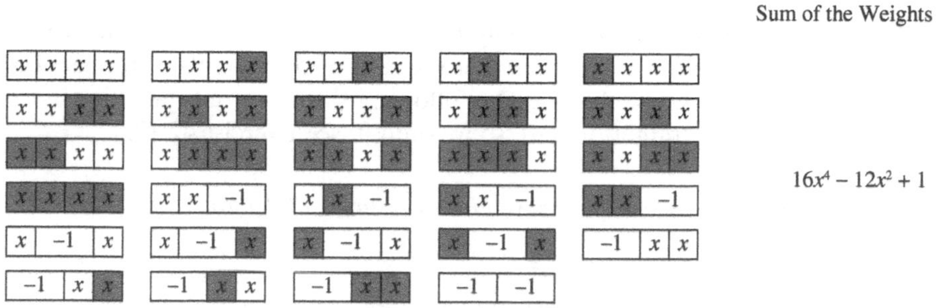

Figure 42.6. Tilings of length 4 using Model II.

Exactly $12 = P_4$ of them begin with a white square; see Figure 42.7.

The remaining 17 tilings do *not* begin with a white square; see Figure 42.8. The sum of their weights is $8x^4 - 8x^2 + 1 = T_4$.

Thus $xU_3 + T_4 = (8x^4 - 4x^2) + (8x^4 - 8x^2 + 1) = 16x^4 - 12x^2 + 1 = U_4$, as expected.

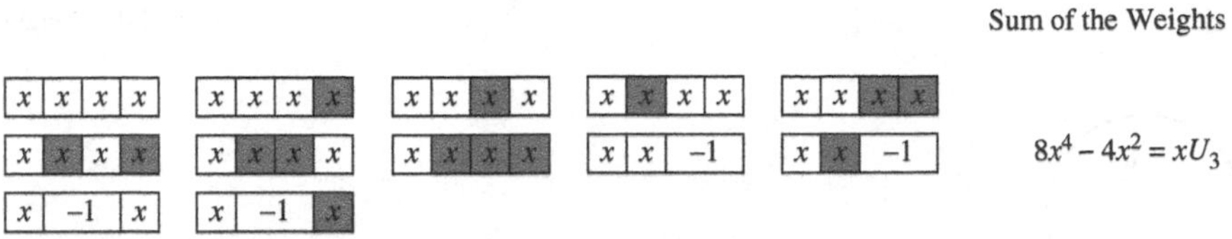

Figure 42.7. Tilings of length 4 beginning with a white square.

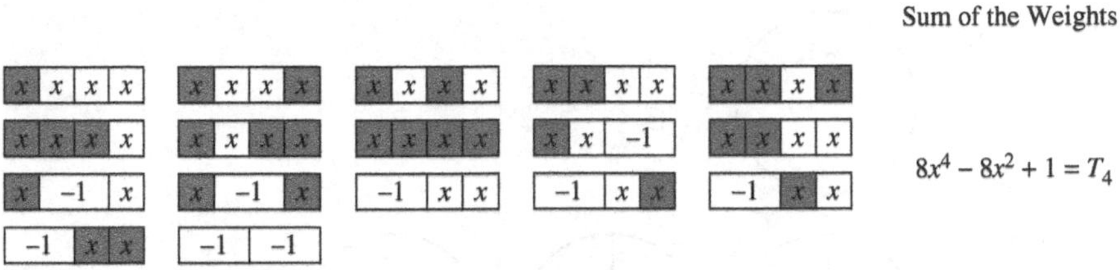

Figure 42.8. Tilings of length 4 *not* beginning with a white square.

It follows from Theorem 42.10 that $T_n = xU_{n-1} - U_{n-2}$; see Exercise 42.15. It would be rewarding to establish this identity independently, using Models II and IV; see Exercise 42.16.

Finally, we turn to a combinatorial model for T_n using circular boards [29, 285]. We used them in Chapter 33 to interpret l_n combinatorially.

42.3 CIRCULAR TILINGS

Model V: Flexible Tiles

In this model, we use (circular) squares and dominoes to tile a circular board with n labeled cells, as we did in Chapter 34; as before, we call such tilings n-bracelets. We let $w(\text{square}) = 2x$ and $w(\text{domino}) = -1$, with one exception: $w(\text{domino}) = 2$ when $n = 2$. We define the weight of the empty tiling to be 2.

Figure 42.9 shows the n-bracelets, where $0 \le n \le 3$. Using the empirical data, we conjecture that the sum of the weights of the n-bracelets is $2T_n$. The next theorem confirms this conjecture. The proof follows an argument similar to the one in Theorem 42.1; see Exercise 42.17.

Theorem 42.11. *The sum of the weights of all n-bracelets is $2T_n$, where $w(square) = 2x$ and $w(domino) = -1$, with one exception: when $n = 2$, $w(domino) = -2$.* ∎

It is easy to confirm that the number of n-bracelets is L_n, where $n \ge 3$; see Exercise 42.18.

For example, there are $4 = L_3$ 3-bracelets (see Figure 42.9) and $7 = L_4$ 4-bracelets.

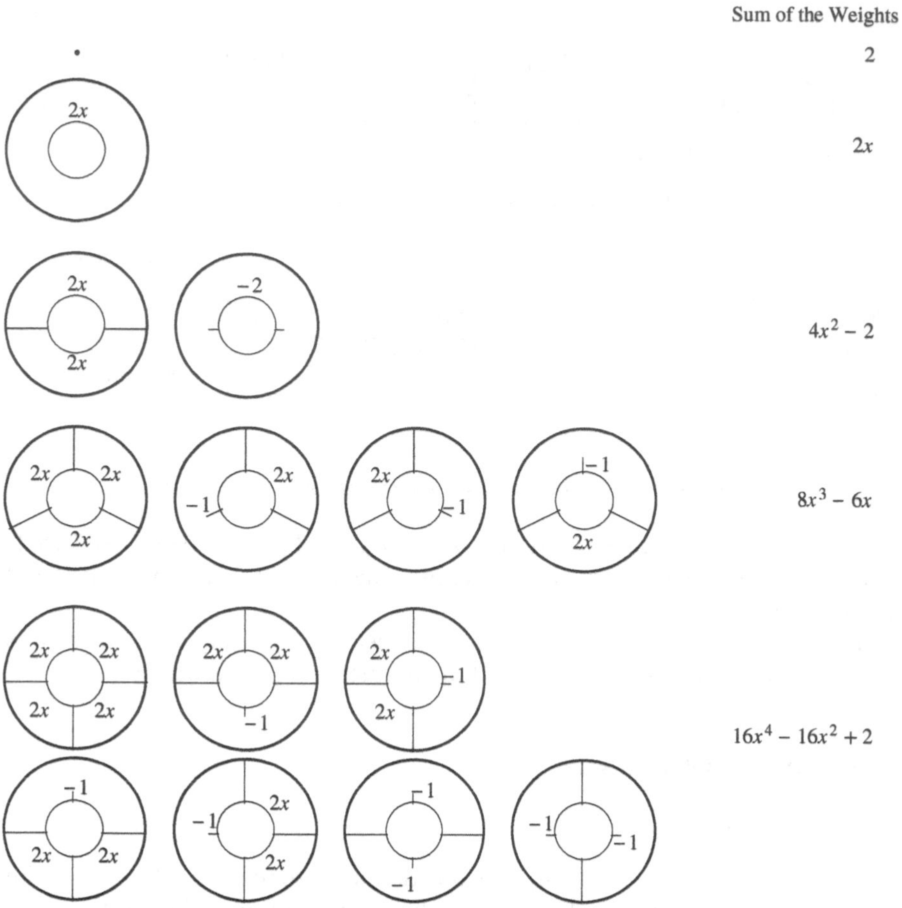

Figure 42.9.

EXERCISES 42

1. Prove Theorem 42.2.

Using Model I, find the sum of the weights of tilings of length 7 with:

2. An even number of dominoes in each.
3. An odd number of dominoes in each.
4. Using Model I, find the number of tilings of length 7.

Using Model I, find the sum of the weights of tilings of length 11 with:

5. An even number of dominoes in each.
6. An odd number of dominoes in each.

Prove each.

7. $U_{m+n} = U_m U_n - U_{m-1} U_{n-1}$, where $m, n \geq 1$.
8. $U_{2n+1} = \sum_{i,j\geq 0} (-1)^{i+j} \binom{n-i}{j} \binom{n-j}{i} (2x)^{2n-2i-2j+1}$.
9. Theorem 42.5.
10. Formula (42.1), using Theorem 42.7.
11. $T_{m+n} = T_m U_n - T_{m-1} U_{n-1}$, where $m, n \geq 1$.
12. Theorem 42.9.
13. Corollary 42.5.
14. $xU_{n-1} + T_n = U_n$, where $n \geq 1$.
15. $T_n = xU_{n-1} - U_{n-2}$, where $n \geq 2$.
16. $T_n = xU_{n-1} - U_{n-2}$, using Models II and IV.
17. Theorem 42.11.
18. The number of n-bracelets in Model V is L_n, where $n \geq 3$.

43

BIVARIATE GIBONACCI FAMILY I

The essence of mathematics lies in its freedom.
–Georg Cantor (1845–1918)

In Chapters 31–33, we investigated univariate Fibonacci and Lucas polynomials. This enabled us to study Pell and Pell–Lucas polynomials, Fibonacci and Lucas numbers, and Pell and Pell–Lucas numbers as dividends. In this chapter, we focus on bivariate Fibonacci polynomials $f_n(x,y)$ and Lucas polynomials $l_n(x,y)$. This investigation yields some valuable results.

43.1 BIVARIATE GIBONACCI POLYNOMIALS

Bivariate gibonacci polynomials in variables x and y satisfy the recurrence

$$s_n(x,y) = xs_{n-1}(x,y) + ys_{n-2}(x,y), \tag{43.1}$$

where $xy \neq 0$, $x^2 + 4y \neq 0$ and $n \geq 2$. When $s_0(x,y) = 0$ and $s_1(x,y) = 1$, $s_n(x,y) = f_n(x,y)$; and when $s_0(x,y) = 2$ and $s_1(x,y) = x$, $s_n(x,y) = l_n(x,y)$. Table 43.1 gives the first ten bivariate Fibonacci and Lucas polynomials.

It follows from recurrence (43.1) that $f_n(x,1) = f_n(x)$ and $l_n(x,1) = l_n(x)$; $f_n(1,1) = F_n$ and $l_n(1,1) = L_n$; $f_n(2x,1) = p_n(x)$, the nth Pell polynomial, and $l_n(2x,1) = q_n(x)$, the nth Pell–Lucas polynomial; $P_n = p_n(1) = f_n(2,1)$ and $2Q_n = q_n(1) = l_n(2,1)$.

Next we develop Binet-like formulas for $f_n(x,y)$ and $l_n(x,y)$.

Fibonacci and Lucas Numbers with Applications, Volume Two. Thomas Koshy.

TABLE 43.1. First 10 Bivariate Fibonacci and Lucas Polynomials

n	$f_n(x,y)$	$l_n(x,y)$
1	1	x
2	x	x^2+2y
3	x^2+y	x^3+3xy
4	x^3+2xy	$x^4+4x^2y+2y^2$
5	$x^4+3x^2y+y^2$	$x^5+5x^3y+5xy^2$
6	$x^5+4x^3y+3xy^2$	$x^6+6x^4y+9x^2y^2+2y^3$
7	$x^6+5x^4y+6x^2y^2+y^3$	$x^7+7x^5y+14x^3y^2+7xy^3$
8	$x^7+6x^5y+10x^3y^2+4xy^3$	$x^8+8x^6y+20x^4y^2+16x^2y^3+2y^4$
9	$x^8+7x^6y+15x^4y^2+10x^2y^3+y^4$	$x^9+9x^7y+27x^5y^2+30x^3y^3+9xy^4$
10	$x^9+8x^7y+21x^5y^2+20x^3y^3+5xy^4$	$x^{10}+10x^8y+35x^6y^2+50x^4y^3+25x^2y^4+2y^5$

Binet-like Formulas

By solving the characteristic equation $t^2-xt-y=0$ of recurrence (43.1), we get $u=u(x,y)=\dfrac{x+\Delta(x,y)}{2}$ and $v=v(x,y)=\dfrac{x-\Delta(x,y)}{2}$, where $t=t(x,y)$ and $\Delta(x,y)=\sqrt{x^2+4y}$. Then $s_n(x,y)=Au^n+Bv^n$, where $A=A(x,y)$ and $B=B(x,y)$ can be determined using the initial conditions. To this end, notice that $u+v=x$, $u-v=\Delta(x,y)$ and $uv=-y$.

Suppose $s_n(x,y)=f_n(x,y)$. Then, solving the linear system

$$A+B=0$$
$$Au+Bv=1,$$

we get $A=\dfrac{1}{u-v}=-B$. Thus $f_n(x,y)=\dfrac{u^n-v^n}{u-v}$. Similarly, $l_n(x,y)=u^n+v^n$.

43.2 BIVARIATE FIBONACCI AND LUCAS IDENTITIES

We can extract a number of properties of bivariate Fibonacci and Lucas polynomials using these explicit formulas; see Exercises 43.7–43.33. In the ⇒ interest of brevity, we denote $f_n(x,y)$ by f_n, and $l_n(x,y)$ by l_n, when there is *no* ambiguity.

1) $f_{-n}=\dfrac{(-1)^{n+1}}{y^n}f_n$ 2) $l_{-n}=\dfrac{(-1)^n}{y^n}l_n$
3) $f_{2n}=f_nl_n$ 4) $f_{n+1}+yf_{n-1}=l_n$
5) $yf_n^2+f_{n+1}^2=f_{2n+1}$ 6) $yl_n^2+l_{n+1}^2=(x^2+4y)f_{2n+1}$
7) $f_{n+1}f_{n-1}-f_n^2=-(-y)^{n-1}$ 8) $l_{n+1}l_{n-1}-l_n^2=(x^2+4y)(-y)^{n-1}$
9) $xf_n+2yf_{n-1}=l_n$ 10) $xl_n+2yl_{n-1}=(x^2+4y)f_n$
11) $xf_n+l_n=2f_{n+1}$ 12) $l_{n+1}+yl_{n-1}=(x^2+4y)f_n$
13) $f_{n+2}+y^2f_{n-2}=(x^2+2y)f_n$ 14) $l_{n+2}+y^2l_{n-2}=(x^2+2y)l_n$
15) $f_{n+2}-y^2f_{n-2}=xl_n$ 16) $l_{n+2}-y^2l_{n-2}=x(x^2+4y)f_n$

17) $l_n^2 + (x^2+4y)f_n^2 = 2l_{2n}$
18) $f_{n+1}^2 - y^2 f_{n-1}^2 = xf_{2n}$
19) $l_n^2 - (x^2+4y)f_n^2 = 4(-y)^n$
20) $f_{m+n} = f_{m+1}f_n + yf_m f_{n-1}$
21) $l_{m+n} = f_{m+1}l_n + yf_m l_{n-1}$
22) $2f_{m+n} = f_m l_n + f_n l_m$.

The next three identities are also a direct application of the Binet-like formulas; we omit their proofs; see Exercises 43.34–43.36. M. Catalani of the University of Torino, Italy, established the first two in 2005 [71, 95]:

$$(x^2+4y)\sum_{k=0}^{n}\binom{n}{k} f_k f_{n-k} = 2^n l_n - 2x^n; \tag{43.2}$$

$$\sum_{k=0}^{n}\binom{n}{k} l_k l_{n-k} = 2^n l_n + 2x^n; \tag{43.3}$$

$$\sum_{k=0}^{n}\binom{n}{k} f_k l_{n-k} = 2^n f_n. \tag{43.4}$$

For example,

$$\begin{aligned}\sum_{k=0}^{5}\binom{5}{k} f_k l_{n-k} &= 5f_1 l_4 + 10f_2 l_3 + 10f_3 l_2 + 5f_4 l_1 + f_5 l_0\\ &= 5(x^4+4x^2y+2y^2) + 10x(x^3+3xy) + 10(x^2+y)(x^2+2y)\\ &\quad + 5(x^3+2xy)x + 2(x^4+3x^2y+y^2)\\ &= 32(x^4+3x^2y+y^2)\\ &= 2^5 f_5.\end{aligned}$$

It follows from identities (43.2)–(43.4) that [287] that

$$(x^2+4)\sum_{k=0}^{n}\binom{n}{k} f_k(x) f_{n-k}(x) = 2^n l_n(x) - 2x^n;$$

$$\sum_{k=0}^{n}\binom{n}{k} l_k(x) l_{n-k}(x) = 2^n l_n(x) + 2x^n;$$

$$\sum_{k=0}^{n}\binom{n}{k} f_k(x) l_{n-k}(x) = 2^n f_n(x);$$

$$(x^2+1)\sum_{k=0}^{n}\binom{n}{k} p_k(x) p_{n-k}(x) = 2^{n-2}[q_n(x) - 2x^n];$$

$$\sum_{k=0}^{n}\binom{n}{k} q_k(x) q_{n-k}(x) = 2^n[q_n(x) + 2x^n];$$

$$\sum_{k=0}^{n}\binom{n}{k} p_k(x) q_{n-k}(x) = 2^n p_n(x);$$

$$5\sum_{k=0}^{n}\binom{n}{k}F_kF_{n-k} = 2^nL_n - 2;$$

$$\sum_{k=0}^{n}\binom{n}{k}L_kL_{n-k} = 2^nL_n + 2;$$

$$\sum_{k=0}^{n}\binom{n}{k}F_kL_{n-k} = 2^nF_n;$$

$$\sum_{k=0}^{n}\binom{n}{k}P_kP_{n-k} = 2^{n-1}(Q_n - 1);$$

$$\sum_{k=0}^{n}\binom{n}{k}Q_kQ_{n-k} = 2^{n-1}(Q_n + 1);$$

$$\sum_{k=0}^{n}\binom{n}{k}P_kQ_{n-k} = 2^{n-1}P_n;$$

see [287] and Chapter 32.

For example, $5\sum_{k=0}^{6} F_kF_{6-k} = 1150 = 2^6L_6 - 2$; $\sum_{k=0}^{6} L_kL_{6-k} = 1154 = 2^6L_6 + 2$; and $\sum_{k=0}^{6} F_kL_{6-k} = 512 = 2^6F_6$.

Bivariate Fibonacci polynomials $f_n(x, y)$ can be computed explicitly using the next theorem. Its proof follows by induction; see Exercise 43.37.

Theorem 43.1. *Let $n \geq 1$. Then*

$$f_{n+1}(x, y) = \sum_{k=0}^{\lfloor n/2 \rfloor}\binom{n-k}{k}x^{n-2k}y^k. \tag{43.5}$$

■

For example, we have

$$\begin{aligned} f_7(x, y) &= \sum_{k=0}^{3}\binom{6-k}{k}x^{6-2k}y^k \\ &= x^6 + 5x^4y + 6x^2y^2 + y^3. \end{aligned}$$

It follows from formula (43.5) that

$$f_{n+1}(x) = \sum_{k=0}^{\lfloor n/2 \rfloor}\binom{n-k}{k}x^{n-2k}; \tag{43.6}$$

$$p_{n+1}(x) = \sum_{k=0}^{\lfloor n/2 \rfloor}\binom{n-k}{k}(2x)^{n-2k};$$

$$F_{n+1} = \sum_{k=0}^{\lfloor n/2 \rfloor} \binom{n-k}{k}; \tag{43.7}$$

$$P_{n+1} = \sum_{k=0}^{\lfloor n/2 \rfloor} \binom{n-k}{k} 2^{n-k}.$$

Formula (43.7) is the well-known *Lucas formula*.
It follows from formula (43.5) that:

1. $f_n(x, y)$ contains a total of $\lfloor (n+1)/2 \rfloor$ terms. Its leading term is x^{n-1}; and the powers of x decrease by 2 and those of y increase by 1 from each term to the next.
2. $f_n(x, y)$ can be found from Pascal's triangle using the entries along its northeast diagonals. For example, $f_7(x, y) = \textcircled{1}x^6 + \textcircled{5}x^4y + \textcircled{6}x^2y^2 + \textcircled{1}y^3$; see Figure 43.1.

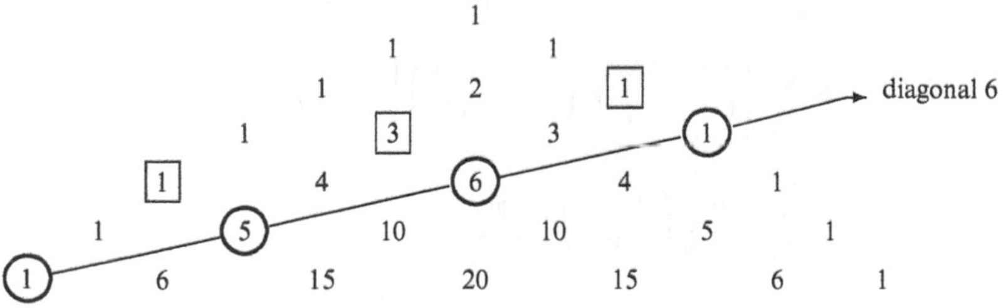

Figure 43.1. Pascal's triangle.

3. $l_n(x, y)$ can be found using the northeast diagonals n and $n-2$. This is true by virtue of identity 4) on page 430.

   For example, using diagonals 6 and 4, we have

$$\begin{aligned} l_6(x, y) &= f_7(x, y) + y f_5(x, y) \\ &= (x^6 + 5x^4y + 6x^2y^2 + y^3) + y(x^4 + 3x^2y + y^2) \\ &= x^6 + 6x^4y + 9x^2y^2 + 2y^3. \end{aligned}$$

Using Theorem 43.1 and identity 4) on page 430, we can develop an explicit formula for $l_n(x, y)$, as the following theorem shows.

Theorem 43.2. *Let n be a positive integer. Then*

$$l_n(x, y) = \sum_{k=0}^{\lfloor n/2 \rfloor} \frac{n}{n-k} \binom{n-k}{k} x^{n-2k} y^k. \tag{43.8}$$

Proof. By Theorem 43.1 and identity 4) on page 430, we have

$$\begin{aligned} l_n(x,y) &= f_{n+1}(x,y) + y f_{n-1}(x,y) \\ &= \sum_{k=0}^{\lfloor n/2 \rfloor} \binom{n-k}{k} x^{n-2k} y^k + y \sum_{k=0}^{\lfloor (n-2)/2 \rfloor} \binom{n-k-2}{k} x^{n-2k-2} y^k. \end{aligned}$$

Suppose n is even. Then

$$\begin{aligned} l_n(x,y) &= \sum_{k=0}^{n/2} \binom{n-k}{k} x^{n-2k} y^k + \sum_{k=0}^{(n-2)/2} \binom{n-k-2}{k} x^{n-2k-2} y^{k+1} \\ &= \sum_{k=0}^{n/2} \binom{n-k}{k} x^{n-2k} y^k + \sum_{k=1}^{n/2} \binom{n-k-1}{k-1} x^{n-2k} y^k \\ &= \sum_{k=0}^{n/2} \binom{n-k}{k} x^{n-2k} y^k + \sum_{k=0}^{n/2} \binom{n-k-1}{k-1} x^{n-2k} y^k \\ &= \sum_{k=0}^{n/2} \left[\binom{n-k}{k} + \binom{n-k-1}{k-1} \right] x^{n-2k} y^k \\ &= \sum_{k=0}^{n/2} \frac{n}{n-k} \binom{n-k}{k} x^{n-2k} y^k. \end{aligned}$$

So the formula works when n is even. Similarly, it works when n is odd. Thus it works for every positive integer n. ■

For example,

$$\begin{aligned} l_5(x,y) &= \sum_{k=0}^{2} \frac{5}{5-k} \binom{5-k}{k} x^{5-2k} y^k \\ &= x^5 + \frac{5}{4}\binom{4}{1} x^3 y + \frac{5}{3}\binom{3}{2} x y^2 \\ &= x^5 + 5x^3 y + 5xy^2. \end{aligned}$$

It follows from Theorem 43.2 that

$$l_n(x) = \sum_{k=0}^{\lfloor n/2 \rfloor} \frac{n}{n-k} \binom{n-k}{k} x^{n-2k};$$

$$q_n(x) = \sum_{k=0}^{\lfloor n/2 \rfloor} \frac{n}{n-k} \binom{n-k}{k} (2x)^{n-2k};$$

$$L_n = \sum_{k=0}^{\lfloor n/2 \rfloor} \frac{n}{n-k}\binom{n-k}{k};$$

$$Q_n = \sum_{k=0}^{\lfloor n/2 \rfloor} \frac{n}{n-k}\binom{n-k}{k}2^{n-2k-1}.$$

For example,

$$q_7(x) = \sum_{k=0}^{3} \frac{7}{7-k}\binom{7-k}{k}(2x)^{7-k} = 128x^7 + 224x^5 + 112x^3 + 14x$$

and $Q_7 = 239$.

The following theorem gives recursive formulas for $f_{2n}(x, y)$ and $f_{2n+1}(x, y)$. We establish them using the strong version of PMI.

Theorem 43.3. *Let n be a positive integer. Then*

$$f_{2n}(x, y) = \sum_{k=1}^{n} \binom{n}{k} x^k y^{n-k} f_k(x, y); \tag{43.9}$$

$$f_{2n+1}(x, y) = \sum_{k=0}^{n} \binom{n}{k} x^k y^{n-k} f_{k+1}(x, y). \tag{43.10}$$

Proof (by PMI). Since

$$\sum_{k=1}^{1} \binom{1}{k} x^k y^{1-k} f_k = x = f_2 \quad \text{and} \quad \sum_{k=1}^{2} \binom{2}{k} x^k y^{2-k} f_k = x^3 + 2xy = f_4,$$

formula (43.9) works when $n = 1$ and $n = 2$. We will establish formula (43.10) in the middle of the induction step.

Assume formula (43.9) works for all positive integers $\leq n$, where $n \geq 2$. Then

$$\begin{aligned} xf_{2n-1} &= f_{2n} - yf_{2n-2} \\ &= \sum_{k=1}^{n} \binom{n}{k} x^k y^{n-k} f_k - y \sum_{k=1}^{n-1} \binom{n-1}{k} x^k y^{n-k-1} f_k \\ &= y \sum_{k=1}^{n} \binom{n}{k} x^k y^{n-k-1} f_k - y \sum_{k=1}^{n-1} \binom{n-1}{k} x^k y^{n-k-1} f_k \\ &= y \sum_{k=1}^{n} \left[\binom{n}{k} - \binom{n-1}{k}\right] x^k y^{n-k-1} f_k \\ &= y \sum_{k=1}^{n} \binom{n-1}{k-1} x^k y^{n-k-1} f_k \end{aligned}$$

$$= \sum_{k=1}^{n} \binom{n-1}{k-1} x^k y^{n-k} f_k$$

$$= \sum_{k=0}^{n-1} \binom{n-1}{k} x^{k+1} y^{n-k-1} f_{k+1}$$

$$= x \sum_{k=0}^{n-1} \binom{n-1}{k} x^k y^{n-k-1} f_{k+1}$$

$$f_{2n+1} = \sum_{k=0}^{n} \binom{n}{k} x^k y^{n-k} f_{k+1}.$$

Consequently, we have

$$\begin{aligned}
f_{2n+2} &= x f_{2n+1} + y f_{2n} \\
&= x \sum_{k=0}^{n} \binom{n}{k} x^k y^{n-k} f_{k+1} + y \sum_{k=1}^{n} \binom{n}{k} x^k y^{n-k} f_k \\
&= \sum_{k=0}^{n} \binom{n}{k} x^{k+1} y^{n-k} f_{k+1} + \sum_{k=1}^{n} \binom{n}{k} x^k y^{n-k+1} f_k \\
&= \sum_{k=1}^{n+1} \binom{n}{k-1} x^k y^{n-k+1} f_k + \sum_{k=1}^{n+1} \binom{n}{k} x^k y^{n-k+1} f_k \\
&= \sum_{k=1}^{n+1} \left[\binom{n}{k-1} + \binom{n}{k}\right] x^k y^{n-k+1} f_k \\
&= \sum_{k=1}^{n+1} \binom{n+1}{k} x^k y^{n-k+1} f_k.
\end{aligned}$$

Thus, by the strong version of PMI, (43.9) works for all positive integers n. ■

For example, we have

$$\begin{aligned}
f_6(x,y) &= \sum_{k=1}^{3} \binom{3}{k} x^k y^{3-k} f_k(x,y) \\
&= 3xy^2 \cdot 1 + 3x^2 y \cdot x + x^3(x^2 + y) \\
&= x^5 + 4x^3 y + 3xy^2; \\
f_7(x,y) &= \sum_{k=0}^{3} \binom{3}{k} x^k y^{3-k} f_{k+1}(x,y) \\
&= y^3 \cdot 1 + 3xy^2 \cdot x + 3x^2 y(x^2 + y) + x^3(x^3 + 2xy) \\
&= x^6 + 5x^4 y + 6x^2 y^2 + y^3.
\end{aligned}$$

It follows from Theorem 43.3 that

$$f_{2n}(x) = \sum_{k=1}^{n} \binom{n}{k} x^k f_k(x);$$

$$f_{2n+1}(x) = \sum_{k=0}^{n} \binom{n}{k} x^k f_{k+1}(x);$$

$$p_{2n}(x) = \sum_{k=1}^{n} \binom{n}{k} (2x)^k p_k(x);$$

$$p_{2n+1}(x) = \sum_{k=0}^{n} \binom{n}{k} (2x)^k p_{k+1}(x);$$

$$F_{2n} = \sum_{k=1}^{n} \binom{n}{k} F_k;$$

$$F_{2n+1} = \sum_{k=0}^{n} \binom{n}{k} F_{k+1};$$

$$P_{2n} = \sum_{k=1}^{n} \binom{n}{k} 2^k P_k(x);$$

$$P_{2n+1} = \sum_{k=0}^{n} \binom{n}{k} 2^k P_{k+1}.$$

For example, $F_{10} = \sum_{k=1}^{5} \binom{5}{k} F_k = 5 \cdot 1 + 10 \cdot 1 + 10 \cdot 2 + 5 \cdot 3 + 1 \cdot 5 = 55.$ Similarly, $F_{11} = \sum_{k=0}^{5} \binom{5}{k} F_{k+1} = 89.$

The following theorem gives a recursive formula for $l_{2n}(x, y)$. The proof is straightforward, so we omit it; see Exercise 43.39.

Theorem 43.4. *Let n be a nonnegative integer. Then*

$$l_{2n}(x, y) = \sum_{k=0}^{n} \binom{n}{k} x^k y^{n-k} l_k(x, y). \tag{43.11}$$

■

For example, we have

$$l_6(x, y) = \sum_{k=0}^{3} \binom{3}{k} x^k y^{3-k} l_k(x, y)$$

$$= y^3 \cdot 2 + 3xy^2 \cdot x + 3x^2y(x^2 + 2y) + x^3(x^3 + 3xy)$$
$$= x^6 + 6x^4y + 9x^2y^2 + 2y^3.$$

It follows from Theorem 43.4 that

$$l_{2n}(x) = \sum_{k=0}^{n} \binom{n}{k} x^k l_k(x);$$

$$q_{2n}(x) = \sum_{k=0}^{n} \binom{n}{k} (2x)^k q_k(x);$$

$$L_{2n} = \sum_{k=0}^{n} \binom{n}{k} L_k;$$

$$Q_{2n} = \sum_{k=0}^{n} \binom{n}{k} 2^k Q_k.$$

For example,

$$L_{10} = \sum_{k=0}^{5} \binom{5}{k} L_k = 1 \cdot 2 + 5 \cdot 1 + 10 \cdot 3 + 10 \cdot 4 + 5 \cdot 7 + 1 \cdot 11 = 123.$$

We can use identity 4) on page 430, coupled with Theorem 43.4, to develop a formula for $l_{2n+1}(x, y)$. In the interest of brevity, we omit its proof; see Exercise 43.40.

Theorem 43.5. *Let n be a nonnegative integer. Then*

$$l_{2n+1}(x, y) = \sum_{k=1}^{n+1} \frac{2n-k+2}{n+1} \binom{n+1}{k} x^k y^{n-k+1} f_k(x, y). \tag{43.12}$$

■

For example, we have

$$l_7(x, y) = \sum_{k=1}^{4} \frac{8-k}{4} \binom{4}{k} x^k y^{4-k} f_k(x, y)$$
$$= 7xy^3 + 9x^3y^2 + 5x^3y(x^2 + y) + x^4(x^3 + 2xy)$$
$$= x^7 + 7x^5y + 14x^3y^2 + 7xy^3.$$

In particular, Theorem 43.5 yields the following results:

$$l_{2n+1}(x) = \sum_{k=1}^{n+1} \frac{2n-k+2}{n+1} \binom{n+1}{k} x^k f_k(x);$$

$$q_{2n+1}(x) = \sum_{k=1}^{n+1} \frac{2n-k+2}{n+1} \binom{n+1}{k} (2x)^k p_k(x);$$

$$L_{2n+1} = \sum_{k=1}^{n+1} \frac{2n-k+2}{n+1}\binom{n+1}{k}F_k;$$

$$Q_{2n+1} = \sum_{k=1}^{n+1} \frac{2n-k+2}{n+1}\binom{n+1}{k}2^{k-1}P_k.$$

For example, $\sum_{k=1}^{6} \frac{12-k}{6}\binom{6}{k}F_k = 199 = L_{11}$.

43.3 CANDIDO'S IDENTITY REVISITED

We can use Candido's identity $\left[x^2 + y^2 + (x+y)^2\right]^2 = 2\left[x^4 + y^4 + (x+y)^4\right]$ to develop an interesting identity for the bivariate Fibonacci family. To this end, we replace x with xs_{n-1}, and y with ys_{n-2}. Then we get the bivariate identity

$$(y^2s_{n-2}^2 + x^2s_{n-1}^2 + s_n^2)^2 = 2(y^4s_{n-2}^4 + x^4s_{n-1}^4 + s_n^4). \qquad (43.13)$$

This result has an interesting geometric interpretation.

A Geometric Bridge

Consider a square $ABCD$, where $AE = x^2s_{n-1}^2$, $EF = y^2s_{n-2}^2$, and $FB = s_n^2$; see Figure 43.2. Then

$$\begin{aligned}\text{Area } ABCD &= (y^2s_{n-2}^2 + x^2s_{n-1}^2 + s_n^2)^2 \\ &= 2(y^4s_{n-2}^4 + x^4s_{n-1}^4 + s_n^4) \\ &= 2(\text{sum of the shaded areas}).\end{aligned}$$

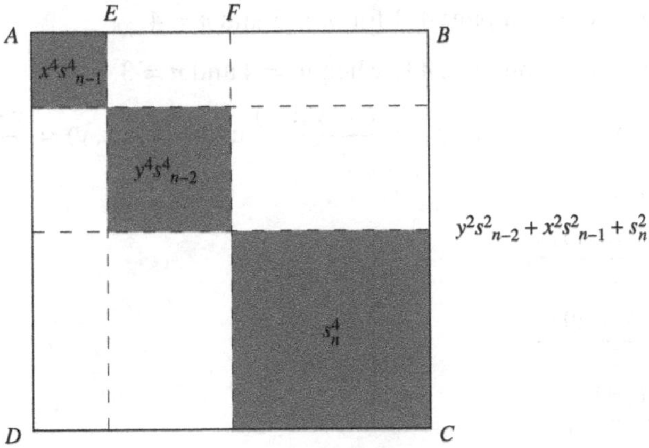

Figure 43.2.

In particular, identity (43.13) yields

$$(y^2 f_{n-2}^2 + x^2 f_{n-1}^2 + f_n^2)^2 = 2(y^4 f_{n-2}^4 + x^4 f_{n-1}^4 + f_n^4);$$
$$(y^2 l_{n-2}^2 + x^2 l_{n-1}^2 + l_n^2)^2 = 2(y^4 l_{n-2}^4 + x^4 l_{n-1}^4 + l_n^4);$$
$$\left[f_{n-2}^2(x) + f_{n-1}^2(x) + f_n^2(x)\right]^2 = 2\left[f_{n-2}^4(x) + f_{n-1}^4(x) + f_n^4(x)\right];$$
$$\left[l_{n-2}^2(x) + l_{n-1}^2(x) + l_n^2(x)\right]^2 = 2\left[l_{n-2}^4(x) + l_{n-1}^4(x) + l_n^4(x)\right];$$
$$\left[p_{n-2}^2(x) + p_{n-1}^2(x) + p_n^2(x)\right]^2 = 2\left[p_{n-2}^4(x) + p_{n-1}^4(x) + p_n^4(x)\right];$$
$$\left[q_{n-2}^2(x) + q_{n-1}^2(x) + q_n^2(x)\right]^2 = 2\left[q_{n-2}^4(x) + q_{n-1}^4(x) + q_n^4(x)\right];$$
$$(F_{n-2}^2 + F_{n-1}^2 + F_n^2)^2 = 2(F_{n-2}^4 + F_{n-1}^4 + F_n^4);$$
$$(L_{n-2}^2 + L_{n-1}^2 + L_n^2)^2 = 2(L_{n-2}^4 + L_{n-1}^4 + L_n^4);$$
$$(P_{n-2}^2 + P_{n-1}^2 + P_n^2)^2 = 2(P_{n-2}^4 + P_{n-1}^4 + P_n^4);$$
$$(Q_{n-2}^2 + Q_{n-1}^2 + Q_n^2)^2 = 2(Q_{n-2}^4 + Q_{n-1}^4 + Q_n^4).$$

For example, $(F_8^2 + F_9^2 + F_{10}^2)^2 = (21^2 + 34^2 + 55^2)^2 = 21{,}362{,}884 = 2(21^4 + 34^4 + 55^4)$. Likewise, $(L_8^2 + L_9^2 + L_{10}^2)^2 = 534{,}256{,}996 = 2(L_8^4 + L_9^4 + L_{10}^4)$.

EXERCISES 43

[*Note*: Throughout the exercises, f_n denotes $f_n(x, y)$ and l_n denotes $l_n(x, y)$.]

1. Using recurrence (43.1), find the polynomials f_{11} and l_{11}.
2. Verify identity 20) on page 431 for $m = 5$ and $n = 4$.
3. Verify identity 21) on page 431 when $m = 4$ and $n = 3$.

Establish each, where $u = u(x, y) = \dfrac{x + \Delta(x, y)}{2}$ and $v = v(x, y) = \dfrac{x - \Delta(x, y)}{2}$.

4. $l_n = u^n + v^n$.
5. $u^n = \dfrac{l_n + \Delta(x, y) f_n}{2}$.
6. $v^n = \dfrac{l_n - \Delta(x, y) f_n}{2}$.
7. $f_{-n} = \dfrac{(-1)^{n+1}}{y^n} f_n$.

8. $l_{-n} = \dfrac{(-1)^n}{y^n} l_n$.
9. $f_{2n} = f_n l_n$.
10. $xf_n + 2yf_{n-1} = l_n$.
11. $f_{n+1} + yf_{n-1} = l_n$.
12. $f_{n+1}^2 + yf_n^2 = f_{2n+1}$.
13. $l_{n+1}^2 + yl_n^2 = (x^2 + 4y)f_{2n+1}$.
14. $f_{n+1}f_{n-1} - f_n^2 = -(-y)^{n-1}$.
15. $l_{n+1}l_{n-1} - l_n^2 = (x^2 + 4y)(-y)^{n-1}$.
16. $xf_n + 2yf_{n-1} = l_n$.
17. $l_{n+1} + yl_{n-1} = (x^2 + 4y)f_n$.
18. $xl_n + 2yl_{n-1} = (x^2 + 4y)f_n$.
19. $xf_n + l_n = 2f_{n+1}$.
20. $f_{n+2} + y^2 f_{n-2} = (x^2 + 2y)f_n$.
21. $l_{n+2} + y^2 l_{n-2} = (x^2 + 2y)l_n$.
22. $f_{n+2} - y^2 f_{n-2} = xl_n$.
23. $l_{n+2} - y^2 l_{n-2} = x(x^2 + 4y)f_n$.
24. $l_n^2 - (x^2 + 4y)f_n^2 = 4(-y)^n$.
25. $l_{2n} = l_n^2 - 2(-y)^n$.
26. $(x^2 + 4y)f_n^2 + l_n^2 = 2l_{2n}$.
27. $f_{n+1}^2 - y^2 f_{n-1}^2 = xf_{2n}$.
28. $l_n^2 - (x^2 + 4y)f_n^2 = 4(-y)^n$.
29. $l_{n+2}^2 - y^4 l_{n-2}^2 = x(x^2 + 2y)(x^2 + 4y)f_{2n}$.
30. $f_m l_n + f_n l_m = 2f_{m+n}$.
31. $F_m L_n + F_n L_m = 2F_{m+n}$.
32. $f_{m+1}f_n + yf_m f_{n-1} = f_{m+n}$.
33. $f_{m+1}l_n + yf_m l_{n-1} = l_{m+n}$.
34. $(x^2 + 4y)\sum_{k=0}^{n} \binom{n}{k} f_k f_{n-k} = 2^n l_n - 2x^n$ (Catalani, [95]).

35. $\sum_{k=0}^{n}\binom{n}{k}l_k l_{n-k} = 2^n l_n + 2x^n$ (Catalani, [95]).

36. $\sum_{k=0}^{n}\binom{n}{k}f_k l_{n-k} = 2^n f_n$.

37. Theorem 43.1.

38. Theorem 43.2 when n is odd.

39. Theorem 43.4.

40. Theorem 43.5.

41. Using Theorem 43.1, find the polynomials f_5 and f_8.

42. Using Theorem 43.4, find the polynomials l_4 and l_8.

43. Using Theorem 43.5, find the polynomials l_5 and l_9.

Evaluate each sum for $n = 4$ and $n = 5$.

44. $\sum_{k=0}^{n}(-1)^k\binom{n}{k}f_k f_{n-k}$.

45. $\sum_{k=0}^{n}(-1)^k\binom{n}{k}l_k l_{n-k}$.

46. $\sum_{k=0}^{n}(-1)^k\binom{n}{k}f_k l_{n-k}$.

Prove each.

47. $\sum_{k=0}^{n}(-1)^k\binom{n}{k}f_k f_{n-k} = \begin{cases} -2(x^2+4y)^{(n-2)/2} & \text{if } n \text{ is even} \\ 0 & \text{otherwise.} \end{cases}$

48. $\sum_{k=0}^{n}(-1)^k\binom{n}{k}l_k l_{n-k} = \begin{cases} 2(x^2+4y)^{n/2} & \text{if } n \text{ is even} \\ 0 & \text{otherwise.} \end{cases}$

49. $\sum_{k=0}^{n}(-1)^k\binom{n}{k}f_k l_{n-k} = \begin{cases} -2(x^2+4y)^{(n-1)/2} & \text{if } n \text{ is odd} \\ 0 & \text{otherwise.} \end{cases}$

50. $\begin{vmatrix} ys_n & xs_{n+1} & s_{n+2} \\ s_{n+2} & ys_n & xs_{n+1} \\ xs_{n+1} & s_{n+2} & ys_n \end{vmatrix} = 2(y^3s_n^3 + x^3s_{n+1}^3)$,
where $s_n = s_n(x, y)$ satisfies the bivariate recurrence (43.1).

44

JACOBSTHAL FAMILY

I went into mathematics and never looked back.
I was struck by the beauty and the appeal
of mathematics. It just captivated me.
–George E. Andrews, 2005

This chapter presents a special member of the bivariate family $\{s_n(x, y)\}$ that we studied in the preceding chapter: the Jacobsthal family. It has charming applications to combinatorics, graph theory, and number theory.

Ernst Erich Jacobsthal (1882–1965) was born in Berlin, Germany. He studied mathematics at the University of Berlin under the well-known mathematicians Ferdinand G. Frobenius, Hermann A. Schwarz, and Issai Schur. In 1906, he received his Ph.D. for his dissertation, *Application of a Formula from the Theory of Quadratic Remainders*, which he completed under the supervision of Frobenius and Schur. Among other results, his dissertation gives a charming proof that every prime $p = 4n + 1$ can be expressed as a sum of two squares.

Fibonacci and Lucas Numbers with Applications, Volume Two. Thomas Koshy.

In 1909, he became a teacher in Berlin at Kaiser Wilhelm High School, and an assistant to Professor E. Lampe at the College of Technology. He became a private senior lecturer at the college and a professor there in 1918.

In 1939, Jacobsthal fled Germany, intending to go to England, where his brother was a professor of archaeology at Oxford. However, he stopped *en route* at Trondheim, Norway, and stayed there as a professor at the Norwegian Institute of Technology until 1943. Later he gave lectures every summer at Free University in Berlin until 1957.

Jacobsthal was a member of the Norwegian Academy of Sciences and a productive researcher, who made significant contributions to algebra, analysis, theory of functions, and theory of numbers.

We begin with a brief introduction to bivariate Jacobsthal and Jacobsthal–Lucas polynomials.

Bivariate Jacobsthal polynomials $J_n(x,y)$ and *Jacobsthal–Lucas polynomials* $j_n(x,y)$ satisfy the same recurrence (43.1). When $s_0(x,y) = 0$ and $s_1(x,y) = 1$, $s_n(x,y) = f_n(x,y) = J_n(x,y)$; and when $s_0(x,y) = 2$ and $s_1(x,y) = x$, $s_n(x,y) = l_n(x,y) = j_n(x,y)$. Consequently, $J_n(y) = J_n(1,y) = f_n(1,y)$ and $j_n(y) = j_n(1,y) = l_n(1,y)$. Then $J_n(2) = J_n(1,2) = f_n(1,2)$ and $j_n(2) = j_n(1,2) = l_n(1,2)$.

For aesthetics and easy readability, we now switch the variable y in both $J_n(y)$ and $j_n(y)$ to x.

44.1 JACOBSTHAL FAMILY

The polynomial $J_n(x)$ is the *Jacobsthal polynomial*, named after Ernst Erich Jacobsthal, who studied them around 1919 [235]. Clearly, $J_n(1) = F_n$. Closely related to $J_n(x)$ is the *Jacobsthal–Lucas polynomial* $j_n(x)$. Obviously, $j_n(1) = L_n$, and hence the name Jacobsthal–Lucas polynomial for $j_n(x)$, where $n \geq 0$. Hoggatt and Bicknell-Johnson, as well as Horadam, studied both Jacobsthal and Jacobsthal–Lucas polynomials extensively [226, 235].

Since $J_n(x) = f_n(1,x)$ and $j_n(x) = l_n(1,x)$, a number of properties of the Jacobsthal family can be deduced from those of bivariate Fibonacci and Lucas polynomials:

1) $J_{-n}(x) = \dfrac{(-1)^{n+1}}{x^n} J_n(x)$
2) $j_{-n}(x) = \dfrac{(-1)^n}{x^n} j_n(x)$
3) $J_{2n}(x) = J_n(x) j_n(x)$
4) $J_{n+1}(x) + xJ_{n-1}(x) = j_n(x)$
5) $xJ_n^2(x) + J_{n+1}^2(x) = J_{2n+1}(x)$
6) $xj_n^2(x) + j_{n+1}^2(x) = (4x+1)J_{2n+1}(x)$
7) $J_{n+1}(x)J_{n-1}(x) - J_n^2(x) = -(-x)^{n-1}$
8) $j_{n+1}(x)j_{n-1}(x) - j_n^2(x) = (4x+1)(-x)^{n-1}$
9) $J_n(x) + 2xJ_{n-1}(x) = j_n(x)$
10) $j_n(x) + 2xj_{n-1}(x) = (4x+1)J_n(x)$
11) $J_n(x) + j_n(x) = 2J_{n+1}(x)$
12) $j_{n+1}(x) + xj_{n-1}(x) = (4x+1)J_n(x)$
13) $J_{n+2}(x) + x^2 J_{n-2}(x) = (2x+1)j_n(x)$
14) $j_{n+2}(x) + x^2 j_{n-2}(x) = (2x+1)j_n(x)$
15) $J_{n+2}(x) - x^2 J_{n-2}(x) = j_n(x)$
16) $j_{n+2}(x) - x^2 j_{n-2}(x) = (4x+1)J_n(x)$
17) $j_n^2(x) + (4x+1)J_n^2(x) = 2j_{2n}(x)$
18) $J_{n+1}^2(x) - x^2 J_{n-1}^2(x) = J_{2n}(x)$
19) $j_n^2(x) - (4x+1)J_n^2(x) = 4(-x)^n$
20) $J_{m+n}(x) = J_{m+1}(x)J_n(x) + xJ_m(x)J_{n-1}(x)$

21) $j_{m+n}(x) = J_{m+1}(x)j_n(x) + xJ_m(x)j_{n-1}(x)$

22) $2J_{m+n}(x) = J_m(x)j_n(x) + J_n(x)j_m(x)$

23) $(4x+1)\sum_{k=0}^{n}\binom{n}{k}J_k(x)J_{n-k}(x) = 2^n j_n - 2$

24) $\sum_{k=0}^{n}\binom{n}{k}j_k(x)j_{n-k}(x) = 2^n j_n(x) + 2$

25) $\sum_{k=0}^{n}\binom{n}{k}J_k(x)j_{n-k}(x) = 2^n J_n(x)$

26) $J_{n+1}(x) = \sum_{k=0}^{\lfloor n/2\rfloor}\binom{n-k}{k}x^k$

27) $j_n(x) = \sum_{k=0}^{\lfloor n/2\rfloor}\frac{n}{n-k}\binom{n-k}{k}x^k$

28) $J_{2n}(x) = \sum_{k=1}^{n}\binom{n}{k}x^{n-k}J_k(x)$

29) $J_{2n+1}(x) = \sum_{k=0}^{n}\binom{n}{k}x^{n-k}J_{k+1}(x)$

30) $j_{2n}(x) = \sum_{k=0}^{n}\binom{n}{k}x^{n-k}j_k(x)$

31) $j_{2n+1}(x) = \sum_{k=1}^{n+1}\frac{2n-k+2}{n+1}\binom{n+1}{k}x^{n-k+1}J_k(x).$

Jacobsthal Recurrence

Jacobsthal polynomials $J_n(x)$ can be defined by the *Jacobsthal recurrence* $J_n(x) = J_{n-1}(x) + xJ_{n-2}(x)$, where $J_0(x) = 0$, $J_1(x) = 1$, and $n \geq 2$. *Jacobsthal–Lucas polynomials* $j_n(x)$ also satisfy the same recurrence, but $j_0(x) = 2$ and $j_1(x) = 1$.

Table 44.1 gives the first ten Jacobsthal and Jacobsthal–Lucas polynomials in x.

TABLE 44.1. First 10 Jacobsthal and Jacobsthal–Lucas Polynomials

n	$J_n(x)$	$j_n(x)$
1	1	1
2	1	$2x+1$
3	$x+1$	$3x+1$
4	$2x+1$	$2x^2+4x+1$
5	x^2+3x+1	$5x^2+5x+1$
6	$3x^2+4x+1$	$2x^3+9x^2+6x+1$
7	x^3+6x^2+5x+1	$7x^3+14x^2+7x+1$
8	$4x^3+10x^2+6x+1$	$2x^4+16x^3+20x^2+8x+1$
9	$x^4+10x^3+15x^2+7x+1$	$9x^4+30x^3+27x^2+9x+1$
10	$5x^4+20x^3+21x^2+8x+1$	$2x^5+25x^4+50x^3+35x^2+10x+1$

The numbers $J_n(2) = J_n$ and $j_n(2) = j_n$ are *Jacobsthal* and *Jacobsthal–Lucas numbers*, respectively. Table 44.2 shows the first ten Jacobsthal and Jacobsthal–Lucas numbers; Table A.3 in the Appendix gives the first 100 such numbers.

TABLE 44.2. First 10 Jacobsthal and Jacobsthal–Lucas Numbers

n	1	2	3	4	5	6	7	8	9	10
J_n	1	1	3	5	11	21	43	85	171	341
j_n	1	5	7	17	31	65	127	257	511	1025

Since $J_n(2) = J_n$ and $j_n(2) = j_n$, the above polynomial identities in turn yield the following for the corresponding Jacobsthal and Jacobsthal–Lucas integer families:

1) $J_{-n} = \frac{(-1)^{n+1}}{2^n} J_n$
2) $j_{-n} = \frac{(-1)^n}{2^n} j_n$
3) $J_{2n} = J_n j_n$
4) $J_{n+1} + 2J_{n-1} = j_n$
5) $2J_n^2 + J_{n+1}^2 = J_{2n+1}$
6) $2j_n^2 + j_{n+1}^2 = 9J_{2n+1}$
7) $J_{n+1}J_{n-1} - J_n^2 = -(-2)^{n-1}$
8) $j_{n+1}j_{n-1} - j_n^2 = 9(-2)^{n-1}$
9) $J_n + 4J_{n-1} = j_n$
10) $j_n + 4j_{n-1} = 9J_n$
11) $J_n + j_n = 2J_{n+1}$
12) $j_{n+1} + 2j_{n-1} = 9J_n$
13) $J_{n+2} + 4J_{n-2} = 5j_n$
14) $j_{n+2} + 4j_{n-2} = 5j_n$
15) $J_{n+2} - 4J_{n-2} = j_n$
16) $j_{n+2} - 4j_{n-2} = 9J_n$
17) $j_n^2 + 9J_n^2 = 2j_{2n}$
18) $J_{n+1}^2 - 4J_{n-1}^2 = J_{2n}$
19) $j_n^2 - 9J_n^2 = 4(-2)^n$
20) $J_{m+n} = J_{m+1}J_n + 2J_m J_{n-1}$
21) $j_{m+n} = J_{m+1}j_n + 2J_m j_{n-1}$
22) $2J_{m+n} = J_m j_n + J_n j_m$
23) $9\sum_{k=0}^{n} \binom{n}{k} J_k J_{n-k} = 2^n j_n - 2$
24) $\sum_{k=0}^{n} \binom{n}{k} j_k j_{n-k} = 2^n j_n + 2$
25) $\sum_{k=0}^{n} \binom{n}{k} J_k j_{n-k} = 2^n J_n$
26) $J_{n+1} = \sum_{k=0}^{\lfloor n/2 \rfloor} \binom{n-k}{k} 2^k$
27) $j_n = \sum_{k=0}^{\lfloor n/2 \rfloor} \frac{n}{n-k} \binom{n-k}{k} 2^k$
28) $J_{2n} = \sum_{k=1}^{n} \binom{n}{k} 2^{n-k} J_k$
29) $J_{2n+1} = \sum_{k=0}^{n} \binom{n}{k} 2^{n-k} J_{k+1}$
30) $j_{2n} = \sum_{k=0}^{n} \binom{n}{k} 2^{n-k} j_k$
31) $j_{2n+1} = \sum_{k=1}^{n+1} \frac{2n-k+2}{n+1} \binom{n+1}{k} 2^{n-k+1} J_k.$

Additional properties are given in the exercises.

It follows from Lucas-like formula 26) that J_{n+1} can be computed using the entries along the northeast diagonal n beginning at row n with weights 2^k, where $0 \le k \le \lfloor n/2 \rfloor$.

The hybrid formula $2J_{m+n} = J_m j_n + J_n j_m$ has two interesting byproducts:

- When $m = n$, it yields $J_{2n} = J_n j_n$; see formula 3).
- A repeated application of this identity yields $J_{2^n} = j_{2^{n-1}} j_{2^{n-2}} \cdots j_2 j_1$. For example, $j_8 j_4 j_2 j_1 = 257 \cdot 17 \cdot 5 \cdot 1 = 21{,}845 = J_{16}$.

Binet-like Formulas

Using Jacobsthal recurrence and the initial conditions, we can derive the Binet-like formulas

$$J_n(x) = \frac{u^n - v^n}{u - v} \quad \text{and} \quad j_n(x) = u^n + v^n,$$

where $u = u(x) = \dfrac{1+\sqrt{4x+1}}{2}$, $v = v(x) = \dfrac{1-\sqrt{4x+1}}{2}$, and $n \ge 0$. Consequently, $J_n = \dfrac{2^n - (-1)^n}{3}$ and $j_n = 2^n + (-1)^n$. When n is even, $J_n = \dfrac{M_n}{3}$, where M_n denotes the nth *Mersenne number* $2^n - 1$ and $n \ge 1$; and when n is odd, $j_n = M_n$.

We can establish an array of properties of Jacobsthal and Jacobsthal–Lucas numbers by using these explicit formulas; see Exercises 44.7–44.97.

Jacobsthal numbers J_n, unlike j_n, satisfy a charming property, as the following example shows. Its proof is simple and straightforward.

Example 44.1. *Prove that the product of every two consecutive Jacobsthal numbers is a triangular number.*

Proof. Let J_nJ_{n+1} be the triangular number $t_k = \dfrac{k(k+1)}{2}$, where $k \ge 1$. Let n be the even integer $2a$. Then

$$\begin{aligned} J_nJ_{n+1} &= \frac{(2^{2a}-1)(2^{2a}+1)}{9} \\ \frac{k(k+1)}{2} &= \frac{(2^{2a}-1)(2^{2a}+1)}{9} \\ 9k^2 + 9k - (2^{4a+2} - 2^{2a+1} - 2) &= 0. \end{aligned}$$

Solving this quadratic equation, we get $k = \dfrac{2^{2a+1}-2}{3} = \dfrac{2^{n+1}-2}{3}$. Thus $J_nJ_{n+1} = t_{(2^{n+1}-2)/3}$.

On the other hand, let n be odd. Then we can show similarly that $J_nJ_{n+1} = t_{(2^{n+1}-1)/3}$; see Exercise 44.98.

Thus

$$J_nJ_{n+1} = \begin{cases} t_{(2^{n+1}-1)/3} & \text{if } n \text{ is odd} \\ t_{(2^{n+1}-2)/3} & \text{otherwise.} \end{cases}$$

■

For example, $J_9J_{10} = 171 \cdot 341 = \dfrac{341 \cdot 342}{2} = t_{341}$; and $J_{10}J_{11} = 341 \cdot 683 = \dfrac{682 \cdot 683}{2} = t_{682}$.

Interestingly, the above piecewise formula for J_nJ_{n+1} can be rewritten in terms of the ceiling function:

$$J_nJ_{n+1} = t_{\lceil (2^{n+1}-2)/3 \rceil};$$

see Exercise 44.99.

An Alternate Formula for J_n

Using the floor and ceiling functions, we can find an alternate explicit formula for J_n:

$$J_n = \begin{cases} \lceil 2^n/3 \rceil & \text{if } n \text{ is odd} \\ \lfloor 2^n/3 \rfloor & \text{otherwise;} \end{cases} \tag{44.1}$$

see Exercise 44.3.

For example, $J_{17} = \lceil 2^{17}/3 \rceil = 43{,}691$, and $J_{20} = \lceil 2^{20}/3 \rceil = 349{,}525$.

Next we compute the number of digits in J_n and j_n.

Number of Digits in J_n

When n is odd, $\frac{2^n}{3} < J_n < \frac{2^n}{3} + 1$; then the number of digits in J_n equals $\lceil 2^n/3 \rceil$. When n is even, $\frac{2^n}{3} - 1 < J_n < \frac{2^n}{3}$; then also the number of digits in J_n equals $\lceil 2^n/3 \rceil$. Thus the same formula works in both cases.

For example, J_{17} is $\lceil 17 \log 2 - \log 3 \rceil = 5$ digits long; similarly, J_{20} is 6 digits long.

Number of Digits in j_n

Since $2^n \equiv 2, 4, 6, \text{or } 8 \pmod{10}$ and $2^n - 1 = j_n < 2^n$, it follows that 2^n and $2^n - 1$ contain the same number of digits. Consequently, j_n consists of $\lceil n \log 2 \rceil$ digits.

For example, j_{17} is $\lceil 17 \log 2 \rceil = 6$ digits long, and j_{20} is $\lceil 20 \log 2 \rceil = 7$ digits long. (Notice that $j_{17} = 131{,}071$ and $j_{20} = 1{,}048{,}577$.)

We can use matrices to generate Jacobsthal and Jacobsthal–Lucas numbers.

Matrix Generators for J_n and j_n

Let $X_n = J_n$ or j_n. Then

$$X_{n-1} = 0X_{n-2} + X_{n-1}$$
$$X_n = 2X_{n-2} + X_{n-1};$$
$$\begin{bmatrix} X_{n-1} \\ X_n \end{bmatrix} = \begin{bmatrix} 0 & 1 \\ 2 & 1 \end{bmatrix} \begin{bmatrix} X_{n-2} \\ X_{n-1} \end{bmatrix}.$$

Let $M = \begin{bmatrix} 0 & 1 \\ 2 & 1 \end{bmatrix}$. Then we can show (see Exercise 44.102) that

$$M^n \begin{bmatrix} X_0 \\ X_1 \end{bmatrix} = \begin{bmatrix} X_n \\ X_{n+1} \end{bmatrix}.$$

Consequently,

$$[1 \quad 0]\, M^n \begin{bmatrix} X_0 \\ X_1 \end{bmatrix} = X_n.$$

In particular,

$$\begin{bmatrix}1 & 0\end{bmatrix} M^n \begin{bmatrix}0\\1\end{bmatrix} = J_n \quad\text{and}\quad \begin{bmatrix}1 & 0\end{bmatrix} M^n \begin{bmatrix}2\\1\end{bmatrix} = j_n.$$

For example, we have

$$\begin{bmatrix}1 & 0\end{bmatrix} M^5 \begin{bmatrix}0\\1\end{bmatrix} = \begin{bmatrix}1 & 0\end{bmatrix} \begin{bmatrix}10 & 11\\22 & 21\end{bmatrix} \begin{bmatrix}0\\1\end{bmatrix} = 11 = J_5;$$

$$\begin{bmatrix}1 & 0\end{bmatrix} M^5 \begin{bmatrix}2\\1\end{bmatrix} = \begin{bmatrix}1 & 0\end{bmatrix} \begin{bmatrix}10 & 11\\22 & 21\end{bmatrix} \begin{bmatrix}2\\1\end{bmatrix} = 31 = j_5.$$

Example 39.1 Revisited

The determinant of the circulant matrix M in Example 39.1 has implications to the extended bivariate family. To see this, we let $a = a(x, y) = xs_{n+1}$ and $b = b(x, y) = ys_n$, where $s_k = s_k(x, y)$. Then $a + b = s_{n+2}$. Consequently,

$$\begin{vmatrix} xs_{n+1} & ys_n & s_{n+2}\\ s_{n+2} & xs_{n+1} & ys_n\\ ys_n & s_{n+2} & xs_{n+1} \end{vmatrix} = 2(a^3 + b^3)$$

$$= 2(x^3 s_{n+1}^3 + y^3 s_n^3).$$

This implies

$$\begin{vmatrix} xs_{n+1} & ys_n & s_{n+2}\\ s_{n+2} & xs_{n+1} & ys_n\\ ys_n & s_{n+2} & xs_{n+1} \end{vmatrix} = 2(s_{n+1}^3 + y^3 s_n^3),$$

where $s_k = s_k(1, y)$. Consequently, we have

$$\begin{vmatrix} J_{n+1}(y) & yJ_n(y) & J_{n+2}(y)\\ J_{n+2}(y) & J_{n+1}(y) & yJ_n(y)\\ yJ_n(y) & J_{n+2}(y) & J_{n+1}(y) \end{vmatrix} = 2[J_{n+1}^3(y) + y^3 J_n^3(y)];$$

$$\begin{vmatrix} J_{n+1} & 2J_n & J_{n+2}\\ J_{n+2} & J_{n+1} & 2J_n\\ 2J_n & J_{n+2} & J_{n+1} \end{vmatrix} = 2(J_{n+1}^3 + 8^3 J_n^3).$$

Similarly,

$$\begin{vmatrix} j_{n+1} & 2j_n & j_{n+2}\\ j_{n+2} & j_{n+1} & 2j_n\\ 2j_n & j_{n+2} & j_{n+1} \end{vmatrix} = 2(j_{n+1}^3 + 8^3 j_n^3).$$

For example, we have

$$\begin{vmatrix} J_6 & 2J_5 & J_7 \\ J_7 & J_6 & 2J_5 \\ 2J_5 & J_7 & J_6 \end{vmatrix} = \begin{vmatrix} 21 & 22 & 43 \\ 43 & 21 & 22 \\ 22 & 43 & 21 \end{vmatrix}$$
$$= 39{,}818$$
$$= 2(J_6^3 + 8^3 J_5^3);$$

and

$$\begin{vmatrix} j_6 & 2j_5 & j_7 \\ j_7 & j_6 & 2j_5 \\ 2j_5 & j_7 & j_6 \end{vmatrix} = \begin{vmatrix} 65 & 62 & 127 \\ 127 & 65 & 62 \\ 62 & 127 & 65 \end{vmatrix}$$
$$= 1{,}025{,}906$$
$$= 2(j_6^3 + 8^3 j_5^3).$$

Candido's Identity Revisited

The bivariate identity (43.13) works for the Jacobsthal family as well:

$$\left(y^2 J_{n-2}^2 + x^2 J_{n-1}^2 + J_n^2\right)^2 = 2(y^4 J_{n-2}^4 + 16x^4 J_{n-1}^4 + J_n^4);$$
$$\left(y^2 j_{n-2}^2 + x^2 j_{n-1}^2 + j_n^2\right)^2 = 2(y^4 j_{n-2}^4 + 16x^4 j_{n-1}^4 + j_n^4);$$
$$\left[y^2 J_{n-2}^2(y) + J_{n-1}^2(y) + J_n^2(y)\right]^2 = 2\left[y^4 J_{n-2}^4(y) + J_{n-1}^4(y) + J_n^4(y)\right];$$
$$\left[y^2 j_{n-2}^2(y) + j_{n-1}^2(y) + j_n^2(y)\right]^2 = 2\left[y^4 j_{n-2}^4(y) + j_{n-1}^4(y) + j_n^4(y)\right];$$
$$\left(4J_{n-2}^2 + J_{n-1}^2 + J_n^2\right)^2 = 2(16J_{n-2}^4 + J_{n-1}^4 + J_n^4);$$
$$\left(4j_{n-2}^2 + j_{n-1}^2 + j_n^2\right)^2 = 2(16j_{n-2}^4 + j_{n-1}^4 + j_n^4).$$

For example, $(4J_7^2 + J_8^2 + J_9^2)^2 = (4 \cdot 43^2 + 85^2 + 171^2)^2 = 1{,}923{,}875{,}044 = 2(16J_7^4 + J_8^4 + J_9^4)$.

Likewise, $(4j_7^2 + j_8^2 + j_9^2)^2 = 153{,}417{,}922{,}596 = 2(16j_7^4 + j_8^4 + j_9^4)$.

44.2 JACOBSTHAL OCCURRENCES

Next we present a few pleasant occurrences of Jacobsthal numbers in totally unrelated contexts. We begin with an occurrence of Jacobsthal numbers as numerators of reduced fractions of partial sums of a very special alternating sum [465].

Example 44.2. *Let S_n denote the nth partial sum (in lowest terms) of the alternating sum $\frac{1}{2} - \frac{1}{4} + \frac{1}{8} - \frac{1}{16} + \frac{1}{32} - \cdots$. Then*

$$S_1 = \frac{\textcircled{1}}{2} \qquad S_2 = \frac{\textcircled{1}}{2}$$

$$S_3 = \frac{\textcircled{3}}{8} \qquad S_4 = \frac{\textcircled{5}}{16}$$

$$S_5 = \frac{\textcircled{11}}{32} \qquad S_6 = \frac{\textcircled{21}}{64}$$

More generally, we can establish that $S_n = \frac{\textcircled{J_n}}{2^n}$, where $n \geq 1$; see Exercise 44.100. ∎

The next example is interesting in its own right. It reveals a close relationship between the number of integers between 2^n and 2^{n+1} that are divisible by 3, and Jacobsthal numbers [465].

Example 44.3. *Find the number of integers between 2^n and 2^{n+1} that are divisible by 3.*

Solution. Let c_n be the number of integers between 2^n and 2^{n+1} divisible by 3.

TABLE 44.3.

n	2^n	2^{n+1}	Integers Between 2^n and 2^{n+1} Divisible by 3	c_n
1	2	4	3	1
2	4	8	6	1
3	8	16	9, 12, 15	3
4	16	32	18, 21, 24, 27, 30	5
5	32	64	33, 36, 39, 42, 45, 48, 51, 54, 57, 60, 63	11
				↑ J_n

Table 44.3 gives the integers between 2^n and 2^{n+1} that are divisible by 3, where $1 \leq n \leq 5$. Here we observe a stunning pattern, so we conjecture that $c_n = J_n$. We will now confirm this.

No. of positive integers $< 2^{n+1}$ divisible by $3 = \lfloor (2^{n+1} - 1)/3 \rfloor$

No. of positive integers $\leq 2^n$ divisible by $3 = \lfloor (2^n - 1)/3 \rfloor$

$$c_n = \lfloor (2^{n+1} - 1)/3 \rfloor - \lfloor (2^n - 1)/3 \rfloor.$$

Suppose n is odd. Then

$$\begin{aligned} c_n &= \frac{2^{n+1} - 1}{3} - \frac{2^n - 2}{3} \\ &= \frac{2^n - (-1)^n}{3} = J_n. \end{aligned}$$

Similarly, $c_n = J_n$ when n is even also. Thus, in both cases, the formula works. ∎

Next we present an occurrence in the analysis of the merge insertion sorting algorithm [273] in computer science.

Example 44.4. *Merge insertion, developed by Lester R. Ford Jr. (1927–2017) and Selmer M. Johnson (1916–1996), is an amalgam of the merge and insertion sorting techniques. A study of the algorithm involves numbers x_n, where $x_{n-1} + x_n = 2^n$ and $x_0 = 1 = x_1$. Then*

$$\begin{aligned}
x_n &= 2^n - x_{n-1} \\
&= 2^n - 2^{n-1} + x_{n-2} \\
&\vdots \\
&= 2^n - 2^{n-1} + 2^{n-2} - \cdots + (-1)^n 2^0 \\
&= 2^n \sum_{k=0}^{n} (-1/2)^k \\
&= \frac{1}{3}[2^{n+1} - (-1)^{n+1}] = J_{n+1}.
\end{aligned}$$

Notice that $J_n + J_{n+1} = 2^n$; see Exercise 44.7. ■

Interestingly, Jacobsthal numbers also occur in the analysis of the *binary gcd algorithm* for computing the gcd (greatest common divisor) of two positive integers [272]; J. Stein discovered this binary algorithm in 1961.

Next we encounter a close relationship between Jacobsthal numbers and compositions of positive integers. We call such compositions *Jacobsthal compositions.*

44.3 JACOBSTHAL COMPOSITIONS

Consider the compositions of positive integers using 1s, and two sorts of 2s, denoted by 2 and 2′ [465]. Table 44.4 shows such compositions of positive

TABLE 44.4. Compositions with 1s, 2s, and 2′s

n	Compositions	Number of Compositions
1	1	1
2	1 + 1, 2, 2′	3
3	1 + 1 + 1, 1 + 2, 1 + 2′, 2 + 1, 2′ + 1	5
4	1 + 1 + 1 + 1, 1 + 1 + 2, 1 + 1 + 2′,	11
	1 + 2 + 1, 1 + 2′ + 1, 2 + 1 + 1, 2′ + 1 + 1,	
	2 + 2, 2 + 2′, 2′ + 2, 2′ + 2′	
		↑ J_{n+1}

integers n, where $1 \leq n \leq 4$. It appears from the table that there are J_{n+1} such compositions. The following theorem establishes its truthfulness.

Theorem 44.1. *There are J_{n+1} compositions of a positive integer n using 1s, 2s, and 2′s.*

Proof. Let a_n denote the number of compositions. Clearly, $a_1 = 1 = J_2$ and $a_2 = 3 = J_3$. Let $n \geq 3$. We will now establish that a_n satisfies the Jacobsthal recurrence using a constructive algorithm. Let x_k denote an arbitrary composition of a positive integer k.

Step 1. Place "1 +" in front of x_{n-1}.

Step 2A. Place "2 +" in front of x_{n-2}.

Step 2B. Place "2′ +" in front of x_{n-2}.

The total number of compositions of n created by the algorithm is $a_{n-1} + 2a_{n-2}$. Since the algorithm is reversible, it follows that $a_n = a_{n-1} + 2a_{n-2}$, where $n \geq 3$. This recurrence, coupled with the initial conditions, implies that $a_n = J_{n+1}$, where $n \geq 1$, as desired. ■

We now illustrate this algorithm with $n = 4$.

Step 1. Placing "1 +" in front of each x_3 yields five compositions of 4: **1** + 1 + 1 + 1, **1** + 1 + 2, **1** + 1 + 2′, **1** + 2 + 1, **1** + 2′ + 1.

Step 2A. Placing "2 +" in front of each x_2 yields three additional compositions: **2** + 1 + 1, **2** + 2, **2** + 2′.

Step 2B. Placing "2′ +" in front of each x_2 yields three more compositions: **2′** + 1 + 1, **2′** + 2, **2′** + 2′.

Combining these steps, we get all $11 = J_5$ compositions of 4.

Next we study compositions with the last summand odd. They were studied extensively by Ralph P. Grimaldi of Rose-Hulman Institute of Technology, Terre Haute, Indiana [200].

Ralph Peter Grimaldi was born in 1943 in New York City. He received his B.S., *summa cum laude*, from the State University of New York at Albany, in 1964, and his M.S. in 1965, both in mathematics. He received his Ph.D. in 1972 from New Mexico State University at Las Cruces, specializing in abstract algebra. He has been on the faculty at Rose-Hulman Institute of Technology since 1972.

Grimaldi is best known for his books *Discrete and Combinatorial Mathematics* (now in its fifth edition) and *Fibonacci and Catalan Numbers*, as well as numerous articles on Fibonacci numbers and other areas of discrete mathematics.

Compositions with Last Summand Odd

Let a_n denote the number of compositions of a positive integer n. Table 44.5 shows such compositions and their counts, where $1 \le n \le 5$. Using the data from the table, we can predict that $a_n = J_n$, where $n \ge 1$.

TABLE 44.5. Compositions with Last Summand Odd

n	Compositions	Number of Compositions
1	1	1
2	1 + 1	1
3	1 + 1 + 1, 2 + 1, 3	3
4	1 + 1 + 1 + 1, 1 + 2 + 1, 2 + 1 + 1, 1 + 3, 3 + 1	5
5	1 + 1 + 1 + 1 + 1, 1 + 1 + 2 + 1, 1 + 2 + 1 + 1, 2 + 1 + 1 + 1, 2 + 2 + 1, 1 + 1 + 3, 1 + 3 + 1, 3 + 1 + 1, 2 + 3, 4 + 1, 5	11
		↑ J_n

The next theorem confirms this observation using a similar constructive algorithm.

Theorem 44.2 (Grimaldi, 2014 [200]). *Let $n \ge 1$. Then $a_n = J_n$.*

Proof. The theorem is trivially true when $n = 1$ and 2. Now let $n \ge 3$. Let x_k denote an arbitrary composition of k with the last summand odd. There are a_{n-1} such compositions of $n - 1$.

Step 1. Placing "1 +" in front of each x_{n-1} results in a_{n-1} compositions of n.

Step 2A. Placing "2 +" in front of each x_{n-2} yields a_{n-2} compositions of n.

Step 2B. Adding 2 to the first summand of x_{n-2} gives an additional a_{n-2} compositions.

These three steps produce a total of $a_{n-1} + 2a_{n-2}$ compositions of n with the desired property. Since the algorithm is reversible, it follows that $a_n = a_{n-1} + 2a_{n-2}$. Thus a_n satisfies the recursive definition of J_n, so $a_n = J_n$, where $n \ge 1$. ■

We now illustrate the algorithm with $n = 5$.

Step 1. Placing "1 +" in front of each x_4 yields $5 = J_4$ compositions of 5: **1** + 1 + 1 + 1, **1** + 1 + 2 + 1, **1** + 2 + 1 + 1, **1** + 1 + 3, **1** + 3 + 1.

Step 2A. Placing "2 +" in front of each x_3 yields $3 = J_3$ additional compositions: **2** + 1 + 1 + 1, **2** + 2 + 1, **2** + 3.

Step 2B. Adding 2 to the first summand of each x_3 yields $3 = J_3$ more compositions: **3** + 1 + 1, **4** + 1, **5**.

Combining these steps yields all $11 = J_5$ compositions of 5.

Next we investigate the frequency of summands in the J_n compositions of n [200].

Frequency of Summands in Compositions

Let $a_{n,k}$ denote the frequency of occurrences of summand k in the J_n compositions of n, where $1 \leq k \leq n$. Table 44.6 shows the values of $a_{n,k}$, where $1 \leq n \leq 8$.

TABLE 44.6. Values of $a_{n,k}$, where $1 \leq n \leq 8$

n \ k	1	2	3	4	5	6	7	8
1	1							
2	2							
3	4	1	1					
4	10	2	2					
5	22	6	(4)	1	1			
6	50	[14]	10	2	2			
7	110	34	22	6	(4)	1	1	
8	242	78	50	[14]	10	2	2	

We now study some interesting properties of $a_{n,k}$. To begin with, suppose we replace the summand k from each of the a_n compositions of n with $k+2$. This results in compositions of $n+2$ with summand $k+2$. This procedure is reversible. Accordingly, we have the following result.

Lemma 44.1 (Grimaldi, 2014 [200]). *Let* $1 \leq k \leq n$. *Then* $a_{n,k} = a_{n+2,k+2}$. ■

For example, $a_{5,3} = (4) = a_{7,5}$, and $a_{6,2} = [14] = a_{8,4}$; see Table 44.6.

It follows from Lemma 44.1 that the values of $a_{n,k}$ can be found from those of $a_{n,1}$ and $a_{n,2}$, where $k \geq 3$:

$$a_{n,k} = \begin{cases} a_{n-2i,1} = a_{n-k+1,1} & \text{if } k = 2i+1 \\ a_{n-2i+2,2} = a_{n-k+2,2} & \text{if } k = 2i. \end{cases}$$

Now we can find formulas for both $a_{n,1}$ and $a_{n,2}$.

To this end, let $s_{n,k}$ denote the number of compositions (from among the J_n compositions) of n that begin with the summand k. For example, $s_{5,1} = 5$, $s_{5,2} = 3$, $s_{5,3} = 1$, $s_{5,4} = 1$, and $s_{5,5} = 1$. Clearly, $\sum_{k=1}^{5} s_{5,k} = J_5$; see Table 44.6.

More generally, we have

$$s_{n,k} = \begin{cases} 0 & \text{if } n < k \\ 0 & \text{if } n = k \text{ is even} \\ 1 & \text{if } n = k \text{ is odd} \\ J_{n-k} & \text{otherwise.} \end{cases}$$

Recurrence for $a_{n,1}$

Using the proof of Theorem 44.2, we can find a recurrence for $a_{n,1}$ in three steps:

Step 1. The quantity $a_{n-1,1} + J_{n-1}$ accounts for the number of 1s obtained by placing "1 +" at the beginning of the J_{n-1} compositions of $n - 1$.

Step 2A. The quantity $a_{n-2,1}$ counts the number of 1s among the J_{n-2} compositions of $n - 2$.

Step 2B. The total number of compositions of $n - 2$ with summand 1 is $a_{n-2,1}$. By adding 2 to the first summand of a composition x_{n-2}, we lose each 1 that appeared as the first summand in x_{n-2}. So the net gain of 1s this way is $a_{n-2,1} - s_{n-2,1}$.

These steps account for all 1s; thus

$$\begin{aligned} a_{n,1} &= (a_{n-1,1} + J_{n-1}) + a_{n-2,1} + (a_{n-2,1} - s_{n-2,1}) \\ &= a_{n-1,1} + 2a_{n-2,1} + J_{n-1} - s_{n-2,1} \\ &= a_{n-1,1} + 2a_{n-2,1} + J_{n-1} - J_{n-3} \\ &= a_{n-1,1} + 2a_{n-2,1} + 2^{n-3}, \quad n \geq 4. \end{aligned} \tag{44.2}$$

Solving the *linear nonhomogeneous recurrence (44.2) with constant coefficients* (LNRWCCs) with $a_{2,1} = 2$ and $a_{3,1} = 4$ (see Exercise 44.103), we get

$$\begin{aligned} a_{n,1} &= \left(\frac{5}{18}\right) 2^n + \left(\frac{2}{9}\right)(-1)^n + \left(\frac{1}{12}\right) n2^n \\ &= \frac{5}{18}(J_n + J_{n+1}) + \frac{2}{9}(J_{n+1} - 2J_n) + \frac{n}{12}(J_n + J_{n+1}) \\ &= \frac{n+6}{12} J_{n+1} + \frac{n-2}{12} J_n, \end{aligned}$$

where $a_{1,1} = 1$ and $n \geq 2$.

For example, $a_{7,1} = \frac{13}{12} J_8 + \frac{5}{12} J_7 = \frac{13}{12} \cdot 85 + \frac{5}{12} \cdot 43 = 110$; see Table 44.6.

Recurrence for $a_{n,2}$

As above, we can obtain a recurrence for $a_{n,2}$ by considering the 2s among the J_n compositions of n:

$$\begin{aligned} a_{n,2} &= a_{n-1,2} + (a_{n-2,2} + J_{n-2}) + (a_{n-2,2} - s_{n-2,2}) \\ &= a_{n-1,2} + 2a_{n-2,2} + J_{n-2} - J_{n-4} \\ &= a_{n-1,2} + 2a_{n-2,2} + 2^{n-4}, \quad n \geq 5. \end{aligned} \tag{44.3}$$

Solving the LNRWCCs (44.3) (see Exercise 44.104), we get

$$\begin{aligned} a_{n,2} &= \left(-\frac{1}{36}\right)2^n + \left(\frac{-2}{9}\right)(-1)^n + \left(\frac{1}{24}\right)n2^n \\ &= -\frac{1}{36}(J_n + J_{n+1}) - \frac{2}{9}(J_{n+1} - 2J_n) + \frac{n}{24}(J_n + J_{n+1}) \\ &= \frac{n-6}{24}J_{n+1} + \frac{n+10}{24}J_n, \end{aligned}$$

where $a_{1,2} = 0 = a_{2,2}$ and $n \geq 3$.

For example, $a_{8,2} = \frac{1}{12}J_9 + \frac{9}{12}J_8 = \frac{1}{12} \cdot 171 + \frac{9}{12} \cdot 85 = 78$; see Table 44.6.

Combining all the results about $a_{n,k}$ yields the following theorem.

Theorem 44.3 (Grimaldi, 2014 [200]). *If k is odd, then $a_{k,k} = 1$; and if $n > k$, then*

$$\begin{aligned} a_{n,k} &= a_{n-k+1,1} \\ &= \frac{n-k+7}{12}J_{n-k+2} + \frac{n-k-1}{12}J_{n-k+1}. \end{aligned}$$

If k is even, then $a_{k+1,k} = 1$; and if $n > k+1$, then

$$\begin{aligned} a_{n,k} &= a_{n-k+2,2} \\ &= \frac{n-k-4}{24}J_{n-k+3} + \frac{n-k+12}{24}J_{n-k+2}. \end{aligned}$$

■

We now turn to a study of the numbers of the plus signs and summands among the compositions of n, which also yields interesting dividends.

Plus Signs in Compositions

Let p_n denote the number of plus signs in the J_n compositions of n. It follows from Table 44.5 that $p_1 = 0, p_2 = 1, p_3 = 3, p_4 = 9$, and $p_5 = 23$. Although these values do not appear to follow a pattern, we do have enough tools at our disposal to develop a recurrence for p_n and hence an explicit formula for it. Let x_k denote an arbitrary composition of k:

Step 1. Placing "1 +" in front of each x_{n-1} contributes $p_{n-1} + J_{n-1}$ plus signs to the J_n compositions of n.

Step 2A. Placing "2 +" in front of each x_{n-2} yields $p_{n-2} + J_{n-2}$ plus signs.

Step 2B. Adding 2 to the first summand of x_{n-2} gives an additional p_{n-2} plus signs.

Combining these counts, we get

$$\begin{aligned} p_n &= (p_{n-1} + J_{n-1}) + (p_{n-2} + J_{n-2}) + p_{n-2} \\ &= p_{n-1} + 2p_{n-2} + J_{n-1} + J_{n-2} \\ &= p_{n-1} + 2p_{n-2} + 2^{n-2}, \quad n \geq 3. \end{aligned}$$

Solving this recurrence (see Exercise 44.105), we get

$$\begin{aligned} p_n &= \left(-\frac{1}{9}\right) 2^n + \frac{1}{9}(-1)^n + \frac{1}{6}n2^n \\ &= -\frac{1}{9}\left[2^n - (-1)^n\right] + \frac{n}{6}2^n \\ &= -\frac{1}{3}J_n + \frac{n}{6}(J_n + J_{n+1}) \\ &= \frac{n}{6}J_{n+1} + \frac{n-2}{6}J_n, \quad n \geq 0. \end{aligned}$$

For example, $p_5 = \frac{5}{6}J_6 + \frac{3}{6}J_5 = \frac{5 \cdot 21}{6} + \frac{3 \cdot 11}{6} = 23.$

Summands in Compositions

We can now find the number of summands s_n in the J_n compositions of n. Since the number of summands in x_k is one more than that of plus signs in it, it follows that

$$\begin{aligned} s_n &= p_n + J_n \\ &= \left(\frac{n}{6}J_{n+1} + \frac{n-2}{6}J_n\right) + J_n \\ &= \frac{n}{6}J_{n+1} + \frac{n+4}{6}J_n, \quad n \geq 1. \end{aligned}$$

For example, $s_5 = \frac{5}{6}J_6 + \frac{9}{6}J_5 = \frac{5 \cdot 21}{6} + \frac{9 \cdot 11}{6} = 34$; see Table 44.6.

We can find several additional occurrences of Jacobsthal numbers using the J_n compositions of n; see Exercises 44.110–44.118.

Compositions with Last Summand Even

Finally, compositions with the last summand even also yield interesting dividends [465]. Using Table 44.7, for example, we can conjecture that the number of such compositions of n is J_{n-1}, where $n \geq 2$; see Exercises 44.119–44.121.

TABLE 44.7. Compositions with Last Summand Even

n	Compositions	Number of Compositions
2	2	1
3	1 + 2	1
4	1 + 1 + 2, 2 + 2, 4	3
5	1 + 1 + 1 + 2, 1 + 2 + 2, 1 + 4, 2 + 1 + 2, 3 + 2	5
6	1 + 1 + 1 + 1 + 2, 1 + 1 + 2 + 2, 1 + 2 + 1 + 2, 1 + 3 + 2, 2 + 1 + 1 + 2, 2 + 2 + 2, 2 + 4, 3 + 1 + 2, 4 + 2, 6	11

↑ J_{n-1}

44.4 TRIANGULAR NUMBERS IN THE FAMILY

Next we identify Jacobsthal and Jacobsthal–Lucas numbers that are also triangular numbers [293]. Congruences, second-order diophantine equations, and linear algebra play a significant role in our discourse. Table 44.8 gives the first fifteen triangular numbers $t_n = \dfrac{n(n+1)}{2}$.

TABLE 44.8. First 15 Triangular Numbers

n	1	2	3	4	5	6	7	8	9	10	11	12	13	14	15
t_n	(1)	[3]	6	10	15	(21)	28	36	45	55	66	78	91	105	120

Quadratic Diophantine Equation $u^2 - Nv^2 = C$

Our goal is to identify all Jacobsthal and Jacobsthal–Lucas numbers that are also triangular. This task hinges on solving the quadratic diophantine equation (QDE) $u^2 - Nv^2 = C$, where N is a nonsquare positive integer and C a positive integer. The solutions of the QDE are closely related to those of the Pell's equation $u^2 - Nv^2 = 1$. So we first need a very brief introduction to solving the QDE [277, 362, 411]. In the interest of brevity, we confine our discussion to solutions (u, v) with $u > 0$.

Let (μ, ν) be the fundamental solution of the Pell's equation $u^2 - Nv^2 = 1$, and (u_0, v_0) a solution of the QDE. Let $u_m + v_m\sqrt{N} = (u_0 + v_0\sqrt{N})(\mu + \nu\sqrt{N})^m$, where m is a positive integer. Then (u_m, v_m) is a solution of the QDE. We then say that the solution (u_m, v_m) is *associated* with the solution (u_0, v_0). Such solutions belong to a class of solutions of the QDE. Suppose (u_0, v_0) has the property that it has the least possible positive value of u among the solutions in the class; then (u_0, v_0) is the *fundamental solution* of the class.

The QDE can have different classes of solutions. Although each class is infinite, the number of distinct classes is finite. Two solutions (u, v) and (u', v') belong to the same class if and only if $uu' \equiv Nvv' \pmod{C}$ and $uv' \equiv u'v \pmod{C}$ [277].

The following theorem provides a mechanism for finding the solutions of the QDE, when it is solvable.

Theorem 44.4. *Let (μ, ν) be the fundamental solution of the Pell's equation $u^2 - Nv^2 = 1$, (u_0, v_0) a fundamental solution of the QDE $u^2 - Nv^2 = C$, and m a positive integer. Then*

i) $0 < u_0 \le \sqrt{\dfrac{(\mu+1)C}{2}}$ *and* $0 < |v_0| \le \nu\sqrt{\dfrac{C}{2(\mu+1)}}$.

[These two inequalities provide computable upper bounds for u_0 and v_0. The number of solutions (u_0, v_0) resulting from these inequalities determines the number of different classes of solutions.]

ii) *Every solution (u_m, v_m) belonging to the class of (u_0, v_0) is given by*

$$u_m + v_m\sqrt{N} = (u_0 + v_0\sqrt{N})(\mu + \nu\sqrt{N})^m. \tag{44.4}$$

iii) *The QDE is* not *solvable if it has no solutions satisfying the inequalities in i).* ■

Recurrence for (u_m, v_m)

Equation (44.4) can be used to derive a recurrence for (u_m, v_m):

$$\begin{aligned} u_{m+1} + v_{m+1}\sqrt{N} &= (u_0 + v_0\sqrt{N})(\mu + \nu\sqrt{N})^{m+1} \\ &= (u_m + v_m\sqrt{N})(\mu + \nu\sqrt{N}) \\ &= (\mu u_m + N\nu v_m) + (\nu u_m + \mu v_m)\sqrt{N}. \end{aligned}$$

Thus we have the following recurrence for (u_m, v_m):

$$\begin{aligned} u_{m+1} &= \mu u_m + N\nu v_m \\ v_{m+1} &= \nu u_m + \mu v_m. \end{aligned} \tag{44.5}$$

These recurrences can be used to develop a second-order recurrence for both u_m and v_m.

A Second-Order Recurrence for (u_m, v_m)

These recurrences can be combined into a matrix equation:

$$\begin{bmatrix} u_{m+1} \\ v_{m+1} \end{bmatrix} = M \begin{bmatrix} u_m \\ v_m \end{bmatrix},$$

where $M = \begin{bmatrix} \mu & N\nu \\ \nu & \mu \end{bmatrix}$.

By the well-known Cayley–Hamilton theorem [13], M satisfies its characteristic equation $|M - \lambda I| = 0$, where I denotes the 2×2 identity matrix; that is, $\lambda^2 - 2\mu\lambda + 1 = 0$. So $M^2 = 2\mu M - I$.

Consequently, we have

$$\begin{aligned}\begin{bmatrix} u_{m+2} \\ v_{m+2} \end{bmatrix} &= M^2 \begin{bmatrix} u_m \\ v_m \end{bmatrix} \\ &= (2\mu M - I) \begin{bmatrix} u_m \\ v_m \end{bmatrix} \\ &= 2\mu \begin{bmatrix} u_{m+1} \\ v_{m+1} \end{bmatrix} - \begin{bmatrix} u_m \\ v_m \end{bmatrix}.\end{aligned}$$

Thus both u_m and v_m satisfy the recurrence

$$r_{m+2} = 2\mu r_{m+1} - r_m, \tag{44.6}$$

where $m \geq 0$.

With these facts at our fingertips, we are now ready to identify all triangular Jacobsthal numbers.

Triangular Jacobsthal Numbers

Clearly, $J_1 = t_1$, $J_2 = t_1$, $J_3 = t_2$, $J_6 = t_6$, and $J_9 = t_{18}$; see Tables 44.2 and 44.8. So there are at least five triangular Jacobsthal numbers.

Next, we show that there are *no* other triangular Jacobsthal numbers. To this end, we use the fact that $8t_k + 1 = (2k + 1)^2$, discovered around 250 A.D. by Diophantus of Alexandria, Egypt. Consequently, J_n is a triangular number if and only if $8J_n + 1$ is the square of an odd integer [277].

Case 1. Suppose J_{2n} is a triangular number, where $n \geq 5$. Then $8J_{2n} + 1 = y^2$ for some odd positive integer y. This yields

$$8 \cdot \frac{2^{2n} - 1}{3} + 1 = y^2$$

$$2w^2 - 3y^2 - 5 = 0 \tag{44.7}$$

$$x^2 - 6y^2 = 10, \tag{44.8}$$

where $w = 2^{n+1}$ and $x = 2w \geq 128$.

Using the theorem, we can solve the QDE (44.8). The fundamental solution of the Pell's equation $x^2 - 6y^2 = 1$ is $(\mu, \nu) = (5, 2)$, and $(\pm 4, \pm 1)$ are solutions of the QDE (44.8). Since $x > 0$, we can safely ignore the solutions $(-4, \pm 1)$. This leaves just two fundamental solutions: $(x_0, y_0) = (4, 1)$ and $(x_0', y_0') = (4, -1)$.

Since $x_0 x_0' - 6 y_0 y_0' = 4 \cdot 4 - 6 \cdot 1 \cdot (-1) \not\equiv 0 \pmod{10}$ and $x_0 y_0' - x_0' y_0 = 4 \cdot (-1) - 4 \cdot 1 \not\equiv 0 \pmod{10}$, it follows that the solutions $(4, 1)$ and $(4, -1)$ belong to two different classes of solutions of the QDE (44.8) [362]; each is the fundamental solution of the corresponding class.

Subcase 1A. Consider the case $(x_0, y_0) = (4, 1)$. Since $(\mu, \nu) = (5, 2)$ and $N = 6$, it follows by equation (44.5) that every solution (x_m, y_m) of (44.8) in the class of (4, 1) is given by

$$x_{m+1} = 5x_m + 12y_m$$
$$y_{m+1} = 2x_m + 5y_m,$$

where $(x_0, y_0) = (4, 1)$.

Consequently, $(x_1, y_1) = (32, 13)$ and $(x_2, y_2) = (316, 129)$ are solutions of QDE (44.8): $32^2 - 6 \cdot 13^2 = 10 = 316^2 - 6 \cdot 129^2$.

Since y_0 and y_1 are odd, and $(\mu, \nu) = (5, 2)$, it follows by PMI that every y_m is odd. Since x_0 is even and $x_{m+1} \equiv x_m \pmod 2$, it also follows that every x_m is even.

Recurrence for (w_m, y_m)

Since $x_m = 2w_m$, the above equations yield the following recurrences for w_m and y_m:

$$w_{m+1} = 5w_m + 6y_m$$
$$y_{m+1} = 4w_m + 5y_m,$$

where $(w_0, y_0) = (2, 1)$ and $m \geq 0$.

Since $w_{m+1} \equiv w_m \pmod 2$ and w_0 is even, it follows that every w_m is even; see Table 44.9.

With $\mu = 5$, recurrence (44.6), which is satisfied by both w_m and y_m, comes in handy when computing the solutions (w_m, y_m) of (44.7); they belong to the class with the fundamental solution (2, 1). Table 44.9 shows the first ten such solutions. Since the y-values do not directly impact the problem at hand, we ignore them.

TABLE 44.9.

m	0	1	2	3	4	5	6	7	8	9
w_m	2	16	158	1564	15482	153256	1517078	15017524	148658162	1471564096
y_m	1	13	129	1277	12641	125133	1238689	12261757	121378881	1201527053

We now show that *no* w_m can be a power of 2. To see this, we use the recurrence $w_{m+2} = 10w_{m+1} - w_m$ to compute the values of $\{w_m \pmod{31}\}_{m\geq 0}$ and $\{w_m \pmod{32}\}_{m\geq 0}$. Both sequences are periodic with periods 32 and 16, respectively; see Table 44.10. But $2^k \pmod{31} = 1, 2, 4, 8$, or 16; and $2^k \pmod{32} = 0$ for $k \geq 5$. It now follows from the table that no w_m satisfies both conditions satisfied by 2^k, where $k \geq 5$; that is, *no* w_m is congruent to 2^k modulo 31 and 32, when $k \geq 5$.

Subcase 1B. Suppose $(x_0', y_0') = (4, -1)$. This solution, coupled with $(\mu, \nu) = (5, 2)$, can be used to generate a different family of solutions (x_m, y_m) of

TABLE 44.10.

m	0	1	2	3	4	5	6	7	8	9	10	11	12	13	14	15
w_m (mod 31)	2	16	3	14	13	23	0	8	18	17	28	15	29	27	24	27
w_m (mod 32)	2	16	30	28	26	8	22	20	18	0	14	12	10	24	6	4
m	16	17	18	19	20	21	22	23	24	25	26	27	28	29	30	31
w_m (mod 31)	29	15	28	17	18	8	0	23	13	14	3	16	2	4	7	4
w_m (mod 32)	2	16	30	28	26	8	22	20	18	0	14	12	10	24	6	4

(44.8) and hence (w_m, y_m). In particular, it follows by recurrence (44.5) that $(x_1, y_1) = (8, 3)$ and $(x_2, y_2) = (76, 31)$ are also solutions of the QDE (44.8): $8^2 - 6 \cdot 3^2 = 10 = 76^2 - 6 \cdot 31$.

TABLE 44.11.

m	0	1	2	3	4	5	6	7	8	9
w_m	2	4	38	376	3722	36844	364718	3610336	35738642	353776084
y_m	−1	3	31	307	3039	30083	297791	2947827	29180479	288856963

Correspondingly, $w_0 = 2$, $w_1 = 4$, and $w_2 = 38$. As in Subcase 1A, w_m satisfies exactly the same recurrence. Table 44.11 shows the first ten solutions (w_m, y_m). Again, we can safely ignore the y-values.

We can now show that *no* w_m can be a power of 2. To see this, notice that the sequences $\{w_m \pmod{31}\}_{m \geq 0}$ and $\{w_m \pmod{32}\}_{m \geq 0}$ are both periodic with period 32; see Table 44.12. It follows from the table that *no* w_m is congruent to 2^k modulo 31 and 32 for any integer $k \geq 5$.

TABLE 44.12.

m	0	1	2	3	4	5	6	7	8	9	10	11	12	13	14	15
w_m (mod 31)	2	4	7	4	2	16	3	14	13	23	0	8	18	17	28	15
w_m (mod 32)	1	2	19	28	5	22	23	16	9	10	27	4	13	30	31	24
m	16	17	18	19	20	21	22	23	24	25	26	27	28	29	30	31
w_m (mod 31)	29	27	24	27	29	15	28	17	18	8	0	23	13	14	3	16
w_m (mod 32)	17	18	3	12	21	6	7	0	25	26	11	20	29	14	15	8

It follows by Subcases 1A and 1B that no w_m can be a power of 2, when $m \geq 5$. Consequently, *no* J_{2n} is a triangular number when $m \geq 5$.

Case 2. Suppose J_{2n+1} is a triangular number, where $n \geq 5$. Then $8J_{2n+1} + 1 = y^2$ for some positive odd integer y. As before, this yields:

$$x^2 - 3y^2 + 11 = 0$$

$$z^2 - 3x^2 = 33, \tag{44.9}$$

where $x = 2^{n+2} \geq 128$, and $z = 3y$ is odd.

The fundamental solution of the Pell's equation $z^2 - 3x^2 = 1$ is $(\mu, \nu) = (2, 1)$. There are two fundamental solutions $(z_0, x_0) = (6, 1)$ and $(z'_0, x'_0) = (6, -1)$ of the QDE (44.9) with the least positive value 6 for z. Since $z_0 z'_0 - 3x_0 x'_0 = 6 \cdot 6 - 3 \cdot 1 \cdot (-1) \not\equiv 0 \pmod{33}$ and $z_0 x'_0 - z'_0 x_0 = 6 \cdot (-1) - 1 \cdot 6 \not\equiv 0 \pmod{33}$, it follows that the solutions (6, 1) and (6, −1) belong to two different classes of solutions of the QDE (44.9) [362]. Using the theorem, we can now find all solutions of (44.9).

Subcase 2A. With the fundamental solution $(z_0, x_0) = (6, 1)$, every solution (z_m, x_m) in its class is given by the recurrence:

$$z_{m+1} = 2z_m + 3x_m$$
$$x_{m+1} = z_m + 2x_m.$$

It now follows that $(z_1, x_1) = (15, 8)$ is also a solution of (44.9). Table 44.13 shows the first ten such solutions (z_m, x_m), where $m \geq 0$.

TABLE 44.13.

m	0	1	2	3	4	5	6	7	8	9
z_m	6	15	54	201	750	2799	10446	38985	145494	542991
x_m	1	8	31	116	433	1616	6031	22508	84001	313496

Recurrence for (y_m, x_m)

Since $z_m = 3y_m$, the above recurrences yield

$$y_{m+1} = 2y_m + x_m$$
$$x_{m+1} = 3y_m + 2x_m.$$

A Second-Order Recurrence for x_m

As before, it follows by (44.6) that x_m satisfies the recurrence

$$x_{m+2} = 4x_{m+1} - x_m, \qquad (44.10)$$

where $m \geq 0$.

It follows from this recurrence that $x_{m+2} \equiv x_m \pmod 2$. Since x_0 is odd and x_1 is even, it follows by induction that $x_m \equiv m + 1 \pmod 2$; so x_m and $m + 1$ have the same parity. Likewise, z_m is always even; see Table 44.13. Since $y_{m+2} \equiv y_m \pmod 2$, it follows that y_m and m have the same parity.

Using recurrence (44.10), we now compute $x_m \pmod 7$ and $x_m \pmod{16}$; see Table 44.14.

TABLE 44.14.

m	0	1	2	3	4	5	6	7	8	9	10	11	12	13	14	15
x_m (mod 7)	1	1	3	4	6	6	4	3	1	1	3	4	6	6	4	3
x_m (mod 16)	1	8	15	4	1	0	15	12	1	8	15	4	1	0	15	12

The sequence $\{x_m \pmod 7\}$ is periodic with period 8: $\underbrace{1\,1\,3\,4\,6\,6\,4\,3}\ \underbrace{1\,1\,3\,4\,6\,6\,4\,3}\cdots$; and so is the sequence $\{x_m \pmod{16}\}$: $\underbrace{1\ 8\ 15\ 4\ 1\ 0\ 15\ 12}\ \underbrace{1\ 8\ 15\ 4\ 1\ 0\ 15\ 12}\cdots$. But $2^k \pmod 7 = 1, 2,$ or 4; and $2^k \pmod{16} = 0$ when $k \geq 5$. Consequently, *no* x_m satisfies both conditions when $k \geq 5$.

Subcase 2B. Consider the solution $(z_0', x_0') = (6, -1)$. Then $(z_1, x_1) = (9, 4)$ and $(z_2, x_2) = (30, 17)$. Since we want $x_m > 0$, we will ignore the solution $(6, -1)$.

TABLE 44.15.

m	1	2	3	4	5	6	7	8	9	10	11	12	13	14	15	16
x_m (mod 127)	4	17	64	112	3	27	105	12	70	14	113	57	115	22	100	124
x_m (mod 128)	4	17	64	111	124	1	8	31	116	49	80	15	108	33	24	63
m	17	18	19	20	21	22	23	24	25	26	27	28	29	30	31	32
x_m (mod 127)	15	63	110	123	1	8	31	116	52	92	62	29	54	60	59	49
x_m (mod 128)	100	81	96	47	92	65	40	95	84	113	112	79	76	97	56	127
m	33	34	35	36	37	38	39	40	41	42	43	44	45	46	47	48
x_m (mod 127)	10	118	81	79	108	99	34	37	114	38	38	114	37	34	99	108
x_m (mod 128)	68	17	0	111	60	1	72	31	52	49	16	15	44	33	88	63
m	49	50	51	52	53	54	55	56	57	58	59	60	61	62	63	64
x_m (mod 127)	79	81	118	10	49	59	60	54	29	62	92	52	116	31	8	1
x_m (mod 128)	36	81	32	47	28	65	104	95	20	113	48	79	12	97	120	127
m	65	66	67	68	69	70	71	72	73	74	75	76	77	78	79	80
x_m (mod 127)	123	110	63	15	124	100	22	115	57	113	14	70	12	105	27	3
x_m (mod 128)	4	17	64	111	124	1	8	31	116	49	80	15	108	33	24	63
m	81	82	83	84	85	86	87	88	89	90	91	92	93	94	95	96
x_m (mod 127)	112	64	17	4	126	119	96	11	75	35	65	98	73	67	68	78
x_m (mod 128)	100	81	96	47	92	65	40	95	84	113	112	79	76	97	56	127
m	97	98	99	100	101	102	103	104	105	106	107	108	109	110	111	112
x_m (mod 127)	117	9	46	48	19	28	93	90	13	89	89	13	90	93	28	19
x_m (mod 128)	68	17	0	111	60	1	72	31	52	49	16	15	44	33	88	63
m	113	114	115	116	117	118	119	120	121	122	123	124	125	126	127	128
x_m (mod 127)	48	46	9	117	78	68	67	73	98	65	35	75	11	96	119	126
x_m (mod 128)	36	81	32	47	28	65	104	95	20	113	48	79	12	97	120	127

Using (44.10), we now compute the sequences $\{x_m \pmod{127}\}$ and $\{x_m \pmod{128}\}$; see Table 44.15. Again, no x_m satisfies the conditions satisfied by $\{2^k \pmod{127}\}_{k\geq 5}$ and $\{2^k \pmod{128}\}_{k\geq 5}$. So *no* x_m in this class can be a power of 2 when $k \geq 5$.

Combining the two subcases 2A and 2B, it follows that *no* J_{2n+1} is a triangular number when $n \geq 5$.

Thus, by Cases 1 and 2, *no* J_n is a triangular number when $n \geq 5$. Consequently, the only triangular Jacobsthal numbers are J_1, J_2, J_3, J_6, and J_9; see Table 44.8.

[*Note*: Since $y_{m+2} \equiv y_m \pmod 2$ and $y_0 = 2$ in both subcases under Case 2, it follows by PMI that y_{2m} is even for $m \geq 0$. But every y-value must be odd. Consequently, we could drop the columns with even values of m from Tables 44.14 and 44.15.]

Next we investigate Jacobsthal–Lucas numbers j_n that are also triangular. To this end, first notice that the sequence $\{t_n \pmod 9\}$ follows an interesting pattern: $\underbrace{1\,3\,6\,1\,6\,3\,1\,9\,9}\ \underbrace{1\,3\,\cdots\,9\,9}\cdots$. Consequently, $t_n \pmod 9$ equals 1, 3, 6, or 9.

Triangular Jacobsthal–Lucas Numbers

It follows from the Binet-like formulas that $9J_n^2 = (2^{2n}+1) - 2(-2)^n = j_{2n} - 2(-2)^n$, so $j_{2n} \equiv 2(-2)^n \pmod 9$. But the sequence $\{2(-2)^n \pmod 9\}$ follows the pattern $\underbrace{5\,8\,2}\ \underbrace{5\,8\,2}\cdots$; so $j_{2n} \pmod 9$ equals 2, 5, or 8. Consequently, *no* Jacobsthal–Lucas number j_{2n} is triangular.

Now, consider the Jacobsthal–Lucas numbers j_{2n+1}, where $n \geq 0$. Since $j_1 = t_1$, we let $n \geq 1$. Then $8j_{2n+1} + 1 = 2^{2(n+2)} - 8$. Since $(2^{n+2}-1)^2 < 8j_{2n+1} + 1 < 2^{2(n+2)}$, and $(2^{n+2}-1)^2$ and $2^{2(n+2)}$ are consecutive squares, it follows that $8j_{2n+1} + 1$ cannot be a square. Consequently, j_{2n+1} *cannot* be triangular when $n \geq 1$.

Thus $j_1 = 1$ is the only triangular Jacobsthal–Lucas number; see Table 44.8.

Next we use Jacobsthal and Jacobsthal–Lucas numbers to generate triangular numbers, and then employ them to construct an interesting class of primitive Pythagorean triples.

Triangular Numbers with Jacobsthal Generators

We develop three interesting properties linking triangular numbers with the Jacobsthal families.

1) Since $2J_{2n} + 1 = J_{2n+1}$, it follows that

$$\begin{aligned}\frac{J_{2n+1}(J_{2n+2}+1)}{2} &= \frac{J_{2n+1}(J_{2n+1}+2J_{2n}+1)}{2}\\ &= \frac{J_{2n+1}(J_{2n+1}+J_{2n+1})}{2}\\ &= J_{2n+1}^2;\end{aligned}$$

and

$$J_{2n}(J_{2n+1}+1) = J_{2n}(2J_{2n}+2)$$
$$= 2J_{2n}(J_{2n}+1).$$

So

$$\frac{J_{2n}(J_{2n+1}+1)}{2} = J_{2n}(J_{2n}+1) = 2t_{J_{2n}}.$$

Thus

$$\frac{J_n(J_{n+1}+1)}{2} = \begin{cases} J_n^2 & \text{if } n \text{ is odd} \\ 2t_{J_n} & \text{otherwise.} \end{cases}$$

For example, $\frac{J_7(J_8+1)}{2} = \frac{43 \cdot 86}{2} = 43^2 = J_7^2$ and $\frac{J_6(J_7+1)}{2} = \frac{21 \cdot 44}{2}$ $= 21 \cdot 22 = 2t_{21} = 2t_{J_6}$.

2) Since $j_{2n} - 1 = 2^{2n} = 2\left[(2^{2n-1}-1)+1\right] = 2(j_{2n-1}+1)$, it follows that

$$\frac{j_{2n-1}(j_{2n}-1)}{2} = j_{2n-1}(j_{2n-1}+1) = 2t_{j_{2n-1}}.$$

For example, $\frac{j_{11}(j_{12}-1)}{2} = \frac{2047 \cdot 4096}{2} = 2047 \cdot 2048 = 2t_{2047} = 2t_{j_{11}}$.

3) Finally, since $j_{2n+1}+1 = 2^{2n+1}$ and $j_{2n+1}+3 = 2j_{2n}$,

$$j_{2n}(j_{2n+1}+1) = j_{2n}2^{2n+1}$$
$$= \left(\frac{j_{2n+1}+3}{2}\right)(j_{2n+1}+1)$$
$$= 2\left(\frac{j_{2n+1}+1}{2}\right)\left(\frac{j_{2n+1}+1}{2}+1\right)$$
$$\frac{j_{2n}(j_{2n+1}+1)}{2} = t_{(j_{2n+1}+1)/2}.$$

For example, $\frac{j_{10}(j_{11}+1)}{2} = \frac{1025(2047+1)}{2} = \frac{1024 \cdot 1025}{2} = t_{1024} = t_{(j_{11}+1)/2}$.

We now employ Jacobsthal and Jacobsthal–Lucas families to generate a class of primitive Pythagorean triples.

Primitive Pythagorean Triples

It is well known that x–y–z is a primitive Pythagorean triple if and only if $x = 2ab$, $y = a^2 - b^2$, and $z = a^2 + b^2$, where a and b are relatively prime with different parity and $a > b$ [277]. Let $a = 2^n$ and $b = 1$. Then $x = 2^{n+1}$, $y = 2^{2n} - 1 = 3J_{2n} = J_3J_{2n}$ and $z = 2^{2n} + 1 = j_{2n}$. So

$$\begin{aligned}(2^{n+1})^2 + (3J_{2n})^2 &= 2^{2n+2} + \left(2^{2n} - 1\right)^2 \\ &= 2^{2n+2} + 2^{4n} - 2^{2n+1} + 1 \\ &= 2^{4n} + 2^{2n+1} + 1 \\ &= \left(2^{2n} + 1\right)^2 \\ &= j_{2n}^2.\end{aligned}$$

Thus 2^{n+1}–J_3J_{2n}–j_{2n} is a Pythagorean triple.

Let p be a prime factor of $3J_{2n}$ and j_{2n}. Then $p|(2^{2n} - 1)$ and $p|(2^{2n} + 1)$; so $p|2$. This is impossible since both $2^{2n} - 1$ and $2^{2n} + 1$ are odd. Consequently, 2^{2n+1}–J_3J_{2n}–j_{2n} is a primitive Pythagorean triple.

The area of the Pythagorean triangle equals $\frac{1}{2} \cdot 2^{n+1} \cdot J_3J_{2n} = 2^nJ_3J_{2n}$; and its perimeter equals

$$\begin{aligned}2^{n+1} + J_3J_{2n} + j_{2n} &= 2^{n+1} + (2^{2n} - 1) + (2^{2n} + 1) \\ &= 2^{n+1} + 2^{2n+1} \\ &= 2 \cdot 2^n(2^n + 1) \\ &= 2(M_n + 1)(M_n + 2) \\ &= 4t_{M_n+1}.\end{aligned}$$

For example, let $n = 5$. Then 2^6–J_3J_{10}–j_{10} = 64–1023–1025 is a primitive Pythagorean triple. The area of the corresponding Pythagorean triangle is $\frac{1}{2} \cdot 2^6 \cdot 1023 = 32{,}736 = 2^5J_3J_{10}$; and its perimeter is $64 + 1023 + 1025 = 2112 = 4 \cdot 528 = 4t_{32}$.

Jacobsthal numbers also appear in the study of formal languages [199], as the next section shows.

44.5 FORMAL LANGUAGES

An *alphabet* Σ is a finite set of symbols. A *word* (or *string*) *over* Σ is a finite sequence of symbols from Σ. The number of symbols in a word is its *length*. The word of length 0 is the *empty word* or *null word*; it is denoted by λ.

The set of all possible words over Σ, denoted by Σ^*, is the *Kleene closure* of Σ, named after the American logician Stephen Kleene. A *language* L over Σ is a subset of Σ^*.

The *concatenation* of two words x and y in L, denoted by xy, is obtained by appending y at the end of x. For example, let $x = x_1x_2 \dots x_m$ and $y = y_1y_2 \cdots y_n$; then $xy = x_1x_2 \dots x_my_1y_2 \cdots y_n$. The *concatenation* of two languages A and B

over Σ, denoted by AB, is defined by $AB = \{ab \mid a \in A \text{ and } b \in B\}$. In particular, $A^2 = \{a_1a_2 \mid a_1, a_2 \in A\}$. More generally, $A^n = \{a_1a_2 \dots a_n \mid a_i \in A\}$. Then $A^* = \bigcup_{n=0}^{\infty} A^n$, where $A^0 = \{\lambda\}$.

Stephen Cole Kleene (1909–1994) was born in Hartford, Connecticut, where his father was a professor of economics at Trinity College. After graduating from Amherst College in 1930, Kleene entered Princeton University, and received his Ph.D. in mathematics in 1934. The eminent logician Alonso Church supervised his dissertation, *A Theory of Positive Integers in Formal Logic*.

After teaching briefly at Princeton, Kleene joined the faculty at the University of Wisconsin, Madison, in 1935. From 1939 to 1940, he was a visiting scholar at the Institute for Advanced Study at Princeton. While there, he laid the foundation for the branch of mathematical logic called *recursive function theory*, which paved the way for theoretical computer science. In 1941, he returned to Amherst College as an associate professor. He spent the next four years with the US Navy, becoming a lieutenant commander.

Kleene returned to the faculty of the University of Wisconsin in 1946. After chairing the Departments of Mathematics and Computer Science, he served as the Dean of the College of Letters and Science, retiring in 1979.

Kleene's research included the theory of algorithms and Church's lambda calculus. His lectures at Wisconsin became three books in mathematical logic. His honors include the Leroy P. Steele Prize from the American Mathematical Society (1983) and the National Medal of Science (1990).

In particular, let $\Sigma = \{0, 1\}$, the *binary alphabet*. Its symbols are the *bits* 0 and 1. Let $L = \{0, 01, 11\}$. Let a_n denote the number of words of length n in L^*, where $n \geq 1$. Table 44.16 shows such words, where $1 \leq n \leq 4$. It appears from the table that $a_n = J_{n+1}$. The following theorem confirms this observation.

TABLE 44.16. Words of Length $1 \leq n \leq 4$

n	Words of Length n	Number of Words
1	0	1
2	00, 01, 11	3
3	000, 010, 110, 001, 011	5
4	0000, 0100, 1100, 0010, 0110	11
	0001, 0101, 1101, 0011, 0111, 1111	

↑
J_{n+1}

Theorem 44.5 (Grimaldi, 2005 [199]). *Let a_n denote the number of words of length n in L^*, where $L = \{0, 01, 11\}$ and $n \geq 1$. Then $a_n = J_{n+1}$.*

Proof. Clearly, $a_1 = 1 = J_2$ and $a_2 = 3 = J_3$. So let $n \geq 3$. It now remains to confirm that a_n satisfies the Jacobsthal recurrence using a constructive algorithm. To this end, let x_k denote an arbitrary word of length k in L^*.

Step 1. Append 0 to the right of x_{n-1}. This yields a_{n-1} words x_n.

Step 2. Append 01 to the right of x_{n-2}. This creates an additional a_{n-2} such words.

Step 3. Append 11 to the right of x_{n-2}. This gives a_{n-2} more words x_n.

Combining these steps produces a total of $a_{n-1} + 2a_{n-2}$ words of length n. Since the algorithm is reversible, it follows that $a_{n-1} + 2a_{n-2} = a_n$. This recurrence, coupled with the initial conditions, yields the desired result. ■

Next we find the numbers of 0s and 1s in the $a_n = J_{n+1}$ words of length n in L^*.

Numbers of 0s and 1s in the a_n Words

Let z_n denote the number of 0s among the a_n words of length n, and w_n that of 1s among them. It follows from Table 44.16 that

$$z_1 = 1,\ z_2 = 3,\ z_3 = 9,\ z_4 = \textcircled{23};\ \text{ and } w_0 = 0,\ w_2 = 3,\ w_3 = 6,\ w_4 = \boxed{21}.$$

We now compute a recurrence and an explicit formula for z_n. Notice that step 1 contributes $z_{n-1} + a_{n-1}$ zeros to z_n; step 2 adds $z_{n-2} + a_{n-2}$ zeros; and step 3 adds an additional z_{n-2} zeros. Thus

$$\begin{aligned} z_n &= (z_{n-1} + a_{n-1}) + (z_{n-2} + a_{n-2}) + z_{n-2} \\ &= (z_{n-1} + 2z_{n-2}) + (J_n + J_{n-1}) \\ &= z_{n-1} + 2z_{n-2} + 2^{n-1}. \end{aligned}$$

Solving this recurrence, we get

$$z_n = \frac{1}{3}[nJ_{n+1} + (n+1)J_n],$$

where $n \geq 1$; see Exercise 44.106.

For example, $z_4 = \frac{1}{3}(4J_5 + 5J_4) = \frac{1}{3}(4 \cdot 11 + 5 \cdot 5) = \textcircled{23}$.

Since $z_n + w_n = na_n = nJ_{n+1}$, we then have

$$\begin{aligned} w_n &= nJ_{n+1} - \frac{1}{3}[nJ_{n+1} + (n+1)J_n] \\ &= \frac{1}{3}[2nJ_{n+1} - (n+1)J_n]. \end{aligned}$$

For example, $w_4 = \frac{1}{3}(8J_5 - 5J_4) = \frac{1}{3}(8 \cdot 11 - 5 \cdot 5) = \boxed{21}$.

Next we investigate the number of runs in the a_n words in L^*.

Runs in the a_n Words

A *run* in a word in L^* is a sequence of consecutive and identical bits that can be preceded and followed by different bits or no bits. For example, the word 01110010 contains five runs: 0, 111, 00, 1, and 0.

Let $a_{n,i}$ denote the number of words of length n that end in i, and $r_{n,i}$ the number of runs among the words of length n that end in i, where $i = 0$ or 1. Let r_n denote the number of runs among the a_n words of length n. Table 44.17 shows the values of $a_{n,i}$, $r_{n,i}$, and r_n, where $1 \leq n \leq 4$.

TABLE 44.17. Values of $a_{n,i}$, $r_{n,i}$, and r_n

n	$a_{n,0}$	$a_{n,1}$	$r_{n,0}$	$r_{n,1}$	r_n
1	1	0	1	0	1
2	1	2	1	3	4
3	3	2	6	4	10
4	5	6	12	14	26

Using the constructive algorithm, we can now find a recurrence and an explicit formula for r_{n+1}. Let $x_{k,i}$ denote an arbitrary word of length k that ends in i.

Step 1A. Appending 0 to $x_{n,0}$ does not create any new runs: $x_{n,0}\underbrace{0} = x_{n+1,0}$. Such words contribute $r_{n,0}$ runs toward r_{n+1}.

Step 1B. Appending 0 to $x_{n,1}$ results in $x_{n+1,0}$: $x_{n,1}\underbrace{0} = x_{n+1,0}$. The number of runs in $x_{n+1,0}$ is one more than the number of runs in $x_{n,1}$. The number of runs in such words $x_{n+1,0}$ is $r_{n,1} + a_{n,1}$.

Step 2A. Appending 01 to $x_{n-1,0}$ results in $x_{n+1,1}$; it ends with a new run of a single 1: $x_{n-1,0}\underbrace{01} = x_{n+1,1}$. The number of runs resulting from such words is $r_{n-1,0} + a_{n-1,0}$.

Step 2B. Appending 01 to $x_{n-1,1}$ results in $x_{n+1,1}$: $x_{n-1,1}\underbrace{01} = x_{n+1,1}$. Such words contribute $r_{n-1,1} + 2a_{n-1,1}$ to the total runs r_{n+1}.

Step 3A. Appending 11 to $x_{n-1,0}$ results in $x_{n+1,1}$: $x_{n-1,0}\underbrace{11} = x_{n+1,1}$. Such words produce an additional $r_{n-1,0} + a_{n-1,0}$ runs.

Step 3B. Appending 11 to $x_{n-1,1}$ creates no new runs in $x_{n+1,1}$: $x_{n-1,1}\underbrace{11} = x_{n+1,1}$. The resulting words contribute $r_{n-1,1}$ runs to the total r_{n+1}.

Since $a_{n,0} = a_{n-1} = J_n$ and $a_{n,1} = 2a_{n-2} = 2J_{n-1}$, combining these steps yields

$$\begin{aligned}
r_{n+1} &= r_{n,0} + (r_{n,1} + a_{n,1}) + (r_{n-1,0} + a_{n-1,0}) + (r_{n-1,1} + 2a_{n-1,1}) \\
&\quad + (r_{n-1,0} + a_{n-1,0}) + r_{n-1,1} \\
&= (r_{n,0} + r_{n,1}) + 2(r_{n-1,0} + r_{n-1,1}) + a_{n,1} + 2a_{n-1,1} + 2a_{n-1,0} \\
&= r_n + 2r_{n-1} + 2J_{n-1} + 4J_{n-2} + 2J_{n-1} \\
&= r_n + 2r_{n-1} + 4(J_{n-1} + J_{n-2}) \\
&= r_n + 2r_{n-1} + 2^n.
\end{aligned}$$

Solving this recurrence, we get

$$r_n = \frac{2n+3}{6}J_{n+1} + \frac{2n-1}{6}J_n, \tag{44.11}$$

where $n \geq 1$; see Exercise 44.107.

For example, $r_4 = \frac{2 \cdot 4 + 3}{6}J_5 + \frac{2 \cdot 4 - 1}{6}J_4 = \frac{11 \cdot 11}{6} + \frac{7 \cdot 5}{6} = 26$; see Table 44.17.

Next we determine the numbers of runs of 0s and 1s.

Runs of 0s and 1s

Let rz_n denote the number of runs of 0s among the a_n words of length n, and rw_n that of 1s. It then follows by the constructive algorithm that

$$\begin{aligned}
rz_n &= (rz_{n-1} + a_{n-1,1}) + (rz_{n-2} + a_{n-2,1}) + rz_{n-2} \\
&= rz_{n-1} + 2rz_{n-2} + 2(J_{n-2} + J_{n-3}) \\
&= rz_{n-1} + 2rz_{n-2} + 2^{n-2},
\end{aligned}$$

where $rz_1 = 1$ and $rz_2 = 2$.

Solving this recurrence, we get

$$rz_n = \frac{n}{6}J_{n+1} + \frac{n+4}{6}J_n, \tag{44.12}$$

where $n \geq 1$; see Exercise 44.108.

For example, $rz_4 = \frac{4}{6}J_5 + \frac{8}{6}J_4 = \frac{2 \cdot 11}{3} + \frac{4 \cdot 5}{3} = 14$; see Table 44.16.

Now we can find rw_n. Since $rz_n + rw_n = r_n$, it follows by equations (44.11) and (44.12) that

$$rw_n = \frac{n+3}{6}J_{n+1} + \frac{n-5}{6}J_n$$

where $n \geq 1$.

For example, $rw_4 = \frac{7}{6}J_5 - \frac{1}{6}J_4 = \frac{7 \cdot 11}{3} - \frac{5}{6} = 12$; see Table 44.16.

Finally, suppose we treat the $a_n = J_{n+1}$ words x_n as binary numbers. We can then compute the decimal value of those binary numbers.

Sum of the a_n Binary Numbers

Let S_n denote the decimal value of the sum of the a_n binary numbers. For example, $S_1 = 0$, $S_2 = 4$, $S_3 = 12$, and $S_4 = 68$; see Table 44.16. Using the constructive algorithm, we now find a recurrence and then an explicit formula for S_n.

Step 1. Appending 0 at the end of x_{n-1} shifts each bit one position to the left. Its effect is multiplying the decimal value of x_{n-1} by 2. The sum of the decimal value of such numbers is $2S_{n-1}$.

Step 2. Appending 01 to the right of x_{n-2} shifts each bit two positions to the left, and then adds a 1. The sum of the values of such numbers is $4S_{n-2} + a_{n-2}$.

Step 3. Appending 11 to the right of x_{n-2} shifts each bit two positions to the left, and then adds a 3. The sum of the values of such numbers is $4S_{n-2} + 3a_{n-2}$.

Thus

$$\begin{aligned} S_n &= 2S_{n-1} + (4S_{n-2} + a_{n-2}) + (4S_{n-2} + 3a_{n-2}) \\ &= 2S_{n-1} + 8S_{n-2} + 4J_{n-1} \\ &= 2S_{n-1} + 8S_{n-2} + \left(\frac{4}{3}\right)2^{n-1} - \left(\frac{4}{3}\right)(-1)^{n-1}. \end{aligned}$$

Solving this recurrence (see Exercise 44.109), we get

$$\begin{aligned} S_n &= \left(\frac{4}{15}\right)4^n + \left(\frac{1}{3}\right)(-2)^n - \left(\frac{1}{3}\right)2^n - \left(\frac{4}{15}\right)(-1)^n \\ &= \left(\frac{4}{15}\right)(J_{2n} + J_{2n+1}) + \left(\frac{1}{3}\right)(J_{n+1}^2 - J_{n+2}J_n) - \left(\frac{1}{3}\right)(J_n + J_{n+1}) - \left(\frac{4}{15}\right)(J_{n+1} - 2J_n) \\ &= \left(\frac{4}{15}\right)(J_{2n} + J_{2n+1}) + \left(\frac{1}{3}\right)(J_{n+1}^2 - J_{n+2}J_n) - \left(\frac{1}{5}\right)(3J_{n+1} - J_n). \end{aligned}$$

For example,

$$\begin{aligned} S_4 &= \left(\frac{4}{15}\right)(J_8 + J_9) + \left(\frac{1}{3}\right)(J_5^2 - J_6J_4) - \left(\frac{1}{5}\right)(3J_5 - J_4) \\ &= \left(\frac{4}{15}\right)(85 + 171) + \left(\frac{1}{3}\right)(11^2 - 21 \cdot 5) - \left(\frac{1}{5}\right)(3 \cdot 11 - 5) = 68, \end{aligned}$$

as expected.

A Ternary Version

Next we pursue a ternary version of the binary case, but with some added restrictions. This appeared in the final round of the 1987 Austrian Olympiad [117, 230]. It is interesting in its own right and has fascinating implications [300].

Let $\Sigma = \{0, 1, 2\}$. The digits 0, 1, and 2 are *ternary digits*. (In the Austrian Olympiad, $\Sigma = \{a, b, c\}$.) Let b_n denote the number of *ternary words* $w_n = x_1x_2 \ldots x_n$ of length n such that $x_1 = 0 = x_n$ and $x_i \neq x_{i+1}$, where $x_i \in \Sigma$ and $1 \leq i \leq n-1$. Clearly, the *reverse* w_n^R of an acceptable word $w_n = 0x_2 \ldots x_{n-1}0$ is also acceptable.

Table 44.18 lists words w_n and the corresponding numbers b_n, where $1 \leq n \leq 6$. Notice that there are *no* ternary words of length 2 that satisfy the given conditions. Although the counts b_n do not seem to follow a pattern, the following theorem establishes a simple formula for b_n using a constructive algorithm.

TABLE 44.18. Ternary Words and Their Counts

n	Ternary Words w_n	b_n
1	0	1
2	.	0
3	010, 020	2
4	0120, 0210	2
5	01210, 02120	6
	01010, 02010	
	01020, 02020	
6	010120, 010210, 020120, 020210, 012120, 021210	10
	012010, 021010	
	012020, 021020	

Theorem 44.6. *Let b_n denote the number of ternary words $w_n = x_1x_2 \ldots x_n$ of length n such that $x_1 = 0 = x_n$ and $x_i \neq x_{i+1}$, where $1 \leq i \leq n-1$. Then $b_n = 2J_{n-2}$, where $n \geq 1$.*

Proof. Since $J_{-1} = 1/2$, $b_1 = 2J_{-1}$, and $b_2 = 2J_0$, $b_3 = 2J_1$, and $b_4 = 2J_2$. Let w_n be an arbitrary ternary word of length $n \geq 4$. We now employ an algorithm to construct words of length n from those of lengths $n-1$ and $n-2$.

Step 1. Replace the last digit $x_{n-1} = 0$ in w_{n-1} with 10 if $x_{n-2} = 2$; otherwise, replace it with 20.

Step 2A. Append 10 at the end of each w_{n-2}.

Step 2B. Append 20 at the end of each w_{n-2}.

Since the algorithm is reversible, it produces all desired ternary words w_n.

Step 1 yields b_{n-1} words w_n. Steps 2A and 2B produce b_{n-2} words each. Thus $b_n = b_{n-1} + 2b_{n-2}$. This recurrence, paired with the initial conditions, gives the desired result. ■

We now illustrate the steps in the proof for the case $n = 6$.

Step 1. There are three words $w_5 = 0x_2x_3x_40$ with $x_4 = 2$; replace each $x_5 = 0$ with 10. The three remaining words have $x_4 = 1$; replace each x_5 with 20:

02120	01020	02020	01210	01010	02010
↓	↓	↓	↓	↓	↓
0212**10**	0102**10**	0202**10**	0121**20**	0101**20**	0201**20**.

Step 2A. Append 10 at the end of each w_4:

0120	0210
↓	↓
0120**10**	0210**10**.

Step 2B. Append 20 at the end of each w_4:

0120	0210
↓	↓
0120**20**	0210**20**.

Clearly, these steps produce the $b_6 = 10$ ternary words.

The proof of Theorem 44.6 will be helpful in our later discussions. But we can establish the theorem using a simpler and shorter alternate technique, based on the one [114] by C. Cooper of Central Missouri State University, Warrensburg, Missouri.

Alternate Proof of Theorem 44.6. Let c_n denote the number of ternary words of length n with $x_1 = 0$ and $x_n = 1$; and d_n the number of such words with $x_1 = 0$ and $x_n = 2$. Then $c_1 = 0 = d_1$, and $c_2 = 1 = d_2$.

Clearly, $b_n = c_{n-1} + d_{n-1}, c_n = b_{n-1} + d_{n-1}$, and $d_n = b_{n-1} + c_{n-1}$, where $n \geq 2$. These equations yield the recurrence $b_{n+1} = b_n + 2b_{n-1}$. Employing the initial conditions, we get $b_n = 2J_{n-2}$, where $n \geq 1$, as expected. ∎

The following result is an immediate consequence of the constructive algorithm.

Corollary 44.1. *There are exactly* $\frac{1}{2}b_n = J_{n-2}$ *ternary words* w_n *that begin with 01 (or end in 10), where* $n \geq 2$. ∎

Consequently, there are J_{n-2} ternary words w_n that begin with (or end in) 02, where $n \geq 2$.

We now have the needed machinery to develop an explicit formula for the number of 0s among the b_n ternary words of length n.

Zeros Among the b_n Ternary Numbers

Let z_n denote the number of 0s among the b_n ternary words w_n of length n. For example, $z_1 = 1, z_2 = 0, z_3 = 4 = z_4, z_5 = 16$, and $z_6 = $ (28); see Table 44.18.

Using the above constructive algorithm, we can easily develop a recurrence for z_n:

$$z_n = z_{n-1} + 2z_{n-2} + \frac{4}{3}\left[2^{n-4} - (-1)^{n-4}\right], \tag{44.13}$$

where $z_1 = 1, z_2 = 0$, and $n \geq 3$; see Exercise 44.122.

For example, $z_6 = z_5 + 2z_4 + 4J_2 = 16 + 2 \cdot 4 + 4 = \textcircled{28}$.

Using the power and techniques of solving recurrences, we can now find an explicit formula for z_n:

$$z_n = \left(\frac{17n+8}{18}\right) J_n - \left(\frac{7n}{18}\right) J_{n+1}, \tag{44.14}$$

where $n \geq 1$; see Exercise 44.123.

For example, $z_{10} = \dfrac{178 \cdot 341}{18} - \dfrac{70 \cdot 683}{18} = 716.$

Nonzero Digits Among the b_n Ternary Numbers

It follows from formula (44.14) that the number of nonzero digits $nonz_n$ among the b_n ternary numbers w_n is given by

$$\begin{aligned} nonz_n &= nb_n - z_n \\ &= n(2J_{n-2}) - \left[\left(\frac{17n+8}{18}\right) J_n - \left(\frac{7n}{18}\right) J_{n+1}\right] \\ &= \left(\frac{7n}{18}\right) J_{n+1} - \left(\frac{17n+8}{18}\right) J_n + 2nJ_{n-2}. \end{aligned} \tag{44.15}$$

For example,

$$nonz_6 = \left(\frac{7 \cdot 6}{18}\right) J_7 - \left(\frac{17 \cdot 6 + 8}{18}\right) J_6 + 12J_4 = \frac{42 \cdot 43}{18} - \frac{110 \cdot 21}{18} + 12 \cdot 5 = 32;$$

see Table 44.18.

Next we compute the cumulative sum of the decimal values of the b_n ternary words when considered as numbers. We will accomplish this using recursion and the constructive algorithm.

Cumulative Sum of the b_n Ternary Numbers

Let S_n denote the cumulative sum of the decimal values of the b_n ternary numbers. It follows from Table 44.18 that $S_1 = 0 = S_2$, $S_3 = 9$, $S_4 = \textcircled{36}$, and $S_5 = \boxed{297}$. Let $w_k = 0x_2x_3 \dots x_{k-1}0$ be an arbitrary ternary number with k digits.

Step 1. Replacing x_n with 10 or 20 shifts $0x_2 \dots x_{n-1}$ by two places to the left. Since there are b_{n-1} ternary numbers with $n-1$ digits, this step contributes

$3S_{n-1} + 2 \cdot 3\left(\frac{1}{2}b_{n-1}\right) + 1 \cdot 3\left(\frac{1}{2}b_{n-1}\right) = 3S_{n-1} + \left(\frac{9}{2}\right)b_{n-1} = 3S_{n-1} + 9J_{n-3}$ to the sum S_n.

Step 2A. Appending 10 at the end of $0x_2 \dots x_{n-3}0$ shifts it by two positions to the left. The contribution resulting from this operation is $3^2S_{n-2} + 1 \cdot 3b_{n-2} = 9S_{n-2} + 6J_{n-4}$.

Step 2B. Appending 20 at the end of $0x_2 \dots x_{n-3}0$ contributes $3^2S_{n-2} + 2 \cdot 3b_{n-2} = 9S_{n-2} + 12J_{n-4}$ to the grand total.

Combining these steps, we get

$$\begin{aligned} S_n &= (3S_{n-1} + 9J_{n-3}) + (9S_{n-2} + 6J_{n-4}) + (9S_{n-2} + 12J_{n-4}) \\ &= 3S_{n-1} + 18S_{n-2} + 9J_{n-3} + 18J_{n-4} \\ &= 3S_{n-1} + 18S_{n-2} + 9J_{n-2}, \end{aligned} \tag{44.16}$$

where $n \geq 4$.

For example, $S_5 = 3S_4 + 18S_3 + 9J_3 = 3 \cdot 36 + 18 \cdot 9 + 9 \cdot 3 = \boxed{297}$.

An Explicit Formula for S_n

It follows from recurrence (44.16) that

$$S_n = 3S_{n-1} + 18S_{n-2} + 3 \cdot 2^{n-2} - 3(-1)^n. \tag{44.17}$$

The particular solution of this recurrence corresponding to the nonhomogeneous part $3 \cdot 2^{n-2}$ has the form $C \cdot 2^n$. Substituting this in the recurrence $S_n = 3S_{n-1} + 18S_{n-2} + 3 \cdot 2^{n-2}$ yields $C = -3/20$. The particular solution of this recurrence corresponding to the nonhomogeneous part $-3(-1)^n$ has the form $D(-1)^n$. Substituting this in the recurrence $S_n = 3S_{n-1} + 18S_{n-2} - 3(-1)^n$ yields $D = 3/14$. Thus

$$S_n = 3S_{n-1} + 18S_{n-2} - \left(\frac{3}{20}\right)2^n + \left(\frac{3}{14}\right)(-1)^n.$$

The roots of the characteristic equation of the homogeneous recurrence $S_n = 3S_{n-1} + 18S_{n-2}$ are -3 and 6. So the general solution of recurrence (44.17) is of the form

$$S_n = A(-3)^n + B \cdot 6^n - \left(\frac{3}{20}\right)2^n + \left(\frac{3}{14}\right)(-1)^n.$$

The initial conditions $S_1 = 0 = S_2$ then yields $A = -\dfrac{14}{140}$ and $B = \dfrac{5}{140}$. Thus

$$S_n = -\left(\frac{14}{140}\right)(-3)^n + \left(\frac{5}{140}\right)6^n - \left(\frac{3}{20}\right)2^n + \left(\frac{3}{14}\right)(-1)^n, \tag{44.18}$$

where $n \geq 1$.

For example, $S_5 = \dfrac{14 \cdot 3^5 + 5 \cdot 6^5}{140} - \dfrac{48 \cdot 14 + 3 \cdot 10}{140} = \boxed{297}$, as expected.

Since $J_n + J_{n+1} = 2^n$ and $J_{n+1} - 2J_n = (-1)^n$, formula (44.18) can be rewritten in terms of Jacobsthal numbers. To this end, for convenience, we let $a = -14/140$, $b = 5/140$, $c = -3/20$, and $d = 3/14$. Then

$$\begin{aligned} S_n &= (3^n b + c)(J_n + J_{n+1}) + (3^n a + d)(J_{n+1} - 2J_n) \\ &= [3^n(a+b) + c + d]J_{n+1} + [3^n(b-2a) + c - 2d]J_n \\ &= \frac{1}{140}\left(-3^{n+2} + 9\right) J_{n+1} - \frac{1}{140}\left[3^n(-33) + 81\right] J_n \\ &= \frac{1}{140}\left(11 \cdot 3^{n+1} - 81\right) J_n - \frac{1}{140}\left(3^{n+2} - 9\right) J_{n+1}. \end{aligned} \tag{44.19}$$

For example,

$$S_4 = \frac{1}{140}\left(11 \cdot 3^5 - 81\right) \cdot 5 - \frac{1}{140}\left(3^6 - 9\right) \cdot 11 = \frac{12{,}960 - 7{,}920}{140} = \textcircled{36},$$

again as expected.

Plus Signs in the Ternary Sums

Suppose we insert a plus sign between every two adjacent digits in the ternary number w_n. Let p_n denote the number of plus signs in the b_n ternary sums. Then

$$\begin{aligned} p_n &= (n-1)b_n \\ &= (n-1)(2J_{n-2}) \\ &= (n-1)(J_n - J_{n-1}) \\ &= (n-1)\left[J_n - \frac{1}{2}(J_{n+1} - J_n)\right] \\ &= \left(\frac{3n-3}{2}\right) J_n - \left(\frac{n-1}{2}\right) J_{n+1}. \end{aligned}$$

For example, $p_6 = \left(\frac{15}{2}\right) J_6 - \left(\frac{5}{2}\right) J_7 = \frac{15 \cdot 21}{2} - \frac{5 \cdot 43}{2} = \boxed{50}$.

Inversions

Next we investigate the number of inversions in words over Σ. To begin with, let $\{a_1, a_2, \ldots, a_n\}$ be a *totally ordered* alphabet with $a_1 < a_2 < \cdots < a_n$. Let $x_1 x_2 \ldots x_k$ be a word of length k over this alphabet. For $1 \le i < j \le k$, call the pair x_i and x_j an *inversion* if $x_i > x_j$.

For example, let $\Sigma = \{0, 1, 2\}$, where $0 < 1 < 2$. Then the word $x_1 x_2 x_3 x_4 x_5 = 01210$ contains four inversions: $x_2 > x_5$, $x_3 > x_4$, $x_3 > x_5$, and $x_4 > x_5$.

Let inv_n count the number of inversions among the b_n ternary words of length n, where $n \geq 3$. Then $inv_3 = 2$, $inv_4 = 5$, $inv_5 = 21$, $inv_6 = 56$, and $inv_7 = 164$. We will now establish that inv_n satisfies the recurrence

$$inv_n = inv_{n-1} + b_{n-1} + \frac{1}{2}\left(nonz_{n-1} - nonz_{n-2}\right)$$
$$+ 2inv_{n-2} + 2b_{n-2} + 2nonz_{n-2} + \frac{1}{2}nonz_{n-2},$$

where $n \geq 3$.

1) From Step 1 of the algorithm, when the last digit x_{n-1} is replaced (by either 10 or 20), a new inversion arises. There are b_{n-1} such inversions.
2) Also from Step 1 of the algorithm, when the last digit x_{n-1} is replaced by 10 (for when $x_{n-2} = 2$), there is a new inversion for each 2 that occurs among the first $n - 2$ digits of the b_{n-1} words that end in 20. There are $\frac{1}{2}\left(nonz_{n-1} - nonz_{n-2}\right)$ such inversions.
3) Steps 2A and 2B of the algorithm each provide b_{n-2} new inversions: 10 in positions $n - 1$ and n for Step 2A; 20 in positions $n - 1$ and n for Step 2B.
4) Each of Steps 2A and 2B of the algorithm provides $nonz_{n-2}$ new inversions with the new 0 now in position n.
5) Finally, from Step 2A, we get $\frac{1}{2}nonz_{n-2}$ new inversions for each of the $\frac{1}{2}nonz_{n-2}$ 2s that occur among the first $n - 2$ positions of the b_n words that end in 10. Each such 2 provides an inversion with the new 1 in position $n - 1$.

Combining these five steps, we get

$$\begin{aligned}
inv_n &= inv_{n-1} + 2inv_{n-2} + b_{n-1} + 2b_{n-2} + \frac{1}{2}nonz_{n-1} + 2nonz_{n-2} \\
&= inv_{n-1} + 2inv_{n-2} + 2J_{n-3} + 4J_{n-4} \\
&\quad + \frac{1}{2}\left\{(n-1)(2J_{n-3}) - \left[\left(\frac{17(n-1)+8}{18}\right)J_{n-1} - \left(\frac{7(n-1)}{18}\right)J_n\right]\right\} \\
&\quad + 2\left\{(n-2)(2J_{n-4}) - \left[\left(\frac{17(n-2)+8}{18}\right)J_{n-2} - \left(\frac{7(n-2)}{18}\right)J_{n-1}\right]\right\} \\
&= inv_{n-1} + 2inv_{n-2} + \frac{1}{3}n(-1)^{n+1} + \frac{1}{12}n(2^n) + \frac{1}{3}(-1)^n - \frac{1}{12}(2^n).
\end{aligned}$$

Consequently, the general solution of the recurrence is of the form

$$inv_n = A(2^n) + B(-1)^n + Cn2^n + Dn^2 2^n + En(-1)^n + Fn^2(-1)^n.$$

Next we determine the coefficients for the particular part of the solution.

1) To find C and D, substitute $inv_n = Cn2^n + Dn^2 2^n$ in the recurrence $inv_n = inv_{n-1} + 2inv_{n-2} + \frac{1}{12}n(2^n) - \frac{1}{12}(2^n)$. After some basic algebra, this gives

$$0 = -\frac{3}{2}C(2^n) - 3Dn(2^n) + \frac{5}{2}D(2^n) + \frac{1}{12}n(2^n) - \frac{1}{12}(2^n).$$

Comparing the coefficients of 2^n and $n2^n$, we get $0 = -\frac{3}{2}C + \frac{5}{2}D - \frac{1}{12}$ and $0 = -3D + \frac{1}{12}$, so $D = \frac{1}{36} = -3C$.

2) To find E and F, substitute $inv_n = En(-1)^n + Fn^2(-1)^n$ in the recurrence $inv_n = inv_{n-1} + 2inv_{n-2} + \frac{1}{3}n(-1)^{n+1} + \frac{1}{3}(-1)^n$. After some simplification, this yields

$$0 = -3E(-1)^n - 6Fn(-1)^n + 7F(-1)^n - \frac{1}{3}n(-1)^n + \frac{1}{3}(-1)^n.$$

Comparing the coefficients of $(-1)^n$ and $n(-1)^n$, we get $0 = -3E + 7F + \frac{1}{3}$ and $0 = -6F - \frac{1}{3}$, so $F = -\frac{1}{18} = 3E$.

Consequently,

$$\begin{aligned} inv_n = A(2^n) + B(-1)^n + \left(-\frac{1}{108}\right)n2^n + \left(\frac{1}{36}\right)n^2 2^n \\ + \left(-\frac{1}{54}\right)n(-1)^n + \left(-\frac{1}{18}\right)n^2(-1)^n. \end{aligned}$$

3) The initial conditions $inv_3 = 2$ and $inv_4 = 5$ yield $A = -\frac{1}{27} = -B$. Thus

$$\begin{aligned} inv_n &= -\frac{1}{27}(2^n) + \frac{1}{27}(-1)^n + \left(-\frac{1}{108}\right)n2^n + \left(\frac{1}{36}\right)n^2 2^n \\ &\quad + \left(-\frac{1}{54}\right)n(-1)^n + \left(-\frac{1}{18}\right)n^2(-1)^n \\ &= -\frac{1}{9}J_n + \frac{n}{36}\left(J_n - J_{n+1}\right) + \frac{n^2}{36}\left(5J_n - J_{n+1}\right) \\ &= \left(\frac{5n^2 + n - 4}{36}\right)J_n - \left(\frac{n^2 + n}{36}\right)J_{n+1}, \end{aligned}$$

where $n \geq 3$.

44.6 A USA OLYMPIAD DELIGHT

The following occurrence of Jacobsthal numbers appears in a fascinating problem. It was proposed by Sam Vandervelde of St. Lawrence University,

Canton, New York, for the 2013 USA Mathematical Olympiad. He presented a beautiful solution to it [495]. Interestingly, two solutions by Kiran S. Kedlaya of the University of California, San Diego, can be found in [495]; both involve matrices.

Example 44.5. *Select n equally spaced points on a circle, where $n \geq 2$. Place a marker A at one of them. Let a_n denote the number of moves needed to advance the marker around the circle exactly twice in the clockwise direction in steps of length 1 or 2; no duplicate moves are allowed. Find a_n.*

Solution. To get an idea about a_n, consider the cases $2 \leq n \leq 4$. (The case $n = 1$ involves duplicate moves.) When $n = 2$, there are exactly three possible moves; see the tree in Figure 44.1, where a (slanted or vertical) bar on an edge indicates a step of length 2; so $a_2 = 3$.

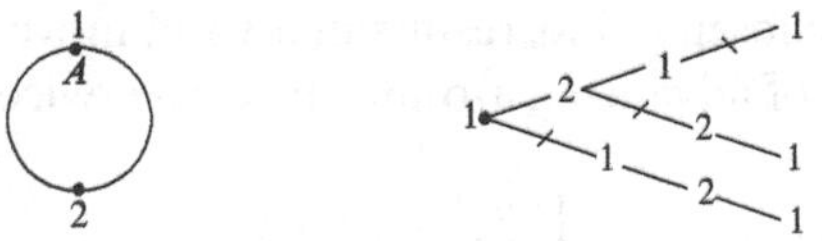

Figure 44.1.

When $n = 3$, there are five possible moves; see Figure 44.2; so $a_3 = 5$.

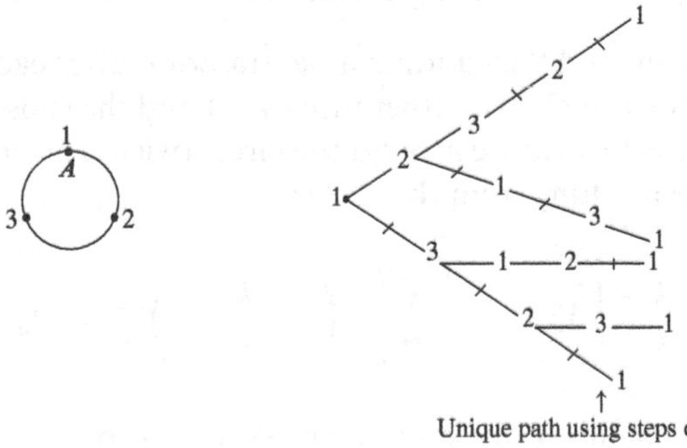

Figure 44.2.

When $n = 4$, there are exactly $11 = a_4$ possible paths; see Figure 44.3 and Table 44.19. The paths are listed for convenience, where $i|j$ indicates a step of length 2 from point i to point j.

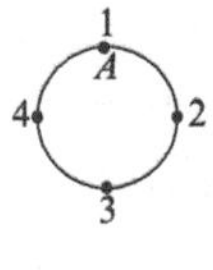

Figure 44.3.

TABLE 44.19. Possible Paths

12341|3|1	12|4|2341	1|34|2|41
1234|2|41	12|4|23|1	1|3|12341
123|1|341	1|34123|1	1|3|12|41
12|41|3|1	1|34|23|1	

Clearly, a pattern slowly emerges: $a_2 = J_3$, $a_3 = J_4$, and $a_4 = J_5$. More generally, we conjecture that $a_n = J_{n+1}$, where $n \geq 2$. We will now confirm it.

Proof. This hinges on the property $\sum_{k=0}^{\lfloor n/2 \rfloor} \binom{n-k}{k} 2^k = J_{n+1}$ from Section 44.1.

To begin with, notice from Figures 44.11–44.13 that the points visited by the marker in its two trips around the circle may not be adjacent. Also, there are two potential moves from any such point to the next, except perhaps at the endpoints.

Suppose we select k of the n points on the circle, where $k \geq 1$. Assume that *no* two of them are adjacent and they do *not* include point A. Although there are 2^k potential ways to advance around the circle, exactly one-half of them involve duplicate moves. So there are only 2^{k-1} possibilities. The k nonadjacent points on the circle, not including A, can be selected in $\binom{n-k}{k}$ ways; adding an extra point behind each of the k chosen points results in a valid move. Correspondingly, the total number of ways of advancing around the circle twice is

$$\sum_{k=1}^{\lfloor n/2 \rfloor} \binom{n-k}{k} 2^{k-1} = \frac{1}{2}\left[\sum_{k=0}^{\lfloor n/2 \rfloor} \binom{n-k}{k} 2^k - 1\right] = \frac{1}{2}(J_{n+1} - 1).$$

On the other hand, suppose the k nonadjacent points *do* include point A. Then there are $\binom{n-k-1}{k-1}$ ways of selecting them; this can be achieved by choosing A, but *not* the next point, and then adding an extra point after each of the $k-1$ selected points. Since we can choose either move at A and the subsequent points, there are exactly 2^k ways to advance around the circle twice. The number of ways of traversing the circle resulting from this case is

$$\sum_{k=1}^{\lfloor n/2 \rfloor} \binom{n-k-1}{k} 2^k = 2 \sum_{k=0}^{\lfloor (n-2)/2 \rfloor} \binom{n-k-2}{k} 2^k = 2J_{n-1}.$$

Now we need to account for one more case. When n is odd, there is a *unique* way of traversing the circle, using steps of length 2; then *no* point is visited twice; see Figures 44.2 and 44.4. The contribution resulting from this case is

$$\begin{cases} 1 & \text{if } n \text{ is odd} \\ 0 & \text{otherwise} \end{cases} = \frac{1}{2}\left[1 - (-1)^n\right].$$

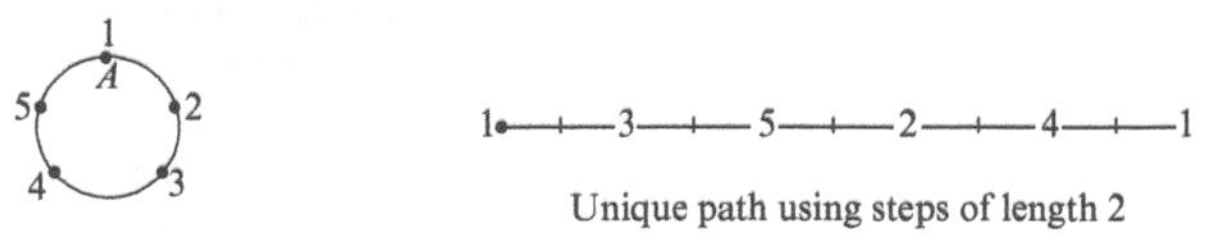

Figure 44.4.

Combining the three cases, we get

$$\begin{aligned} a_n &= \frac{1}{2}(J_{n+1} - 1) + 2J_{n-1} + \frac{1}{2}\left[1 - (-1)^n\right] \\ &= \frac{1}{2}J_{n+1} + \frac{1}{2}\left[4J_{n-1} - (-1)^n\right] \\ &= \frac{1}{2}J_{n+1} + \frac{1}{2}J_{n+1} = J_{n+1}, \end{aligned}$$

as expected. ■

In particular, let $n = 5$. Then there are $J_6 = 21$ ways of advancing the marker around the circle twice. It would be a good exercise to list them all.

Next we present a stunning occurrence of Jacobsthal numbers in an unexpected setting.

44.7 A STORY OF 1, 2, 7, 42, 429, ...

It is unlikely that many mathematicians will have encountered the sequence $1, 2, 7, 42, 429, 7436, 218348, \ldots$. Its terms may not appear to follow a pattern, yet they do. To our great surprise, the sequence is directly related to Jacobsthal numbers J_m, as we will see shortly.

To narrate the story of this rich but hidden relationship, we first introduce a very special class of square matrices, discovered in the early 1980s by David P. Robbins and Howard Rumsey Jr. of the Institute for Defense Analyses, Princeton, New Jersey.

Alternating Sign Matrices

An *alternating sign matrix* (ASM) is an $n \times n$ matrix of 0s, 1s, and -1s such that:

- the sum of the entries in each row and column is 1; and
- the signs of the nonzero entries in each row and column alternate.

For example, there are seven 3×3 ASMs; see Table 44.20.

TABLE 44.20. The Seven 3 × 3 ASMs

$\begin{bmatrix} \textcircled{1} & 0 & 0 \\ 0 & 1 & 0 \\ 0 & 0 & 1 \end{bmatrix}$	$\begin{bmatrix} \textcircled{1} & 0 & 0 \\ 0 & 0 & 1 \\ 0 & 1 & 0 \end{bmatrix}$	$\begin{bmatrix} 0 & \boxed{1} & 0 \\ 1 & 0 & 0 \\ 0 & 0 & 1 \end{bmatrix}$	$\begin{bmatrix} 0 & \boxed{1} & 0 \\ 0 & 0 & 1 \\ 1 & 0 & 0 \end{bmatrix}$	$\begin{bmatrix} 0 & \boxed{1} & 0 \\ 1 & -1 & 1 \\ 0 & 1 & 0 \end{bmatrix}$
$\begin{bmatrix} 0 & 0 & \boxed{\boxed{1}} \\ 1 & 0 & 0 \\ 0 & 1 & 0 \end{bmatrix}$	$\begin{bmatrix} 0 & 0 & \boxed{\boxed{1}} \\ 0 & 1 & 0 \\ 1 & 0 & 0 \end{bmatrix}$			

Intellectual curiosity led William H. Mills and his colleagues, Robbins and Rumsey, to investigate the number of $n \times n$ ASMs $A(n)$. Their extensive computer search yielded the sequence $1, 2, 7, 42, 429, 7436, 218348, \ldots$. It is astonishing that the same sequence appeared earlier in the study of an unrelated problem in the theory of plane partitions by George E. Andrews of the Pennsylvania State University [48].

The relentless pursuit of an explicit formula for $A(n)$ by Robbins, Rumsey, and Mills led to the development of an interesting triangular array T; see Figure 44.5.

George E. Andrews was born in Oregon in 1938, and received his B.S. and M.A. simultaneously from Oregon State University in 1960. After a year as a Fulbright Scholar at the University of Cambridge, England, he received his Ph.D. in 1964 from the University of Pennsylvania. He joined the faculty at Pennsylvania State University, becoming full professor at 32, the same age at which the Indian mathematical genius Srinivasa Ramanujan died. In 1981, Andrews became the Evan Pugh Professor of Mathematics for his excellence in research and teaching.

Andrews is considered the world's leading authority on the theory of partitions and has made significant contributions to number theory, combinatorics, and theoretical physics. An expert on q-series, he is the author of *q-series: Their Development and Application in Analysis, Number Theory, Combinatorics, Physics and Computer Algebra*, published by The American Mathematical Society (AMS) in 1986. An author of more than 300 articles, he served as thesis advisor to more than 40 Ph.D. and master's degree recipients.

Andrews has a long-term interest in the work of Ramanujan, whose last notebook he unearthed in 1976. He is currently working with Bruce Bernt of the University of Illinois, Urbana-Champaign, to explicate the brilliant and sometimes enigmatic ideas in this notebook.

Andrews was elected to the American Academy of Arts and Sciences in 1997, and the National Academy of Sciences in 2003. He was awarded an honorary professorship at Nankai University in 2008, and became a Society for Industrial and Applied Mathematics Fellow the following year. He received honorary degrees from the Universities of Parma (Italy), Florida, Waterloo (Canada), Illinois, and SASTRA University (India). He was the President of the AMS during 2009–2011, becoming a Fellow of the AMS in 2012.

$$
\begin{array}{ccccccccccccc}
 & & & & & & 1 & & & & & & \\
 & & & & & 1 & & 1 & & & & & \\
 & & & & \boxed{2} & & \boxed{3} & & \boxed{2} & & & & \\
 & & & 7 & & 14 & & 14 & & 7 & & & \\
 & & 42 & & 105 & & 135 & & 105 & & 42 & & \\
 & \boxed{429} & & 1287 & & 2002 & & 2002 & & 1287 & & 429 & \\
7436 & & 26026 & & 47320 & & 56784 & & 47320 & & 26026 & & 7436
\end{array}
$$

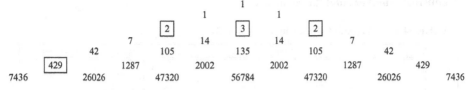

Figure 44.5. Number of $n \times n$ ASMs: triangular array T.

The element $t_{n,k}$ of array T gives the number of $n \times n$ ASMs with a 1 at the top of column k, where we have defined $t_{1,1} = 1$.

For example, $t_{3,1} = \boxed{2}$, $t_{3,2} = \boxed{3}$, and $t_{3,3} = \boxed{2}$; see Figure 44.5.

Like Pascal's triangle, this array is also symmetric about the vertical line through the middle: $t_{n,k} = t_{n,n-k+1}$. This is so, since the vertical reflection of an ASM is again an ASM.

In addition, array T satisfies two interesting properties:

- the sum of the elements in row n is $A(n)$; and
- the leading entry in row n is $A(n-1)$.

For example, the sum of the elements in row 4 is 42, the same as the leading element in row 5.

The ASM Conjecture

Using these two properties, Robbins, Rumsey, and Williams conjectured an explicit formula for A_n in terms of factorials:

$$
\begin{aligned}
A_n &= \prod_{i=0}^{n-1} \frac{(3i+1)!}{(n+i)!} \\
&= \frac{1!4!7!\cdots(3n-2)!}{n!(n+1)!\cdots(2n-1)!}.
\end{aligned}
$$

In 1995, three distinct proofs confirmed the validity of this conjecture; the first by D. Zeilberger of Rutgers University, New Brunswick, New Jersey; G. Kuperberg of the University of California, Berkeley; and again by Zeilberger.

For example,

$$
\begin{aligned}
A_5 &= \prod_{i=0}^{4} \frac{(3i+1)!}{(5+i)!} = \frac{1!4!7!10!13!}{5!6!7!8!9!} \\
&= \boxed{429};
\end{aligned}
$$

see Figure 44.5.

We now pursue the close bond between the sequence and Jacobsthal numbers [176].

Jacobsthal Numbers and the Sequence

TABLE 44.21. Values of $A(n)$, where $1 \leq n \leq 21$

n	$A(n)$	Least (prime) factor of $A(n)$
(1)	1	[1]
2	2	2
(3)	7	[7]
4	42	2
(5)	429	[3]
6	7,436	2
7	218,348	2
8	10,850,216	2
9	911,835,460	2
10	129,534,272,700	2
(11)	31,095,744,852,375	[3]
12	12,611,311,859,677,500	2
13	8,639,383,518,297,652,500	2
14	9,995,541,355,448,167,482,000	2
15	19,529,076,234,661,277,104,897,200	2
16	64,427,185,703,425,689,356,896,743,840	2
17	358,869,201,916,137,601,447,486,156,417,296	2
18	33,748,606,392,587,505,622,695,144,915,229,925,456	2
19	53,580,350,833,984,348,888,878,646,149,709,092,313,244	2
20	14,360,389,347,155,382,009,131,556,826,370,512,043,768,827,212	2
(21)	64,971,294,999,808,427,895,847,904,380,524,143,538,858,551,437,757	[7]

Table 44.21 shows the values of $A(n)$ and the least prime factor of $A(n)$, where $1 \leq n \leq 21$. A close examination of the table and additional computations reveal an astonishing pattern: $A(n)$ is odd when $n = 1, 3, 5, 11, 21, 43, 87$, and 171; that is, $A(J_m)$ is odd, where $2 \leq m \leq 9$.

More generally, we establish that $A(n)$ is odd if and only if n is a (nonnegative) Jacobsthal number [176]. The proof hinges on several results and hence is unusually long. We omit their proofs in the interest of brevity, but focus on the proof of the main result; see [277] for a proof of Lemma 44.2 and [176] for the others.

Lemma 44.2 (Legendre's formula).† *The largest exponent e of a prime p in the prime factorization of $N!$ is given by* $e = \sum_{k\geq 1} \left\lfloor \frac{N}{p^k} \right\rfloor$. ■

For example, when $N = 12$ and $p = 2$, $e = \sum_{k\geq 1} \left\lfloor \frac{12}{2^k} \right\rfloor = \frac{12}{2} + \frac{12}{4} + \frac{12}{8} = 10.$

This lemma plays a pivotal role in the proof of the principal result. To see this, let $N^*(J_{m+1}) = \prod_{i=0}^{J_{m+1}-1} (3i+1)!$ and $D^*(J_{m+1}) = \prod_{i=0}^{J_{m+1}-1} (n+i)!$. We can show

†Adrien-Marie Legendre (1752–1893).

that the largest exponent of 2 in the prime factorization of $N^*(J_{m+1})$ and that in $D^*(J_{m+1})$ are equal; this would mean that $2 \nmid A(J_{m+1})$; that is, $A(J_{m+1})$ is odd.

It follows by Lemma 44.2 that the largest exponent of 2 in $N(n) = \prod_{i=0}^{n-1}(3i+1)!$ is given by

$$N^*(n) = \sum_{i=0}^{n-1}\sum_{k\geq 1}\left\lfloor \frac{3i+1}{2^k}\right\rfloor = \sum_{k\geq 1} N_k^*(n),$$

where

$$N_k^*(n) = \sum_{i=0}^{n-1}\left\lfloor \frac{3i+1}{2^k}\right\rfloor.$$

Similarly, the largest exponent of 2 in $D(n) = \prod_{i=0}^{n-1}(n+i)!$ is given by

$$D^*(n) = \sum_{k\geq 1} D_k^*(n),$$

where

$$D_k^*(n) = \sum_{i=0}^{n-1}\left\lfloor \frac{n+i}{2^k}\right\rfloor.$$

Lemma 44.3 (Frey and Sellers, 2000 [176]). *The least value of i for which*

$$\left\lfloor \frac{3i+1}{2^k}\right\rfloor = m$$

is

$$\begin{cases} \frac{m}{3}2^k & \text{if } m \equiv 0 \pmod 3 \\ \frac{m-1}{3}2^k + J_k & \text{if } m \equiv 1 \pmod 3 \\ \frac{m-2}{3}2^k + J_{k+1} & \text{otherwise,} \end{cases}$$

where m and k are positive integers and $k \geq 2$. ■

The next lemma follows from this; see Exercises 44.125 and 44.126.

Lemma 44.4 (Frey and Sellers, 2000 [176]). *Let k be a positive integer. Then*

$$\sum_{v=0}^{2^k-1}\left\lfloor \frac{3v+1}{2^k}\right\rfloor = 2^k.$$ ■

Lemmas 44.3 and 44.4 yield the following theorem.

Theorem 44.7 (Frey and Sellers, 2000 [176]). *Let $n = 2^k q + r$, where $q \geq 0$ and $0 \leq r < 2^k$. Then*

$$N_k^*(n) = \frac{n-r}{2^{k+1}}[3(n-r) - 2^k] + \omega(n),$$

where

$$\omega(n) = \begin{cases} 3qr & \text{if } 0 \leq r \leq J_k \\ 3qr + (r - J_k) & \text{if } J_k < r \leq J_{k+1} \\ (3q+2)r - 2^k & \text{if } J_{k+1} < r < 2^k. \end{cases}$$

■

The next theorem gives a similar formula for $D_k^*(n)$.

Theorem 44.8 (Frey and Sellers, 2000 [176]). *Let $n = 2^k q + r$, where $q \geq 0$ and $0 \leq r < 2^k$. Then*

$$D_k^*(n) = \begin{cases} \dfrac{n-r}{2^{k+1}}[3(n+r) - 2^k] & \text{if } 0 \leq r \leq 2^{k-1} \\ \dfrac{n+r-2^k}{2^{k+1}}[3(n-r) - 2^{k+1}] & \text{if } 2^{k-1} < r < 2^k. \end{cases}$$

■

With Theorems 44.7 and 44.8 at our fingertips, we can establish the oddness of $A(J_m)$. Again, in the interest of brevity, we omit most of the algebraic steps.

Theorem 44.9 (Frey and Sellers, 2000 [176]). *$A(J_m)$ is odd, where $m \geq 1$.*

Proof. Clearly, $A(J_1) = A(1)$ is odd. So it suffices to prove that $A(J_{m+1})$ is odd, where $m \geq 1$.

We will show that $N_k^*(J_{m+1}) = D_k^*(J_{m+1})$ for all k. This implies that $N^*(J_{m+1}) = D^*(J_{m+1})$. Then the largest exponent of 2 in $A(J_{m+1})$ is zero; this implies the oddity of $A(J_{m+1})$.

We split the proof into two cases depending on the parities of m and k.

Case 1. Let $m \equiv k \pmod 2$. Then $2^k(J_{m-k+1} - 1) + J_{k+1} = J_{m+1}$; see Exercise 44.127. Using Theorems 44.7 and 44.8, and $q = J_{m-k+1} - 1$ and $r = J_{k+1}$, we can compute both $N_k^*(J_{m+1})$ and $D_k^*(J_{m+1})$:

$$\begin{aligned} N_k^*(J_{m+1}) &= \frac{J_{m+1} - J_{k+1}}{2^{k+1}}\left[3(J_{m+1} - J_{k+1}) - 2^k\right] + 3(J_{m-k+1} - 1) + (J_{k+1} - J_k) \\ &= \frac{1}{3}\left[2^{2m-k+1} - 2^m + (-1)^k 2^{m-k+1}\right]; \end{aligned}$$

see Exercise 44.128.

Since $2^{k-1} < r = J_{k+1} < 2^k$,

$$\begin{aligned} D_k^*(J_{m+1}) &= \frac{J_{m+1} + J_{k+1} - 2^k}{2^{k+1}}\left[3(J_{m+1} - J_{k+1}) + 2^{k+1}\right] \\ &= \frac{1}{3}\left[2^{2m-k+1} - 2^m + (-1)^k 2^{m-k+1}\right]; \end{aligned}$$

see Exercise 44.129. Thus $N_k^*(J_{m+1}) = D_k^*(J_{m+1})$.

Case 2. Let $m \not\equiv k \pmod 2$. Then $2^k J_{m-k+1} + J_k = J_{m+1}$; see Exercise 44.130. With $q = J_{m-k+1}$ and $r = J_k$, Theorems 44.7 and 44.8 now yield the following values of $N_k^*(J_{m+1})$ and $D_k^*(J_{m+1})$:

$$\begin{aligned}
N_k^*(J_{m+1}) &= \frac{J_{m+1} - J_k}{2^{k+1}}\left[3(J_{m+1} - J_k) - 2^k\right] + 3J_{m-k+1}J_k \\
&= \frac{1}{3}\left[2^{2m-k+1} - 2^m - (-1)^k 2^{m-k+1} + (-1)^k\right]; \\
D_k^*(J_{m+1}) &= \frac{J_{m+1} - J_k}{2^{k+1}}\left[3(J_{m+1} + J_k) - 2^k\right] \\
&= \frac{1}{3}\left[2^{2m-k+1} - 2^m - (-1)^k 2^{m-k+1} + (-1)^k\right];
\end{aligned}$$

see Exercises 44.131 and 44.132. Again, $N_k^*(J_{m+1}) = D_k^*(J_{m+1})$.

Thus $N_k^*(J_{m+1}) = D_k^*(J_{m+1})$ for all values of k, so $N^*(J_{m+1}) = D^*(J_{m+1})$. Consequently, $A(J_{m+1})$ is odd, as desired. ■

For example, let $n = 11 = J_6 = 2^k q + r$, where $(k, q, r) = (1, 5, 1)$, $(2, 2, 3)$, or (3,1,3). Then

$$\begin{aligned}
N_1^*(11) &= \frac{11-1}{2^2}[3(11-1) - 2] + 3 \cdot 5 \cdot 1 - \textcircled{85}; \\
D_1^*(11) &= \frac{11-1}{2^2}[3(11+1) - 2] = \textcircled{85}; \\
N_2^*(11) &= \frac{11-3}{2^3}[3(11-3) - 2^2] + 3 \cdot 2 \cdot 3 + (3-1) = \boxed{40}; \\
D_2^*(11) &= \frac{11+3-2^2}{2^3}[3(11-3) + 2^3] = \boxed{40}; \\
N_3^*(11) &= \frac{11-3}{2^4}[3(11-3) - 2^3] + 3 \cdot 1 \cdot 3 = \boxed{\boxed{17}}; \\
D_3^*(11) &= \frac{11-3}{2^4}[3(11+3) - 2^3] = \boxed{\boxed{17}}.
\end{aligned}$$

Thus $N_k^*(11) = D_k^*(J_1 1)$ for all k; so $A(11) = A(J_6)$ is odd, as expected.

Fortunately, the converse of Theorem 44.9 is also true, as the next theorem confirms.

Theorem 44.10 (Frey and Sellers, 2000 [176]). *If n is* not *a Jacobsthal number, then $A(J_m)$ is even.* ■

As we might expect, the proof takes a lot of machinery. The essence of the proof lies in showing that there is some value of k for which $N_k^*(n) > D_k^*(n)$, when n is *not* a Jacobsthal number. For the sake of brevity, we omit the proof; see [176].

For example, $n = 10$ is *not* a Jacobsthal number. Let $10 = 2^k q + r$, where $(k, q, r) = (1, 5, 0), (2, 2, 2)$ or $(3,1,2)$. Then

$$N_1^*(10) = \frac{10-0}{2^2}[3(10-0) - 2] + 3 \cdot 5 \cdot 0 = 70;$$

$$D_1^*(10) = \frac{10-0}{2^2}[3(10+0) - 2] = 70;$$

$$N_2^*(10) = \frac{10-2}{2^3}[3(10-2) - 2^2] + 3 \cdot 2 \cdot 2 + (2-1) = \textcircled{33};$$

$$D_2^*(10) = \frac{10-2}{2^3}[3(10+2) - 2^2] = \boxed{32};$$

$$N_3^*(10) = \frac{10-2}{2^4}[3(10-2) - 2^3] + 3 \cdot 1 \cdot 2 = 14;$$

$$D_3^*(10) = \frac{10-2}{2^4}[3(10+2) - 2^3] = 14.$$

Since $N_k^*(10) \neq D_k^*(10)$ for some value of k, it follows that $A(10) = 129{,}534{,}272{,}700$ is even.

Finally, Table 44.22 shows the values of $N_1^*(n)$, where $1 \le n \le 15$ [176]. The sequence $\{N_1^*(n)\}$ occurs in [465] as sequence A001859.

TABLE 44.22. Values of $N_1^*(n)$, where $1 \le n \le 15$

n	(1)	2	(3)	4	(5)	6	7	8	9	10	(11)	12	13	14	15
$N_1^*(n)$	0	2	5	10	16	24	33	44	56	70	85	102	120	140	161

44.8 CONVOLUTIONS

The next example illustrates the power of generating functions and partial fractions in the study of convolutions of Fibonacci and Jacobsthal numbers.

Example 44.6. *Develop a formula for the sum* $S_n = \sum_{k=0}^{n} F_k J_{n-k}$.

Solution. We have

$$f(x) = \sum_{n=0}^{\infty} F_n x^n = \frac{x}{1-x-x^2} \quad \text{and} \quad J(x) = \sum_{n=0}^{\infty} J_n x^n = \frac{x}{1-x-2x^2}$$

(see Exercise 44.139). Then

$$\sqrt{5} f(x) = \frac{\sqrt{5}x}{(1-\alpha x)(1-\beta x)} = \frac{1}{1-\alpha x} - \frac{1}{1-\beta x}$$

$$3J(x) = \frac{3x}{(1-2x)(1+x)} = \frac{1}{1-2x} - \frac{1}{1+x}$$

$$\begin{aligned}
3\sqrt{5}f(x)J(x) &= \left(\frac{1}{1-\alpha x} - \frac{1}{1-\beta x}\right)\left(\frac{1}{1-2x} - \frac{1}{1+x}\right) \\
&= \frac{1}{(1-\alpha x)(1-2x)} - \frac{1}{(1-\alpha x)(1+x)} \\
&\quad - \frac{1}{(1-\beta x)(1-2x)} + \frac{1}{(1-\beta x)(1+x)} \\
&= \left[\frac{-\alpha}{\beta^2(1-\alpha x)} + \frac{2}{\beta^2(1-2x)}\right] - \left[\frac{\alpha}{\alpha^2(1-\alpha x)} + \frac{1}{\alpha^2(1+x)}\right] \\
&\quad - \left[\frac{-\beta}{\alpha^2(1-\beta x)} + \frac{2}{\alpha^2(1-2x)}\right] + \left[\frac{\beta}{\beta^2(1-\beta x)} + \frac{1}{\beta^2(1+x)}\right] \\
&= \frac{-\alpha}{1-\alpha x}\left(\frac{1}{\beta^2} + \frac{1}{\alpha^2}\right) + \frac{2}{1-2x}\left(\frac{1}{\beta^2} - \frac{1}{\alpha^2}\right) + \frac{1}{1+x}\left(\frac{1}{\beta^2} - \frac{1}{\alpha^2}\right) \\
&\quad + \frac{\beta}{1-\beta x}\left(\frac{1}{\beta^2} + \frac{1}{\alpha^2}\right) \\
&= \frac{2\sqrt{5}}{1-2x} + \frac{\sqrt{5}}{1+x} - \frac{3\alpha}{1-\alpha x} + \frac{3\beta}{1-\beta x}.
\end{aligned}$$

Equating the coefficients of x^n from both sides, we get

$$\begin{aligned}
3\sqrt{5}S_n &= [2^{n+1} - (-1)^{n+1}]\sqrt{5} - 3\sqrt{5}F_{n+1} \\
S_n &= J_{n+1} - F_{n+1}.
\end{aligned}$$

■

In particular,

$$\begin{aligned}
\sum_{k=0}^{7} F_k J_{7-k} &= F_1J_6 + F_2J_5 + F_3J_4 + F_4J_3 + F_5J_2 + F_6J_1 \\
&= 64 = 85 - 21 \\
&= J_8 - F_8, \quad \text{as expected.}
\end{aligned}$$

Similarly, we can show (see Exercises 44.141–44.143) that

$$\begin{aligned}
\sum_{k=0}^{n} F_k j_{n-k} &= j_{n+1} - L_{n+1} \\
\sum_{k=0}^{n} L_k J_{n-k} &= j_{n+1} - L_{n+1} \qquad (44.20)
\end{aligned}$$

$$\sum_{k=0}^{n} L_k j_{n-k} = 9J_{n+1} - 5F_{n+1}.$$

M. Griffiths of the University of Essex and A. Bramham of the University of Oxford, United Kingdom, discovered formula (44.20) [198].

EXERCISES 44

1. Derive the Binet-like formula for $J_n(x)$.
2. Derive the Binet-like formula for $j_n(x)$.
3. Derive formula (44.1).
4. Using formula (44.1), compute J_{15} and J_{18}.
5. Compute the number of digits in J_{14} and J_{19}.
6. Compute the number of digits in j_{15} and j_{21}.

Establish the following Jacobsthal and Jacobsthal–Lucas properties.

7. $J_n + J_{n+1} = 2^n$.
8. $J_n < 2^n$.
9. Every J_n is odd.
10. $J_n + J_{n+1} + J_{n+2} + J_{n+3} = 5 \cdot 2^n$.
11. $J_{n+2} = J_n + 2^n$.
12. $J_n = 2J_{n-1} - (-1)^n$.
13. $J_n + J_{n+3} = 3 \cdot 2^n$.
14. $J_{3n} \equiv 0 \pmod{3}$.
15. $J_n \equiv (-1)^n \pmod{4}$, $n \geq 2$.
16. $J_{5n} \equiv 0 \pmod{11}$.
17. $J_{n+k}J_{n-k} - J_n^2 = (-1)^{n-k+1}2^{n-k}J_k^2$.
18. $(J_n, J_{n+1}) = 1$.
19. If $m|n$, then $J_m|J_n$.
20. $(J_m, J_n) = J_{(m,n)}$.
21. $J_{n+k} - J_{n-k} = 2^{n-k}J_{2k}$.
22. $\lim_{n\to\infty} \frac{J_{n+1}}{J_n} = 2$.
23. $J_{n+1}^2 - J_n^2 = 2^{n+1}J_{n-1}$.
24. $J_{n+1}^2 - 4J_{n-1}^2 = J_{2n}$.

25. $3J_{n-k} + 2J_{n-k-1} = J_{n-k+2}$.
26. $J_n = J_{m+1}J_{n-m} + 2J_mJ_{n-m-1}$.
27. $2^{n-1}J_{m-n} = (-1)^n(J_mJ_{n-1} - J_{m-1}J_n)$.
28. $3(J_{n+k}^2 - J_{n-k}^2) = 4^{n-k}J_{4k} + (-2)^{n-k+1}J_{2k}$.
29. $3(J_n^2 + J_{n+1}^2) = 2J_{2n+1} + 2^{n+1}J_{n-1}$.
30. $J_{n+1} = \sum_{j=0}^{\lfloor n/2 \rfloor} \binom{n-j}{j} 2^j$.
31. $2\sum_{i=0}^{n} J_i = J_{n+2} - 1$.
32. $3\sum_{i=0}^{n} J_{2i} = J_{n+2} - n - 1$.
33. $3\sum_{i=0}^{n} J_{2i+1} = J_{2n+3} + n$.
34. $\sum_{i=0}^{n} \binom{n}{i} J_i = 3^{n-1}$.
35. $\sum_{i=0}^{n} \binom{n}{i} J_{i+k} = 2^k 3^{n-1}$.
36. $9\sum_{i=0}^{n} J_i^2 = J_{2n+2} - (-1)^n J_{n+2} + n$.
37. $J_n = \sum_{i=0}^{n-1} (-1)^i \binom{n}{i} 3^{n-i-1}$.
38. $9\sum_{i=0}^{n} \binom{n}{i} J_{i+k}^2 = 4^k \cdot 5^n + 2^n - (-1)^{n+k} 2^k$.
39. $9\sum_{i=0}^{n} (-1)^i \binom{n}{i} J_{i+k}^2 = [(-1)^n 4^k + (-2)^{k+1}]3^n$.
40. $J_n = \sum_{k=1}^{n} \binom{n}{k} (-1)^{n+k} 3^{k-1}$ (Sloane, [465]).
41. $J_{n+1} = \sum_{k=0}^{n} (-1)^{n-k} 2^k$ (Sloane, [465]).
42. $J_{n+1} = \lceil 2^n/3 \rceil + \lfloor 2^n/3 \rfloor$ (Sloane, [465]).
43. Redo Exercise 44.42 using the binomial theorem.
44. $J_n = \lceil 2^{n+1}/3 \rceil - \lceil 2^n/3 \rceil$ (Sloane, [465]).

45. $J_n = \lfloor 2^{n+1}/3 \rceil - \lfloor 2^n/3 \rceil$ (Sloane, [465]).
46. $J_n + J_{n+2k+1} = 2^n J_{2k+1}$ (Sloane, [465]).
47. $8J_{n-1}J_{n-2} + 1 = J_n^2$ (Sloane, [465]).
48. $J_n = 2^k J_{n-k} + (-1)^{n+k} J_k$ (Sloane, [465]).
49. $2^n J_{-n} = (-1)^{n+1} J_n$ (Sloane, [465]).
50. $J_{n+1}^2 + 2J_n^2 = J_{2n+1}$.
51. $j_{n+1}^2 - 2j_n^2 = j_{2n+1}$.
52. Every j_n is odd.
53. $j_n + j_{n+1} = 3 \cdot 2^n$.
54. $j_n + j_{n+3} = 9 \cdot 2^n$.
55. $j_n \equiv (-1)^n \pmod 4, \quad n \geq 2$.
56. $j_{4n+2} \equiv 5 \pmod{10}$.
57. $j_n + j_{n+1} + j_{n+2} + j_{n+3} = 15 \cdot 2^n$.
58. $j_n = j_{n-2} + 3 \cdot 2^{n-2}$.
59. $j_n = 2j_{n-1} + 3(-1)^n$.
60. $j_{n+k} - j_{n-k} = 3 \cdot 2^{n-k} J_{2k}$.
61. $j_{n+k} j_{n-k} - j_n^2 = 9(-2)^{n-k} J_k^2$.
62. $(j_n, j_{n+1}) = 1$.
63. $j_{2n} = 2j_{2n-1} + 3$.
64. $j_n^2 = j_{2n} + 2(-1)^n$.
65. $j_{n+1}^2 - j_n^2 = 3 \cdot 2^{n+1} j_{n-1}$.
66. $j_{n+1}^2 - 4j_{n-1}^2 = 9J_{2n}$.
67. $j_{n+1}^2 + 2j_n^2 = 3(j_{2n+1} + 2)$.
68. $j_n = \sum_{i=0}^{n-1} (-1)^i \binom{n}{i} 3^{n-i} + (-1)^n$.
69. $(j_{n-1}^2 + 4j_{n-2}^2 + j_n^2)^2 = 2(j_{n-1}^4 + 16j_{n-2}^4 + j_n^4)$.
70. $j_n = \sum_{r=0}^{\lfloor n/2 \rfloor} \frac{n}{n-r} \binom{n-r}{r} 2^r$.
71. $2 \sum_{i=0}^{n} j_i = j_{n+2} - 1$.
72. $\sum_{i=0}^{n} j_{2i} = J_{n+2} + n + 1$.

73. $\sum_{i=0}^{n} j_{2i+1} = 2J_{n+2} - n - 1.$

74. $\sum_{i=0}^{n} \binom{n}{i} j_{i+k} = 2^k 3^n.$

75. $\sum_{i=0}^{n} \binom{n}{i} j_{i+k}^2 = 4^k \cdot 5^n + (-1)^{n+k} 2^{k+1} + 2^n.$

76. $j_n + J_n = 2J_{n+1}$ (Horadam, [233]).

77. $j_n - J_n = 4J_{n-1}$ (Horadam, [233]).

78. $j_n + 3J_n = 2^{n+1}$ (Horadam, [233]).

79. $j_n = J_{n+1} + 2J_{n-1}$ (Horadam, [233]).

80. $9J_n = j_{n+1} + 2j_{n-1}$ (Horadam, [233]).

81. $j_{n+1}^2 - 4j_{n-1}^2 = 9J_{2n}.$

82. $j_n^2 - J_n^2 = 8J_{n-1}J_{n+1}.$

83. $9J_n^2 = j_{2n} - 2(-2)^n.$

84. $9J_n^2 = j_n^2 - 4(-2)^n.$

85. $(J_n, j_n) = 1.$

86. $j_n^2 + 9J_n^2 = 2j_{2n}$ (Horadam, [233]).

87. $j_n^2 - 9J_n^2 = 4(-2)^n$ (Horadam, [233]).

88. $j_n^2 - J_n^2 = 8J_{n-1}J_{n+1}.$

89. $J_m j_n + J_n j_m = 2J_{m+n}$ (Horadam, [233]).

90. $J_m j_n - J_n j_m = (-1)^n 2^{n+1} J_{m-n}$ (Horadam, [233]).

91. $j_m j_n + 9J_m J_n = 2j_{m+n}$ (Horadam, [233]).

92. $j_m j_n - 9J_m J_n = (-1)^n 2^{n+1} j_{m-n}$ (Horadam, [233]).

93. $j_{m+n} = j_m J_{n+1} + 2j_{m-1} J_n.$

94. $2^{n-1} j_{m-n} = j_m J_{n-1} - j_{m-1} J_n.$

95. $9J_{m+n}^2 + j_{m-n}^2 = 4^{m-n} j_{4n} - 6(-2)^{m-n} J_{2n} + 2.$

96. $9J_{m+n}^2 - j_{m-n}^2 = 3 \cdot 4^{m-n} J_{4n} - 2(-2)^{m-n} j_{2n}.$

97. $\lim_{n\to\infty} \frac{j_n}{J_n} = 3.$

98. $J_n J_{n+1} = t_{(2^{n+1}-1)/3}$, where n is odd.

99. $J_n J_{n+1} = t_{\lceil (2^{n+1}-2)/3 \rceil}.$

100. Let S_n denote the nth partial sum (in lowest terms) of the alternating sum $\sum_{n=1}^{\infty} \frac{(-1)^{n-1}}{2^n}$. Then $2^n S_n = J_n$.

101. Let c_n denote the number of integers between 2^n and 2^{n+1} that are divisible by 3, where n is even. Then $c_n = J_n$.

102. $M^n \begin{bmatrix} X_0 \\ X_1 \end{bmatrix} = \begin{bmatrix} X_n \\ X_{n+1} \end{bmatrix}$, where $M = \begin{bmatrix} 0 & 1 \\ 2 & 1 \end{bmatrix}$, and $X_n = J_n$ or j_n.

Solve each recurrence.

103. $x_n = x_{n-1} + 2x_{n-2} + 2^{n-3}$, where $x_2 = 2, x_3 = 4$, and $n \geq 4$ (Grimaldi, [199]).

104. $x_n = x_{n-1} + 2x_{n-2} + 2^{n-4}$, where $x_3 = 1, x_4 = 2$, and $n \geq 5$ (Grimaldi, [199]).

105. $x_n = x_{n-1} + 2x_{n-2} + 2^{n-2}$, where $x_1 = 0, x_2 = 1$, and $n \geq 3$ (Grimaldi, [199]).

106. $z_n = z_{n-1} + 2z_{n-2} + 2^{n-1}$, where $z_1 = 1, z_2 = 3$, and $n \geq 3$ (Grimaldi, [199]).

107. $z_{n+1} = z_n + 2z_{n-1} + 2^n$, where $z_1 = 1, z_2 = 4$, and $n \geq 3$ (Grimaldi, [199]).

108. $z_n = z_{n-1} + 2z_{n-2} + 2^{n-2}$, where $z_1 = 1, z_2 = 2$, and $n \geq 3$ (Grimaldi, [199]).

109. $x_n = 2x_{n-1} + 8x_{n-2} + \left(\frac{4}{3}\right) 2^{n-1} - \left(\frac{4}{3}\right)(-1)^{n-1}$, where $x_1 = 0$, $x_2 = 4$, and $n \geq 3$ (Grimaldi, [199]).

110. Find a recurrence satisfied by the even number of summands e_n in the J_n compositions of a positive integer n (Grimaldi, [200]).

111. Find an explicit formula for the even number of summands e_n in the J_n compositions of a positive integer n (Grimaldi, [200]).

112. Find an explicit formula for the odd number of summands o_n in the J_n compositions of a positive integer n (Grimaldi, [200]).

113. Let $starte_n$ denote the number of compositions of n that begin with an even summand. Find a recurrence satisfied by $starte_n$ (Grimaldi, [200]).

114. Find an explicit formula for $starte_n$ (Grimaldi, [200]).

115. Let $starto_n$ denote the number of compositions of n that begin with an odd summand. Find an explicit formula for $starto_n$ (Grimaldi, [200]).

116. Let es_n denote the sum of all even summands of the J_n compositions of n. Find a recurrence for es_n (Grimaldi, [200]).

117. Find an explicit formula for the sum of all even summands es_n of the J_n compositions of n (Grimaldi, [200]).

118. Find an explicit formula for the sum of all odd summands os_n of the J_n compositions of n (Grimaldi, [200]).

Let b_n denote the number of compositions of a positive integer n with the last summand even, where $n \geq 2$.

119. Prove that $b_n = J_{n-1}$.

120. Find the number of plus signs p_n in the b_n compositions of n.
121. Find the number of summands s_n in the b_n compositions of n.

Let z_n denote the number of 0s among the $b_n = 2J_{n-2}$ ternary words $x_1x_2 \dots x_{n-1}x_n$, where $x_1 = 0 = x_n$, $x_i \neq x_{i+1}$, and $1 \leq i \leq n-1$.

122. Find a recurrence satisfied by z_n.
123. Solve the recurrence in Exercise 44.122 for z_n.

Prove each.

124. The number of $n \times n$ ASMs with a 1 in row 1 is $A(n-1)$.
125. Lemma 44.3 (Frey and Sellers, [176]).
126. Lemma 44.4 (Frey and Sellers, [176]).
127. Let m and k be positive integers, and $m \equiv k \pmod 2$. Then $2^k(J_{m-k+1} - 1) + J_{k+1} = J_{m+1}$ (Frey and Sellers, [176]).
128. $N^*(J_{m+1}) = \frac{1}{3}[2^{2m-k+1} - 2^m + (-1)^k 2^{m-k+1}]$, where $m \equiv k \pmod 2$, $q = J_{m-k+1} - 1$ and $r = J_{k+1}$ (Frey and Sellers, [176]).
129. $D^*(J_{m+1}) = \frac{1}{3}[2^{2m-k+1} - 2^m + (-1)^k 2^{m-k+1}]$, where $m \equiv k \pmod 2$, $q = J_{m-k+1} - 1$ and $r = J_{k+1}$ (Frey and Sellers, [176]).
130. Suppose $m \not\equiv k \pmod 2$. Then $2^k J_{m-k+1} + J_k = J_{m+1}$ (Frey and Sellers, [176]).
131. $N^*(J_{m+1}) = \frac{1}{3}[2^{2m-k+1} - 2^m - (-1)^k 2^{m-k+1} + (-1)^k]$, where $m \not\equiv k \pmod 2$, $q = J_{m-k+1}$ and $r = J_k$ (Frey and Sellers, [176]).
132. $D^*(J_{m+1}) = \frac{1}{3}[2^{2m-k+1} - 2^m - (-1)^k 2^{m-k+1} + (-1)^k]$, where $m \not\equiv k \pmod 2$, $q = J_{m-k+1}$ and $r = J_k$ (Frey and Sellers, [176]).
133. Using Theorem 44.9, confirm that $A(5)$ is odd.
134. Using Theorem 44.9, confirm that $A(21)$ is odd.
135. Using Theorem 44.10, confirm that $A(6)$ is even.
136. Using Theorem 44.10, confirm that $A(12)$ is odd.

Find a generating function for each sequence.

137. $\{J_n(x)\}$.
138. $\{j_n(x)\}$.
139. $\{J_n\}$.
140. $\{j_n\}$.

Prove each.

141. $\sum_{k=0}^{n} F_k j_{n-k} = j_{n+1} - L_{n+1}$.

142. $\sum_{k=0}^{n} L_k J_{n-k} = j_{n+1} - L_{n+1}$ (Griffiths and Bramham, [198]).

143. $\sum_{k=0}^{n} L_k j_{n-k} = 9J_{n+1} - 5F_{n+1}$.

Develop a formula for each sum S_n.

144. $\sum_{k=0}^{n} P_k J_{n-k}$.

145. $\sum_{k=0}^{n} P_k j_{n-k}$.

146. $\sum_{k=0}^{n} Q_k J_{n-k}$.

147. $\sum_{k=0}^{n} Q_k j_{n-k}$.

Prove each.

148. Let p be a prime > 3. Then $J_{n+p-1} \equiv J_n \pmod{p}$ (Griffiths and Bramham, [198]).
149. Let $k \geq 0$. Then $4^{3^k} \equiv 3^{k+1} + 1 \pmod{3^{k+2}}$ (Griffiths and Bramham, [198]).
150. Let n be a nonnegative even integer. Then $J_{n+2\cdot 3^k} \equiv J_n + 3^k \pmod{3^{k+1}}$ (Griffiths and Bramham, [198]).

45

JACOBSTHAL TILINGS AND GRAPHS

I have no particular talent.
I am only inquisitive.
–Albert Einstein (1879–1955)

This chapter presents combinatorial and graph-theoretic models for the Jacobsthal family. They offer delightful opportunities for investigating the beauty and ubiquity of the family. We begin with tilings of a $1 \times n$ board.

45.1 $1 \times n$ TILINGS

Model I

Suppose we would like to tile a $1 \times n$ board with square tiles and dominoes. Let the weights be w(square) = 1 and w(domino) = x. The weight of a tiling is the product of the weights of all tiles in the tiling. The weight of the empty tiling of length 0 is defined as 1.

Figure 45.1 shows the possible tilings of length n and the sum of the weights of such tilings, where $0 \le n \le 4$.

Suppose S_n denotes the sum of the weights of tilings of length n. We now establish that $S_n = J_{n+1}(x)$. First, notice that $S_0 = 1 = J_1(x)$ and $S_1 = 1 = J_2(x)$.

Now consider a tiling of an arbitrary length n, where $n \ge 2$. Suppose it ends in a square; the sum of the weights of such tilings is $S_{n-1} \cdot 1 = S_{n-1}$. On the other hand, suppose it ends in a domino; the sum of the weights of such tilings

Fibonacci and Lucas Numbers with Applications, Volume Two. Thomas Koshy.

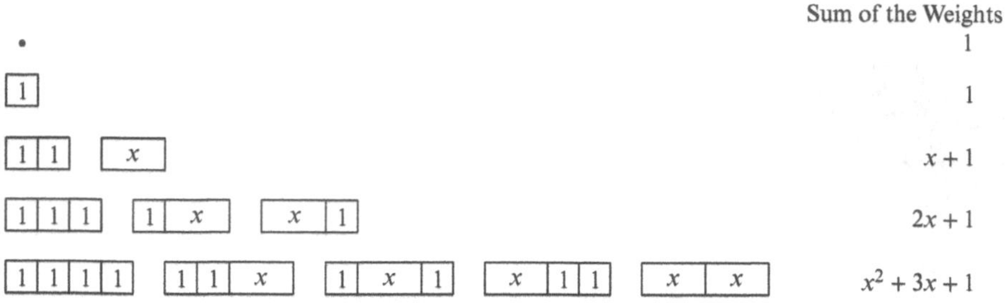

Figure 45.1.

is $S_{n-2} \cdot x = xS_{n-2}$. Consequently, $S_n = S_{n-1} + xS_{n-2}$, the Jacobsthal recurrence. By virtue of the initial conditions, this implies that $S_n = J_{n+1}(x)$, as desired.

Thus we have the following result.

Theorem 45.1. *Suppose the weight of a square is* 1 *and that of a domino is* x. *Then the sum of the weights of tilings of length* n *is* $J_{n+1}(x)$, *where* $n \geq 0$. ■

Since $J_{n+1}(1) = F_{n+1}$, it follows that there are exactly F_{n+1} tilings of length n. Using the concept of breakability, we can show that

$$J_{m+n+1}(x) = J_{m+1}(x)J_{n+1}(x) + xJ_m(x)J_n(x); \tag{45.1}$$

see Exercise 45.1. This can be confirmed algebraically using the Binet-like formula for $J_n(x)$.

In particular, formula (45.1) yields

$$J_{2n}(x) = J_n(x)[J_{n+1}(x) + xJ_n(x)] \qquad J_{2n+1}(x) = J_{n+1}^2(x) + xJ_n^2(x)$$
$$J_{m+n+1} = J_{m+1}J_{n+1} + J_mJ_n \qquad J_{2n} = J_nj_n$$
$$J_{2n+1} = J_{n+1}^2 + 2J_n^2 \qquad F_{m+n+1} = F_{m+1}F_{n+1} + F_mF_n$$
$$F_{2n} = F_nL_n \qquad F_{2n+1} = F_{n+1}^2 + F_n^2.$$

For example, $J_{10}(x) = J_5(x)[J_6(x) + xJ_4(x)] = (x^2 + 3x + 1)[(3x^2 + 4x + 1) + x(2x + 1)] = 5x^4 + 20x^3 + 21x^2 + 8x + 1$; see Table 45.1 So $J_{10} = 5 \cdot 2^4 + 20 \cdot 2^3 + 21 \cdot 2^2 + 8 \cdot 2 + 1 = 341$; see Table 45.2.

Next we show how Jacobsthal polynomials can be extracted from Pascal's triangle with appropriate weights.

Lucas-like Formula for $J_n(x)$

Suppose a tiling has exactly k dominoes; the weight of such a tiling is x^k. Such a tiling has $n - 2k$ squares; so $0 \leq k \leq \lfloor n/2 \rfloor$. Since there is a total of $n - k$ tiles in the tiling, the k dominoes can be placed in $\binom{n-k}{k}$ different ways;

the weight of a such a tiling is $\binom{n-k}{k}x^k$. Consequently, the cumulative sum of the weights of tilings of length n equals $\sum_{k=0}^{\lfloor n/2 \rfloor} \binom{n-k}{k}x^k$. Thus, by Theorem 45.1,

$$J_{n+1}(x) = \sum_{k=0}^{\lfloor n/2 \rfloor} \binom{n-k}{k}x^k.$$

Horadam established this result algebraically in 1997 [235].

In particular, this yields $J_{n+1} = \sum_{k=0}^{\lfloor n/2 \rfloor} \binom{n-k}{k}2^k$ and $F_{n+1} = \sum_{k=0}^{\lfloor n/2 \rfloor} \binom{n-k}{k}$. For example, $J_6(x) = \sum_{k=0}^{2} \binom{5-k}{k}x^k = 3x^2 + 4x + 1$; see Table 45.1.

Consequently, we can compute Jacobsthal numbers using the northeast diagonals of Pascal's triangle with weights 2^k, where $k \geq 0$; see Figure 45.2.

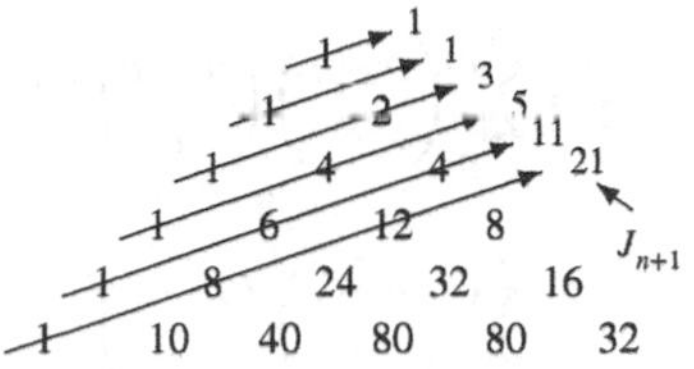

Figure 45.2. Weighted Pascal's triangle.

The median square in a tiling of length $2n + 1$ [31] can be used to develop an interesting summation formula for $J_{2n+2}(x)$.

A Double-Summation Formula for $J_{2n+2}(x)$

Consider a tiling of length $2n + 1$. By Theorem 45.1, the sum of the weights of all such tilings is J_{2n+2}.

Since the length of the tiling is odd, it contains an odd number of squares. So there must a *median square* M with an equal number of squares on either side:

$$\underbrace{\quad \cdots \quad}_{i \text{ dominoes}} \boxed{M} \underbrace{\quad \cdots \quad}_{j \text{ dominoes}}$$

$\uparrow$ median square

Suppose there are i dominoes to the left of M and j dominoes to its right, where $0 \leq i, j \leq n$. Then the tiling contains $2n + 1 - (2i + 2j)$ squares; so M has

$n-i-j$ squares on either side. Consequently, there are $i+(n-i-j)=n-j$ tiles to the left of M and $j+(n-i-j)=n-i$ tiles to its right. So the i dominoes can be placed in $\binom{n-j}{i}$ different ways and the j dominoes in $\binom{n-i}{j}$ different ways. By Theorem 45.1, the sum of the weights of all tilings of length $2n+1$ equals

$$\sum_{i,j\geq 0}\binom{n-j}{i}x^i\cdot 1\cdot\binom{n-i}{j}x^j=\sum_{i,j\geq 0}\binom{n-i}{j}\binom{n-j}{i}x^{i+j}.$$

Thus

$$J_{2n+2}(x)=\sum_{i,j=0}^{n}\binom{n-i}{j}\binom{n-j}{i}x^{i+j}. \tag{45.2}$$

■

For example, $J_8(x)=\sum_{i,j=0}^{3}\binom{3-i}{j}\binom{3-j}{i}(2x)^{i+j}=4x^3+10x^2+6x+1.$

Formula (45.2) yields

$$J_{2n+2}=\sum_{i,j=0}^{n}\binom{n-i}{j}\binom{n-j}{i}2^{i+j};$$

$$F_{2n+2}=\sum_{i,j=0}^{n}\binom{n-i}{j}\binom{n-j}{i}.$$

Next we present a related model for $J_n(x)$.

Model II: A Related Model for $J_n(x)$

Suppose dominoes come in two colors, black and white. As in Model I, w(square) = 1; but w(domino) = $x/2$. The sum of the weights of such tilings of length n is again $J_{n+1}(x)$, where $n\geq 0$.

Next we present two models for Jacobsthal–Lucas polynomials.

Model A: A Combinatorial Model for $j_n(x)$

Suppose we let w(square) = 1, except that the weight of the initial square in a tiling is $2x+1$; and w(domino) = x.

Figure 45.3 shows the tilings of length n and the sum of their weights, where $0\leq n\leq 4$ and $w=2x+1$. The next result follows as in Theorem 45.1; see Exercise 45.2.

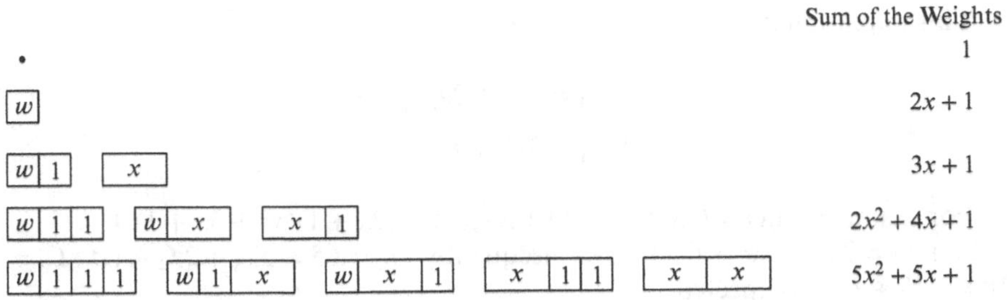

Figure 45.3.

Theorem 45.2. *The sum of the weights of tilings of length n in Model A is $j_{n+1}(x)$, where $n \geq 0$.* ■

Suppose we let $x = 1$ in Model A; then the sum of the weights of tilings of length n is L_{n+1}.

The concept of breakability, coupled with Theorem 45.2, can be applied to confirm the addition formula for $j_n(x)$:

$$j_{m+n+1}(x) = j_{m+1}(x)J_{n+1}(x) + xj_m(x)J_n(x). \tag{45.3}$$

This can be established algebraically also using the Binet-like formula for $j_n(x)$. It follows from formula (45.3) that

$$j_{2n}(x) = j_{n+1}(x)J_n(x) + xj_n(x)J_{n-1}(x) \qquad j_{2n+1}(x) = j_{n+1}(x)J_{n+1}(x) + xj_n(x)J_n(x)$$

$$j_{m+n+1} = j_{m+1}J_{n+1} + j_mJ_n \qquad L_{m+n+1} = L_{m+1}F_{n+1} + L_mF_n$$

$$L_{m+n+1} = L_{m+1}F_{n+1} + L_mF_n \qquad L_{n+1} = F_{n+2} + F_n.$$

For example, $j_9(x) = j_6(x)J_4(x) + xj_5(x)J_3(x) = (2x^3 + 9x^2 + 6x + 1)(2x + 1) + x(5x^2 + 5x + 1)(x + 1) = 9x^4 + 30x^3 + 27x^2 + 9x + 1$; see Table 45.1. So $j_9 = j_6J_4 + j_5J_3 = 511$ and $L_9 = 76 = L_6F_4 + L_5F_3 = 55 + 21 = F_{10} + F_8$.

Another Jacobsthal Hybridity

We can employ Model A to develop a hybrid Jacobsthal polynomial identity. To this end, consider a tiling of length n. By Theorem 45.2, the sum of the weights of such tilings is $j_{n+1}(x)$.

Suppose the tiling begins with a square. Then, by Theorem 45.1, the sum of the weights of such tilings is $(2x + 1)J_n(x)$. On the other hand, suppose it begins with a domino. The sum of the weights of such tilings is $xJ_{n-1}(x)$. So the sum of the weights of tilings of length n is $(2x + 1)J_n(x) + xJ_{n-1}(x)$.

Thus

$$j_{n+1}(x) = (2x + 1)J_n(x) + xJ_{n-1}(x). \tag{45.4}$$

This implies that

$$j_{n+1} = 5J_n + 2J_{n-1};$$
$$L_{n+1} = 3F_n + F_{n-1}.$$

For example, $j_6(x) = (2x+1)J_5(x) + xJ_4(x) = (2x+1)(x^2+3x+1) + x(2x+1) = 2x^3+9x^2+6x+1$. Consequently, $j_6 = 65 = 5J_5 + 2J_4$ and $L_6 = 18 = 3F_5 + F_4$, as expected.

Model A can be used to extract Jacobsthal–Lucas polynomials from Pascal's triangle with appropriate weights:

$$j_{n+1}(x) = (2x+1)\sum_{k=0}^{\lfloor (n-1)/2\rfloor}\binom{n-k-1}{k}x^k + \sum_{k=0}^{\lfloor (n-2)/2\rfloor}\binom{n-k-2}{k}x^{k+1}. \quad (45.5)$$

In the interest of brevity, we omit the proof; see Exercise 45.4.

It follows by formula (45.5) that $j_{n+1}(x)$ can be computed using the adjacent northeast diagonals $n-1$ and $n-2$ in Pascal's triangle; diagonal $n-1$ uses the weights $(2x+1)x^k$, where $0 \le k \le \lfloor (n-1)/2\rfloor$; and diagonal $n-2$ uses the weights x^{k+1}, where $0 \le k \le \lfloor (n-2)/2\rfloor$.

It also follows from formula (45.5) that

$$\begin{aligned} j_{n+1} &= 5\sum_{k=0}^{\lfloor (n-1)/2\rfloor}\binom{n-k-1}{k}2^k + \sum_{k=0}^{\lfloor (n-2)/2\rfloor}\binom{n-k-2}{k}2^{k+1} \\ &= 5J_n + 2J_{n-1}; \\ L_{n+1} &= 3\sum_{k=0}^{\lfloor (n-1)/2\rfloor}\binom{n-k-1}{k} + \sum_{k=0}^{\lfloor (n-2)/2\rfloor}\binom{n-k-2}{k} \\ &= 3F_n + F_{n-1}, \end{aligned}$$

as we found earlier.

Finally, we present a model for $j_n(x)$ using colored dominoes.

Model B: A Related Model for $j_n(x)$

Suppose the weight of a square is $w = 2x+1$ if it is the first square in a tiling; otherwise it is 1. The weight of a domino, black or white, is $x/2$. Then the sum of the weights of such tilings of length n is $j_{n+1}(x)$, where $n \ge 0$.

Next we study a slightly different combinatorial model for Jacobsthal numbers. As a dividend, we encounter an unexpected occurrence of Fibonacci numbers.

45.2 $2 \times n$ TILINGS

Suppose we would like to tile a $2 \times n$ board using 2×1 tiles (dominoes) and 2×2 tiles; it appeared as an exercise in [202] in 1997. Figure 45.4 shows the possible tilings and the number of tilings of the board, where $0 \le n \le 4$.

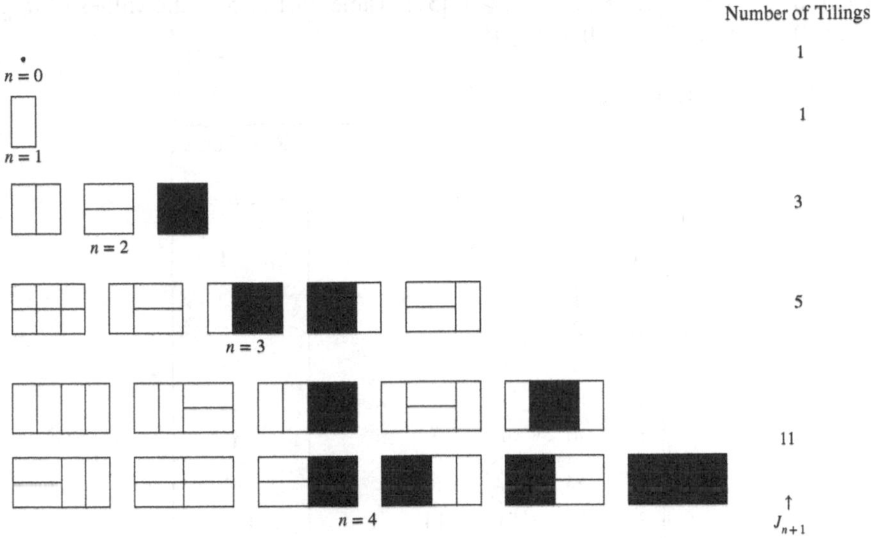

Figure 45.4. Tilings of a $2 \times n$ board, where $0 \le n \le 4$.

Using the empirical data, we can conjecture that a $2 \times n$ board can be tiled in J_{n+1} different ways. The following theorem confirms this observation.

Theorem 45.3. *A $2 \times n$ board can be tiled with 1×2 and 2×2 tiles in J_{n+1} different ways.*

Proof. Let S_n denote the number of tilings of the board. Clearly, $S_0 = 1 = J_1$ and $S_1 = 1 = J_2$.

Let T be an arbitrary tiling of a $2 \times n$ board, where $n \ge 2$. Suppose it ends in a vertical domino: $\underbrace{\text{subtiling}}_{\text{length } n-1}$ 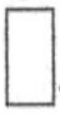. There are S_{n-1} such tilings.

On the other hand, suppose T does *not* end in a vertical domino; there are two such possibilities: $\underbrace{\text{subtiling}}_{\text{length } n-2}$ 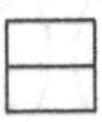 or $\underbrace{\text{subtiling}}_{\text{length } n-2}$ . There are $2S_{n-2}$ such tilings.

So the total number of tilings of length n equals $S_{n-1} + 2S_{n-2}$. Thus $S_n = S_{n-1} + 2S_{n-2}$, where $S_0 = J_1, S_1 = J_2$, and $n \ge 2$. Consequently, $S_n = J_{n+1}$, as desired. ■

Let $a_{n,k}$ denote the number of tilings of the board with exactly k 2×2 tiles each. For example, $a_{4,0} = 5, a_{4,1} = 5$, and $a_{4,2} = 1$; see Figure 45.3.

It follows from the proof of Theorem 45.3 that $a_{n,k}$ satisfies the recurrence

$$a_{n,k} = a_{n-1,k} + a_{n-2,k} + a_{n-2,k-1}, \tag{45.6}$$

where $a_{0,0} = 1 = a_{1,0}, n \geq 2$, and $k \geq 0$ [51]. Table 45.1 shows the values of $a_{n,k}$, where $0 \leq k \leq \lfloor n/2 \rfloor$ and $0 \leq n \leq 10$.

TABLE 45.1.

n \ k	0	1	2	3	4	5	Row Sums
0	1						1
1	1						1
2	2	1					3
3	3	2					5
4	5	5	1				11
5	8	10 →	3				21
6	13	20	9	1			43
7	21	38	(22)	4			85
8	34	71	51	14	1		171
9	55	130	111	[40]	5		341
10	89	235	233	105	20	1	683
	↑ F_{n+1}						↑ J_{n+1}

For example, $a_{7,2} = \textcircled{22} = 9 + 3 + 10 = a_{6,2} + a_{5,2} + a_{4,1}$; see Table 45.1. As expected, the nth row sum is J_{n+1}.

Interestingly, $a_{n,0} = F_{n+1}$; this can be confirmed easily; see Exercise 45.5.

There is an explicit formula for $a_{n,k}$ [465]:

$$a_{n,k} = \sum_{i=k}^{\lfloor n/2 \rfloor} \binom{n-i}{i}\binom{i}{k};$$

see Exercise 45.6.

For example, $a_{7,2} = \sum_{i=2}^{3} \binom{7-i}{i}\binom{i}{2} = \binom{5}{2}\binom{2}{2} + \binom{4}{3}\binom{3}{2} = \textcircled{22}$; see Table 45.1.

The entries $a_{n,k}$ in Table 45.1 exhibit several interesting properties. The following theorem reveals some of them [51]. We prove one of them and leave the others as routine exercises; see Exercises 45.7–45.10.

Theorem 45.4 (Brigham, Chin, and Grimaldi, 1999 [51]). *Let $n \geq 0$. Then*

1) $a_{2n,n} = 1$, *where* $n \geq 1$;
2) $a_{2n+1,n} = n + 1$;
3) $a_{2n,n-1} = \left(\frac{1}{2}\right) n^2 + \left(\frac{3}{2}\right) n$, *where* $n \geq 1$;
4) $a_{2n+3,n} = \frac{n}{6}(n^2 + 12n + 29) + 3$;
5) $a_{2n,n-2} = \frac{n}{24}(n^3 + 14n^2 + 11n - 26)$, *where* $n \geq 2$.

Proof. We will prove part (4). Using parts (1)–(3), it follows from recurrence (45.6) that

$$\begin{aligned} a_{2n+3,n} &= a_{2n+2,n} + a_{2n+1,n} + a_{2n+1,n-1} \\ &= \left[\left(\frac{1}{2}\right)(n+1)^2 + \left(\frac{3}{2}\right)(n+1)\right] + (n+1) + a_{2n+1,n-1}. \end{aligned}$$

Letting $x_n = a_{2n+3,n}$, this gives the *nonhomogeneous recurrence with constant coefficients* (NHRWCCs) $x_n = x_{n-1} + \left(\frac{1}{2}\right) n^2 + \left(\frac{7}{2}\right) n + 3$.

The particular solution of the recurrence has the form $n(An^2 + Bn + C)$. Substituting this in the recurrence yields $A = \frac{1}{6}$, $B = 2$, and $C = \frac{29}{6}$. So the general solution is of the form $x_n = D + n\left(\frac{n^2}{6} + 2n + \frac{29}{6}\right)$. Since $x_0 = a_{3,0} = 3 = D$, we get $a_{2n+3,n} = x_n = \frac{n}{6}(n^2 + 12n + 29) + 3$. ■

For example, $a_{9,3} = a_{2\cdot 3+} = \frac{3}{6}(9 + 12 \cdot 3 + 29) + 3 = \boxed{40}$; see Table 45.1.

It follows from part (3) of Theorem 45.4 that $a_{2n+2,n} = \frac{1}{2}(n+1)(n+4)$. For example, $a_{8,3} = \frac{1}{2} 4 \cdot 7 = 14$; see Table 45.1.

The next theorem gives an unexpected dividend, a link between $a_{n,1}$, and Fibonacci and Lucas numbers.

Theorem 45.5 (Brigham, Chin, and Grimaldi, 1999 [51]). *Let* $n \geq 0$. *Then* $a_{n,1} = \frac{1}{5}(nL_n - F_n)$.

Proof. Since $a_{n,0} = F_{n+1}$, it follows by recurrence (45.6) that

$$\begin{aligned} a_{n+2,1} &= a_{n+1,1} + a_{n,1} + a_{n,0} \\ &= a_{n+1,1} + a_{n,1} + \frac{\alpha^{n+1} - \beta^{n+1}}{\alpha - \beta}. \end{aligned}$$

The general solution of this recurrence has the form $a_{n,1} = A\alpha^n + B\beta^n + Cn\alpha^n + Dn\beta^n$. Since $\alpha + 2 = \sqrt{5}\alpha$, substituting $Cn\alpha^n$ in the recurrence yields

$$\begin{aligned}[(n+2)\alpha^2 - (n+1)\alpha - n]C &= \frac{\alpha}{\sqrt{5}}\\ (\alpha+2)C &= \frac{\alpha}{\sqrt{5}}\\ C &= \frac{1}{5}.\end{aligned}$$

Likewise, $D = \dfrac{1}{5}$.

Then

$$\begin{aligned}a_{n,1} &= A\alpha^n + B\beta^n + \frac{n}{5}(\alpha^n + \beta^n)\\ &= A\alpha^n + B\beta^n + \frac{n}{5}L_n.\end{aligned}$$

Using the initial conditions $a_{0,1} = 0 = a_{1,1}$, we then get $A = -\dfrac{1}{5(\alpha - \beta)} = -B$. Thus

$$a_{n,1} = \frac{1}{5}(nL_n - F_n),$$

as desired. ■

For example, $a_{10,1} = \frac{1}{5}(10L_{10} - F_{10}) = \dfrac{10 \cdot 123 - 55}{5} = 235$; see Table 45.1.

The next theorem gives another unexpected gift, a close link between the triangular array $(a_{n,k})$ and Pell numbers P_n. To see this, we turn to the southeast diagonals of the array. The diagonal sums D_n appear to be Pell numbers; see Figure 45.5. The following theorem does indeed confirm this observation.

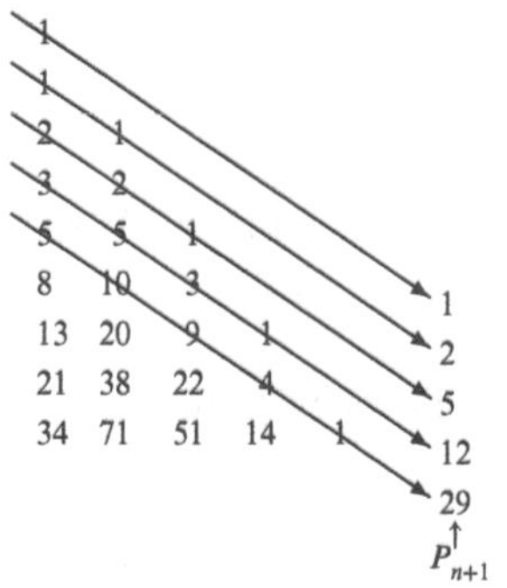

Figure 45.5. Southeast diagonal sums.

Theorem 45.6 (Brigham, Chin, and Grimaldi, 1999 [51]). *Let D_n denote the nth southeast diagonal sum of the array $(a_{n,k})$, where $n \geq 0$. Then $D_n = P_{n+1}$.*

Proof. Since $D_0 = 1 = P_1$ and $D_1 = 2 = P_2$, it suffices to show that D_n satisfies the Pell recurrence, where $n \geq 2$. We accomplish this using recurrence (45.6).

Since $a_{2n-1,n} = 0 = a_{2n-3,n-1} = a_{2n-2,n}$, we have

$$\begin{aligned}
D_n &= \sum_{i=0}^{n} a_{n+i,i} \\
&= (a_{n-1,0} + a_{n-2,0}) + \sum_{i=1}^{n} a_{n+i,i} \\
&= (a_{n-1,0} + a_{n-2,0}) + \sum_{i=1}^{n} (a_{n+i-1,i} + a_{n+i-2,i} + a_{n+i-2,i-1}) \\
&= \left(a_{n-1,0} + \sum_{i=1}^{n-1} a_{n+i-1,i} + 0 \right) + \left(a_{n-2,0} + \sum_{i=1}^{n-2} a_{n+i-2,i} + 0 + 0 \right) + \sum_{i=1}^{n} a_{n+i-2,i-1} \\
&= \sum_{i=0}^{n-1} a_{n-1+i,i} + \sum_{i=0}^{n-2} a_{n-2+i,i} + \sum_{i=0}^{n-1} a_{n-1+i,i} \\
&= D_{n-1} + D_{n-2} + D_{n-1} \\
&= 2D_{n-1} + D_{n-2}.
\end{aligned}$$

Consequently, $D_n = P_{n+1}$, as desired. ■

For example, $D_5 = \sum_{i=0}^{5} a_{5+i,i} = 8 + 20 + 22 + 14 + 5 + 1 = 70 = P_6$.

The triangular array yields another spectacular dividend. To see this, consider the northeast diagonal sums $E_n = \sum_{i=0}^{\lfloor n/3 \rfloor} a_{n-i,i}$, where $n \geq 0$; see Figure 45.6. They are the well-known *tribonacci numbers* T_n [51, 465], defined by the recurrence $T_n = T_{n-1} + T_{n-2} + T_{n-3}$, where $T_1 = 1 = T_2$, $T_3 = 2$, and $n \geq 4$ (see Chapter 49). The following theorem establishes this observation.

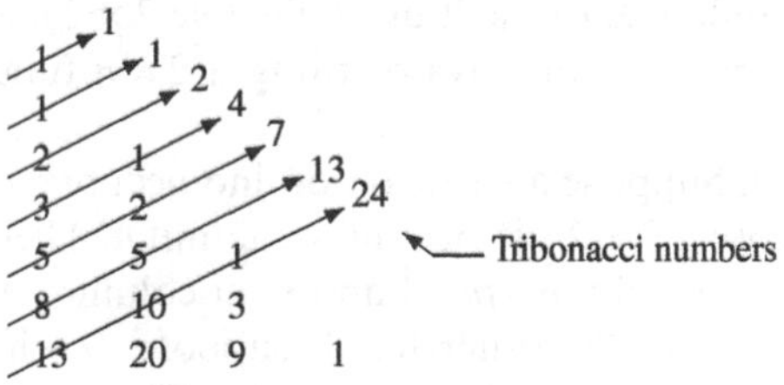

Figure 45.6. Northeast diagonal sums.

Theorem 45.7 (Brigham, Chin, and Grimaldi, 1999 [51]). *Let E_n denote the nth northeast diagonal sum of the array $(a_{n,k})$, where $n \geq 0$. Then $D_n = T_{n+1}$.*

Proof. Since $E_0 = 1 = T_1$, $E_1 = 1 = T_2$, and $E_2 = 2 = T_3$, it remains to show that E_n satisfies the tribonacci recurrence when $n \geq 3$. Since the upper limit of the formula for E_n is $\lfloor n/3 \rfloor$, we split the proof into three cases: $n \equiv 0 \pmod 3$, $n \equiv 1 \pmod 3$, and $n \equiv 2 \pmod 3$.

Suppose $n = 3k + 2$. Using recurrence (45.6), we then have

$$\begin{aligned}
&E_{n-1} + E_{n-2} + E_{n-3}\\
&\quad= E_{3k+1} + E_{3k} + E_{3k-1}\\
&\quad= \sum_{i=0}^{k} a_{3k+1-i,i} + \sum_{i=0}^{k} a_{3k-i,i} + \sum_{i=0}^{k-1} a_{3k-1-i,i}\\
&\quad= (a_{3k+1,0} + a_{3k,0}) + (a_{3k,1} + a_{3k-1,1} + a_{3k-1}, 0)\\
&\qquad + (a_{3k-1,2} + a_{3k-2,2} + a_{3k-2}, 1) + \cdots + (a_{2k+1,k} + a_{2k,k} + a_{2k,k-1})\\
&\quad= \sum_{i=0}^{k} a_{3k+2-i,i}\\
&\quad= E_{3k+2}\\
&\quad= E_n.
\end{aligned}$$

The cases $n = 3k$ and $n = 3k + 1$ follow similarly. Thus E_n satisfies the tribonacci recurrence in each case, so $E_n = T_{n+1}$, where $n \geq 0$. ■

For example, $E_7 = \sum_{i=0}^{\lfloor 7/3 \rfloor} a_{7-i,i} = a_{7,0} + a_{6,1} + a_{5,2} = 21 + 20 + 3 = 44 = T_8$.

The next section provides a fascinating combinatorial model for Jacobsthal–Lucas numbers [51].

45.3 2 × n TUBULAR TILINGS

Consider a $2 \times n$ board such that the right edge of column n is glued with the left edge of column 1, where the columns are labeled and $n \geq 1$. This results in a tubular board. We would like to tile it using flexible 2×2 and 2×1 tiles.

Let b_n denote the number of ways of tiling a $2 \times n$ tubular board. Clearly, $b_1 = 1$.

Let $n \geq 2$ and even. Suppose a vertical domino occupies column n. There are J_n such tilings. Suppose a 2×2 tile occupies column n. Then there are two possibilities: the tile occupies columns $n - 1$ and n, or columns n and 1. The number of such tilings is $2J_{n-1}$. On the other hand, suppose two horizontal dominoes occupy columns $n - 1$ and n, or columns n and 1. The number of such tilings is also $2J_{n-1}$.

Finally, the nth column can be occupied by two horizontal dominoes in two different ways: either the top tile is in columns n and 1, and the lower tile in columns $n-1$ and n, or the top tile is in columns $n-1$ and n, and the lower tile in columns n and 1.

Combining all the counts, we get $b_n = J_n + 4J_{n-1} + 2$, where $n \geq 2$ and is even.

On the other hand, let $n \geq 2$ and is odd. Then it follows from the above argument that $b_n = J_n + 4J_{n-1}$.

Thus

$$b_n = \begin{cases} 1 & \text{if } n = 1 \\ J_n + 4J_{n-1} & \text{if } n \geq 2 \text{ and is odd} \\ J_n + 4J_{n-1} + 2 & \text{if } n \geq 2 \text{ and is even.} \end{cases} \tag{45.7}$$

This formula yields the following theorem, where $b_0 = 4$.

Theorem 45.8 (Brigham, Chin, and Grimaldi, 1999 [51]). *Let b_n denote the number of tilings of a $2 \times n$ tubular board, where $n \geq 0$. Then*

1) $b_n = \begin{cases} j_n & \textit{if } n \textit{ is odd} \\ j_n + 2 & \textit{otherwise;} \end{cases}$

2) $b_n = b_{n-1} + 2b_{n-2} - 2$, *where* $b_0 = 4, b_1 = 1$, *and* $n \geq 2$;

3) $b_n = 2^n + 2(-1)^n + 1$, *where* $n \geq 0$.

Proof.

1) Clearly, $b_1 = 1 = j_1$. Now let $n \geq 2$ and odd. Then, by formula (45.7),

$$\begin{aligned} b_n &= \frac{2^n + 1}{3} + \frac{4(2^{n-1} - 1)}{3} \\ &= 2^n + (-1)^n \\ &= j_n. \end{aligned}$$

On the other hand, let n be even. Then, again by formula (45.7),

$$\begin{aligned} b_n &= \frac{2^n - 1}{3} + \frac{4(2^{n-1} + 1)}{3} + 2 \\ &= [2^n + (-1)^n] + 2 \\ &= j_n + 2. \end{aligned}$$

We leave the other two parts are routine exercises; see Exercises 45.11 and 45.12. ■

Next we study tubular tilings with exactly k square tiles.

Tubular Tilings with Exactly k Square Tiles

Let $b_{n,k}$ denote the number of tubular tilings with exactly k square tiles. Table 45.2 shows the values of $b_{n,k}$, where $1 \le n \le 10$ and $0 \le k \le \lfloor n/2 \rfloor$.

TABLE 45.2. Values of $b_{n,k}$, where $1 \le n \le 10$

n \ k	0	1	2	3	4	5	b_n
1	1						1
2	5	2					7
3	(4)	3					7
4	[9]	8	2				19
5	11	15	5				31
6	20	30	15	2			67
7	29	56	35	7			127
8	49	104	80	24	2		259
9	76	189	171	66	9		511
10	125	340	355	170	35	2	1027

The triangular array $B = b_{n,k}$ contains several interesting patterns. To begin with, column 0 reveals two interesting patterns:

$$\begin{array}{llll} b_{3,0} = \textcircled{4} & = 5 + 1 - 2 & = b_{2,0} + b_{1,0} - 2 \\ b_{5,0} = 11 & = 9 + 4 - 2 & = b_{4,0} + b_{3,0} - 2 \\ b_{7,0} = 29 & = 20 + 11 - 2 & = b_{6,0} + b_{5,0} - 2 \\ \quad\uparrow & \quad\vdots & \\ \quad L_n & & \end{array} \qquad \begin{array}{llll} b_{4,0} = \boxed{9} & = 4 + 5 & = b_{3,0} + b_{2,0} \\ b_{6,0} = 20 & = 11 + 9 & = b_{5,0} + b_{4,0} \\ b_{8,0} = 49 & = 4 + 5 & = b_{7,0} + b_{6,0} \\ \quad\uparrow & \quad\vdots & \\ \quad L_n + 2 & & \end{array}$$

More generally, consider $b_{n,0}$, where n is odd. Then $b_{n-1,0}$ counts the number of tubular tilings with a vertical domino in column n; $b_{n-2,0}$ counts those with two vertical dominoes in columns $n-1$ and n. But $b_{n-1,0}$ includes 2 tilings with two horizontal dominoes: the top tile in columns $n-1$ and n, and the bottom tile in columns n and 1; and the top tile in columns n and 1, and the bottom tile in columns $n-1$ and n. So we must discount 2 from the total count. Thus

$$b_n = b_{n-1,0} + b_{n-2,0} - 2, \tag{45.8}$$

where n is odd.

Likewise,

$$b_n = b_{n-1,0} + b_{n-2,0}, \tag{45.9}$$

where n is even.

These two recurrences, together with PMI, yield the following result; see Exercise 45.13.

Theorem 45.9 (Brigham, Chin, and Grimaldi, 1999 [51]).

$$b_{n,0} = \begin{cases} L_n & \text{if } n \text{ is odd} \\ L_n + 2 & \text{otherwise.} \end{cases}$$ ∎

Earlier, we found that $a_{n,0} = F_{n+1}$. Theorem 45.9 can now be invoked to establish a link between $a_{n,0}$ and $b_{n,0}$, as the next theorem shows. Its proof is simple and straightforward; so we omit it; see Exercise 45.14.

Theorem 45.10 (Brigham, Chin, and Grimaldi, 1999 [51]).

$$b_{n,0} = \begin{cases} a_{n-1,0} + 2a_{n-2,0} & \text{if } n \text{ is odd} \\ a_{n-1,0} + 2a_{n-2,0} + 2 & \text{otherwise.} \end{cases}$$ ∎

For example,

$$a_{6,0} + 2a_{5,0} = 13 + 2 \cdot 8 = 29 = b_{7,0};$$
$$a_{7,0} + 2a_{6,0} + 2 = 21 + 2 \cdot 13 = 49 = b_{8,0}.$$

Using Theorem 45.10, we can derive a recurrence for $b_{n,1}$, similar to equations (45.8) and (45.9). When n is odd,

$$\begin{aligned} b_{n,1} &= b_{n-1,1} + b_{n-2,1} + b_{n-2,0} \\ &= b_{n-1,1} + b_{n-2,1} + L_{n-2}; \end{aligned}$$

and when n is even,

$$\begin{aligned} b_{n,1} &= b_{n-1,1} + b_{n-2,1} + b_{n-2,0} - 2 \\ &= b_{n-1,1} + b_{n-2,1} + (L_{n-2} + 2) - 2 \\ &= b_{n-1,1} + b_{n-2,1} + L_{n-2}. \end{aligned}$$

Combining the two cases, we get

$$b_{n,1} = b_{n-1,1} + b_{n-2,1} + L_{n-2}, \tag{45.10}$$

for all $n \geq 2$.

This recurrence comes in handy in extracting an explicit formula for $b_{n,1}$. To this end, take a close look at the elements of column 1 in Table 45.2. They display an interesting pattern:

$$\begin{aligned}
b_{1,1} &= 0 = 1F_0\\
b_{2,1} &= 2 = 2F_1\\
b_{3,1} &= 3 = 3F_2\\
b_{4,1} &= 8 = 4F_3\\
b_{5,1} &= 15 = 5F_4\\
&\vdots
\end{aligned}$$

More generally, we have the following result.

Theorem 45.11 (Brigham, Chin, and Grimaldi, 1999 [51]). *Let* $n \geq 1$. *Then* $b_{n,1} = nF_{n-1}$. ■

The proof follows by recurrence (45.10) and PMI; so we omit it; see Exercise 45.15.

45.4 3 × *n* TILINGS

Suppose we would like to tile a $3 \times n$ board using 1×1 and 2×2 tiles. The *weight* of a 1×1 tile is 1 and that of a 2×2 tile is $x/2$. As usual, the *weight of a tiling* is the product of the weights of all tiles in the tiling. We define the weight of the empty tiling to be 1.

Figure 45.7 shows the tilings of a $3 \times n$ board and the sum of the weights of the tilings, where $0 \leq n \leq 4$. Using the experimental data, we conjecture that the sum of the weights of tilings of a $3 \times n$ board is $J_{n+1}(x)$. The next theorem confirms this observation.

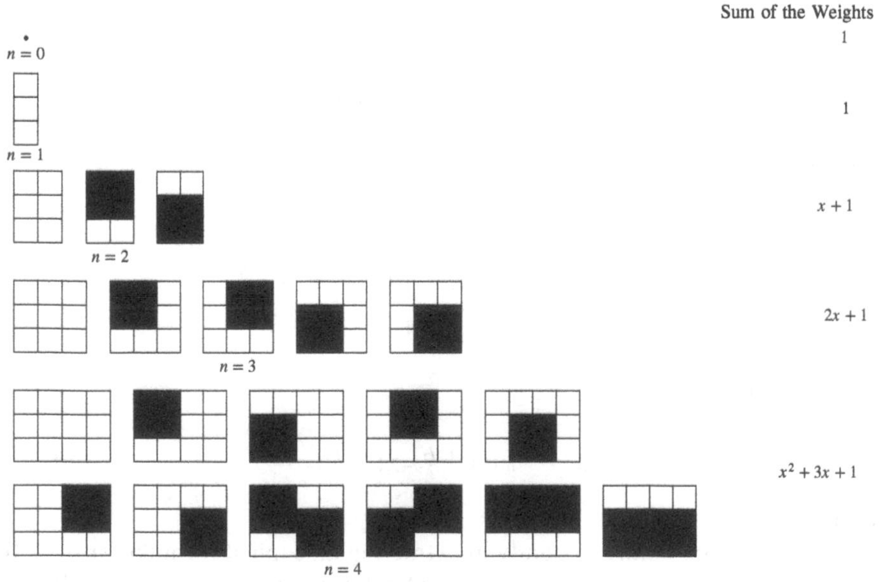

Figure 45.7. Tilings of a $3 \times n$ board, where $0 \leq n \leq 4$.

Theorem 45.12. *The sum of the weights of tilings of a $3 \times n$ board using 1×1 and 2×2 tiles is $J_{n+1}(x)$, where $w(1 \times 1$ tile$) = 1$, $w(2 \times 2$ tile$) = x/2$, and $w($empty tiling$) = 1$.*

Proof. Let $S_n(x)$ denote the sum of the weights of tilings of the board. Clearly, $S_0(x) = 1 = J_1(x)$ and $S_1(x) = 1 = J_2(x)$. We now confirm that $S_n(x)$ satisfies the Jacobsthal recurrence. To this end, consider an arbitrary tiling T of the board.

Suppose the last column of tiling T consists of unit squares:

$$T = \underbrace{3 \times (n-1) \text{ tiling}}$$ 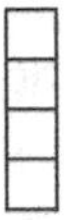.

The sum of the weights of such tilings is $S_{n-1}(x) \cdot 1 = S_{n-1}(x)$.

On the other hand, suppose the last column of T contains a 2×2 tile. There are two such possibilities: $\underbrace{3 \times (n-2) \text{ tiling}}$  and $\underbrace{3 \times (n-2) \text{ tiling}}$ . The sum of the weights of such tilings is $(x/2)S_{n-2}(x) + (x/2)S_{n-2}(x) = xS_{n-2}(x)$.

Combining the two cases, $S_n(x) = S_{n-1}(x) + xS_{n-2}(x)$, where $S_0(x) = J_1(x)$ and $S_1(x) = J_2(x)$ and $n \geq 2$. Thus $S_n(x) = J_{n+1}(x)$, as desired. ■

The next corollary is an immediate consequence of Theorem 45.12.

Corollary 45.1. *Let $n \geq 0$. Then the following hold.*

1) *The number of tilings of a $3 \times n$ board using 1×1 and 2×2 tilings is F_{n+1}.*
2) *Let $w(1 \times 1$ tile$) = 1 = w(2 \times 2$ tile$)$. Then the sum of the weights of tilings of a $3 \times n$ board is J_{n+1} [209, 465].* ■

Next we employ Model A to establish the Lucas-like explicit formula for Jacobsthal polynomials.

Theorem 45.13. *Let $n \geq 0$. Then $J_{n+1}(x) = \sum_{k=0}^{\lfloor n/2 \rfloor} \binom{n-k}{k} x^k$.*

Proof. Let T be an arbitrary tiling of a $3 \times n$ board. By Theorem 45.12, the sum of the weights of its tilings is $J_{n+1}(x)$.

We now compute this sum in a different way. To this end, suppose tiling T contains k 2×2 tiles in the top two rows of the board. Then the top row contains $(n-2k)$ 1×1 tiles. So the top row contains $(n-2k) + k = n-k$ tiles. Consequently, the k 2×2 tiles can be placed among those $(n-k)$ positions in $\binom{n-k}{k}$ different ways; that is, there are $\binom{n-k}{k}$ tilings T with k 2×2 tiles in their top two rows.

Sliding down by one row a 2×2 tile from T also results in a tiling with exactly k 2×2 tiles. So there are $\binom{n-k}{k} \cdot 2^k$ tilings with exactly k 2×2 tiles; and the sum of their weights is $\binom{n-k}{k} 2^k (x/2)^k = \binom{n-k}{k} x^k$. Thus the sum of the weights of all tilings T is $\sum_{k=0}^{\lfloor n/2 \rfloor} \binom{n-k}{k} x^k$. Equating the two sums yields the desired result. ■

For example, let $n = 4$. There is exactly one tiling with $k = 0$ 2×2 tiles; six tilings with $k = 1$ 2×2 tiles; and four tilings with $k = 2$ 2×2 tiles. The sum of their weights is $1 + 6(x/2) + 4(x/2)^2 = x^2 + 3x + 1 = \sum_{k=0}^{2} \binom{4-k}{k} x^k$.

TABLE 45.3. Jacobsthal Polynomials

$i \backslash j$	0	1	2	3	4	5	Sum of the Weights
0	1						1
1	1						1
2	1	x					$x + 1$
3	1	$2x$					$2x + 1$
4	1	$3x$	x^2				$x^2 + 3x + 1$
5	1	$4x$	$3x^2$				$3x^2 + 4x + 1$
6	1	$5x$	$6x^2$	x^3			$x^3 + 6x^2 + 5x + 1$

↑ $J_{n+1}(x)$

Suppose $t_{n,k} = t_{n,k}(x)$ denotes the sum of the weights of tilings of a $3 \times n$ board with exactly k 2×2 tiles, where $0 \le k \le \lfloor n/2 \rfloor$. Table 45.3 shows the possible values of $t_{n,k}$, where $0 \le n \le 6$; see Figure 45.7.

It follows from the proof of Theorem 45.13 that $t_{n,k} = \binom{n-k}{k} x^k$ and hence $\sum_{k=0}^{\lfloor n/2 \rfloor} t_{n,k} = J_{n+1}(x)$, as expected.

In particular, let $x = 2$. Then the row sums yield the Jacobsthal numbers; see Table 45.4.

Here is an interesting recurrence satisfied by $t_{n,k}$: $t_{n,k} = x t_{n-2,k-1} + t_{n-1,k}$; see Exercise 45.16. For example, $t_{6,2} = 6x^2 = x(3x) + 3x^2 = x t_{4,1} + t_{5,2}$; see Tables 45.3 and 45.4.

TABLE 45.4. Jacobsthal Numbers

$i \backslash j$	0	1	2	3	4	5	Row Sums
0	1						1
1	1						1
2	1	2					3
3	1	4					5
4	1	6	4				11
5	1	8	12				21
6	1	10	24	8			43

↑ J_{n+1}

Breakability

We can extend the concept of breakability to the tilings of a $3 \times n$ board. To this end, first we label the columns 1 through n from left to right. A tiling is *unbreakable* at column i if a 2×2 tile occupies columns i and $i + 1$; otherwise, it is *breakable* at column i.

For example, consider the tiling in Figure 45.8. It is breakable at columns 1, 3, 5, 7, 8, and 10, but unbreakable at columns 2, 4, 6, and 9.

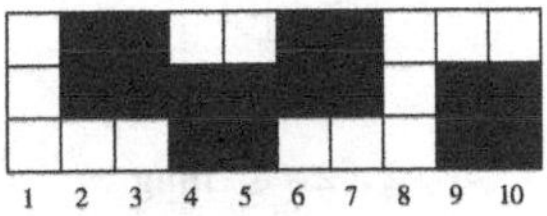

Figure 45.8.

The concept of breakability comes in handy in developing an *addition formula* for Jacobsthal polynomials, as the following theorem shows.

Theorem 45.14. *Let* $m, n \geq 1$. *Then* $J_{m+n}(x) = J_{m+1}(x)J_n(x) + xJ_m(x)J_{n-1}(x)$.

Proof. Consider a $3 \times (m + n - 1)$ board. By Theorem 45.12, the sum of the weights of its tilings is $J_{m+n}(x)$.

Let T be an arbitrary tiling of the board. Suppose it is breakable at column m: $\underbrace{\text{subtiling}}_{\text{length } m}\underbrace{\text{subtiling}}_{\text{length } n-1}$. The sum of the weights of such tilings is $J_{m+1}(x)J_n(x)$.

On the other hand, suppose T is *not* breakable at column m:

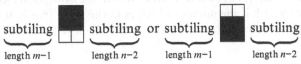

$\underbrace{\text{subtiling}}_{\text{length } m-1}$  $\underbrace{\text{subtiling}}_{\text{length } n-2}$ or $\underbrace{\text{subtiling}}_{\text{length } m-1}$  $\underbrace{\text{subtiling}}_{\text{length } n-2}$

The sum of the weights of such tilings is $J_m(x)(x/2)J_{n-1}(x) + J_m(x)(x/2)J_{n-1}(x) = xJ_m(x)J_{n-1}(x)$.

Combining the two cases, the sum of the weights of the tilings of the board is $J_{m+1}(x)J_n(x) + xJ_m(x)J_{n-1}(x)$. The addition formula now follows by equating the two sums. ■

It follows from Theorem 45.14 that $F_{m+n} = F_{m+1}F_n + F_mF_{n-1}$ and $J_{m+n} = J_{m+1}J_n + 2J_mJ_{n-1}$.

Finally, we add an interesting tidbit. Let a_n denote the number of tilings of a $2 \times n$ board with 2×1 and 2×2 tiles, and b_n the number of tilings of a $3 \times n$ board with 1×1 and 2×2 tiles. Recall that $a_n = J_{n+1} = b_n$. In addition, let c_n denote the number of ways of filling a $2 \times 2 \times n$ hole with $2 \times 2 \times 1$ bricks. Then $c_n = J_{n+1}$ gives a three-dimensional interpretation of Jacobsthal numbers [465]. Thus $a_n = b_n = c_n = J_{n+1}$. Griffiths illustrated this short chain of equalities using a bijective proof "without words" in 2009 [197]; see Figure 45.9.

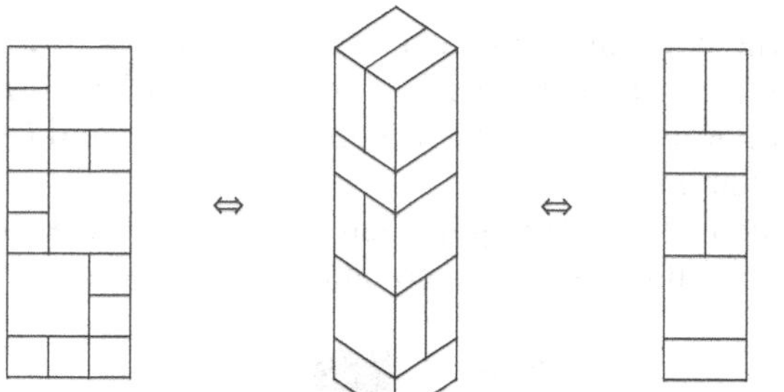

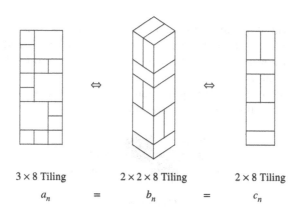

3×8 Tiling $\qquad$ $2 \times 2 \times 8$ Tiling $\qquad$ 2×8 Tiling

$a_n \quad = \quad b_n \quad = \quad c_n$

Figure 45.9. A bijective proof without words.

Next we construct graph-theoretic models for the Jacobsthal family.

45.5 GRAPH-THEORETIC MODELS

Consider a graph with n vertices $v_1, v_2, \ldots, v_n$. An edge from v_i to v_j is denoted by v_i–v_j, i–j, or by the "word" ij for brevity.

A *path* from v_i to v_j in a connected graph is a sequence v_i–e_i–v_{i+1}–$\cdots$–v_{j-1}–e_{j-1}–v_j of vertices v_k and directed edges e_k, where edge e_k is incident with vertices v_k and v_{k+1}. The path is *closed* if its endpoints are the same; otherwise, it is *open*. The *length* ℓ of a path is the number of edges in the path; that is, it takes ℓ steps to reach from one endpoint to the other.

Model A

Consider the complete graph K_3 with three vertices v_1, v_2, and v_3 [276]; see Figure 45.10.

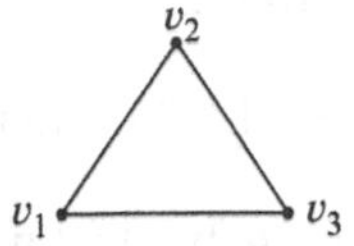

Figure 45.10. A complete graph K_3.

Its adjacency matrix A is given by

$$A = \begin{bmatrix} 0 & 1 & 1 \\ 1 & 0 & 1 \\ 1 & 1 & 0 \end{bmatrix}.$$

It is a *binary matrix*, meaning each entry a_{ij} is a *bit* (0 or 1). It is also *circulant*, meaning each row (column) can be obtained from the previous row (column) by cyclically shifting it by one element.

The following theorem gives an explicit formula for A^n.

Theorem 45.15. *Let $n \geq 1$. Then*

$$A^n = \begin{bmatrix} a & J_n & J_n \\ J_n & a & J_n \\ J_n & J_n & a \end{bmatrix},$$

where $a = \begin{cases} J_n - 1 & \text{if } n \text{ is odd} \\ J_n + 1 & \text{otherwise.} \end{cases}$ ■

The proof follows by PMI; so we omit it; see Exercise 45.17.

Theorem 45.15 has several interesting byproducts. For example, since $|A| = 2$, it follows that $|A^n| = 2^n$. This also follows from Theorem 45.15; see Exercise 45.18. The next three corollaries reveal some additional ones.

Corollary 45.2.

1) *Let n be odd. Then there are exactly $J_n - 1$ closed paths of length n from v_i to itself, where $1 \leq i \leq 3$; otherwise, there are $J_n + 1$ such paths.*
2) *There are exactly J_n paths of length n from v_i to v_j, where $1 \leq i, j \leq 3$ and $i \neq j$.* ■

For example, there are six closed paths of length 4 from v_1 to itself:

$$12121 \quad 13121 \quad 12131 \quad 13131 \quad 12321 \quad 13231;$$

and five paths of length 4 from v_1 to v_3:

$$12123 \quad 12313 \quad 12323 \quad 13123 \quad 13213;$$

where we have denoted the edge $v_i{-}v_j$ by the "word" ij for brevity.

Let $a_{11}^{(n)}$ denote the leading element in A^n. The values of $a_{11}^{(n)}$ manifest an interesting pattern:

$$\begin{aligned}
a_{11}^{(1)} &= 0 = 2 \cdot 0\\
a_{11}^{(2)} &= 2 = 2 \cdot 1\\
a_{11}^{(3)} &= 0 = 2 \cdot 1\\
a_{11}^{(4)} &= 0 = 2 \cdot 3\\
a_{11}^{(5)} &= 0 = 2 \cdot 5\\
a_{11}^{(5)} &= 0 = 2 \cdot \underset{\substack{\uparrow\\ J_{n-1}}}{11}.
\end{aligned}$$

More generally, we have the following result. The proof is straightforward; see Exercise 45.19.

Corollary 45.3. *Let* $n \geq 1$. *Then* $a_{11}^{(n)} = 2J_{n-1}$. ■

The next result follows from this; see Exercise 45.20.

Corollary 45.4. *Let* $n \geq 1$. *Then* $\text{trace}(A^n) = 6J_{n-1}$. ■

Next we present another model for Jacobsthal numbers using a path graph with three vertices v_1, v_2, and v_3.

Model B

Consider the *path graph* P_3 with a loop at the middle vertex v_2; see Figure 45.11 [465]. Its adjacency matrix A is the binary matrix

$$A = \begin{bmatrix} 0 & 1 & 0\\ 1 & 1 & 1\\ 0 & 1 & 0 \end{bmatrix}.$$

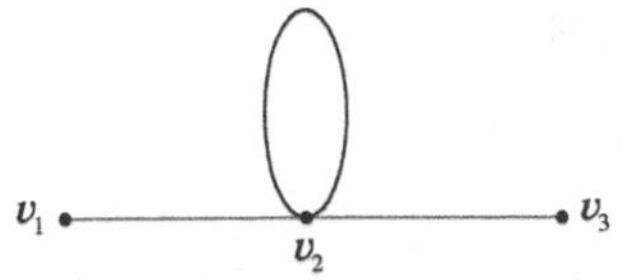

Figure 45.11. Path graph P_3 with a loop at v_2.

The first six powers of A show a spectacular pattern:

$$A = \begin{bmatrix} \boxed{0} & 1 & 0 \\ \boxed{\boxed{1}} & 1 & 1 \\ 0 & 1 & 0 \end{bmatrix} \quad A^2 = \begin{bmatrix} \boxed{1} & 1 & 1 \\ \boxed{\boxed{1}} & 3 & 1 \\ 1 & 1 & 1 \end{bmatrix} \quad A^3 = \begin{bmatrix} \boxed{1} & 3 & 1 \\ \boxed{\boxed{3}} & 5 & 3 \\ 1 & 3 & 1 \end{bmatrix}$$

$$A^4 = \begin{bmatrix} \boxed{3} & 5 & 3 \\ \boxed{\boxed{5}} & 11 & 5 \\ 3 & 5 & 5 \end{bmatrix} \quad A^5 = \begin{bmatrix} \boxed{5} & 11 & 5 \\ \boxed{\boxed{11}} & 21 & 11 \\ 5 & 11 & 5 \end{bmatrix} \quad A^6 = \begin{bmatrix} \boxed{11} & 21 & 11 \\ \boxed{\boxed{21}} & 43 & 21 \\ 11 & 21 & 11 \end{bmatrix}.$$

Using the pattern, we conjecture that

$$A^n = \begin{bmatrix} J_{n-1} & J_n & J_{n-1} \\ J_n & J_{n+1} & J_n \\ J_{n-1} & J_n & J_{n-1} \end{bmatrix}.$$

The next theorem confirms it. The proof is simple and straightforward; so we omit it; see Exercise 45.21.

Theorem 45.16. *Let A be the adjacency matrix of the graph in Figure 45.11 and $n \geq 1$. Then*

$$A^n = \begin{bmatrix} \boxed{J_{n-1}} & J_n & J_{n-1} \\ \boxed{\boxed{J_n}} & J_{n+1} & J_n \\ J_{n-1} & J_n & J_{n-1} \end{bmatrix}.$$

■

It follows from Theorem 45.16 that every element of A^n is a Jacobsthal number, a charming property we did not encounter earlier. Consequently, the number of paths of length n from v_i to v_j is a Jacobsthal number, where $1 \leq i \leq j \leq 3$.

For example, there are $11 = J_5$ paths of length 4 from v_2 to itself:

21212 21222 21232 22122 22212 22222
22232 22322 23212 23222 23232.

The next section presents three intriguing occurrences of the Jacobsthal family in the study of digraphs [291, 465]. A *digraph* is a *di*rected *graph*, meaning every edge is directed.

45.6 DIGRAPH MODELS

Model D

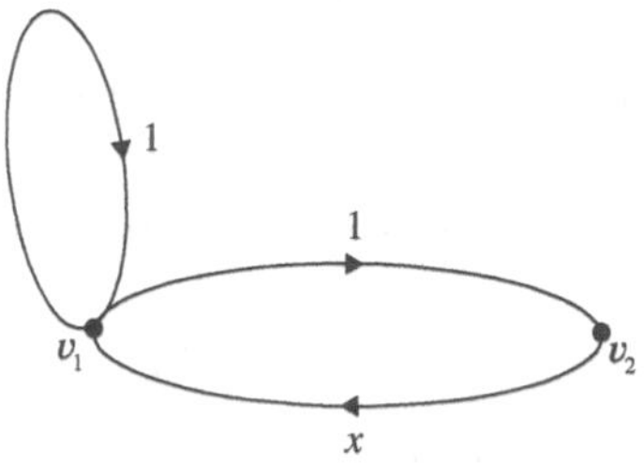

Figure 45.12. Weighted digraph.

Consider the digraph with two vertices v_1 and v_2 in Figure 45.12. It has a loop at v_1, an edge from v_1 to v_2, and an edge from v_2 to v_1. These edges are assigned the *weights* 1, 1, and x, respectively, where $x \geq 1$. The *weight of a directed path* is the product of the weights of all edges along the path.

The *weighted adjacency matrix* of the digraph is given by

$$A = A(x) = \begin{bmatrix} 1 & 1 \\ x & 0 \end{bmatrix}.$$

The next result follows by induction, so we omit its proof for brevity.

Theorem 45.17. *Let $n \geq 1$. Then*

$$A^n = \begin{bmatrix} J_{n+1}(x) & J_n(x) \\ xJ_n(x) & xJ_{n-1}(x) \end{bmatrix}.$$

■

Let $|D|$ denote the determinant of the square matrix D. It then follows from Theorem 45.17 that

$$|A^n| = \begin{vmatrix} J_{n+1}(x) & J_n(x) \\ xJ_n(x) & xJ_{n-1}(x) \end{vmatrix}$$

$$= x[J_{n+1}(x)J_{n-1}(x) - J_n^2(x)].$$

Since $|A^n| = |A|^n = (-x)^n$, it then follows that $x[J_{n+1}(x)J_{n-1}(x) - J_n^2(x)] = (-x)^n$. This yields the Cassini-like formula we encountered in Section 44.1.

In particular, $J_{n+1}J_{n-1} - J_n^2 = -(-2)^{n-1}$, where $n \geq 1$. Since $J_n(1) = F_n$, Cassini's formula for Fibonacci numbers also follows from it: $F_{n+1}F_{n-1} - F_n^2 = (-1)^n$ [287].

Let $m, n \geq 1$. It follows by Theorem 45.17 that

$$A^{m+(n-1)} = \begin{bmatrix} J_{m+n}(x) & J_{m+n-1}(x) \\ xJ_{m+n-1}(x) & xJ_{m+n-2}(x) \end{bmatrix}. \tag{45.11}$$

But

$$\begin{aligned} A^{m+(n-1)} &= A^m \cdot A^{n-1} \\ &= \begin{bmatrix} J_{m+1}(x) & J_m(x) \\ xJ_m(x) & xJ_{m-1}(x) \end{bmatrix} \begin{bmatrix} J_n(x) & J_{n-1}(x) \\ xJ_{n-1}(x) & xJ_{n-2}(x) \end{bmatrix} \\ &= \begin{bmatrix} J_{m+1}(x)J_n(x) + xJ_m(x)J_{n-1}(x) & * \\ * & * \end{bmatrix}. \end{aligned} \tag{45.12}$$

Equating the corresponding elements in equations (45.11) and (45.12), we get the *addition formula* for Jacobsthal polynomials:

$$J_{m+n}(x) = J_{m+1}(x)J_n(x) + xJ_m(x)J_{n-1}(x).$$

In particular, this implies

$$\begin{aligned} F_{m+n} &= F_{m+1}F_n + F_mF_{n-1}; \\ J_{m+n} &= J_{m+1}J_n + 2J_mJ_{n-1}. \end{aligned}$$

We can use the weighted adjacency matrix to find the sum of the weights of paths of length n from vertex v_i to vertex v_j, as the following theorem shows [276].

Theorem 45.18. *Let $A^n = (a_{ij}^{(n)})$, where $n \geq 1$. Then the element $(a_{ij}^{(n)})$ gives the sum of the weights of the edges of the paths of length n from vertex v_i to vertex v_j, where $1 \leq i, j \leq 2$.* ■

For example,

$$A^4 = \begin{bmatrix} J_5(x) & J_4(x) \\ xJ_4(x) & xJ_3(x) \end{bmatrix} = \begin{bmatrix} x^2 + 3x + 1 & 2x + 1 \\ 2x^2 + x & x^2 + x \end{bmatrix}.$$

So the sum of the weights of the edges of closed paths of length 4 at v_1 is $x^2 + 3x + 1$. Table 45.5 shows the corresponding paths, their weights, and the cumulative sum.

Since $J_k(1) = F_k$, the next corollary follows from Theorem 45.18.

Corollary 45.5. *The element $(a_{ij}^{(n)})$ of the matrix $A^n(1)$ gives the number of paths of length n from vertex v_i to vertex v_j, where $1 \leq i, j \leq 2$.* ■

TABLE 45.5. Closed Paths of Length 4 at v_1

Closed Paths of Length 4	Weights of Paths
11111	1
11121	x
11211	x
12111	x
12121	x^2
Cumulative sum	$x^2 + 3x + 1$

For example, there are $(a_{ij}^{(n)})(1) = 5$ closed paths of length 4 at v_1; see Table 45.5.

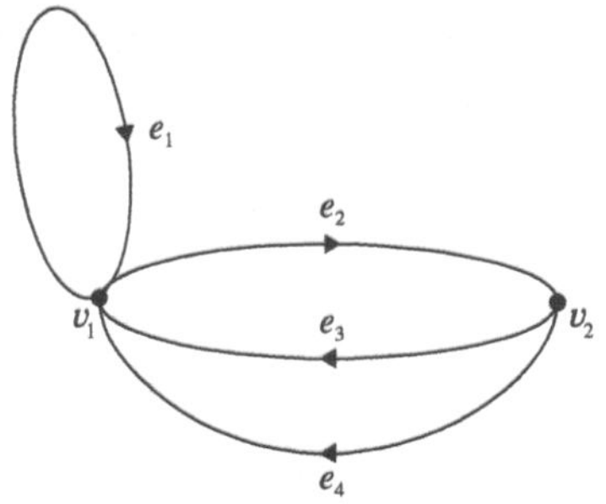

Figure 45.13. A digraph with four edges.

Suppose we interpret the weight of a directed edge v_i–v_j to mean the number of directed edges from v_i to v_j. Let $x = 2$. Then Figure 45.13 shows the corresponding digraph with four edges e_1, e_2, e_3, and e_4.

It then follows from Theorem 45.17 that

$$A^n = \begin{bmatrix} J_{n+1} & J_n \\ 2J_n & 2J_{n-1} \end{bmatrix}.$$

Consequently, for example, there are exactly J_{n+1} closed paths of length n at v_1 and $2J_{n-1}$ such paths at v_2.

Since

$$2J_n = \begin{cases} J_{n+1} + 1 & \text{if } n \text{ is odd} \\ J_{n+1} - 1 & \text{otherwise,} \end{cases}$$

we can rewrite A^n as follows. If n is odd, then

$$A^n = \begin{bmatrix} J_{n+1} & J_n \\ J_{n+1} + 1 & J_n - 1 \end{bmatrix};$$

otherwise,

$$A^n = \begin{bmatrix} J_{n+1} & J_n \\ J_{n+1} - 1 & J_n + 1 \end{bmatrix}.$$

For example, there are $J_5 = 11$ closed paths of length 4 at v_1; and $J_4 + 1 = 6$ closed paths of length 4 at v_2; see Table 45.6.

TABLE 45.6. Closed Paths of Length 4 at v_1 and v_2

Closed Paths of Length 4 at v_1		Closed Paths of Length 4 at v_2
$e_1e_1e_1e_1$	$e_2e_3e_1e_1$	$e_3e_1e_1e_2$
$e_1e_1e_2e_3$	$e_2e_3e_2e_3$	$e_3e_2e_3e_2$
$e_1e_1e_2e_4$	$e_2e_3e_2e_4$	$e_3e_2e_4e_2$
$e_1e_2e_3e_1$	$e_2e_4e_1e_1$	$e_4e_1e_1e_2$
$e_1e_2e_4e_1$	$e_2e_4e_2e_3$	$e_4e_2e_3e_2$
	$e_2e_4e_2e_4$	$e_4e_2e_4e_2$

When n is odd, $J_n = \frac{1}{3}(j_n + 2)$; otherwise, $J_n = \frac{1}{3}(j_n - 2)$. Consequently, the matrix A^n can be rewritten in terms of Jacobsthal–Lucas numbers:

$$A^n = \begin{cases} \dfrac{1}{3}\begin{bmatrix} j_{n+1} - 2 & j_n + 2 \\ j_{n+1} + 1 & j_n - 1 \end{bmatrix} & \text{if } n \equiv 1 \pmod 2 \\[2ex] \dfrac{1}{3}\begin{bmatrix} j_{n+1} + 2 & j_n - 2 \\ j_{n+1} - 1 & j_n + 1 \end{bmatrix} & \text{otherwise.} \end{cases}$$

For example,

$$A^5 = \frac{1}{3}\begin{bmatrix} j_6 - 2 & j_5 + 2 \\ j_6 + 1 & j_5 - 1 \end{bmatrix} = \begin{bmatrix} 21 & 11 \\ 22 & 10 \end{bmatrix};$$

$$A^6 = \frac{1}{3}\begin{bmatrix} j_7 + 2 & j_6 - 2 \\ j_7 - 1 & j_6 + 1 \end{bmatrix} = \begin{bmatrix} 43 & 21 \\ 42 & 22 \end{bmatrix}.$$

Model E

Consider the digraph in Figure 45.14 with three vertices v_1, v_2, and v_3, and six directed edges [279, 281]. Table 45.7 shows all possible paths of length 4 from v_i to v_j, where $1 \le i,j \le 3$. Notice that, by symmetry, a path from v_j to v_i is the same as the one from v_i to v_j in the reverse order.

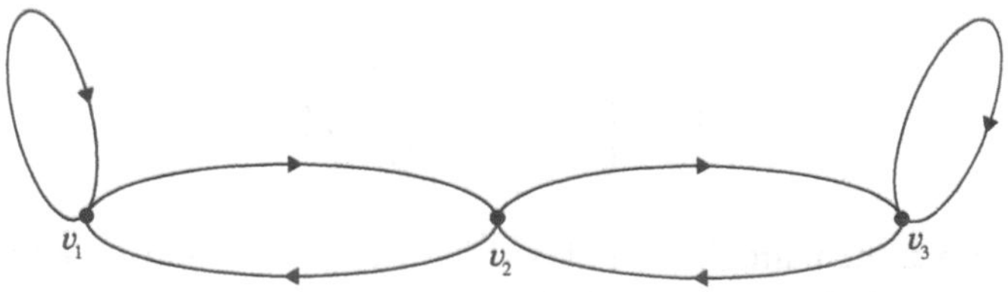

Figure 45.14. A digraph model for the Jacobsthal family.

TABLE 45.7. Directed Paths of Length 4

Paths from v_1 to v_1	Paths from v_1 to v_2	Paths from v_1 to v_3	Paths from v_2 to v_2	Paths from v_2 to v_3	Paths from v_3 to v_3
11111	11112	11123	21112	21123	32123
11121	11212	11233	21212	21233	32323
11211	11232	12123	21232	23233	32333
12111	12112	12323	23332	23333	33233
12121	12332	12333	23212		33323
12321			23232		33333

Adjacency Matrix

The adjacency matrix M of the digraph is the circulant matrix

$$M = \begin{bmatrix} 1 & 1 & 0 \\ 1 & 0 & 1 \\ 0 & 1 & 1 \end{bmatrix}.$$

Let $|M|$ denote the determinant of matrix M. Then $|M| = -2$; so $|M^n| = |M|^n = (-2)^n$, where $n \geq 1$.

The adjacency matrix M can be used to compute the number of directed paths of a given length n between any two vertices, as the next theorem confirms; its proof follows by induction [277].

Theorem 45.19. *Let M be the adjacency matrix of a connected digraph with vertices $v_1, v_2, \ldots, v_k$, and n a positive integer. Then the ij-th entry of the matrix M^n gives the number of directed paths of length n from v_i to v_j, where $1 \leq i,j \leq n$.* ∎

For example,

$$M^4 = \begin{bmatrix} 6 & 5 & 5 \\ 5 & 6 & 5 \\ 5 & 5 & 6 \end{bmatrix}.$$

So there are six directed paths from v_1 to itself, and five from v_1 to v_3; see Table 45.7.

Although we do not yet know the entries of matrix M^n, we will find them a bit later using an indirect route. The next theorem paves the way.

Theorem 45.20. *Let n be a positive integer. If n is odd, M^n has the form*

$$M^n = \begin{bmatrix} a+1 & a+1 & a \\ a+1 & a & a+1 \\ a & a+1 & a+1 \end{bmatrix},$$

where a is a nonnegative integer; otherwise, M^n is of the form

$$M^n = \begin{bmatrix} a+1 & a & a \\ a & a+1 & a \\ a & a & a+1 \end{bmatrix},$$

where a is a positive integer. ■

The proofs of both cases follow by induction; so in the interest of brevity, we omit them.

Let n be odd. Then, by Theorem 45.20,

$$|M^n| = \begin{vmatrix} a+1 & a+1 & a \\ a+1 & a & a+1 \\ a & a+1 & a+1 \end{vmatrix}$$
$$= -(3a+2).$$

Consequently, $|M^n| = -(3a+2) = (-2)^n$ and hence $3a+2 = 2^n$.

For example, let $n = 7$. Then

$$M^7 = \begin{bmatrix} 43 & 43 & 42 \\ 43 & 42 & 43 \\ 42 & 43 & 43 \end{bmatrix}.$$

So $a = 42$ and hence $3a+2 = 128 = 2^7$, as desired.

On the other hand, let n be even. Then

$$|M^n| = \begin{vmatrix} a+1 & a & a \\ a & a+1 & a \\ a & a & a+1 \end{vmatrix}$$
$$= 3a+1.$$

So $|M^n| = 3a+1 = (-2)^n = 2^n$.

For example, let $n = 10$. Then

$$M^{10} = \begin{bmatrix} 342 & 341 & 341 \\ 341 & 342 & 341 \\ 341 & 341 & 342 \end{bmatrix}.$$

Consequently, $a = 341$ and hence $3a + 1 = 1024 = 2^{10}$, as expected.

Digraph and J_n

When n is odd, $a = \dfrac{2^n - 2}{3} = J_n - 1$; and when n is even, $a = \dfrac{2^n - 1}{3} = J_n$. Theorem 45.20 then offers a spectacular interpretation of J_n, as the next corollary reveals.

Corollary 45.6. *Let n be a positive integer. If n is odd, then*

$$M^n = \begin{bmatrix} J_n & J_n & J_n - 1 \\ J_n & J_n - 1 & J_n \\ J_n - 1 & J_n & J_n \end{bmatrix};$$

otherwise,

$$M^n = \begin{bmatrix} J_n + 1 & J_n & J_n \\ J_n & J_n + 1 & J_n \\ J_n & J_n & J_n + 1 \end{bmatrix}.$$

■

It now follows that there are exactly J_n paths of length n from v_1 to itself if n is odd, and $J_n + 1$ such paths if n is even.

Let m and n be odd. Then, by Corollary 45.6,

$$\begin{aligned} M^{m+n} &= M^m \cdot M^n \\ &= \begin{bmatrix} J_m & J_m & J_m - 1 \\ J_m & J_m - 1 & J_m \\ J_m - 1 & J_m & J_m \end{bmatrix} \begin{bmatrix} J_n & J_n & J_n - 1 \\ J_n & J_n - 1 & J_n \\ J_n - 1 & J_n & J_n \end{bmatrix} \\ &= \begin{bmatrix} 2J_mJ_n + (J_m - 1)(J_n - 1) & * & * \\ * & * & * \\ * & * & * \end{bmatrix}, \end{aligned} \tag{45.13}$$

where an asterisk denotes some element of the matrix.

Since $m+n$ is even, again by Corollary 45.6, we have

$$M^{m+n} = \begin{bmatrix} J_{m+n}+1 & J_{m+n} & J_{m+n} \\ J_{m+n} & J_{m+n}+1 & J_{m+n} \\ J_{m+n} & J_{m+n} & J_{m+n}+1 \end{bmatrix}. \tag{45.14}$$

It follows from equations (45.13) and (45.14) that

$$J_{m+n} = 3J_mJ_n - J_m - J_n, \tag{45.15}$$

where both m and n are odd.

For example, $J_{12} = J_{7+5} = 3J_7J_5 - J_7 - J_5 = 3 \cdot 43 \cdot 11 - 43 - 11 = 1365$.

In particular, equation (45.15) implies

$$J_{2n} = 3J_n^2 - 2J_n.$$

Consequently, $\dfrac{J_{2n}+2J_n}{3}$ is a Jacobsthal square, where n is odd.

For example,

$$\frac{J_{14}+2J_7}{3} = \frac{5461+2\cdot 43}{3} = 1849 = 43^2 = J_7^2.$$

On the other hand, let n be even. Then $J_{2n} = 3J_n^2 + 2J_n$; so $\dfrac{J_{2n}-2J_n}{3}$ is a Jacobsthal square.

Digraph and j_n

Since $j_n = 2^n + (-1)^n$, Corollary 45.6 can be expressed in terms of Jacobsthal–Lucas numbers also, as the following corollary shows.

Corollary 45.7. *If n is odd, then*

$$M^n = \frac{1}{3}\begin{bmatrix} j_n+2 & j_n+2 & j_n-1 \\ j_n+2 & j_n-1 & j_n+2 \\ j_n-1 & j_n+2 & j_n+2 \end{bmatrix};$$

otherwise,

$$M^n = \frac{1}{3}\begin{bmatrix} j_n+1 & j_n-2 & j_n-2 \\ j_n-2 & j_n+1 & j_n-2 \\ j_n-2 & j_n-2 & j_n+1 \end{bmatrix}.$$

■

This corollary has an interesting relationship with *Mersenne numbers* $M_p = 2^p - 1$, where p is an odd prime p. Then $\frac{1}{3}(j_p + 2) = \frac{1}{3}(M_p + 2)$ gives the number of closed paths of length p from v_1 to itself.

Since $|M^n| = (-2)^n$, the two corollaries yield the following apparently "charming" Jacobsthal and Jacobsthal–Lucas identities:

Let n be a positive integer. Then

$$\begin{vmatrix} J_n & J_n & J_n - 1 \\ J_n & J_n - 1 & J_n \\ J_n - 1 & J_n & J_n \end{vmatrix} = -2^n = \frac{1}{3}\begin{vmatrix} j_n + 2 & j_n + 2 & j_n - 1 \\ j_n + 2 & j_n - 1 & j_n + 2 \\ j_n - 1 & j_n + 2 & j_n + 2 \end{vmatrix}$$

if n is odd; and

$$\begin{vmatrix} J_n + 1 & J_n & J_n \\ J_n & J_n + 1 & J_n \\ J_n & J_n & J_n + 1 \end{vmatrix} = 2^n = \frac{1}{3}\begin{vmatrix} j_n + 1 & j_n - 2 & j_n - 2 \\ j_n - 2 & j_n + 1 & j_n - 2 \\ j_n - 2 & j_n - 2 & j_n + 1 \end{vmatrix}$$

if n is even. These are in fact the Binet-like formulas in disguise.

For example,

$$\begin{vmatrix} J_9 & J_9 & J_9 - 1 \\ J_9 & J_9 - 1 & J_9 \\ J_9 - 1 & J_9 & J_9 \end{vmatrix} = \begin{vmatrix} 171 & 171 & 170 \\ 171 & 170 & 171 \\ 170 & 171 & 171 \end{vmatrix} = -2^9;$$

$$\begin{vmatrix} J_{12} + 1 & J_{12} & J_{12} \\ J_{12} & J_{12} + 1 & J_{12} \\ J_{12} & J_{12} & J_{12} + 1 \end{vmatrix} = \begin{vmatrix} 1366 & 1365 & 1365 \\ 1365 & 1366 & 1365 \\ 1365 & 1365 & 1366 \end{vmatrix} = 2^{12}.$$

Corollaries 45.6 and 45.7 have interesting byproducts; they can be used to extract addition formulas for the Jacobsthal family, as the next corollary shows.

Corollary 45.8. *If* $r \equiv 0 \equiv s \pmod 2$, *then*

$$J_{r+s} = (J_r + 1)J_s + J_r(J_s + 1) + J_rJ_s$$

$$3j_{r+s} = (j_r + 1)(j_s + 1) + 2(j_r - 2)(j_s - 2) - 3.$$

If $r \equiv 1 \equiv s \pmod 2$, *then*

$$J_{r+s} = J_rJ_s + J_r(J_s - 1) + (J_r - 1)J_s$$

$$3j_{r+s} = 2(j_r + 2)(j_s + 2) + (j_r - 1)(j_s - 1) - 3.$$

If $r \equiv 1 \pmod 2$ *and* $s \equiv 0 \pmod 2$, *then*

$$J_{r+s} = J_r(J_s + 1) + J_r J_s + (J_r - 1)J_s$$

$$3j_{r+s} = 2(j_r + 2)(j_s - 2) + (j_r - 1)(j_s + 1) + 3.$$

■

In particular, this corollary yields the following properties:
If $r \equiv 0 \pmod 2$, then

$$J_{2r} = 2(J_r + 1)J_r + J_r^2$$

$$3j_{2r} = (j_r + 1)^2 + 2(j_r - 2)^2 - 3.$$

If $r \equiv 1 \pmod 2$, then

$$J_{2r} = 2J_r(J_r - 1) + J_r^2$$

$$3j_{2r} = 2(j_r + 2)^2 + (j_r - 1)^2 - 3.$$

Model F
Consider the digraph in Figure 45.15 with five vertices and 11 edges.

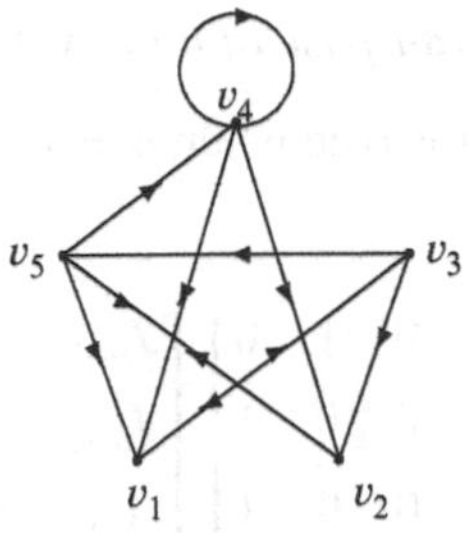

Figure 45.15.

Its adjacency matrix A is also a binary matrix, and it enjoys the special property that its square is also a binary matrix:

$$A = \begin{bmatrix} \boxed{0} & 0 & 0 & 1 & 0 \\ 0 & 0 & 0 & 0 & 1 \\ 1 & 1 & 0 & 0 & 1 \\ 1 & 1 & 0 & 1 & 0 \\ \textcircled{1} & 1 & 0 & 1 & 0 \end{bmatrix} \quad \text{and} \quad A^2 = \begin{bmatrix} \boxed{1} & 1 & 0 & 0 & 1 \\ 1 & 1 & 0 & 1 & 0 \\ 1 & 1 & 1 & 1 & 1 \\ 1 & 1 & 1 & 1 & 1 \\ \textcircled{1} & 1 & 1 & 1 & 1 \end{bmatrix}.$$

An interesting observation: A has $11 = J_5$ nonzero elements, and A^2 has $21 = J_6$ nonzero elements.

Although not obvious by now, the next four powers of A exhibit an interesting pattern:

$$A^3 = \begin{bmatrix} \boxed{1} & 1 & 1 & 1 & 1 \\ 1 & 1 & 1 & 1 & 1 \\ 3 & 3 & 1 & 2 & 2 \\ 3 & 3 & 1 & 2 & 2 \\ \textcircled{3} & 3 & 3 & 2 & 2 \end{bmatrix}; \qquad A^4 = \begin{bmatrix} \boxed{3} & 3 & 1 & 2 & 2 \\ 3 & 3 & 1 & 2 & 2 \\ 5 & 5 & 3 & 4 & 4 \\ 5 & 5 & 3 & 4 & 4 \\ \textcircled{5} & 5 & 3 & 4 & 4 \end{bmatrix};$$

$$A^5 = \begin{bmatrix} \boxed{5} & 5 & 3 & 4 & 4 \\ 5 & 5 & 3 & 4 & 4 \\ 11 & 11 & 5 & 8 & 8 \\ 11 & 11 & 5 & 8 & 8 \\ \textcircled{11} & 11 & 5 & 8 & 8 \end{bmatrix}; \qquad A^6 = \begin{bmatrix} \boxed{11} & 11 & 5 & 8 & 8 \\ 11 & 11 & 5 & 8 & 8 \\ 21 & 21 & 11 & 16 & 16 \\ 21 & 21 & 11 & 16 & 16 \\ \textcircled{21} & 21 & 11 & 16 & 16 \end{bmatrix}.$$

More generally, we have the following result. We confirm it using PMI.

Theorem 45.21. *Let $C_1^{(n)}$ denote column 1 of A^n. Then $C_1^{(n)} = [J_{n-1}, J_{n-1}, J_n, J_n, \textcircled{J_n}\,]^{\mathrm{T}}$, where X^{T} denotes the transpose of matrix X.*

Proof. Clearly, the statement is true when $n = 1$. Suppose, it is true for an arbitrary integer $n \geq 1$. Then

$$A^{n+1} = \begin{bmatrix} 0 & 0 & 0 & 1 & 0 \\ 0 & 0 & 0 & 0 & 1 \\ 1 & 1 & 0 & 0 & 1 \\ 1 & 1 & 0 & 1 & 0 \\ 1 & 1 & 0 & 1 & 0 \end{bmatrix} \begin{bmatrix} J_{n-1} & * & * & * & * \\ J_{n-1} & * & * & * & * \\ J_n & * & * & * & * \\ J_n & * & * & * & * \\ J_n & * & * & * & * \end{bmatrix}$$

$$= \begin{bmatrix} J_n & * & * & * & * \\ J_n & * & * & * & * \\ J_{n+1} & * & * & * & * \\ J_{n+1} & * & * & * & * \\ J_{n+1} & * & * & * & * \end{bmatrix},$$

$$\uparrow$$
$$C_1^{(n+1)}$$

where $*$ indicates an element we are *not* interested in. So the statement is true for $n+1$ also.

Thus, by PMI, the result is true for every positive integer, as desired. ■

The next two corollaries are immediate consequences of Theorem 45.21.

Corollary 45.9. *Let $a_{ij}^{(n)}$ denote the ij-th element of A^n. Then $a_{31}^{(n)} = a_{41}^{(n)} = a_{51}^{(n)} = J_n$, where $n \geq 1$.* ■

For example, $a_{41}^{(5)} = 11 = J_5$ and $a_{51}^{(6)} = \textcircled{21} = J_6$.

Corollary 45.10.

1) *There are J_{n-1} directed paths of length n from v_1 to v_1, and v_2 to v_1.*
2) *There are J_n directed paths of length n from v_3, v_4, and v_5 to v_1.* ■

For example, There are $5 = J_4$ directed paths of length 5 from v_2 to v_1:

251352 252542 254132 254252;

and $11 = J_5$ directed paths of length 5 from v_3 to v_2:

313132 313252 313542 325132 325252 325442
351352 352542 354132 354252 354442.

EXERCISES 45

1. Formula (45.1).
2. Theorem 45.2.
3. Formula (45.3).
4. Formula (45.5).

Let $a_{n,k}$ denote the number of tilings of a $2 \times n$ board with exactly k 2×2 tiles. Prove each (Brigham *et al.*, [51]).

5. $a_{n,0} = F_{n+1}$, where $n \geq 0$.
6. $a_{n,k} = \sum_{i=k}^{\lfloor n/2 \rfloor} \binom{n-i}{i}\binom{i}{k}$ (Sloane, [465]).
7. $a_{2n,n} = 1$.
8. $a_{2n+1,n} = n+1$.
9. $a_{2n,n-1} = \left(\frac{1}{2}\right)n^2 + \left(\frac{3}{2}\right)n$, where $n \geq 1$.
10. $a_{2n,n-2} = \frac{n}{24}(n^3 + 14n^2 + 11n - 26)$, where $n \geq 2$.

11. $b_n = b_{n-1} + 2b_{n-2} - 2$, using part (1) of Theorem 45.8, where $b_0 = 4, b_1 = 1$, and $n \geq 2$.
12. Solve the recurrence $b_n = b_{n-1} + 2b_{n-2} - 2$, where $b_0 = 4, b_1 = 1$, and $n \geq 2$.
13. Prove Theorem 45.9.
14. Prove Theorem 45.10.
15. Prove Theorem 45.11.
16. $t_{n,k} = xt_{n-2,k-1} + t_{n-1,k}$.
17. Theorem 45.15.
18. $|A^n| = 2^n$, using Theorem 45.15.
19. Corollary 45.3.
20. Corollary 45.4.
21. Theorem 45.16.
22. Theorem 45.17.
23. Let A denote the adjacency matrix of the graph in Figure 45.11. Then $|A^n| = 0$.
24. $2J_n - 1 = J_{n+1}$, if n is odd.
25. $2J_n + 1 = J_{n+1}$, if n is even.

Find the paths of length 5 in the digraph in Figure 45.14 from:

26. Vertex v_1 to itself.
27. Vertex v_1 to vertex v_3.

Prove each, where $M = \begin{bmatrix} 1 & 1 & 0 \\ 1 & 0 & 1 \\ 0 & 1 & 1 \end{bmatrix}$.

28. If n is odd, then $M^n = \begin{bmatrix} a+1 & a+1 & a \\ a+1 & a & a+1 \\ a & a+1 & a+1 \end{bmatrix}$, where a is a nonnegative integer.
29. If n is even, then $M^n = \begin{bmatrix} a+1 & a & a \\ a & a+1 & a \\ a & a & a+1 \end{bmatrix}$, where a is a positive integer.

Find the number of paths of length 11 in the digraph in Figure 45.14 from:

30. Vertex v_1 to vertex v_1.
31. Vertex v_3 to vertex v_1.

Find the number of paths of length 12 in the digraph in Figure 45.14 from:

32. Vertex v_1 to vertex v_1.
33. Vertex v_3 to vertex v_2.

Prove each, where $r \equiv 0 \equiv s \pmod 2$.

34. $J_{r+s} = (J_r + 1)J_s + J_r(J_s + 1) + J_rJ_s$.
35. $3j_{r+s} = (j_r + 1)(j_s + 1) + 2(j_r - 2)(j_s - 2) - 3$.

Prove each, where $r \equiv 1 \equiv s \pmod 2$.

36. $J_{r+s} = J_rJ_s + J_r(J_s - 1) + (J_r - 1)J_s$.
37. $3j_{r+s} = 2(j_r + 2)(j_s + 2) + (j_r - 1)(j_s - 1) - 3$.

Prove each, where $r \equiv 1 \pmod 2$ and $s \equiv 0 \pmod 2$.

38. $J_{r+s} = J_r(J_s + 1) + J_rJ_s + (J_r - 1)J_s$.
39. $3j_{r+s} = 2(j_r + 2)(j_s - 2) + (j_r - 1)(j_s + 1) + 3$.

46

BIVARIATE TILING MODELS

Mathematicians stand on each other's shoulders.
–Carl Friedrich Gauss (1777–1855)

In Chapter 43, we studied bivariate Fibonacci and Lucas polynomials as subfamilies of the gibonacci family $\{s_n(x,y)\}$, where $s_n(x,y) = xs_{n-1}(x,y) + ys_{n-2}(x,y)$, $x^2 + 4y \neq 0$, and $n \geq 2$. Recall that when $s_0(x,y) = 0$ and $s_1(x,y) = 1$, $s_n(x,y) = f_n(x,y)$ and $s_n(x,1) = f_n(x)$; and when $s_0(x,y) = 2$ and $s_1(x,y) = x$, $s_n(x,y) = l_n(x,y)$ and $l_n(x,1) = l_n(x)$. We also have $f_n(2x,y) = p_n(x,y)$ and $l_n(2x,y) = q_n(x,y)$; $f_n(x,y) = J_n(x,y)$ and $f_n(1,y) = J_n(y)$; and $l_n(x,y) = j_n(x,y)$ and $l_n(1,y) = l_n(y) = j_n(y)$.

In this chapter, we present combinatorial models for $f_n(x,y)$ and $l_n(x,y)$, and their subfamilies; they are in fact adaptations of earlier versions.

46.1 A MODEL FOR $f_n(x,y)$

Model I

Suppose we assign a *weight* x to a square (tile) and y to a domino: w(square tile) = x and w(domino) = y. The *weight* of a tiling is the product of the weights of all its tiles. The weight of the empty tiling is defined as 1. Figure 46.1 shows the weights of the tiles, the weights of the tilings, and the sum of the weights of the tilings of a $1 \times n$ board, where $0 \leq n \leq 5$.

Fibonacci and Lucas Numbers with Applications, Volume Two. Thomas Koshy.

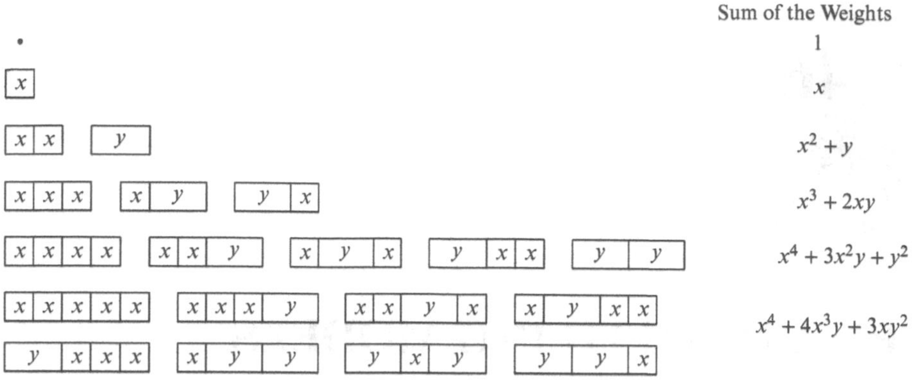

Sum of the Weights
1
x
x^2 + y
x^3 + 2xy
x^4 + 3x^2y + y^2
x^4 + 4x^3y + 3xy^2

Figure 46.1.

It appears from the figure that the sum of the weights of tilings of a $1 \times n$ board is $f_{n+1}(x, y) = J_{n+1}(x, y)$, where $n \geq 0$. The following theorem confirms this observation.

Theorem 46.1. *The sum of the weights of tilings of a* $1 \times n$ *board is* $f_{n+1}(x, y)$, *where w(square tile)* $= x$, *w(domino)* $= y$, *w(empty tiling)* $= 1$, *and* $n \geq 0$.

Proof. Let $w_n = w_n(x, y)$ denote the sum of the weights of tilings of the board. Then $w_0 = 1 = f_1(x, y)$ and $w_1 = x = f_2(x, y)$.

Now consider an arbitrary tiling of a $1 \times n$ board, where $n \geq 2$. Suppose it ends in a square: $\underbrace{\cdots}_{1\times(n-1)\text{ board}}\ \boxed{x}$. The sum of the weights of such tilings is xw_{n-1}.

On the other hand, suppose the tiling ends in a domino: $\underbrace{\cdots}_{1\times(n-2)\text{ board}}\ \boxed{\ y\ }$.

The sum of the weights of such tilings is yw_{n-2}.

Thus $w_n = xw_{n-1} + yw_{n-2}$. This, coupled with the two initial conditions, implies that $w_n = f_{n+1}(x, y)$, as desired. ■

Several interesting byproducts follow from this theorem:

1) $w_n(x, 1) = f_{n+1}(x)$

2) $w_n(1, 1) = F_{n+1}$

3) $w_n(2x, y) = f_{n+1}(2x, y) = p_{n+1}(x, y)$

4) $w_n(2x, 1) = f_{n+1}(2x, 1) = p_{n+1}(x)$

5) $w_n(2, 1) = f_{n+1}(2, 1) = P_{n+1}$

6) $w_n(1, y) = J_{n+1}(1, y) = J_{n+1}(y)$

7) $w_n(1, 2) = J_{n+1}(1, 2) = J_{n+1}$.

Since $w_n(1, 1) = F_{n+1}$, it follows that there are F_{n+1} tilings of a $1 \times n$ board; see Figure 46.1.

Using this model, we now establish Theorem 43.1 combinatorially. The proof pivots on the number of dominoes in tilings of length n.

Theorem 46.2. *Let $n \geq 1$. Then*

$$f_{n+1}(x,y) = \sum_{k=0}^{\lfloor n/2 \rfloor} \binom{n-k}{k} x^{n-2k} y^k.$$

Prove this combinatorially.

Proof. By Theorem 46.1, the sum of the weights of tilings of a $1 \times n$ board is $f_{n+1}(x,y)$. Consider a tiling with exactly k dominoes. It has $n-2k$ squares, where $0 \leq k \leq \lfloor n/2 \rfloor$. The weight of such a tiling is $x^{n-2k}y^k$.

Since there are k dominoes and $n-2k$ squares, the tiling consists of $k+(n-2k) = n-k$ tiles. Consequently, the k dominoes can be placed in $\binom{n-k}{k}$ different ways in the tiling. So the sum of the weights of tilings with exactly k dominoes is $\binom{n-k}{k} x^{n-2k}y^k$.

Since $0 \leq k \leq \lfloor n/2 \rfloor$, it follows that the sum of the weights of all tilings of the board is $\sum_{k=0}^{\lfloor n/2 \rfloor} \binom{n-k}{k} x^{n-2k}y^k = f_{n+1}(x,y)$. Thus $f_n(x,y) = \sum_{k=0}^{\lfloor (n-1)/2 \rfloor} \binom{n-k-1}{k} x^{n-2k-1}y^k$, as desired. ■

In particular, we have

$$p_n(x,y) = \sum_{k=0}^{\lfloor (n-1)/2 \rfloor} \binom{n-k-1}{k} (2x)^{n-2k-1} y^k;$$

$$J_n(x,y) = \sum_{k=0}^{\lfloor (n-1)/2 \rfloor} \binom{n-k-1}{k} x^{n-2k-1} (2y)^k;$$

$$F_n = \sum_{k=0}^{\lfloor (n-1)/2 \rfloor} \binom{n-k-1}{k};$$

$$P_n = \sum_{k=0}^{\lfloor (n-1)/2 \rfloor} \binom{n-k-1}{k} 2^{n-2k-1};$$

$$J_n = \sum_{k=0}^{\lfloor (n-1)/2 \rfloor} \binom{n-k-1}{k} 2^k.$$

46.2 BREAKABILITY

The next theorem establishes combinatorially the addition formula for $f_n(x,y)$; see identity (20) in Section 44.1. To this end, we introduce the concept of *breakability*. A tiling is *unbreakable* at cell k if a domino occupies cells k and

$k+1$; otherwise, it is *breakable* at cell k. Thus a tiling is breakable at cell k if and only if a domino does *not* cover cells k and $k+1$.

For example, consider the two tilings in Figure 46.2. The one in Figure 46.2(a) is breakable at cell 3, whereas the one in Figure 46.2(b) is *not*.

(a) (b)

Figure 46.2.

Theorem 46.3. *Let $m, n \geq 0$. Then*

$$f_{m+n}(x,y) = f_{m+1}(x,y)f_n(x,y) + yf_m(x,y)f_{n-1}(x,y).$$

Proof. Consider a $1 \times (m+n)$ board. By Theorem 46.1, the sum of the weights of its tilings is f_{m+n+1}.

Now consider an arbitrary tiling of the board. Suppose it is breakable at cell m, so the tiling can be split up into two sub-tilings, one of length m and the other of length n:

Since the sum of the weights of tilings of length m is $f_{m+1}(x,y)$ and that of tilings of length n is $f_{n+1}(x,y)$, it follows that the sum of the weights of tilings that are breakable at cell m is $f_{m+1}(x,y)f_{n+1}(x,y)$.

On the other hand, suppose the tiling is *not* breakable at cell m, so a domino occupies cells m and $m+1$. Consequently, the tiling is breakable at cell $m-1$:

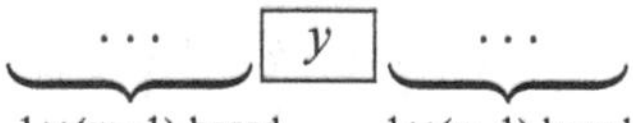

The sum of the weights of such tilings is $yf_m(x,y)f_n(x,y)$.

Combining the two cases, the sum of the weights of all tilings of the board equals $f_{m+1}(x,y)f_{n+1}(x,y) + yf_m(x,y)f_n(x,y) = f_{m+n+1}(x,y)$. Thus $f_{m+n}(x,y) = f_{m+1}(x,y)f_n(x,y) + yf_m(x,y)f_{n-1}(x,y)$, as desired. ∎

This addition formula yields the following special cases:

$$f_{m+n+1}(x,y) = f_{m+1}(x,y)f_{n+1}(x,y) + yf_m(x,y)f_n(x,y);$$

$$p_{m+n}(x,y) = p_{m+1}(x,y)p_n(x,y) + yp_m(x,y)p_{n-1}(x,y);$$

$$f_{m+n}(x) = f_{m+1}(x)f_n(x) + f_m(x)f_{n-1}(x);$$

$$p_{m+n}(x) = p_{m+1}(x)p_n(x) + p_m(x)p_{n-1}(x);$$

$$J_{m+n}(y) = J_{m+1}(y)J_n(y) + yJ_m(y)J_{n-1}(y);$$

$$F_{m+n} = F_{m+1}F_n + F_mF_{n-1};$$

$$P_{m+n} = P_{m+1}P_n + P_mP_{n-1};$$
$$J_{m+n} = J_{m+1}J_n + 2J_mJ_{n-1}.$$

In particular, we have

$$f_{2n+1}(x,y) = f_{n+1}^2(x,y) + yf_n^2(x,y) \qquad p_{2n+1}(x,y) = p_{n+1}^2(x,y) + yp_n^2(x,y)$$
$$J_{2n+1}(x,y) = J_{n+1}^2(x,y) + yJ_n^2(x,y) \qquad f_{2n}(x,y) = f_n(x,y)l_n(x,y)$$
$$p_{2n}(x,y) = p_n(x,y)q_n(x,y) \qquad J_{2n}(x,y) = J_n(x,y)j_n(x,y).$$

We can take advantage of this model to develop an explicit double-summation formula for $f_{2n+2}(x,y)$, as the following theorem shows.

Theorem 46.4. *Let $n \geq 0$. Then*

$$f_{2n+2}(x,y) = \sum_{i,j=0}^{n} \binom{n-j}{i}\binom{n-i}{j} x^{2n-2i-2j+1}y^{i+j}. \tag{46.1}$$

Proof. Consider a $1 \times (2n+1)$ board. By Theorem 46.1, the sum of the weights of tilings of the board is $f_{2n+2}(x,y)$.

We now compute this sum in a different way. Consider an arbitrary tiling of the board. Since its length is odd, it must contain an odd number of square tiles. So it must contain a special square tile M, called the *median square tile*, with an equal number of square tiles on either side. Suppose there are i dominoes to the left of M and j dominoes to its right:

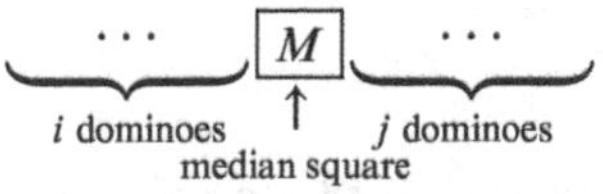

Since there are $(2n+1) - (2i+2j)$ square tiles in the tiling, it follows that there are $n-i-j$ square tiles on either side of M. So there are $(n-i-j)+i = n-j$ tiles to the left of M and $(n-i-j)+j = n-i$ tiles to its right.

The i dominoes to the left of M can be placed in $\binom{n-j}{i}$ different ways and the j dominoes to its right in $\binom{n-i}{j}$ different ways. So the weight of the tiling is $\binom{n-j}{i}x^{n-i-j}y^i \cdot x \cdot \binom{n-i}{j}x^{n-i-j}y^j = \binom{n-i}{j}\binom{n-j}{i}x^{2n-2i-2j}y^{i+j}$.

The sum of the weights of all tilings of the board is $\sum_{i,j\geq 0}\binom{n-j}{i}\binom{n-i}{j}x^{2n-2i-2j+1}y^{i+j}$. Thus

$$f_{2n+2}(x,y) = \sum_{i,j=0}^{n} \binom{n-j}{i}\binom{n-i}{j} x^{2n-2i-2j+1}y^{i+j}.$$ ■

For example, we have, as expected,

$$f_8(x,y) = \sum_{i,j=0}^{3} \binom{3-j}{i}\binom{3-i}{j} x^{7-2i-2j}y^{i+j} = x^7 + 6x^5y + 10x^3y^2 + 4xy^3.$$

It follows from formula (46.1) that

$$p_{2n+2}(x, y) = \sum_{i,j=0}^{n} \binom{n-j}{i}\binom{n-i}{j}(2x)^{2n-2i-2j+1}y^{i+j};$$

$$f_{2n+2}(x) = \sum_{i,j=0}^{n} \binom{n-j}{i}\binom{n-i}{j}x^{2n-2i-2j+1};$$

$$p_{2n+2}(x) = \sum_{i,j=0}^{n} \binom{n-j}{i}\binom{n-i}{j}(2x)^{2n-2i-2j+1};$$

$$J_{2n+2}(y) = \sum_{i,j=0}^{n} \binom{n-j}{i}\binom{n-i}{j}y^{i+j};$$

$$F_{2n+2} = \sum_{i,j=0}^{n} \binom{n-j}{i}\binom{n-i}{j};$$

$$P_{2n+2} = \sum_{i,j=0}^{n} \binom{n-j}{i}\binom{n-i}{j}2^{2n-2i-2j+1};$$

$$J_{2n+2} = \sum_{i,j=0}^{n} \binom{n-j}{i}\binom{n-i}{j}2^{i+j}.$$

For example,

$$P_{10} = \sum_{i,j=0}^{4} \binom{4-j}{i}\binom{4-i}{j}2^{9-2i-2j} = 2^9 + 8 \cdot 2^7 + 21 \cdot 2^5 + 20 \cdot 2^3 + 5 \cdot 2 = 2378;$$

and

$$J_{10} = \sum_{i,j=0}^{4} \binom{4-j}{i}\binom{4-i}{j}2^{i+j} = 1 + 8 \cdot 2 + 21 \cdot 2^2 + 20 \cdot 2^3 + 5 \cdot 2^7 = 341.$$

46.3 COLORED TILINGS

Model II

Suppose square tiles come in two colors, black and white. Let w(square tile) = x, w(domino) = y, and w(empty tiling) = 1. Figure 46.3 shows the resulting colored tilings of a $1 \times n$ board and the sum of the weights of tilings, where $0 \le n \le 4$.

From these experimental data, we can conjecture that the sum of the weights of such colored tilings of a $1 \times n$ board is $p_{n+1}(x, y)$, where $n \ge 0$. The next theorem confirms this observation. We omit its proof for the sake of brevity; see Exercise 46.12.

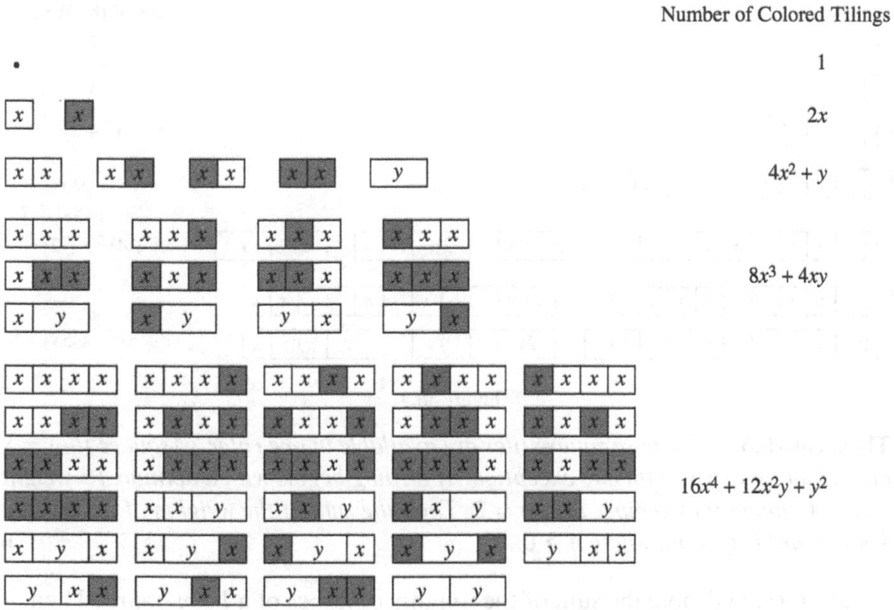

Figure 46.3.

Theorem 46.5. *Suppose square tiles are available in two colors, black and white; w(square tile) = x, w(domino) = y, and w(empty tiling) = 1. Then the sum of the weights of such colored tilings of a $1 \times n$ board is $p_{n+1}(x, y)$, where $n \geq 0$.* ■

In particular, this theorem implies that there are P_{n+1} such colored tilings of the board; see Exercise 46.13. In addition, we can use it to establish the bivariate Pell recurrence. Its development hinges on the initial tile of a tiling; see Exercise 46.14.

Next we present a combinatorial model for the bivariate Lucas polynomial $l_n(x, y)$.

46.4 A MODEL FOR $l_n(x, y)$

Model III

Once again, our objective is to tile a $1 \times n$ board with square tiles and dominoes. Suppose square tiles are all white, w(square tile) = x, and w(domino) = y, with one exception: If the tiling begins with a domino, its weight is $2y$. The weight of the empty tiling is 2.

Figure 46.4 shows the tilings, weights of tiles, and the sum of the weights of tilings of a $1 \times n$ board, where $0 \leq n \leq 5$.

We can conjecture from these data that the sum of the weights of tilings in this model is $l_n(x, y)$; see Table 43.1. The next theorem confirms this observation; we leave the proof as an exercise; see Exercise 46.15.

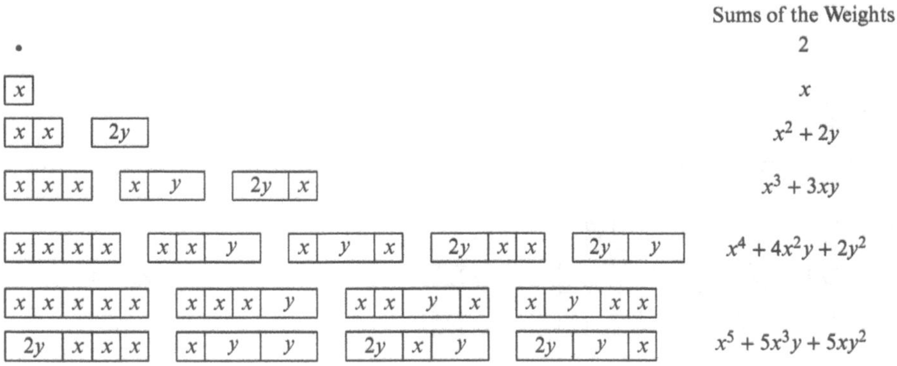

Figure 46.4.

Theorem 46.6. *Suppose square tiles are available in one color, w(square tile) = x and w(domino) = y, with one exception: If a tiling begins with a domino, its weight is $2y$. Assume that w(empty tiling) = 2. Then the sum of the weights of tilings of a $1 \times n$ board is $l_n(x, y)$, where $n \geq 0$.* ■

Let $w_n(x, y)$ denote the sum of the weights of tilings of a $1 \times n$ board. Then, it follows by Theorem 46.6 that

1) $w_n(1, 1) = l_n(1, 1) = L_n$
2) $w_n(2x, y) = l_n(2x, y) = q_n(x, y)$
3) $w_n(2, 1) = l_n(2, 1) = q_n(1, 1) = 2Q_n$
4) $w_n(1, y) = l_n(1, y) = j_n(y)$
5) $w_n(1, 2) = l_n(1, 2) = j_n(2) = j_n$.

Using Model III and the concept of breakability, we can derive the addition formula for $l_{m+n}(x, y)$ [see identity (21) in Section 43.2], as the next theorem shows.

Theorem 46.7. *Let $m, n \geq 0$. Then*

$$l_{m+n}(x, y) = f_{m+1}(x, y)l_n(x, y) + yf_m(x, y)l_{n-1}(x, y).$$

Proof. Consider a $1 \times (m + n)$ board. By Theorem 46.6, the sum of the weights of tilings of the board is $l_{m+n}(x, y)$.

Now consider an arbitrary tiling of the board. Suppose it is breakable at cell m into two sub-tilings, one of length m and the other of length n: $\underbrace{\cdots}_{1\times m \text{ board}}\underbrace{\cdots}_{1\times n \text{ board}}$. By Theorem 46.6, the sum of the weights of tilings of length m is $l_m(x, y)$; and by Theorem 46.1, that of tilings of length n is $f_{n+1}(x, y)$. So the sum of the weights of tilings that are breakable at cell m is $l_m(x, y)f_{n+1}(x, y)$.

On the other hand, suppose the tiling is *not* breakable at cell m, so a domino occupies cells m and $m + 1$: $\underbrace{\cdots}_{1\times(m-1)\text{ board}}\boxed{\;y\;}\underbrace{\cdots}_{1\times(n-1)\text{ board}}$. By Theorem 46.6, the sum of the weights of tilings of length $m-1$ is $l_{m-1}(x,y)$; and by Theorem 46.1, that of tilings of length $n-1$ is $f_n(x,y)$. So the sum of the weights of such tilings is $l_{m-1}(x,y)\cdot y\cdot f_n(x,y) = yf_n(x,y)l_{m-1}(x,y)$.

Thus $l_{m+n}(x,y) = f_{n+1}(x,y)l_m(x,y) + yf_n(x,y)l_{m-1}(x,y) = f_{m+1}(x,y)l_n(x,y) + yf_m(x,y)l_{n-1}(x,y)$, as desired. ■

In particular, this theorem yields

$$\begin{aligned}
l_{m+n}(x) &= f_{m+1}(x)l_n(x) + f_m(x)l_{n-1}(x);\\
q_{m+n}(x,y) &= p_{m+1}(x,y)q_n(x,y) + yp_m(x,y)q_{n-1}(x,y);\\
q_{m+n}(x) &= p_{m+1}(x)q_n(x) + p_m(x)q_{n-1}(x);\\
j_{m+n}(y) &= J_{m+1}(y)j_n(y) + yJ_m(y)j_{n-1}(y);\\
L_{m+n} &= F_{m+1}L_n + F_mL_{n-1};\\
Q_{m+n} &= P_{m+1}Q_n + P_mQ_{n-1};\\
j_{m+n} &= J_{m+1}j_n + 2J_mj_{n-1}.
\end{aligned}$$

In particular, we have

$$\begin{aligned}
l_{2n+1}(x,y) &= f_{2n+2}(x,y) + yf_{2n}(x,y);\\
q_{2n+1}(x,y) &= p_{2n+2}(x,y) + yp_{2n}(x,y);\\
j_{2n+1}(x,y) &= J_{2n+2}(x,y) + yJ_{2n}(x,y).
\end{aligned}$$

Next we construct a different model for $q_n(x,y)$ using colored tiles.

46.5 COLORED TILINGS REVISITED

Model IV

Suppose square tiles come in two colors, black and white. Let w(square tile) $= x$ and w(domino) $= y$, with one exception: If a tiling begins with a domino, its weight is $2y$. The weight of the empty tiling remains 2.

Figure 46.5 shows the tilings and the sum of the weights of tilings of a $1\times n$ board, where $0\le n\le 4$. It appears from the figure that the sum of the weights is $q_n(x,y)$. The following theorem confirms this observation. Again, we omit its proof in the interest of brevity.

Theorem 46.8. *Suppose square tiles are available in two colors, black and white. Let w(square tile) $= x$ and w(domino) $= y$, with one exception: If a tiling begins with a domino, then its weight is $2y$. The weight of the empty tiling is 2. Then the sum of the weights of tilings of a $1\times n$ board is $q_n(x,y)$, where $n\ge 0$.* ■

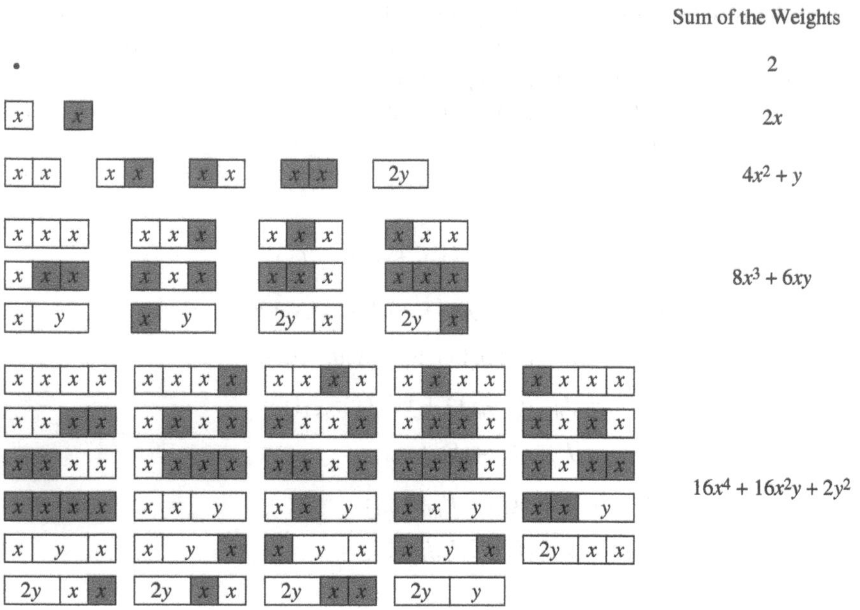

Figure 46.5.

Let $w_n(x, y)$ denote the sum of the weights of tilings of a $1 \times n$ board in Model IV. Then by Theorem 46.8, we have

1) $w_n(x, 1) = q_n(x, 1) = q_n(x)$
2) $w_n(1, 1) = q_n(1, 1) = q_n(1) = 2Q_n$
3) $w_n(x/2, y) = q_n(x/2, y) = l_n(x, y)$
4) $w_n(1/2, 1) = q_n(1/2, 1) = l_n(1, 1) = L_n$
5) $w_n(1/2, y) = q_n(1/2, y) = l_n(1, y) = j_n(y)$
6) $w_n(1/2, 2) = q_n(1/2, 2) = l_n(1, 2) = j_n(2) = j_n$.

Interesting Byproducts

We can use this theorem, coupled with Theorem 46.5, to extract some interesting facts:

1) First we find the sum of the weights of tilings that begin with a domino in Model IV:

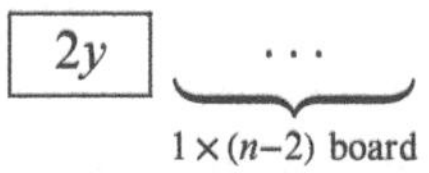

By Theorem 46.5, we find that the sum of the weights of tilings of length $n - 2$ is $p_{n-1}(x, y)$. So the sum of the weights of such tilings of a $1 \times n$ board is $2yp_{n-1}(x, y)$. Consequently, there are P_{n-1} tilings beginning with a domino.

For example, the sum of the weights of such tilings of length 4 is $2yp_3(x,y) = 2y(4x^2+y) = 8x^2y + 2y^2$. There are $P_3 = 5$ tilings beginning with a domino; see Figure 46.5.

2) Next we find the sum of the weights of tilings beginning with a square:

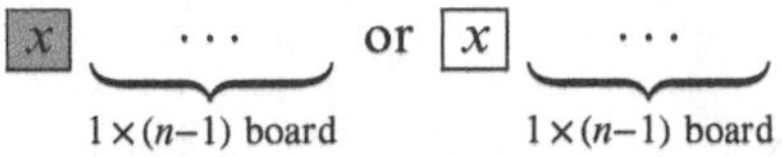

Using Theorem 46.5, we find that the sum of the weights of such tilings is $2xp_n(x,y)$. There are $2P_n$ such tilings.

For example, the sum of the weights of tilings of a 1×4 board that begin with a square is $16x^4 + 4x^2y = 2x(8x^3 + 4xy) = 2xp_4(x,y)$. There are $24 = 2P_4$ such tilings.

3) Thus, by Theorem 46.8, the sum of the weights of tilings of a $1 \times n$ board is given by the bivariate identity

$$2xp_n(x,y) + 2yP_{n-1}(x,y) = q_n(x,y). \tag{46.2}$$

[This is a special case of bivariate identity (9) in Section 43.2.]

For example, $2xp_5(x,y) + 2yp_4(x,y) = 2x(16x^4 + 12x^2 + y^2) + 2y(8x^3 + 4xy) = 32x^5 + 24x^3y + 2xy^2 = q_5(x,y)$.

4) It follows by parts 1) and 3) that there are exactly $2P_n + P_{n-1} = P_{n+1}$ tilings of a $1 \times n$ board in Model IV. For example, there are $29 = P_5$ tilings of a 1×4 board; see Figure 46.5.

5) Identity (46.1) implies that $2P_n + 2P_{n-1} = 2Q_n$; that is, $P_n + P_{n-1} = Q_n$.

Next we investigate tilings of a circular board.

46.6 CIRCULAR TILINGS AGAIN

Model V

Consider a circular board with n labeled cells, ordered in the counterclockwise direction. We would like to tile the board with square tiles and dominoes. (Although square tiles and dominoes are *not* circular, we will keep the same terminology for convenience and consistency.) Let w(square tile) $= x$, w(domino) $= y$, and w(empty tiling) $= 2$, with one exception: w(domino) $= 2y$ when $n = 2$.

Figure 46.6 shows the tilings and the sum of their weights of a $1 \times n$ bracelet, where $0 \le n \le 4$.

Using these data, we can conjecture that the sum of the weights of the tilings in this model is $l_n(x,y)$, where $n \ge 0$. The following theorem confirms this observation.

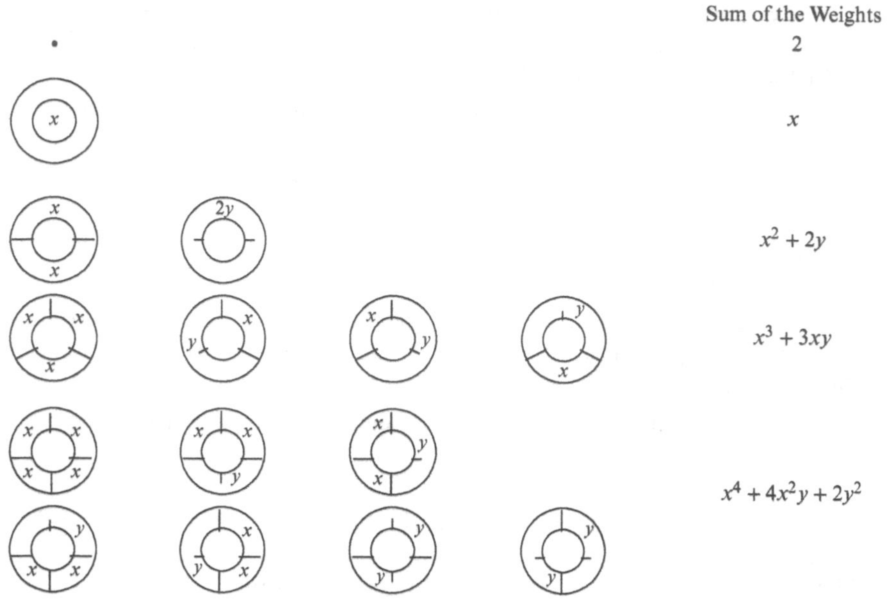

Figure 46.6.

Theorem 46.9. *Suppose* $w(\textit{square tile}) = x$, $w(\textit{domino}) = y$, *and* $w(\textit{empty tiling}) = 2$, *with one exception:* $w(\textit{domino}) = 2y$ *when* $n = 2$. *Then the sum of the weights of tilings of a bracelet with n cells is* $l_n(x, y)$, *where* $n \geq 0$.

Proof. Let $S_n(x, y)$ be the sum of the weights of tilings of a bracelet with n cells. Then $S_0(x, y) = 2 = l_0(x, y)$, $S_1(x, y) = x = l_1(x, y)$, $S_2(x, y) = x^2 + 2y = l_2(x, y)$.

Now consider an arbitrary tiling of the bracelet, where $n \geq 3$. Suppose it begins with a square tile. We can use the remaining $n - 1$ cells to form a bracelet; by definition, the sum of the weights of such sub-bracelets is $S_{n-1}(x, y)$. So the sum of the weights of such bracelets is $xS_{n-1}(x, y)$.

On the other hand, suppose the tiling begins with a domino. The sum of the weights of such bracelets is $yS_{n-2}(x, y)$.

Combining the two cases, $S_n(x, y) = xS_{n-1}(x, y) + yS_{n-2}(x, y)$. This recurrence, together with the initial conditions, yields the desired result. ■

This theorem yields

1) $S_n(x, 1) = l_n(x, 1) = l_n(x)$
2) $S_n(1, 1) = l_n(1, 1) = l_n(1) = L_n$
3) $S_n(2x, y) = l_n(2x, y) = q_n(x, y)$
4) $S_n(2x, 1) = l_n(2x, 1) = q_n(x, 1) = q_n(x)$
5) $S_n(2, 1) = l_n(2, 1) = q_n(1) = 2Q_n$
6) $S_n(1, y) = l_n(1, y) = j_n(y)$
7) $S_n(1, 2) = l_n(1, 2) = j_n(2) = j_n$.

Theorem 46.9 has another delightful consequence. We can use it to develop combinatorially the explicit formula for $l_n(x, y)$ in Theorem 46.2, as the next theorem shows. The essence of its proof hinges on counting the tilings with exactly k dominoes, as in Theorem 46.2.

Theorem 46.10. *Let $n \geq 0$. Then*

$$l_n(x,y) = \sum_{k=0}^{\lfloor n/2 \rfloor} \frac{n}{n-k}\binom{n-k}{k} x^{n-2k}y^k.$$

Proof. Consider a bracelet with n cells. By Theorem 46.9, the sum of the weights of its tilings is $l_n(x,y)$.

We now find the sum in a different way. To this end, consider an arbitrary tiling. Assume it has exactly k dominoes. So the tiling contains $n-2k$ squares and a total of $n-k$ tiles, where $0 \leq k \leq \lfloor n/2 \rfloor$. The weight of such a tiling is $x^{n-2k}y^k$.

Suppose a domino occupies cells n and 1:

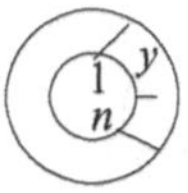

There are $n-k-1$ tiles and $k-1$ dominoes left in the tiling. So the $k-1$ dominoes can be placed in $\binom{n-k-1}{k-1}$ ways; that is, there are $\binom{n-k-1}{k-1}$ tilings with a domino in cells n and 1.

On the other hand, suppose a domino does not occupy cells n and 1. Then the bracelet can be considered a linear board with exactly k dominoes:

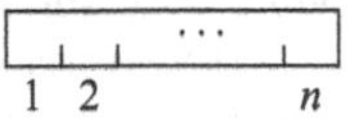

There are $\binom{n-k}{k}$ such tilings.

Combining the two cases, we find there is a total of $\binom{n-k-1}{k-1} + \binom{n-k}{k} = \frac{n}{n-k}\binom{n-k}{k}$ tilings, each with exactly k dominoes. Since each has weight $x^{n-2k}y^k$, the sum of the weights of all tilings of the bracelet is $\sum_{k=0}^{\lfloor n/2 \rfloor} \frac{n}{n-k}\binom{n-k}{k} x^{n-2k}y^k$. This yields the desired result. ■

It follows by this theorem that there are $l_n(1,1) = L_n$ tilings of a bracelet with n cells.

We now present one more circular tiling model.

Circular Tilings Revisited

Model VI

Suppose w(square tile) $= 2x$, w(domino) $= y$, and w(empty tiling) $= 2$, but with one exception: when $n = 2$, w(domino) $= 2y$. Figure 46.7 shows such tilings and the sum of their weights of a bracelet with n cells, where $0 \leq n \leq 4$.

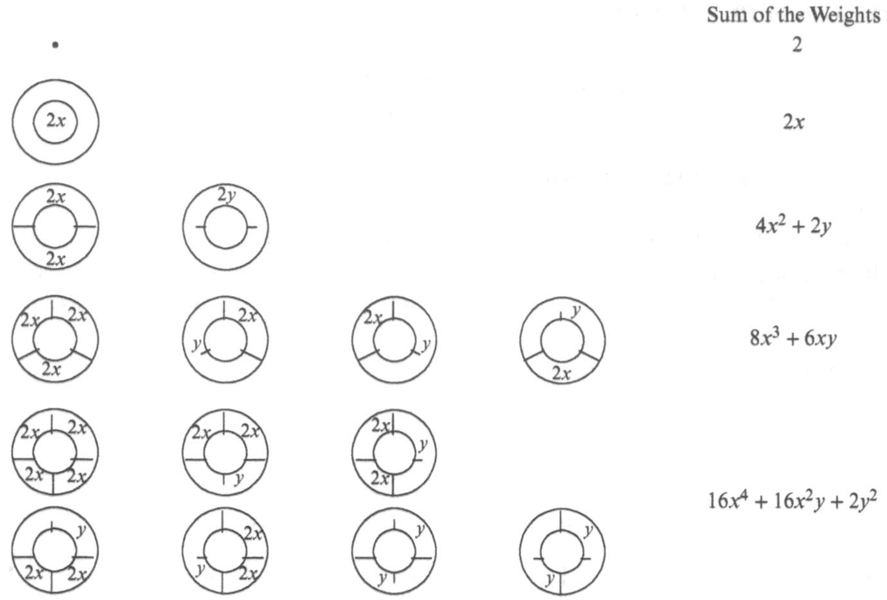

Figure 46.7.

The following theorem confirms that the sum of the weights of tilings of a bracelet with n cells in such a model is $q_n(x, y)$. The proof follows the same argument as in Theorem 46.9, so we omit it for the sake of brevity; see Exercise 46.17.

Theorem 46.11. *Suppose w(square tile) = x, w(domino) = y, w(empty tiling) = 2, with one exception: when n = 2, w(domino) = 2y. Then the sum of the weights of tilings of a bracelet with n cells in such a model is $q_n(x, y)$, where $n \geq 0$.* ∎

EXERCISES 46

Prove each, using Theorem 46.3.

1. $f_{2n+1}(x, y) = f_{n+1}^2(x, y) + y f_n^2(x, y)$.
2. $f_{2n}(x, y) = f_n(x, y) l_n(x, y)$.
3. $p_{2n}(x, y) = p_n(x, y) q_n(x, y)$.
4. $J_{2n}(x, y) = J_n(x, y) j_n(x, y)$.
5. $p_{2n+1}(x, y) = p_{n+1}^2(x, y) + y p_n^2(x, y)$.
6. $J_{2n+1}(x, y) = J_{n+1}^2(x, y) + y J_n^2(x, y)$.

Find each, using Theorem 46.4.

7. $f_9(x, y)$.
8. $p_8(x, y)$.
9. $J_8(y)$.
10. P_{10}.
11. J_{10}.

Establish each.

12. Theorem 46.5.
13. The number of colored tilings in Theorem 46.5 is P_{n+1}, where $n \geq 0$.
14. The bivariate Pell recurrence, using Model II.
15. Theorem 46.6.
16. Theorem 46.8.
17. Theorem 46.11.

47

VIETA POLYNOMIALS

I, who do not profess to be a mathematician,
but who, whenever there is leisure, delight
in mathematical studies.
–François Vieta, 1595

In our investigation of Fibonacci and Lucas polynomials, we encountered the numbers

$$a_{nk} = \binom{n-k-1}{k} \quad \text{and} \quad b_{nk} = \frac{n}{n-k}\binom{n-k}{k}.$$

Recall that a_{nk} and b_{nk} are the coefficients of x^k in $f_n(x)$ and $l_n(x)$, respectively; see Tables 31.2 and 31.3. We came across the arrays (a_{nk}) and (b_{nk}) while studying the topological indices of paraffins C_nH_{2n+2} and cycloparaffins C_nH_{2n} (see Tables 3.8 and 3.9 in Volume One [287]), and also while studying the sums and differences of the roots of the equation $x^2 - px - q = 0$ (see Tables 12.1 and 12.2 in [287]). N. Robbins of San Francisco University attributes the discovery of array (b_{nk}) to the French mathematician François Vieta (1540–1603) [236, 405].

In this chapter, we use the numbers a_{nk} and b_{nk} to study two related classes of polynomials. We call them *Vieta polynomials* $V_n(x)$ and *Vieta–Lucas polynomials* $v_n(x)$, respectively. Jacobsthal, Robbins, and Shannon and Horadam studied them in depth [236, 405]. Again, we *omit* the argument from the functional notation when there is *no* ambiguity.

Fibonacci and Lucas Numbers with Applications, Volume Two. Thomas Koshy.

47.1 VIETA POLYNOMIALS

Vieta polynomials V_n and *Vieta–Lucas polynomials* v_n satisfy the recurrence

$$h_n = xh_{n-1} - h_{n-2},$$

where $n \geq 2$. When $h_0 = 0$ and $h_1 = 1$, $h_n = V_n$; and when $h_0 = 2$ and $h_1 = x$, $h_n = v_n$. Table 47.1 shows the Vieta and Vieta–Lucas polynomials V_n and v_n, where $1 \leq n \leq 10$.

TABLE 47.1. First 10 Vieta and Vieta–Lucas Polynomials

n	V_n	v_n
1	1	x
2	x	$x^2 - 2$
3	$x^2 - 1$	$x^3 - 3x$
4	$x^3 - 2x$	$x^4 - 4x^2 + 2$
5	$x^4 - 3x^2 + 1$	$x^5 - 5x^3 + 5x$
6	$x^5 - 4x^3 + 3x$	$x^6 - 6x^4 + 9x^2 - 2$
7	$x^6 - 5x^4 + 6x^2 - 1$	$x^7 - 7x^5 + 14x^3 - 7x$
8	$x^7 - 6x^5 + 10x^3 - 4x$	$x^8 - 8x^6 + 20x^4 - 16x^2 + 2$
9	$x^8 - 7x^6 + 15x^4 - 10x^2 + 1$	$x^9 - 9x^7 + 27x^5 - 30x^3 + 9x$
10	$x^9 - 8x^7 + 21x^5 - 20x^3 + 5x$	$x^{10} - 10x^8 + 35x^6 - 50x^4 + 25x^2 - 2$

Binet-like Formulas

The roots of the characteristic equation $z^2 - xz + 1 = 0$ of the Vieta recurrence are

$$r = \frac{x + \sqrt{x^2 - 4}}{2} \quad \text{and} \quad s = \frac{x - \sqrt{x^2 - 4}}{2},$$

where $r + s = x$, $rs = 1$, and $r - s = \sqrt{x^2 - 4}$. Consequently, the Binet-like formulas for V_n and v_n are

$$V_n = \frac{r^n - s^n}{r - s} \quad \text{and} \quad v_n = r^n + s^n,$$

respectively.

As we can predict, V_n and f_n are closely related; and so are v_n and l_n. We can now establish such links. This, in turn, helps us extract properties of Vieta and Vieta–Lucas polynomials from the corresponding properties of Fibonacci and Lucas polynomials [289].

Links Between V_n and f_n, and v_n and l_n

Let $i = \sqrt{-1}$. It follows by the Fibonacci recurrence that

$$i^{n-1} f_n = (ix)(i^{n-2} f_{n-1}) - (i^{n-3} f_{n-2}),$$

so that $i^{n-1} f_n$ satisfies the Vieta recurrence. Since $V_1(ix) = 1 = f_1(x)$ and $V_2(ix) = ix = i^{2-1} f_2$, it follows that

$$V_n(ix) = i^{n-1} f_n(x). \tag{47.1}$$

Likewise, $i^n l_n$ satisfies the Vieta recurrence. Since $v_1(ix) = ix = il_1(x)$ and $v_2(ix) = -(x^2 + 2) = i^2 l_2$, it follows that

$$v_n(ix) = i^n l_n(x). \tag{47.2}$$

For example, $V_6(ix) = (ix)^5 - 4(ix)^3 + 3(ix) = i^5(x^5 + 4x^3 + 3x) = i^5 f_6(x)$. Similarly, $v_6(ix) = i^6(x^6 + 6x^4 + 9x^2 + 2) = i^6 l_6(x)$.

As in the case of Fibonacci and Lucas polynomials, we can develop an array of properties for Vieta and Vieta–Lucas polynomials. They can be established independently or extracted from their Fibonacci and Lucas counterparts; see Exercises 47.1–47.47. The next example uses the latter technique.

Example 47.1. *Prove that $v_{n+1} v_{n-1} - v_n^2 = x^2 - 4$.*

Proof. Since $v_n = i^n l_n(-ix)$, by the Cassini-like formula for Lucas polynomials, we have

$$\begin{aligned}
v_{n+1} v_{n-1} - v_n^2 &= i^{n+1} l_{n+1}(-ix) \cdot i^{n-1} l_{n-1}(-ix) - i^{2n} l_n^2(-ix) \\
&= (-1)^n [l_{n+1}(-ix) l_{n-1}(-ix) - l_n^2(-ix)] \\
&= (-1)^n \cdot (-1)^{n-1}(-x^2 + 4) \\
&= x^2 - 4.
\end{aligned}$$

■

For example,

$$\begin{aligned}
v_6 v_4 - v_5^2 &= (x^6 - 6x^4 + 9x^2 - 2)(x^4 - 4x^2 + 2) - (x^5 - 5x^3 + 5x)^2 \\
&= x^2 - 4.
\end{aligned}$$

The next example gives another explicit formula for V_{n+1}.

Example 47.2. *Prove that*

$$V_{n+1} = \sum_{k=0}^{\lfloor n/2 \rfloor} (-1)^k \binom{n-k}{k} x^{n-2k}. \tag{47.3}$$

Proof. Since $f_{n+1} = \sum_{k=0}^{\lfloor n/2 \rfloor} \binom{n-k}{k} x^{n-2k}$, we have

$$\begin{aligned} V_{n+1} &= i^n f_{n+1}(-ix) \\ &= i^n \sum_{k=0}^{\lfloor n/2 \rfloor} \binom{n-k}{k} (-ix)^{n-2k} \\ &= \sum_{k=0}^{\lfloor n/2 \rfloor} (-1)^k \binom{n-k}{k} x^{n-2k}, \end{aligned}$$

as desired. ■

For example,

$$\begin{aligned} V_7 &= \sum_{k=0}^{3} (-1)^k \binom{6-k}{k} x^{6-2k} \\ &= \binom{6}{0}x^6 - \binom{5}{1}x^4 - \binom{4}{2}x^2 - \binom{3}{3}x^0 \\ &= x^6 - 5x^4 + 6x^2 - 1; \end{aligned}$$

see Table 47.1.

Similarly, we can show that

$$v_n = \sum_{k=0}^{\lfloor n/2 \rfloor} (-1)^k \frac{n}{n-k} \binom{n-k}{k} x^{n-2k}; \tag{47.4}$$

see Exercise 47.44.

For example,

$$\begin{aligned} v_7 &= \sum_{k=0}^{3} (-1)^k \frac{7}{7-k} \binom{7-k}{k} x^{7-2k} \\ &= \binom{7}{0}x^7 - \frac{7}{6}\binom{6}{1}x^5 + \frac{7}{5}\binom{5}{2}x^3 - \frac{7}{4}\binom{4}{3}x \\ &= x^7 - 7x^5 + 14x^3 - 7x. \end{aligned}$$

It follows by formula (47.3) that we can compute V_{n+1} by using the diagonal entries $(-1)^k \binom{n-k}{k}$ with weights x^{n-2k} in the modified Pascal's triangle in Figure 47.1, where $0 \le k \le \lfloor n/2 \rfloor$.

In Chapter 31, we studied a number of occurrences of the charming identity $l_n^2 - (x^2+4)f_n^2 = 4(-1)^n$ in different contexts. As we might predict,

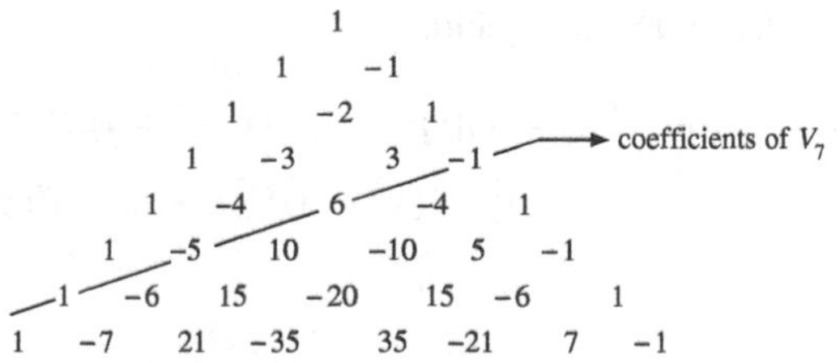

Figure 47.1. A modified Pascal's triangle.

it has an equally charming counterpart for Vieta polynomials, as the next example shows.

Example 47.3. *Prove that* $v_n^2 - (x^2-4)V_n^2 = 4$.

Proof. It follows by the identity $l_n^2 - (x^2+4)f_n^2 = 4(-1)^n$ that

$$l_n^2(-ix) + (x^2-4)f_n^2(-ix) = 4(-1)^n$$

$$\left(\frac{v_n}{i^n}\right)^2 + (x^2-4)\left(\frac{V_n}{i^n}\right)^2 = 4(-1)^n$$

$$v_n^2 - (x^2-4)V_n^2 = 4. \qquad \blacksquare$$

For a specific case, let $n = 5$. Then

$$\begin{aligned} v_5^2 - (x^2-4)V_5^2 &= (x^5-5x^3+5x)^2 - (x^2-4)(x^4-3x^2+1)^2 \\ &= 4. \end{aligned}$$

It follows from Example 47.3 that (v_n, V_n) is a solution of the Pell's equation $u^2 - (x^2-4)v^2 = 4$, where $x \geq 3$. In particular, let $x = 5$. Since $v_5(5) = 2525$ and $V_5(5) = 551$, it follows that $(2525, 551)$ is a solution of the equation $u^2 - 21v^2 = 4$: $2525^2 - 21 \cdot 551^2 = 4$.

Identity (31.51) Revisited

In the next example, we derive the Vieta counterpart of the identity

$$l_n^4 - [(x^2+4)f_n^2]^2 - 8(-1)^n(x^2+4)f_n^2 = 16 \tag{47.5}$$

that we established in Example 31.10. Recall that it is an application of identity (31.51).

Example 47.4. *Derive the Vieta counterpart of identity (47.5).*

Solution. Recall that $V_n(x) = i^{n-1}f_n(-ix)$ and $v_n(x) = i^n l_n(-ix)$. Replacing x with $-ix$ in identity (47.5), we get

$$l_n^4(-ix) - [(x^2-4)f_n^2(-ix)]^2 + 8(-1)^n(x^2-4)f_n^2(-ix) = 16.$$

Multiplying both sides by i^{4n}, this yields

$$[i^n l_n^4(-ix)]^4 - \{(x^2-4)[i^{n-1} f_n(-ix)]^2\}^2 - 8(-1)^n(x^2-4)[i^{n-1} f_n(-ix)]^2 = 16$$
$$v_n^4 - [(x^2-4)V_n^2]^2 - 8(-1)^n(x^2-4)V_n^2 = 16. \tag{47.6}$$

This is the desired result. ∎

The next theorem presents an elegant result satisfied by Vieta–Lucas polynomials: Composition is a binary commutative operation on the set of Vieta–Lucas polynomials. It was discovered by Jacobsthal, according to Horadam [236].

Theorem 47.1 (Jacobsthal, 1955 [236]). $v_m \circ v_n = v_{mn} = v_n \circ v_m$.

Proof. Since $rs = 1$, it follows by the Binet-like formula for v_n that $v_n(x) = r^n + r^{-n}$, so

$$\begin{aligned} v_{mn}(x) &= r^{mn} + r^{-mn} \\ &= v_m(r^n + r^{-n}) \\ &= v_m(v_n(x)). \end{aligned}$$

Similarly, $v_{mn}(x) = v_n(v_m(x))$. Thus $v_m(v_n(x)) = v_{mn}(x) = v_n(v_m(x))$, as desired. ∎

For example,

$$\begin{aligned} v_3(v_2(x)) &= v_3(x^2-2) \\ &= (x^2-2)^3 - 3(x^2-2) \\ &= x^6 - 6x^4 + 9x^2 - 2 = V_6(x) \\ &= (x^3-3x)^2 - 2 \\ &= v_2(v_3(x)). \end{aligned}$$

But

$$\begin{aligned} V_3(V_4(x)) &= x^6 - 4x^4 + 4x^2 - 1 \\ &\neq x^6 - 3x^4 + x^2 + 1 \\ &= V_4(V_3(x)). \end{aligned}$$

So composition is *not* a binary operation on the set of Vieta polynomials.

Generalized Cassini-like Formulas

The Cassini-like formulas $V_{n+1}V_{n-1} - V_n^2 = -1$ and $v_{n+1}v_{n-1} - v_n^2 = x^2 - 1$ can be generalized. In the interest of brevity, we do the former case and leave the latter case as an exercise; see Exercise 47.41.

It follows by identity (32.8) that

$$\frac{V_m}{i^{m-1}} \cdot \frac{V_{m+n+k}}{i^{m+n+k-1}} - \frac{V_{m+k}}{i^{m+k-1}} \cdot \frac{V_{m+n}}{i^{m+n-1}} = (-1)^{m+1}\frac{V_n}{i^{n-1}} \cdot \frac{V_k}{i^{k-1}}$$

$$V_{m+k}V_{m+n} - V_m V_{m+n+k} = V_n V_k. \tag{47.7}$$

Likewise,

$$v_m v_{m+n+k} - v_{m+k}v_{m+n} = (x^2-4)V_n V_k. \tag{47.8}$$

It follows from identities (47.7) and (47.8) that

$$V_{n+k}V_{n-k} - V_n^2 = -V_k^2; \tag{47.9}$$

$$v_{n+k}v_{n-k} - v_n^2 = (x^2-4)V_k^2; \tag{47.10}$$

$$V_m V_{n+1} - V_{m+1}V_n = V_{m-n}; \tag{47.11}$$

$$v_m v_{n+1} - v_{m+1}v_n = (4-x^2)V_{m-n}. \tag{47.12}$$

For example, let $x = 3, n = 5$, and $k = 2$. Then

$$\begin{aligned}
V_7(11)V_3(11) - V_5^2(11) &= 1{,}699{,}081 \cdot 120 - 14{,}279^2\\
&= -121\\
&= -V_2^2(11);\\
v_7(11)v_3(11) - v_5^2(11) &= 18{,}378{,}371 \cdot 1298 - 154{,}451^2\\
&= 14{,}157 = 9 \cdot 13 \cdot 11^2\\
&= 9 \cdot 13V_2^2(11).
\end{aligned}$$

Identities (47.9) and (47.10) imply that

$$V_{n+1}^2 - V_n^2 = V_{2n+1};$$

$$v_{n+1}^2 + (x^2-4)V_n^2 = xv_{2n+1}.$$

Addition Formulas

As in the case of Fibonacci and Lucas polynomials, we can easily establish the following addition formulas for Vieta and Vieta–Lucas polynomials:

$$\begin{aligned}
v_{m+n} &= v_{m+1}V_n - v_mV_{n-1};\\
v_{m-n} &= v_mV_{n+1} - v_{m+1}V_n;\\
2V_{m+n} &= V_mv_n + V_nv_m;\\
2V_{m-n} &= V_mv_n - V_nv_m;\\
2v_{m+n} &= v_mv_n + (x^2-4)V_mV_n;\\
2v_{m-n} &= v_mv_n - (x^2-4)V_mV_n;
\end{aligned}$$

see Exercises 47.33 and 47.38.

It follows from these identities that

$$\begin{aligned}
V_nv_n &= V_{2n};\\
v_n^2 + (x^2-4)V_n^2 &= 2v_{2n};\\
v_n^2 - (x^2-4)V_n^2 &= 4.
\end{aligned}$$

Additional Bridges

The next theorem establishes links between V_n and f_{2n}, and v_n and l_{2n}.

Theorem 47.2. *Let $n \geq 0$. Then*

$$xV_n(x^2+2) = f_{2n}; \tag{47.13}$$

$$v_n(x^2+2) = l_{2n}. \tag{47.14}$$

Proof. We can establish identity (47.13) using PMI. Clearly, it is true when $n = 0$ and $n = 1$.

Now assume that it is true for all nonnegative integers $< n$. Since $f_{2n} = (x^2+2)f_{2n-2} - f_{2n-4}$, by the Vieta recurrence we have

$$\begin{aligned}
V_n(x^2+2) &= (x^2+2)V_{n-1}(x^2+2) - V_{n-2}(x^2+2)\\
&= (x^2+2)\cdot\frac{1}{x}f_{2n-2} - \frac{1}{x}f_{2n-4}\\
xV_n(x^2+2) &= f_{2n}.
\end{aligned}$$

So the given result is true for n also.

Thus, by PMI, it is true for all $n \geq 0$.

We can similarly establish identity (47.14); see Exercise 47.45. ■

For example, $V_6(x^2+2) = (x^2+2)^5 - 4(x^2+2)^3 + 3(x^2+2)$; so $xV_6(x^2+2) = x(x^{10} + 10x^8 + 36x^6 + 56x^4 + 35x^4 + 35x^2 + 6) = f_{12}$. Likewise, $v_3(x^2+2) = (x^2+2)^3 - 3(x^2+2) = x^6 + 6x^4 + 9x^2 + 2 = l_6$.

It follows from Theorem 47.2 that

$$\sqrt{x} f_{2n}(1/\sqrt{x}) = V_n\left(\frac{2x+1}{x}\right) \quad \text{and} \quad l_{2n}(1/\sqrt{x}) = v_n\left(\frac{2x+1}{x}\right).$$

The next corollary [460] follows from Theorem 47.2.

Corollary 47.1 (Shannon and Horadam, 1999 [460]). *Let $n \geq 0$. Then $V_n(3) = F_{2n}$ and $v_n(3) = L_{2n}$.* ■

Since $v_n' = nV_n$ (see Exercise 47.75), it follows that $v_n'(3) = nF_{2n}$, where the prime denotes differentiation with respect to x. For example, $v_5'(x) = 5(x^4 - 3x^2 + 1)$; so $v_n'(3) = 3 \cdot 55 = 5F_{10}$.

The following corollary also follows from Theorem 47.2, where the prime denotes differentiation with respect to x.

Corollary 47.2. *Let $n \geq 0$. Then*

$$V_n(x^2+2) + 2x^2 V_n'(x^2+2) = f_{2n}';$$
$$2x^2 v_n'(x^2+2) = l_{2n}'.$$ ■

It follows by Corollary 47.2 that

$$2V_n'(3) = f_{2n}'(1) - F_{2n}$$
$$2v_n'(3) = l_{2n}'(1).$$

So $l_{2n}'(1) = 2nF_{2n}$.

For example,

$$\begin{aligned} 2V_5'(3) &= f_{10}'(1) - F_{10} \\ &= 2(4 \cdot 3^3 - 6 \cdot 3) - 55 \\ V_5'(3) &= 90. \end{aligned}$$

Likewise, $2v_5'(3) = 550 = l_{10}'(1) = 10F_{10}$.

In the next example, we use Theorem 47.2 to establish two hybrid identities.

Example 47.5. *Prove each:*

$$f_{2n+2} - f_{2n-2} = xl_{2n}; \tag{47.15}$$
$$l_{2n+2} - l_{2n-2} = x(x^2+4)f_{2n}. \tag{47.16}$$

Proof.

1) Since $v_n(x) = V_{n+1}(x) - V_{n-1}(x)$ (see Exercise 47.14), it follows by Theorem 47.2 that

$$\begin{aligned} v_n(x^2+2) &= V_{n+1}(x^2+2) - V_{n-1}(x^2+2) \\ l_{2n} &= \frac{1}{x}f_{2n+2} - \frac{1}{x}f_{2n-2}. \end{aligned}$$

This yields identity (47.15).

2) Using the identity $v_{n+1}(x) - v_{n-1}(x) = (x^2-4)V_n(x)$ (see Exercise 47.17), identity (47.16) follows similarly. ■

Interestingly, the Vieta and Pell families are also closely related. We now investigate a few links between them.

Vieta–Pell Links

Since $p_n(x) = f_n(2x)$ and $q_n(x) = l_n(2x)$, it follows by identities (47.1) and (47.2) that

$$p_n(x) = (-i)^{n-1}V_n(2ix), \tag{47.17}$$

$$q_n(x) = (-i)^n v_n(2ix). \tag{47.18}$$

For example,

$$\begin{aligned} p_5(x) &= V_5(2ix) \\ &= (2ix)^4 - 3(2ix)^2 + 1 \\ &= 16x^4 + 12x^2 + 1; \\ q_6(x) &= -v_6(2ix) \\ &= -[(2ix)^6 - 6(2ix)^4 + 9(2ix)^2 - 2] \\ &= 64x^6 + 96x^4 + 36x^2 + 2. \end{aligned}$$

Theorem 47.2 yields the following results.

Corollary 47.3.

$$p_{2n}(x) = 2xV_n(4x^2+2); \tag{47.19}$$

$$q_{2n}(x) = v_n(4x^2+2). \tag{47.20}$$

■

For example,

$$\begin{aligned} p_6(x) &= 2xV_3(4x^2+2) \\ &= 2x[(4x^2+2)^2-1] \\ &= 32x^5+32x^3+6x. \end{aligned}$$

Similarly, $q_6(x) = v_3(4x^2+2) = 64x^6+96x^4+36x^2+2$.

The next two results follow from Corollary 47.3.

Corollary 47.4. *Let $n \ge 0$. Then $P_{2n} = 2V_n(6)$ and $2Q_{2n} = v_n(6)$.* ■

For example,

$$\begin{aligned} P_{10} &= 2V_5(6) = 2(6^4 - 3\cdot 6^2 + 1) = 2378 \\ Q_{10} &= \frac{1}{2}v_5(6) = \frac{1}{2}(6^5 - 5\cdot 6^3 + 5\cdot 6) = 3363. \end{aligned}$$

Since $v_n' = nV_n$ and $P_{2n} = 2V_n(6)$, it follows that $v_n'(6) = \dfrac{nP_{2n}}{2}$. For example, $v_n'(6) = 5(6^4 - 3\cdot 6^2 + 1) = 5945 = \dfrac{5\cdot 2378}{2} = \dfrac{5P_{10}}{2}$.

Corollary 47.3 also implies the following results.

Corollary 47.5. *Let $n \ge 0$. Then*

$$\begin{aligned} 2V_n(4x^2+2) + 16x^2V_n'(4x^2+2) &= p_{2n}'(x); \\ 8xv_n'(4x^2+2) &= q_{2n}'(x). \end{aligned}$$ ■

It follows by this corollary that

$$\begin{aligned} 2V_n(6) + 16V_n'(6) &= p_{2n}'(1) \\ 8v_n'(6) &= q_{2n}'(1). \end{aligned}$$

For example, $2V_5(6) + 16V_5'(6) = 15{,}626 = p_{10}'(1)$ and $8v_5'(6) = 47{,}560 = q_{10}'(1)$.

The next theorem presents two charming identities involving Vieta polynomials.

Theorem 47.3. *Let $n \ge 0$. Then*

$$v_n(x^2-2) - (x^2-4)V_n^2(x) = 2; \tag{47.21}$$

$$v_n(x^2-2) - v_n^2(x) = -2. \tag{47.22}$$

Proof. We have $v_n(-x) = (-1)^n v_n(x)$, and $V_n(x) = i^{n-1} f_n(-ix)$. Since $v_n(x^2+2) = l_{2n}(x)$ by Theorem 47.1, $v_n(-x^2+2) = l_{2n}(-ix)$; that is, $(-1)^n v_n(x^2-2) = l_{2n}(-ix)$. Since $l_{2n}(u) = (u^2+4)f_n^2(u) + 2(-1)^n$ by Exercise 31.49, we then have

$$(-1)^n v_n(x^2-2) = -(x^2-4)\frac{V_n^2(x)}{i^{2n-2}} + 2(-1)^n.$$

This yields identity (47.21).

Identity (47.22) follows similarly; see Exercise 47.46. ■

For example,

$$\begin{aligned} v_4(x^2-2) - (x^2-4)V_4^2(x) &= (x^2-2)^4 - 4(x^2-2)^2 + 2 - (x^2-4)(x^3-2x)^2 \\ &= 2; \\ v_4(x^2-2) - v_4^2(x) &= (x^2-2)^4 - 4(x^2-2)^2 + 2 - (x^4-4x^2+2)^2 \\ &= -2. \end{aligned}$$

Horadam attributes identity (47.22) to Jacobsthal [236].

Both identities (47.21) and (47.22) imply that $v_n(2) = 2$; see Exercise 47.47. For example, $v_7(2) = 2^7 - 7\cdot 2^5 + 14\cdot 2^3 - 7\cdot 2 = 2$.

Theorem 47.3 has interesting Pell consequences, as the next two corollaries show.

Corollary 47.6. *Let $n \geq 0$. Then*

$$q_{2n}(x) - 4(x^2+1)p_n^2(x) = 2(-1)^n; \qquad (47.23)$$

$$q_{2n}(x) - q_n^2(x) = 2(-1)^{n+1}. \qquad (47.24)$$

Proof. We have $v_n(-x) = (-1)^n v_n(x)$, $V_n(ix) = i^{n-1} p_n(x/2)$, and $q_{2n}(x) = v_n(4x^2+2)$. It then follows by identity (47.21) that

$$\begin{aligned} v_n(-x^2-2) + (x^2+4)V_n^2(ix) &= 2 \\ (-1)^n v_n(x^2+2) + (x^2+4)\cdot i^{2n-2} p_n^2(x/2) &= 2 \\ q_{2n}(x/2) - (x^2+4)p_n^2(x/2) &= 2(-1)^n \\ q_{2n}(x) - 4(x^2+1)p_n^2(x) &= 2(-1)^n. \end{aligned}$$

Identity (47.24) can be established similarly; see Exercise 47.48. ■

For example,

$$\begin{aligned} q_6(x) - 4(x^2+1)p_3^2(x) &= (64x^6 + 96x^4 + 36x^2 + 2) - 4(x^2+1)(4x^2+1)^2 \\ &= -2. \end{aligned}$$

Similarly, $q_6(x) - q_3^2(x) = (64x^6 + 96x^4 + 36x^2 + 2) - (8x^3 + 6x)^2 = 2$.

It follows from Corollary 47.6 that $Q_{2n} = 4P_n^2 + (-1)^n = 2Q_n^2 - (-1)^n$; see Exercises 47.49 and 49.50.

The next corollary follows from identities (47.23) and (47.24).

Corollary 47.7. *Let $n \geq 0$. Then $q_n^2 - 4(x^2+1)p_n^2 = 4(-1)^n$.* ■

This corollary has a magnificent byproduct. It implies that $Q_n^2 - 2P_n^2 = (-1)^n$. Consequently, (Q_n, P_n) is a solution of the Pell's equation $u^2 - 2v^2 = (-1)^n$; its converse is also true [285].

Vieta–Jacobsthal Links

Polynomials $V_n(x)$ and $J_n(x)$ are closely related, and so are $v_n(x)$ and $j_n(x)$. The next theorem establishes such links [236, 460].

Theorem 47.4. *Let $n \geq 0$. Then*

$$V_{n+1}(x) = x^n J_{n+1}(-1/x^2); \tag{47.25}$$

$$v_n(x) = x^n j_n(-1/x^2). \tag{47.26}$$

Proof. Recall from identity (26) in Section 44.1 that $J_{n+1}(x) = \sum_{k=0}^{\lfloor n/2 \rfloor} \binom{n-k}{k} x^k$. Consequently, by formula (47.3), we then have

$$\begin{aligned} V_{n+1}(x) &= x^n \sum_{k=0}^{\lfloor n/2 \rfloor} \binom{n-k}{k} \left(-\frac{1}{x^2}\right)^k \\ &= x^n J_{n+1}(-1/x^2). \end{aligned}$$

This yields identity (47.25).

Identity (47.26) follows similarly; see Exercise 47.52. ■

For example,

$$\begin{aligned} J_5(x) &= x^2 + 3x + 1 \\ x^4 J_5(-1/x^2) &= x^4 \left[\left(-\frac{1}{x^2}\right)^2 + 3\left(-\frac{1}{x^2}\right) + 1\right] \end{aligned}$$

$$= x^4 - 3x^2 + 1$$
$$= V_5(x).$$

Similarly, $x^4 j_5(-1/x^2) = x^5 - 5x^3 + 5x = v_5(x)$.

Jacobsthal–Fibonacci Links

It follows from identities (47.1) and (47.25) that

$$x^n J_{n+1}(-1/x^2) = i^n f_{n+1}(-ix).$$

Replacing x with $i/\sqrt{x}$, this yields

$$J_{n+1}(x) = x^{n/2} f_{n+1}(1/\sqrt{x}). \tag{47.27}$$

Likewise,

$$j_n(x) = x^{n/2} l_n(1/\sqrt{x}); \tag{47.28}$$

see Exercise 47.53.

For example,

$$\begin{aligned} J_6(x) &= x^{5/2} f_6(1/\sqrt{x}) \\ &= x^{5/2}\left(\frac{1}{x^2\sqrt{x}} + \frac{4}{x\sqrt{x}} + \frac{3}{\sqrt{x}}\right) \\ &= 3x^2 + 4x + 1. \end{aligned}$$

Likewise, $j_6(x) = x^3 l_6(1/\sqrt{x} = 2x^3 + 9x^2 + 6x + 1$.

It follows from identity (47.27) that $J_{n+1}(1) = F_{n+1}$ and $J_{n+1} = 2^{n/2} f_{n+1}(1/\sqrt{2})$. Similarly, $j_n(1) = L_n$ and $j_n = 2^{n/2} l_n(1/\sqrt{2})$.

It follows from identities (47.13) and (47.27) that

$$\begin{aligned} J_{2n}(x) &= x^{(2n-1)/2} \cdot \frac{1}{\sqrt{x}} V_n\left(\frac{2x+1}{x}\right) \\ &= x^{n-1} V_n\left(\frac{2x+1}{x}\right). \end{aligned} \tag{47.29}$$

Likewise,

$$j_{2n}(x) = x^n v_n\left(\frac{2x+1}{x}\right); \tag{47.30}$$

see Exercise 47.54.

For example,

$$\begin{aligned} J_{10}(x) &= x^4 V_5\left(\frac{2x+1}{x}\right) \\ &= x^4\left[\left(\frac{2x+1}{x}\right)^4 - 3\left(\frac{2x+1}{x}\right)^2 + 1\right] \\ &= 5x^4 + 20x^3 + 21x^2 + 8x + 1. \end{aligned}$$

Similarly, $j_8(x) = x^4 v_4\left(\frac{2x+1}{x}\right) = 2x^4 + 16x^3 + 20x^3 + 8x + 1.$

Identities (47.29) and (47.30) imply that $J_{2n} = 2^{n-1}V_n(5/2)$ and $j_{2n} = 2^n v_n(5/2)$, respectively.

47.2 AURIFEUILLE'S IDENTITY

In 1879, the French mathematician Léon-François-Antoine Aurifeuille (1822–1882) discovered an elegant factorization of L_{5n}, where n is odd [52, 357]:

$$L_{5n} = L_n(L_{2n} + 5F_n + 3)(L_{2n} - 5F_n + 3). \quad (47.31)$$

For example, $L_3[(L_6 + 3)^2 - 25F_3^2] = 4[(18 + 3)^2 - 25 \cdot 2^2] = 1{,}364 = L_{15}$.
Identity (47.31) has indeed a beautiful polynomial extension:

$$l_{5n} = l_n\left\{[l_{2n} - 3(-1)^n]^2 + 5(-1)^n\Delta^2 f_n^2\right\}. \quad (47.32)$$

Using the Binet-like formulas, and the identities $l_{2n} = l_n^2 - 2(-1)^n$ and $l_{2n} = \Delta^2 f_n^2 + 2(-1)^n$, we now confirm its validity.

Proof. We have

$$\begin{aligned} l_{5n} &= \alpha^{5n} + \beta^{5n} \\ &= (\alpha^n + \beta^n)\left(\alpha^{4n} - \alpha^{3n}\beta^n + \alpha^{2n}\beta^{2n} - \alpha^n\beta^{3n} + \beta^{4n}\right) \\ &= l_n[l_{4n} + 1 - (\alpha\beta)^n(\alpha^{2n} + \beta^{2n})] \\ &= l_n[(l_{2n}^2 - 2) + 1 - (-1)^n l_{2n}] \\ &= l_n\left\{[l_{2n}^2 - 6(-1)^n l_{2n} + 9] + 5(-1)^n[l_{2n} - 2(-1)^n]\right\} \\ &= l_n\left\{[l_{2n} - 3(-1)^n]^2 + 5(-1)^n\Delta^2 f_n^2\right\}, \end{aligned}$$

as desired. ■

For example,

$$\begin{aligned} & l_3[(l_6 + 3)^2 - 5(x^2 + 4)f_3^2] \\ &\quad = (x^3 + 3x)[(x^6 + 6x^4 + 9x^2 + 5)^2 - 5(x^2 + 4)(x^2 + 1)^2] \\ &\quad = x^{15} + 15x^{13} + 90x^{11} + 275x^9 + 450x^7 + 378x^5 + 140x^3 + 15x \\ &\quad = l_{15}. \end{aligned}$$

Case 1. Suppose n is odd. Then identity (47.32) implies that

$$q_{5n} = q_n(q_{2n} + 2\sqrt{5}Dp_n + 3)(q_{2n} - 2\sqrt{5}Dp_n + 3);$$
$$Q_{5n} = Q_n(2Q_{2n} + 2\sqrt{10}P_n + 3)(2Q_{2n} - 2\sqrt{10}P_n + 3).$$

For example,

$$\begin{aligned} Q_{25} &= Q_5[(2Q_{10} + 3)^2 - 40P_5^2] \\ &= 41[(2 \cdot 3363 + 3)^2 - 40 \cdot 29^2] \\ &= 1{,}855{,}077{,}841. \end{aligned}$$

Since the Fibonacci and Jacobsthal families are related by the links (47.27) and (47.28), identity (47.32) has Jacobsthal implications as well. It follows by identity (47.32) that

$$l_{5n}(1/\sqrt{x}) = l_n(1/\sqrt{x})\left[\left(l_{2n}(1/\sqrt{x}) + 3\right)^2 - \frac{5(4x+1)}{x} f_n^2(1/\sqrt{x})\right]$$
$$x^{-5n/2} j_{5n}(x) = x^{-n/2} j_n(x)\left[\left(x^{-n} j_{2n}(x) + 3\right)^2 - \frac{5(4x+1)}{x} \cdot x^{-(n-1)} J_n^2(x)\right]$$
$$j_{5n}(x) = j_n(x)\left[\left(j_{2n}(x) + 3x^n\right)^2 - 5(4x+1)x^n J_n^2(x)\right]$$
$$j_{5n} = j_n\left[\left(j_{2n} + 3 \cdot 2^n\right)^2 - 45 \cdot 2^n J_n^2\right].$$

For example,

$$\begin{aligned} j_5\left[\left(j_{10} + 3 \cdot 2^5\right)^2 - 45 \cdot 2^5 J_5^2\right] &= 31\left[(1025 + 3 \cdot 32)^2 - 45 \cdot 32 \cdot 11^2\right] \\ &= 33{,}554{,}431 \\ &= j_{25}. \end{aligned}$$

Case 2. Suppose n is even. Then identity (47.32) implies that

$$L_{5n} = L_n\left[(L_{2n} - 3)^2 + 25F_n^2\right]; \tag{47.33}$$
$$q_{5n} = q_n\left[(q_n - 3)^2 + 20D^2p_n^2\right];$$
$$Q_{5n} = Q_n\left[(2Q_n - 3)^2 + 40P_n^2\right];$$
$$j_{5n}(x) = j_n(x)\left[\left(j_{2n}(x) - 3x^n\right)^2 + 5(4x+1)x^n J_n^2(x)\right];$$
$$j_{5n} = j_n\left[\left(j_{2n} - 3 \cdot 2^n\right)^2 + 45 \cdot 2^n J_n^2\right].$$

For example, $j_4\left[(j_8 - 3\cdot 2^4)^2 + 45\cdot 2^4 J_4^2\right] = 17\left[(257-48)^2 + 45\cdot 16\cdot 5^2\right] = 1{,}048{,}577 = j_{20}$.

Dresel discovered identity (47.33) using a totally different technique [134].

Next we derive the Jacobsthal counterpart of identity (47.5).

Example 47.6. *Derive the Jacobsthal counterpart of identity (47.5).*

Solution. We have $J_n(x) = x^{(n-1)/2} f_n(1/\sqrt{x})$ and $j_n(x) = x^{n/2} l_n(1/\sqrt{x})$. Replacing x with $1/\sqrt{x}$ in identity (47.5), and multiplying the resulting equation by x^{2n}, we find that

$$[x^{n/2} l_n(1/\sqrt{x})]^4 - \{(4x+1)[x^{(n-1)/2} f_n(1/\sqrt{x})]^2\}^2$$
$$- 8(-x)^n(4x+1)[x^{(n-1)/2} f_n(1/\sqrt{x})]^2 = 16x^{2n}$$
$$j_n^4 - [(4x+1)J_n^2(x)]^2 - 8(-x)^n(4x+1)J_n^2(x) = 16x^{2n}. \qquad (47.34)$$

This is the desired identity. ■

In particular,

$$\begin{aligned} j_n^4(x) &- (4x+1)^2 J_n^4(x) - 8(-x)^n(4x+1)J_n^2(x) \\ &= (3x+1)^4 - (4x+1)^2(x+1)^4 + 8x^3(4x+1)(x+1)^2 \\ &= 16x^6. \end{aligned}$$

Letting $x = 2$, identity (47.34) yields

$$j_n^4 - 81J_n^4 - 72(-2)^n J_n^2 = 4^{n+2}. \qquad (47.35)$$

For example, $j_5^4 - 81J_5^4 - 72(-2)^5 J_5^2 = 31^4 - 81\cdot 11^4 + 72\cdot 32\cdot 11^2 = 4^7$.

Theorem 47.4, coupled with Theorem 47.2, yields the next theorem; it provides a link between $f_{2n}(x)$ and $J_n(x)$, and $l_{2n}(x)$ and $j_n(x)$.

Theorem 47.5. *Let $n \geq 0$. Then*

$$f_{2n}(x) = x(x^2+2)^{n-1} J_n\left(-\frac{1}{(x^2+2)^2}\right); \qquad (47.36)$$

$$l_{2n}(x) = (x^2+2)^n j_n\left(-\frac{1}{(x^2+2)^2}\right). \qquad (47.37)$$

Proof. By Theorems 47.4 and 47.2, we have

$$xV_n(x^2+2) = x(x^2+2)^{n-1}J_n\left(-\frac{1}{(x^2+2)^2}\right)$$
$$f_{2n}(x) = x(x^2+2)^{n-1}J_n\left(-\frac{1}{(x^2+2)^2}\right).$$

Identity (47.37) follows similarly; see Exercise 47.55. ■

For example, let $n = 4$. Then

$$\begin{aligned} x(x^2+2)^3J_4\left(-\frac{1}{(x^2+2)^2}\right) &= x(x^2+2)^3\left[\frac{-2}{(x^2+2)^2}+1\right] \\ &= x^7+6x^5+10x^3+4x \\ &= f_8(x). \end{aligned}$$

Likewise, $(x^2+2)^4j_4\left(-\dfrac{1}{(x^2+2)^2}\right) = x^8+8x^6+20x^4+16x^2+2 = l_8(x)$.

Theorem 47.5 has consequences to the Pell family, as the following corollary shows.

Corollary 47.8. *Let $n \geq 0$. Then*

$$p_{2n}(x) = 2x(4x^2+2)^{n-1}J_n\left(-\frac{1}{(4x^2+2)^2}\right);$$
$$q_{2n}(x) = (4x^2+2)^n j_n\left(-\frac{1}{(4x^2+2)^2}\right).$$ ■

The next corollary follows from Theorem 47.5 and Corollary 47.8.

Corollary 47.9. *Let $n \geq 0$. Then*

$$\begin{aligned} F_{2n} &= 3^{n-1}J_n(-1/9) & L_{2n} &= 3^n j_n(-1/9) \\ P_{2n} &= 2\cdot 6^{n-1}J_n(-1/36) & 2Q_{2n} &= 6^n j_n(-1/36). \end{aligned}$$ ■

For example,

$$\begin{aligned} 3^5j_5(-1/9) &= 3^5[5(-1/9)^2+5(-1/9)+1] \\ &= 123 = L_{10}; \\ 2\cdot 6^4J_5(-1/36) &= 2\cdot 6^4[(-1/36)^2+3(-1/36)+1] \\ &= 2{,}378 = P_{10}. \end{aligned}$$

Theorems 47.3 and 47.4 together yield the next two results. Their proofs are straightforward, so we omit them.

Theorem 47.6. *Let $n \geq 0$. Then*

$$(x^2-2)^n j_n\left(-\frac{1}{(x^2-2)^2}\right) - x^{2n-2}(x^2-4)J_n^2\left(-\frac{1}{x^2}\right) = 2;$$

$$(x^2-2)^n j_n\left(-\frac{1}{(x^2-2)^2}\right) - x^{2n} j_n^2\left(-\frac{1}{x^2}\right) = -2.$$

■

For example,

$$\begin{aligned}(x^2-2)^4 j_4\left(-\frac{1}{(x^2-2)^2}\right) &- x^6(x^2-4)J_4^2\left(-\frac{1}{x^2}\right)\\ &= 2 - 4(x^2-2)^2 + (x^2-2)^4 - x^2(x^2-4)(x^2-2)^2\\ &= 2.\end{aligned}$$

The next result follows from Theorem 47.6.

Corollary 47.10. *Let $n \geq 0$. Then*

$$x^{2n} j_n^2(-1/x^2) - x^{2n-2}(x^2-4)J_n^2(-1/x^2) = 4. \tag{47.38}$$

■

Identity (47.38) implies

$$j_n^2(x) - (4x+1)J_n^2(x) = 4(-x)^n;$$

see Exercise 47.56.

Since $j_n(2) = j_n$ and $J_n(2) = J_n$, identity (47.38) yields the following result, linking Jacobsthal and Jacobsthal–Lucas numbers.

Corollary 47.11. *Let $n \geq 0$. Then*

$$j_n^2 - 9J_n^2 = 4(-2)^n. \tag{47.39}$$

For example, $j_5^2 - 9J_5^2 = 31^2 - 9 \cdot 11^2 = -2^7$. ■

Identity (47.39) has a delightful byproduct. Since $3J_{2n} = 4^n - 1$ and $j_{2n} = 4^n + 1$, it implies that $3J_{2n} - 2^{n+1} - j_{2n} = (4^n - 1) - 2^{n+1} - (4^n + 1)$ is a Pythagorean triple; clearly, it is primitive. The area of the Pythagorean triangle is $3 \cdot 2^n J_{2n} = 2^n(4^n - 1)$.

For example, $(3J_{10})^2 + 2^{12} = (3 \cdot 341)^2 + 2^{12} = 1{,}050{,}625 = 1025^2 = j_{10}^2$; so $(1023, 64, 1025)$ is a primitive Pythagorean triple, and the area of the Pythagorean triangle is 32,736.

Next we investigate the close relationship between Vieta and Chebyshev polynomials.

47.3 VIETA–CHEBYSHEV BRIDGES

Recall that Chebyshev polynomials $T_n(x)$ and $U_n(x)$ satisfy the recurrence $z_n = 2xz_{n-1} - z_{n-2}$, where $n \geq 1$. Since $v_0(x) = 2 = 2T_0(x/2)$ and $v_1(x) = x = 2T_1(x/2)$, and $V_0(x) = 0 = U_{-1}(x/2)$ and $V_1(x) = 1 = U_0(x/2)$, it follows that

$$v_n(x) = 2T_n(x/2), \tag{47.40}$$

$$V_n(x) = U_{n-1}(x/2), \tag{47.41}$$

where $n \geq 1$. Using these links, we can translate Vieta identities into Chebyshev ones and vice versa.

For example, the Vieta identity $v_n^2 - (x^2 - 4)V_n^2 = 4$ can be translated into a Chebyshev identity:

$$4T_n^2(x/2) - (x^2 - 4)U_{n-1}^2(x/2) = 4$$
$$T_n^2(x) - (x^2 - 1)U_{n-1}^2(x) = 1,$$

which we studied in Section 41.5; see Exercise 47.67 also.

Likewise, identity (47.34) has a Chebyshev counterpart:

$$T_n^4 - (x^2 - 1)^2 U_{n-1}^4 - 2(-1)^n(x^2 - 1)U_{n-1}^2 = 1. \tag{47.42}$$

In the interest of brevity, we omit its proof; see Exercise 47.68.

It follows by identities (47.1) and (47.41) that $i^{-n}U_n(ix/2) = f_{n+1}$. Likewise, $2i^{-n}T_n(ix/2) = l_n$.

For example, $2i^{-5}T_5(ix/2) = 2i^{-1}[16(ix/2)^5 - 20(ix/2)^3 + 5(ix/2)] = x^5 + 5x^3 + 5x = l_5$.

It follows by Theorem 47.2 that

$$f_{2n}(x) = xU_{n-1}\left(\frac{x^2+2}{2}\right), \tag{47.43}$$

$$l_{2n}(x) = 2T_n\left(\frac{x^2+2}{2}\right). \tag{47.44}$$

For example,

$$\begin{aligned} f_8(x) &= xU_3\left(\frac{x^2+2}{2}\right) \\ &= x\left[8\left(\frac{x^2+2}{2}\right)^3 - 4\left(\frac{x^2+2}{2}\right)\right] \\ &= x^7 + 6x^5 + 10x^3 + 4x. \end{aligned}$$

Similarly, $l_8(x) = 2T_4\left(\frac{x^2+2}{2}\right) = x^8 + 8x^6 + 20x^4 + 16x^2 + 2.$

See Exercises 47.63–47.89 for some additional properties linking the Vieta and Chebyshev families.

Next we investigate some properties bridging Jacobsthal and Chebyshev polynomials.

47.4 JACOBSTHAL–CHEBYSHEV LINKS

Identities (47.25) and (47.41) imply that

$$x^n J_{n+1}(-1/x^2) = U_n(x/2).$$

Consequently,

$$\left(\frac{i}{\sqrt{x}}\right)^n J_{n+1}(x) = U_n\left(\frac{i}{2\sqrt{x}}\right)$$

$$J_{n+1}(x) = (-i\sqrt{x})^n U_n\left(\frac{i}{2\sqrt{x}}\right). \tag{47.45}$$

Similarly,

$$j_n(x) = 2(-i\sqrt{x})^n T_n\left(\frac{i}{2\sqrt{x}}\right); \tag{47.46}$$

see Exercise 47.72.

For example,

$$\begin{aligned} j_5(x) &= 2(-i\sqrt{x})^5 T_5\left(\frac{i}{2\sqrt{x}}\right) \\ &= -2ix^2\sqrt{x}\left[16\left(\frac{i}{2\sqrt{x}}\right)^5 - 20\left(\frac{i}{2\sqrt{x}}\right)^3 + 5\left(\frac{i}{2\sqrt{x}}\right)\right] \\ &= 5x^2 + 5x + 1. \end{aligned}$$

Similarly, $J_6(x) = (-i\sqrt{x})^5 U_5\left(\frac{i}{2\sqrt{x}}\right) = 3x^2 + 4x + 1.$

It follows by identity (47.45) that $F_{n+1} = (-i)^n U_n(i/2)$ and $J_{n+1} = (-\sqrt{2}i)^n U_n(i/2\sqrt{2})$. Similarly, $L_n = 2(-i)^n T_n(i/2)$ and $j_n = 2(-\sqrt{2}i)^n T_n(i/2\sqrt{2})$.

For example, $j_7 = 2(-\sqrt{2}i)^7 T_7(i/2\sqrt{2}) = 127$ and $J_7 = (-\sqrt{2}i)^6 U_6(i/2\sqrt{2}) = 43$.

Next we focus on a charming gibonacci identity [282]. We find its Vieta counterpart, and use the counterpart to extract the corresponding Jacobsthal, Chebyshev, and Pell identities.

47.5 TWO CHARMING VIETA IDENTITIES

Consider the gibonacci identity

$$g_{n+k}^3 - (-1)^k l_k g_n^3 + (-1)^k g_{n-k}^3 = \begin{cases} f_k f_{2k} g_{3n} & \text{if } g_n = f_n \\ (x^2+4) f_k f_{2k} g_{3n} & \text{if } g_n = l_n. \end{cases} \tag{47.47}$$

Vieta Counterparts

The next theorem gives the equally beautiful counterpart for Vieta polynomials. The proof is short and neat, and hinges on identities (47.1) and (47.2).

Theorem 47.7.

$$h_{n+k}^3 - v_k h_n^3 + h_{n-k}^3 = \begin{cases} h_k h_{2k} h_{3n} & \text{if } h_n = V_n \\ (x^2-4) V_k V_{2k} h_{3n} & \text{if } h_n = v_n. \end{cases} \tag{47.48}$$

Proof. Suppose $h_n = V_n$ and $g_n = f_n$. By identity (47.47), we have

$$f_{n+k}^3 - (-1)^k l_k f_n^3 + (-1)^k f_{n-k}^3 = f_k f_{2k} f_{3n}.$$

Now replace x with $-ix$ and multiply the resulting equation with i^{3n+3k}, where $i = \sqrt{-1}$. Since $V_n(x) = i^{n-1} f_n(-ix)$, this yields

$$-iV_{n+k}^3 + iv_k V_n^3 - iV_{n-k}^3 = -iV_k V_{2k} V_{3n}$$

$$V_{n+k}^3 - v_k V_n^3 + V_{n-k}^3 = V_k V_{2k} V_{3n}. \tag{47.49}$$

On the other hand, let $h_n = v_n$. Again, by identity (47.47), we have

$$l_{n+k}^3 - (-1)^k l_k l_n^3 + (-1)^k l_{n-k}^3 = (x^2+4) f_k f_{2k} l_{3n}.$$

Since $v_n(x) = i^n l_n(-ix)$, as before, this yields

$$v_{n+k}^3 - v_k v_n^3 + v_{n-k}^3 = (x^2-4) V_k V_{2k} v_{3n}. \tag{47.50}$$

Combining identities (47.49) and (47.50), we get the desired result. ∎

For example,

$$\begin{aligned} V_5^3 - v_2 V_3^3 + V_1^3 &= (x^4 - 3x^2 + 1)^3 - (x^2-2)(x^2-1)^3 + 1 \\ &= x^{12} - 9x^{10} + 29x^8 - 40x^6 + 21x^4 - 2x^2 \\ &= x(x^3 - 2x)(x^8 - 7x^6 + 15x^4 - 10x^2 + 1) \\ &= V_2 V_4 V_9. \end{aligned}$$

It follows by identity (47.48) that

$$h_{n+1}^3 - xh_n^3 + h_{n-1}^3 = \begin{cases} xh_{3n} & \text{if } h_n = V_n \\ x(x^2-4)h_{3n} & \text{if } h_n = v_n. \end{cases}$$

Since $xV_n(x^2+2) = f_{2n}$ and $v_n(x^2+2) = l_{2n}$, it also follows by identity (47.48) that

$$h_{2n+2}^3 - (x^2+2)h_{2n}^3 + h_{2n-2}^3 = \begin{cases} x^2(x^2+2)h_{6n} & \text{if } h_n = f_n \\ x^2(x^2+2)(x^2+4)h_{6n} & \text{if } h_n = l_n; \end{cases}$$

see Exercises 47.73 and 47.74 [288].

Next we extract the Jacobsthal counterparts from identity (47.48).

Jacobsthal Counterparts

We have $J_n(x) = (-i\sqrt{x})^{n-1}V_n(i/\sqrt{x})$ and $j_n(x) = (-i\sqrt{x})^n v_n(i/\sqrt{x})$. Now replace x with $i/\sqrt{x}$ in equation (47.48), and multiply the resulting equation with $(-i\sqrt{x})^{3n+3k}$. We then get

$$z_{n+k}^3 - (-x)^k j_k(x) z_n^3 + (-1)^k x^{3k} z_{n-k}^3$$
$$= \begin{cases} J_k(x)J_{2k}(x)z_{3n} & \text{if } z_n(x) = J_n(x) \\ (4x+1)J_k(x)J_{2k}(x)z_{3n} & \text{if } z_n(x) = j_n(x). \end{cases} \qquad (47.51)$$

For example,

$$\begin{aligned} j_5^3(x) - x^2 j_2(x) j_3^3(x) + x^6 j_1^3(x) &= (5x^2+5x+1)^3 - x^2(2x+1)(3x+1)^3 + x^6 \\ &= 72x^6 + 294x^5 + 405x^4 + 264x^3 + 89x^2 + 15x + 1 \\ &= (4x+1)\cdot 1\cdot(2x+1)(9x^4+30x^3+27x^2+9x+1) \\ &= (4x+1)J_2(x)J_4(x)j_9(x). \end{aligned}$$

Likewise, $J_5^3(x) - x^2 j_2(x) J_3^3(x) + x^6 J_1^3(x) = (2x+1)(x^4+10x^3+15x^2+7x+1) = J_2(x)J_4(x)J_9(x)$.

In particular, identity (47.51) implies that

$$J_{n+k}^3 - (-2)^k j_k J_n^3 + (-8)^k J_{n-k}^3 = J_k J_{2k} J_{3n};$$
$$j_{n+k}^3 - (-2)^k j_k j_n^3 + (-8)^k j_{n-k}^3 = 9J_k J_{2k} j_{3n}.$$

For example,

$$\begin{aligned} J_{10}^3 - (-2)^3 j_3 J_7^3 + (-8)^3 J_4^3 &= 341^3 + 8\cdot 7\cdot 43^3 - 512\cdot 5^3 \\ &= 44{,}040{,}213 \\ &= 3\cdot 21\cdot 699{,}051 = J_3 J_6 J_{21}. \end{aligned}$$

Likewise, $j_{10}^3 - (-2)^3 j_3 j_7^3 + (-8)^3 j_4^3 = 1{,}189{,}084{,}617 = 9 \cdot 3 \cdot 21 \cdot 2{,}097{,}151 = 9J_3 J_6 j_{21}$.

Next we find the Chebyshev counterparts of identity (47.48).

Chebyshev Counterparts

Since $U_n(x) = V_{n+1}(2x)$ and $2T_n(x) = v_n(2x)$, it follows from identity (47.48) that

$$z_{n+k}^3 - 2T_k z_n^3 + z_{n-k}^3 = \begin{cases} z_{k-1} z_{2k-1} z_{3n+2} & \text{if } z_n = U_n \\ (x^2-1) U_{k-1} U_{2k-1} z_{3n} & \text{if } z_n = T_n. \end{cases} \tag{47.52}$$

Next we find the Pell counterparts of identity (47.48).

Pell Counterparts

Since $p_n(x) = f_n(2x)$ and $q_n(x) = l_{2n}(x)$, it is easy to find the Pell counterparts directly from identity (47.48) [282]:

$$z_{n+k}^3 - (-1)^k q_k z_n^3 + (-1)^k z_{n-k}^3 = \begin{cases} z_k z_{2k} z_{3n} & \text{if } z_n = p_n \\ 4(x^2+1) p_k p_{2k} z_{3n} & \text{if } z_n = q_n. \end{cases} \tag{47.53}$$

Obviously, they can be obtained from identity (47.48) using the conversion formulas $p_n(x) = (-i)^{n-1} V_n(2ix)$ and $q_n(x) = (-i)^n v_n(2ix)$.

Using these techniques, we can transform gibonacci polynomial identities to Vieta, Pell, Jacobsthal, and Chebyshev polynomial identities. For example, Fibonacci enthusiasts may want to find the Vieta, Pell, Jacobsthal, and Chebyshev counterparts of the following gibonacci identities, where $g_n = f_n$ or l_n [288]:

$$g_{n+3}^2 = (x^2+1) g_{n+2}^2 + (x^2+1) g_{n+1}^2 - g_n^2 \tag{47.54}$$

$$g_{n+4}^3 = (x^3+2x) g_{n+3}^3 + (x^4+3x^2+2) g_{n+2}^3 - (x^3+2x) g_{n+1}^3 - g_n^3 \tag{47.55}$$

$$\begin{aligned} g_{n+5}^4 &= (x^4+3x^2+1) g_{n+4}^4 + (x^6+5x^4+7x^2+2) g_{n+3}^4 \\ &\quad - (x^6+5x^4+7x^2+2) g_{n+2}^4 - (x^4+3x^2+1) g_{n+1}^4 + g_n^4. \end{aligned} \tag{47.56}$$

Next we construct combinatorial models for Vieta polynomials.

47.6 TILING MODELS FOR V_n

Model I

Suppose we would like to tile a $1 \times n$ board with square tiles and dominoes. We assign the weight x to each square, and -1 to each domino. As usual, the weight of a tiling is the product of the weights of all tiles in the tiling. The weight of the empty tiling is defined to be 1.

Figure 47.2 shows the Vieta tilings of a $1 \times n$ board, where $0 \le n \le 5$.

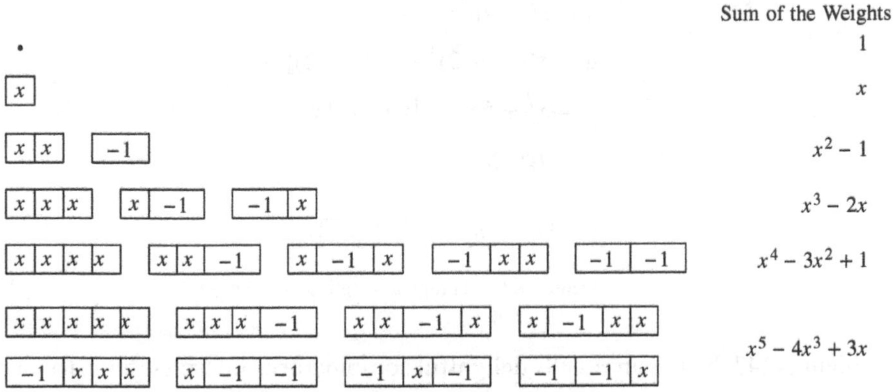

Sum of the Weights
1
x
x² − 1
x³ − 2x
x⁴ − 3x² + 1
x⁵ − 4x³ + 3x

Figure 47.2. Tiling models for V_n using Model I.

Based on the experimental data, we can predict that the sum of the weights of such tilings is V_{n+1}, where $n \ge 0$. The following theorem confirms this observation; the proof runs as in Theorem 43.1.

Theorem 47.8. *The sum of the weights of tilings of a $1 \times n$ board is V_{n+1}, where $w(square) = x$, $w(domino) = -1$, and $n \ge 0$.*

Proof. Clearly, the theorem is true when $n = 0$ and $n = 1$. Assume it is true for all nonnegative integers $< n$.

Now consider an arbitrary tiling T of length n. Suppose it ends in a square: $\underbrace{\text{subtiling}}_{\text{length } n-1}\boxed{x}$. By the inductive hypothesis, the sum of the weights of such tilings T is xV_n.

On the other hand, suppose tiling T ends in a domino: $\underbrace{\text{subtiling}}_{\text{length } n-2}\boxed{\;-1\;}$; the sum of the weights of such tilings is $-V_{n-1}$.

So the sum of the weights of tilings of a $1 \times n$ board is $xV_n - V_{n-1} = V_{n+1}$.

Thus, by PMI, the theorem is true for all $n \ge 0$. ∎

Since $V_{n+1}(2x) = U_n(x)$, it follows by Theorem 47.8 that the sum of the weights of tilings of a $1 \times n$ board is U_n, where w(square) $= 2x$.

Identities (47.13) and (47.29) Revisited

In light of Theorem 47.8, we find that the identity $xV_n(x^2+2) = f_{2n}(x)$ has a beautiful combinatorial interpretation. To see this, suppose we let w(square) $= u = x^2 + 2$. Let w be the sum of the weights of tilings of a board of length $n - 1$. Then $xw = f_{2n}(x)$.

For example, consider the tilings of length 3 in Figure 47.3. Then

$$\begin{aligned} w &= u^3 - 2u \\ xw &= x[(x^2+2)^3 - 2(x^2+2)] \\ &= x^7 + 6x^5 + 10x^3 + 4x \\ &= f_8(x). \end{aligned}$$

u | u | u		u | −1		−1 | u

Figure 47.3. Tilings of length 3.

Identity (47.29) has an equally delightful combinatorial interpretation. To this end, we let w(square) $= u = \dfrac{2x+1}{x}$. The sum of the weights w of length $n - 1$ is $V_n(u)$. Then $x^{n-1}w = J_{2n}(x)$.

For example, consider the tilings of length 4 in Figure 47.4. Then

$$\begin{aligned} w &= u^4 - 3u^2 + 1 \\ x^4 w &= x^4\left[\left(\frac{2x+1}{x}\right)^4 - 3\left(\frac{2x+1}{x}\right)^2 + 1\right] \\ &= 5x^4 + 20x^3 + 21x^2 + 8x + 1 \\ &= J_{10}(x). \end{aligned}$$

u | u | u | u	u | u | −1	u | −1 | u	−1 | u | u	−1 | −1

Figure 47.4. Tilings of length 4.

Using standard techniques, we can establish interesting Vieta properties. The next theorem establishes one such result.

Theorem 47.9. *Let $n \geq 0$. Then*

$$V_{2n+1} = \sum_{k=0}^{n} (-1)^{n-k} \binom{n}{k} x^k V_{k+1}. \tag{47.57}$$

Proof. Consider a board of length $2n$. By Theorem 47.8, the sum of the weights of its tilings is V_{2n+1}.

Since the length of the board is even, every tiling must contain an even number of squares. So each tiling must contain at least n tiles.

Now consider an arbitrary tiling T with k squares (and hence $n-k$ dominoes) among the first n tiles. There are $\binom{n}{k}$ such tilings.

The first n tiles partitions T into two subtilings:

$$\underbrace{k \text{ squares, } n-k \text{ dominoes}}_{n \text{ tiles}}\ \underbrace{\text{subtiling}}_{\text{length } k}.$$

The sum of the weights of such tilings is $(-1)^{n-k}\binom{n}{k}x^k V_{k+1}$. So the sum of the weights of all tilings is $\sum_{k=0}^{n}(-1)^{n-k}\binom{n}{k}x^k V_{k+1}$.

This, coupled with the original sum, gives the desired result. ■

For example,

$$\begin{aligned}\sum_{k=0}^{3}(-1)^{3-k}\binom{3}{k}x^k V_{k+1} &= -V_1 + 3xV_2 - 3x^2V_3 + x^3V_4\\ &= -1 + 3x\cdot x - 3x^2(x^2-1) + x^3(x^3-2x)\\ &= x^6 - 5x^4 + 6x^2 - 1\\ &= V_7.\end{aligned}$$

Using identity (47.13), we find that formula (47.57) gives an explicit formula for f_{4n+2}:

$$\begin{aligned}V_{2n+1}(x^2+2) &= \sum_{k=0}^{n}(-1)^{n-k}\binom{n}{k}(x^2+2)^k V_{k+1}(x^2+2)\\ f_{4n+2} &= \sum_{k=0}^{n}(-1)^{n-k}\binom{n}{k}(x^2+2)^k xV_{k+1}(x^2+2)\\ &= \sum_{k=0}^{n}(-1)^{n-k}\binom{n}{k}(x^2+2)^k f_{2k+2}.\end{aligned}$$

For example,

$$\begin{aligned}\sum_{k=0}^{2}(-1)^{2-k}\binom{2}{k}(x^2+2)^k f_{2k+2} &= f_2 - 2(x^2+2)f_4 + (x^2+2)^2 f_6\\ &= x - 2(x^2+2)(x^3+2x) + (x^2+2)^2(x^5+4x^3+3x)\\ &= x^9 + 8x^7 + 21x^5 + 20x^3 + 5x\\ &= f_{10}.\end{aligned}$$

Next we present a slightly modified model for V_n.

Model II

Suppose square tiles come in two colors, black and white. Suppose w(black square) = $x/2$ = w(white square), and w(domino) = -1.

Figure 47.5 shows such tilings of board of length n, where $0 \leq n \leq 3$. As in Theorem 43.5, we can confirm that the sum of the weights of such tilings of length n is V_{n+1}, where $n \geq 0$.

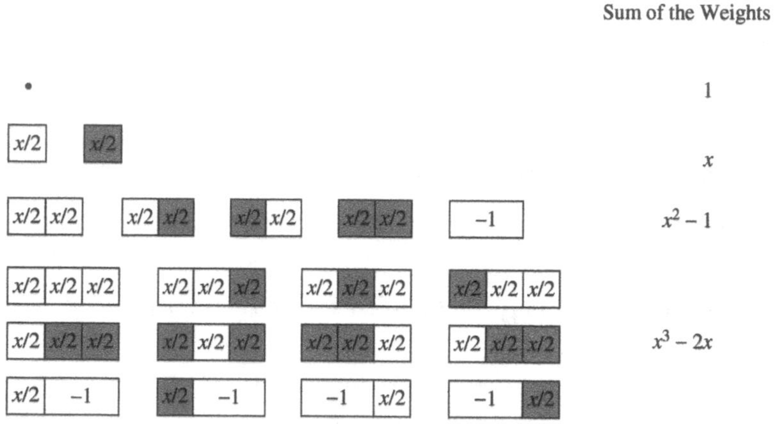

Figure 47.5. Tiling models for V_n using Model II.

Next we briefly study a different domino model for V_n.

Model III: Domino Tilings

Suppose we would like to tile a $2 \times n$ board with dominoes. The weight of a vertical domino is x and that of a horizontal domino is $i = \sqrt{-1}$. The weight of the empty tiling is 1.

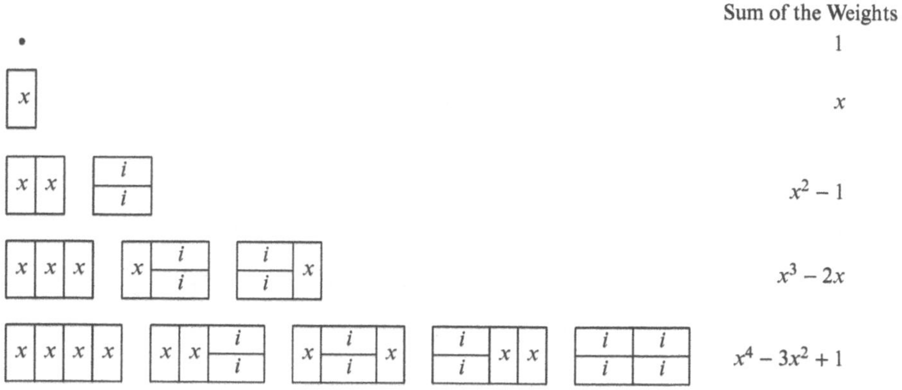

Figure 47.6. Domino tilings of a $2 \times n$ board.

Figure 47.6 shows such tilings of a $2 \times n$ board and the sum of their weights, where $0 \leq n \leq 4$.

Using the data from Figure 47.6, we conjecture that the sum of the weights of domino tilings of a $2 \times n$ board is V_{n+1}, where $n \geq 0$. The next theorem confirms this conjecture.

Theorem 47.10. *The sum of the weights of domino tilings of a $2 \times n$ board is V_{n+1}, where $n \geq 0$.*

Proof. Clearly, the theorem is true when $n = 0$ and $n = 1$. Now assume that it is true for all nonnegative integers $< n$.

Consider an arbitrary tiling T of a $2 \times n$ board. Suppose it begins with a vertical domino: $T = \boxed{x}\,\underbrace{\text{subtiling}}_{\text{length } n-1}$. By the hypothesis, the sum of the weights of such tilings is xV_n.

On the other hand, suppose T begins with a horizontal domino. Since horizontal dominoes appear in pairs, the tiling is of the form $\begin{array}{|c|}\hline i \\ \hline i \\ \hline \end{array}\,\underbrace{\text{subtiling}}_{\text{length } n-2}$.

The sum of the weights of such tilings is $-V_{n-1}$. Thus the sum of the weights of all tilings of the board is $xV_n - V_{n-1} = V_{n+1}$.

Consequently, by PMI, the statement is true for all $n \geq 0$, as desired. ■

In particular, let w(vertical domino) $= x^2 + 2$. Then the sum of the weights of the corresponding tilings of a $2 \times n$ board is $V_{n+1}(x^2 + 2)$. By identity (47.15), it equals $\frac{1}{x} f_{2n+2}$.

For example, Figure 47.7 shows such domino tilings of a 2×5 board. The sum of their weights is

$$\begin{aligned} V_6(x^2 + 2) &= (x^2 + 2)^5 - 4(x^2 + 2)^3 + 3(x^2 + 2) \\ &= x^{10} + 10x^8 + 36x^6 + 56x^4 + 35x^2 + 6 \\ &= \frac{1}{x} f_{12}. \end{aligned}$$

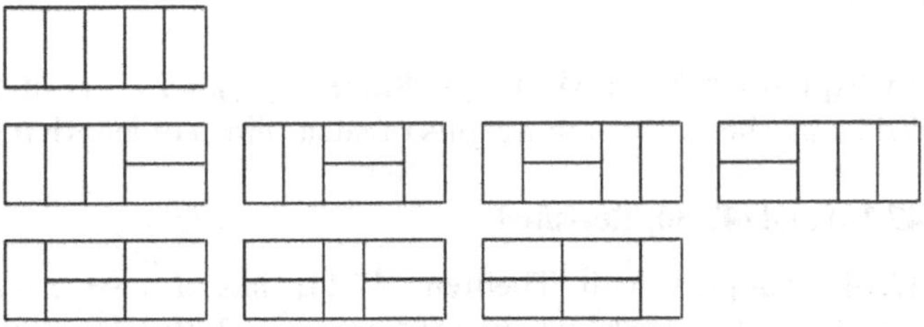

Figure 47.7. Domino tilings of a 2×5 board.

On the other hand, suppose w(vertical domino) = $2x$. Since $V_{n+1}(2x) = U_n(x)$, it follows by Theorem 47.10 that the sum of the weights of domino tilings of a $2 \times n$ board is U_n.

Using Model III, we can confirm a number of properties of V_n; see Exercises 47.98–47.100.

Next we present tiling models for Vieta–Lucas polynomials.

47.7 TILING MODELS FOR $v_n(x)$

Model IV

In this model, the weight of a square is x and that of a domino is -1, with one exception: If a tiling ends in a domino, then w(domino) = -2. The weight of the empty tiling is 2.

Figure 47.8 shows such tilings of a board of length n, where $0 \leq n \leq 5$.

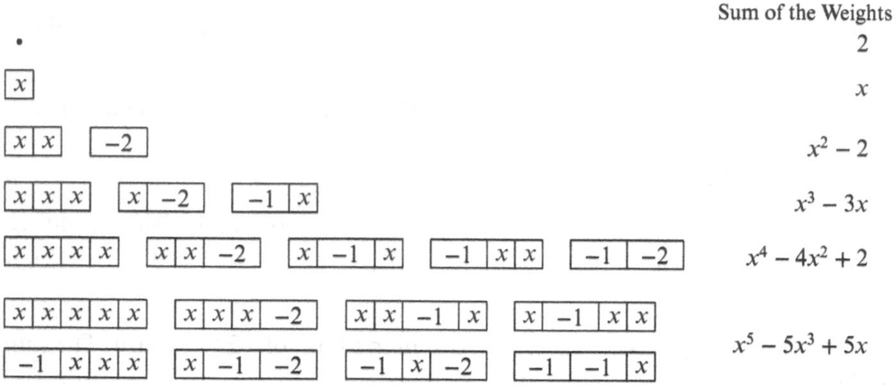

Sum of the Weights
2
x
x² − 2
x³ − 3x
x⁴ − 4x² + 2
x⁵ − 5x³ + 5x

Figure 47.8. Tiling models for v_n using Model IV.

It appears from Figure 47.8 that the sum of the weights of such tilings of length n is v_n, where $n \geq 1$. The next theorem validates this observation. In the interest of brevity, we omit its proof; see Exercise 47.90.

Theorem 47.11. *The sum of the weights of tilings of length n, using Model IV, is* v_n, *where* $n \geq 0$. ■

Suppose w(square) = $2x$ in Model IV. Since $v_n(2x) = 2T_n(x)$, it follows by Theorem 47.11 that the sum of the weights of such tilings of length n is $2T_n$.

Identities (47.14) and (47.30) Revisited

Identity (47.14), coupled with Theorem 47.11, has a nice combinatorial interpretation. To see this, we let w(square) = $u = x^2 + 2$. By Theorem 47.11, the sum of the weights w of such tilings of length n is $v_n(x^2 + 2) = l_{2n}$.

For example, consider the tilings of length 4 in Figure 47.9. Then

$$\begin{aligned} w &= u^4 - 4u^2 + 2 \\ &= (x^2+2)^4 - 4(x^2+2)^2 + 2 \\ &= x^8 + 8x^6 + 20x^4 + 16x^2 + 2 \\ &= l_8. \end{aligned}$$

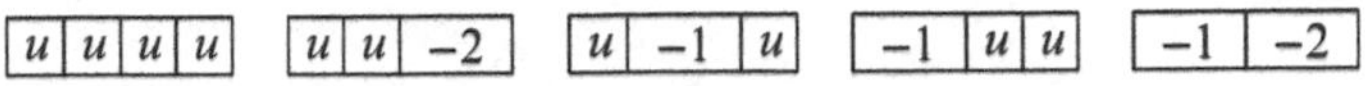

Figure 47.9. Tilings of length 4.

Identity (47.30), together with Theorem 47.11, has an equally charming interpretation. To see this, let w denote the sum of the weights of a board of length n, where w(square) $= u = \dfrac{2x+1}{x}$. Then $x^n w = j_{2n}(x)$.

For example, consider the tilings in Figure 47.8. Then

$$\begin{aligned} x^4 w &= x^4(u^4 - 4u^2 + 2) \\ &= x^4\left[\left(\frac{2x+1}{x}\right)^4 - 4\left(\frac{2x+1}{x}\right)^2 + 2\right] \\ &= 2x^4 + 16x^3 + 20x^2 + 8x + 1 \\ &= j_8(x). \end{aligned}$$

The next theorem gives an expansion of v_{2n} in terms of v_0 through v_n. The proof runs as in Theorem 47.9; so we omit it; see Exercise 47.101.

Theorem 47.12. *Let $n \geq 0$. Then*

$$v_{2n} = \sum_{k=0}^{n} (-1)^{n-k}\binom{n}{k} x^k v_k. \tag{47.58}$$

■

For example,

$$\begin{aligned} \sum_{k=0}^{3} (-1)^{3-k}\binom{3}{k} x^k v_k &= -v_0 + 3xv_1 - 3x^2v_2 + x^3v_3 \\ &= -2 + 3x^2 - 3x^2(x^2-2) + x^3(x^3-3x) \\ &= x^6 - 6x^4 + 9x^2 - 2 \\ &= v_6. \end{aligned}$$

Using identity (47.14), formula (47.58) can be rewritten in two ways:

$$\begin{aligned} l_{4n} &= \sum_{k=0}^{n}(-1)^{n-k}\binom{n}{k}(x^2+2)^k v_k(x^2+2) \\ &= \sum_{k=0}^{n}(-1)^{n-k}\binom{n}{k}(x^2+2)^k l_{2k}. \end{aligned}$$

For example,

$$\begin{aligned} \sum_{k=0}^{2}(-1)^{2-k}\binom{2}{k}(x^2+2)^k l_{2k} &= l_0 - 2(x^2+2)l_2 + (x^2+2)^2 l_4 \\ &= 2 - 2(x^2+2)^2 + (x^2+2)^2(x^4+4x^2+2) \\ &= x^8 + 8x^6 + 20x^4 + 16x^2 + 2 \\ &= l_8. \end{aligned}$$

Next we construct a combinatorial model for v_n using circular boards.

Model V: Circular Tilings

Consider a circular board with n cells, labeled 1 through n in the counter-clockwise direction, as in Figure 33.26. We would like to tile the board with (curved) square tiles and (curved) dominoes. The weight of a square (tile) is x and that of a domino is -1, with one exception: when $n = 2$, the weight of the domino is -2. The weight of the empty tiling is 2.

Figure 47.10 shows the tilings of a circular board with n cells, where $0 \le n \le 4$, and the sum of their weights.

Based on the experimental data, we conjecture that the sum of the weights of an n-bracelet is v_n. The following theorem confirms this conjecture.

Theorem 47.13. *The sum of the weights of n-bracelets, using Model V, is v_n, where $n \ge 0$.*

Proof. Clearly, the theorem is true when $n = 0$ and $n = 1$. Assume it is true for all nonnegative integers $< n$.

Consider an arbitrary n-bracelet. Suppose a square occupies cell n. Remove the square tile and glue the ends of the remaining subtiling to form an $(n-1)$-bracelet. By the inductive hypothesis, the sum of the weights of $(n-1)$-bracelets is v_{n-1}. So the sum of the weights of such n-bracelets is xv_{n-1}.

On the other hand, suppose a domino occupies cell n. (Then it covers cells n and 1, or cells n and $n-1$.) Remove the domino, and glue the ends of the

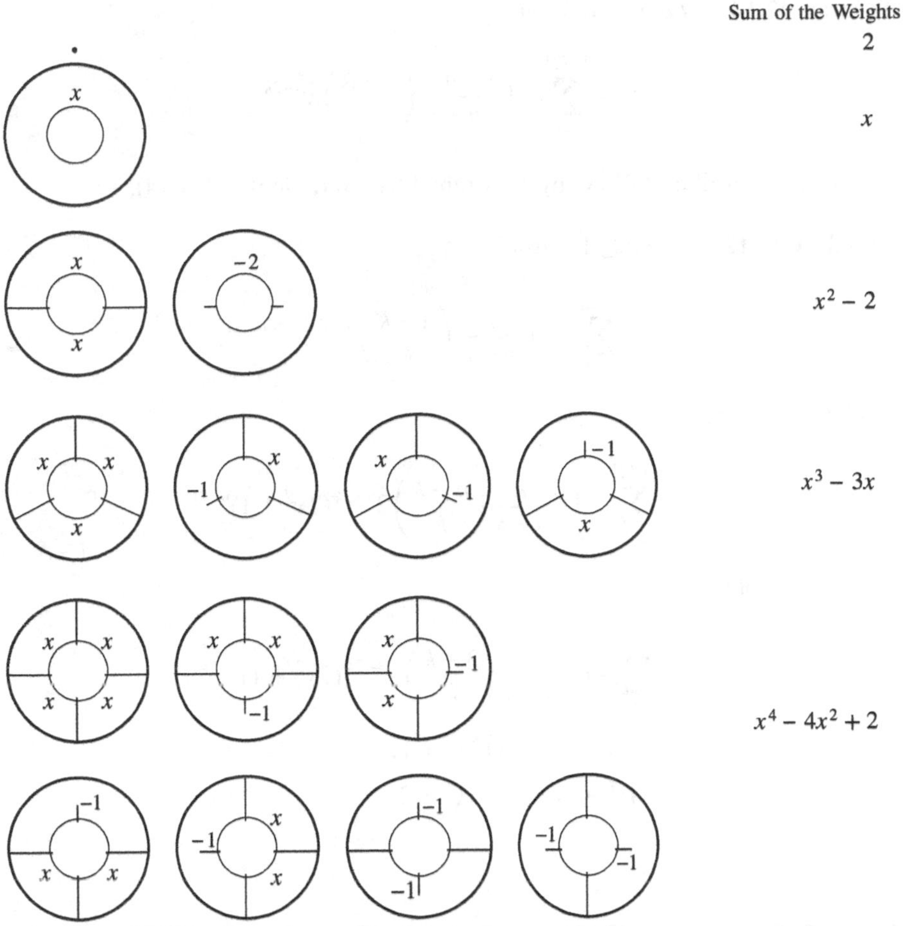

Figure 47.10. Circular tilings.

remaining subtiling to form an $(n-2)$-bracelet. By the hypothesis, the sum of the weights of $(n-2)$-bracelets is v_{n-2}. So the sum of the weights of such n-bracelets is $-v_{n-2}$.

Hence, the sum of the weights of all tilings of the circular board is $xv_{n-1} - v_{n-2} = v_n$. So the given statement is true for n also. Thus, by PMI, the theorem is true for all integers $n \geq 0$. ■

In particular, suppose w(square) = $2x$. Since $v_n(2x) = 2T_n(x)$, it follows by Theorem 47.13 that the sum of the weights of such n-bracelets is $2T_n$.

The next theorem gives an explicit formula for v_n. We leave its proof as an exercise; see Exercise 47.102.

Theorem 47.14. *Let $n \geq 1$. Then*

$$v_n = \sum_{k=0}^{\lfloor n/2 \rfloor} (-1)^k \frac{n}{n-k} \binom{n-k}{k} x^{n-2k}. \qquad \blacksquare$$

The next corollary follows by Theorem 47.14 and identity (47.14).

Corollary 47.12. *Let $n \geq 1$. Then*

$$l_{2n} = \sum_{k=0}^{\lfloor n/2 \rfloor} (-1)^k \frac{n}{n-k} \binom{n-k}{k} (x^2+2)^{n-2k}. \qquad \blacksquare$$

Consequently,

$$q_{2n} = \sum_{k=0}^{\lfloor n/2 \rfloor} (-1)^k \frac{n}{n-k} \binom{n-k}{k} 2^{n-2k} (2x^2+1)^{n-2k}.$$

For example,

$$\begin{aligned} q_6 &= \sum_{k=0}^{1} (-1)^k \frac{3}{3-k} \binom{3-k}{k} 2^{3-2k} (2x^2+1)^{3-2k} \\ &= 8(2x^2+1)^3 - 6(2x^2+1) \\ &= 64x^6 + 96x^4 + 36x^2 + 2. \end{aligned}$$

EXERCISES 47

1. Compute $v_n(0)$.

Prove each, where $i = \sqrt{-1}$ and the prime denotes differentiation with respect to x.

2. $V_n(x) = i^{n-1} f_n(-ix)$.
3. $v_n(x) = i^n l_n(-ix)$.
4. $V_{-n}(x) = -V_n(x)$.
5. $v_{-n}(x) = v_n(x)$.
6. $V_n(-x) = (-1)^{n+1} V_n(x)$.
7. $v_n(-x) = (-1)^n v_n(x)$.

8. $V_{n+1}V_{n-1} - V_n^2 = -1.$

9. $(x^2-4)(V_n^2 + V_{n+1}^2) = xV_{2n+1} - 4.$

10. $v_n^2 + v_{n+1}^2 = xv_{2n+1} + 4.$

11. $V_{2n} = V_n v_n.$

12. $v_{2n} = v_n^2 - 2.$

13. $v_{2n} = (x^2-4)V_n^2 + 2.$

14. $V_{n+1} - V_{n-1} = v_n.$

15. $V_{n+1}^2 - V_{n-1}^2 = xV_{2n}.$

16. $2V_{n+1} - xV_n = v_n.$

17. $v_{n+1} - v_{n-1} = (x^2-4)V_n.$

18. $v_{n+1}^2 - v_{n-1}^2 = x(x^2-4)V_{2n}.$

19. $2v_{n+1} - xv_n = (x^2-4)V_n.$

20. $V_{n+1}^2 - V_n^2 = V_{2n+1}.$

21. $v_{n+1}^2 + (x^2-4)V_n^2 = xv_{2n+1}.$

22. $(x^2-4)V_{n+1}^2 + v_n^2 = xv_{2n+1}.$

23. $V_{n+2} + V_{n-2} = (x^2-2)V_n.$

24. $V_{n+2} - V_{n-2} = xV_n.$

25. $V_{n+2}^2 - V_{n-2}^2 = x(x^2-2)V_{2n}.$

26. $V_{n+k}^2 - V_{n-k}^2 = V_{2n}V_{2k}.$

27. $v_{n+2} + v_{n-2} = (x^2-2)v_n.$

28. $v_{n+2} - v_{n-2} = x(x^2-4)V_n.$

29. $v_{n+2}^2 - v_{n-2}^2 = x(x^2-2)(x^2-4)V_{2n}.$

30. $v_{n+k}^2 - v_{n-k}^2 = (x^2-4)V_{2n}V_{2k}.$

31. $v_n V_{n-1} + V_n v_{n-1} = 2V_{2n-1}.$

32. $v_n V_{n-1} - V_n v_{n-1} = 2.$

33. $v_{m+n} = v_{m+1}V_n - v_m V_{n-1}.$

34. $v_{m-n} = v_m V_{n+1} - v_{m+1}V_n.$

35. $2V_{m+n} = V_m v_n + V_n v_m.$

36. $2V_{m-n} = V_m v_n - V_n v_m.$

37. $2v_{m+n} = v_m v_n + (x^2-4)V_m V_n$.
38. $2v_{m-n} = v_m v_n - (x^2-4)V_m V_n$.
39. $(x^2-4)V_n^3 = V_{3n} - 3V_n$.
40. $v_n^3 = v_{3n} + 3v_n$.
41. $v_m v_{m+n+k} - v_{m+k} v_{m+n} = (x^2-4)V_n V_k$.
42. $V_n(x)V_{n-1}(-x) + V_n(-x)V_{n-1}(-x) = 0$ (Shannon and Horadam, [460]).
43. $v_n(x)v_{n-1}(-x) + v_n(-x)v_{n-1}(-x) = 0$ (Horadam, [236]).
44. Formula (47.4).
45. $v_n(x^2+2) = l_{2n}$.
46. $v_n(x^2-2) - v_n^2(x) = -2$ (Jacobsthal, as per [236]).
47. $v_n(2) = 2$.
48. $q_{2n}(x) - q_n^2(x) = 2(-1)^{n+1}$.
49. $Q_{2n} = 4P_n^2 + (-1)^n$.
50. $Q_{2n} = 2Q_n^2 - (-1)^n$.
51. $J_n(x) = x^{(n-1)/2} f_n(1/\sqrt{x})$.
52. $v_n(x) = x^n j_n(-1/x^2)$.
53. $j_n(x) = x^{n/2} l_n(1/\sqrt{x})$.
54. $j_{2n}(x) = x^n v_n\left(\dfrac{2x+1}{x}\right)$.
55. $l_{2n}(x) = (x^2+2)^n j_n\left(-\dfrac{1}{(x^2+2)^2}\right)$.
56. $j_n^2(x) - (4x+1)J_n^2(x) = 4(-x)^n$.
57. $U_n(x) = i^n f_{n+1}(-2ix)$.
58. $2T_n(x) = i^n l_n(-2ix)$.
59. $F_{2n} = U_{n-1}(3/2)$.
60. $L_{2n} = 2T_n(3/2)$.
61. $2T_{n+1}(x) = U_{n+1}(x) - U_{n-1}(x)$.
62. $T_{n+1}(x) - T_{n-1}(x) = 2(x^2-1)U_{n-1}(x)$.
63. $p_{2n}(x) = 2xU_{n-1}(2x^2+1)$.
64. $q_{2n}(x) = 2T_n(2x^2+1)$.

65. $P_{2n} = 2U_{n-1}(3)$.

66. $Q_{2n} = T_n(3)$.

67. $T_n(2x^2 - 1) - 2(x^2 - 1)U_{n-1}^2(x) = 1$.

68. $T_n^4 - (x^2 - 1)^2 U_{n-1}^4 - 2(-1)^n(x^2 - 1)U_{n-1}^2 = 1$.

69. $T_n(2x^2 - 1) - 2T_n^2(x) = -1$.

70. $T_n^2(x) - (x^2 - 1)U_{n-1}^2(x) = 1$.

71. $J_{n+1} = (\sqrt{2}i)^n U_n\left(\dfrac{-i}{2\sqrt{2}}\right)$.

72. $j_n(x) = 2(-i\sqrt{x})^n T_n\left(\dfrac{i}{2\sqrt{x}}\right)$.

73. $f_{2n+2}^3 - (x^2 + 2)f_{2n}^3 + f_{2n-2}^3 = x^2(x^2 + 2)f_{6n}$.

74. $l_{2n+2}^3 - (x^2 + 2)l_{2n}^3 + l_{2n-2}^3 = x^2(x^2 + 2)(x^2 + 4)l_{6n}$.

75. $v_n' = nV_n$.

76. $n\displaystyle\int_0^x V_n(y)dy = v_n(x) - v_n(0)$.

77. $n\displaystyle\int_0^3 V_n(y)dy = L_{2n} - v_n(0)$.

78. $V_{n+3}^2 = (x^2 - 1)V_{n+2}^2 - (x^2 - 1)V_{n+1}^2 + V_n^2$.

79. $v_{n+3}^2 = (x^2 - 1)v_{n+2}^2 - (x^2 - 1)v_{n+1}^2 + v_n^2$.

80. $J_{n+3}^2(x) = (x + 1)J_{n+2}^2(x) + x(x + 1)J_{n+1}^2(x) - x^3 J_n^2(x)$.

81. $j_{n+3}^2(x) = (x + 1)j_{n+2}^2(x) + x(x + 1)j_{n+1}^2(x) - x^3 j_n^2(x)$.

82. $U_{n+3}^2 = (4x^2 - 1)U_{n+2}^2(x) - (4x^2 - 1)U_{n+1}^2(x) - U_n^2(x)$.

83. $T_{n+3}^2 = (4x^2 - 1)T_{n+2}^2(x) - (4x^2 - 1)T_{n+1}^2(x) - T_n^2(x)$.

84. $V_{n+4}^3 = (x^3 - 2x)V_{n+3}^3 - (x^4 - 3x^2 + 2)V_{n+2}^3 + (x^3 - 2x)V_{n+1}^3 - V_n^3$.

85. $v_{n+4}^3 = (x^3 - 2x)v_{n+3}^3 - (x^4 - 3x^2 + 2)v_{n+2}^3 + (x^3 - 2x)v_{n+1}^3 - v_n^3$.

86. $J_{n+4}^3(x) = (2x + 1)J_{n+3}^3(x) + x(2x^2 + 3x + 1)J_{n+2}^3(x) - x^3(2x + 1)J_{n+1}^3(x) - x^6 J_n^3(x)$.

87. $j_{n+4}^3(x) = (2x + 1)j_{n+3}^3(x) + x(2x^2 + 3x + 1)j_{n+2}^3(x) - x^3(2x + 1)j_{n+1}^3(x) - x^6 j_n^3(x)$.

88. $U_{n+4}^3 = 4(2x^3 - x)U_{n+3}^3 - (16x^4 - 12x^2 + 2)U_{n+2}^3 + 4(2x^3 - x)U_{n+1}^3 - U_n^3.$

89. $T_{n+4}^3 = 4(2x^3 - x)T_{n+3}^3 - (16x^4 - 12x^2 + 2)T_{n+2}^3 + 4(2x^3 - x)T_{n+1}^3 - T_n^3.$

Using combinatorial tilings, prove each.

90. Theorem 47.11.

91. $V_{n+1} = \sum_{k=0}^{\lfloor n/2 \rfloor} (-1)^k \binom{n-k}{k} x^{n-2k}.$

92. $f_{2n+2} = x \sum_{k=0}^{\lfloor n/2 \rfloor} (-1)^k \binom{n-k}{k} (x^2+2)^{n-2k}.$

93. $V_{n+2} = \sum_{\substack{i,j \ge 0 \\ i+j \le n}} (-1)^{i+j} \binom{n-i}{j} \binom{n-j}{i} x^{2n-2i-2j+1}.$

94. $f_{2n+4} = x \sum_{\substack{i,j \ge 0 \\ i+j \le n}} (-1)^{i+j} \binom{n-i}{j} \binom{n-j}{i} (x^2+2)^{2n-2i-2j+1}.$

95. $V_{n+2} + V_{n-2} = (x^2 - 2)V_n.$

96. $V_{n+1}^2 - V_{n-1}^2 = xV_{2n}.$

97. $V_{m+n} = V_{m+1}V_n - V_m V_{n-1}.$

Using Model III, establish each.

98. Formula (47.3).

99. $V_{m+n} = V_{m+1}V_n - V_m V_{n-1}.$

100. $V_{n+2} = \sum_{\substack{i,j \ge 0 \\ i+j \le n}} (-1)^{i+j} \binom{n-i}{j} \binom{n-j}{i} x^{2n-2i-2j+1}.$

Using Model IV or V, prove each.

101. $v_{2n} = \sum_{k=0}^{n} (-1)^{n-k} \binom{n}{k} x^k v_k.$

102. Theorem 47.14.

103. $v_{m+n} = v_{m+1}V_n - v_m V_{n-1}.$

104. $V_{2n} = V_n v_n.$

105. $v_{n+1} + v_{n-1} = (x^2 - 4)V_n.$

Find a generating function for each.

106. $V_n(x).$

107. $v_n(x).$

48

BIVARIATE GIBONACCI FAMILY II

To ask the right question is harder than to answer it.
–Georg Cantor (1845–1918)

In Chapter 43, we introduced an extended bivariate family $\{s_n(x, y)\}$, which includes the well-known Fibonacci, Pell, and Jacobsthal families, and their Lucas counterparts as subfamilies. In Chapters 43–45, we explored some interesting properties of the extended family, using Candido's identity. In this chapter, we explore some additional properties of the extended family and their special cases applicable to the subfamilies.

48.1 BIVARIATE IDENTITIES

We begin our study with a variant of Candido's identity:

$$x^4 + y^4 + (x + y)^4 = 2(x^2 + xy + y^2)^2. \tag{48.1}$$

This variant also has an interesting geometric interpretation. To see this, consider the square $ABCD$ in Figure 48.1, where $AE = x^2$, $EF = y^2$, and $FB = xy$. Clearly, the area of the square is $(x^2 + xy + y^2)^2$.

Now place a copy of square $ABCD$ to its right to complete a rectangle; see Figure 48.2. The area of the resulting rectangle $AGHD$ is $2(x^2 + xy + y^2)^2$.

Fibonacci and Lucas Numbers with Applications, Volume Two. Thomas Koshy.

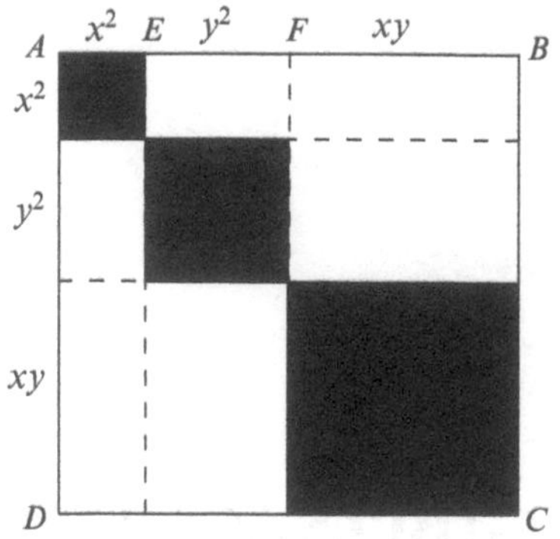

Figure 48.1.

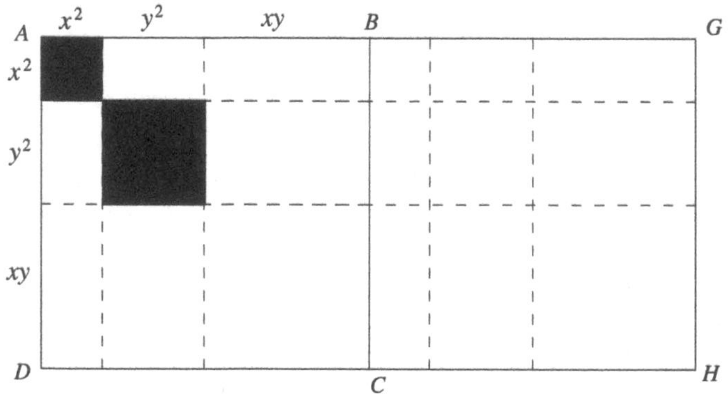

Figure 48.2.

We now compute this area in a different way. The unshaded area is $x^4 + y^4 + 4x^3y + 6x^2y^2 + 4xy^3 = (x + y)^4$. So the area $AGHD$ equals $x^4 + y^4 + (x + y)^4$. Thus $x^4 + y^4 + (x + y)^4 = 2(x^2 + xy + y^2)^2$.

Identity (48.1) has an abundance of interesting bivariate Fibonacci–Pell–Jacobsthal implications. To see this, we replace x with xs_n, and y with ys_{n-1} in the identity. It then yields the bivariate identity

$$x^4s_n^4 + y^4s_{n-1}^4 + s_{n+1}^4 = 2(x^2s_n^2 + xys_{n-1}s_n + y^2s_{n-1}^2)^2. \tag{48.2}$$

In particular, this identity yields

$$a_{n-1}^4 + x^4a_n^4 + a_{n+1}^4 = 2(a_{n-1}^2 + xa_{n-1}a_n + x^2a_n^2)^2, \tag{48.3}$$

$$b_{n-1}^4 + 16x^4b_n^4 + b_{n+1}^4 = 2(b_{n-1}^2 + 2xb_{n-1}b_n + 4x^2b_n^2)^2, \tag{48.4}$$

$$16y^4c_{n-1}^4 + c_n^4 + c_{n+1}^4 = 2(4y^2c_{n-1}^2 + 2yc_{n-1}c_n + c_n^2)^2, \tag{48.5}$$

where $a_n = a_n(x)$, $b_n = b_n(x)$, and $c_n = c_n(y)$ satisfy the Fibonacci, Pell, and Jacobsthal recurrences, respectively.

It follows from these polynomial identities that

$$f_{n-1}^4(x) + x^4 f_n^4(x) + f_{n+1}^4(x) = 2[f_{n-1}^2(x) + x f_{n-1}(x) f_n(x) + x^2 f_n^2(x)]^2; \quad (48.6)$$

$$p_{n-1}^4(x) + 16x^4 p_n^4(x) + p_{n+1}^4(x) = 2[p_{n-1}^2(x) + 2x p_{n-1}(x) p_n(x) + 4x^2 p_n^2(x)]^2; \quad (48.7)$$

$$16y^4 J_{n-1}^4(y) + J_n^4(y) + J_{n+1}^4(y) = 2[4y^2 J_{n-1}^2(y) + 2y J_{n-1}(y) J_n(y) + J_n^2(y)]^2; \quad (48.8)$$

$$F_{n-1}^4 + F_n^4 + F_{n+1}^4 = 2(F_{n-1}^2 + F_{n-1}F_n + F_n^2)^2; \quad (48.9)$$

$$P_{n-1}^4 + 16P_n^4 + P_{n+1}^4 = 2(P_{n-1}^2 + 2P_{n-1}P_n + 4P_n^2)^2; \quad (48.10)$$

$$16J_{n-1}^4 + J_n^4 + J_{n+1}^4 = 2(4J_{n-1}^2 + 2J_{n-1}J_n + J_n^2)^2. \quad (48.11)$$

Similar results apply for their Lucas counterparts also. For example, $P_4^4 + 16P_5^4 + P_6^4 = 12^4 + 16 \cdot 29^4 + 70^4 = 35{,}347{,}232 = 2(P_4^2 + 2P_4P_5 + P_5^2)^2$. Similarly, $16j_5^4 + j_6^4 + j_7^4 = 292{,}771{,}602 = 2(4j_5^2 + 2j_5j_6 + j_6^2)^2$.

Using Fibonacci recurrence and Cassini's formula, we can rewrite identity (48.6) in a different way:

$$\begin{aligned} f_{n-1}^4(x) + x^4 f_n^4(x) + f_{n+1}^4(x) &= 2\{f_{n-1}(x)[f_{n-1}(x) + x f_n(x)] + x^2 f_n^2(x)\}^2 \\ &= 2[f_{n+1}(x) f_{n-1}(x) + x^2 f_n^2(x)]^2 \\ &= \{[f_n^2(x) + (-1)^n] + x^2 f_n^2(x)\}^2 \\ &= 2[(x^2+1) f_n^2(x) + (-1)^n]^2. \end{aligned} \quad (48.12)$$

In particular,

$$F_{n-1}^4 + F_n^4 + F_{n+1}^4 = 2[2F_n^2 + (-1)^n]^2. \quad (48.13)$$

J.A.H. Hunter of Toronto, Canada, discovered this identity in 1966 [249]. It has similar counterparts to the other subfamilies; see Exercises 48.1–48.5.

Identities (48.7) and (48.8) also can be rewritten using Pell and Jacobsthal recurrences and the corresponding Cassini's formulas; see Exercises 48.6–48.10.

We can rewrite identity (48.9) in yet another way, as C.C. Yalavigi of Government College, Mercara, India, did in 1969 [512]:

$$\begin{aligned} F_{n-1}^4 + F_n^4 + F_{n+1}^4 &= 2[(F_{n-1} + F_n)^2 - F_nF_{n-1}]^2 \\ &= 2(F_{n+1}^2 - F_nF_{n-1})^2. \end{aligned}$$

Next we employ some algebraic identities to develop bivariate identities for $s_n = s_n(x, y)$.

48.2 ADDITIONAL BIVARIATE IDENTITIES

Consider the identity $(x+y)^3 - x^3 - y^3 = 3xy(x+y)$. Replacing x with xs_n and y with ys_{n-1}, this yields

$$
\begin{aligned}
(xs_n + ys_{n-1})^3 - x^3 s_n^3 - y^3 s_{n-1}^3 &= 3 \cdot xs_n \cdot ys_{n-1}(xs_n + ys_{n-1}) \\
s_{n+1}^3 - x^3 s_n^3 - y^3 s_{n-1}^3 &= 3xys_{n+1}s_n s_{n-1}. \qquad (48.14)
\end{aligned}
$$

In particular, we have

$$
\begin{aligned}
f_{n+1}^3(x) - x^3 f_n^3(x) - f_{n-1}^3(x) &= 3xf_{n+1}(x)f_n(x)f_{n-1}(x); \\
p_{n+1}^3(x) - 8x^3 p_n^3(x) - p_{n-1}^3(x) &= 6xp_{n+1}(x)p_n(x)p_{n-1}(x); \\
J_{n+1}^3(y) - J_n^3(y) - 8y^3 J_{n-1}^3(y) &= 6yJ_{n+1}(y)J_n(y)J_{n-1}(y); \\
F_{n+1}^3 - F_n^3 - F_{n-1}^3 &= 3F_{n+1}F_nF_{n-1}; \\
L_{n+1}^3 - L_n^3 - L_{n-1}^3 &= 3L_{n+1}L_nL_{n-1}; \\
P_{n+1}^3 - 8P_n^3 - P_{n-1}^3 &= 6P_{n+1}P_nP_{n-1}; \\
Q_{n+1}^3 - 8Q_n^3 - Q_{n-1}^3 &= 6Q_{n+1}Q_nQ_{n-1}; \\
J_{n+1}^3 - J_n^3 - 8J_{n-1}^3 &= 6J_{n+1}J_nJ_{n-1}; \\
j_{n+1}^3 - j_n^3 - 8j_{n-1}^3 &= 6j_{n+1}j_nj_{n-1}.
\end{aligned}
$$

For example, we have

$$
\begin{aligned}
f_5^3(x) - x^3 f_4^3(x) - f_3^3(x) &= (x^4+3x^2+1)^3 - x^3(x^3+2x)^3 - (x^2+1)^3 \\
&= 3x^2(x^8+6x^6+12x^4+9x^2+2) \\
&= 3x(x^4+3x^2+1)(x^3+2x)(x^2+1) \\
&= 3xf_5(x)f_4(x)f_3(x); \\
F_5^3 - F_4^3 - F_3^3 = 90 &= 3F_5F_4F_3; \\
P_5^3 - 8P_4^3 - P_3^3 = 10,440 &= 6P_5P_4P_3.
\end{aligned}
$$

We can use the following identities from Section 35.1 to extract additional bivariate identities and their special cases; see Exercises 48.16–48.33:

$$(x+y)^5 - x^5 - y^5 = 5xy(x+y)(x^2+xy+y^2); \qquad (48.15)$$

$$(x+y)^7 - x^7 - y^7 = 7xy(x+y)(x^2+xy+y^2)^2. \qquad (48.16)$$

Catalani studied the next example in 2005 [97]. The featured proof, based on the one given by Seiffert in the following year, shows how we can use a formula for $f_{2n}(x)$ to develop a bivariate one for $f_{2n}(x, y)$ [453].

Example 48.1 (Catalani, 2005 [97]). *Let $xy \neq 0$ and $x^2 + 4y \neq 0$. Prove that*

$$f_{2n}(x, y) = x \sum_{k=0}^{n-1} \binom{2n-k-1}{k} (x^2 + 4y)^{n-k-1}(-y)^k. \tag{48.17}$$

Proof. For convenience, we split the proof into four parts.

1) To begin with, we establish an interesting link between $f_n(x, y)$ and $f_n(x/\sqrt{y})$. To this end, let $\Delta(x, y) = \sqrt{x^2 + 4y}$, $\Delta = \Delta(x) = \Delta(x, 1)$, $u = u(x, y) = \dfrac{x + \Delta(x, y)}{2}$, $v = v(x, y) = \dfrac{x - \Delta(x, y)}{2}$, $a = a(x) = \dfrac{x + \Delta}{2}$, and $b = b(x) = \dfrac{x - \Delta}{2}$. Then

$$\begin{aligned} a(x/\sqrt{y}) &= \frac{x + \sqrt{x^2 + 4y}}{2\sqrt{y}} \\ &= \frac{u}{\sqrt{y}}. \end{aligned}$$

Similarly, $b(x/\sqrt{y}) = \dfrac{v}{\sqrt{y}}$.

Using the Binet-like formula, we then have

$$\begin{aligned} \sqrt{x^2 + 4} f_n(x) &= a^n - b^n \\ \sqrt{x^2 + 4y} f_n(x\sqrt{y}) &= \frac{u^n - v^n}{y^{(n-1)/2}} \\ y^{(n-1)/2} f_n(x/\sqrt{y}) &= f_n(x, y). \end{aligned} \tag{48.18}$$

2) Let $i = \sqrt{-1}$. Then

$$\begin{aligned} a(i\Delta) &= \frac{i\Delta + \sqrt{-\Delta^2 + 4}}{2} \\ &= \frac{i(x + \Delta)}{2}; \\ b(i\Delta) &= \frac{i(x - \Delta)}{2}. \end{aligned}$$

Consequently,

$$\begin{aligned} f_k(i\Delta) &= \frac{i^k}{ix}(a^k - b^k) \\ &= \frac{i^{k-1}\Delta}{x} f_k(x) \\ f_k(x) &= \frac{x}{i^{k-1}\Delta} f_k(i\Delta). \end{aligned} \tag{48.19}$$

3) By the Lucas-like formula for $f_n(x)$, we have

$$f_{2n}(x) = \sum_{k=0}^{n-1} \binom{2n-k-1}{k} x^{2n-2k-1}$$

$$f_{2n}(i\Delta) = \sum_{k=0}^{n-1} \binom{2n-k-1}{k} (i\Delta)^{2n-2k-1}.$$

Using formula (48.19), we then have

$$\begin{aligned} f_{2n}(x) &= \frac{x}{i^{2n-1}\Delta} \sum_{k=0}^{n-1} \binom{2n-k-1}{k} (i\Delta)^{2n-2k-1} \\ &= x \sum_{k=0}^{n-1} \binom{2n-k-1}{k} (-1)^k \Delta^{2n-2k-2}. \end{aligned} \tag{48.20}$$

4) Since $\Delta(x/\sqrt{y}) = \sqrt{\dfrac{x^2+4y}{y}}$, it follows from (48.18) that

$$\begin{aligned} f_{2n}(x, y) &= y^{(2n-1)/2} f_{2n}(x\sqrt{y}) \\ &= y^{(2n-1)/2} \cdot \frac{x}{\sqrt{y}} \sum_{k=0}^{n-1} \binom{2n-k-1}{k} (-1)^k \left(\frac{x^2+4y}{y}\right)^{n-k-1} \\ &= x \sum_{k=0}^{n-1} \binom{2n-k-1}{k} (x^2+4y)^{n-k-1}(-y)^k, \end{aligned}$$

as desired. ■

For example, we have

$$\begin{aligned} f_6(x, y) &= x \sum_{k=0}^{2} \binom{5-k}{k} (x^2+4y)^{2-k}(-y)^k \\ &= x[(x^2+4y)^2 + 4(x^2+4y)(-y) + 3(-y)^2] \\ &= x^5 + 4x^3y + 3xy^2; \end{aligned}$$

see Table 43.1.

Formula (48.17) yields several interesting byproducts:

$$p_{2n}(x,y) = x\sum_{k=0}^{n-1}\binom{2n-k-1}{k}2^{2n-2k-1}(x^2+y)^{n-k-1}(-y)^k;$$

$$f_{2n}(x) = x\sum_{k=0}^{n-1}(-1)^k\binom{2n-k-1}{k}(x^2+4)^{n-k-1};$$

$$p_{2n}(x) = x\sum_{k=0}^{n-1}(-1)^k\binom{2n-k-1}{k}2^{2n-2k-1}(x^2+1)^{n-k-1};$$

$$J_{2n}(y) = \sum_{k=0}^{n-1}(-1)^k\binom{2n-k-1}{k}(4y+1)^{n-k-1}(-y)^k;$$

$$F_{2n} = \sum_{k=0}^{n-1}(-1)^k\binom{2n-k-1}{k}5^{n-k-1};$$

$$P_{2n} = \sum_{k=0}^{n-1}(-1)^k\binom{2n-k-1}{k}8^{n-k-1};$$

$$J_{2n} = \sum_{k=0}^{n-1}\binom{2n-k-1}{k}(-2)^k\cdot 9^{n-k-1}.$$

For example,

$$\begin{aligned}p_6(x,y) &= x\sum_{k=0}^{2}\binom{5-k}{k}2^{5-2k}(x^2+y)^{2-k}(-y)^k\\ &= x\left[2^5(x^2+y)^2 - 4\cdot 2^3(x^2+y)y + 3\cdot 2y^2\right]\\ &= 32x^5+32x^3y+6xy^2.\end{aligned}$$

Likewise, $J_6 = \sum_{k=0}^{2}\binom{5-k}{k}(-2)^k\cdot 9^{2-k} = 21$; see Table 44.2.

Example 48.1 has an interesting consequence, as the next example reveals.

Example 48.2. *Find an explicit formula and a recurrence for the sum* $a_n = \sum_{k=0}^{n-1}\binom{2n-k-1}{k}(-3)^{n-k-1}$.

Solution. By equation (48.20), we have

$$\begin{aligned} f_{2n}(i) &= i\sum_{k=0}^{n-1}\binom{2n-k-1}{k}(-1)^k 3^{n-k-1} \\ &= i(-1)^{n-1}a_n \\ a_n &= i(-1)^n f_{2n}(i). \end{aligned}$$

Using the Binet-like formula, this yields

$$a_n = \frac{i(-1)^n}{\sqrt{3}}\left[\left(\frac{\sqrt{3}+i}{2}\right)^{2n} - \left(\frac{\sqrt{3}-i}{2}\right)^{2n}\right].$$

Since $\cos \pi/6 = \sqrt{3}/2$ and $\sin \pi/6 = 1/2$, it follows by the well-known *Euler's formula* $e^{it} = \cos t + i \sin t$, that

$$\begin{aligned} a_n &= \frac{i(-1)^n}{\sqrt{3}}\left(e^{n\pi i/3} - e^{-n\pi i/3}\right) \\ &= \frac{i(-1)^n}{\sqrt{3}} \cdot 2i\sin(n\pi/3) \\ &= \frac{2}{\sqrt{3}}(-1)^{n-1}\sin(n\pi/3). \end{aligned} \tag{48.21}$$

This is the desired explicit formula for a_n, where $n \geq 0$.

Next we find a recurrence for a_n using the addition formula for the sine function. Using formula (48.21), we have

$$\begin{aligned} a_{n-1} + a_{n-2} &= \frac{2}{\sqrt{3}}(-1)^{n-2}\left[\sin\frac{(n-1)\pi}{3} - \sin\frac{(n-2)\pi}{3}\right] \\ &= \frac{2}{\sqrt{3}}(-1)^{n-2}\left[\sin\frac{n\pi}{3}\left(\cos\frac{\pi}{3} - \cos\frac{2\pi}{3}\right) - \cos\frac{n\pi}{3}\left(\sin\frac{\pi}{3} - \sin\frac{2\pi}{3}\right)\right] \\ &= \frac{2}{\sqrt{3}}(-1)^{n-2}\sin\left(\frac{n\pi}{3}\right) \\ &= -a_n. \end{aligned}$$

So $a_n = a_{n-1} + a_{n-2}$, where $n \geq 2$. ■

48.3 A BIVARIATE LUCAS COUNTERPART

Using the techniques in Example 48.1, we can develop a similar formula for $l_{2n}(x, y)$:

$$l_{2n}(x,y) = \sum_{k=0}^{n} \frac{2n}{2n-k}\binom{2n-k}{k}(-1)^k(x^2+4y)^{n-k}y^k, \qquad (48.22)$$

where $n \geq 1$; see Exercises 48.34–48.38.

For example,

$$\begin{aligned} l_6(x,y) &= \sum_{k=0}^{3} \frac{6}{6-k}\binom{6-k}{k}(-1)^k(x^2+4y)^{3-k}y^k \\ &= (x^2+4y)^3 - 6(x^2+4y)^2y + 9(x^2+4y)y^2 - 2y^3 \\ &= x^6 + 6x^4y + 9x^2y^2 + 2y^3; \end{aligned}$$

see Table 43.1.

Interesting Dividends

Formula (48.22) has interesting consequences:

$$q_{2n}(x,y) = \sum_{k=0}^{n} \frac{2n}{2n-k}\binom{2n-k}{k}(-1)^k\, 4^{n-k}(x^2+y)^{n-k}y^k;$$

$$l_{2n}(x) = \sum_{k=0}^{n} \frac{2n}{2n-k}\binom{2n-k}{k}(-1)^k(x^2+y)^{n-k};$$

$$q_{2n}(x) = \sum_{k=0}^{n} \frac{2n}{2n-k}\binom{2n-k}{k}(-1)^k\, 4^{n-k}(x^2+1)^{n-k};$$

$$j_{2n}(y) = \sum_{k=0}^{n} \frac{2n}{2n-k}\binom{2n-k}{k}(-1)^k(4y+1)^{n-k}y^k;$$

$$L_{2n} = \sum_{k=0}^{n} \frac{2n}{2n-k}\binom{2n-k}{k}(-1)^k\, 5^{n-k};$$

$$Q_{2n} = \sum_{k=0}^{n} \frac{2n}{2n-k}\binom{2n-k}{k}(-1)^k\, 2^{3n-3k-1};$$

$$j_{2n} = \sum_{k=0}^{n} \frac{2n}{2n-k}\binom{2n-k}{k}(-1)^k\, 5^{n-k}.$$

As examples, we have

$$\begin{aligned}
j_6(y) &= \sum_{k=0}^{3} \frac{6}{6-k}\binom{6-k}{k}(-1)^k(4y+1)^{3-k}y^k \\
&= (4y+1)^3 - 6y(4y+1)^2 + 9y^2(4y+1) - 2y^3 \\
&= 2y^3 + 9y^2 + 6y + 1; \\
Q_8 &= \sum_{k=0}^{4} \frac{8}{8-k}\binom{8-k}{k}(-1)^k\, 2^{11-3k} = 577;
\end{aligned}$$

see Tables 44.1 and 31.6.

We can use the techniques used in Example 48.1, coupled with the recurrence for $f_{2n}(x)$ in Theorem 43.3, to develop a double-summation formula for it; we accomplish this in four small steps.

48.4 A SUMMATION FORMULA FOR $f_{2n}(x,y)$

1) Recall from Example 48.1 that

$$f_k(x,y) = y^{(k-1)/2} f_k(x/\sqrt{y})$$
$$f_k(x) = \frac{x}{i^{k-1}\Delta} f_k(i\Delta).$$

2) By the Lucas-like formula, we have

$$f_k(x) = \sum_{j=0}^{\lfloor (k-1)/2 \rfloor} \binom{k-j-1}{j} x^{k-2j-1}$$

$$f_k(i\Delta) = \sum_{j=0}^{\lfloor (k-1)/2 \rfloor} \binom{k-j-1}{j} (i\Delta)^{k-2j-1}.$$

3) By invoking Theorem 43.3, we get

$$\begin{aligned}
f_{2n}(x) &= \sum_{k=1}^{n} \binom{n}{k} x^k f_k(x) \\
&= \sum_{k=1}^{n} \binom{n}{k} x^k \sum_{j=0}^{\lfloor (k-1)/2 \rfloor} \binom{k-j-1}{j} x^{k-2j-1}
\end{aligned}$$

$$= \frac{x}{i^{2n-1}\Delta} \sum_{k=1}^{n} \binom{n}{k} (i\Delta)^k \sum_{j=0}^{\lfloor (k-1)/2 \rfloor} \binom{k-j-1}{j} (i\Delta)^{k-2j-1}$$

$$= \frac{x}{i^{2n-1}} \sum_{k=1}^{n} \binom{n}{k} \sum_{j=0}^{\lfloor (k-1)/2 \rfloor} \binom{k-j-1}{j} i^{2k-2j-1} \Delta^{2k-2j-2}$$

$$= x \sum_{k=1}^{n} \binom{n}{k} \sum_{j=0}^{\lfloor (k-1)/2 \rfloor} \binom{k-j-1}{j} (-1)^{n+k+j} (x^2+4)^{k-j-1}.$$

4) Since $f_k(x, y) = y^{(k-1)/2} f_k(x/\sqrt{y})$, this yields

$$\begin{aligned} f_{2n}(x, y) &= y^{(2n-1)/2} f_{2n}(x/\sqrt{y}) \\ &= y^{(2n-1)/2} \cdot \frac{x}{\sqrt{y}} \sum_{k=1}^{n} \binom{n}{k} \sum_{j=0}^{\lfloor (k-1)/2 \rfloor} \binom{k-j-1}{j} (-1)^{n+k+j} \left(\frac{x^2+4y}{y} \right)^{k-j-1} \\ &= x \sum_{k=1}^{n} \binom{n}{k} \sum_{j=0}^{\lfloor (k-1)/2 \rfloor} \binom{k-j-1}{j} (-1)^{n+k+j} (x^2+4y)^{k-j-1} y^{n-k+j}. \end{aligned} \quad (48.23)$$

This is the desired summation formula for $f_{2n}(x, y)$.

For example, we have

$$\begin{aligned} f_6(x, y) &= x \sum_{k=1}^{3} \binom{3}{k} \sum_{j=0}^{\lfloor (k-1)/2 \rfloor} \binom{k-j-1}{j} (-1)^{k+j+1} (x^2+4y)^{k-j-1} y^{3-k+j} \\ &= x[3y^2 - 3y(x^2+4y) + (x^2+4y)^2 - (x^2+4y)y] \\ &= x^5 + 4x^3y + 3xy^2; \end{aligned}$$

see Table 43.1.

As can be expected, formula (48.23) also has several byproducts:

$$p_{2n}(x, y) = 2x \sum_{k=1}^{n} \binom{n}{k} \sum_{j=0}^{\lfloor (k-1)/2 \rfloor} \binom{k-j-1}{j} (-1)^{n+k+j}\, 4^{k-j-1} (x^2+y)^{k-j-1} y^{n-k+j};$$

$$f_{2n}(x) = \frac{x}{2} \sum_{k=1}^{n} \binom{n}{k} \sum_{j=0}^{\lfloor (k-1)/2 \rfloor} \binom{k-j-1}{j} (-1)^{n+k+j}\, 4^{k-j-1} (x^2+1)^{k-j-1};$$

$$p_{2n}(x) = 2 \sum_{k=1}^{n} \binom{n}{k} \sum_{j=0}^{\lfloor (k-1)/2 \rfloor} \binom{k-j-1}{j} (-1)^{n+k+j}\, 4^{k-j-1} (x^2+1)^{k-j-1};$$

$$F_{2n} = \frac{1}{2}\sum_{k=1}^{n}\binom{n}{k}\sum_{j=0}^{\lfloor (k-1)/2\rfloor}\binom{k-j-1}{j}(-1)^{n+k+j}\,8^{k-j-1};$$

$$P_{2n} = 2\sum_{k=1}^{n}\binom{n}{k}\sum_{j=0}^{\lfloor (k-1)/2\rfloor}\binom{k-j-1}{j}(-1)^{n+k+j}\,8^{k-j-1};$$

$$J_{2n} = \sum_{k=1}^{n}\binom{n}{k}\sum_{j=0}^{\lfloor (k-1)/2\rfloor}\binom{k-j-1}{j}(-1)^{n+k+j}\,2^{n-k+j}\,9^{k-j-1}.$$

For example, we have

$$\begin{aligned} J_6(y) &= \sum_{k=1}^{3}\binom{3}{k}\sum_{j=0}^{1}\binom{k-j-1}{j}(-1)^{k+j+1}\,(4y+1)^{k-j-1}y^{3-k+j} \\ &= 3y^2 - 3y(4y+1) + (4y+1)^2 - (4y+1)y \\ &= 3y^2 + 4y + 1; \\ J_6 &= J_6(2) = 21; \end{aligned}$$

see Tables 44.1 and 44.2.

48.5 A SUMMATION FORMULA FOR $l_{2n}(x, y)$

We can use the above ideas, together with Theorem 43.5, to derive a double-summation formula for $l_{2n}(x, y)$:

$$l_{2n}(x,y) = \sum_{k=1}^{n}\binom{n}{k}\sum_{j=0}^{\lfloor k/2\rfloor}\frac{k}{k-j}\binom{k-j}{j}(-1)^{n+k+j}(x^2+4y)^{k-j}y^{n-k+j} + 2(-y)^n; \tag{48.24}$$

see Exercise 48.39.

As an example, we have

$$\begin{aligned} l_6(x,y) &= \sum_{k=1}^{3}\binom{3}{k}\sum_{j=0}^{1}\frac{k}{k-j}\binom{k-j}{j}(-1)^{k+j+1}(x^2+4y)^{k-j}y^{3-k+j} + 2(-y)^3 \\ &= 9(x^2+4y)y^2 - 6(x^2+4y)^2y + (x^2+4y)^3 - 2y^3 \\ &= x^6 + 6x^4y + 9x^2y^2 + 2y^3; \end{aligned}$$

see Table 43.1. Similarly, $l_4(x, y) = x^4 + 4x^2y + 2y^2$.

Formula (48.24) also has interesting consequences:

$$q_{2n}(x,y) = \sum_{k=1}^{n} \binom{n}{k} \sum_{j=0}^{\lfloor k/2 \rfloor} \frac{k}{k-j}\binom{k-j}{j}(-1)^{n+k+j}\,4^{k-j}(x^2+y)^{k-j}y^{n-k+j} + 2(-y)^n;$$

$$l_{2n}(x) = \sum_{k=1}^{n} \binom{n}{k} \sum_{j=0}^{\lfloor k/2 \rfloor} \frac{k}{k-j}\binom{k-j}{j}(-1)^{n+k+j}(x^2+4)^{k-j} + 2(-1)^n;$$

$$q_{2n}(x) = \sum_{k=1}^{n} \binom{n}{k} \sum_{j=0}^{\lfloor k/2 \rfloor} \frac{k}{k-j}\binom{k-j}{j}(-1)^{n+k+j}\,4^{k-j}(x^2+1)^{k-j} + 2(-1)^n;$$

$$j_{2n}(y) = \sum_{k=1}^{n} \binom{n}{k} \sum_{j=0}^{\lfloor k/2 \rfloor} \frac{k}{k-j}\binom{k-j}{j}(-1)^{n+k+j}(4y+1)^{k-j}y^{n-k+j} + 2(-y)^n;$$

$$L_{2n} = \sum_{k=1}^{n} \binom{n}{k} \sum_{j=0}^{\lfloor k/2 \rfloor} \frac{k}{k-j}\binom{k-j}{j}(-1)^{n+k+j}\,5^{k-j} + 2(-1)^n;$$

$$Q_{2n} = \sum_{k=1}^{n} \binom{n}{k} \sum_{j=0}^{\lfloor k/2 \rfloor} \frac{k}{k-j}\binom{k-j}{j}(-1)^{n+k+j}\,2^{3k-3j-1} + (-1)^n;$$

$$j_{2n} = \sum_{k=1}^{n} \binom{n}{k} \sum_{j=0}^{\lfloor k/2 \rfloor} \frac{k}{k-j}\binom{k-j}{j}(-1)^{n+k+j}\,2^{n-k+j}\,9^{k-j} + 2(-2)^n.$$

For example, we have

$$\begin{aligned} q_4(x) &= \sum_{k=1}^{2} \binom{2}{k} \sum_{j=0}^{\lfloor k/2 \rfloor} \frac{k}{k-j}\binom{k-j}{j}(-1)^{k+j}\,4^{k-j}(x^2+1)^{k-j} + 2 \\ &= -8(x^2+1) + 16(x^2+1)^2 - 8(x^2+1) + 2 \\ &= 16x^4 + 16x^2 + 2; \end{aligned}$$

similarly, $j_4(y) = 2y^2 + 4y + 1$; see Tables 31.5 and 44.1.

48.6 BIVARIATE FIBONACCI LINKS

We can use Binet's formulas, coupled with the binomial theorem and a bit of differential calculus, to develop interesting relationships among the bivariate Fibonacci polynomials. To this end, recall that $f_n(x,y) = \dfrac{u^n - v^n}{u - v}$, where $u = u(x,y) = \dfrac{x + \Delta(x,y)}{2}$, $v = v(x,y) = \dfrac{x - \Delta(x,y)}{2}$, $\Delta(x,y) = \sqrt{x^2+4y}$, $u + v = x$, $uv = -y$, $xy \neq 0$, and $\Delta(x,y) \neq 0$.

Since $x+2+\Delta(x,y) = 2(u+1)$ and $x+2-\Delta(x,y) = 2(v+1)$, it follows that $(x+2)^2 - \Delta^2(x,y) = 4(uv+1) = 4(x-y+1)$; and $x^2+4y = \Delta^2(x,y) = (x+2)^2 + 4(y-x-1)$. But we also have $x^2+4y = \Delta^2(x,y) = (2u-x)^2 = 4u^2 - 4ux + x^2$; so $ux+y = u^2$. Similarly, $vx+y = v^2$.

Using the binomial theorem, we have

$$\begin{aligned}\sum_{k=1}^{n}\binom{n}{k} ks^{k-1}t^{n-k} &= \frac{d}{ds}\left[\sum_{k=0}^{n}\binom{n}{k} s^k t^{n-k}\right]\\ &= \frac{d}{ds}(s+t)^n\\ &= n(s+t)^{n-1}. \end{aligned} \tag{48.25}$$

By Binet's formula, we then have

$$\begin{aligned}\sum_{k=1}^{n}\binom{n}{k} kf_{k-1}(x,y) &= \sum_{k=1}^{n}\binom{n}{k} k\cdot\frac{u^{k-1}-v^{k-1}}{u-v}\\ &= n\cdot\frac{(u+1)^{k-1}-(v+1)^{k-1}}{u-v}\\ &= nf_{n-1}(x+2, y-x-1). \end{aligned} \tag{48.26}$$

For instance, let $f_k = f_k(x,y)$ and $n = 3$. Then

$$\begin{aligned}\sum_{k=1}^{5}\binom{n}{k} kf_{k-1} &= 5f_0 + 20f_1 + 30f_2 + 20f_3 + 5f_4\\ &= 0 + 20 + 30x + 20(x^2+y) + 5(x^3+2xy)\\ &= 5(x^3 + 4x^2 + 2xy + 6x + 4y + 4)\\ &= 5\left[(x+2)^3 + 2(x+2)(y-x-1)\right]\\ &= 5f_4(x+2, y-x-1). \end{aligned}$$

With these tools, we can develop another Fibonacci delight:

$$\begin{aligned}\sum_{k=1}^{n}\binom{n}{k} kx^{k-1}y^{n-k} f_k(x,y) &= \sum_{k=1}^{n}\binom{n}{k} kx^{k-1}y^{n-k}\cdot\frac{u^k-v^k}{u-v}\\ &= \frac{1}{u-v}\sum_{k=1}^{n}\binom{n}{k} k\left[u(ux)^{k-1} - v(vx)^{k-1}\right]y^{n-k}\\ &= \frac{n}{u-v}\left[u(ux+y)^{n-1} - v(vx+y)^{n-1}\right]\\ &= n\cdot\frac{u^{2n-1}-v^{2n-1}}{u-v}\\ &= nf_{2n-1}(x,y). \end{aligned} \tag{48.27}$$

For example,

$$\sum_{k=1}^{3} \binom{3}{k} kx^{k-1}y^{3-k} f_k(x,y) = 3(x^4 + 3x^2y + y^2)$$

$$= 3f_5(x,y).$$

Bivariate formulas (48.26) and (48.27) yield the following byproducts:

$$\sum_{k=1}^{n} \binom{n}{k} kp_{k-1}(x,y) = np_{n-1}(2x+2, y-2x-1);$$

$$\sum_{k=1}^{n} \binom{n}{k} kf_{k-1}(x,y) = nf_{n-1}(x+2, -x);$$

$$\sum_{k=1}^{n} \binom{n}{k} kp_{k-1}(x) = np_{n-1}(2x+2, -2x);$$

$$\sum_{k=1}^{n} \binom{n}{k} kJ_{k-1}(y) = nf_{n-1}(3, y-2);$$

$$\sum_{k=1}^{n} \binom{n}{k} kF_{k-1} = nf_{n-1}(3, -1);$$

$$\sum_{k=1}^{n} \binom{n}{k} kP_{k-1} = np_{n-1}(4, -2);$$

$$\sum_{k=1}^{n} \binom{n}{k} kJ_{k-1} = nf_{n-1}(3, 0);$$

$$\sum_{k=1}^{n} \binom{n}{k} k(2x)^{k-1}y^{n-k} p_k(x,y) = np_{2n-1}(x,y);$$

$$\sum_{k=1}^{n} \binom{n}{k} kx^{k-1} f_k(x) = nf_{2n-1}(x);$$

$$\sum_{k=1}^{n} \binom{n}{k} k(2x)^{k-1} p_k(x) = np_{2n-1}(x);$$

$$\sum_{k=1}^{n} \binom{n}{k} ky^{n-k} J_k(y) = nJ_{2n-1}(y);$$

$$\sum_{k=1}^{n} \binom{n}{k} kF_k = nF_{2n-1};$$

$$\sum_{k=1}^{n}\binom{n}{k}k2^{k-1}P_k = nP_{2n-1};$$

$$\sum_{k=1}^{n}\binom{n}{k}k2^{n-k}J_k = nJ_{2n-1}.$$

48.7 BIVARIATE LUCAS LINKS

Interestingly, properties (48.26) and (48.27) work for bivariate Lucas polynomials also:

$$\sum_{k=1}^{n}\binom{n}{k}kl_{k-1}(x,y) = nl_{n-1}(x+2, y-x-1); \tag{48.28}$$

$$\sum_{k=1}^{n}\binom{n}{k}kx^{k-1}y^{n-k}l_k(x,y) = nl_{2n-1}(x,y). \tag{48.29}$$

We omit their proofs for the sake of brevity; see Exercises 48.40 and 48.41.

For example,

$$\begin{aligned}\sum_{k=1}^{5}\binom{5}{k}kl_{k-1}(x,y) &= 5(x^4+4x^3+4x^2y+6x^2+12xy+2y^2+4x+12y+2)\\ &= 5[(x+2)^2+4(x+2)^2(y-x-1)+2(y-x-1)^2]\\ &= 5l_4(x+2, y-x-1).\end{aligned}$$

Similarly, $\sum_{k=1}^{3}\binom{3}{k}kx^{k-1}y^{3-k}l_{k-1}(x,y) = 3(x^5+5x^3y+5xy^2) = 3l_5(x,y)$.

Formulas (48.28) and (48.29) also yield interesting special cases. For example, we have

$$\sum_{k=1}^{n}\binom{n}{k}k(2x)^{k-1}y^{n-k}q_k(x,y) = nq_{2n-1}(x,y);$$

$$\sum_{k=1}^{n}\binom{n}{k}kx^{k-1}l_k(x) = nl_{2n-1}(x);$$

$$\sum_{k=1}^{n}\binom{n}{k}k(2x)^{k-1}q_k(x) = nq_{2n-1}(x);$$

$$\sum_{k=1}^{n}\binom{n}{k}ky^{n-k}j_k(y) = nj_{2n-1}(y);$$

$$\sum_{k=1}^{n} \binom{n}{k} kL_k = nL_{2n-1};$$

$$\sum_{k=1}^{n} \binom{n}{k} k \cdot 2^{k-1} Q_k = nQ_{2n-1};$$

$$\sum_{k=1}^{n} \binom{n}{k} k \cdot 2^{n-k} j_k = nj_{2n-1}.$$

For example, $\sum_{k=1}^{4} \binom{4}{k} k \cdot 2^{k-1} Q_k = 956 = 4 \cdot 239 = 4Q_7$; and $\sum_{k=1}^{5} \binom{5}{k} k \cdot 2^{5-k} j_k = 2555 = 5 \cdot 511 = 5j_9$.

EXERCISES 48

Establish the following identities.

1. $L_{n-1}^4 + L_n^4 + L_{n+1}^4 = 2[2L_n^2 - 5(-1)^n]^2$.
2. $P_{n-1}^4 + 16P_n^4 + P_{n+1}^4 = 2[5P_n^2 + (-1)^n]^2$.
3. $Q_{n-1}^4 + 16Q_n^4 + Q_{n+1}^4 = 2[5Q_n^2 - 2(-1)^n]^2$.
4. $16J_{n-1}^4 + J_n^4 + J_{n+1}^4 = 2[3J_n^2 + (-2)^n]^2$.
5. $16j_{n-1}^4 + j_n^4 + j_{n+1}^4 = 18[j_n^2 - 3(-2)^n]^2$.
6. $l_{n-1}^4(x) + x^4 l_n^4(x) + l_{n+1}^4(x) = 2[(x^2+1)l_n^2(x) - (x^2+4)(-1)^n]^2$.
7. $p_{n-1}^4(x) + 16x^4 p_n^4(x) + p_{n+1}^4(x) = 2[(4x^2+1)p_n^2(x) + (-1)^n]^2$.
8. $q_{n-1}^4(x) + 16x^4 q_n^4(x) + q_{n+1}^4(x) = 2[(4x^2+1)q_n^2(x) - 4(x^2+1)(-1)^n]^2$.
9. $16y^4 J_{n-1}^4(y) + J_n^4(y) + J_{n+1}^4(y) = 2[(2y+1)J_n^2(y) + (-2y)^n]^2$.
10. $16y^4 j_{n-1}^4(y) + j_n^4(y) + j_{n+1}^4(y) = 2[(2y+1)j_n^2(y) - (8y+1)(-2y)^n]^2$.

Suppose $a_n(x)$, $b_n(x)$, and $c_n(y)$ satisfy the Fibonacci, Pell, and Jacobsthal recurrences in one variable, respectively. Prove each.

11. $a_{n-1}^4(x) + x^4 a_n^4(x) + a_{n+1}^4(x) = 2[a_{n+1}^2(x) - xa_n(x)a_{n-1}(x)]^2$.
12. $b_{n-1}^4(x) + 16x^4 b_n^4(x) + b_{n+1}^4(x) = 2[b_{n+1}^2(x) - 2xb_n(x)b_{n-1}(x)]^2$.
13. $16y^4 c_{n-1}^4(x) + c_n^4(x) + c_{n+1}^4(x) = 2[c_{n+1}^2(x) - 2yc_n(x)c_{n-1}(x)]^2$.

Establish the following bivariate identities, where $s_n = s_n(x, y)$.

14. $s_{n+1}^5 - x^5 s_n^5 - y^5 s_{n-1}^5 = 5xys_{n+1}s_n s_{n-1}(x^2 s_n^2 + xys_n s_{n-1} + y^2 s_{n-1}^2)$.
15. $s_{n+1}^7 - x^7 s_n^5 - y^7 s_{n-1}^7 = 7xys_{n+1}s_n s_{n-1}(x^2 s_n^2 + xys_n s_{n-1} + y^2 s_{n-1}^2)^2$.

Establish the following Fibonacci, Pell, and Jacobsthal identities.

16. $f_{n+1}^5(x) - x^5 f_n^5(x) - f_{n-1}^5(x)$
$= 5xf_{n+1}(x)f_n(x)f_{n-1}(x)[x^2 f_n^2(x) + xf_n(x)f_{n-1}(x) + f_{n-1}^2(x)]$.
17. $p_{n+1}^5(x) - 32x^5 p_n^5(x) - p_{n-1}^5(x)$
$= 10xp_{n+1}(x)p_n(x)p_{n-1}(x)[4x^2 p_n^2(x) + 2xp_n(x)p_{n-1}(x) + p_{n-1}^2(x)]$.
18. $J_{n+1}^5(y) - J_n^5(y) - 32y^5 J_{n-1}^5(y)$
$= 10yJ_{n+1}(y)J_n(y)J_{n-1}(y)[J_n^2(y) + 2yJ_n(y)J_{n-1}(y) + 4y^2 J_{n-1}^2(y)]$.
19. $F_{n+1}^5 - F_n^5 - F_{n-1}^5 = 5F_{n+1}F_nF_{n-1}[2F_n^2 + (-1)^n]$ (Carlitz, [83, 88]).
20. $L_{n+1}^5 - L_n^5 - L_{n-1}^5 = 5L_{n+1}L_nL_{n-1}[2L_n^2 - 5(-1)^n]$ (Carlitz, [83, 88]).
21. $P_{n+1}^5 - 32P_n^5 - P_{n-1}^5 = 10P_{n+1}P_nP_{n-1}[5P_n^2 + (-1)^n]$.
22. $Q_{n+1}^5 - 32Q_n^5 - Q_{n-1}^5 = 10Q_{n+1}Q_nQ_{n-1}[5Q_n^2 - 2(-1)^n]$.
23. $J_{n+1}^5 - J_n^5 - 32J_{n-1}^5 = 10J_{n+1}J_nJ_{n-1}[3J_n^2 + (-2)^n]$.
24. $j_{n+1}^5 - j_n^5 - 32j_{n-1}^5 = 30j_{n+1}j_nj_{n-1}[j_n^2 - 3(-2)^n]$.
25. $f_{n+1}^7(x) - x^7 f_n^7(x) - f_{n-1}^7(x)$
$= 7xf_{n+1}(x)f_n(x)f_{n-1}(x)[x^2 f_n^2(x) + xf_n(x)f_{n-1}(x) + f_{n-1}^2(x)]^2$.
26. $p_{n+1}^7(x) - 128x^7 p_n^7(x) - p_{n-1}^7(x)$
$= 14xp_{n+1}(x)p_n(x)p_{n-1}(x)[4x^2 p_n^2(x) + 2xp_n(x)p_{n-1}(x) + p_{n-1}^2(x)]^2$.
27. $J_{n+1}^7(y) - J_n^7(y) - 128y^7 J_{n-1}^7(y)$
$= 14yJ_{n+1}(y)J_n(y)J_{n-1}(y)[J_n^2(y) + 2yJ_n(y)J_{n-1}(y) + 4y^2 J_{n-1}^2(y)]^2$.
28. $F_{n+1}^7 - F_n^7 - F_{n-1}^7 = 7F_{n+1}F_nF_{n-1}[2F_n^2 + (-1)^n]^2$ (Carlitz, [83, 88]).
29. $L_{n+1}^7 - L_n^7 - L_{n-1}^7 = 7L_{n+1}L_nL_{n-1}[2L_n^2 - 5(-1)^n]^2$ (Carlitz, [83, 88]).
30. $P_{n+1}^7 - 128P_n^7 - P_{n-1}^7 = 14P_{n+1}P_nP_{n-1}[5P_n^2 + (-1)^n]^2$.
31. $Q_{n+1}^7 - 128Q_n^7 - Q_{n-1}^7 = 14Q_{n+1}Q_nQ_{n-1}[5Q_n^2 - 2(-1)^n]^2$.
32. $J_{n+1}^7 - J_n^7 - 128J_{n-1}^7 = 14J_{n+1}J_nJ_{n-1}[3J_n^2 + (-2)^n]^2$.
33. $j_{n+1}^7 - j_n^7 - 128j_{n-1}^7 = 126j_{n+1}j_nj_{n-1}[j_n^2 - 3(-2)^n]^2$.
34. $l_n(x, y) = y^{n/2} l_n(x/\sqrt{y})$.
35. $l_{2n}(x) = (-1)^n l_{2n}(i\Delta)$.

36. $l_{2n}(i\Delta) = \sum_{k=0}^{n} \frac{2n}{2n-k}\binom{2n-k}{k}(-1)^{n-k}\Delta^{2n-2k}$.

37. $l_{2n}(x) = \sum_{k=0}^{n} \frac{2n}{2n-k}\binom{2n-k}{k}(-1)^{n-k}(x^2+4)^{n-k}$.

38. $l_{2n}(x,y) = \sum_{k=0}^{n} \frac{2n}{2n-k}\binom{2n-k}{k}(-1)^{n-k}(x^2+4y)^{n-k}y^k$.

39. $l_{2n}(x,y) = \sum_{k=1}^{n}\binom{n}{k}\sum_{j=0}^{\lfloor k/2\rfloor}\frac{k}{k-j}\binom{k-j}{j}(-1)^{n+k+j}(x^2+4y)^{k-j}y^{n-k+j} + 2(-y)^n$.

40. $\sum_{k=1}^{n}\binom{n}{k}kl_{k-1}(x,y) = nl_{n-1}(x+2, y-x-1)$ (Catalani, [95]).

41. $\sum_{k=1}^{n}\binom{n}{k}kx^{k-1}y^{n-k}l_k(x,y) = nl_{2n-1}(x,y)$ (Catalani, [95]).

49

TRIBONACCI POLYNOMIALS

> Theorems are fun when you are the prover,
> but then the pleasure fades. What keeps us
> going are the unsolved problems.
> –Carl Pomerance, 2000

Recall that in the case of both Fibonacci and Lucas numbers, there are two initial conditions, and each succeeding element is the sum of its two immediate predecessors. However, suppose we are given three initial conditions, and add three immediate predecessors to compute their successor to construct a number sequence. Such a sequence is the *tribonacci sequence*, originally studied by M. Feinberg in 1963, when he was a 14-year-old ninth-grader at Susquehanna Township Junior High School in Pennsylvania [159, 290].

This leads us to the definition of tribonacci numbers.

49.1 TRIBONACCI NUMBERS

The nth *tribonacci number* T_n is defined by the recurrence

$$T_n = T_{n-1} + T_{n-2} + T_{n-3}, \tag{49.1}$$

where $T_1 = 1 = T_2$, $T_3 = 2$, and $n \geq 4$.

Appendix A.4 gives the first 100 tribonacci numbers. Just as the ratios of consecutive Fibonacci and Lucas numbers converge to the golden ratio α, the tribonacci ratios T_{n+1}/T_n converge to the irrational number 1.83928675521416....

A Tribonacci Array

We are able to compute Fibonacci (and hence Lucas) numbers by using the rising diagonals of Pascal's triangle. Likewise, we can compute tribonacci numbers by adding up the elements on the rising diagonals of a similar triangular array, a *tribonacci array*; see Figure 49.1. Every element $t(i,j)$ of the array is defined recursively, as follows, where $i, j \geq 0$:

$$t(i,j) = \begin{cases} 0 & \text{if } i > j \text{ or } j < 0 \\ 1 & \text{if } i = j \\ t(i-1, j-1) + t(i-1, j) + t(i-2, j-1) & \text{if } i \geq 2. \end{cases}$$

It follows by the recurrence that every element of the array can be obtained by adding the three immediate neighbors in the two preceding rows. For example, (63) = 25 + 25 + 13. The rising diagonal sums are indeed tribonacci numbers.

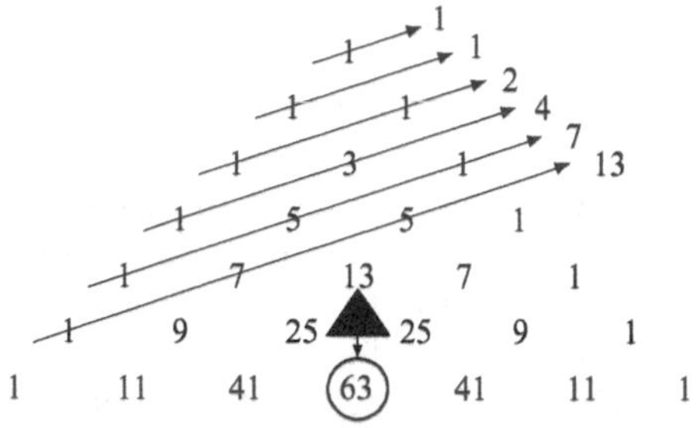

Figure 49.1. Tribonacci array.

There is another triangular array that yields all tribonacci numbers. To construct this array, first we find the trinomial expansions of $(1 + x + x^2)^n$ for several values of $n \geq 0$:

$$(1 + x + x^2)^0 = 1$$
$$(1 + x + x^2)^1 = 1 + x + x^2$$
$$(1 + x + x^2)^2 = 1 + 2x + 3x^2 + 2x^3 + x^4$$
$$(1 + x + x^2)^3 = 1 + 3x + 6x^2 + 7x^3 + 6x^4 + 3x^5 + x^6$$
$$(1 + x + x^2)^4 = 1 + 4x + 10x^2 + 16x^3 + 19x^4 + 16x^5 + 10x^6 + 4x^7 + x^8.$$

TABLE 49.1.

1												
1	1	1										
1	2	3	2	1								
1	3	6 →	7 →	6 ↓	3	1						
1	4	10	16	(19)	16	10	4	1				
1	5	15	30	45	51	45	30	15	5	1		
1	6	21	50	90	126	141	126	90	50	21	6	1

Now we arrange the coefficients in the various expansions to form a left-justified triangular array; see Table 49.1.

Obviously, every row is symmetric. With the exception of the first two rows, every row can be obtained from the preceding row. For example, (19) = 6 + 7 + 6; see the arrows in the table. The rising diagonal sums of this array also yield the trinomial numbers.

Recall from Chapter 4 that the number of compositions of a positive integer n with summands 1 and 2 is F_{n+1}. Suppose we permit the numbers 1, 2, and 3 as summands. What can we say about the number of such compositions C_n?

49.2 COMPOSITIONS WITH SUMMANDS 1, 2, and 3

We begin our investigation with such compositions of integers 1 through 5; see Table 49.2. It appears from the table that $C_n = T_{n+1}$, where $n \geq 1$. Fortunately, this is indeed the case; see Exercise 49.2.

TABLE 49.2.

n	Compositions	C_n
1	1	1
2	1 + 1, 2	2
3	1 + 1 + 1, 1 + 2, 2 + 1, 3	4
4	1 + 1 + 1 + 1, 1 + 1 + 2, 1 + 2 + 1, 2 + 1 + 1, 2 + 2, 1 + 3, 3 + 1	7
5	1 + 1 + 1 + 1 + 1, 1 + 1 + 1 + 2, 1 + 1 + 3, 1 + 1 + 2 + 1, 1 + 2 + 1 + 1, 2 + 1 + 1 + 1, 1 + 3 + 1, 3 + 1 + 1, 2 + 3, 3 + 2, 1 + 2 + 2, 2 + 1 + 2, 2 + 2 + 1	13
		↑ T_{n+1}

Theorem 49.1. *The number of compositions of a positive integer n, using the summands 1, 2, and 3, is* T_{n+1}. ■

Next we explore an explicit formula for the number of additions a_n needed to compute T_n recursively. For example, it takes four additions to compute T_4; that is, $a_4 = 2$.

Using recurrence (49.1), we can define a_n recursively:

$$a_1 = a_2 = a_3 = 0$$
$$a_n = a_{n-1} + a_{n-2} + a_{n-3} + 2, \quad n \geq 4.$$

Letting $b_n = a_n + 1$, we can rewrite this definition:

$$b_1 = b_2 = b_3 = 1$$
$$b_n = b_{n-1} + b_{n-2} + b_{n-3}.$$

The first 12 elements of the sequence $\{b_n\}$ are 1, 1, 1, 3, 5, 9, 17, 31, 57, 105, 193, and 355.

Here we make an interesting observation: Suppose we write these numbers in a row, except the first three; then we write the first ten tribonacci numbers, except the first, in a row right below; now we add the two rows:

$$\begin{array}{rrrrrrrrrr} & 3 & 5 & 9 & 17 & 31 & 57 & 105 & 193 & 355 \\ + & 1 & 2 & 4 & 7 & 13 & 24 & 44 & 81 & 149 \\ \hline & 4 & 7 & 13 & 24 & 44 & 81 & 149 & 274 & 504 \end{array} \leftarrow T_n$$

See an intriguing pattern? The resulting sums are tribonacci numbers T_n, where $n \geq 4$.

So we conjecture that $b_n + T_{n-2} = T_n$; that is, $b_n = T_n - T_{n-2} = T_{n-1} + T_{n-3}$, where $n \geq 4$. Thus we predict that $a_n = T_{n-1} + T_{n-3} - 1$, where $n \geq 4$.

The next theorem confirms this using strong induction.

Theorem 49.2. *Let a_n be the number of additions needed to compute T_n recursively. Then $a_n = T_{n-1} + T_{n-3} - 1$, where $n \geq 4$.*

Proof. Since $T_3 + T_1 - 1 = 2 + 1 - 1 = a_4$, the formula works when $n = 4$.

Now assume it works for all positive integers $k \leq n$, where $n \geq 4$. Using the tribonacci recurrence, we then have:

$$\begin{aligned} a_{n+1} &= a_n + a_{n-1} + a_{n-2} + 2 \\ &= (T_{n-1} + T_{n-3} - 1) + (T_{n-2} + T_{n-4} - 1) + (T_{n-3} + T_{n-5} - 1) + 2 \\ &= (T_{n-1} + T_{n-2} + T_{n-3}) + (T_{n-3} + T_{n-4} + T_{n-5}) - 1 \\ &= T_n + T_{n-2} - 1. \end{aligned}$$

Thus, by the strong version of PMI, the formula works for every $n \geq 4$. ■

It follows by the theorem that

$$a_n = \begin{cases} 0 & \text{if } 1 \le n \le 3 \\ T_{n-1} + T_{n-3} - 1 & \text{otherwise.} \end{cases}$$

For example, it takes $a_6 = T_5 + T_3 - 1 = 7 + 2 - 1 = 8$ additions to compute T_6 recursively.

We can represent the recursive computation of T_n pictorially by a complete 3-ary rooted tree; see Figure 49.2. Each internal vertex of the tree (see the dots in the figure) represents two additions, so $a_n = 2 \times$ (number of internal vertices of the tree rooted at T_n). For example, the tree in Figure 49.2 has four internal vertices, so it takes eight additions to compute T_6 recursively, as we just discovered.

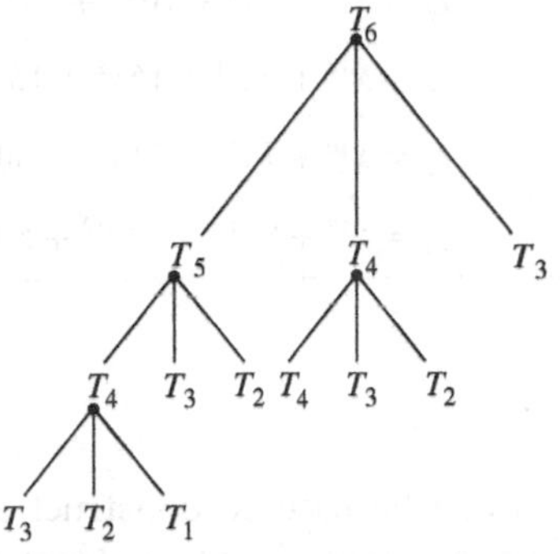

Figure 49.2. A complete 3-ary rooted tree.

Next we develop a generating function for tribonacci numbers.

A Generating Function for $\{T_n\}$

Let $g(x) = \sum_{n=0}^{\infty} T_n x^n$. Since $T_0 = 0$, it follows by the tribonacci recurrence that

$$\begin{aligned}(1 - x - x^2 - x^3)g(x) &= T_1 x \\ \frac{x}{1 - x - x^2 - x^3} &= g(x) \\ &= x + x^2 + 2x^3 + 4x^4 + \cdots .\end{aligned}$$

Now we introduce a family of integer polynomials, which generalizes tribonacci numbers.

49.3 TRIBONACCI POLYNOMIALS

In 1973, Hoggatt and Bicknell introduced a family of integer polynomials called *tribonacci polynomials* t_n [225]. They are defined recursively by the recurrence

$$t_n(x) = x^2 t_{n-1}(x) + x t_{n-2}(x) + t_{n-3}(x),$$

where $t_0(x) = 0$, $t_1(x) = 1$, and $t_2(x) = x^2$. Notice that $t_n(1) = T_n$, the nth tribonacci number. Once again, we omit the argument from the functional notation when there is *no* confusion. Table 49.3 gives the first 10 tribonacci polynomials.

TABLE 49.3. First 10 Tribonacci Polynomials

$t_1 = 1$	$t_6 = x^{10} + 4x^7 + 6x^4 + 2x$
$t_2 = x^2$	$t_7 = x^{12} + 5x^9 + 10x^6 + 7x^3 + 1$
$t_3 = x^4 + x$	$t_8 = x^{14} + 6x^{11} + 15x^8 + 16x^5 + 6x^2$
$t_4 = x^6 + 2x^3 + 1$	$t_9 = x^{16} + 7x^{13} + 21x^{10} + 30x^7 + 19x^4 + 3x$
$t_5 = x^8 + 3x^5 + 3x^2$	$t_{10} = x^{18} + 8x^{15} + 28x^{12} + 50x^9 + 45x^6 + 16x^3 + 1$

Tribonacci Array

The tribonacci coefficients can be used to construct a left-justified array; see Table 49.4. As expected, the row sums yield tribonacci coefficients.

TABLE 49.4. Tribonacci Array

n \ j								Row Sums
1	1							1
2	1							1
3	1	1						2
4	1	2	1					4
5	1	3	3					7
6	1	4	6	2				13
7	1	5	10	7	1			24
8	1	6	15	16	6			44
9	1	7	21	(30)	19	3		81
10	1	8	28	50	45	16	1	149

↑ T_n

Let $T(n,j)$ denote the element in row n and column j of this array, where $n > j \geq 0$. It satisfies the recurrence

$$T(n, j) = T(n-1, j) + T(n-2, j-1) + T(n-3, j-2),$$

where $n \geq 4$. For example, (30) $= 4 + 10 + 16$; see the arrows in the table.

Interestingly, each row of the array in Table 49.4 is a rising diagonal of the triangular array of coefficients in the trinomial expansion of $(x + y + z)^n$, where $n \geq 0$; see the left-justified trinomial coefficient array in Table 49.5.

TABLE 49.5.

1												
1	1	1										
1	2	3	2	1								
1	3	6	7	6	3	1						
1	4	10	16	19	16	10	4	1				
1	5	15	30	45	51	45	30	15	5	1		
1	6	21	50	90	126	141	126	90	50	21	6	1

→ coefficients of t_6

Obviously, every row is symmetric. With the exception of the first two rows, every row can be obtained from the preceding row. We can obtain the tribonacci array from this array by lowering each column one level more than the preceding column. Consequently, the rising diagonal sums of the trinomial coefficient array also yield the tribonacci numbers. Therefore, the sum of every rising diagonal is a tribonacci number.

A Tribonacci Formula

We can develop an explicit formula for the tribonacci polynomial:

$$t_{n+1} = \sum_{j=0}^{\lfloor 2n/3 \rfloor} T(n+1, j) x^{2n-3j}.$$

For example,

$$\begin{aligned} t_5 &= \sum_{j=0}^{2} T(5,j) x^{8-3j} \\ &= T(5,0)x^8 + T(5,1)x^5 + T(5,2)x^2 \\ &= x^8 + 3x^5 + 3x^2. \end{aligned}$$

Next we construct a combinatorial model for tribonacci polynomials [290].

49.4 A COMBINATORIAL MODEL

Suppose we would like to tile a $1 \times n$ board with 1×1 tiles, 1×2 tiles, and 1×3 tiles. We call such tiles squares, dominoes, and *triminoes*, respectively.

We now assign a *weight* to each tile. The weight of a square is x^2, that of a domino is x, and that of a trimino is 1. The *weight of a tiling* is the product of the weights of tiles in the tiling. The weight of the empty tiling is defined as 1.

Figure 49.3 shows such tilings of a $1 \times n$ board, where $0 \le n \le 4$. Using the empirical data, we conjecture that the sum of the weights of tilings of such a board is t_{n+1}. The following theorem confirms this observation [290].

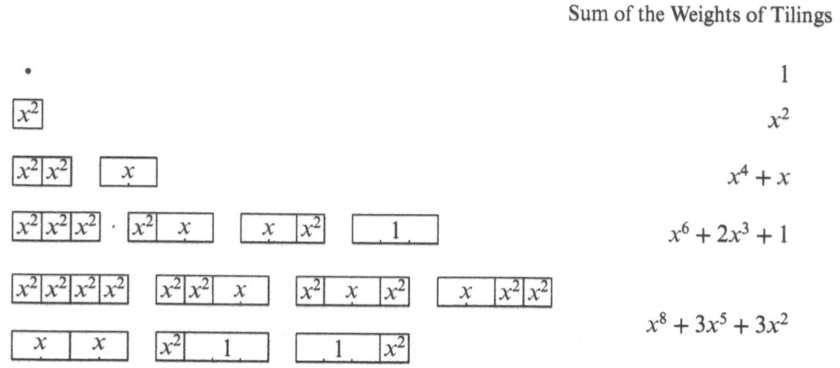

Figure 49.3. Tribonacci tilings of length 4.

Theorem 49.3. *The sum of the weights of tilings of a* $1 \times n$ *board is* t_{n+1}, *where* $n \ge 0$.

Proof. Let $a_n = a_n(x)$ denote the sum of the weights of tilings of the board. Clearly, $a_0 = t_1$, $a_1 = t_2$, and $a_2 = t_3$. Assume that the result is true for all boards of length $< n$, where $n \ge 3$.

Consider an arbitrary tiling T of a board of length n. Suppose T ends in a square: $\underbrace{\text{subtiling}}_{\text{length } n-1}\boxed{x^2}$. By the inductive hypothesis, the sum of the weights of such tilings is $x^2 a_{n-1}$.

Suppose T ends in a domino: $\underbrace{\text{subtiling}}_{\text{length } n-2}\boxed{\quad x \quad}$. The sum of the weights of such tilings is $x a_{n-2}$.

Similarly, the sum of the weights of tilings ending in a trimino is a_{n-3}.

Combining the three cases, we get $a_n = x^2 a_{n-1} + x a_{n-2} + a_{n-3}$. This recurrence, coupled with the initial conditions, implies that $a_n = t_{n+1}$, as desired. ∎

The next result is a direct consequence of this theorem [290].

Corollary 49.1. *The number of tribonacci tilings of a* $1 \times n$ *board is* T_{n+1}, *where* $n \geq 0$. ∎

Next we establish a summation formula for t_{n+1} [290].

Theorem 49.4. $$t_{n+1} = \sum_{\substack{i,j \geq 0 \\ i+2j \leq \lfloor 2n/3 \rfloor \\ 2i+3j \leq n}} \binom{n-i-2j}{i+j} \binom{i+j}{i} x^{2n-3i-6j}.$$

Proof. Consider a $1 \times n$ board. By Theorem 49.2, the sum of the weights of tilings of the board is t_{n+1}.

Now consider an arbitrary tiling T of the board. Suppose it contains i dominoes and j triminoes. Then the tiling contains $n - 2i - 3j$ squares and hence a total of $(n - 2i - 3j) + i + j = n - i - 2j$ tiles. The weight of such a tiling is $x^{2n-3i-6j}$. The $i + j$ nonsquares can be placed among the $n - i - 2j$ tiles in $\binom{n-i-2j}{i+j}$ ways; and the i dominoes among the $i + j$ nonsquares in $\binom{i+j}{i}$ ways. Consequently, there are $\binom{n-i-2j}{i+j}\binom{i+j}{i}$ such tilings T. So the sum of the weights of all tilings of the board is

$$\sum_{\substack{i,j \geq 0 \\ i+2j \leq \lfloor 2n/3 \rfloor \\ 2i+3j \leq n}} \binom{n-i-2j}{i+j} \binom{i+j}{i} x^{2n-3i-6j}.$$

The bounds $i + 2j \leq \lfloor 2n/3 \rfloor$ and $2i + 3j \leq n$ follow from the conditions $n - i - 2j \geq i + j$ and $2n \geq 3i + 6j$.

This, together with the original sum, yields the given result. ∎

We now illustrate this formula with $n = 9$. Table 49.6 shows the possible values of i and j, and corresponding values of $\binom{n-i-2j}{i+j}\binom{i+j}{i} x^{18-3i-6j}$. It follows from the last column that $t_{10} = x^{18} + 8x^{15} + 28x^{12} + 50x^9 + 45x^6 + 16x^3 + 1$, as expected.

The following result is an immediate consequence of Theorem 49.3 [290].

Corollary 49.2. *Let* $n \geq 0$. *Then* $$T_{n+1} = \sum_{\substack{i,j \geq 0 \\ i+2j \leq \lfloor 2n/3 \rfloor \\ 2i+3j \leq n}} \binom{n-i-2j}{i+j} \binom{i+j}{i}.$$ ∎

TABLE 49.6. Tribonacci Weights of Length 9

i	j	$i+2j \le \lfloor 2n/3 \rfloor$?	$2i+3j \le n$?	$\binom{n-i-2j}{i+j}\binom{i+j}{i}x^{18-3i-6j}$
0	0	yes	yes	x^{18}
0	1	yes	yes	$7x^{12}$
0	2	yes	yes	$10x^6$
0	3	yes	yes	1
1	0	yes	yes	$8x^{15}$
1	1	yes	yes	$30x^9$
1	2	yes	yes	$12x^3$
2	0	yes	yes	$21x^{12}$
2	1	yes	yes	$30x^6$
3	0	yes	yes	$20x^9$
3	1	yes	yes	$4x^3$
4	0	yes	yes	$5x^6$

↑ Sum $= t_{10}$

The next theorem gives a summation formula for t_{2n+2} [290].

Theorem 49.5. *Let s denote the number of squares, d the number of dominoes, and t the number of triminoes in a tiling. Then $t_{2n+2} = A(x) + B(x)$, where*

$$A(x) = \sum_{\substack{s \text{ odd}; t \text{ even} \\ 0 \le i \le d \le n \\ 0 \le j \le t \le \lfloor (2n+1)/3 \rfloor \\ s+2d+3t=2n+1}} \binom{(s-1)/2+i+j}{i+j}\binom{i+j}{i}\binom{(s-1)/2+d+t-i-j}{d+t-i-j}\binom{d+t-i-j}{d-i}x^{2s+d};$$

$$B(x) = \sum_{\substack{s \text{ even}; t \text{ odd} \\ 0 \le a \le s; 0 \le b \le d \\ 0 \le t \le \lfloor (2n+1)/3 \rfloor \\ s+2d+3t=2n+1}} \binom{a+b+(t-1)/2}{b+(t-1)/2}\binom{b+(t-1)/2}{b}\binom{s+d-a-b+(t-1)/2}{d-b+(t-1)/2}\binom{d-b+(t-1)/2}{d-b}x^{2s+d}.$$

Proof. Consider a $1 \times (2n+1)$ board. By Theorem 49.3, the sum of the weights of its tribonacci tilings is t_{2n+2}.

Let T be an arbitrary tiling of the board. Then $2n+1 = s+2d+3t$. Consequently, $s+t$ is odd, and hence s and t have opposite parity.

Suppose s is odd. Then T contains an odd number of squares and hence contains a median square M. Suppose there are i dominoes and j triminoes to the left of M; then it has $d-i$ dominoes and $t-j$ triminoes to its right:

$\underbrace{(s-1)/2 \text{ squares, } i \text{ dominoes, } j \text{ triminoes}}$ $\boxed{x^2}$ $\underbrace{(s-1)/2 \text{ squares, } d-i \text{ dominoes, } t-j \text{ triminoes.}}$
↑ M

The weight of such a tiling is x^{2s+d}.

There are $(s-1)/2+i+j$ tiles to the left of M. So the $i+j$ nonsquares can be placed among them in $\binom{(s-1)/2+i+j}{i+j}$ different ways, and the i dominoes among the $i+j$ nonsquares in $\binom{i+j}{i}$ ways. So the $(s-1)/2+i+j$ tiles can be placed in $\binom{(s-1)/2+i+j}{i+j}\binom{i+j}{i}$ different ways. Similarly, the $d+t-i-j$ nonsquares to the right of M can be placed in $\binom{(s-1)/2+d+t-i-j}{d+t-i-j}\binom{d+t-i-j}{d-i}$ different ways. The sum of the weights of all such tilings is given by

$$A(x) = \sum_{\substack{s\ \mathrm{odd};t\ \mathrm{even}\\ 0\le i\le d\le n\\ 0\le j\le t\le\lfloor(2n+1)/3\rfloor\\ s+2d+3t=2n+1}} \binom{(s-1)/2+i+j}{i+j}\binom{i+j}{i}\binom{(s-1)/2+d+t-i-j}{d+t-i-j}\binom{d+t-i-j}{d-i}x^{2s+d}.$$

The bounds on the indices follow from the conditions $i,j,s,d,t \ge 0; 2i, 2d, 3t \le 2n+1; d-i \ge 0$; and $t-j \ge 0$.

On the other hand, suppose t is odd. Then tiling T contains a median trimino M. Suppose there are a squares and b dominoes to the left of M, and hence $s-a$ squares and $d-b$ dominoes to the right of M:

$$\underbrace{a \text{ squares, } b \text{ dominoes, } (t-1)/2 \text{ triminoes}}\ \boxed{1}\ \underbrace{s-a \text{ squares, } d-b \text{ dominoes, } (t-1)/2 \text{ triminoes.}}$$

$\uparrow$ M

The weight of such a tiling is also x^{2s+d}.

Since there are $a+b+(t-1)/2$ tiles to the left of M, the $b+(t-1)/2$ nonsquares can be placed among them in $\binom{a+b+(t-1)/2}{b+(t-1)/2}\binom{b+(t-1)/2}{b}$ different ways; and the $d-b+(t-1)/2$ nonsquares among the $s+d-a-b+(t-1)/2$ tiles in $\binom{s+d-a-b+(t-1)/2}{d-b+(t-1)/2}\binom{d-b+(t-1)/2}{d-b}$ different ways. So there are

$$\binom{a+b+(t-1)/2}{b+(t-1)/2}\binom{b+(t-1)/2}{b}\binom{s+d-a-b+(t-1)/2}{d-b+(t-1)/2}\binom{d-b+(t-1)/2}{d-b}$$

such tilings T. The sum of the weights of all such tilings is given by

$$B(x) = \sum_{\substack{s\ \mathrm{even};t\ \mathrm{odd}\\ 0\le a\le s;0\le b\le d\\ 0\le t\le\lfloor(2n+1)/3\rfloor\\ s+2d+3t=2n+1}} \binom{a+b+(t-1)/2}{b+(t-1)/2}\binom{b+(t-1)/2}{b}\binom{s+d-a-b+(t-1)/2}{d-b+(t-1)/2}\binom{d-b+(t-1)/2}{d-b}x^{2s+d}.$$

The bounds on the indices follow from the conditions $a, b, s, d, t \ge 0; s+2d+3t = 2n+1; s-a, d-b \ge 0$; and $2d, 3t \le 2n+1$. Thus $t_{2n+2} = A(x)+B(x)$, as desired. ■

Example 49.1 We now illustrate this theorem for $n = 2$. The sum of the weights of tribonacci tilings of a 1×5 board is $t_6 = x^{10} + 4x^7 + 6x^4 + 2x$.

Table 49.7 shows the possible values of i, j, s, d, t, $C = \binom{(s-1)/2 + i + j}{i+j}\binom{i+j}{i}$, $D = \binom{(s-1)/2 + d + t - i - j}{d+t-i-j}\binom{d+t-i-j}{d-i}$, and CDx^{2s+d}, where s is odd. It follows from the table that $A(x) = x^{10} + 4x^7 + 3x^4$. Figure 49.4 shows the tilings of the board.

TABLE 49.7.

i	j	s	d	t	C	D	CDx^{2s+d}
0	0	1	2	0	1	1	x^4
0	1	1	2	1	1	1	x^4
1	0	1	2	0	1	1	x^4
0	0	3	1	0	1	2	$2x^7$
1	0	3	1	0	2	1	$2x^7$
0	0	5	0	0	1	1	x^{10}

$$A(x) \overset{\uparrow}{=} x^{10} + 4x^7 + 3x^4$$

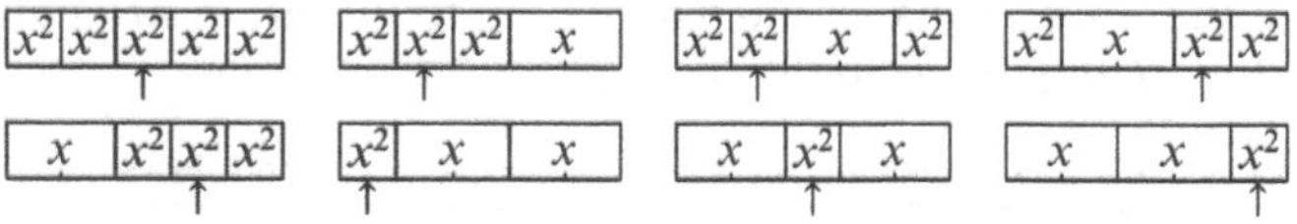

Figure 49.4. Tilings with an odd number of squares.

Table 49.8 shows possible values of a, b, s, d, t,
$E = \binom{a + b + (t-1)/2}{b + (t-1)/2}\binom{b + (t-1)/2}{b}$,
$F = \binom{s + d - a - b + (t-1)/2}{d - b + (t-1)/2}\binom{d - b + (t-1)/2}{d-b}$, and EFx^{2s+d}, and Figure 49.5 the corresponding tilings of the board. It follows from Table 49.8 that $B(x) = 3x^4 + 2x$.

TABLE 49.8.

a	b	s	d	t	E	F	EFx^{2s+d}
0	0	0	1	1	1	1	x
0	1	0	1	1	1	1	x
0	0	2	0	1	1	1	x^4
1	0	2	0	1	1	1	x^4
2	0	2	0	1	1	1	x^4

$$B(x) \overset{\uparrow}{=} 3x^4 + 2x$$

Thus $t_6 = A(x) + B(x) = x^{10} + 4x^7 + 6x^4 + 2x$. ■

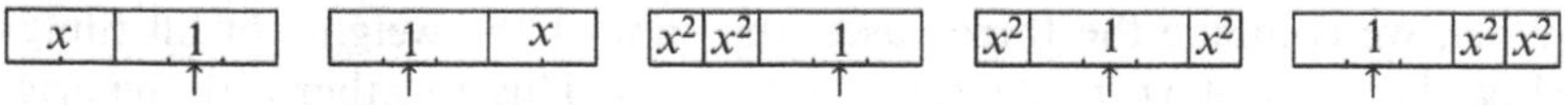

Figure 49.5. Tilings with an odd number of triminoes.

For a 1×9 board, we have

$$A(x) = x^{18} + 8x^{15} + 21x^{12} + 20x^9 + 15x^6 + 12x^3$$

$$B(x) = 7x^{12} + 30x^9 + 30x^6 + 4x^3 + 1$$

$$A(x) + B(x) = t_{10}.$$

The computations of $A(x)$ and $B(x)$ involve 24 and 30 cases, respectively. In the interest of brevity, we leave the steps for the curious-minded to complete.

Theorem 49.5 yields the following result [290].

Corollary 49.3. *$T_{2n+2} = A + B$, where*

$$A = \sum_{\substack{s\ \text{odd};t\ \text{even}\\ 0\le i\le d\le n\\ 0\le j\le t\le \lfloor (2n+1)/3\rfloor\\ s+2d+3t=2n+1}} \binom{(s-1)/2+i+j}{i+j}\binom{i+j}{i}\binom{(s-1)/2+d+t-i-j}{d+t-i-j}\binom{d+t-i-j}{d-i};$$

$$B = \sum_{\substack{s\ \text{even};t\ \text{odd}\\ 0\le a\le s;0\le b\le d\\ 0\le t\le \lfloor (2n+1)/3\rfloor\\ s+2d+3t=2n+1}} \binom{a+b+(t-1)/2}{b+(t-1)/2}\binom{b+(t-1)/2}{b}\binom{s+d-a-b+(t-1)/2}{d-b+(t-1)/2}\binom{d-b+(t-1)/2}{d-b}.$$

■

Using the concept of breakability [31, 285], we can find an addition formula for tribonacci polynomials, as the next theorem shows [290].

Theorem 49.6 (Addition formula). *Let $m, n \ge 1$. Then*

$$t_{m+n} = t_{m+1}t_n + xt_m t_{n-1} + t_m t_{n-2} + t_{m-1}t_{n-1}.$$

Proof. Consider a $1 \times (m+n-1)$ board. The sum of the weights of its tilings is t_{m+n}.

Let T be an arbitrary tiling of the board. Suppose it is breakable at cell m: $\underbrace{\text{subtiling}}_{\text{length } m}\underbrace{\text{subtiling}}_{\text{length } n-1}$. The sum of the weights of such tilings is $t_{m+1}t_n$.

On the other hand, suppose tiling T is *not* breakable at cell m. Then a domino or trimino must occupy cell m.

Suppose a domino occupies cell m: $\underbrace{\text{subtiling}}_{\text{length } m-1}$ $\boxed{x}_{m}$ $\underbrace{\text{subtiling}}_{\text{length } n-2}$. The sum of the weights of such tilings is $xt_m t_{n-1}$.

Suppose a trimino occupies cell m. Then there are two possibilities: $\underbrace{\text{subtiling}}_{\text{length } m-2}$ $\boxed{1}_{m}$ $\underbrace{\text{subtiling}}_{\text{length } n-2}$ or $\underbrace{\text{subtiling}}_{\text{length } m-1}$ $\boxed{1}_{m}$ $\underbrace{\text{subtiling}}_{\text{length } n-3}$.

The sum of the weights of such tilings is $t_{m-1}t_{n-1} + t_m t_{n-2}$.

When we combine the three cases, the sum of the weights of all tilings of the board is $t_{m+1}t_n + xt_mt_{n-1} + t_mt_{n-2} + t_{m-1}t_{n-1}$. This, together with the original sum, gives the desired result. ■

For example,

$$\begin{aligned} t_7 = t_{4+3} &= t_5t_3 + xt_4t_2 + t_4t_1 + t_3t_2 \\ &= (x^8 + 3x^5 + 3x^2)(x^4 + x) + x(x^6 + 2x^3 + 1)x^2 + (x^6 + 2x^3 + 1) \cdot 1 + (x^4 + x)x^2 \\ &= x^{12} + 5x^9 + 10x^6 + 7x^3 + 1; \end{aligned}$$

see Table 49.3.

The following corollary is a consequence of Theorem 49.6.

Corollary 49.4. *Let $n \geq 1$. Then*

1) $T_{m+n} = T_{m+1}T_n + T_mT_{n-1} + T_mT_{n-2} + T_{m-1}T_{n-1}$
2) $t_{2n} = t_{n+1}t_n + xt_nt_{n-1} + t_nt_{n-2} + t_{n-1}^2$
3) $T_{2n} = T_{n+1}T_n + T_nT_{n-1} + T_nT_{n-2} + T_{n-1}^2$. ■

49.5 TRIBONACCI POLYNOMIALS AND THE Q-MATRIX

In Chapter 32, we experienced the beauty and power of the Q-matrix for extracting properties of Fibonacci polynomials [287]. Likewise, we can find such a matrix for tribonacci polynomials [223]:

$$Q = \begin{bmatrix} x^2 & 1 & 0 \\ x & 0 & 1 \\ 1 & 0 & 0 \end{bmatrix}.$$

Using PMI, we can show that

$$Q^n = \begin{bmatrix} t_{n+1} & t_n & t_{n-1} \\ xt_n + t_{n-1} & xt_{n-1} + t_{n-2} & xt_{n-2} + t_{n-3} \\ t_n & t_{n-1} & t_{n-2} \end{bmatrix}; \tag{49.2}$$

see Exercise 49.8.

Since $|Q| = 1$, it follows that $|Q^n| = 1$; thus

$$\begin{vmatrix} t_{n+1} & t_n & t_{n-1} \\ xt_n + t_{n-1} & xt_{n-1} + t_{n-2} & xt_{n-2} + t_{n-3} \\ t_n & t_{n-1} & t_{n-2} \end{vmatrix} = 1.$$

Now multiply row 1 by x^2 and add to row 2; then exchange rows 1 and 2. This yields the *tribonacci polynomial identity*:

$$\begin{vmatrix} t_{n+2} & t_{n+1} & t_n \\ t_{n+1} & t_n & t_{n-1} \\ t_n & t_{n-1} & t_{n-2} \end{vmatrix} = -1.$$

In particular, this yields the identity

$$\begin{vmatrix} T_{n+2} & T_{n+1} & T_n \\ T_{n+1} & T_n & T_{n-1} \\ T_n & T_{n-1} & T_{n-2} \end{vmatrix} = -1.$$

For example,

$$\begin{vmatrix} T_7 & T_6 & T_5 \\ T_6 & T_5 & T_4 \\ T_5 & T_4 & T_3 \end{vmatrix} = \begin{vmatrix} 24 & 13 & 7 \\ 13 & 7 & 4 \\ 7 & 4 & 2 \end{vmatrix} = -1.$$

Next we construct a graph-theoretic model for tribonacci polynomials [290].

49.6 TRIBONACCI WALKS

We can represent the matrix $Q = (q_{ij})_{3\times3}$ by a *weighted digraph* with three vertices v_1, v_2, and v_3, and five weighted edges. We will denote the edge from v_i to v_j by v_i–v_j, or the "word" ij when there is no confusion. The *weight of the edge* ij is q_{ij}; see Figure 49.6. The *weight of a walk* is the product of the weights of the edges along the walk. The Q-matrix is the weighted adjacency matrix of the digraph.

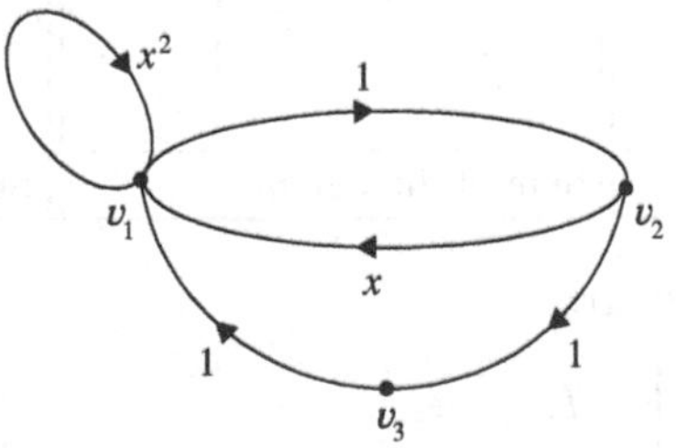

Figure 49.6. A tribonacci digraph.

It follows from equation (49.2) that the sum of the weights of closed walks of length n originating at v_1 is t_{n+1}, and that of length n from v_1 to v_2 is t_n.

In particular,

$$Q^4 = \begin{bmatrix} t_5 & t_4 & t_3 \\ xt_4 + t_3 & xt_3 + t_2 & xt_2 + t_1 \\ t_4 & t_3 & t_2 \end{bmatrix}.$$

Table 49.9 shows all walks of length 4, their weights, and the sums of weights of walks from v_i to v_j, where $1 \le i,j \le 3$.

TABLE 49.9.

Walks from v_1 to v_1	Weight	Walks from v_1 to v_2	Weight	Walks from v_1 to v_3	Weight
11111	x^8	11112	x^6	11123	x^4
11121	x^5	11212	x^3	12123	x
11211	x^5	12112	x^3		
12111	x^5	12312	1		
11231	x^2				
12121	x^2				
12311	x^2				
Sum of the weights	t_5	Sum of the weights	t_4	Sum of the weights	t_3
Walks from v_2 to v_1	Weight	Walks from v_2 to v_2	Weight	Walks from v_2 to v_3	Weight
21111	x^7	21112	x^5	21123	x^3
21121	x^4	21212	x^2	23123	1
21211	x^4	23112	x^2		
23111	x^4				
21231	x				
23121	x				
Sum of the weights	$xt_4 + t_3$	Sum of the weights	$xt_3 + t_2$	Sum of the weights	$xt_2 + 1$
Walks from v_3 to v_1	Weight	Walks from v_3 to v_2	Weight	Walks from v_3 to v_3	Weight
31111	x^6	31112	x^4	31123	x^2
31121	x^3	31212	x		
31211	x^3				
31231	1				
Sum of the weights	t_4	Sum of the weights	t_3	Sum of the weights	t_2

The elements of the matrix

$$Q^n(1) = \begin{bmatrix} T_{n+1} & T_n & T_{n-1} \\ T_n + T_{n-1} & T_{n-1} + T_{n-2} & T_{n-2} + T_{n-3} \\ T_n & T_{n-1} & T_{n-2} \end{bmatrix}$$

give the number of walks of length n from v_i to v_j, where $1 \le i,j \le 3$. For example, there are exactly $T_5 = 7$ walks of length 4 from v_1 to itself; see Table 49.9.

Using the following result [281, 282], we can compute the sum of the weights of walks of length $\leq n$ from v_i to v_j [290].

Theorem 49.7. *Let A be the weighted adjacency matrix of a weighted graph with vertices $v_1, v_2, \ldots, v_k$, and n a positive integer. Then the ij-th entry of the matrix $A + A^2 + \cdots + A^n$ gives the sum of the weights of the walks of length $\leq n$ from v_i to v_j.* ■

This yields the following byproduct [290].

Corollary 49.5. *The ij-th entry in $\sum_{k=1}^{n} Q^k(x)$ gives the sum of the weights of walks of length $\leq n$ from v_i to v_j.* ■

For example, $\sum_{k=1}^{n} t_{k+1}$ gives the sum of the weights of walks of length $\leq n$ from v_1 to v_1.

It also follows from this corollary that the *ij*th entry in $\sum_{k=1}^{n} T_{k+1}$ gives the number of walks of length $\leq n$ from v_1 to v_1. For example, there are exactly $T_2 + T_3 + T_4 = T_5 = 7$ walks of length $\leq n$ from v_1 to v_1.

49.7 A BIJECTION BETWEEN THE TWO MODELS

Interestingly, there is a bijection between the set of tribonacci tilings of a $1 \times n$ board and the set of closed walks of length n originating at v_1. To see this, consider a closed weighted walk of length n. Replace the edge 11 with a square, the walk 121 with a domino, and the walk 1231 with a trimino. This results in a tribonacci tiling of length n. This algorithm is clearly reversible, so it establishes the desired bijection [290].

TABLE 49.10. Closed Walks of Length 4 and the Corresponding Tribonacci Tilings

Walks from v_1 to v_1	Corresponding Tribonacci Tilings	Weights of Tilings
11111	x^2 x^2 x^2 x^2	x^8
11121	x^2 x^2 x	x^5
11211	x^2 x x^2	x^5
12111	x x^2 x^2	x^5
11231	x x	x^2
12121	x^2 1	x^2
12311	1 x^2	x^2
Sum of the weights	t_5	t_5

To illustrate this algorithm, consider the seven walks of length 4 from v_1 to v_1. Table 49.10 shows the corresponding tribonacci tilings and their weights.

EXERCISES 49

1. Find the compositions of the integer 6 with summands 1, 2, and 3.
2. Prove Theorem 49.1.

$$
\begin{array}{ccccccccccc}
 & & & & & 1 & & & & & \\
 & & & & 1 & & 1 & & & & \\
 & & & 1 & & 3 & & 1 & & & \\
 & & 1 & & 5 & & 5 & & 1 & & \\
 & 1 & & 7 & & 13 & & 7 & & 1 & \\
1 & & 9 & & 25 & & 25 & & 9 & & 1
\end{array}
$$

Figure 49.7. Array B.

The Pascal-like array B in Figure 49.7 can be used to generate all tribonacci numbers.

3. Let $B(n,j)$ denote the element in row n and column j of the array. Define $B(n,j)$ recursively.
4. Prove that the sum of the elements on the nth rising diagonal is T_n; that is, $\sum_{j=0}^{\lfloor (n-1)/2 \rfloor} B(n,j) = T_n$.
5. Find $t_{-1}(x)$.
6. Find $t_{11}(x)$ and $t_{12}(x)$.
7. Find Q^2 and Q^3.
8. Establish the formula for Q^n.
9. In 1973, Hoggatt and Bicknell defined the nth *quadranacci number* T_n^* by $T_n^* = T_{n-1}^* + T_{n-2}^* + T_{n-3}^* + T_{n-4}^*$, where $T_1^* = 1 = T_2^*$, $T_3^* = 2$, $T_4^* = 4$, and $n \geq 5$ [225]. Compute T_5^*, T_6^*, and T_7^*.
10. In 1973, Hoggatt and Bicknell also introduced a new family of polynomials t_n^*, called *quadranacci polynomials*. They are defined by $t_n^*(x) = x^3 t_{n-1}^*(x) + x^2 t_{n-2}^*(x) + x t_{n-3}^*(x) + t_{n-4}^*(x)$, where $t_{-2}^*(x) = t_{-1}^*(x) = t_0^*(x) = 0$, $t_1^*(x) = 1$, and $n \geq 5$ [225]. Find $t_3^*(x)$, $t_4^*(x)$, $t_5^*(x)$, and $t_6^*(x)$.
11. Find $t_n^*(1)$.

Quadranacci polynomials can be generated by the Q-matrix

$$
Q = \begin{bmatrix} x^3 & 1 & 0 & 0 \\ x^2 & 0 & 1 & 0 \\ x & 0 & 0 & 1 \\ 1 & 0 & 0 & 0 \end{bmatrix} \quad \text{(Hoggatt and Bicknell, [225]).}
$$

12. Find Q^2 and Q^3.
13. Find Q^n (Hoggatt and Bicknell, [225]).
14. Find $|Q^n|$ (Hoggatt and Bicknell, [225]).
15. Prove that

$$\begin{vmatrix} T^*_{n+3} & T^*_{n+2} & T^*_{n+1} & T^*_{n} \\ T^*_{n+2} & T^*_{n+1} & T^*_{n} & T^*_{n-1} \\ T^*_{n+1} & T^*_{n} & T^*_{n-1} & T^*_{n-2} \\ T^*_{n} & T^*_{n-1} & T^*_{n-2} & T^*_{n-3} \end{vmatrix} = (-1)^{n+1} \text{ (Hoggatt and Bicknell, [225]).}$$

Find the weight of each walk.

16. 112121
17. 121231
18. 1231231
19. 1123121

Find the closed walk corresponding to each tribonacci tiling.

20. 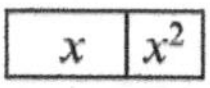

21. 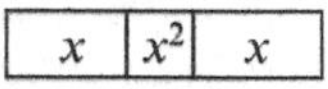

22. 

23. 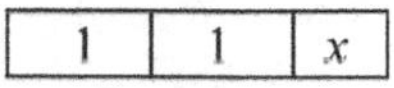

Find the tribonacci tiling corresponding to each closed walk.

24. 121121
25. 1212311
26. 1231121
27. 12123111
28. Find the sum of the weights of walks of length ≤ 3 from v_1 to v_1.
29. Find the sum of the weights of walks of length 4 from v_2 to v_2.

A

APPENDIX

APPENDIX A.1 THE FIRST 100 FIBONACCI AND LUCAS NUMBERS

n	F_n	L_n
1	1	1
2	1	3
3	2	4
4	3	7
5	5	11
6	8	18
7	13	29
8	21	47
9	34	76
10	55	123
11	89	199
12	144	322
13	233	521
14	377	843
15	610	1,364
16	987	2,207
17	1,597	3,571
18	2,584	5,778
19	4,181	9,349
20	6,765	15,127

(*Continued*)

n	F_n	L_n
21	10,946	24,476
22	17,711	39,603
23	28,657	64,079
24	46,368	103,682
25	75,025	167,761
26	121,393	271,443
27	196,418	439,204
28	317,811	710,647
29	514,229	1,149,851
30	832,040	1,860,498
31	1,346,269	3,010,349
32	2,178,309	4,870,847
33	3,524,578	7,881,196
34	5,702,887	12,752,043
35	9,227,465	20,633,239
36	14,930,352	33,385,282
37	24,157,817	54,018,521
38	39,088,169	87,403,803
39	63,245,986	141,422,324
40	102,334,155	228,826,127
41	165,580,141	370,248,451
42	267,914,296	599,074,578
43	433,494,437	969,323,029
44	701,408,733	1,568,397,607
45	1,134,903,170	2,537,720,636
46	1,836,311,903	4,106,118,243
47	2,971,215,073	6,643,838,879
48	4,807,526,976	10,749,957,122
49	7,778,742,049	17,393,796,001
50	12,586,269,025	28,143,753,123
51	20,365,011,074	45,537,549,124
52	32,951,280,099	73,681,302,247
53	53,316,291,173	119,218,851,371
54	86,267,571,272	192,900,153,618
55	139,583,862,445	312,119,004,989
56	225,851,433,717	505,019,158,607
57	365,435,296,162	817,138,163,596
58	591,286,729,879	1,322,157,322,203
59	956,722,026,041	2,139,295,485,799
60	1,548,008,755,920	3,461,452,808,002

(*Continued*)

n	F_n	L_n
61	2,504,730,781,961	5,600,748,293,801
62	4,052,739,537,881	9,062,201,101,803
63	6,557,470,319,842	14,662,949,395,604
64	10,610,209,857,723	23,725,150,497,407
65	17,167,680,177,565	38,388,099,893,011
66	27,777,890,035,288	62,113,250,390,418
67	44,945,570,212,853	100,501,350,283,429
68	72,723,460,248,141	162,614,600,673,847
69	117,669,030,460,994	263,115,950,957,276
70	190,392,490,709,135	425,730,551,631,123
71	308,061,521,170,129	688,846,502,588,399
72	498,454,011,879,264	1,114,577,054,219,522
73	806,515,533,049,393	1,803,423,556,807,921
74	1,304,969,544,928,657	2,918,000,611,027,443
75	2,111,485,077,978,050	4,721,424,167,835,364
76	3,416,454,622,906,707	7,639,424,778,862,807
77	5,527,939,700,884,757	12,360,848,946,698,171
78	8,944,394,323,791,464	20,000,273,725,560,978
79	14,472,334,024,676,221	32,361,122,672,259,149
80	23,416,728,348,467,685	52,361,396,397,820,127
81	37,889,062,373,143,906	84,722,519,070,079,276
82	61,305,790,721,611,591	137,083,915,467,899,403
83	99,194,853,094,755,497	221,806,434,537,978,679
84	160,500,643,816,367,088	358,890,350,005,878,082
85	259,695,496,911,122,585	580,696,784,543,856,761
86	420,196,140,727,489,673	939,587,134,549,734,843
87	679,891,637,638,612,258	1,520,283,919,093,591,604
88	1,100,087,778,366,101,931	2,459,871,053,643,326,447
89	1,779,979,416,004,714,189	3,980,154,972,736,918,051
90	2,880,067,194,370,816,120	6,440,026,026,380,244,498
91	4,660,046,610,375,530,309	10,420,180,999,117,162,549
92	7,540,113,804,746,346,429	16,860,207,025,497,407,047
93	12,200,160,415,121,876,738	27,280,388,024,614,569,596
94	19,740,274,219,868,223,167	44,140,595,050,111,976,643
95	31,940,434,634,990,099,905	71,420,983,074,726,546,239
96	51,680,708,854,858,323,072	115,561,578,124,838,522,882
97	83,621,143,489,848,422,977	186,982,561,199,565,069,121
98	135,301,852,344,706,746,049	302,544,139,324,403,592,003
99	218,922,995,834,555,169,026	489,526,700,523,968,661,124
100	354,224,848,179,261,915,075	792,070,839,848,372,253,127

APPENDIX A.2 THE FIRST 100 PELL AND PELL–LUCAS NUMBERS

n	P_n	Q_n
1	1	1
2	2	3
3	5	7
4	12	17
5	29	41
6	70	99
7	169	239
8	408	577
9	985	1,393
10	2,378	3,363
11	5,741	8,119
12	13,860	19,601
13	33,461	47,321
14	80,782	114,243
15	195,025	275,807
16	470,832	665,857
17	1,136,689	1,607,521
18	2,744,210	3,880,899
19	6,625,109	9,369,319
20	15,994,428	22,619,537
21	38,613,965	54,608,393
22	93,222,358	131,836,323
23	225,058,681	318,281,039
24	543,339,720	768,398,401

25	1,311,738,121	1,855,077,841
26	3,166,815,962	4,478,554,083
27	7,645,370,045	10,812,186,007
28	18,457,556,052	26,102,926,097
29	44,560,482,149	63,018,038,201
30	107,578,520,350	152,139,002,499
31	259,717,522,849	367,296,043,199
32	627,013,566,048	886,731,088,897
33	1,513,744,654,945	2,140,758,220,993
34	3,654,502,875,938	5,168,247,530,883
35	8,822,750,406,821	12,477,253,282,759
36	21,300,003,689,580	30,122,754,096,401
37	51,422,757,785,981	72,722,761,475,561
38	124,145,519,261,542	175,568,277,047,523
39	299,713,796,309,065	423,859,315,570,607
40	723,573,111,879,672	1,023,286,908,188,737
41	1,746,860,020,068,409	2,470,433,131,948,081
42	4,217,293,152,016,490	5,964,153,172,084,899
43	10,181,446,324,101,389	14,398,739,476,117,879
44	24,580,185,800,219,268	34,761,632,124,320,657
45	59,341,817,924,539,925	83,922,003,724,759,193
46	143 263 821 649 299 118	202,605,639,573,839,043
47	345,869,461,223,138,161	489,133,282,872,437,279
48	835,002,744,095,575,440	1,180,872,205,318,713,601
49	2,015,874,949,414,289,041	2,850,877,693,509,864,481
50	4,866,752,642,924,153,522	6,882,627,592,338,442,563

(*Continued*)

n	P_n	Q_n
51	11,749,380,235,262,596,085	16,616,132,878,186,749,607
52	28,365,513,113,449,345,692	40,114,893,348,711,941,777
53	68,480,406,462,161,287,469	96,845,919,575,610,633,161
54	165,326,326,037,771,920,630	233,806,732,499,933,208,099
55	399,133,058,537,705,128,729	564,459,384,575,477,049,359
56	963,592,443,113,182,178,088	1,362,725,501,650,887,306,817
57	2,326,317,944,764,069,48,4905	3,289,910,387,877,251,662,993
58	5,616,228,332,641,321,147,898	7,942,546,277,405,390,632,803
59	13,558,774,610,046,711,780,701	19,175,002,942,688,032,928,599
60	32,733,777,552,734,744,709,300	46,292,552,162,781,456,490,001
61	79,026,329,715,516,201,199,301	111,760,107,268,250,945,908,601
62	190,786,436,983,767,147,107,902	269,812,766,699,283,348,307,203
63	460,599,203,683,050,495,415,105	651,385,640,666,817,642,523,007
64	1,111,984,844,349,868,137,938,112	1,572,584,048,032,918,633,353,217
65	2,684,568,892,382,786,771,291,329	3,796,553,736,732,654,909,229,441
66	6,481,122,629,115,441,680,520,770	9,165,691,521,498,228,451,812,099
67	15,646,814,150,613,670,132,332,869	22,127,936,779,729,111,812,853,639
68	37,774,750,930,342,781,945,186,508	53,421,565,080,956,452,077,519,377
69	91,196,316,011,299,234,022,705,885	128,971,066,941,642,015,967,892,393
70	220,167,382,952,941,249,990,598,278	311,363,698,964,240,484,013,304,163
71	531,531,081,917,181,734,0039,02,441	751,698,464,870,122,983,994,500,719
72	1,283,229,546,787,304,717,998,403,160	1,814,760,628,704,486,452,002,305,601
73	3,097,990,175,491,791,170,000,708,761	4,381,219,722,279,095,887,999,111,921
74	7,479,209,897,770,887,057,999,820,682	10,577,200,073,262,678,228,000,529,443
75	18,056,409,971,033,565,286,000,350,125	25,535,619,868,804,452,344,000,170,807

76	43,592,029,839,838,017,630,000,520,932	61,648,439,810,871,582,916,000,871,057
77	105,240,469,650,709,600,546,001,391,989	148,832,499,490,547,618,176,001,912,921
78	254,072,969,141,257,218,722,003,304,910	359,313,438,791,966,819,268,004,696,899
79	613,386,407,933,224,037,990,008,001,809	867,459,377,074,481,256,712,011,306,719
80	1,480,845,785,007,705,294,702,019,308,528	2,094,232,192,940,929,332,692,027,310,337
81	3,575,077,977,948,634,627,394,046,618,865	5,055,923,762,956,339,922,096,065,927,393
82	8,631,001,740,904,974,549,490,112,546,258	12,206,079,718,853,609,176,884,159,165,123
83	20,837,081,459,758,583,726,374,271,711,381	29,468,083,200,663,558,275,864,384,257,639
84	50,305,164,660,422,142,002,238,655,969,020	71,142,246,120,180,725,728,612,927,680,401
85	121,447,410,780,602,867,730,851,583,649,421	171,752,575,441,025,009,733,090,239,618,441
86	293,199,986,221,627,877,463,941,823,267,862	414,647,397,002,230,745,194,793,406,917,283
87	707,847,383,223,858,622,658,735,230,185,145	1,001,047,369,445,486,500,122,677,053,453,007
88	1,708,894,752,669,345,122,781,412,283,638,152	2,416,742,135,893,203,745,440,147,513,823,297
89	4,125,636,888,562,548,868,221,559,797,461,449	5,834,531,641,231,893,991,002,972,081,099,601
90	9,960,168,529,794,442,859,224,531,878,561,050	14,085,805,418,356,991,727,446,091,676,022,499
91	24,045,973,948,151,434,586,670,623,554,583,549	34,006,142,477,945,877,445,895,155,433,144,599
92	58,052,116,426,097,312,032,565,778,987,728,148	82,098,090,374,248,746,619,236,402,542,311,697
93	140,150,206,800,346,058,651,802,181,530,039,845	198,202,323,226,443,370,684,367,960,517,767,993
94	338,352,530,026,789,429,336,170,142,047,807,838	478,502,736,827,135,487,987,972,323,577,847,683
95	816,855,266,853,924,917,324,142,465,625,655,521	1,155,207,796,880,714,346,660,312,607,673,463,359
96	1,972,063,063,734,639,263,984,455,073,299,118,880	2,788,918,330,588,564,181,308,597,538,924,774,401
97	4,760,981,394,323,203,445,293,052,612,223,893,281	6,733,044,458,057,842,709,277,507,685,523,012,161
98	11,494,025,852,381,046,154,570,560,297,746,905,442	16,255,007,246,704,249,599,863,612,909,970,798,723
99	27,749,033,099,085,295,754,434,173,207,717,704,165	39,243,058,951,466,341,909,004,733,505,464,609,607
100	66,992,092,050,551,637,663,438,906,713,182,313,772	94,741,125,149,636,933,417,873,079,920,900,017,937

APPENDIX A.3 THE FIRST 100 JACOBSTHAL AND JACOBSTHAL–LUCAS NUMBERS

n	J_n	j_n
1	1	1
2	1	5
3	3	7
4	5	17
5	11	31
6	21	65
7	43	127
8	85	257
9	171	511
10	341	1,025
11	683	2,047
12	1,365	4,097
13	2,731	8,191
14	5,461	16,385
15	10,923	32,767
16	21,845	65,537
17	43,691	131,071
18	87,381	262,145
19	174,763	524,287
20	349,525	1,048,577
21	699,051	2,097,151
22	1,398,101	4,194,305
23	2,796,203	8,388,607
24	5,592,405	16,777,217

25	11,184,811	33,554,431
26	22,369,621	67,108,865
27	44,739,243	134,217,727
28	89,478,485	268,435,457
29	178,956,971	536,870,911
30	357,913,941	1,073,741,825
31	715,827,883	2,147,483,647
32	1,431,655,765	4,294,967,297
33	2,863,311,531	8,589,934,591
34	5,726,623,061	17,179,869,185
35	11,453,246,123	34,359,738,367
36	22,906,492,245	68,719,476,737
37	45,812,984,491	137,438,953,471
38	91,625,968,981	274,877,906,945
39	183,251,937,963	549,755,813,887
40	366,503,875,925	1,099,511,627,777
41	733,007,751,851	2,199,023,255,551
42	1,466,015,503,701	4,398,046,511,105
43	2,932,031,007,403	8,796,093,022,207
44	5,864,062,014,805	17,592,186,044,417
45	11,728,124,029,611	35,184,372,088,831
46	23,456,248,059,221	70,368,744,177,665
47	46,912,496,118,443	140,737,488,355,327
48	93,824,992,236,885	281,474,976,710,657
49	187,649,984,473,771	562,949,953,421,311
50	375,299,968,947,541	1,125,899,906,842,625

(*Continued*)

n	J_n	j_n
51	750,599,937,895,083	2,251,799,813,685,247
52	1,501,199,875,790,165	4,503,599,627,370,497
53	3,002,399,751,580,331	9,007,199,254,740,991
54	6,004,799,503,160,661	18,014,398,509,481,985
55	12,009,599,006,321,323	36,028,797,018,963,967
56	24,019,198,012,642,645	72,057,594,037,927,937
57	48,038,396,025,285,291	144,115,188,075,855,871
58	96,076,792,050,570,581	288,230,376,151,711,745
59	192,153,584,101,141,163	576,460,752,303,423,487
60	384,307,168,202,282,325	1,152,921,504,606,846,977
61	768,614,336,404,564,651	2,305,843,009,213,693,951
62	1,537,228,672,809,129,301	4,611,686,018,427,387,905
63	3,074,457,345,618,258,603	9,223,372,036,854,775,807
64	6,148,914,691,236,517,205	18,446,744,073,709,551,617
65	12,297,829,382,473,034,411	36,893,488,147,419,103,231
66	24,595,658,764,946,068,821	73,786,976,294,838,206,465
67	49,191,317,529,892,137,643	147,573,952,589,676,412,927
68	98,382,635,059,784,275,285	295,147,905,179,352,825,857
69	196,765,270,119,568,550,571	590,295,810,358,705,651,711
70	393,530,540,239,137,101,141	1,180,591,620,717,411,303,425
71	787,061,080,478,274,202,283	2,361,183,241,434,822,606,847
72	1,574,122,160,956,548,404,565	4,722,366,482,869,645,213,697
73	3,148,244,321,913,096,809,131	9,444,732,965,739,290,427,391
74	6,296,488,643,826,193,618,261	18,889,465,931,478,580,854,785
75	12,592,977,287,652,387,236,523	37,778,931,862,957,161,709,567

76	25,185,954,575,304,774,473,045	75,557,863,725,914,323,419,137
77	50,371,909,150,609,548,946,091	151,115,727,451,828,646,838,271
78	100,743,818,301,219,097,892,181	302,231,454,903,657,293,676,545
79	201,487,636,602,438,195,784,363	604,462,909,807,314,587,353,087
80	402,975,273,204,876,391,568,725	1,208,925,819,614,629,174,706,177
81	805,950,546,409,752,783,137,451	2,417,851,639,229,258,349,412,351
82	1,611,901,092,819,505,566,274,901	4,835,703,278,458,516,698,824,705
83	3,223,802,185,639,011,132,549,803	9,671,406,556,917,033,397,649,407
84	6,447,604,371,278,022,265,099,605	19,342,813,113,834,066,795,298,817
85	12,895,208,742,556,044,530,199,211	38,685,626,227,668,133,590,597,631
86	25,790,417,485,112,089,060,398,421	77,371,252,455,336,267,181,195,265
87	51,580,834,970,224,178,120,796,843	154,742,504,910,672,534,362,390,527
88	103,161,669,940,448,356,241,593,685	309,485,009,821,345,068,724,781,057
89	206,323,339,880,896,712,483,187,371	618,970,019,642,690,137,449,562,111
90	412,646,679,761,793,424,966,374,741	1,237,940,039,285,380,274,899,124,225
91	825,293,359,523,586,849,932,749,483	2,475,880,078,570,760,549,798,248,447
92	1,650,586,719,047,173,699,865,498,965	4,951,760,157,141,521,099,596,496,897
93	3,301,173,438,094,347,399,730,997,931	9,903,520,314,283,042,199,192,993,791
94	6,602,346,876,188,694,799,461,995,861	19,807,040,628,566,084,398,385,987,585
95	13,204,693,752,377,389,598,923,991,723	39,614,081,257,132,168,796,771,975,167
96	26,409,387,504,754,779,197,847,983,445	79,228,162,514,264,337,593,543,950,337
97	52,818,775,009,509,558,395,695,966,891	158,456,325,028,528,675,187,087,900,671
98	105,637,550,019,019,116,791,391,933,781	316,912,650,057,057,350,374,175,801,345
99	211,275,100,038,038,233,582,783,867,563	633,825,300,114,114,700,748,351,602,687
100	422,550,200,076,076,467,165,567,735,125	1,267,650,600,228,229,401,496,703,205,377

APPENDIX A.4 THE FIRST 100 TRIBONACCI NUMBERS

n	T_n	n	T_n
1	1	51	10,562,230,626,642
2	1	52	19,426,970,897,100
3	2	53	35,731,770,264,967
4	4	54	65,720,971,788,709
5	7	55	120,879,712,950,776
6	13	56	222,332,455,004,452
7	24	57	408,933,139,743,937
8	44	58	752,145,307,699,165
9	81	59	1,383,410,902,447,554
10	149	60	2,544,489,349,890,656
11	274	61	4,680,045,560,037,375
12	504	62	8,607,945,812,375,585
13	927	63	15,832,480,722,303,616
14	1,705	64	29,120,472,094,716,576
15	3,136	65	53,560,898,629,395,777
16	5,768	66	98,513,851,446,415,969
17	10,609	67	181,195,222,170,528,322
18	19,513	68	333,269,972,246,340,068
19	35,890	69	612,979,045,863,284,359
20	66,012	70	1,127,444,240,280,152,749
21	121,415	71	2,073,693,258,389,777,176
22	223,317	72	3,814,116,544,533,214,284
23	410,744	73	7,015,254,043,203,144,209
24	755,476	74	12,903,063,846,126,135,669

25	1,389,537	75	23,732,434,433,862,494,162
26	2,555,757	76	43,650,752,323,191,774,040
27	4,700,770	77	80,286,250,603,180,403,871
28	8,646,064	78	147,669,437,360,234,672,073
29	15,902,591	79	271,606,440,286,606,849,984
30	29,249,425	80	499,562,128,250,021,925,928
31	53,798,080	81	918,838,005,896,863,447,985
32	98,950,096	82	1,690,006,574,433,492,223,897
33	181,997,601	83	3,108,406,708,580,377,597,810
34	334,745,777	84	5,717,251,288,910,733,269,692
35	615,693,474	85	10,515,664,571,924,603,091,399
36	1,132,436,852	86	19,341,322,569,415,713,958,901
37	2,082,876,103	87	35,574,238,430,251,050,319,992
38	3,831,006,429	88	65,431,225,571,591,367,370,292
39	7,046,319,384	89	120,346,786,571,258,131,649,185
40	12,960,201,916	90	221,352,250,573,100,549,339,469
41	23,837,527,729	91	407,130,262,715,950,048,358,946
42	43,844,049,029	92	748,829,299,860,308,729,347,600
43	80,641,778,674	93	1,377,311,813,149,359,327,046,015
44	148,323,355,432	94	2,533,271,375,725,618,104,752,561
45	272,809,183,135	95	4,659,412,488,735,286,161,146,176
46	501,774,317,241	96	8,569,995,677,610,263,592,944,752
47	922,906,855,808	97	15,762,679,542,071,167,858,843,489
48	1,697,490,356,184	98	28,992,087,708,416,717,612,934,417
49	3,122,171,529,233	99	53,324,762,928,098,149,064,722,658
50	5,742,568,741,225	100	98,079,530,178,586,034,536,500,564

ABBREVIATIONS

Abbreviation	Meaning
gcd	greatest common divisor
lcm	least common multiple
PMI	principle of mathematical induction
RHS	right-hand side
LHS	left-hand side
LNHRWCCs	linear nonhomogeneous recurrence with constant coefficients
NHRWCCs	nonhomogeneous recurrence with constant coefficients
QDE	quadratic diophantine equation
ISCF	infinite simple continued fraction

Fibonacci and Lucas Numbers with Applications, Volume Two. Thomas Koshy.

BIBLIOGRAPHY

[1] M. Abramowitz and I.A. Stegun, *Handbook of Mathematical Functions*, Dover, New York, 1972.

[2] G.L. Alexanderson, Jarden Products, *The Fibonacci Quarterly*, 2 (1964), 235–236.

[3] G.L. Alexanderson, Problem B-123, *The Fibonacci Quarterly*, 5 (1967), 288.

[4] Br. U. Alfred, Problem H-8, *The Fibonacci Quarterly*, 1(1) (1963), 48.

[5] Br. U. Alfred, Problem B-59, *The Fibonacci Quarterly*, 3 (1965), 74.

[6] E.G. Alptekin, Solution to Problem B-1011, *The Fibonacci Quarterly*, 44 (2006), 372–373.

[7] R. André-Jeannin, Problem H-450, *The Fibonacci Quarterly*, 29 (1991), 89.

[8] R. André-Jeannin, Problem B-745, *The Fibonacci Quarterly*, 31 (1993), 277.

[9] R. André-Jeannin, Problem B-749, *The Fibonacci Quarterly*, 31 (1993), 371.

[10] R. André-Jeannin, Problem B-761, *The Fibonacci Quarterly*, 32 (1994), 180.

[11] R. André-Jeannin, Problem B-766, *The Fibonacci Quarterly*, 32 (1994), 373.

[12] R.H. Anglin, Problem B-160, *The Fibonacci Quarterly*, 7 (1969), 218.

[13] H. Anton, *Elementary Linear Algebra*, 11th edition, John Wiley & Sons, Inc., New York, 2014.

[14] D. Antzoulakos, Solution to Problem B-604, *The Fibonacci Quarterly*, 26 (1988), 373.

[15] R.A. Askey, Fibonacci and Related Sequences, *The Mathematics Teacher*, 97 (2004), 116–119.

[16] J.H. Avila, Solution to Problem B-19, *The Fibonacci Quarterly*, 2 (1964), 75–76.

[17] M.K. Azarian, Problem B-1133, *The Fibonacci Quarterly*, 51 (2013), 275.

[18] W.W.R. Ball, *A Short Account of the History of Mathematics*, Dover, New York, 1960.

[19] S.-J. Bang, Problem B-746, *The Fibonacci Quarterly*, 31 (1993), 278.

[20] E.J. Barbeau, *Pell's Equation*, Springer, New York, 2003.

[21] W.C. Barley, Problem B-234, *The Fibonacci Quarterly*, 10 (1972), 330.

[22] S.L. Basin, Problem B-23, *The Fibonacci Quarterly*, 1(3) (1963), 76.

[23] S.L. Basin, Problem B-42, *The Fibonacci Quarterly*, 2 (1964), 155.

[24] M. Bataille, Problem 90.G, *The Mathematical Gazette*, 90 (2006), 354.

[25] M. Bataille, Solution to Problem 90.G, *The Mathematical Gazette*, 91 (2007), 160–161.

[26] M. Bataille, Problem 957, *The College Mathematics Journal*, 42 (2011), 329.

[27] D.M. Bătineţu-Giurgiu and N. Stanciu, Problem B-1142, *The Fibonacci Quarterly*, 52 (2014), 81.

[28] B.D. Beasley, Solution to Problem B-988, *The Fibonacci Quarterly*, 43 (2005), 280–281.

[29] A.T. Benjamin and J.J. Quinn, Recounting Fibonacci and Lucas Identities, *The College Mathematics Journal*, 30 (1999), 359–366.

[30] A.T. Benjamin and J.J. Quinn, The Fibonacci Numbers – Exposed More Discretely, *Mathematics Magazine*, 76 (2003), 182–192.

[31] A.T. Benjamin and J.J. Quinn, *Proofs That Really Count*, Math. Association of America, Washington, D.C., 2003.

[32] G.E. Bergum, W.J. Wagner, and V.E. Hoggatt, Jr., Chebyshev Polynomials and Related Sequences, *The Fibonacci Quarterly*, 13 (1975), 19–24.

[33] K.S. Bhanu and M.N. Deshpande, Problem 92.H, *The Mathematical Gazette*, 92 (2008), 356–357.

[34] K.S. Bhanu and M.N. Deshpande, Solution to Problem 92.H, *The Mathematical Gazette*, 93 (2009), 162–163.

[35] M. Bicknell, Solution to Problem B-12, *The Fibonacci Quarterly*, 1(4) (1963), 78.

[36] M. Bicknell, Solution to Problem B-13, *The Fibonacci Quarterly*, 1(4) (1963), 79.

[37] M. Bicknell, A Primer for the Fibonacci Numbers: Part VII, *The Fibonacci Quarterly*, 8 (1970), 407–420.

[38] M. Bicknell, In Memoriam, *The Fibonacci Quarterly*, 18 (1980), 289.

[39] M. Bicknell-Johnson, The Fibonacci Association: Historical Snapshots, *The Fibonacci Quarterly*, 50 (2012), 290–293.

[40] M. Bicknell-Johnson, Private Communication, 2014.

[41] M. Bicknell and V.E. Hoggatt, Jr., Fibonacci Matrices and Lambda Functions, *The Fibonacci Quarterly*, 1(2) (1963), 47–52.

[42] D. Block, Curiosum 330, *Scripta Mathematica*, 19 (1953), 191.

[43] J. Booth and T. Walker, Solution to Problem B-1140, *The Fibonacci Quarterly*, 52 (2014), 372–373.

[44] W.G. Brady, Problem H-273, *The Fibonacci Quarterly*, 15 (1977), 185.

[45] T.A. Brennan, Problem H-5, *The Fibonacci Quarterly*, 1(1) (1963), 47.

[46] T.A. Brennan, Fibonacci Powers and Pascal's Triangle in a Matrix – Part I, *The Fibonacci Quarterly*, 2 (1964), 93–103.

[47] D. M. Bressoud, *Proofs and Confirmations: the Story of the Alternating Sign Matrix Conjecture*, Cambridge University Press, New York, 1999.

[48] D. Bressoud and J. Propp, How the Alternating Sign Matrix Conjecture Was Solved, *Notices of the AMS*, 46(6) (1999), 637–646.

[49] C.A. Bridger, Solution to Problem H-118, *The Fibonacci Quarterly*, 11 (1973), 74–75.

[50] R.C. Brigham, R.M. Caron, P.Z. Chinn, and R.P. Grimaldi, A Tiling Scheme for the Fibonacci Numbers, *Journal of Recreational Mathematics*, 28 (1996–1997), 10–16.

[51] R.C. Brigham, P.Z. Chinn, and R.P. Grimaldi, Tilings and Patterns of Enumeration, *Congressus Numerantium*, 137 (1999), 207–219.

[52] M. Brooke, Fibonacci Formulas, *The Fibonacci Quarterly*, 1(2) (1963), 60.

[53] Br. A. Brousseau, Problem H-92, *The Fibonacci Quarterly*, 6 (1968), 145.

[54] Br. A. Brousseau, Linear Recursion Relations, *The Fibonacci Quarterly*, 7 (1969), 99–104, 106.

[55] Br. A. Brousseau, Summation of Infinite Series, *The Fibonacci Quarterly*, 7 (1969), 143–148.

[56] Br. A. Brousseau, Fibonacci Numbers and Geometry, *The Fibonacci Quarterly*, 10 (1972), 303–318, 323.

[57] J.L. Brown, Solution to Problem H-31, *The Fibonacci Quarterly*, 2 (1964), 306–308.

[58] J.L. Brown, Solution to Problem H-30, *The Fibonacci Quarterly*, 3 (1965), 117–120.

[59] J.L. Brown, Problem H-71, *The Fibonacci Quarterly*, 3 (1965), 299.

[60] J.L. Brown, Solution to Problem B-142, *The Fibonacci Quarterly*, 7 (1969), 220.

[61] P.S. Bruckman, Solution to Problem H-410, *The Fibonacci Quarterly*, 27 (1989), 474.

[62] P.S. Bruckman, Solution to Problem H-444, *The Fibonacci Quarterly*, 30 (1992), 93–96.

[63] P.S. Bruckman, Solution to Problem H-466, *The Fibonacci Quarterly*, 30 (1992), 188–189.

[64] P.S. Bruckman, Solution to Problem H-450, *The Fibonacci Quarterly*, 30 (1992), 191–192.

[65] P.S. Bruckman, Solution to Problem B-696, *The Fibonacci Quarterly*, 30 (1992), 279.

[66] P.S. Bruckman, Solution to Problem B-700, *The Fibonacci Quarterly*, 30 (1992), 372.

[67] P.S. Bruckman, Solution to Problem H-460, *The Fibonacci Quarterly*, 31 (1993), 190–192.

[68] P.S. Bruckman, Solution to Problem B-847, *The Fibonacci Quarterly*, 36 (1998), 472.

[69] P.S. Bruckman, Solution to Problem B-895, *The Fibonacci Quarterly*, 39 (2001), 87–88.

[70] P.S. Bruckman, Solution to Problem B-928, *The Fibonacci Quarterly*, 40 (2002), 375–376.

[71] P.S. Bruckman, Solution to Problem 990, *The Fibonacci Quarterly*, 43 (2004), 282.

[72] P.S. Bruckman, Problem B-1003, *The Fibonacci Quarterly*, 43 (2005), 278.

[73] P.S. Bruckman and H. Kwong, Solution to Problem B-990, *The Fibonacci Quarterly*, 43 (2005), 282.

[74] J.H. Butchart, Problem B-124, *The Fibonacci Quarterly*, 5 (1967), 464.

[75] P.F. Byrd, Problem B-12, *The Fibonacci Quarterly*, 1(2) (1963), 86.

[76] N.D. Cahill, J.R. D'Errico, D.A. Narayan, and J.Y. Narayan, Fibonacci Determinants, *The College Mathematics Journal*, 33 (2002), 221–225.

[77] P.J. Cameron, *Combinatorics: Topics, Techniques and Algorithms*, Cambridge University Press, New York, 1994.

[78] L. Carlitz, Solution to Problem H-10, *The Fibonacci Quarterly*, 1(4) (1963), 49.

[79] L. Carlitz, Problem B-19, *The Fibonacci Quarterly*, 1(3) (1963), 75.

[80] L. Carlitz, Problem H-97, *The Fibonacci Quarterly*, 4 (1966), 332.

[81] L. Carlitz, Problem H-106, *The Fibonacci Quarterly*, 5 (1967), 70.

[82] L. Carlitz, Problem H-80, *The Fibonacci Quarterly*, 5 (1967), 442.

[83] L. Carlitz, Problem H-112, *The Fibonacci Quarterly*, 5 (1967), 71.

[84] L. Carlitz, Solution to Problem H-78, *The Fibonacci Quarterly*, 5 (1967), 438–440.

[85] L. Carlitz, Solution to Problem H-92, *The Fibonacci Quarterly*, 6 (1968), 145.

[86] L. Carlitz, Solution to Problem H-85, *The Fibonacci Quarterly*, 6 (1968), 56–57.

[87] L. Carlitz, Problem B-135, *The Fibonacci Quarterly*, 6 (1968), 90.

[88] L. Carlitz, Solution to Problem H-112, *The Fibonacci Quarterly*, 7 (1969), 61–62.

[89] L. Carlitz, Problem B-185, *The Fibonacci Quarterly*, 8 (1970), 325.

[90] L. Carlitz, Problem B-186, *The Fibonacci Quarterly*, 8 (1970), 326.

[91] L. Carlitz, Solution to Problem B-169, *The Fibonacci Quarterly*, 8 (1970), 330–331.

[92] L. Carlitz and J.A.H. Hunter, Sums of Powers of Fibonacci and Lucas Numbers, *The Fibonacci Quarterly*, 7 (1969), 467–473.

[93] M. Catalani, Problem B-949, *The Fibonacci Quarterly*, 41 (2003), 378.

[94] M. Catalani, Problem H-608, *The Fibonacci Quarterly*, 42 (2004), 92.

[95] M. Catalani, Problem B-990, *The Fibonacci Quarterly*, 42 (2004), 371.

[96] M. Catalani, Problem B-995, *The Fibonacci Quarterly*, 43 (2005), 86.

[97] M. Catalani, Problem H-630, *The Fibonacci Quarterly*, 43 (2005), 284.

[98] B.Y. Chen, Solution to Problem 43.7, *Mathematical Spectrum*, 44 (2011–2012), 46–47.

[99] W. Cheves, Problem B-192, *The Fibonacci Quarterly*, 8 (1970), 443.

[100] C.A. Church, Jr., Problem B-46, *The Fibonacci Quarterly*, 2 (1964), 231.

[101] C.A. Church, Jr., Problem B-54, *The Fibonacci Quarterly*, 2 (1964), 324.

[102] P.Z. Chinn, Curriculum Vitae, http://users.humboldt.edu/phyllis/vita.html.

[103] R.J. Clarke, Letter to the Editor, *Mathematical Spectrum*, 48 (2014–2015), 39–40.

[104] S. Clary and P.D. Hemenway, On Sums of Cubes of Fibonacci Numbers, *Applications of Fibonacci Numbers*, Vol. 5 (ed. G.E. Bergum *et al.*), 123–136, Kluwer, Dordrecht, 1993.

[105] CMC 328, Carleton College, Northfield, Minnesota, *Mathematics Magazine*, 80 (2007), 395–396.

[106] D.I.A. Cohen, *Basic Techniques of Combinatorial Theory*, John Wiley & Sons, Inc., New York, 1978.

[107] I. Cook, The Euclidean Algorithm and Fibonacci, *The Mathematical Gazette*, 74 (1990), 47–48.

[108] C.K. Cook, Solution to Problem B-976, *The Fibonacci Quarterly*, 43 (2005), 86–87.

[109] C.K. Cook, Solution to Problem B-912, *The Fibonacci Quarterly*, 39 (2001), 469.

[110] C.K. Cook, Solution to Problem B-962, *The Fibonacci Quarterly*, 42 (2004), 183.

[111] C.K. Cook, Problem H-636, *The Fibonacci Quarterly*, 44 (2006), 91.

[112] C.K. Cook, Solution to Problem B-1012, *The Fibonacci Quarterly*, 44 (2006), 373–374.

[113] C.K. Cook, Solution to Problem H-636, *The Fibonacci Quarterly*, 45 (2007), 188.

[114] C. Cooper, Alternate Solution, *Crux Mathematicorum*, 15 (1989), 265.

[115] R. Courant, *Differential and Integral Calculus*, Vol. 1, John Wiley & Sons, Inc., New York, 1937.

[116] T. Cross, Solution to Problem 1809, *Crux Mathematicorum*, 20 (1994), 19–20.

[117] Crux, 1987 Austrian Olympiad, *Crux Mathematicorum*, 15 (1989), 264.

[118] A. Cusumano, Problem 2792, *Journal of Recreational Mathematics*, 36 (2007), 66.

[119] K.B. Davenport, Solution to Problem H-608, *The Fibonacci Quarterly*, 43 (2005), 95–96.

[120] M. DeNobili, Problem B-844, *The Fibonacci Quarterly*, 36 (1998), 85.

[121] T.P. Dence, Problem B-129, *The Fibonacci Quarterly*, 5 (1967), 465.

[122] J.E. Desmond, Solution to Problem B-83, *The Fibonacci Quarterly*, 4 (1966), 375.

[123] J.E. Desmond, Problem B-178, *The Fibonacci Quarterly*, 8 (1970), 105.

[124] M.N. Deshpande, An Unexpected Encounter With the Fibonacci Numbers, *The Fibonacci Quarterly*, 32 (1994), 108–109.

[125] M.N. Deshpande, Problem B-911, *The Fibonacci Quarterly*, 39 (2001), 85.

[126] J.L. Diáz-Barrero, Problem B-989, *The Fibonacci Quarterly*, 42 (2004), 371.

[127] J.L. Diáz-Barrero, Problem B-1011, *The Fibonacci Quarterly*, 44 (2006), 85.

[128] J.L. Diáz-Barrero, Solution to Problem H-635, *The Fibonacci Quarterly*, 45 (2007), 96.

[129] J.L. Diáz-Barrero, Problem B-1140, *The Fibonacci Quarterly*, 51 (2011), 368.

[130] L.E. Dickson, *History of the Theory of Numbers*, Vol. 1, Chelsea, New York, 1966.

[131] C.R. Diminnie, Problem 1909, *Crux Mathematicorum*, 20 (1994), 17.

[132] D. Doster, Problem 1809, *Crux Mathematicorum*, 19 (1993), 16.

[133] L.A.G. Dresel, Solution to Problem B-731, *The Fibonacci Quarterly*, 31 (1993), 181.

[134] L.A.G. Dresel, Transformations of Fibonacci–Lucas Identities, *Applications of Fibonacci Numbers*, Vol. 5 (ed. G.E. Bergum *et al.*), 169–184, Kluwer, Dordrecht, 1993.

[135] L.A.G. Dresel, Solution to Problem B-734, *The Fibonacci Quarterly*, 32 (1994), 183–184.

[136] L.A.G. Dresel, Solution to Problem B-738, *The Fibonacci Quarterly*, 32 (1994), 470.

[137] L.A.G. Dresel, Solution to Problem B-766, *The Fibonacci Quarterly*, 33 (1995), 467.

[138] L.A.G. Dresel, Solution to Problem B-811, *The Fibonacci Quarterly*, 35 (1997), 185.

[139] L.A.G. Dresel, Solution to Problem B-843, *The Fibonacci Quarterly*, 36 (1998), 468.

[140] L.A.G. Dresel, Solution to Problem B-942, *The Fibonacci Quarterly*, 41 (2003), 183–184.

[141] A.A. Dubrulle and E.W. Weisstein, Hessenberg Matrix, http://mathworld.wolfram.com/Hessenberg Matrix.html.

[142] R.L. Duncan, An Application of Uniform Distributions to the Fibonacci Numbers, *The Fibonacci Quarterly*, 5 (1967), 137–140.

[143] S. Edwards, Solution to Problem B-858, *The Fibonacci Quarterly*, 37 (1999), 183–184.

[144] S. Edwards, Problem B-961, *The Fibonacci Quarterly*, 41 (2003), 374.

[145] S. Edwards, Problem B-962, *The Fibonacci Quarterly*, 41 (2003), 374.

[146] S. Edwards, Problem B-980, *The Fibonacci Quarterly*, 42 (2004), 182.

[147] S. Edwards, Problem B-988, *The Fibonacci Quarterly*, 42 (2004), 370.

[148] S. Edwards, Solution to Problem B-986, *The Fibonacci Quarterly*, 43 (2005), 279.

[149] S. Edwards, Problem B-1154, *The Fibonacci Quarterly*, 52 (2014), 275.

[150] R. Euler, Problem B-688, *The Fibonacci Quarterly*, 29 (1991), 181.

[151] R. Euler, Problem B-758, *The Fibonacci Quarterly*, 32 (1994), 86.

[152] R. Euler, Problem B-811, *The Fibonacci Quarterly*, 34 (1996), 182.

[153] R. Euler, Problem B-912, *The Fibonacci Quarterly*, 39 (2001), 85.

[154] R. Euler, Problem B-992, *The Fibonacci Quarterly*, 43 (2005), 85.

[155] D. Everman *et al.*, Problem E1396, *American Mathematical Monthly*, 67 (1960), 81–82.

[156] H. Eves, *An Introduction to the History of Mathematics*, 3rd edition, Holt, Rinehart and Winston, New York, 1969.

[157] S. Fairgrieve and H.W. Gould, Product Difference Fibonacci Identities of Simson, Gelin–Cesàro, Taguiri, and Generalizations, *The Fibonacci Quarterly*, 43 (2005), 137–141.

[158] F.J. Faase, On the Number of Specific Spanning Subgraphs of the Graphs $G \times P_n$, *Ars Combinatoria*, 49 (1998), 129–154.

[159] M. Feinberg, Fibonacci–Tribonacci, *The Fibonacci Quarterly*, 1(3) (1963), 70–74.

[160] H.H. Ferns, Solution to Problem B-42, *The Fibonacci Quarterly*, 2 (1964), 329.

[161] H.H. Ferns, Problem B-115, *The Fibonacci Quarterly*, 5 (1967), 202.

[162] P. Filipponi, Problem B-720, *The Fibonacci Quarterly*, 30 (1992), 275.

[163] P. Filipponi, Solution to Problem B-744, *The Fibonacci Quarterly*, 32 (1994), 472.

[164] P. Filipponi, Problem B-769, *The Fibonacci Quarterly*, 32 (1994), 373.

[165] R. Finkelstein, Problem B-143, *The Fibonacci Quarterly*, 6 (1968), 288.

[166] R. Flórez, R.A. Higuita, and A. Mukherjee, Star of David and Other Patterns in the Hosoya-like Polynomials Triangles, arXiv:1706.04247v1 (2017).

[167] H.T. Freitag, Solution to Problem B-178, *The Fibonacci Quarterly*, 8 (1970), 544–545.

[168] H.T. Freitag, Problem B-696, *The Fibonacci Quarterly*, 29 (1991), 277.

[169] H.T. Freitag, Problem B-700, *The Fibonacci Quarterly*, 29 (1991), 371.

[170] H.T. Freitag, Problem B-701, *The Fibonacci Quarterly*, 29 (1991), 371.

[171] H.T. Freitag, Solution to Problem B-701, *The Fibonacci Quarterly*, 30 (1992), 372–373.

[172] H.T. Freitag, Solution to Problem B-718, *The Fibonacci Quarterly*, 30 (1992), 275.

[173] H.T. Freitag, Problem B-736, *The Fibonacci Quarterly*, 31 (1993), 181.

[174] H.T. Freitag, Problem B-744, *The Fibonacci Quarterly*, 31 (1993), 277.

[175] H.T. Freitag, Problem B-748, *The Fibonacci Quarterly*, 31 (1993), 371.

[176] D.D. Frey and J.A. Sellers, Jacobsthal Numbers and Alternating Sign Matrices, *Journal of Integer Sequences*, 3 (2000), article 00.2.3.

[177] J.A. Fuchs and J. Erbacher, Solution to Problem H-8, *The Fibonacci Quarterly*, 1(3) (1963), 51–52.

[178] O. Furdui, Problem H-634, *The Fibonacci Quarterly*, 43 (2005), 378.

[179] S.E. Ganis, Notes on the Fibonacci Sequence, *American Mathematical Monthly*, 66 (1959), 129–130.

[180] C. Georghiou, Solution to Problem 669, *The Fibonacci Quarterly*, 29 (1991), 185.

[181] C. Georghiou, Solution to Problem B-708, *The Fibonacci Quarterly*, 31 (1993), 88–89.

[182] R.E. Giudici, Solution to Problem E1846, *American Mathematical Monthly*, 74 (1966), 592–593.

[183] J. Gill and G. Miller, Newton's Method and Ratios of Fibonacci Numbers, *The Fibonacci Quarterly*, 19 (1981), 1–4.

[184] J. Ginsburg, A Relationship Between Cubes of Fibonacci Numbers, *Scripta Mathematica*, (1953), 242.

[185] J.R. Goggins, Problem B-714, *The Fibonacci Quarterly*, 30 (1992), 182.

[186] M. Goldberg, Solution to Problem H-19, *The Fibonacci Quarterly*, 2 (1964), 130–131.

[187] M. Goldberg and M. Kaplan, Problem 1758, *Mathematics Magazine*, 79 (2006), 393.

[188] H.W. Gould, Generating Functions for Products of Powers of Fibonacci Numbers, *The Fibonacci Quarterly*, 1(2) (1963), 1–16.

[189] H.W. Gould, Problem B-7, *The Fibonacci Quarterly*, 1(3) (1963), 80.

[190] H.W. Gould, Problem H-85, *The Fibonacci Quarterly*, 4 (1966), 148.

[191] H.W. Gould, The Bracket Function and the Fontené–Ward Generalized Binomial Coefficients With Application to Fibonomial Coefficients, *The Fibonacci Quarterly*, 7 (1969), 23–40.

[192] H.W. Gould, A New Greatest Common Divisor Property of the Binomial Coefficients, *The Fibonacci Quarterly*, 10 (1972), 579–584, 628.

[193] R.L. Graham, Problem B-9, *The Fibonacci Quarterly*, 1(2) (1963), 85.

[194] R.L. Graham, Problem H-10, *The Fibonacci Quarterly*, 1(2) (1963), 53.

[195] R.L. Graham, Problem H-45, *The Fibonacci Quarterly*, 2 (1964), 205.

[196] M. Grau-Sánchez and J.L. Diáz-Barrero, Solution to Problem H-631, *The Fibonacci Quarterly*, 44 (2006), 287–288.

[197] M. Griffiths, Bijective Proof Without Words, *The College Mathematics Journal*, 41 (2009), 100.

[198] M. Griffiths and A. Bramham, The Jacobsthal Numbers: Two Results and Two Questions, *The Fibonacci Quarterly*, 53 (2015), 147–151.

[199] R.P. Grimaldi, Binary Strings and the Jacobsthal Numbers, *Congressus Numerantium*, 174 (2005), 3–22.

[200] R.P. Grimaldi, Compositions With the Last Summand Odd, *Ars Combinatoria*, 113A (2014), 299–319.

[201] R.P. Grimaldi and S. Heubach, Binary Strings Without Odd Runs of Zeros, *Ars Combinatoria*, 75 (2005), 241–255.

[202] J. Gruska, *Foundations of Computing*, International Thomson Computer Press, Boston, Mass., 1997.

[203] A.R. Gugheri, Letter to the Editor, *Mathematical Spectrum*, 43 (2010–2011), 42.

[204] A.R. Gugheri, Periodicity of the Fibonacci Sequence, *Mathematical Spectrum*, 45 (2012–2013), 37–38.

[205] A.R. Gugheri, Summing a Finite Series of Fibonacci Numbers, *Mathematical Spectrum*, 45 (2012–2013), 43–44.

[206] A.K. Gupta, Solution to Problem B-171, *The Fibonacci Quarterly*, 8 (1970), 333–334.

[207] R.K. Guy and R.J. Nowakowski, Problem 10316, *American Mathematical Monthly*, 100 (1993), 589.

[208] R.J. Hendel, Solution to Problem B-692, *The Fibonacci Quarterly*, 30 (1992), 185.

[209] S. Heubach, Tiling an m-by-n Area With Squares up to k-by-k ($m \leq k$), *Congressus Numerantium*, 140 (1999), 43–64.

[210] A.P. Hillman and V.E. Hoggatt, Jr., A Proof of Gould's Pascal Hexagon Conjecture, *The Fibonacci Quarterly*, 10 (1972), 565–568, 598.

[211] H.J. Hindin, Problem B-929, *The Fibonacci Quarterly*, 39 (2001), 468.

[212] V.E. Hoggatt, Jr., Problem H-31, *The Fibonacci Quarterly*, 2 (1964), 49–50.

[213] V.E. Hoggatt, Jr., Problem H-72, *The Fibonacci Quarterly*, 3 (1965), 300.

[214] V.E. Hoggatt, Jr., Problem H-73, *The Fibonacci Quarterly*, 3 (1965), 300.

[215] V.E. Hoggatt, Jr., Problem H-78, *The Fibonacci Quarterly*, 4 (1966), 56–57.

[216] V.E. Hoggatt, Jr., Problem H-82, *The Fibonacci Quarterly*, 4 (1966), 57.

[217] V.E. Hoggatt, Jr., Problem H-88, *The Fibonacci Quarterly*, 4 (1966), 146.

[218] V.E. Hoggatt, Jr., Problem B-109, *The Fibonacci Quarterly*, 5 (1967), 107.

[219] V.E. Hoggatt, Jr., Problem B-149, *The Fibonacci Quarterly*, 6 (1968), 400.
[220] V.E. Hoggatt, Jr., Problem B-150, *The Fibonacci Quarterly*, 6 (1968), 400.
[221] V.E. Hoggatt, Jr., *Fibonacci and Lucas Numbers*, Houghton-Mifflin, Boston, Mass., 1969.
[222] V.E. Hoggatt, Jr., Problem B-208, *The Fibonacci Quarterly*, 9 (1971), 217.
[223] V.E. Hoggatt, Jr. and M. Bicknell, Some New Fibonacci Identities, *The Fibonacci Quarterly*, 2 (1964), 29–32.
[224] V.E. Hoggatt, Jr. and M. Bicknell, Roots of Fibonacci Polynomials, *The Fibonacci Quarterly*, 11 (1973), 271–274.
[225] V.E. Hoggatt, Jr. and M. Bicknell, Generalized Fibonacci Polynomials, *The Fibonacci Quarterly*, 11 (1973), 457–465.
[226] V. E. Hoggatt, Jr. and M. Bicknell-Johnson, Convolution Arrays for Jacobsthal and Fibonacci Polynomials, *The Fibonacci Quarterly*, 16 (1978), 385–402.
[227] V. E. Hoggatt, Jr. and W. Hansell, The Hidden Hexagon Squares, *The Fibonacci Quarterly*, 9 (1971), 120, 133.
[228] V.E. Hoggatt, Jr. and I.D. Ruggles, A Primer for the Fibonacci Sequence: Part III, *The Fibonacci Quarterly*, 1(3) (1963), 61–65.
[229] J.E. Homer, Solution to Problem B-138, *The Fibonacci Quarterly*, 7 (1969), 110.
[230] R. Honsberger, *From Erdös to Kiev*, Math. Association of America, Washington, D.C., 1996.
[231] R. Honsberger, *Mathematical Delights*, Math. Association of America, Washington, D.C., 2004.
[232] A.F. Horadam, Pell Identities, *The Fibonacci Quarterly*, 9 (1971), 245–263.
[233] A.F. Horadam, Jacobsthal Representation Numbers, *The Fibonacci Quarterly*, 34 (1996), 40–54.
[234] A. F. Horadam, Jacobsthal and Pell Curves, *The Fibonacci Quarterly*, 26 (1988), 77–83.
[235] A.F. Horadam, Jacobsthal Representation Polynomials, *The Fibonacci Quarterly*, 35 (1997), 137–148.
[236] A.F. Horadam, Vieta Polynomials, *The Fibonacci Quarterly*, 40 (2002), 223–232.
[237] A.F. Horadam and Bro. J.M. Mahon, Pell and Pell–Lucas Polynomials, *The Fibonacci Quarterly*, 23 (1985), 7–20.
[238] W.W. Horner, Problem B-146, *The Fibonacci Quarterly*, 6 (1968), 289.
[239] F.T. Howard, The Sum of the Squares of Two Generalized Fibonacci Numbers, *The Fibonacci Quarterly*, 41 (2003), 80–84.
[240] https://en.wikipedia.org/wiki/Carl-Gustav-Jacob-Jacobi.
[241] https://en.wikipedia.org/wiki/Ernst-Jacobsthal.
[242] https://en.wikipedia.org/wiki/Joseph-Raphson.
[243] https://en.wikipedia.org/wiki/Stephen-Cole-Kleene.
[244] https://en.wikipedia.org/wiki/François-Viète.
[245] https://en.wikipedia.org/wiki/Phyllis-Chinn.
[246] D. Huang, Fibonacci Identities, Matrices, and Graphs, *The Mathematics Teacher*, 98 (2005), 400–403.
[247] J.A.H. Hunter, Problem H-30, *The Fibonacci Quarterly*, 2 (1964), 49.
[248] J.A.H. Hunter, Solution to Problem H-30, *The Fibonacci Quarterly*, 2 (1964), 305–306.
[249] J.A.H. Hunter, Problem H-79, *The Fibonacci Quarterly*, 4 (1966), 57.
[250] J.A.H. Hunter, Problem H-80, *The Fibonacci Quarterly*, 4 (1966), 57.

[251] J.A.H. Hunter, Problem H-95, *The Fibonacci Quarterly*, 4 (1966), 258.
[252] J.A.H. Hunter, Problem H-124, *The Fibonacci Quarterly*, 5 (1967), 252.
[253] V. Ivanoff, Problem H-107, *The Fibonacci Quarterly*, 5 (1967), 70.
[254] R.F. Jackson, Solution to Problem E1708, *American Mathematical Monthly*, 72 (1965), 671–672.
[255] W.D. Jackson, Problem B-142, *The Fibonacci Quarterly*, 6 (1968), 288.
[256] N. Jensen, Solution to Problem B-744, *The Fibonacci Quarterly*, 32 (1994), 472.
[257] N. Jensen, Solution to Problem H-492, *The Fibonacci Quarterly*, 34 (1996), 91–96.
[258] S. Jerbic, Problem H-63, *The Fibonacci Quarterly*, 3 (1965), 116.
[259] B. Johnson, Problem B-960, *The Fibonacci Quarterly*, 41 (2003), 182.
[260] L.B.W. Jolley, *Summation of Series*, 2nd edition, Dover, New York, 1961.
[261] H. Kappus, Solution to Problem B-722, *The Fibonacci Quarterly*, 31 (1993), 375.
[262] H. Kappus, Solution to Problem B-758, *The Fibonacci Quarterly*, 33 (1995), 186.
[263] M.J. Karameh, Problem B-976, *The Fibonacci Quarterly*, 42 (2004), 181.
[264] M.A. Khan, Letter to the Editor, *Mathematical Spectrum*, 42 (2009–2010), 94.
[265] M.A. Khan, Problem 43.11, *Mathematical Spectrum*, 43 (2010–2011), 136.
[266] M.A. Khan, Letter to the Editor, *Mathematical Spectrum*, 46 (2013–2014), 92.
[267] M.A. Khan, Summing Fibonacci Numbers, *Mathematical Spectrum*, 48 (2015–2016), 42.
[268] B.W. King, Solution to Problem B-145, *The Fibonacci Quarterly*, 7 (1969), 222–223.
[269] B.W. King, Problem B-184, *The Fibonacci Quarterly*, 8 (1970), 325.
[270] M.S. Klamkin, Comment on the Solution to Problem B-769, *The Fibonacci Quarterly*, 33 (1995), 469.
[271] K. Klinger and R. Hess, Solution to Problem 2792, *Journal of Recreational Mathematics*, 37 (2008), 86.
[272] D.E. Knuth, *The Art of Computer Programming*, Vol. 2, 2nd edition, Addison-Wesley, Reading, Mass., 1981.
[273] D.E. Knuth, *The Art of Computer Programming*, Vol. 3, Addison-Wesley, Reading, Mass., 1973.
[274] B. Kolman and D. Hill, *Elementary Linear Algebra*, 9th edition, Pearson, Upper Saddle River, New Jersey, 2007.
[275] N. Komanda, Solution to Problem 10203, *American Mathematical Monthly*, 101 (1994), 279–280.
[276] T. Koshy, *Discrete Mathematics with Applications*, Academic Press, Burlington, Mass., 2007.
[277] T. Koshy, *Elementary Number Theory with Applications*, 2nd edition, Academic Press, Burlington, Mass., 2007.
[278] T. Koshy, *Catalan Numbers with Applications*, Oxford University Press, New York, 2009.
[279] T. Koshy, *Triangular Arrays with Applications*, Oxford University Press, New York, 2011.
[280] T. Koshy, Pell Walks, *The Mathematical Gazette*, 97 (2013), 27–35.
[281] T. Koshy, Graph-Theoretic Models for the Fibonacci Family, *The Mathematical Gazette*, 98 (2014), 256–265.
[282] T. Koshy, Polynomial Extensions of the Lucas and Ginsburg Identities, *The Fibonacci Quarterly*, 52 (2014), 141–147.
[283] T. Koshy, Graph-Theoretic Models for the Univariate Fibonacci Family, *The Fibonacci Quarterly*, 53 (2015), 135–146.

[284] T. Koshy, Fibonacci Polynomials, Lucas Polynomials, and Operators, *Mathematical Spectrum*, 47 (2014–2015), 122–124.

[285] T. Koshy, *Pell and Pell–Lucas Numbers with Applications*, Springer, New York, 2014.

[286] T. Koshy, Fibonacci Walks, *The Mathematical Scientist*, 40 (2015), 128–133.

[287] T. Koshy, *Fibonacci and Lucas Numbers with Applications*, Vol. 1, 2nd edition, John Wiley & Sons, Inc., New York, 2017.

[288] T. Koshy, Gibonomial Coefficients With Interesting Byproducts, *The Fibonacci Quarterly*, 53 (2015), 340–348.

[289] T. Koshy, Vieta Polynomials and Their Close Relatives, *The Fibonacci Quarterly*, 54 (2016), 141–148.

[290] T. Koshy, Combinatorial Models for Tribonacci Polynomials, *The Mathematical Scientist*, 41 (2016), 101–107.

[291] T. Koshy, A Digraph Model for the Jacobsthal Family, *The Mathematical Scientist*, 42 (2017), 151–156.

[292] T. Koshy, Differences of Gibonacci Polynomial Products of Orders 2, 3, and 4, *The Fibonacci Quarterly*, (to appear).

[293] T. Koshy and Z. Gao, Triangular Numbers in the Jacobsthal Family, *Integers*, 12 (2012), article A64.

[294] T. Koshy and Z. Gao, A Fibonacci Curiosity, *Mathematical Spectrum*, 48 (2015–2016), 13–15.

[295] T. Koshy and Z. Gao, Variations of a Charming Putnam Problem, *Mathematical Spectrum*, 48 (2015–2016), 100–104.

[296] T. Koshy and Z. Gao, Polynomial Extensions of a Diminnie Delight, *The Fibonacci Quarterly*, 55 (2017), 13–20.

[297] T. Koshy and Z. Gao, Polynomial Extensions of a Putnam Delight, *Mathematics Magazine*, (to appear).

[298] T. Koshy and M. Griffiths, A Graph-Theoretic Approach to Jacobsthal Polynomials, *The Mathematical Scientist*, 42 (2017), 143–150.

[299] T. Koshy and M. Griffiths, Some Gibonacci Convolutions With Dividends, *The Fibonacci Quarterly*, (to appear).

[300] T. Koshy and R.P. Grimaldi, Ternary Words and Jacobsthal Numbers, *The Fibonacci Quarterly*, 55 (2017), 129–136.

[301] J.J. Kostal, Problem B-708, *The Fibonacci Quarterly*, 30 (1992), 85.

[302] M. Krebs and N.C. Martinez, The Combinatorial Trace Method in Action, *The College Mathematics Journal*, 44 (2013), 32–36.

[303] H.V. Krishna, Solution to Problem H-109, *The Fibonacci Quarterly*, 7 (1969), 59.

[304] H.V. Krishna, Problem B-227, *The Fibonacci Quarterly*, 10 (1972), 218.

[305] H.K. Krishnaprian and J.J. Kostal, Solution to Problem B-749, *The Fibonacci Quarterly*, 33 (1995), 88–89.

[306] L. Kupiers, Problem B-702, *The Fibonacci Quarterly*, 30 (1992), 371.

[307] L. Kupiers, Problem B-717, *The Fibonacci Quarterly*, 30 (1992), 183.

[308] H. Kwong, Solution to Problem B-761, *The Fibonacci Quarterly*, 33 (1995), 373–374.

[309] H. Kwong, Solution to Problem B-844, *The Fibonacci Quarterly*, 36 (1998), 469.

[310] H. Kwong, Solution to Problem B-995, *The Fibonacci Quarterly*, 43 (2006), 375–376.

[311] H. Kwong, Solution to Problem H-640, *The Fibonacci Quarterly*, 45 (2007), 285.

[312] H. Kwong, Solution to Problem B-1147, *The Fibonacci Quarterly*, 53 (2015), 182–183.

[313] H. Kwong, Solution to Problem B-1155, *The Fibonacci Quarterly*, 53 (2015), 278.

[314] W. Lang, Problem B-858, *The Fibonacci Quarterly*, 36 (1998), 373–374.
[315] G. Ledin, Jr., Problem H-109, *The Fibonacci Quarterly*, 5 (1967), 70.
[316] G. Ledin, Jr., Problem H-117, *The Fibonacci Quarterly*, 5 (1967), 162.
[317] G. Ledin, Jr., Problem H-118, *The Fibonacci Quarterly*, 5 (1967), 162.
[318] G. Ledin, Jr., Problem H-173, *The Fibonacci Quarterly*, 8 (1970), 383.
[319] H.T. Leonard, Jr. and V.E. Hoggatt, Jr., Problem H-141, *The Fibonacci Quarterly*, 6 (1968), 252.
[320] K.E. Lewis, Solution to Problem B-961, *The Fibonacci Quarterly*, 42 (2005), 182.
[321] K.E. Lewis, Solution to Problem B-980, *The Fibonacci Quarterly*, 43 (2005), 90.
[322] C. Libis, Problem B-978, *The Fibonacci Quarterly*, 42 (2004), 181.
[323] D. Lind, Solution to Problem B-59, *The Fibonacci Quarterly*, 3 (1965), 237–238.
[324] D. Lind, Problem H-93, *The Fibonacci Quarterly*, 4 (1966), 252.
[325] D. Lind, Problem H-128, *The Fibonacci Quarterly*, 6 (1968), 51.
[326] D. Lind, Problem B-134, *The Fibonacci Quarterly*, 6 (1968), 90.
[327] D. Lind, Problem B-138, *The Fibonacci Quarterly*, 6 (1968), 185.
[328] D. Lind, Problem B-140, *The Fibonacci Quarterly*, 6 (1968), 185.
[329] D. Lind, Problem B-145, *The Fibonacci Quarterly*, 6 (1968), 289.
[330] D. Lind, Problem B-165, *The Fibonacci Quarterly*, 7 (1969), 219.
[331] D. Lind, Problem H-140, *The Fibonacci Quarterly*, 8 (1970), 81.
[332] K.M. Lindberg, Solution to Problem B-1154, *The Fibonacci Quarterly*, 53 (2015), 277.
[333] E.H. Lockwood, A Side-light on Pascal's Triangle, *The Mathematical Gazette*, 51 (1967), 243–244.
[334] C.T. Long, Discovering Fibonacci Identities, *The Fibonacci Quarterly*, 24 (1986), 160–167.
[335] C.T. Long, Solution to Problem B-746, *The Fibonacci Quarterly*, 33 (1995), 87.
[336] C.T. Long and J.H. Jordan, A Limited Arithmetic on Simple Continued Fractions, *The Fibonacci Quarterly*, 5 (1967), 113–128.
[337] G. Lord, Solution to Problem B-714, *The Fibonacci Quarterly*, 31 (1993), 278.
[338] N. Lord, A Fibonacci Curiosity, *Mathematical Spectrum*, 48 (2016), 87–88.
[339] C.C. MacDuffee, *The Theory of Matrices*, Chelsea, New York, 1956.
[340] M.K. Mahanthappa, Arithmetic Sequences and Fibonacci Quadratics, *The Fibonacci Quarterly*, 29 (1991), 343–346.
[341] Br. J. Mahon, Problem B-986, *The Fibonacci Quarterly*, 42 (2004), 370.
[342] Br. J. Mahon, Problem B-1012, *The Fibonacci Quarterly*, 44 (2006), 86.
[343] Br. J. Mahon and A.F. Horadam, Inverse Trigonometrical Summation Formulas Involving Pell Polynomials, *The Fibonacci Quarterly*, 23 (1985), 319–324.
[344] P. Mana, Solution to Problem B-123, *The Fibonacci Quarterly*, 6 (1968), 191.
[345] P. Mana, Problem B-152, *The Fibonacci Quarterly*, 6 (1968), 401.
[346] P. Mana, Problem B-163, *The Fibonacci Quarterly*, 7 (1969), 219.
[347] P. Mana, Problem B-171, *The Fibonacci Quarterly*, 7 (1969), 332.
[348] P. Mana, Problem B-194, *The Fibonacci Quarterly*, 8 (1970), 444.
[349] D.C.B. Marsh, Solution to Problem E1911, *American Mathematical Monthly*, 75 (1968), 81.
[350] R.H. McNutt, Problem B-843, *The Fibonacci Quarterly*, 36 (1998), 85.
[351] D.G. Mead, Problem B-67, *The Fibonacci Quarterly*, 3 (1965), 153.

[352] R.S. Melham, Sums of Certain Products of Fibonacci and Lucas Numbers, *The Fibonacci Quarterly*, 37 (1999), 248–251.

[353] R.S. Melham, Some Analogs of the Identity $F_n^2 + F_{n+1}^2 = F_{2n+1}$, *The Fibonacci Quarterly*, 37 (1999), 305–311.

[354] R.S. Melham, Families of Identities Involving Sums of Powers of the Fibonacci and Lucas Numbers, *The Fibonacci Quarterly*, 37 (1999), 315–319.

[355] R.S. Melham, A Fibonacci Identity in the Spirit of Simson and Gelin–Cesàro, *The Fibonacci Quarterly*, 41 (2003), 142–143.

[356] R.S. Melham, A Three-Variable Identity Involving Cubes of Fibonacci Numbers, *The Fibonacci Quarterly*, 41 (2003), 220–223.

[357] R.S. Melham, Ye Olde Fibonacci Curiosity Shoppe Revisited, *The Fibonacci Quarterly*, 42 (2004), 155–160.

[358] R.S. Melham and A.G. Shannon, Inverse Trigonometric and Hyperbolic Summation Formulas Involving Generalized Fibonacci Numbers, *The Fibonacci Quarterly*, 33 (1995), 32–40.

[359] J.W. Milsom, Solution to Problem B-227, *The Fibonacci Quarterly*, 11 (1973), 107–108.

[360] L.J. Mordell, *Diophantine Equations*, Academic Press, New York, 1969.

[361] J. Morgado, Some Remarks on an Identity of Catalan Concerning the Fibonacci Numbers, *Portugaliae Mathematica*, 39 (1980), 341–348.

[362] T. Nagell, *Introduction to Number Theory*, 2nd edition, Chelsea, New York, 1964.

[363] K. Nkwanta and L.W. Shapiro, Pell Walks and Riordan Matrices, *The Fibonacci Quarterly*, 43 (2005), 170–180.

[364] J.J. O'Connor and E.F. Robertson, *Adrien-Marie Legendre*, MacTutor History of Mathematics Archive, University of St. Andrews, Scotland, 1999.

[365] J.J. O'Connor and E.F. Robertson, *François Viète*, MacTutor History of Mathematics Archive, University of St. Andrews, Scotland, 2000.

[366] J.J. O'Connor and E.F. Robertson, *Ernst Jacobsthal*, MacTutor History of Mathematics Archive, University of St. Andrews, Scotland, 2008.

[367] J.J. O'Connor and E.F. Robertson, *Stephen Cole Kleene*, MacTutor History of Mathematics Archive, University of St. Andrews, Scotland, 2008.

[368] J.J. O'Connor and E.F. Robertson, *Eugène Charles Catalan*, MacTutor History of Mathematics Archive, University of St. Andrews, Scotland, 2012.

[369] H. Ohtsuka, Problem H-724, *The Fibonacci Quarterly*, 50 (2012), 281.

[370] H. Ohtsuka, Solution to Problem H-724, *The Fibonacci Quarterly*, 52 (2012), 186–187.

[371] H. Ohtsuka, Problem B-1147, *The Fibonacci Quarterly*, 52 (2012), 178.

[372] H. Ohtsuka, Problem B-1155, *The Fibonacci Quarterly*, 53 (2015), 278.

[373] T.J. Osler and A. Hilburn, An Unusual Proof that F_m Divides F_{mn} Using Hyperbolic Functions, *The Mathematical Gazette*, 91 (2007), 510–512.

[374] G.C. Padilla, Problem B-172, *The Fibonacci Quarterly*, 7 (1969), 545.

[375] G.C. Padilla, Problem B-173, *The Fibonacci Quarterly*, 7 (1969), 545.

[376] F.D. Parker, Solution to Problem B-9, *The Fibonacci Quarterly*, 1(4) (1963), 76.

[377] J.M. Patel, Problem H-631, *The Fibonacci Quarterly*, 43 (2005), 377.

[378] J.M. Patel, Problem H-635, *The Fibonacci Quarterly*, 44 (2006), 91.

[379] J.M. Patel, Problem H-640, *The Fibonacci Quarterly*, 44 (2006), 188.

[380] C.B.A. Peck, Solution to Problem H-103, *The Fibonacci Quarterly*, 6 (1968), 353–354.

[381] C.B.A. Peck, Solution to Problem H-117, *The Fibonacci Quarterly*, 7 (1969), 62–63.

[382] C.B.A. Peck, Solution to Problem H-120, *The Fibonacci Quarterly*, 7 (1969), 173.
[383] C.B.A. Peck, Solution to Problem H-129, *The Fibonacci Quarterly*, 7 (1969), 284–285.
[384] P. Philipponi, Solution to Problem B-744, *The Fibonacci Quarterly*, 32 (1994), 472.
[385] J. Pla, Problem B-967, *The Fibonacci Quarterly*, 41 (2003), 466.
[386] Ā. Plaza and S. Falcón, Problem B-1153, *The Fibonacci Quarterly*, 52 (2012), 275.
[387] Ā. Plaza and M. Woltermann, Solution to Problem 957, *The College Mathematics Journal*, 43 (2012), 339–340.
[388] B. Prielipp, Solution to Problem B-703, *The Fibonacci Quarterly*, 31 (1993), 84.
[389] B. Prielipp and G. Lord, Solution to Problem B-717, *The Fibonacci Quarterly*, 31 (1993), 280–281.
[390] S. Rabinowitz, Problem H-129, *The Fibonacci Quarterly*, 6 (1968), 51–52.
[391] S. Rabinowitz, Solution to Problem B-115, *The Fibonacci Quarterly*, 6 (1968), 93.
[392] S. Rabinowitz, Problem B-842, *The Fibonacci Quarterly*, 36 (1998), 85.
[393] S. Rabinowitz, Problem B-831, *The Fibonacci Quarterly*, 36 (1998), 90.
[394] S. Rabinowitz, Solution to Problem B-844, *The Fibonacci Quarterly*, 36 (1998), 469.
[395] S. Rabinowitz, Problem B-942, *The Fibonacci Quarterly*, 40 (2002), 372.
[396] S. Rabinowitz, Problem B-964, *The Fibonacci Quarterly*, 41 (2003), 375.
[397] A. Ralston, *A First Course in Numerical Analysis*, McGraw-Hill, New York, 1965.
[398] J.F. Ramaley, Problem E1708, *American Mathematical Monthly*, 71 (1964), 680.
[399] K.R. Rebman, The Sequence 1 5 16 45 121 320 …, *The Fibonacci Quarterly*, 13 (1975), 51–55.
[400] K.-G. Recke, Problem B-153, *The Fibonacci Quarterly*, 6 (1968), 401.
[401] D. Redmond, Solution to Problem B-749, *The Fibonacci Quarterly*, 33 (1995), 88–89.
[402] D. Redmond, Solution to Problem B-771, *The Fibonacci Quarterly*, 33 (1995), 470–471.
[403] T.J. Rivlin, *Chebyshev Polynomials*, 2nd edition, John Wiley & Sons, Inc., New York, 1990.
[404] D.P. Robbins, The Story of 1, 2, 7, 42, 429, 7436, …, *The Mathematical Intelligencer*, 3 (1991), 12–19.
[405] N. Robbins, Vieta's Triangular Array and a Related Family of Polynomials, *International Journal of Mathematics and Mathematical Sciences*, 14 (1991), 239–244.
[406] M.A. Rose, Solution to Problem B-908, *The Fibonacci Quarterly*, 39 (2001), 376.
[407] N. Routledge, Fibonacci and Lucas Numbers, *Mathematical Spectrum*, 38 (2005–2006), 36.
[408] R. Roy, *Sources in the Development of Mathematics*, Cambridge University Press, New York, 2011.
[409] F. Scheid, *Theory and Problems of Numerical Analysis*, Schaum's Outline Series, McGraw-Hill, New York, 1968.
[410] D. Schepler, Solution to Problem 10316, *American Mathematical Monthly*, 103 (1996), 905.
[411] S.J. Schlicker, Numbers Simultaneously Polygonal and Centered Polygonal, *Mathematics Magazine*, 84 (2011), 339–350.
[412] J. Seibert, Solution to Problem B-978, *The Fibonacci Quarterly*, 43 (2005), 88–89.
[413] H.-J. Seiffert, Problem H-410, *The Fibonacci Quarterly*, 25 (1987), 186.
[414] H.-J. Seiffert, Problem B-604, *The Fibonacci Quarterly*, 25 (1987), 370.
[415] H.-J. Seiffert, Problem H-444, *The Fibonacci Quarterly*, 28 (1990), 283.

[416] H.-J. Seiffert, Problem B-703, *The Fibonacci Quarterly*, 29 (1991), 372.
[417] H.-J. Seiffert, Problem H-460, *The Fibonacci Quarterly*, 29 (1991), 377.
[418] H.-J. Seiffert, Solution to Problem B-705, *The Fibonacci Quarterly*, 29 (1991), 372.
[419] H.-J. Seiffert, Solution to Problem B-688, *The Fibonacci Quarterly*, 30 (1992), 183.
[420] H.-J. Seiffert, Solution to Problem B-702, *The Fibonacci Quarterly*, 30 (1992), 374–375.
[421] H.-J. Seiffert, Solution to Problem B-722, *The Fibonacci Quarterly*, 30 (1992), 276.
[422] H.-J. Seiffert, Problem B-731, *The Fibonacci Quarterly*, 31 (1993), 82.
[423] H.-J. Seiffert, Problem B-771, *The Fibonacci Quarterly*, 32 (1994), 374.
[424] H.-J. Seiffert, Problem H-492, *The Fibonacci Quarterly*, 32 (1994), 473–474.
[425] H.-J. Seiffert, Solution to Problem B-738, *The Fibonacci Quarterly*, 32 (1994), 377.
[426] H.-J. Seiffert, Problem H-508, *The Fibonacci Quarterly*, 34 (1996), 89–90, 188.
[427] H.-J. Seiffert, Problem H-510, *The Fibonacci Quarterly*, 34 (1996), 187.
[428] H.-J. Seiffert, Problem H-518, *The Fibonacci Quarterly*, 34 (1996), 473–474.
[429] H.-J. Seiffert, Solution to Problem H-508, *The Fibonacci Quarterly*, 35 (1997), 188–190.
[430] H.-J. Seiffert, Solution to Problem H-510, *The Fibonacci Quarterly*, 35 (1997), 191–192.
[431] H.-J. Seiffert, Problem H-542, *The Fibonacci Quarterly*, 36 (1998), 379.
[432] H.-J. Seiffert, Solution to Problem H-518, *The Fibonacci Quarterly*, 36 (1998), 92–94.
[433] H.-J. Seiffert, Solution to Problem B-858, *The Fibonacci Quarterly*, 37 (1999), 183–184.
[434] H.-J. Seiffert, Solution to Problem B-844, *The Fibonacci Quarterly*, 36 (1998), 469.
[435] H.-J. Seiffert, Solution to Problem H-542, *The Fibonacci Quarterly*, 37 (1999), 381–382.
[436] H.-J. Seiffert, Problem H-570, *The Fibonacci Quarterly*, 39 (2001), 91.
[437] H.-J. Seiffert, Problem B-928, *The Fibonacci Quarterly*, 39 (2001), 468.
[438] H.-J. Seiffert, Problem H-586, *The Fibonacci Quarterly*, 40 (2002), 379.
[439] H.-J. Seiffert, Solution to Problem B-928, *The Fibonacci Quarterly*, 40 (2002), 375–376.
[440] H.-J. Seiffert, Solution to Problem B-929, *The Fibonacci Quarterly*, 40 (2002), 376–378.
[441] H.-J. Seiffert, Solution to Problem H-586, *The Fibonacci Quarterly*, 41 (2003), 189–191.
[442] H.-J. Seiffert, Solution to Problem B-960, *The Fibonacci Quarterly*, 42 (2004), 90–91.
[443] H.-J. Seiffert, Solution to Problem B-964, *The Fibonacci Quarterly*, 42 (2004), 184.
[444] H.-J. Seiffert, Solution to Problem B-967, *The Fibonacci Quarterly*, 42 (2004), 279.
[445] H.-J. Seiffert, Problem H-617, *The Fibonacci Quarterly*, 42 (2004), 377.
[446] H.-J. Seiffert, Problem H-626, *The Fibonacci Quarterly*, 43 (2005), 188.
[447] H.-J. Seiffert, Solution to Problem B-989, *The Fibonacci Quarterly*, 43 (2005), 281.
[448] H.-J. Seiffert, Solution to Problem H-570, *The Fibonacci Quarterly*, 43 (2005), 284–286.
[449] H.-J. Seiffert, Solution to Problem H-617, *The Fibonacci Quarterly*, 43 (2005), 378–379.
[450] H.-J. Seiffert, Solution to Problem B-1003, *The Fibonacci Quarterly*, 44 (2006), 185–186.
[451] H.-J. Seiffert, Problem H-639, *The Fibonacci Quarterly*, 44 (2006), 187.

[452] H.-J. Seiffert, Solution to Problem H-626, *The Fibonacci Quarterly*, 44 (2006), 189–191.
[453] H.-J. Seiffert, Solution to Problem H-630, *The Fibonacci Quarterly*, 44 (2006), 285–287.
[454] H.-J. Seiffert, Solution to Problem H-634, *The Fibonacci Quarterly*, 45 (2007), 95–96.
[455] H.-J. Seiffert, Solution to Problem H-639, *The Fibonacci Quarterly*, 45 (2007), 191–192.
[456] J.A. Sellers, Solution to Problem B-992, *The Fibonacci Quarterly*, 43 (2005), 374.
[457] J. Shallit, Problem B-423, *The Fibonacci Quarterly*, 18 (1980), 85.
[458] A.G. Shannon, Solution to Problem B-138, *The Fibonacci Quarterly*, 7 (1969), 110.
[459] A.G. Shannon, Solution to Problem H-273, *The Fibonacci Quarterly*, 16 (1978), 568–569.
[460] A.G. Shannon and A.F. Horadam, Some Relationships Among Vieta, Morgan–Voyce and Jacobsthal Polynomials, *Applications of Fibonacci Numbers* (ed. F.T. Howard), 307–323, Kluwer, Dordrecht, 1999.
[461] L.W. Shapiro, A Combinatorial Proof of a Chebyshev Polynomial Identity, *Discrete Mathematics*, 34 (1981), 203–206.
[462] E.M. Sheuer, Solution to Problem E1396, *American Mathematical Monthly*, 67 (1960), 694.
[463] A. Sinefakopoulos, Solution to Problem 1909, *Crux Mathematicorum*, 20 (1994), 295–296.
[464] S. Singh, Solution to Problem B-796, *The Fibonacci Quarterly*, 34 (1996), 469.
[465] N.J.A. Sloane, *The Online Encyclopedia of Integer Sequences*, http://oeis.org; A001045.
[466] P.A. Smith, Problem E3210, *American Mathematical Monthly*, 94 (1987), 457.
[467] G.W. Smith, Problem B-847, *The Fibonacci Quarterly*, 36 (1998), 86.
[468] L. Somer, Solution to Problem B-842, *The Fibonacci Quarterly*, 36 (1998), 377–378.
[469] M.Z. Spievy, Fibonacci Identities via the Determinant Sum Property, *The College Mathematics Journal*, 37 (2006), 286–289.
[470] W. Square, Problem H-83, *The Fibonacci Quarterly*, 4 (1966), 57.
[471] P. Stănică, Solution to Problem B-911, *The Fibonacci Quarterly*, 39 (2001), 468.
[472] I. Strazdins, Problem B-895, *The Fibonacci Quarterly*, 38 (2000), 181.
[473] I. Strazdins, Problem B-908, *The Fibonacci Quarterly*, 38 (2000), 468.
[474] Students in the 1987 Mathematical Olympiad Program, U.S. Military Academy, West Point, New York, Solution to Problem E3210, *American Mathematical Monthly*, 95 (1988), 879–880.
[475] M.N.S. Swamy, Solution to Problem B-67, *The Fibonacci Quarterly*, 3 (1965), 326.
[476] M.N.S. Swamy, Problem B-83, *The Fibonacci Quarterly*, 4 (1966), 90.
[477] M.N.S. Swamy, Problem B-84, *The Fibonacci Quarterly*, 4 (1966), 90.
[478] M.N.S. Swamy, Problem E1846, *American Mathematical Monthly*, 73 (1966), 81.
[479] M.N.S. Swamy, Problem H-120, *The Fibonacci Quarterly*, 5 (1967), 252.
[480] M.N.S. Swamy, Problem H-79, *The Fibonacci Quarterly*, 5 (1967), 441.
[481] M.N.S. Swamy, Solution to Problem H-83, *The Fibonacci Quarterly*, 6 (1968), 54–55.
[482] M.N.S. Swamy, Solution to Problem H-93, *The Fibonacci Quarterly*, 6 (1968), 145–148.
[483] M.N.S. Swamy, Solution to Problem H-95, *The Fibonacci Quarterly*, 6 (1968), 148–150.
[484] M.N.S. Swamy, Pythagoreans and All That Stuff, *The Fibonacci Quarterly*, 6 (1968), 259.

[485] M.N.S. Swamy, Solution to Problem H-101, *The Fibonacci Quarterly*, 6 (1968), 259–260.

[486] M.N.S. Swamy, Solution to Problem B-134, *The Fibonacci Quarterly*, 6 (1968), 404.

[487] M.N.S. Swamy, Problem H-150, *The Fibonacci Quarterly*, 7 (1969), 57.

[488] M.N.S. Swamy, Problem H-155, *The Fibonacci Quarterly*, 7 (1969), 170.

[489] M.N.S. Swamy, Problem H-158, *The Fibonacci Quarterly*, 7 (1969), 277.

[490] M.N.S. Swamy, Solution to Problem H-155, *The Fibonacci Quarterly*, 8 (1970), 497–498.

[491] M.N.S. Swamy, Problem B-796, *The Fibonacci Quarterly*, 33 (1995), 466.

[492] R.F. Torretto and J.A. Fuchs, Generalized Binomial Coefficients, *The Fibonacci Quarterly*, 2 (1964), 296–302.

[493] B.R. Toskey, Problem E1911, *American Mathematical Monthly*, 73 (1966), 775.

[494] C.W. Trigg, Problem 750, *Mathematics Magazine*, 43 (1970), 48.

[495] 2013 USAMO Problems, Committee on the American Mathematics Competitions, Math. Association of America, Washington, D.C., 2016.

[496] H.L. Umansky, Problem B-233, *The Fibonacci Quarterly*, 10 (1972), 329–330.

[497] H.L. Umansky and M.H. Tallman, Problem H-101, *The Fibonacci Quarterly*, 4 (1966), 333.

[498] S. Vajda, Some Summation Formulas for Binomial Coefficients, *The Mathematical Gazette*, 34 (1950), 211–212.

[499] I. Vidav, Problem 10203, *American Mathematical Monthly*, 99 (1992), 265.

[500] W.J. Wagner, Two Explicit Expressions for $\cos nx$, *The Mathematics Teacher*, 67 (1974), 234–237.

[501] C.R. Wall, Problem H-19, *The Fibonacci Quarterly*, 1(3) (1963), 46.

[502] C.R. Wall, Solution to Problem H-45, *The Fibonacci Quarterly*, 3 (1965), 127–128.

[503] C.R. Wall, Solution to Problem B-135, *The Fibonacci Quarterly*, 6 (1968), 405–406.

[504] J.D. Watson, Solution to Problem B-1133, *The Fibonacci Quarterly*, 52 (2014), 277–278.

[505] William Lowell Putnam Mathematical Competition, *Mathematics Magazine*, 81 (2008), 72–77.

[506] G. Wulczyn, Problem B-423, *The Fibonacci Quarterly*, 19 (1981), 92.

[507] G. Wulczyn, Problem B-669, *The Fibonacci Quarterly*, 28 (1990), 183.

[508] G. Wulczyn, Solution to Problem B-669, *The Fibonacci Quarterly*, 29 (1991), 185.

[509] G. Wulczyn, Problem B-692, *The Fibonacci Quarterly*, 29 (1991), 182.

[510] www.britannica.com/EBchecked/topics/584793/Brook-Taylor.

[511] www.pnas.org/cgi/doi/10.1073/pnas.0501311102.

[512] C.C. Yalavigi, Problem B-169, *The Fibonacci Quarterly*, 7 (1969), 332.

[513] M. Yodder, Solution to Problem B-143, *The Fibonacci Quarterly*, 7 (1969), 221.

[514] M. Yodder, Solution to Problem B-165, *The Fibonacci Quarterly*, 8 (1970), 112.

[515] D. Zeitlin, Problem H-17, *The Fibonacci Quarterly*, 1(4) (1963), 51.

[516] D. Zeitlin, Problem H-103, *The Fibonacci Quarterly*, 5 (1967), 69.

[517] D. Zeitlin, Solution to Problem H-97, *The Fibonacci Quarterly*, 6 (1968), 256–257.

[518] D. Zeitlin, Solution to Problem H-106, *The Fibonacci Quarterly*, 6 (1968), 358.

[519] D. Zeitlin, Solution to Problem B-163, *The Fibonacci Quarterly*, 8 (1970), 110.

[520] D. Zeitlin, Solution to Problem H-141, *The Fibonacci Quarterly*, 8 (1970), 272–275.

[521] Z. Zhusheng, Problem 43.7, *Mathematical Spectrum*, 43 (2010–2011), 92.

SOLUTIONS TO ODD-NUMBERED EXERCISES

EXERCISES 31

1. $a_{n-1,k} + a_{n-2,k-1} = \binom{n-k-2}{k} + \binom{n-k-3}{k-1} = \binom{n-k-1}{k} = a_{n,k}.$

3. Since $b_{n,k}$ is the coefficient of x^{n-2k}, $b_{n,k} = \frac{n}{n-k}\binom{n-k}{k}$. Then $b_{n-1,k} + b_{n-2,k-1} = \frac{n-1}{n-k-1}\binom{n-k-1}{k} + \frac{n-2}{n-k-1}\binom{n-k-1}{k-1} = \frac{n}{n-k}\binom{n-k}{k} = b_{n,k}.$

5. Since $b_{n,k} = \frac{n}{n-k}\binom{n-k}{k}$ and $x_n = \sum_{k=0}^{\lfloor (n-1)/3 \rfloor} b_{n-k,k}$, the given result follows.

7. The general solution of the recurrence is $g_n = A\alpha^n + B\beta^n$, where $A = A(x)$ and $B = B(x)$. The initial conditions $g_0 = 2$ and $g_1 = x$ yield the equations $A + B = 2$ and $A\alpha + B\beta = x$. Solving them, we get $A = 1 = B$. So $l_n = \alpha^n + \beta^n$.

9. $\alpha l_n + l_{n-1} = \alpha(\alpha^n + \beta^n) + (\alpha^{n-1} + \beta^{n-1}) = \alpha^{n-1}(\alpha^2 + 1) - \beta^{n-1} + \beta^{n-1} = \alpha^{n-1} \cdot \alpha(\alpha - \beta) = \alpha^n \Delta$.

11. Using Exercise 31.9, LHS $= \alpha^m \cdot \Delta\alpha^n = \alpha^n(\Delta\alpha^m) =$ RHS.

Fibonacci and Lucas Numbers with Applications, Volume Two. Thomas Koshy.

13. Using Exercises 31.8 and 31.9, LHS = $(\alpha l_n + l_{n-1}) - x(\alpha f_n + f_{n-1}) = \Delta\alpha^n - x\alpha^n = \alpha^n(\Delta - x) = \alpha^n \cdot 2\beta = -2\alpha^{n-1}$.

15. $\Delta(\text{LHS}) = (\alpha^{n+1} - \beta^{n+1}) + (\alpha^{n-1} - \beta^{n-1}) = (\alpha^{n+1} + \alpha^{n-1}) - (\beta^{n+1} + \beta^{n-1}) = \alpha^n(\alpha - \beta) - \beta^n(\beta - \alpha) = \Delta l_n$. So LHS = l_n.

17. LHS = $(xf_n + f_{n-1}) + f_{n-1} = f_{n+1} + f_{n-1} = l_n$ = RHS.

19. LHS = $(l_n + xf_n)(l_n - xf_n) = 2f_{n+1} \cdot 2f_{n-1} = 4f_{n+1}f_{n-1}$ = RHS.

21. LHS = $(xf_{n+1} + f_n) - (f_n - xf_{n-1}) = x(f_{n+1} + f_{n-1}) = xl_n$ = RHS.

23. LHS = $(l_{n+2} + l_{n-2})(l_{n+2} - l_{n-2}) = x(x^2+4)f_n \cdot (x^2+2)l_n$ = RHS.

25. Adding the equations $l_{n+1}^2 - l_n^2 = xl_{2n+1} - 4(-1)^n$, $l_{n+2}^2 - l_{n+1}^2 = xl_{2n+3} + 4(-1)^n$, and $l_{n+3}^2 - l_{n+2}^2 = xl_{2n+5} - 4(-1)^n$, we get LHS = $x(l_{2n+5} + l_{2n+1}) + xl_{2n+3} - 4(-1)^n = (x^3 + 2x)l_{2n+3} + xl_{2n+3} - 4(-1)^n$ = RHS.

27. $\Delta^2(\text{LHS}) = (\alpha^{n+1} - \beta^{n+1})^2 + (\alpha^n - \beta^n)^2 = \alpha^{2n+1}(\alpha - \beta) + \beta^{2n+1}(\beta - \alpha) = \Delta(\alpha^{2n+1} - \beta^{2n+1})$. So LHS = f_{2n+1} = RHS.

29. LHS = $(l_{n+2} - xl_{n+1})^2 + (xl_{n+2} + l_{n+1})^2 = (x^2+1)(l_{n+2}^2 + l_{n+1}^2) = (x^2+1)(x^2+4)f_{2n+3}$.

31. LHS = $(x^2+1)f_n + xf_{n-1} - (f_n - xf_{n-1}) = x(xf_n + 2f_{n-1}) = xl_n$ = RHS.

33. LHS = $(l_{n+1} + l_{n-1})(l_{n+1} - l_{n-1}) = (x^2+4)f_n \cdot xl_n$ = RHS.

35. LHS = $(xl_{n+1} + l_n) - (l_n - xl_{n-1}) = x(l_{n+1} + l_{n-1}) = x \cdot (x^2+4)f_n$ = RHS.

37. Let k be odd. Then $\Delta(\text{LHS}) = (\alpha^{n+k} - \beta^{n+k}) - (\alpha^{n-k} - \beta^{n-k}) = \alpha^n(\alpha^k + \beta^k) - \beta^n(\beta^k + \alpha^k) = (\alpha^n - \beta^n)(\alpha^k + \beta^k)$. So LHS = $f_n l_k$. The case k even follows similarly.

39. Let k be odd. Then LHS = $\alpha^n(\alpha^k - \beta^k) + \beta^n(\beta^k - \alpha^k) = (\alpha^n - \beta^n)(\alpha^k - \beta^k) = (x^2+4)f_k l_n$. Similarly, when k is even, LHS = $l_k l_n$.

41. Follows by Exercises 31.39 and 31.40.

43. LHS = $\Delta(\alpha^n - \beta^n) + x(\alpha^n + \beta^n) = \alpha^n(\Delta + x) + \beta^n(x - \Delta) = \alpha^n(2\alpha) + \beta^n(2\beta) = 2l_{n+1}$.

45. LHS = $\dfrac{f_{n+1}}{l_{n+1}} = \dfrac{2f_{n+1}}{2l_{n+1}} = \dfrac{xf_n + l_n}{(x^2+4)f_n + xl_n} = \dfrac{xR_n + 1}{(x^2+4)R_n + 1}$.

47. LHS = $(\alpha^{2n+2} - \alpha^{2n}) + (\beta^{2n+2} - \beta^{2n}) - 4(-1)^n = \alpha^{2n+1}(\alpha + \beta) + \beta^{2n+1}(\beta + \alpha) - 4(-1)^n = x(\alpha^{2n+1} + \beta^{2n+1}) - 4(-1)^n$ = RHS.

49. RHS = $(\alpha^n - \beta^n)^2 + 2(-1)^n = \alpha^{2n} + \beta^{2n} = l_{2n}$ = LHS.

51. RHS = $l_{2n} + [l_{2n} + 2(-1)^n] = l_{2n} + l_n^2$ = LHS.

53. LHS = $(\alpha^n + \beta^n)(\alpha^{n+1} + \beta^{n+1}) = (\alpha^{2n+1} + \beta^{2n+1}) + (\alpha\beta)^n(\alpha + \beta) = l_{2n+1} + (-1)^n x$ = RHS.

55. LHS = $(\alpha^{n+2} - \beta^{n+2})(\alpha^{n-1} - \beta^{n-1}) = l_{2n+1} - (-1)^n\left(\frac{\alpha^2}{\beta} + \frac{\beta^2}{\alpha}\right) =$ $l_{2n+1} + (-1)^n l_3$ = RHS.
57. RHS = $f_n^2[l_n^2 - 4(-1)^n] = f_n^2 \cdot (x^2+4)f_n^2$ = LHS.
59. Δ(LHS) = $(\alpha^{2n} - \beta^{2n}) - (\alpha^{n-1} - \beta^{n-1})(\alpha^{n+1} + \beta^{n+1}) =$ $-\alpha^{n-1}\beta^{n+1} + \alpha^{n+1}\beta^{n-1} = (-1)^n\left(\frac{\alpha}{\beta} - \frac{\beta}{\alpha}\right) = (-1)^{n+1}(\alpha^2 - \beta^2)$. So LHS = $(-1)^{n+1}x$ = RHS.
61. Δ(LHS) = $(\alpha^{2n} - \beta^{2n}) - (\alpha^{n+1} - \beta^{n+1})(\alpha^{n-1} + \beta^{n-1}) = (-1)^n\left(\frac{\alpha}{\beta} - \frac{\beta}{\alpha}\right) =$ $(-1)^n(\alpha^2 - \beta^2)$. So LHS = $(-1)^n x$ = RHS.
63. LHS = $(\alpha^{n+1} + \beta^{n+1})(\alpha^{n-1} + \beta^{n-1}) - (\alpha^n + \beta^n)^2 =$ $\alpha^{n+1}\beta^{n-1} + \alpha^{n-1}\beta^{n+1} - 2(-1)^n = (-1)^{n+1}\left(\frac{1}{\beta^2} + \frac{1}{\alpha^2}\right) - 2(-1)^n =$ $(-1)^{n+1}(l_2 + 2) = (-1)^{n+1}(x^2+4)$ = RHS.
65. Δ(LHS) = $(\alpha^n + \beta^n)(\alpha^{n-1} - \beta^{n-1}) + (\alpha^n - \beta^n)(\alpha^{n-1} + \beta^{n-1}) =$ $2(\alpha^{2n-1} - \beta^{2n-1}) = 2\Delta f_{2n-1}$; so LHS = RHS.
67. Adding the equations $l_{2n+2}l_{2n} - l_{2n+1}^2 = (x^2+4)$ (Exercise 31.63) and $l_{2n+1}^2 - (x^2+4)f_{2n+1}^2 = -4$ (Exercise 31.56) yields the given identity.
69. Using the Germain identity $4a^4 + b^4 = [a^2 + (a+b)^2][a^2 + (a-b)^2]$ with $a = l_{n+1}$ and $b = l_{n-1}$, LHS = $[l_{n+1}^2 + (l_{n+1} + l_{n-1})^2][l_{n+1}^2 + (l_{n+1} - l_{n-1})^2] =$ $(l_{n+1}^2 + \Delta^2 f_n^2)(l_{n+1}^2 + x^2 l_n^2)$ = RHS.
71. Follows by subtracting equation (31.7) from equation (31.6).
73. Δ^2(LHS) = $(\alpha^m - \beta^m)(\alpha^{m+n+k} - \beta^{m+n+k}) - (\alpha^{m+k} - \beta^{m+k})(\alpha^{m+n} - \beta^{m+n}) =$ $(-1)^m\beta^k(\alpha^n - \beta^n) - (-1)^m\alpha^k(\alpha^n - \beta^n) = (-1)^{m+1}(\alpha^n - \beta^n)(\alpha^k - \beta^k)$. So LHS = $(-1)^{m+1}f_n f_k$ = RHS.
75. LHS = $(\alpha^{n+1} + \beta^{n+1})(\alpha^{n+3} + \beta^{n+3}) - (\alpha^n - \beta^n)(\alpha^{n+4} - \beta^{n+4}) =$ $(-1)^n(\alpha^4 + \beta^4 + \alpha^3\beta + \alpha\beta^3) = (-1)^n(\alpha + \beta)(\alpha^3 + \beta^3) = (-1)^n l_1 l_3$ = RHS.
77. Follows from Exercise 31.74 by letting $m = n - k$ and $n = k$.
79. The identity follows by letting $m = 1$ and $k = n$.
81. Δ(RHS) = $(\alpha^{2n} - \beta^{2n})(\alpha^n + \beta^n) - (-1)^n(\alpha^n - \beta^n) = \alpha^{3n} - \beta^{3n}$, so RHS = LHS.
83. Using the identity $x^5 - y^5 = (x-y)(x^4 + x^3y + x^2y^2 + xy^3 + y^4)$, $\Delta f_{5n} = (\alpha^n)^5 - (\beta^n)^5 = (\alpha^n - \beta^n)[\alpha^{4n} + \beta^{4n} + (-1)^n(\alpha^{2n} + \beta^{2n}) + 1]$. So $f_{5n} = [l_{4n} + (-1)^n l_{2n} + 1]f_n$.
85. $\alpha^{4n} + \beta^{4n} = (\alpha^n + \beta^n)^4 - 4(\alpha\beta)^n(\alpha^{2n} + \beta^{2n}) - 6$. So $l_{4n} =$ $l_n^4 - 4(-1)^n[(\alpha^n + \beta^n)^2 - 2(-1)^n] - 6 = l_n^4 - 4(-1)^n l_n^2 + 2$.

87. $\alpha^{6n}+\beta^{6n}=(\alpha^n+\beta^n)4^6-6(-1)^n(\alpha^{4n}+\beta^{4n})-15(\alpha^{2n}+\beta^{2n})-20(-1)^n$. This implies $l_{6n}=l_n^6-6(-1)^n[l_n^4-4(-1)^n l_n^2+2]-15[l_n^2-2(-1)^n]-20(-1)^n=l_n^6-6(-1)^n l_n^4+9l_n^2-2(-1)^n$.

89. Using Exercises 31.48 and 31.85, $\Delta^4 f_n^4=[l_n^4-2(-1)^n l_n^2+2]-4(-1)^n[l_n^2-2(-1)^n]+6=l_n^4-6(-1)^n l_n^2+16$.

91. Since $\Delta^2(f_{n+1}^2+f_{n-1}^2)=(\alpha^{n+1}-\beta^{n+1})^2+(\alpha^{n-1}-\beta^{n-1})^2=(x^2+2)l_{2n}+4(-1)^n$, and $f_{n+1}^2-f_{n-1}^2=xf_{2n}$ by Exercise 31.38, the given result follows.

93. Follows by Exercises 31.33 and 31.92.

95. Since $a_n+\Delta b_n=\sum\limits_{k=0}^{n}\binom{n}{k}x^{n-k}\Delta^k=\sum\limits_{k\geq 0}^{n}\binom{n}{2k}x^{n-2k}(x^2+4)^k+\Delta\sum\limits_{k\geq 0}^{n}\binom{n}{2k+1}x^{n-2k-1}(x^2+4)^k$, the given result follows by equations (31.31) and (31.32).

97. The formula works when $n=1$ and $n=2$. Assume it works for all positive integers $\leq n$, where $n\geq 2$. Then $l_{n+1}=xl_n+l_{n-1}=$
$x\sum\limits_{k=0}^{\lfloor n/2\rfloor}\frac{n}{n-k}\binom{n-k}{k}x^{n-2k}+\sum\limits_{k=0}^{\lfloor (n-1)/2\rfloor}\frac{n-1}{n-k-1}\binom{n-k-1}{k}x^{n-2k-1}$.
Let n be even, say, $n=2m$. Then

$$
\begin{aligned}
l_{n+1}&=\sum_{k=0}^{m}\frac{2m}{2m-k}\binom{2m-k}{k}x^{2m-2k+1}+\sum_{k=0}^{m-1}\frac{2m-1}{2m-k-1}\binom{2m-k-1}{k}x^{2m-2k-1}\\
&=\sum_{k=0}^{m}\frac{2m}{2m-k}\binom{2m-k}{k}x^{2m-2k+1}+\sum_{k=0}^{m}\frac{2m-1}{2m-k}\binom{2m-k}{k-1}x^{2m-2k+1}\\
&=\sum_{k=0}^{m}\left[\frac{2m}{2m-k}\binom{2m-k}{k}+\frac{2m-1}{2m-k}\binom{2m-k}{k-1}\right]x^{2m-2k+1}\\
&=\sum_{k=0}^{m}\frac{2m+1}{2m+1-k}\binom{2m+1-k}{k}x^{2m+1-2k}=l_{2m+1}.
\end{aligned}
$$

Similarly, the formula works when n is odd. Thus, by PMI, it works for all $n\geq 1$.

99. We have $2^n\Delta f_n=(x+\Delta)^n-(x-\Delta)^n=\sum\limits_{k=0}^{n}\binom{n}{k}x^{n-k}\Delta^k[1-(-1)^k]=\sum\limits_{k\text{ odd}}\binom{n}{k}x^{n-k}2\Delta^k$. So $f_n=\frac{1}{2^{n-1}}\sum\limits_{k=0}^{\lfloor (n-1)/2\rfloor}\binom{n}{2k+1}(x^2+4)^k x^{n-2k-1}$.

101. It follows from the expansions of α^{2n} and β^{2n} in Example 31.3 that $f_{2n} = \sum_{k=0}^{\lfloor n/2\rfloor} \binom{n}{2k} f_{2k}x^{2k} + \sum_{k=0}^{\lfloor (n-1)/2\rfloor} \binom{n}{2k+1} f_{2k+1}x^{2k+1}$. From the expansions of α^n and β^n, we get
$$-f_n = \sum_{k=0}^{\lfloor n/2\rfloor} \binom{n}{2k} f_{2k}x^{n-2k} - \sum_{k=0}^{\lfloor (n-1)/2\rfloor} \binom{n}{2k+1} f_{2k+1}x^{n-2k+1}.$$
Adding these two equations yields the given formula.

103. Since $P_n = \frac{\gamma^n}{2\sqrt{2}}[1-(\delta/\gamma)]^n$, where $|\delta| < |\gamma|$, the given result follows.

105. When n is sufficiently large, $P_n \approx \frac{\gamma^n}{2\sqrt{2}}$. So $\log P_n \approx n\log\gamma - 1.5\log 2$. So number of digits in $P_n = 1 +$ characteristic of $\log P_n = \lceil \log P_n \rceil = \lceil n\log\gamma - 1.5\log 2\rceil$.

107. $(\gamma-\delta)P_{-n} = \gamma^{-n} - \delta^{-n} = \frac{1}{\gamma^n} - \frac{1}{\delta^n} = (-1)^{n-1}(\gamma-\delta)P_n$. This yields the desired result.

109. Since $f_{n+1}(2x)f_{n-1}(2x) - f_n^2(2x) = (-1)^n$, the given result follows.

111. $\Delta^3 f_n^3 = (\alpha^n - \beta^n)^3 = (\alpha^{3n} - \beta^{3n}) - 3(-1)^n(\alpha^n - \beta^n)$. This yields the given result.

113. Since $p_n(x) = f_n(2x)$, this follows from Exercise 31.111.

115.
$$\begin{aligned}\Delta^{2m+1} f_n^{2m+1} = (\alpha^n - \beta^n)^{2m+1} &= \sum_{k=0}^{2m+1} (-1)^k \binom{2m+1}{k} \alpha^{(2m+1)n-nk}\beta^{nk}\\ &= \sum_{k=0}^{m} (-1)^k \binom{2m+1}{k} \left(\alpha^{2mn-nk+n}\beta^{nk} - \alpha^{nk}\beta^{2mn-nk+n}\right)\\ &= \sum_{k=0}^{m} (-1)^k \binom{2m+1}{k} (\alpha\beta)^{nk}\left[\alpha^{2n(m-k)+n} - \beta^{2n(m-k)+n}\right].\end{aligned}$$
The given result now follows.

117. LHS $= (\alpha^{n+m} + \beta^{n+m}) - (\alpha\beta)^m(\alpha^{n-m} + \beta^{n-m}) = \alpha^{n+m} + \beta^{n+m} - \alpha^n\beta^m - \alpha^m\beta^n = (\alpha^m - \beta^m)(\alpha^n - \beta^n) =$ RHS.

119. By identity (31.51), $l_{2mn}^2 \equiv 4 \pmod{\Delta^2 f_{2mn}^2}$. Since $f_{2mn} = f_{mn}l_{mn}$, this yields the given congruence.

121. Since $q_i(x) = l_i(2x)$, the given formula follows from Exercise 31.120.

123. This follows from Exercise 31.121.

125. LHS $= (l_n^2 + l_{n+1}^2)^k(l_{n+2}^2 + l_{n+3}^2)^k = \Delta^{2k}(f_n^2 + f_{n+1}^2)^k \cdot \Delta^{2k}(f_{n+2}^2 + f_{n+3}^2)^k =$ RHS.

127. Since $\alpha^n\beta^{2m-n} = \beta^{2m-2n}$ and $\beta^n\alpha^{2m-n} = \alpha^{2m-2n}$, and n is even, $l_m^2 - l_{m-n}^2 = \alpha^{2m} + \beta^{2m} - \alpha^{2m-2n} - \beta^{2m-2n} = \alpha^{2m} + \beta^{2m} - \beta^n\alpha^{2m-n} - \alpha^n\beta^{2m-n} = (\alpha^n - \beta^n)(\alpha^{2m-n} - \beta^{2m-n}) = (x^2+4)f_n f_{2m-n}$. But by identity (31.65), $l_m^2 - l_{m-n}^2 = (x^2+4)(f_m^2 - f_{m-n}^2)$. Combining these two equations, we get the desired result.

129. Since $xl_n = x(f_{n+1} + f_{n-1}) = (xf_{n+1} + f_n) + (xf_{n-1} - f_n) = f_{n+2} - f_{n-2}$, it follows that $xf_{n+1} < xl_n < f_{n+2}$.

131. Clearly, the inequality works when $n = 3$ and $n = 4$. Now assume it works for all positive integers $\leq n$, where $n \geq 4$. Let $f = x^2 + 2$, so $f_4 = xf$.

Since $\dfrac{2xf_{n-1}f_n}{f_3^2 f^{n-2}} \leq \dfrac{2xf_3 f^{(n-4)/2} \cdot f_3 f^{(n-3)/2}}{f_3^2 f^{n-2}} = \dfrac{2x}{f\sqrt{f}}$, we have

$$\begin{aligned}
f_{n+1}^2 &= (xf_n + f_{n-1})^2 = x^2 f_n^2 + f_{n-1}^2 + 2xf_n f_{n-1} \\
&\leq x^2 f_3^2 f_4^{n-3} + f_3^2 f_4^{n-4} + 2xf_n f_{n-1} \leq f_3^2 f^{n-2}\left(\frac{x^2}{f} + \frac{1}{f^2} + \frac{2x}{f\sqrt{f}}\right) \\
&= f_3^2 f^{n-2}\left(\frac{x}{\sqrt{f}} + \frac{1}{f}\right)^2 \leq f_3^2 f^{n-2}.
\end{aligned}$$

So $x^{n-2}f_{n+1}^2 \leq f_3^2 f_4^{n-2}$. Thus by the strong version of PMI, the inequality holds for every $n \geq 3$.

133. Follows by the identity $t_n + t_{n-1} = n^2$.

135. Follows by the identity $t_n^2 + t_{n-1}^2 = t_{n^2}$.

137. Let A_n denote the LHS of the formula and $B_n = g_n g_{n+1}$. Then $B_n - B_{n-1} = xg_n^2 = A_n - A_{n-1}$. Consequently, $A_n - B_n = A_{n-1} - B_{n-1} = A_0 - B_0 = 0 - g_0 g_1 = -g_0 g_1$. Thus $A_n = B_n - g_0 g_1 = g_n g_{n+1} - g_0 g_1$.

139. Follows from Exercise 31.137 with $g_k = l_k$.

141. Follows from Exercise 31.140.

143.
$$\begin{aligned}
\text{LHS} &= (\alpha^m + \beta^m)(\alpha^n + \beta^n) - (\alpha^{m+c} - \beta^{m+c})(\alpha^{n+c} - \beta^{n+c}) \\
&= \alpha^{m+n} + \beta^{m+n} - \alpha^{m+n+2c} - \beta^{m+n+2c} + \alpha^m\beta^n + \alpha^n\beta^m \\
&\quad + \alpha^{m+c}\beta^{n+c} + \alpha^{n+c}\beta^{m+c} \\
&= \alpha^{m+n} + \beta^{m+n} - \alpha^{m+n+2c} - \beta^{m+n+2c} + [1 + (-1)^c](\alpha^m\beta^n + \alpha^n\beta^m).
\end{aligned}$$

Let c be odd. Then LHS $= \alpha^{m+n} + \beta^{m+n} - \alpha^{m+n+2c} - \beta^{m+n+2c}$ $= (\alpha^{-c} + \beta^{-c})(\alpha^{m+n+c} + \beta^{m+n+c}) = l_{-c}l_{m+n+c}$.

On the other hand, let c be even. Then

$$\begin{aligned}
\text{LHS} &= \alpha^{m+n} + \beta^{m+n} - \alpha^{m+n+2c} - \beta^{m+n+2c} + 2\alpha^m\beta^n + 2\alpha^n\beta^m \\
&= \alpha^{m+n} + \beta^{m+n} - \alpha^{m+n+c}\beta^{-c} - \beta^{m+n+c}\alpha^{-c} + 2(\alpha\beta)^n(\alpha^{m-n} + \beta^{m-n}) \\
&= (\alpha^{-c} - \beta^{-c})\left(\alpha^{m+n+c} - \beta^{m+n+c}\right) + 2(-1)^n(\alpha^{m-n} + \beta^{m-n}) \\
&= (x^2+4)f_{-c}f_{m+n+c} + 2(-1)^n l_{m-n}.
\end{aligned}$$

145. LHS $= (\alpha^m + \beta^m)(\alpha^n + \beta^n) - (\alpha^{m-c} - \beta^{m-c})(\alpha^{n+c} - \beta^{n+c})$
$= \alpha^m\beta^n + \alpha^n\beta^m + \alpha^{m-c}\beta^{n+c} + \alpha^{n+c}\beta^{m-c}$
$= (\alpha\beta)^m \left(\alpha^{n-m} + \beta^{n-m} + \alpha^{-c}\beta^{n-m+c} + \alpha^{n-m+c}\beta^{-c}\right)$
$= (\alpha\beta)^m(\alpha^{-c} + \beta^{-c})\left(\alpha^{n-m+c} + \beta^{n-m+c}\right) = (-1)^m l_{-c} l_{n-m+c}.$

147. Follows from Exercise 31.144 by letting $c = 0$.

149. Since $\dfrac{xg_{2k}}{g_{2k+1}g_{2k-1}} = \dfrac{1}{g_{2k-1}} - \dfrac{1}{g_{2k+1}}$, it follows that $\sum_{k=1}^{n} \dfrac{xg_{2k}}{g_{2k+1}g_{2k-1}} = \dfrac{1}{g_1} - \dfrac{1}{g_{2n+1}}$. So $\dfrac{1}{g_1} - \sum_{k=1}^{n} \dfrac{xg_{2k}}{g_{2k+1}g_{2k-1}} = \dfrac{1}{g_{2n+1}}$. Likewise, $\dfrac{1}{g_2} - \sum_{k=1}^{n} \dfrac{xg_{2k+1}}{g_{2k+2}g_{2k}} = \dfrac{1}{g_{2n+2}}$. Combining these two equations, we get the desired result.

151. It follows from Exercise 31.149 that
$\left(\dfrac{1}{l_1} - \sum_{k=1}^{n} \dfrac{xl_{2k}}{l_{2k+1}l_{2k-1}}\right)\left(\dfrac{1}{l_2} - \sum_{k=1}^{n} \dfrac{xl_{2k+1}}{l_{2k+2}l_{2k}}\right) = \dfrac{1}{l_{2n+1}l_{2n+2}}$. Since $l_1 = x$ and $l_2 = x^2 + 2$, the given result follows from this.

153. It follows from the generating function $\dfrac{2 - xz}{1 - xz - z^2} = \sum_{n=0}^{\infty} l_n z^n$ that

$\dfrac{2 - 2xz}{1 - 2xz - z^2} = \sum_{n=0}^{\infty} q_n z^n$.

155. Since $l_{2n+3} = (x^2 + 2)l_{2n+1} - l_{2n-1}$, $z_{n+2} = 3[(x^2 + 2)l_{2n+1} - l_{2n-1}] + (x + 1) = (x^2 + 2)(3l_{2n+1} + x + 1) - (3l_{2n-1} + x + 1) - (x + 1)(x^2 + 2) + 2(x + 1) = (x^2 + 2)z_{n+1} - z_n - (x + 1)x^2$, where $z_0 = 1 - 2x$ and $z_1 = 4x + 1$.

157. Since $l_{3n} = l_n^3 - 3(-1)^n l_n$, $\frac{1}{2}l_{3^{n+1}} = 4\left(\dfrac{l_{3^n}^3}{8}\right) + 3\left(\dfrac{l_{3^n}}{2}\right)$. So

$z_{n+1} = 4z_n^3 + 3z_n$, where $z_0 = \dfrac{x}{2}$.

159. $z_{k+1} = \alpha^{kn+n} + \beta^{kn+n} = (\alpha^n + \beta^n)(\alpha^{kn} + \beta^{kn}) - (\alpha^n\beta^{kn} + \beta^{kn}\alpha^n) = l_n z_k - (-1)^n[\alpha^{(k-1)n} + \beta^{(k-1)n}] = l_n z_k - (-1)^n z_{k-1}$, where $z_0 = 2$ and $z_1 = l_n$.

161. Since $l_{3n} = l_n^3 - 3(-1)^n l_n$, $l_{3^{n+1}} = l_{3^n}^3 + 3l_{3^n}$. So $z_{n+1} = z_n^3 + 3z_n$, where $z_0 = 2$.

163. Since $l_{2n} = l_n^2 - 2(-1)^n$, it follows that $z_{n+1} = z_n^2 - 2$, where $z_0 = x$.

165. It follows by Exercise 31.87 that $z_{n+1} = z_n^6 - 6z_n^4 + 9z_n^2 - 2$, where $z_0 = x$.

167. $\dfrac{l_{n+1}}{\alpha^{n-1}} + \dfrac{l_n}{\alpha^n} = \dfrac{\alpha^{n+1} + \beta^{n+1}}{\alpha^{n-1}} + \dfrac{\alpha^n + \beta^n}{\alpha^n} = \alpha^2 - \dfrac{\beta^n}{\alpha^n} + 1 + \dfrac{\beta^n}{\alpha^n} = \alpha^2 + 1.$

169. By Exercise 31.168, $x + \sum_{i=1}^{n} \frac{(-1)^{i+1}}{f_i f_{i+1}} = \frac{f_{n+2}}{f_{n+1}}$. So $x + \sum_{i=1}^{\infty} \frac{(-1)^{i+1}}{f_i f_{i+1}} = \lim_{n\to\infty} \frac{f_{n+2}}{f_{n+1}} = \alpha$.

171. $\prod_{k=1}^{n} \left(x + \frac{g_{k-1}}{g_k}\right) = \prod_{k=1}^{n} \frac{g_{k+1}}{g_k} = \frac{g_{n+1}}{g_1}$.

173. It follows by Exercise 31.170 that $\prod_{k=1}^{\infty} \frac{x l_{2k} l_{2k+2} + l_{2k-1} l_{2k+2}}{x l_{2k} l_{2k+2} + l_{2k} l_{2k+1}} = \frac{l_3}{l_2} \cdot \lim_{n\to\infty} \frac{l_{2n+2}}{l_{2n+3}} = \frac{x^3+3x}{x^2+2} \cdot \frac{1}{\alpha} = \frac{x^3+3x}{(x^2+2)\alpha}$.

175. It follows from Example 31.14 that the given infinite product equals $2\alpha(1) - 1 = \sqrt{5}$.

177. Since $(u^2+x)^n = \sum_{k=0}^{n} \binom{n}{k} u^{2(n-k)} x^k$, we have

$$(u^2+x)^n = \sum_{k=0}^{\lfloor n/2 \rfloor} \binom{n}{2k} u^{2(n-2k)} x^{2k} + \sum_{k=0}^{\lfloor (n-1)/2 \rfloor} \binom{n}{2k+1} u^{2(n-2k-1)} x^{2k+1} \tag{1}$$

$$\frac{(u^2+\alpha)^n - (u^2+\beta)^n}{\Delta} = \sum_{k=0}^{\lfloor n/2 \rfloor} \binom{n}{2k} u^{2(n-2k)} f_{2k} + \sum_{k=0}^{\lfloor (n-1)/2 \rfloor} \binom{n}{2k+1} u^{2(n-2k-1)} f_{2k+1}.$$

Now let $u = \alpha\sqrt{x}$ and then $u = \beta\sqrt{x}$; add the two results. Since $\alpha^2 x + \alpha = \alpha^3, \beta^2 x + \beta = \beta^3, \alpha^2 x + \beta = (x^2-1)\alpha + 2x$, and $\beta^2 x + \alpha = (x^2-1)\beta + 2x$, the resulting sum yields

$$f_{3n} - A + B = \sum_{k=0}^{\lfloor n/2 \rfloor} \binom{n}{2k} l_{2(n-2k)} f_{2k} + \sum_{k=0}^{\lfloor (n-1)/2 \rfloor} \binom{n}{2k+1} l_{2(n-2k-1)} f_{2k+1}. \tag{2}$$

Using equation (1), we have

$$\frac{(y^2-\alpha x)^n - (y^2-\beta x)^n}{\Delta} = \sum_{k=0}^{\lfloor n/2 \rfloor} \binom{n}{2k} y^{2(n-2k)} x^{2k} f_{2k} - \sum_{k=0}^{\lfloor (n-1)/2 \rfloor} \binom{n}{2k+1} y^{2(n-2k-1)} x^{2k+1} f_{2k+1}.$$

Letting $y = \alpha$ and $y = \beta$, and adding the two sums, we get

$$\frac{(2\beta x - x^2 + 1)^n - (2\alpha x - x^2 + 1)^n}{\Delta}$$
$$= \sum_{k=0}^{\lfloor n/2\rfloor} \binom{n}{2k} l_{2(n-2k)} x^{2k} f_{2k} - \sum_{k=0}^{\lfloor (n-1)/2\rfloor} \binom{n}{2k+1} l_{2(n-2k-1)} x^{2k+1} f_{2k+1}. \quad (3)$$

The given result now following by adding equations (2) and (3).

179. The formula is true when $n = 0$. Now assume it is true for an arbitrary integer $n \geq 0$. Then $x_{n+1} = 3x_n + \lceil 2\sqrt{2}x_n \rceil = \lceil (3 + 2\sqrt{2})x_n \rceil = \lceil \gamma^2 x_n \rceil =$ $\lceil \gamma^2 P_{2n+2} \rceil = \left\lceil \frac{\gamma^2}{\gamma - \delta}(\gamma^{2n+2} - \delta^{2n+2}) \right\rceil$
$= \left\lceil \frac{1}{\gamma - \delta}\left[(\gamma^{2n+4} - \delta^{2n+4}) + (\delta^{2n+4} - \delta^{2n})\right] \right\rceil$
$= P_{2n+4} + \left\lceil \frac{\delta^{2n+2}}{\gamma - \delta}(\delta^2 - \gamma^2) \right\rceil = P_{2n+4} + \lceil -2\delta^{2n+2} \rceil.$
Since $\delta^2 = 3 - 2\sqrt{2}, \delta^{2n} = (3 - 2\sqrt{2})^n$; so $-2\delta^{2n+2} = -2(3 - 2\sqrt{2})^{n+1}$ and hence $-1 < -2\delta^{2n+2} < 0$. Consequently, $\lceil -2\delta^{2n+2} \rceil = 0$. This implies $x_{n+1} = P_{2n+4} + 0 = P_{2n+4}$; so the formula works for $n + 1$ as well. Thus, by PMI, it works for all $n \geq 0$, as desired.

181. The formula works when $n = 0$. Assume it is true for an arbitrary $n \geq 0$. Then $x_{n+1} = (x^2 + 2)x_n + \lceil x\Delta x_n \rceil = \lceil (x^2 + x\Delta + 2)x_n \rceil = \lceil 2\alpha^2 x_n \rceil =$ $\lceil 2\alpha^2 \cdot 2^{n-1} l_{2n+3} \rceil = \lceil 2^n \alpha^2 (\alpha^{2n+3} + \beta^{2n+3}) \rceil =$ $\lceil 2^n [(\alpha^{2n+5} + \beta^{2n+5}) - (\beta^{2n+5} - \beta^{2n+1})] \rceil = 2^n l_{2n+5} + \lceil 2^n \beta^{2n+3}(\alpha^2 - \beta^2) \rceil =$ $2^n l_{2n+5} + \lceil 2^n x\Delta\beta^{2n+3} \rceil$. Since $x\Delta\beta^2 = x\Delta \cdot \frac{(x-\Delta)^2}{4} = \frac{x\Delta}{4} \cdot \frac{(x^2 - \Delta^2)^2}{(x+\Delta)^2} =$ $\frac{x\Delta}{4} \cdot \frac{(-4)^2}{(x+\Delta)^2} = \frac{4x\Delta}{(x+\Delta)^2} < 1$ and $\beta^2 = \left(\frac{x-\Delta}{2}\right)^2 = \left[\frac{\Delta^2 - x^2}{2(\Delta + x)}\right]^2 =$ $\frac{4}{(\Delta + x)^2} \leq \frac{4}{(1+\sqrt{5})^2} < \frac{1}{2}$, it follows that $x\Delta\beta^2 \cdot (\beta^2)^n < (1/2)^n$; that is, $2^n x\Delta\beta^{2n+2} < 1$. Since $-1 < \beta < 0$, this implies $-1 < 2^n x\Delta\beta^{2n+3} < 0$; so $\lceil 2^n x\Delta\beta^{2n+3} \rceil = 0$. Thus $x_{n+1} = 2^n l_{2n+5}$. Hence the formula works for $n + 1$ as well. So, by PMI, it works for all $n \geq 0$, as desired.

183. Letting $\theta = \sqrt{x^2 + 1}, \gamma = x + \theta$ and $\delta = x - \theta$. The formula works when $n = 0$. Assume it works for an arbitrary $n \geq 0$. Then $x_{n+1} =$ $\lfloor (2x^2 + 1 + 2x\theta)x_n \rfloor = \lfloor \gamma^2 x_n \rfloor = \lfloor \gamma^2 q_{2n+2} \rfloor = \lfloor \gamma^2(\gamma^{2n+2} + \delta^{2n+2}) \rfloor =$ $\lfloor (\gamma^{2n+4} + \delta^{2n+4}) - (\delta^{2n+4} - \delta^{2n}) \rfloor = q_{2n+4} + \lfloor 4x\theta\delta^{2n+2} \rfloor.$

Since $\delta^2 = (x^2 - \theta^2)^2/(x+\theta)^2 = 1/(x+\theta)^2 < 1$ and $4x\theta\delta^2 = 4x\theta \cdot (x^2-\theta^2)^2/(x+\theta)^2 = 4x\theta/(x+\theta)^2 < 1$, it follows that $0 < 4x\theta\delta^{2n+2} < 1$ and hence $\lfloor 4x\theta\delta^{2n+2} \rfloor = 0$.
Thus $x_{n+1} = q_{2n+4}$. So the formula works for $n+1$ also. Thus, by PMI, it works for all $n \geq 0$.

EXERCISES 32

1. The result is clearly true when $n = 1$. Assume it is true for an arbitrary positive integer k. Then $Q^{k+1} = Q \cdot Q^k = \begin{bmatrix} x & 1 \\ 1 & 0 \end{bmatrix} \begin{bmatrix} f_{k+1} & f_k \\ f_k & f_{k-1} \end{bmatrix} = \begin{bmatrix} f_{k+2} & f_{k+1} \\ f_{k+1} & f_k \end{bmatrix}$. So, by PMI, the result is true for all positive integers n.
3. It follows by Exercise 32.2 that $p_{n+1}p_{n-1} - p_n^2 = |M|^n = (-1)^n$.
5. Since $xf_{2j} = f_{2j+1} - f_{2j-1}$, it follows that $x\sum_{j=1}^{n} f_{2j} = \sum_{j=1}^{n}(f_{2j+1} - f_{2j-1}) = f_{2n+1} - 1$.
7. By the addition formula, $f_n = f_{k+(n-k)} = f_{k+1}f_{n-k} + f_k f_{n-k-1}$.
9. Since $\deg(f_{mn}) = mn - 1$, $\deg(f_m f_n) = \deg(f_m) + \deg(f_n) = m + n - 2$, and $mn > m + n - 1$, it follows that $f_{mn} > f_m f_n$.
11. $x(\text{RHS}) = (\alpha^{m+1} - \beta^{m+1})(\alpha^{n+1} + \beta^{n+1}) - (\alpha^{m-1} - \beta^{m-1})(\alpha^{n-1} + \beta^{n-1}) = (\alpha^{m+n} + \beta^{m+n})(\alpha^2 - \beta^2) = x\Delta l_{m+n}$. So RHS = LHS.
13. Since $f_{m+n} = f_{m+1}f_n + f_m f_{n-1}$ and $f_{m-n} = (-1)^n(f_m f_{n-1} - f_{m-1}f_n)$, $f_{m+n} - (-1)^n f_{m-n} = f_n(f_{m+1} + f_{m-1}) = f_n l_m$.
15. It follows by Exercise 32.14 that $l_{m-n} = f_{m+1}l_{-n} + f_m l_{-(n+1)} = (-1)^n f_{m+1} l_n + (-1)^{n+1} f_m l_{n+1} = (-1)^n(f_{m+1}l_n - f_m l_{n+1})$.
17. By Exercise 32.16, $l_{n+4k} - l_n = l_{(n+2k)+2k} - (-1)^{2k} l_{(n+2k)-2k} = (x^2+4)f_{n+2k}f_{2k}$.
19. It follows by Exercise 32.16 that $l_{m+n} + l_{m-n} = (x^2+4)f_m f_n$, where n is odd. On the other hand, let n be even. Then $l_{m+n} + l_{m-n} = (\alpha^{m+n} + \alpha^{m-n}) + (\beta^{m+n} + \beta^{m-n}) = \alpha^m(\alpha^n + \alpha^{-n}) + \beta^m(\beta^n + \beta^{-n}) = \alpha^m[\alpha^n + (-\beta)^n] + \beta^m[\beta^n + (-\alpha)^n] = \alpha^m(\alpha^n + \beta^n) + \beta^m(\beta^n + \alpha^n) = (\alpha^m + \beta^m)(\alpha^n + \beta^n) = l_m l_n$.
21. Follows by Exercises 32.19 and 32.20.
23. Using Exercise 32.19 and the identity $l_k^2 = l_{2k} + 2(-1)^k$, LHS = $[l_{2m+2n} + 2(-1)^{m+n}] + [l_{2m-2n} + 2(-1)^{m-n}] = (l_{2m+2n} + l_{2m-2n}) + 4(-1)^{m+n} = l_{2m}l_{2n} + 4(-1)^{m+n}$ = RHS.

25. Since $a+b$ is odd, RHS $= (\alpha^{a+b}+\beta^{a+b})(\alpha^{a-b}+\beta^{a-b})+4(-1)^a =$
$\alpha^{2a}+\beta^{2a}+\alpha^{a+b}\beta^{a-b}+\alpha^{a-b}\beta^{a+b}+4(-1)^a = \alpha^{2a}+\beta^{2a}+(-1)^a[\alpha^{2b}(-1)^b+$
$\beta^{2b}(-1)^b]+4(-1)^a = \alpha^{2a}+\beta^{2a}+(-1)^{a+b}(\alpha^{2b}+\beta^{2b})+4(-1)^a =$
$l_{2a}-l_{2b}+4(-1)^a = l_{2a}-l_{2b}+2[(-1)^a-(-1)^b] = (\alpha^a+\beta^a)^2-(\alpha^b+\beta^b)^2 =$
LHS.

27. Follows from Exercise 32.26.

29. Since $f_{m+1}+f_{m-1}=l_m$, by the Cassini-like and addition formulas, we have RHS $= f_{2^n-1}\left(f_{2^n+1}+f_{2^n-1}\right) = f_{2^n-1}f_{2^n+1}+f^2_{2^n-1} = f_{2^n-1}f_{2^n+1}+$ $\left(f_{2^n}f_{2^n-2}+1\right) = f_{2^n-1}f_{2^n+1}+f_{2^n}f_{2^n-2})+1 = f_{2^n+(2^n-1)}+1 = f_{2^{n+1}-1}+1.$

31. Using identity (32.33), $\begin{bmatrix} f_{4n} \\ l_{4n} \end{bmatrix} = \dfrac{1}{2}\begin{bmatrix} l_n & f_n \\ \Delta^2 f_n & l_n \end{bmatrix}\begin{bmatrix} f_{3n} \\ l_{3n} \end{bmatrix} =$

$$\frac{1}{2}\begin{bmatrix} l_n & f_n \\ \Delta^2 f_n & l_n \end{bmatrix}\cdot\frac{1}{4}\begin{bmatrix} \Delta^2 f_n^3+3f_nl_n^2 \\ l_n^3+3\Delta^2 f_n^2 l_n \end{bmatrix} = \frac{1}{8}\begin{bmatrix} 4\Delta^2 f_n^3 l_n^3+4f_nl_n^3 \\ l_n^4+6\Delta^2 f_n^2 l_n^2+\Delta^4 f_n^4 \end{bmatrix}.$$

This yields the desired result.

33. See identity (31.54).

35. Using the identity $l_{2m}+(-1)^{m+n}l_{2n} = l_{m+n}l_{m-n}$, LHS $= (l_{2a+2b}+\nu l_{2c})(l_{2c}+$
$\nu l_{2a-2b}) = l_{2c}(l_{2a+2b}+l_{2a-2b})+\nu l^2_{2c}+\nu l_{2a+2b}l_{2a-2b} = l_{2c}(l_{2a}l_{2b})+\nu l^2_{2c}+$
$\nu(l_{4a}+l_{4b}) = l_{2a}l_{2b}l_{2c}+\nu l^2_{2c}+\nu[l^2_{2a}+l^2_{2b}-2(-1)^{2a}-2(-1)^{2b}] =$ RHS.

37. The result is true when $n=0$. Assume it works for an arbitrary integer $n\ge 0$. Then $au_{m+1}u_{n+2}-cu_mu_{n+1} = -(bx+cu_n)u_{m+1}+(au_{m+2}+$ $bxu_{m+1})u_{n+1} = au_{m+2}u_{n+1}-cu_{m+1}u_n = au_{m+(n+1)+1}$. So the formula also works for $n+1$. Thus, by PMI, it works for all $m,n\ge 0$.

39. Since $l_{2^k}-2=\Delta^2 f^2_{2^{k-1}}$ and $l_{2^k}+2=l^2_{2^{k-1}}$ when $k\ge 2$, we have $\Delta^2 f^2_{2^{k-1}}\left(l^2_{2^{k-1}}-1\right) = (l_{2^k}-2)(l_{2^k}+1) = l^2_{2^k}-l_{2^k}-2 = l^2_{2^k}-l^2_{2^{k-1}}$.
Consequently,

$$\frac{f^2_{2^{k-1}}}{l^2_{2^k}-1} = \frac{l^2_{2^k}-l^2_{2^{k-1}}}{\Delta^2(l^2_{2^{k-1}}-1)(l^2_{2^k}-1)} = \frac{1}{\Delta^2}\left(\frac{1}{l^2_{2^{k-1}}-1}-\frac{1}{l^2_{2^{k-1}}-1}\right)$$

$$\sum_{k=1}^{\infty}\frac{f^2_{2^{k-1}}}{l^2_{2^k}-1} = \frac{f_1^2}{l_2^2-1}+\frac{1}{\Delta^2}\sum_{k=2}^{\infty}\left(\frac{1}{l^2_{2^{k-1}}-1}-\frac{1}{l^2_{2^{k-1}}-1}\right)$$

$$= \frac{1}{l_2^2-1}+\frac{1}{\Delta^2}\cdot\frac{1}{l_2^2-1} = \frac{x^2+5}{(x+1)(x+3)(x^2+4)}.$$

41. Let $S_n = \sum_{k=1}^{n} \left(\frac{1}{l_{2k}l_{2k+1}} + \frac{1}{l_{2k}l_{2k+2}} - \frac{1}{l_{2k+1}l_{2k+2}} \right)$. Then

$$S_n = \sum_{k=1}^{n} \frac{l_{2k+1} + (l_{2k+2} - l_{2k})}{l_{2k}l_{2k+1}l_{2k+2}} = \sum_{k=1}^{n} \frac{(x+1)l_{2k+1}}{l_{2k}l_{2k+1}l_{2k+2}}$$

$$= \sum_{k=1}^{n} \frac{x+1}{l_{2k}l_{2k+2}} = (x+1) \sum_{k=1}^{n} \left(\frac{1}{l_{2k}l_{2k+1}} - \frac{1}{l_{2k+1}l_{2k+2}} \right)$$

$$= (x+1) \sum_{i=2}^{n} \frac{(-1)^i}{l_i l_{i+1}} = \frac{x+1}{\Delta^2} \sum_{i=2}^{n} \left(\frac{l_i}{l_{i+1}} - \frac{l_{i-1}}{l_i} \right) = \frac{x+1}{\Delta^2} \left(\frac{l_n}{l_{n+1}} - \frac{l_1}{l_2} \right)$$

$$\text{Given sum} = \lim_{n\to\infty} S_n = \frac{x+1}{\Delta^2} \left(\frac{1}{\alpha} - \frac{x}{x^2+2} \right) = \frac{(x+1)(x^2 - \alpha x + 2)}{(x^2+2)(x^2+4)\alpha}.$$

EXERCISES 33

1. Since $f_{k+1} + f_{k-1} = l_k$ and $f_k l_k = f_{2k}$, it follows by formula (33.1) that

$$f_{2n}^2 = x \sum_{k=0}^{n-1} f_{2k+1}(f_{2k} + f_{2k+2}) = x \sum_{k=0}^{n-1} f_{2k+1}l_{2k+1} = x \sum_{k=0}^{n-1} f_{4k+2} = x \sum_{k=1}^{n} f_{4k-2}.$$

3. Adding equations (33.4) and (33.6), we get $\sum_{k=1}^{n} (F_{4k-3} + F_{4k-2}) =$ $F_{2n}^2 + F_{2n}F_{2n-1}$; that is, $\sum_{k=1}^{n} F_{4k-1} = F_{2n+1}F_{2n}$.

5. Consider a $1 \times 2n$ board. The sum of the weights of its tilings is f_{2n+1}. Let T be an arbitrary tiling of the board. Suppose it is breakable at cell n: $\underbrace{\text{subtiling}}_{\text{length } n}\underbrace{\text{subtiling}}_{\text{length } n}$. The sum of the weights of such tilings is f_{n+1}^2. Suppose T is not breakable at cell n: $\underbrace{\text{subtiling}}_{\text{length } n-1}\boxed{1}\underbrace{\text{subtiling}}_{\text{length } n-1}$. The sum of the weights of such tilings is f_n^2. So the sum of the weights of tilings of the board equals $f_{n+1}^2 + f_n^2$. Thus $f_{2n+1} = f_{n+1}^2 + f_n^2$.

7. Consider a board of length $m - 1$. The sum of the weights of all its tilings is f_m. Let T be an arbitrary tiling of the board. Suppose it is breakable at cell n: $\underbrace{\text{subtiling}}_{\text{length } n}\ \underbrace{\text{subtiling}}_{\text{length } m-n-1}$. The sum of the weights of such

tilings is $f_{n+1}f_{m-n}$. On the other hand, suppose T is unbreakable at cell n: $\underbrace{\text{subtiling}}_{\text{length } n-1}\boxed{\ 1\ }\underbrace{\text{subtiling}}_{\text{length } m-n-2}$. The sum of the weights of such tilings is $f_n f_{m-n-1}$.

Thus the sum of the weights of all tilings is $f_{n+1}f_{m-n} + f_n f_{m-n-1} = f_m$.

9. By Exercise 32.8, $f_{2m} > xf_m^2$. So $f_{3m} = f_{2m+m} > xf_{2m} \cdot f_m > x \cdot xf_m^2 \cdot f_m = x^2 f_m^3$. Thus the given statement is true when $n = 2$ and $n = 3$. Now assume it is true for all integers $< n$, where $n \geq 3$. Then $f_{nm} = f_{(n-1)m+m} > xf_{(n-1)m} \cdot f_m > x \cdot x^{n-2} f_m^{n-1} \cdot f_m = x^{n-1} f_m^n$. Thus the statement is true for all integers ≥ 2.

11. Let s_n denote the sum of the weights of tilings of a $1 \times n$ board. Then $s_0 = 1 = \frac{1}{2}q_0$ and $s_1 = x = \frac{1}{2}q_1$. Let T be an arbitrary tiling of the board, where $n \geq 2$. Suppose it ends in a square: $\underbrace{\text{subtiling}}_{\text{length } n-1}\boxed{2x}$. The sum of the weights of such tilings is $2xs_{n-1}$. On the other hand, suppose T ends in a domino: $\underbrace{\text{subtiling}}_{\text{length } n-2}\boxed{\ 1\ }$. The sum of the weights of such tilings is s_{n-2}. So the sum of the weights of all tilings of the board is $2xs_{n-1} + s_{n-2} = s_n$. Thus s_n satisfies the Pell recurrence with $s_0 = \frac{1}{2}q_0$ and $s_1 = \frac{1}{2}q_1$. So $s_n = \frac{1}{2}q_n$, as desired.

13. Let s_n denote the sum of the weights of colored tilings of a $1 \times n$ board. Then $s_0 = 1 = p_1$ and $s_1 = 2x = p_2$. Let T be an arbitrary tiling of the board, where $n \geq 2$. Suppose it ends in a square: $\underbrace{\text{subtiling}}_{\text{length } n-1}\boxed{x}$ or $\underbrace{\text{subtiling}}_{\text{length } n-1}\boxed{x}$. The sum of the weights of such tilings is $2xs_{n-1}$. On the other hand, suppose T ends in a domino: $\underbrace{\text{subtiling}}_{\text{length } n-2}\boxed{\ 1\ }$. The sum of the weights of such tilings is s_{n-2}. Thus the sum of the weights of all tilings of the board is $s_n = 2xs_{n-1} + s_{n-2}$. This recurrence, coupled with the initial conditions, implies that $s_n = p_{n+1}$, where $n \geq 0$.

15. Consider a $1 \times 2n$ board. The sum of the weights of its tilings is f_{2n+1}. One of the tilings consists of dominoes. So the sum of the weights of its tilings that contain at least one square is $f_{2n+1} - 1$.
Since the length of the board is even, each tiling must contain an even number of squares. So the last square must occur at an even-numbered cell, say, $2k$. Such a tiling has the form $\underbrace{\text{subtiling}}_{\text{length } 2k-1}\square\underbrace{\text{subtiling}}_{\text{all dominoes}}$. Since square tiles

occur in x colors, the sum of the weights of such tilings is xf_{2k}. So the sum of the weights of all tilings containing at least one square is $\sum_{k=1}^{n} xf_{2k}$. The given formula now follows by equating the two sums.

17. Consider a circular tiling with n cells. By Theorem 33.5, the sum of the weights of its tilings is q_n.
Suppose an arbitrary tiling T contains exactly k dominoes. Then T contains $n-2k$ square tiles. Suppose a domino occupies cells n and 1. Since there are $n-k-1$ tiles covering cells 2 through $n-1$, the remaining $k-1$ cells can be placed among them in $\binom{n-k-1}{k-1}$ different ways. There are $\binom{n-k-1}{k-1}$ such bracelets.
On the other hand, suppose a domino does not occupy cells n and 1. Then T can be considered a tiling of length n with k dominoes. This time T contains $n-k$ tiles. Consequently, there are $\binom{n-k}{k}$ such tilings.
Thus there are $\binom{n-k-1}{k-1} + \binom{n-k}{k} = \frac{n}{n-k}\binom{n-k}{k}$ bracelets, each with exactly k dominoes. The sum of the weights of all tilings is
$q_n = \sum_{k=0}^{\lfloor n/2 \rfloor} \frac{n}{n-k}\binom{n-k}{k}(2x)^{n-2k}$, as desired.

19. Consider a $2 \times n$ board. The sum of the weights of all tilings of the board is f_{n+1}.
Let T be an arbitrary tiling. Clearly, horizontal dominoes in the tiling must occur in pairs, one below the other. Assume there are k such pairs; so the tiling must contain $n-2k$ vertical dominoes. Since the location of a horizontal domino uniquely determines the position of its companion, there are exactly $(n-2k)+k = n-k$ tiling positions. So the k horizontal pairs can be placed in $\binom{n-k}{k}$ different ways; such tilings contribute a total of $\binom{n-k}{k}x^{n-2k} \cdot 1^{2k} = \binom{n-k}{k}x^{n-2k}$ to the sum of the weights.
Since $0 \le k \le n$, it follows that the sum of all weights of tilings of the board is $\sum_{k=0}^{\lfloor n/2 \rfloor}\binom{n-k}{k}x^{n-2k}$. Combining this with the initial total yields the desired formula.

21. Consider a $2 \times 2n$ board. The sum of the weights of all tilings of the board is f_{2n+1}.
Let T be an arbitrary tiling. Suppose it is unbreakable at cell n. The sum of the weights of such tilings is $f_n \cdot 1 \cdot f_n = f_n^2$. On the other hand, suppose it is unbreakable at cell n. The sum of the weights of such tilings is $f_{n+1} \cdot f_{n+1} = f_{n+1}^2$.
Thus the sum of the weights of all tilings is $f_{2n+1} = f_n^2 + f_{n+1}^2$.

EXERCISES 34

1. 111111, 111121, 111211, 112111, 121111, 112121, 121121, 121211.
3. 211111, 211121, 211211, 212111, 212121.
5. $f_6 = x^5 + 4x^3 + 3x$.
7. $f_5 = x^4 + 3x^2 + 1$.
9. $|Q - \lambda I| = \begin{vmatrix} x - \lambda & 1 \\ 1 & -\lambda \end{vmatrix} = -\lambda(x - \lambda) - 1$. The solutions of the equation $|Q - \lambda I| = 0$ are $\alpha(x)$ and $\beta(x)$; so they are the eigenvalues of Q.
11. The sum of the weights of closed paths of length $2n$ originating at v_1 is f_{2n+1}.
    Let P be such an arbitrary path. Suppose it lands at v_1 after n steps. The sum of the weights of such paths is f_{n+1}^2. On the other hand, suppose P ends at v_2 after n steps. The sum of the weights of such paths is f_n^2. Thus the sum of the weights of all closed paths of length $2n$ originating at v_1 is $f_{n+1}^2 + f_n^2 = f_{2n+1}$, as desired.
13. The sum of the weights of closed paths of length $2n$ is l_{2n}.
    Case 1. Consider the closed paths P of length $2n$ originating at v_1. Path P can end at v_1 or v_2 after n steps. The sum of the weights of such paths P ending at v_1 is f_{n+1}^2 and that of such paths P ending at v_2 is f_n^2.
    Case 2. Now consider the closed paths P of length $2n$ originating at v_2. Such paths also can end at v_1 or v_2 after n steps. The sum of the weights of such paths P ending at v_1 is f_n^2 and that of such paths P ending at v_2 is f_{n-1}^2.
    Thus the grand total of the sums of all closed paths of length $2n$ is $f_{n+1}^2 + 2f_n^2 + f_{n-1}^2 = f_{n+1}^2 + 2f_n^2 + f_{n-1}(f_{n+1} - xf_n) = f_{n+1}(f_{n+1} + f_{n-1}) + f_n(2f_n - xf_{n-1}) = f_{n+1}l_n + f_nl_{n-1}$.
    Equating the two totals gives the desired result.
15.

x	1	x	1

17. 112111
19. 112111211
21. x(sum of the weights of all closed paths of even length $\leq 2n - 1$) $= l_{2n} - 2$.
23. x(sum of the weights of all closed paths of even length ≤ 6 from v_1 to v_2) $=$
    $x \sum_{k=1}^{3} f_{2k} = f_7 - 1 = x^6 + 5x^4 + 6x^2$.
25. Answer $= x \sum_{k=1}^{4} f_{2k-1} = f_8 = x^7 + 6x^5 + 10x^3 + 4x$.
27. Answer $= x \sum_{k=1}^{4} f_{2k} = f_9 - 1 = x^8 + 7x^6 + 15x^4 + 10x^2$.

EXERCISES 35

1. $g_{n-1}u^2 - (g_{n+1} - g_n)u - g_{n+1} = 0$, so $(u+1)(g_{n-1}u - g_{n+1}) = 0$. Thus $u = -1, \dfrac{g_{n+1}}{g_{n-1}}$.
3. $g_{2n-1} = xg_{2n-2} + g_{2n-3} = x(xg_{2n-3} + g_{2n-4}) + g_{2n-3} = (x^2+1)g_{2n-3} + (g_{2n-3} - g_{2n-5}) = (x^2+2)g_{2n-3} - g_{2n-5}$.
5. Using Theorem 35.1 and the Fibonacci addition formula, RHS = $(af_{n-2} + bf_{n-1})f_{m+1} + (af_{n-3} + bf_{n-2})f_m = a(f_{n-2}f_{m+1} + f_{n-3}f_m) + b(f_{n-1}f_{m+1} + f_{n-2}f_m) = af_{n+m-2} + bf_{n+m-1} = g_{n+m}$ = LHS.
7. $c(x) - d(x) = [a + (ax-b)\beta] - [a + (ax-b)\alpha] = (b - ax)\sqrt{x^2+4}$.
9. Letting $h = 1$ and $k = -1$, the result follows.
11. When $g_n = f_n$, $\mu = 1$. It then follows from Exercise 35.10 that $f_n^2 = (x^2+2)f_{n-1}^2 - f_{n-2}^2 - 2(-1)^n$.
13. When $g_n = q_n$, $\mu = -4(x^2+1)$. It then follows from Exercise 35.10 that $q_n^2 = 2(2x^2+1)q_{n-1}^2 - q_{n-2}^2 + 8(-1)^n(x^2+1)$.
15. Follows from identity (35.4) by letting $h = -k$.
17. $g_{n+3}^3 = (xg_{n+2} + g_{n+1})^3 = x^3g_{n+2}^3 + 3x^2g_{n+2}^2g_{n+1} + 3xg_{n+2}g_{n+1}^2 + g_{n+1}^3$.
19.
$$\begin{aligned}
g_n^2 g_{n+3} &= (g_{n+2} - xg_{n+1})^2(xg_{n+2} + g_{n+1})\\
g_n^2 g_{n+3} - g_{n+1}^3 &= xg_{n+2}^3 + (1-2x^2)g_{n+2}^2g_{n+1} + (x^3-2x)g_{n+2}g_{n+1}^2 + (x^2-1)g_{n+1}^3\\
&= xg_{n+2}^3 + x^2g_{n+1}^2(xg_{n+2} + g_{n+1}) - g_{n+1}^2(xg_{n+2} + g_{n+1})\\
&\quad - xg_{n+2}g_{n+1}^2 + (1-2x^2)g_{n+2}^2g_{n+1}\\
&= xg_{n+2}^3 + (x^2-1)g_{n+3}g_{n+1}^2 + (1-x^2)g_{n+2}^2g_{n+1}\\
&\quad - xg_{n+3}g_{n+2}g_{n+1}\\
&= xg_{n+2}^3 + (x^2-1)g_{n+1}[g_{n+2}^2 + \mu(-1)^{n+2}] + (1-x^2)g_{n+2}^2g_{n+1}\\
&\quad - xg_{n+2}[g_{n+2}^2 + \mu(-1)^{n+2}]\\
&= \mu(-1)^{n+1}[xg_{n+2} + (1-x^2)g_{n+1}] = \mu(-1)^{n+1}(g_{n+3} - x^2g_{n+1}).
\end{aligned}$$
21. Letting $a = n$ and $b = n-2$, identity (35.43) yields $f_{n+1}f_{n-3} - f_nf_{n-2} = (-1)^n(x^2+1)$; that is, $f_{n+1}f_{n-3} - [f_{n-1}^2 + (-1)^{n-1}] = (-1)^n(x^2+1)$. This gives the desired result.
23. By Theorem 35.1, $g_{-n} = af_{-(n+2)} + bf_{-(n+1)} = a(-1)^{n+1}f_{n+2} + b(-1)^nf_{n+1} = (-1)^{n+1}(af_{n+2} - bf_{n+1})$.
25. By Theorem 35.1 and Exercise 31.20, LHS = $(af_n + bf_{n+1}) + (af_{n-4} + bf_{n-3}) = a(f_n + f_{n-4}) + b(f_{n+1} + f_{n-3}) = (x^2+2)(af_{n-2} + bf_{n-1}) = (x^2+2)g_n$ = RHS.
27. Using Theorem 35.1 and the Fibonacci addition formula, RHS = $(af_{m-2} + bf_{m-1})f_{n+1} + (af_{m-3} + bf_{m-2})f_n = a(f_{m-2}f_{n+1} + f_{m-3}f_n) + b(f_{m-1}f_{n+1} + f_{m-2}f_n) = af_{m+n-2} + bf_{m+n-1} = g_{m+n}$ = LHS.
29. Follows by Exercises 35.27 and 35.28.

31. Follows by Exercises 35.29 and 35.30.

33. Using the identity $r^4 + s^4 + (r+s)^4 = 2(r^2 + rs + s^2)^2$ with $r = xg_n$ and $s = g_{n-1}$, $g_{n+1}^4 + x^4 g_n^4 + g_{n-1}^4 = 2(x^2 g_n^2 + xg_n g_{n-1} + g_{n-1}^2)^2 = 2[xg_n(xg_n + g_{n-1}) + g_{n-1}^2]^2 = 2(xg_{n+1}g_n + g_{n-1}^2)^2 = \text{RHS}$.

35. Using Theorem 35.1 and Exercises 31.15 and 31.27, LHS $= (af_{n-1} + bf_n)^2 + (af_{n-2} + bf_{n-1})^2 = a^2(f_{n-1}^2 + f_{n-2}^2) + b^2(f_n^2 + f_{n-1}^2) + 2abf_{n-1}(f_n + f_{n-2}) = a^2 f_{2n-3} + b^2 f_{2n-1} + 2abf_{2n-2} = \text{RHS}$.

37. By Exercise 35.36, $H_n = (F_{n+1}^2 + F_{n+2}^2)^2 = F_{2n+3}^2$.

39. LHS $= (g_{n+2} - xg_{n+1})^2(g_{n+2} + xg_{n+1})^2 + 4x^2 g_{n+2}^2 g_{n+1}^2 = (g_{n+2}^2 - x^2 g_{n+1}^2)^2 + 4x^2 g_{n+2}^2 g_{n+1}^2 = (g_{n+2}^2 + x^2 g_{n+1}^2)^2 = \text{RHS}$.

41. Since $L_{n+2}^2 + L_{n+1}^2 = 5F_{2n+3}$, the result follows from Exercise 35.39.

43. LHS $= \sum_{k=0}^{n} (g_{2k+2} - g_{2k}) = g_{2n+2} - g_0 = g_{2n+2} - (b - ax) = g_{2n+2} + ax - b =$ RHS.

45. $\Delta^2(\text{LHS}) = (\alpha^{2n} - \beta^{2n})^2 - (\alpha^{2n-1} - \beta^{2n-1})^2 - x(\alpha^{2n} - \beta^{2n})(\alpha^{2n-1} - \beta^{2n-1}) + (x^2+4) = \alpha^{4n-2}(\alpha^2 - \alpha x - 1) + \beta^{4n-2}(\beta^2 - \beta x - 1) - 4 - x^2 + x^2 + 4 = 0$. So LHS = 0 = RHS.

47. Since $p + q = r$ and $q + r = s$, $(rs - pq)^2 = [r(q+r) - q(r-q)]^2 = (r^2 + q^2)^2 = (r^2 - q^2)^2 + (2rq)^2 = (ps)^2 + (2qr)^2$.

49. Since $L_{n+2}L_{n+3} - L_n L_{n+1} = L_{n+2}(L_{n+2} + L_{n+1}) - (L_{n+2} - L_{n+1})L_{n+1} = L_{n+2}^2 + L_{n+1}^2 = 5F_{2n+3}$, the given identity follows by Exercise 35.47. (It also follows from Exercise 35.39.)

51. It follows by Exercise 35.15 that $L_{n+k}L_{n-k} - L_n^2 = (-1)^{n-k}5F_k^2$. This implies $L_{n+1}L_{n-1} + 5(-1)^n = L_n^2$ and $L_{n+2}L_{n-2} - 5(-1)^n = L_n^2$. Adding these two equations yields the desired result. (This also follows by Exercise 35.50.)

53. This follows by squaring the identities in Exercises 35.51 and 35.52, and then adding the resulting equations.

55. When $g_n = l_n$, $\mu = -(x^2+4)$. So the given identity follows from Exercise 35.54.

57. Follows from identity (35.31) by letting $k = 2$.

59. When $g_n = l_n$, $\mu = -(x^2+4)$. So the given identity follows from Exercise 35.57 with $k = 2$.

61. Let A_k = LHS and $B_k = g_{n+2k+1}$. Then $B_k - B_{k-1} = g_{n+2k+1} - g_{n+2k-1} = xg_{n+2k} = A_k - A_{k-1}$. So $A_k - B_k = A_1 - B_1 = xg_{n+2} - g_{n+3} = -g_{n+1}$. Thus $A_k = B_k - g_{n+1}$, as desired.

63. Since $f_0 = 0$, the formula follows from Exercise 35.62.

65. When $g_n = l_n$, $\mu = -(x^2+4)$. So the given identity follows from identity (35.37).

67. Follows from Exercise 35.65.

69. When $g_n = l_n$, $\mu = -(x^2+4)$. So the given identity follows from identity (35.46) by letting $h = 3$ and $k = 2$.

71. Follows from Exercise 35.69.

73. Follows from Exercise 35.69.

75. When $g_k = f_k$, equation (35.47) yields $f_{n+1}^3 + xf_n^3 - f_{n-1}^3 = xf_{n+1}^2 f_n + xf_n^3 + xf_n f_{n-1} l_n = xf_n(f_{n+1}^2 + f_n^2) + xf_{n-1}f_{2n} = x(f_n f_{2n+1} + f_{n-1}f_{2n}) = xf_{3n}$, as desired.

77. Let $g_k = f_k$. Then, using Lemma 35.3, equation (35.49) yields

$$\begin{aligned}
&f_{n+2}^3 - (x^2+2)f_n^3 + f_{n-2}^3 \\
&= (x^2+2)f_{n+2}f_n \cdot xl_n - f_{n+2}^2 f_n + xf_{n+2}f_{n+1}f_n + f_n f_{n-2} f_{n-4} \\
&= x(x^2+2)f_{n+2}f_{2n} - f_{n+2}^2 f_n + xf_{n+2}f_{n+1}f_n + f_n f_{n-2} f_{n-4} \\
&= x(x^2+2)f_{2n}(xf_{n+1} + f_n) - f_{n+2}^2 f_n + xf_{n+2}f_{n+1}f_n + f_n f_{n-2} f_{n-4} \\
&= x^2(x^2+2)f_{2n}f_{n+1} + x(x^2+2)f_n(xf_{2n-1} + f_{2n-2}) \\
&\quad - f_{n+2}^2 f_n + xf_{n+2}f_{n+1}f_n + f_n f_{n-2} f_{n-4} \\
&= x^2(x^2+2)(f_{n+1}f_{2n} + f_n f_{2n-1}) + f_n[x(x^2+2)f_{2n-2} \\
&\quad - f_{n+2}^2 + xf_{n+2}f_{n+1} + f_{n-2}f_{n-4}] \\
&= x^2(x^2+2)f_{3n} + 0 = x^2(x^2+2)f_{3n}.
\end{aligned}$$

79. Using the identities $f_{n+k} = f_n f_{k-1} + f_{n+1} f_k$ and $(-1)^k f_{n-k} = f_n f_{k+1} - f_{n+1} f_k$, and Lemma 35.4, we have

$$\begin{aligned}
f_{n+k}^3 + (-1)^{3k} f_{n-k}^3 &= (f_n f_{k-1} + f_{n+1} f_k)^3 + (f_n f_{k+1} - f_{n+1} f_k)^3 \\
&= f_n^3(f_{k+1}^3 + f_{k-1}^3) + 3f_n f_{n+1} f_{k-1} f_k (f_n f_{k-1} + f_{n+1} f_k) \\
&\quad -3 f_n f_{n+1} f_k f_{k+1}(f_n f_{k+1} - f_{n+1} f_k)
\end{aligned}$$

$$\begin{aligned}
f_{n+k}^3 + (-1)^k f_{n-k}^3 &= f_n^3(f_{k+1} + f_{k-1})(f_{k+1}^2 - f_{k+1}f_{k-1} + f_{k-1}^2) \\
&\quad - 3f_n^2 f_{n+1} f_k (f_{k+1}^2 - f_{k-1}^2) + 3f_n f_{n+1}^2 f_k^2 (f_{k+1} + f_{k-1}) \\
&= f_n^3 l_k[(f_{k+1} - f_{k-1})^2 + f_{k+1} f_{k-1}] - 3f_n^2 f_{n+1} f_k (f_{k+1}^2 - f_{k-1}^2) \\
&\quad + 3 f_n f_{n+1}^2 f_k^2 l_k \\
&= f_n^3 l_k (x^2 f_k^2 + f_{k+1} f_{k-1}) - 3 f_n^2 f_{n+1} f_k \cdot l_k \cdot x f_k + 3 f_n f_{n+1}^2 f_k^2 l_k \\
&= f_n^3 l_k[(x^2+1) f_k^2 + (-1)^k] + 3 f_n f_{n+1} f_k^2 l_k (f_{n+1} - x f_n) \\
&= f_n^3 l_k[(x^2+1) f_k^2 + (-1)^k] + 3 f_{n-1} f_n f_{n+1} f_k^2 l_k \\
&= f_k^2 l_k[(x^2+1) f_n^3 + 3 f_{n-1} f_n f_{n+1}] + (-1)^k f_n^3 l_k \\
&= f_k f_{2k} f_{3n} + (-1)^k f_n^3 l_k.
\end{aligned}$$

This yields the desired result.

81. $g_{2n} = xg_{2n-1} + g_{2n-2} = x(xg_{2n-2} + g_{2n-3}) + g_{2n-2} = (x^2+1)g_{2n-2} + (g_{2n-2} - g_{2n-4}) = (x^2+2)g_{2n-2} - g_{2n-4}$.

83. $g_{n+3}^2 = (xg_{n+2} + g_{n+1})^2 = x^2g_{n+2}^2 + g_{n+1}^2 + xg_{n+1}(xg_{n+1} + g_n) + g_{n+2}(g_{n+2} - g_n) = (x^2+1)(g_{n+2}^2 + g_{n+1}^2) - g_n(g_{n+2} - xg_{n+1}) = (x^2+1)(g_{n+2}^2 + g_{n+1}^2) - g_n^2$.

85. Using Exercise 35.83, $p_{n+3}^2 = (4x^2+1)p_{n+2}^2 + (4x^2+1)p_{n+1}^2 - p_n^2$.

87. $Q_{n+3}^2 = 5Q_{n+2}^2 + 5Q_{n+1}^2 - Q_n^2$.

89. Using the recurrence $F_n^2 = 2F_{n-1}^2 + 2F_{n-2}^2 - F_{n-3}^2$, it follows that

$$\frac{t-t^2}{1-2t-2t^2+t^3} = \sum_{n=0}^{\infty} F_n^2 t^n.$$

91.
$$\begin{aligned}
z_{n+4} = x_{n+4}y_{n+4} &= (px_{n+3} + qx_{n+2})(ry_{n+3} + sy_{n+2}) \\
&= prz_{n+3} + qsz_{n+2} + psx_{n+3}y_{n+2} + qrx_{n+2}y_{n+3} \\
&= prz_{n+3} + qsz_{n+2} + psy_{n+2}(px_{n+2} + qx_{n+1}) + qrx_{n+2}(ry_{n+2} + sy_{n+1}) \\
&= prz_{n+3} + (p^2s + qr^2 + qs)z_{n+2} + pqsx_{n+1}(ry_{n+1} + sy_n) \\
&\quad + qrsx_{n+2}\left(\frac{y_{n+2} - sy_n}{r}\right) \\
&= prz_{n+3} + (p^2s + qr^2 + 2qs)z_{n+2} + pqrsz_{n+1} + pqs^2x_{n+1}y_n \\
&\quad - qs^2y_n(px_{n+1} + qx_n) \\
&= prz_{n+3} + (p^2s + qr^2 + 2qs)z_{n+2} + pqrsz_{n+1} - q^2s^2z_n.
\end{aligned}$$

93. Let $z_n(t) = p_n^2(t)$. It then follows from Exercise 35.82 that $z_{n+4} = 4x^2z_{n+3} + 2(4x^2+1)z_{n+2} + 4x^2z_{n+1} - z_n$.

95. Let $B_k = L_kP_k$. Then, by Exercise 35.94, $9\sum_{k=1}^{n} B_k = B_{n+4} - B_{n+3} - 8B_{n+2} - 10B_{n+1} + (10B_1 + 8B_2 + B_3 - B_4) = (L_{n+2}P_{n+2} - L_{n+1}P_{n+1} + L_nP_n - L_{n-1}P_{n-1}) - 6 = 3(L_{n+1}P_n + L_nP_{n+1}) - 6$. This gives the desired result.

97. Let $D_k = L_kQ_k$. Then, by Exercise 35.94, $9\sum_{k=1}^{n} D_k = D_{n+4} - D_{n+3} - 8D_{n+2} - 10D_{n+1} + (10D_1 + 8D_2 + D_3 - D_4) = (L_{n+2}Q_{n+2} - L_{n+1}Q_{n+1} + L_nQ_n - L_{n-1}Q_{n-1}) - 9 = 3(L_{n+1}Q_n + L_nQ_{n+1}) - 9$. The given result follows from this.

99. Follows from formula 35.6.

101. It follows by formula 35.6 that LHS $= \lim_{n\to\infty} \dfrac{f_{2n}}{f_{2n+1}} = \dfrac{1}{\alpha^2}$.

103. Follows by the formula $\sum_{k=1}^{n} \dfrac{4x^2}{p_{2k}p_{2k+2}} = \dfrac{p_{2n}}{p_{2n+2}}$.

105. We have $\sum_{k=1}^{n} \frac{x}{g_{2k}g_{2k+2}} + \sum_{k=1}^{n} \frac{x}{g_{2k-1}g_{2k+1}} = \sum_{k=1}^{2n} \frac{x}{g_k g_{k+2}} =$
$\sum_{k=1}^{2n} \left(\frac{1}{g_k g_{k+1}} - \frac{1}{g_{k+1}g_{k+2}} \right) = \frac{1}{g_1 g_2} - \frac{1}{g_{2n+1}g_{2n+2}}$. So LHS =
$\mu x \left(\frac{1}{g_1 g_2} - \frac{1}{g_{2n+1}g_{2n+2}} \right) - \sum_{k=1}^{n} \frac{\mu x^2}{g_{2k}g_{2k+2}} = \mu x \left(\frac{1}{g_1 g_2} - \frac{1}{g_{2n+1}g_{2n+2}} \right)$
$- \left(\frac{g_{2n}}{g_{2n+2}} - \frac{g_0}{g_2} \right) = \frac{\mu x + g_0 g_1}{g_1 g_2} - \frac{g_{2n}g_{2n+1} + \mu x}{g_{2n+1}g_{2n+2}} =$ RHS.

107. Follows from formula 35.10.

109. Follows from Exercise 35.107.

EXERCISES 36

1. LHS $= \sum_{k=0}^{n} (g_{2k+1} - g_{2k-1}) = g_{2n+1} - g_{-1} =$ RHS.

3. $\sum_{k=0}^{n} l_k = \sum_{k=0}^{n} (\alpha^k + \beta^k) = \frac{1 - \alpha^{n+1}}{1 - \alpha} + \frac{1 - \beta^{n+1}}{1 - \beta}$. So $x \sum_{k=0}^{n} l_k = (\alpha^{n+1} + \beta^{n+1}) +$ $(\alpha^n + \beta^n) + (\alpha + \beta) - 2 = l_{n+1} + l_n + x - 2$.

5. Let A_n denote the LHS and $B_n = g_{2n+2}$. Then $B_n - B_{n-1} = xg_{2n+1} =$ $A_n - A_{n-1}$. Then $A_n - B_n = A_{n-1} - B_{n-1} = A_0 - B_0 = xg_1 - g_2 = -g_0$. Thus $A_n = B_n - g_0$, as desired.

7. Follows by Exercise 36.5.

9. LHS $= x \sum_{k=1}^{n} (f_{2k}^2 + f_{2k-1}^2) = x \sum_{k=1}^{2n} f_k^2 = f_{2n} f_{2n+1}$.

11. $\Delta(\text{LHS}) = \sum_{k=0}^{n} [(\alpha x)^k - (\beta x)^k] = \frac{1 - (\alpha x)^{n+1}}{1 - \alpha x} - \frac{1 - (\beta x)^{n+1}}{1 - \beta x} =$
$\frac{\Delta(x - x^{n+1} f_{n+1} - x^{n+2} f_n)}{1 - 2x^2}$. This yields the given result.

13. Follows by PMI.

15. Follows by PMI.

17. $xR_n = x \sum_{k=0}^{n} f_{2k+1} = f_{2n+2}$, by Exercise 36.6.

19. $S_n = \sum_{j=0}^{n}(j+1)f_{2n-2j+1} = \sum_{i=0}^{n}\sum_{j=i}^{n} f_{2n-2j+1}$. By Exercise 36.9, we then have

$$x^2S_n = x\sum_{i=0}^{n}\left(\sum_{j=i}^{n} xf_{2n-2j+1}\right) = x\sum_{i=0}^{n}\sum_{j=0}^{n-i} xf_{2(n-i)-2j+1} = x\sum_{i=0}^{n} f_{2n-2i+2}$$
$$= \sum_{i=0}^{n}(f_{2n-2i+3} - f_{2n-2i+1}) = f_{2n+3} - f_1 = f_{2n+3} - 1.$$

21. Follows by Exercise 36.7.

23. $S_n = \sum_{j=0}^{n}(j+1)l_{2n-2j+1} = \sum_{i=0}^{n}\sum_{j=i}^{n} l_{2n-2j+1}$. By Exercise 36.9, we then have

$$x^2S_n = x\sum_{i=0}^{n}\left(\sum_{j=i}^{n} xl_{2n-2j+1}\right) = x\sum_{i=0}^{n}\sum_{j=0}^{n-i} xl_{2(n-i)-2j+1} = x\sum_{i=0}^{n}(l_{2n-2i+2} - 2)$$
$$= \sum_{i=0}^{n}(l_{2n-2i+3} - l_{2n-2i+1}) - 2(n+1)x = l_{2n+3} - l_1 - 2(n+1)x$$
$$= l_{2n+3} - (2n+3)x.$$

25. Follows by Exercise 36.1.

27. $S_n = \sum_{j=0}^{n}(j+1)f_{2n-2j} = \sum_{i=0}^{n}\sum_{j=i}^{n} f_{2n-2j} = \sum_{i=0}^{n}\sum_{j=0}^{n-i} f_{2(n-i)-2j}$. Then

$$x^2S_n = x\sum_{i=0}^{n}\left[\sum_{j=0}^{n-i} xf_{2(n-i)-2j}\right] = x\sum_{i=0}^{n}[f_{2(n-i)+1} - 1]$$
$$= \sum_{i=0}^{n} xf_{2n-2i+1} - (n+1)x = \sum_{i=0}^{n}(f_{2n-2i+2} - f_{2n-2(i+1)+2}) - (n+1)x$$
$$= f_{2n+2} - f_0 - (n+1)x = f_{2n+2} - (n+1)x.$$

29. Follows by Exercise 36.8.

31. $S_n = \sum_{j=0}^{n}(j+1)l_{2n-2j} = \sum_{i=0}^{n}\sum_{j=i}^{n} l_{2n-2j} = \sum_{i=0}^{n}\sum_{j=0}^{n-i} l_{2(n-i)-2j}$. Then

$$x^2S_n = x\sum_{i=0}^{n}\left[\sum_{j=0}^{n-i} xl_{2(n-i)-2j}\right] = x\sum_{i=0}^{n}[l_{2(n-i)+1} + x]$$
$$= \sum_{i=0}^{n} xl_{2n-2i+1} + (n+1)x^2 = \sum_{i=0}^{n}(l_{2n-2(i-1)} - l_{2n-2i}) + (n+1)x^2$$
$$= l_{2n+2} - l_0 + (n+1)x^2 = l_{2n+2} + (n+1)x^2 - 2.$$

33. The formula works when $n = 0$ and $n = 1$. Suppose it works for an arbitrary integer $n \geq 0$.
Let $n+1$ be odd. Then, by Exercises 31.32, 31.46, and 31.50, we have

$$x\sum_{k=1}^{n+1} l_{2k-1} = x\sum_{k=1}^{n} l_{2k-1} + xl_{2n+1} = (x^2+4)f_n^2 + xl_{2n+1}$$
$$= (2l_{2n} - l_n^2) + xl_{2n+1} = (l_{2n} + l_{2n+2}) - l_n^2$$
$$= (x^2+4)f_{2n+1} - l_n^2 = l_{n+1}^2.$$

On the other hand, let $n+1$ be even. Then, by Exercises 31.48 and 31.49,

$$\begin{aligned} x\sum_{k=1}^{n+1} l_{2k-1} &= x\sum_{k=1}^{n} l_{2k-1} + xl_{2n+1} = l_n^2 + xl_{2n+1} \\ &= (l_{2n} + xl_{2n+1}) + 2(-1)^n = l_{2n+2} + 2(-1)^n \\ &= (x^2+4)f_{n+1}^2. \end{aligned}$$

So the formula works for $n+1$ also. Thus, by PMI, it works for all $n \ge 0$.

35. $\text{LHS} = \sum_{k=0}^{n}(l_{2k+1} - l_{2k-1}) = l_{2n+1} - l_{-1} = l_{2n+1} + x = \text{RHS}.$

37. By the Catalan-like identity $g_{k+r}g_{k-r} - g_r^2 = (-1)^{k+r+1}\mu f_r^2$ and Exercise 36.62,

$$\begin{aligned} \text{LHS} &= x\sum_{k=0}^{n} g_k^2 - \mu x^3 \sum_{k=0}^{n}(-1)^k \\ &= (g_n g_{n+1} - g_{-1}g_0) - \mu x^3 \begin{cases} 0 & \text{if } n \text{ is odd} \\ 1 & \text{otherwise.} \end{cases} \end{aligned}$$

This gives the desired result.

39. $$\begin{aligned} (x^2+4)\sum_{k=0}^{n} f_k^2 &= \sum_{k=0}^{n}(\alpha^k - \beta^k)^2 = \sum_{k=0}^{n} l_{2k} - 2\sum_{k=0}^{n}(-1)^k \\ &= \left(\frac{1}{x}l_{2n+1} + 1\right) - 2\begin{cases} 0 & \text{if } n \text{ is odd} \\ 1 & \text{otherwise.} \end{cases} \end{aligned}$$

This yields the given formula.

41. The formula works when $n = 1$ and $n = 2$. Suppose it works for $n-1$. Since $x\sum_{k=1}^{n} f_k^2 = f_n f_{n+1}$,

$$\begin{aligned} x^2\sum_{k=1}^{2n-1}(2n-k)f_k^2 &= x^2\sum_{k=1}^{2n-3}(2n-k-2)f_k^2 + 2x^2\sum_{k=1}^{2n-2} f_k^2 + x^2 f_{2n-1}^2 \\ &= f_{2n-2}^2 + 2xf_{2n-2}f_{2n-1} + x^2 f_{2n-1}^2 = (xf_{2n-1} + f_{2n-2})^2 = f_{2n}^2. \end{aligned}$$

Thus, by PMI, the formula works for all $n \ge 1$.

43. $$\begin{aligned} \text{LHS} &= \sum_{k=0}^{n} l_{k+1}(xl_k) = \sum_{k=0}^{n} l_{k+1}(l_{k+1} - l_{k-1}) = \sum_{k=0}^{n} l_{k+1}^2 - \sum_{k=0}^{n}[l_k^2 - (-1)^k\Delta^2] \\ &= \sum_{k=0}^{n}\left(l_{k+1}^2 - l_k^2\right) + \Delta^2\sum_{k=0}^{n}(-1)^k = (l_{n+1}^2 - 4) + \Delta^2 \begin{cases} 0 & \text{if } n \text{ is odd} \\ 1 & \text{otherwise.} \end{cases} \\ &= \text{RHS}. \end{aligned}$$

45. $$\begin{aligned} \text{LHS} &= x\sum_{k=1}^{n}\left[f_{k+1}^2 - (-1)^k\right] = x\left(\sum_{k=1}^{n+1} f_k^2 - 1\right) - x\sum_{k=1}^{n}(-1)^k \\ &= f_{n+1}f_{n+2} - x - x\sum_{k=1}^{n}(-1)^k = \text{RHS}. \end{aligned}$$

47. $$\begin{aligned}\text{LHS} &= x\sum_{k=1}^{n}\left[l_{k+1}^2-(-1)^k\Delta^2\right] = x\sum_{k=1}^{n} l_{k+1}^2 - x\Delta^2\sum_{k=1}^{n}(-1)^k\\ &= x\sum_{k=0}^{n+1} l_k^2 - x^2 - x\Delta^2\sum_{k=1}^{n}(-1)^k\\ &= l_{2n+3} + \begin{cases}0 & \text{if } n+1 \text{ is odd}\\ 2x & \text{otherwise}\end{cases} - x^2 - x\Delta^2\begin{cases}-1 & \text{if } n \text{ is odd}\\ 0 & \text{otherwise}\end{cases}\\ &= \begin{cases} l_{2n+3}+2x-x^2+x(x^2+4) & \text{if } n \text{ is odd}\\ l_{2n+3}+0-x^2-0 & \text{otherwise}\end{cases} = \text{RHS}.\end{aligned}$$

49. Using formula (36.18), $4\sum_{k=1}^{n} kP_k = (2n-1)P_{n+1}+2n-2P_n-P_{n-1}+2 = (2n-1)(P_{n+1}+P_n)-(P_n+P_{n-1})+2 = (2n-1)Q_{n+1}-Q_n+2 = 2nQ_{n+1}-(Q_{n+1}+Q_n)+2 = 2nQ_{n+1}-2P_n+2$. This yields the desired result.

51. Follows by replacing x with $2x$ in Exercise 36.50.

53. Using Exercise 36.44, LHS $= (n-1)L_{n+1}+(n-2)L_n-L_{n-1}+4 = n(L_{n+1}+L_n)-[(L_{n+1}+L_n)+(L_n+L_{n-1})]+4 = nL_{n+2}-L_{n+3}+4$.

55. Using Exercise 36.46, $4\sum_{k=1}^{n} kQ_k = (2n-1)Q_{n+1}+(2n-2)Q_n-Q_{n-1}+2 = 2n(Q_{n+1}+Q_n)-[(Q_{n+1}+Q_n)+(Q_n+Q_{n-1})]+2 = 2n\cdot 2P_{n+1}-(2P_{n+1}+2P_n)+2 = 4nP_{n+1}-2Q_{n+1}+2$. This gives the desired result.

57. Let $C_n = x^2\sum_{k=1}^{n} kl_k$ and $D_n = x^2\sum_{k=1}^{n}(n-k+1)l_k$. By formulas (36.2) and (36.19), we then have

$$\begin{aligned}C_n+D_n &= (n+1)x^2\sum_{k=1}^{n} l_k = (n+1)x(l_n+l_{n+1}+x-2)\\ D_n &= [(n+1)x(l_n+l_{n+1}+x-2)]\\ &\quad -[(nx-1)l_{n+1}+(nx-2)l_n-l_{n-1}+4]\\ &= (x+1)l_{n+1}+(x+2)l_n+l_{n-1}+(n+1)(x^2-2x)-4.\end{aligned}$$

59. Since $l_1 = x$ and $l_0 = 2$, the given result follows from formula (36.20).

61. Follows from Exercise 36.51 by letting $x = 1$.

63. By Example 36.3 and formula (36.20),

$$\begin{aligned}2x^2\sum_{k=0}^{n} kg_{2k+1} &= x^2\sum_{k=0}^{n}(2k+1)g_{2k+1} - x^2\sum_{k=0}^{n} g_{2k+1}\\ &= [(2n+1)xg_{2n+2}-2g_{2n+1}+2g_1-xg_0]-x(g_{2n+2}-g_0)\\ &= 2nxg_{2n+2}-2g_{2n+1}+2g_1.\end{aligned}$$

This yields the given result.

65. Follows from Exercise 36.55.

67. Follows from Exercise 36.57.

69. Follows from Exercise 36.60.

71. Follows from Exercise 36.61.

73. $\text{LHS} = \Delta(\text{LHS}) = \alpha^{a+2b} - \beta^{a+2b} - \alpha^a + \beta^a = \begin{cases} \Delta l_b f_{a+b} & \text{if } b \text{ is odd} \\ \Delta f_b l_{a+b} & \text{otherwise.} \end{cases}$

This gives the desired result.

75. Since $\alpha^2 = \alpha x + 1$ and $\beta^2 = \beta x + 1$,

$$\begin{aligned}
\Delta^2 \sum_{k=1}^{n} f_k f_{k+c} &= \sum_{k=1}^{n} (\alpha^k - \beta^k)(\alpha^{k+c} - \beta^{k+c}) \\
&= \sum_{k=1}^{n} [\alpha^{2k+c} + \beta^{2k+c} - (-1)^k \alpha^c - (-1)^k \beta^c] \\
&= \alpha^{c+2}\left(\frac{\alpha^{2n} - 1}{\alpha^2 - 1}\right) + \beta^{c+2}\left(\frac{\beta^{2n} - 1}{\beta^2 - 1}\right) - l_c \sum_{k=1}^{n} (-1)^k \\
x\Delta^2 \sum_{k=1}^{n} f_k f_{k+c} &= \alpha^c(\alpha^{2n+1} - \alpha) + \beta^c(\beta^{2n+1} - \beta) - x l_c \sum_{k=1}^{n} (-1)^k \\
&= (\alpha^{2n+c+1} + \beta^{2n+c+1}) - (\alpha^{c+1} + \beta^{c+1}) - x l_c \sum_{k=1}^{n} (-1)^k \\
&= l_{2n+c+1} - l_{c+1} - x l_c \begin{cases} -1 & \text{if } n \text{ is odd} \\ 0 & \text{otherwise} \end{cases} \\
&= \begin{cases} l_{2n+c+1} - l_{c-1} & \text{if } n \text{ is odd} \\ l_{2n+c+1} - l_{c+1} & \text{otherwise} \end{cases} \\
&= \begin{cases} l_{2n+c+1} + l_{c+1} - \Delta^2 f_c & \text{if } n \text{ is odd} \\ \Delta^2 f_n f_{n+c+1} & \text{otherwise} \end{cases} \\
&= \begin{cases} \Delta^2 f_n f_{n+c+1} - \Delta^2 f_c & \text{if } n \text{ is odd} \\ \Delta^2 f_n f_{n+c+1} & \text{otherwise.} \end{cases}
\end{aligned}$$

The given result follows from this.

77. $\text{RHS} = (f_{n+1} - f_{n-1})(f_{n+1} + f_{n-1}) = x f_n \cdot l_n = x f_{2n} = \text{LHS}.$

79. Follows from Exercise 36.70.

81. Follows from Exercise 36.71.

83. Follows from Exercise 36.74.

85. $\Delta(\text{LHS}) = \sum_{k=0}^{n} \binom{n}{k} [(\alpha x)^k - (\beta x)^k] = (1 + \alpha x)^n - (1 + \beta x)^n = \alpha^{2n} - \beta^{2n} = \Delta f_{2n}$, so LHS = RHS.

87. $\text{LHS} = \sum_{k=0}^{n} \binom{n}{k} x^{n-k}[(-\alpha)^k + (-\beta)^k] = (x - \alpha)^n + (x - \beta)^n = \beta^n + \alpha^n = l_n =$ RHS.

89. The formula works when $n = 1$. Assume it works for an arbitrary integer $n \geq 1$. Then $\sum_{i=1}^{n+1} x(x^2+1)^{n+1-i} f_{2i+k-3} + (x^2+1)^{n+1} f_k =$
$(x^2+1)\left[\sum_{i=1}^{n} x(x^2+1)^{n-i} f_{2i+k-3} + (x^2+1)^n f_k\right] + x f_{2n+k-1} =$
$(x^2+1) f_{2n+k} + x f_{2n+k-1} = f_{2(n+1)+k}$. So the formula works for $n+1$. Thus, by PMI, it works for all $n \geq 1$.

91. $$\Delta^2 \sum_{k=1}^{n} f_{2k-1}^2 = \sum_{k=1}^{n} \left(\alpha^{2k-1} - \beta^{2k-1}\right)^2 = \sum_{k=1}^{n} \left(\alpha^{4k-2} - \beta^{4k-2} + 2\right)$$
$$= \beta^2 \sum_{k=0}^{n} \alpha^{4k} + \alpha^2 \sum_{k=0}^{n} \beta^{4k} + 2n - l_2$$
$$= \beta^2 \cdot \frac{\alpha^{4n+4} - 1}{\alpha^4 - 1} + \alpha^2 \cdot \frac{\beta^{4n+4} - 1}{\beta^4 - 1} + 2n - l_2$$
$$= \frac{\beta^2}{\alpha^2} \cdot \frac{\alpha^{4n+4} - 1}{\alpha^2 - \beta^2} + \frac{\alpha^2}{\beta^2} \cdot \frac{\beta^{4n+4} - 1}{\beta^2 - \alpha^2} + 2n - l_2$$
$$= \frac{\beta^4(\alpha^{4n+4} - 1)}{x\Delta} - \frac{\alpha^4(\beta^{4n+4} - 1)}{x\Delta} + 2n - l_2$$
$$= \frac{\alpha^{4n} - \beta^{4n} + \alpha^4 - \beta^4}{x\Delta} + 2n - l_2$$
LHS $= f_{4n} + f_4 + 2nx - xl_2 = f_{4n} + 2nx =$ RHS.

93. Using Exercises 36.83 and 36.84, we have $x\Delta^2 \sum_{k=1}^{j} f_{2k-1}^2 = f_{4j} + 2jx$. Then
$$x^2\Delta^2 \sum_{j=1}^{i} \sum_{k=1}^{j} f_{2k-1}^2 = x \sum_{j=1}^{i} f_{4j} + 2x^2 \sum_{j=1}^{i} j = f_{2i} f_{2i+2} + i(i+1)x^2$$
$$x^3\Delta^4 \sum_{i=1}^{n-1} \sum_{j=1}^{i} \sum_{k=1}^{j} f_{2k-1}^2 = x\Delta^2 \sum_{i=1}^{n-1} f_{2i} f_{2i+2} + x^3\Delta^2 \sum_{i=1}^{n-1} (i^2 + i)$$
$$= f_{4n} - nx(x^2+2) + x^3\Delta^2 \left[\frac{(n-1)n(2n-1)}{6} + \frac{(n-1)n}{2}\right]$$
$$= f_{4n} - nx(x^2+2) + (n-1)n(n+1)x^3\Delta^2/3.$$
This gives the desired result.

95. $$\Delta(\text{LHS}) = \sum_{k=0}^{n} \binom{n}{k} \left(\alpha^{4mk} - \beta^{4mk}\right) = (1 + \alpha^{4m})^n - (1 + \beta^{4m})^n$$
$$= \alpha^{2mn}(\alpha^{2m} + \beta^{2m})^n - \beta^{2mn}(\beta^{2m} + \alpha^{2m})^n$$
$$= (\alpha^{2m} + \beta^{2m})^n \left(\alpha^{2mn} - \beta^{2mn}\right).$$
This yields the given formula.

97. By the given hint, $a(\alpha) + a(\beta) = b(\alpha) + b(\beta)$. Then

$$\text{LHS} = \sum_{k=0}^{n} \binom{n}{k}\binom{n+k}{k}\left[(\alpha-1)^{n-k} + (\beta-1)^{n-k}\right]$$
$$= \sum_{k=0}^{n} \binom{n}{k}\binom{n+k}{k}\left[(-\beta)^{n-k} + (-\alpha)^{n-k}\right]$$
$$= \sum_{k=0}^{n} (-1)^{n-k}\binom{n}{k}\binom{n+k}{k} L_{n-k} = \text{RHS}.$$

99. By the given hint, $a(\alpha^2) + a(\beta^2) = b(\alpha^2) + b(\beta^2)$. Then

$$\sum_{k=0}^{n} \binom{n}{k}^2 (\alpha^{2k} + \beta^{2k}) = \sum_{k=0}^{n} \binom{n}{k}\binom{n+k}{k}\left[(\alpha^2-1)^{n-k} + (\beta^2-1)^{n-k}\right]$$
$$\sum_{k=0}^{n} \binom{n}{k}^2 L_{2k} = \sum_{k=0}^{n} \binom{n}{k}\binom{n+k}{k}(\alpha^{n-k} + \beta^{n-k}) = \sum_{k=0}^{n} \binom{n}{k}\binom{n+k}{k} L_{n-k}.$$

101. Notice that $F_{3n+3}^4 - F_{3n}^4 = (F_{3n+3} - F_{3n})(F_{3n+3} + F_{3n})(F_{3n+3}^2 + F_{3n}^2) = 2F_{3n+1} \cdot 2F_{3n+2} \cdot 2F_{6n+3}$; so $F_{3n+3}^4 = F_{3n}^4 + 8F_{3n+1}F_{3n+2}F_{6n+3}$. The given formula works when $n = 0$ and $n = 1$. Assume it works for $n-1$. Then $8\sum_{k=0}^{n} F_{3k+1}F_{3k+2}F_{6k+3} = F_{3n}^4 + 8F_{3n+1}F_{3n+2}F_{6n+3} = F_{3n+3}^4$. So the formula works for n also. Thus, by PMI, it works for every $n \ge 0$.

103. Let $a(t) = e^{\alpha xt} + e^{\beta xt} = b(t)$. Then $\left(e^{2\alpha xt} + e^{2\beta xt}\right) + 2e^{x^2 t} =$
$$\sum_{n=0}^{\infty}\left[x^n \sum_{k=0}^{n}\binom{n}{k} l_k l_{n-k}\right]\frac{t^n}{n!};\text{ that is, } \sum_{n=0}^{\infty} x^n(2^n l_n + 2x^n)\frac{t^n}{n!}$$
$$= \sum_{n=0}^{\infty}\left[x^n \sum_{k=0}^{n}\binom{n}{k} l_k l_{n-k}\right]\frac{t^n}{n!}.$$ The given formula follows from this.

105. Let $a(t) = \dfrac{e^{\alpha^2 t} - e^{\beta^2 t}}{\alpha - \beta}$ and $b(t) = e^{-t}$. Then $\dfrac{(e^{\alpha^2 t} - e^{\beta^2 t})e^{-t}}{\alpha-\beta} =$
$$\sum_{n=0}^{\infty}\left[\sum_{k=0}^{n}(-1)^{n-k}\binom{n}{k} f_{2k}\right]\frac{t^n}{n!}.\text{ But } \frac{(e^{\alpha^2 t} - e^{\beta^2 t})e^{-t}}{\alpha-\beta} = \frac{e^{(\alpha^2-1)t} - e^{(\beta^2-1)t}}{\alpha-\beta} =$$
$$\frac{e^{\alpha xt} - e^{\beta xt}}{\alpha-\beta} = \sum_{n=0}^{\infty} x^n f_n \frac{t^n}{n!}.\text{ Thus } \sum_{n=0}^{\infty} x^n f_n \frac{t^n}{n!} = \sum_{n=0}^{\infty}\left[\sum_{k=0}^{n}(-1)^{n-k}\binom{n}{k} f_{2k}\right]\frac{t^n}{n!}.$$
This gives the desired result.

107. With $a(t) = e^{\alpha t} + e^{\beta t}$ and $b(t) = e^{-xt}$, $a(t)b(t) = e^{(\alpha-x)t} + e^{(\beta-x)t} =$
$$e^{-\alpha t} + e^{-\beta t} = \sum_{n=0}^{\infty}(-1)^n l_n \frac{t^n}{n!}.\text{ But } a(t)b(t) = \sum_{n=0}^{\infty}\left[\sum_{k=0}^{n}(-1)^{n-k}\binom{n}{k} l_k x^{n-k}\right].$$
$$\text{Thus } \sum_{n=0}^{\infty}\left[\sum_{k=0}^{n}(-1)^{n-k}\binom{n}{k} l_k x^{n-k}\right] = \sum_{n=0}^{\infty}(-1)^n l_n \frac{t^n}{n!}.$$ This implies the given result.

109. We have $e^{\alpha xt}+e^{\beta xt}=\sum_{n=0}^{\infty} l_n x^n \frac{t^n}{n!}$. Let $a(t)=\frac{d^r}{dt^r}e^{\alpha xt}+e^{\beta xt}$ and $b(t)=e^t$. Then $a(t)=\sum_{n=0}^{\infty} l_{n+r}x^{n+r}\frac{t^n}{n!}$. So $a(t)b(t)=\sum_{n=0}^{\infty}\left[\sum_{k=0}^{n}\binom{n}{k} l_{k+r}x^{k+r}\right]\frac{t^n}{n!}$. We also have $a(t)b(t)=x^r\left[\alpha^r e^{(\alpha x+1)t}+\beta^r e^{(\beta x+1)t}\right]=x^r\left(\alpha^r e^{\alpha^2 t}+\beta^r e^{\beta^2 t}\right)=x^r\sum_{n=0}^{\infty} l_{2n+r}\frac{t^n}{n!}$. The given formula now follows by equating the two sums for $a(t)b(t)$.

111. Let $a(t)=\frac{d^r}{dt^r}(e^{\alpha t}+e^{\beta t})=\sum_{n=0}^{\infty} l_{n+r}\frac{t^n}{n!}=b(t)$. Then $a(t)b(t)=\sum_{n=0}^{\infty}\left[\sum_{k=0}^{n}\binom{n}{k} l_{k+r}l_{n-k+r}\right]\frac{t^n}{n!}$. Since $\frac{d^r}{dt^r}(e^{\alpha t}+e^{\beta t})=\alpha^r e^{\alpha t}+\beta^r e^{\beta t}$, we also have $a(t)b(t)=(\alpha^r e^{\alpha t}+\beta^r e^{\beta t})^2=\alpha^{2r}e^{2\alpha t}+\beta^{2r}e^{2\beta t}+2(-1)^r e^{xt}=\sum_{k=0}^{n}\left[2^n(\alpha^{n+2r}+\beta^{n+2r})+2(-1)^r x^n\right]\frac{t^n}{n!}=\sum_{k=0}^{n}\left[2^n l_{n+2r}+2(-1)^r x^n\right]\frac{t^n}{n!}$. The given result now follows by equating the two values of $a(t)b(t)$.

113. We have $\frac{e^{\alpha^2 t}-e^{\beta^2 t}}{\Delta}=\sum_{n=0}^{\infty} f_{2n}\frac{t^n}{n!}$. Then $a(t)=\frac{d^r}{dt^r}\left(\frac{e^{\alpha^2 t}-e^{\beta^2 t}}{\Delta}\right)$ $=\frac{1}{\Delta}(\alpha^{2r}e^{\alpha^2 t}-\beta^{2r}e^{\beta^2 t})=\sum_{n=0}^{\infty} f_{2n+2r}\frac{t^n}{n!}$. With $b(t)=e^{-t}$, $a(t)b(t)=\sum_{n=0}^{\infty}\left[\sum_{k=0}^{n}(-1)^{n-k}\binom{n}{k} f_{2k+2r}\right]\frac{t^n}{n!}$. We also have $\Delta a(t)b(t)=\alpha^{2r}e^{(\alpha^2-1)t}-\beta^{2r}e^{(\beta^2-1)t}=\alpha^{2r}e^{\alpha xt}-\beta^{2r}e^{\beta xt}$. So $a(t)b(t)=\sum_{n=0}^{\infty} f_{n+2r}x^n\frac{t^n}{n!}$. The desired formula now follows by equating the two values of $a(t)b(t)$.

115.
$$\frac{xg_n}{g_{n-1}g_{n+1}}=\frac{1}{g_{n-1}}-\frac{1}{g_{n+1}}=\left(\frac{1}{g_{n-1}}-\frac{1}{g_n}\right)+\left(\frac{1}{g_n}-\frac{1}{g_{n+1}}\right)$$
$$\sum_{n=2}^{m}\frac{xg_n}{g_{n-1}g_{n+1}}=\sum_{n=2}^{m}\left[\left(\frac{1}{g_{n-1}}-\frac{1}{g_n}\right)+\left(\frac{1}{g_n}-\frac{1}{g_{n+1}}\right)\right]$$
$$=\left(\frac{1}{g_1}-\frac{1}{g_m}\right)+\left(\frac{1}{g_2}-\frac{1}{g_{m+1}}\right)$$
$$\sum_{n=2}^{\infty}\frac{xg_n}{g_{n-1}g_{n+1}}=\frac{1}{a}+\frac{1}{b}-0-0=\text{RHS}.$$

117.
$$\begin{aligned}g_{n+6}&=xg_{n+5}+g_{n+4}=x(xg_{n+4}+g_{n+3})+g_{n+4}=(x^2+1)g_{n+4}+xg_{n+3}\\&=(x^2+1)g_{n+4}+x(xg_{n+2}+g_{n+1})\\&=(x^2+1)g_{n+4}+x^2g_{n+2}+(g_{n+2}-g_n)\\&=(x^2+1)[(x^2+2)g_{n+2}-g_n]+(x^2+1)g_{n+2}-g_n=\text{RHS}.\end{aligned}$$

119. Let $r = \alpha$, $s = \beta$, $A = \dfrac{1}{\alpha - \beta} = -B$, and $k = 4$. Then $S_n = F_n$, $1 + \dfrac{r}{k} = 2\sqrt{5} - 1$, $1 - \dfrac{r}{k} = \dfrac{7 - \sqrt{5}}{8}$, $1 + \dfrac{k}{s} = -(2\sqrt{5} + 1)$, and $1 - \dfrac{s}{k} = \dfrac{7 + \sqrt{5}}{8}$. Then, by equation (36.39), $\displaystyle\sum_{n=1}^{\infty} \frac{2n+1}{n(n+1)4^n} F_n = -\frac{2\sqrt{5} - 1}{\sqrt{5}} l_n \left(\frac{7 - \sqrt{5}}{8}\right) - \frac{2\sqrt{5} + 1}{\sqrt{5}} l_n \left(\frac{7 + \sqrt{5}}{8}\right) = 2l_n \frac{16}{11} + \frac{1}{\sqrt{5}} l_n \frac{27 - 7\sqrt{5}}{22}$.

121.
$$\begin{aligned}
\sum_{n=1}^{k} \frac{1}{f_{2n+1} f_{2n} f_{2n-1}} &= \sum_{n=1}^{k} \frac{f_{2n+1} f_{2n-1} - f_{2n}^2}{f_{2n+1} f_{2n} f_{2n-1}} = \sum_{n=1}^{k} \left(\frac{1}{f_{2n}} - \frac{f_{2n}}{f_{2n+1} f_{2n-1}} \right) \\
&= \sum_{n=1}^{k} \frac{1}{f_{2n}} - \frac{1}{x} \sum_{n=1}^{k} \frac{f_{2n+1} - f_{2n-1}}{f_{2n+1} f_{2n-1}} \\
&= \sum_{n=1}^{k} \frac{1}{f_{2n}} + \frac{1}{x} \sum_{n=1}^{k} \left(\frac{1}{f_{2n+1}} - \frac{1}{f_{2n-1}} \right) \\
&= \sum_{n=1}^{k} \frac{1}{f_{2n}} + \frac{1}{x} \left(\frac{1}{f_{2k+1}} - 1 \right).
\end{aligned}$$

So $\displaystyle\sum_{n=1}^{\infty} \frac{1}{f_{2n+1} f_{2n} f_{2n-1}} = \sum_{n=1}^{\infty} \frac{1}{f_{2n}} + \frac{1}{x}(0 - 1)$.

This gives the desired result.

123. The series $\sum\limits_{n=0}^{\infty} n[(\alpha x)^n + (\beta x)^n]$ converges to $\dfrac{\alpha x}{(1 - \alpha x)^2} + \dfrac{\beta x}{(1 - \beta x)^2} = \dfrac{x(1 + 4x - x^2)}{(1 - x - x^2)^2}$, when $|x| < \dfrac{1}{\alpha}$. Thus $\sum\limits_{n=0}^{\infty} nL_n x^n = \dfrac{x(1 + 4x - x^2)}{(1 - x - x^2)^2}$. When $x = 2/5$, this implies $\displaystyle\sum_{n=0}^{\infty} \frac{n2^n}{5^n} L_n = \frac{610}{121}$.

125. As in Exercise 36.115, $\sum\limits_{n=0}^{\infty} n[(\gamma x)^n + (\delta x)^n] = \dfrac{\gamma x}{(1 - \gamma x)^2} + \dfrac{\delta x}{(1 - \delta x)^2}$; that is, $\sum\limits_{n=0}^{\infty} nQ_n x^n = \dfrac{1 + 2x - x^2}{(1 - 2x - x^2)^2}$. Letting $x = 2/5$, this yields $\displaystyle\sum_{n=0}^{\infty} \frac{n2^n}{5^n} Q_n = 410$.

EXERCISES 37

1. Since $x^2 f_n^2 + 4 f_{n+1} f_{n-1} = x^2 f_n^2 + 4[f_n^2 + (-1)^n] = (x^2 + 4) f_n^2 + 4(-1)^n = l_n^2$, the given result follows.

3. Since $l_{n+k}l_{n-k} - l_n^2 = (-1)^{n+k}[l_k^2 - 4(-1)^k]$ and $l_k^2 - (x^2+4)f_k^2 = 4(-1)^k$, $l_{n+k}l_{n-k} - l_n^2 = (-1)^{n+k}[l_k^2 - 4(-1)^k] = (-1)^{n+k}l_k^2 - 4(-1)^n$. This gives the desired result.
5. LHS $= (\alpha^{n+k+1} + \beta^{n+k+1})^2 + (\alpha^{n-k} + \beta^{n-k})^2 = \alpha^{2n+2k+2} + \beta^{2n+2k+2} + \alpha^{2n-2k} + \beta^{2n-2k} = (\alpha^{2n+1} - \beta^{2n+1})(\alpha^{2k+1} - \beta^{2k+1}) = (x^2+4)f_{2n+1}f_{2k+1}$.
7. Using the identities $l_{n+1}^2 + l_n^2 = \Delta^2 f_{2n+1}$ and $l_{n+1}^2 - l_{n-1}^2 = x\Delta^2 f_{2n}$, LHS − RHS $= x^2(l_{n+3}^2 + l_{n+2}^2) + x^2(l_{n+2}^2 + l_{n+1}^2) + (l_{n+2}^2 - l_n^2) - (l_{n+4}^2 - l_{n+2}^2)$ $= x\Delta^2[(xf_{2n+5} - f_{2n+6}) + (xf_{2n+3} + f_{2n+2}] = x\Delta^2(-f_{2n+4} + f_{2n+4}) = 0$. So LHS = RHS.
9. Since $l_{a+b} = f_{a+1}l_b + f_a l_{b-1}$, $l_{a+b+2} = l_{(a+2)+b} = f_{a+3}l_b + f_{a+2}l_{b-1}$. Then $l_{a+b} + l_{a+b+2} = (f_{a+1} + f_{a+3})l_b + (f_a + f_{a+2})l_{b-1}$; that is, $(x^2+4)f_{a+b+1} = l_{a+2}l_b + l_{a+1}l_{b-1} = l_a l_b + l_{a+1}l_{b+1}$.
11. LHS $= (f_{n+1} - f_{n-1})(f_{n+1} + f_{n-1}) = xf_n \cdot l_n = xf_{2n}$ = RHS.
13. $$\begin{aligned}\Delta^3 f_{n+1}^3 &= \alpha^{3n+3} - 3\alpha^{2n+2}\beta^{n+1} + 3\alpha^{n+1}\beta^{2n+2} - \beta^{3n+3}\\ &= \alpha^{3n}(f_3\alpha + x) + 3(-1)^n\alpha^{n+1} - 3(-1)^n\beta^{n+1} - \beta^{3n}(f_3\beta + x);\\ \Delta^3 f_n^3 &= \alpha^{3n} - 3\alpha^{2n}\beta^n + 3\alpha^n\beta^{2n} - \beta^{3n} = \alpha^{3n} - 3(-1)^n\alpha^n + 3(-1)^n\beta^n - \beta^{3n};\\ \Delta^3 f_{n-1}^3 &= \alpha^{3n-3} - 3\alpha^{2n-2}\beta^{n-1} + 3\alpha^{n-1}\beta^{2n-2} - \beta^{3n-3}\\ &= -\alpha^{3n}(f_3\beta + x) + 3(-1)^n\alpha^{n-1} - 3(-1)^n\beta^{n-1} + \beta^{3n}(f_3\alpha + x).\\ \Delta^3(f_{n+1}^3 + xf_n^3 - f_{n-1}^3) &= \alpha^{3n}(f_3x + 3x) - \beta^{3n}(f_3x + 3x)\\ &\quad + 3(-1)^n\alpha^n\left(\alpha - x - \frac{1}{\alpha}\right) - 3(-1)^n\beta^n\left(\beta - x - \frac{1}{\beta}\right)\\ &= x\Delta^2(\alpha^{3n} - \beta^{3n}) + 0 - 0.\end{aligned}$$
This yields the given identity.
15. The FL equation is homogeneous of degree 2 in n. Since it is true when $n = 0$ and $n = \pm k$ $(\neq 0)$, the result is true by Dresel's theorem.
17. Since $2 = 2(-1)^{2n}$, the FL equation is homogeneous of degree 4 in n. It is true when $n = 0, \pm 1$. So the result follows by Dresel's theorem.
19. The FL equation is homogeneous of degree 2 in n. Since it is true when $n = 0, \pm 1$, the result is true by Dresel's theorem.
21. The FL equation is homogeneous of degree 4 in n. When $n = 0, -4$, LHS = 225 = RHS; when $n = -1, -3$, LHS = 49 = RHS; and when $n = -2$, LHS = 16 = RHS. So, the result follows by Dresel's theorem.
23. $$\begin{aligned}\text{LHS} &= (g_{n+1} - g_{n-1})(g_{n+1}^2 + g_{n+1}g_{n-1} + g_{n-1}^2) - x^3g_n^3\\ &= xg_n(g_{n+1}^2 + g_{n+1}g_{n-1} + g_{n-1}^2) - x^3g_n^3\\ &= xg_ng_{n+1}(xg_n + g_{n-1}) + xg_{n+1}g_ng_{n-1} + xg_ng_{n-1}(g_{n+1} - xg_n) - x^3g_n^3\\ &= 3xg_{n+1}g_ng_{n-1} + x^2g_n^2(g_{n+1} - g_{n-1} - x^3g_n^3 = 3xg_{n+1}g_ng_{n-1}\\ &\quad + x^3g_n^3 - x^3g_n^3 = \text{RHS}.\end{aligned}$$

25. Suppose $g_k = f_k$. Then, by Exercise 37.13, $f_{n+2}^3 = x^3 f_{n+1}^3 + f_n^3 + 3xf_{n+2}f_{n+1}f_n$. By identity (38.14), $xf_{3n} = f_{n+1}^3 + xf_n^3 - f_{n-1}^3$. The given identity follows from these two equations.

27. $$\begin{aligned} \text{LHS} &= (xg_n + g_{n-1})^2 - 4xg_ng_{n-1} = (xg_n - g_{n-1})^2 \\ &= (xg_n - g_{n-1})[x(xg_{n-1} + g_{n-2}) - g_{n-1}] \\ &= x(x^2-1)g_ng_{n-1} + x^2g_ng_{n-2} - g_{n-1}(x^2g_{n-1} + xg_{n-2} - g_{n-1}) \\ &= x(x^2-1)g_ng_{n-1} + x^2g_ng_{n-2} - x^2g_{n-1}^2 + g_{n-1}g_{n-3} \\ &= x(x^2-1)g_ng_{n-1} + x^2(xg_{n-1} + g_{n-2})g_{n-2} - x^2g_{n-1}^2 + g_{n-1}g_{n-3} \\ &= x(x^2-1)g_ng_{n-1} + x^3g_{n-1}g_{n-2} + x^2g_{n-2}^2 - x^2g_{n-1}^2 + g_{n-1}g_{n-3} \\ &= x(x^2-1)g_ng_{n-1} + x^2g_{n-2}^2 + x^2g_{n-1}(xg_{n-2} - g_{n-1}) + g_{n-1}g_{n-3} \\ &= x(x^2-1)g_ng_{n-1} + x^2g_{n-2}^2 - (x^2-1)g_{n-1}g_{n-3} \\ &= x^2g_{n-2}^2 + (x^2-1)g_{n-1}(xg_n - g_{n-3}) = \text{RHS}. \end{aligned}$$

29. $$\begin{aligned} \text{LHS} &= \begin{bmatrix} x & 1 \\ 1 & 0 \end{bmatrix}^2 - x\begin{bmatrix} x & 1 \\ 1 & 0 \end{bmatrix} - \begin{bmatrix} 1 & 0 \\ 0 & 1 \end{bmatrix} = \begin{bmatrix} x^2+1 & x \\ x & 1 \end{bmatrix} - \begin{bmatrix} x^2+1 & x \\ x & 1 \end{bmatrix} \\ &= \begin{bmatrix} 0 & 0 \\ 0 & 0 \end{bmatrix} = \text{RHS}. \end{aligned}$$

31. $$\begin{aligned} \text{Let } B &= 3x^2g_{n+3}^2g_{n+2} = 2x^2g_{n+3}^2(xg_{n+2}) + x^2g_{n+3}^2g_{n+2} \\ &= 2xg_{n+3}^2(g_{n+3} - g_{n+1}) + x^2g_{n+2}(xg_{n+2} + g_{n+1})^2 \\ &= 2xg_{n+3}^3 - 2xg_{n+3}^2g_{n+1} + x^2g_{n+2}(x^2g_{n+2}^2 + 2xg_{n+2}g_{n+1} + g_{n+1}^2) \\ &= (2xg_{n+3}^3 + x^4g_{n+2}^3) + 2x^3g_{n+2}^2g_{n+1} + x^2g_{n+2}g_{n+1}^2 \\ &\quad - 2\,xg_{n+1}(xg_{n+2} + g_{n+1})^2 \\ &= 2xg_{n+3}^3 + x^4g_{n+2}^3 - 3x^2g_{n+2}g_{n+1}^2 - 2xg_{n+1}^3 \\ &= 2xg_{n+3}^3 + x^4g_{n+2}^3 - 2xg_{n+1}^3 - 3x^2g_{n+1}^2(xg_{n+1} + g_n) \\ &= 2xg_{n+3}^3 + x^4g_{n+2}^3 - 2xg_{n+1}^3 - 3x^3g_{n+1}^3 - 3x^2g_{n+1}^2g_n. \end{aligned}$$

$$\begin{aligned} \text{Let } C &= 3xg_{n+2}^2(xg_{n+2} + g_{n+1}) = 3x^2g_{n+2}^3 + xg_{n+2}^2g_{n+1} + 2xg_{n+2}^2g_{n+1} \\ &= 3x^2g_{n+2}^3 + g_{n+2}^2(g_{n+2} - g_n) + 2xg_{n+1}(xg_{n+1} + g_n)^2 \\ &= (3x^2+1)g_{n+2}^3 - g_{n+2}^2g_n + 2xg_{n+1}(xg_{n+1} + g_n)^2 \\ &= (3x^2+1)g_{n+2}^3 - g_n(xg_{n+1} + g_n)^2 \\ &\quad + 2xG_{n+1}(x^2g_{n+1}^2 + 2xg_{n+1}g_n + g_n^2) \\ &= (3x^2+1)g_{n+2}^3 - g_n(x^2g_{n+1}^2 + 2xg_{n+1}g_n + g_n^2) + 2x^3g_{n+1}^3 \\ &\quad + 4x^2g_{n+1}^2g_n + 2xg_{n+1}g_n^2 \\ &= (3x^2+1)g_{n+2}^3 + 2x^3g_{n+1}^3 - g_n^3 + 3x^2g_{n+1}^2g_n. \end{aligned}$$

$$\begin{aligned} \text{Then } g_{n+4}^3 &= (xg_{n+3} + g_{n+2})^3 = x^3g_{n+3}^3 + g_{n+2}^3 + B + C \\ &= (x^3+2x)g_{n+3}^3 + (x^4+3x^2+2)g_{n+2}^3 - (x^3+2x)g_{n+1}^3 - g_n^3. \end{aligned}$$

33. $\text{RHS} = \frac{xf_{n-1}^*}{f_r^* f_{n-r-1}^*} f_{r+1} + \frac{xf_{n-1}^*}{f_{r-1}^* f_{n-r}^*} f_{n-r-1} = \frac{xf_{n-1}^*}{f_r^* f_{n-r}^*}(f_{n-r}f_{r+1} + f_{n-r-1}f_r)$
$= \frac{xf_{n-1}^* f_n}{f_r^* f_{n-r}^*} = \text{LHS}.$

35. Follows by adding the recurrences in Exercises 37.33 and 37.34.

37. Follows by the identity $(x+y)^5 - x^5 - y^5 = 5xy(x+y)(x^2+xy+y^2)$.

39. $b_{n+4} = 2x^2 b_{n+3} + (5x^2+2)b_{n+2} + 2x^2 b_{n+1} - b_n$, where $b_0 = 0, b_1 = 2x$, $b_2 = 4x^3 + 2x, b_3 = 8x^5 + 14x^3 + 6x$, and $n \geq 0$.

41. $d_{n+4} = 2x^2 d_{n+3} + (5x^2+2)d_{n+2} + 2x^2 d_{n+1} - d_n$, where $d_0 = 4, d_1 = 2x^2$, $d_2 = 4x^4 + 10x^2 + 4, d_3 = 8x^6 + 30x^4 + 18x^2$, and $n \geq 0$.

43. $B_{n+4} = 2B_{n+3} + 7B_{n+2} + 2B_{n+1} - B_n$, where $B_0 = 0, B_1 = 1, B_2 = 3, B_3 = 14$, and $n \geq 0$.

45. $D_{n+4} = 2D_{n+3} + 7D_{n+2} + 2D_{n+1} - D_n$, where $D_0 = 0, D_1 = 1, D_2 = 9$, $D_3 = 28$, and $n \geq 0$.

47. It follows from the identity (37.51) that

$$\begin{aligned}
g_{n+4}^2 &= x^2 g_{n+3}^2 + 2x^2 g_{n+2}^2 + 2(xg_{n+1} + g_n)^2 + x^2 g_{n+1}^2 - g_n^2 \\
&= x^2 g_{n+3}^2 + 2x^2 g_{n+2}^2 + 3x^2 g_{n+1}^2 + 4xg_{n+1}g_n + g_n^2 \\
&= x^2 g_{n+3}^2 + 2x^2 g_{n+2}^2 + 4xg_{n+1}(xg_{n+1} + g_n) - x^2 g_{n+1}^2 + g_n^2 \\
&= x^2 g_{n+3}^2 + 2x^2 g_{n+2}^2 + 4xg_{n+1}g_{n+2} - x^2 g_{n+1}^2 + g_n^2 \\
&= x^2 g_{n+3}^2 + 2x(xg_{n+2} + 2g_{n+1})g_{n+2} - x^2 g_{n+1}^2 + g_n^2 \\
&= x^2 g_{n+3}^2 + 2(g_{n+3} + g_{n+1})(g_{n+3} - g_{n+1}) - x^2 g_{n+1}^2 + g_n^2 \\
&= x^2 g_{n+3}^2 + 2(g_{n+3}^2 - g_{n+1}^2) - x^2 g_{n+1}^2 + g_n^2 = (x^2+2)g_{n+3}^2 \\
&\quad -(x^2+2)g_{n+1}^2 + g_n^2.
\end{aligned}$$

49. $\Delta(\text{RHS}) = (\alpha^{2n} - \beta^{2n})(\alpha^n + \beta^n) - (-1)^n(\alpha^n - \beta^n) = \alpha^{3n} - \beta^{3n}$, so RHS = LHS.

51. $l_{3n} = \alpha^{3n} + \beta^{3n} = [\alpha^{3n} + \beta^{3n} + (\alpha\beta)^n(\alpha^n + \beta^n)] - (-1)^n l_n$
$= (\alpha^n + \beta^n)(\alpha^{2n} + \beta^{2n}) - (-1)^n l_n = [l_{2n} - (-1)^n] l_n$.

53. $\text{LHS} = (\alpha^{n+3k} + \beta^{n+3k})(\alpha^{3k} + \beta^{3k}) - (\alpha^{n+6k} + \beta^{n+6k}) = (-1)^k(\alpha^n + \beta^n) =$ RHS.

55. Using the identity $x^5 + y^5 = (x+y)(x^4 - x^3y + x^2y^2 - xy^3 + y^4)$, $l_{5n} = (\alpha^n)^5 + (\beta^n)^5 = (\alpha^n + \beta^n)(\alpha^{4n} - \alpha^{3n}\beta^n + \alpha^{2n}\beta^{2n} - \alpha^n\beta^{3n} + \beta^{4n}) =$ $[l_{4n} - (-1)^n l_{2n} + 1] l_n$.

57. $l_{6n} = \alpha^{6n} + \beta^{6n} = (\alpha^{2n} + \beta^{2n})(\alpha^{4n} - \alpha^{2n}\beta^{2n} + \beta^{4n}) = (L_{4n} - 1)l_{2n}$.

59. Since $l_{3n} = l_{(n-1)+2n+1} = l_{n-1}f_{2n} + l_n f_{2n+1}, xl_{3n} = l_{n-1}(xf_{2n}) + xl_n f_{2n+1} =$ $l_{n-1}(f_{n+1}^2 - f_{n-1}^2) + xl_n(f_{n+1}^2 + f_n^2) = f_{n+1}^2(xl_n + l_{n-1}) + xl_n f_n^2 - l_{n-1}f_{n-1}^2 =$ RHS.

61. By the Cassini-like formula, LHS $-(x^4-1)l_n^4$
$= 2(x^2l_n^2 + l_{n+1}l_{n-1})^2 - (x^4-1)l_n^4 = (x^4+1)l_n^4 + 2l_{n+1}^2l_{n-1}^2 + 4x^2l_n^2l_{n+1}l_{n-1}$
$= l_n^4 + x^2l_n^2(x^2l_n^2 + 4l_{n+1}l_{n-1}) + 2l_{n+1}^2l_{n-1}^2$
$= l_n^4 + x^2l_n^2[(l_{n+1}-l_{n-1})^2 + 4l_{n+1}l_{n-1}] + 2l_{n+1}^2l_{n-1}^2$
$= l_n^4 + (l_{n+1}-l_{n-1})^2(l_{n+1}+l_{n-1})^2 + 2l_{n+1}^2l_{n-1}^2$
$= l_n^4 + (l_{n+1}^2 - l_{n-1}^2)^2 + 2l_{n+1}^2l_{n-1}^2$
$= l_n^4 + l_{n+1}^4 + l_{n-1}^4$. So LHS = RHS.

63. The desired recurrence is $z_n = 5z_{n-1} + 15z_{n-2} - 15z_{n-3} - 5z_{n-4} + z_{n-5}$, where $z_n = F_n^4$. The corresponding FL equation is homogeneous of degree 4 in n. When $n = 0, 1, 2, 3$, and 4, LHS = RHS. Therefore, by Dresel's theorem, the FL equation is an identity and hence the result follows.

EXERCISES 38

1. Follows by Exercise 36.108.

3. Let $x = 1$ and $y = 0$ in Theorem 38.2. Then $z = \sqrt{5}$ and $\Delta(xy/z) = 2$. Clearly, $f_n = 0$ if and only if n is odd. Let $n = 2m+1$. Then
$\text{LHS} = \sum_{k=0}^{m}\binom{2m+1}{k} f_{2m+1-2k}(1)f_{2m+1-2k}(0) = \sum_{k=0}^{m}\binom{2m+1}{k}F_{2m+1-2k}$
$= \sum_{k=0}^{m}\binom{2m+1}{m-k}F_{2k+1}$; and RHS $= (\sqrt{5})^{n-1}f_n(0) = (\sqrt{5})^{2m}f_{2m+1}(0) = 5^m$.
Thus $\sum_{k=0}^{n}\binom{2n+1}{n-k}F_{2k+1} = 5^n$, as desired.

5. Let $x = 1 = y$ in identity (38.13). Then $z = 2\sqrt{3}$, $4xy/z = 2\sqrt{3}/3$, $\Delta(4xy/z) = 4\sqrt{3}/3$, $\alpha(4xy/z) = \sqrt{3}$, and $\beta(4xy/z) = -2\sqrt{3}/6$. It then follows by identity (38.13) that $\sum_{k=0}^{\lfloor n/2\rfloor}\binom{n}{k}P_{n-2k}^2 = (2\sqrt{3})^{n-1}f_n(2\sqrt{3}/3) = 2^{n-1}[3^n - (-1)^n] = \text{RHS}$.

7. The characteristic roots are $z = \dfrac{xy-4 \pm \sqrt{(xy-4)^2 + 4(x+y)^2}}{2}$. Let $\Delta(x) = \sqrt{x^2+4}$ and $t = \dfrac{xy-4}{x+y}$. Then $\Delta(t) = \dfrac{1}{x+y}\sqrt{(xy-4)^2+4(x+y)^2}$ and $z = \dfrac{xy-4 \pm (x+y)\Delta(t)}{2}$. So the general solution of the recurrence is $A_n = A\left[\dfrac{xy-4+(x+y)\Delta(t)}{2}\right]^n + B\left[\dfrac{xy-4-(x+y)\Delta(t)}{2}\right]^n$,

where A and B are unknowns to be determined. Using the two initial conditions, we get $A = \dfrac{1}{(x+y)\Delta(t)} = -B$. Since $2\alpha(t) = t + \Delta(t)$ and $2\beta(t) = t - \Delta(t)$, it follows that $A_n = \dfrac{(x+y)^n}{x+y}\left[\dfrac{\alpha^n(t) - \beta^n(t)}{\Delta(t)}\right] = (x+y)^{n-1} f_n(t) = (x+y)^{n-1} f_n\left(\dfrac{xy-4}{x+y}\right)$.

9. $\alpha(3i) = \alpha^2 i$ and $\beta(3i) = \beta^2 i$. So
$$f_n(3i) = \frac{\alpha^n(3i) - \beta^n(3i)}{\alpha(3i) - \beta(3i)} = \frac{(\alpha^{2n} - \beta^{2n})i^n}{\sqrt{5}i} = i^{n-1}F_{2n}.$$

11. $\alpha(\sqrt{5}) = \alpha^2$ and $\beta(\sqrt{5}) = -\beta^2$. So
$$f_{2n-1}(\sqrt{5}) = \frac{\alpha^{2n-1}(\sqrt{5}) - \beta^{2n-1}(\sqrt{5})}{\Delta(\sqrt{5})} = \frac{\alpha^{4n-2} + \beta^{4n-2}}{3} = \frac{1}{3}L_{4n-2}.$$

13. Replacing x with $-x$ and letting $y = x+1$, formula (38.21) becomes
$$f_n(-x)f_n(x+1) = \sum_{k=0}^{n-1} \frac{n}{k+1}\binom{n+k}{2k+1} f_{k+1}[-(x^2+x+4)];\text{ that is,}$$
$$f_n(x)f_n(x+1) = n\sum_{k=0}^{n-1} \frac{(-1)^{n-k+1}}{k+1}\binom{n+k}{2k+1} f_{k+1}(x^2+x+4).$$

15. Follows from formula (38.21) by letting $x = 1$.

17. Follows from formula (38.23) by letting $x = 2$.

19. By Exercise 38.19, $f_{2n-1}(\sqrt{5}) = \frac{1}{3}L_{4n-2}$. Since $\alpha(4/\sqrt{5}) = \sqrt{5}$, $\beta(4/\sqrt{5}) = -1/\sqrt{5}$, and $\Delta(4/\sqrt{5}) = 6/\sqrt{5}$,
$$f_{2n-1}(4/\sqrt{5}) = \frac{(\sqrt{5})^{2n-1} - (-1/\sqrt{5})^{2n-1}}{6/\sqrt{5}} = \frac{5^{2n-1}+1}{6\cdot 5^{n-1}}.$$
Letting $x = \sqrt{5}$ and replacing n with $2n-1$ in formula (38.22), we get
$$\frac{1}{3}L_{4n-2}\cdot\frac{5^{2n-1}+1}{6\cdot 5^{n-1}} = \sum_{k=0}^{n-1}\frac{2n-1}{2k+1}\binom{2n+2k-1}{4k+1}\frac{81^k}{5^k}$$
$$l_{4n-2} = \frac{18(2n-1)}{5^{2n-1}+1}\sum_{k=0}^{n-1}\frac{1}{2k+1}\binom{2n+2k-1}{4k+1}81^k 5^{n-k-1}.$$

21. By Exercise 38.9, $f_n(3i) = i^{n-1}F_{2n}$. Formula (38.9) then yields $(-1)^{n-1}F_{2n}^2 = n\sum_{k=0}^{n-1}\dfrac{(-1)^{n-k+1}}{k+1}\dbinom{n+k}{2k+1}(-5)^k$; that is, $F_{2n}^2 = n\sum_{k=0}^{n-1}\dfrac{1}{k+1}\dbinom{n+k}{2k+1}5^k$.

23. By Exercise 38.9, $f_n(3i) = i^{n-1}F_{2n}$. Then, by formula (38.24) $(-1)^{n-1}F_{2n}^2 = n\sum_{k=0}^{n-1}\frac{1}{k+1}\binom{n+k}{2k+1}\frac{(-9)^{k+1}-(-4)^{k+1}}{-5}$. This yields the desired result.

25. Since $\alpha(4) = \alpha^3, \beta(4) = \beta^3$, and $\Delta(4) = 2\sqrt{5}$, $f_j(4) = \frac{\alpha^{3j}-\beta^{3j}}{2\sqrt{5}} = \frac{1}{2}F_{3j}$, the result follows by letting $x = 4$ in formula (38.25).

27. Follows by letting $x = 2$ in formula (38.25).

29. Since $\alpha(i) = \frac{\sqrt{3}+i}{2} = e^{\pi i/6}, \beta(i) = -\frac{\sqrt{3}-i}{2} = -e^{-\pi i/6}$, and $\Delta(i) = \sqrt{3}$, $f_j(i) = \frac{e^{j\pi i/6} - (-e^{-\pi i/6})^j}{\sqrt{3}}$. In particular, $f_{2n-1}(i) = \frac{e^{(2n-1)\pi i/6} + e^{-(2n-1)\pi i/6}}{\sqrt{3}} = \frac{2}{\sqrt{3}}\cos(2n-1)\pi/6$. Consequently, by formula (38.23), $\frac{4}{3}\cos^2(2n-1)\pi/6 = (2n-1)\sum_{k=0}^{2n-2}\frac{(-1)^k}{k+1}\binom{2n+k-1}{2k+1}3^k$. This gives the desired formula.

31. From Exercise 38.29, $f_{2n-1}(i) = \frac{2}{\sqrt{3}}\cos(2n-1)\pi/6$. Now replace x with i, and n with $2n-1$ in formula (38.24). We then get $\frac{4}{3}\cos^2(2n-1)\pi/6 = (2n-1)\sum_{k=0}^{2n-2}\frac{1}{k+1}\binom{2n+k-1}{2k+1}\frac{(-1)^{k+1}-(-4)^{k+1}}{3}$. This gives the desired result.

33. Let $x = 4$ and $y = 2$. From Exercise 38.25, $f_k(4) = \frac{1}{2}F_{3k}$. The given result now follows from formula (38.45).

35. Follows by letting $y = x$.

37. Follows from formula (38.51) by letting $x = 1$.

39. Follows from formula (38.51) by letting $x = 2$.

41. Follows from formula (38.53) by letting $x = 2$.

43. Follows from formula (38.54).

45. Follows from formula (38.55) by letting $x = 2$.

47. Follows from formula (38.65) by letting $x = 2$.

49. Follows by Exercise 31.32.

51. $$P_5 = \frac{1}{8}\sum_{\substack{0\le k\le 5\\ 2k\not\equiv 1 \pmod 4}}(-1)^{\lfloor (2-k)/2\rfloor}\binom{11}{2k+1} = 29 \text{ and}$$
$$P_6 = \frac{1}{8}\sum_{\substack{0\le k\le 6\\ 2k\not\equiv 2 \pmod 4}}(-1)^{\lfloor (5-2k)/4\rfloor}\binom{13}{2k+1} = 70.$$

53. Follows by the strong version of PMI.
55. Follows by the strong version of PMI and the Legendre recurrence. $\left(\text{It also follows from the fact that } P_n(x) = \frac{1}{2^n n!}\frac{d^n}{dx^n}(x^2-1)^n.\right)$
57. Using Exercise 38.56, $A_n = \frac{(2n)!}{2^n(n!)^2} = \frac{n+1}{2^n}\cdot\frac{(2n)!}{n!(n+1)!} = \frac{n+1}{2^n}C_n$.
59. Clearly, $P_1'(1) = t_1$ and $P_2'(1) = t_2$. Assume $P_k'(1) = t_k$ for $k = n-1$, where $n \geq 2$. Then, by Exercises 38.53 and 38.58, $P_n'(1) = P_{n-1}'(1) + nP_{n-1}(1) = t_{n-1} + n = t_n$. So, by PMI, the result is true for all $n \geq 1$.
61. Since $f_j(4) = \frac{1}{2}F_{3j}$ by Exercise 38.8, the given result follows from formula (38.109).
63. Let $x = 3i$. Then $\Delta = \sqrt{5}i$ and $f_n(3i) = i^{n-1}F_{2n}$; see Exercise 38.9. The given result now follows from formula (38.109).
65. Let $x = \sqrt{5}$. Since $\alpha(\sqrt{5}) = \alpha^2, \beta(\sqrt{5}) = -\beta^2$, and $\Delta = 3$, it follows that $l_{2n+1}(\sqrt{5}) = \sqrt{5}F_{4n+2}$. Then, by formula (38.110), $\sqrt{5}F_{4n+2} = \sqrt{5}\sum_{k=0}^{2n} P_k(3/2)P_{n-k}(3/2)$. This yields the given result.
67. Let $x = i$. Since $\alpha(i) = e^{\pi i/3}, \beta(i) = -e^{-\pi i/3}$, and $\Delta(i) = \sqrt{3}$, $f_{2n}(i) = \frac{2i}{\sqrt{3}}\sin 2n\pi/3$. So, by formula (38.109), $\frac{2i}{\sqrt{3}}\sin 2n\pi/3 = \frac{i}{\sqrt{3}}\sum_{k=0}^{2n-1} P_k(\sqrt{3}/2)P_{2n-k-1}(\sqrt{3}/2)$. The given formula now follows.

EXERCISES 39

1. $a+b+c+d = x(l_{n+3}+l_{n+1}) + (l_{n+2}+l_n) = x(\Delta^2 f_{n+2}) + \Delta^2 f_{n+1} = (x^2+4)f_{n+3}$.
3. $a-b+c-d = x(l_{n+3}+l_{n+1}) - (l_{n+2}+l_n) = x(\Delta^2 f_{n+2}) - \Delta^2 f_{n+1} = \Delta^2(xf_{n+2} - f_{n+1})$.
5. Using Exercise 31.49, LHS $= \Delta^2[f_n^2 + (-1)^n] = [l_{2n} - 2(-1)^n] + \Delta^2(-1)^n = l_{2n} + (x^2+2)(-1)^n =$ RHS.
7. LHS $= F_{n+1}f_{n-1} + [L_{2n} + 3(-1)^n] = F_n^2 + L_{2n} + 4(-1)^n$
$= f_n^2 + L_{2n} + (L_n^2 - 5F_n^2) = L_n^2 + L_{2n} - 4F_n^2 =$ RHS.
9. LHS $= 2l_n^2 + (l_n^2 + \Delta^2 f_n^2) = 2l_n^2 + 2l_{2n} =$ RHS.

11. Let D denote the given determinant. By the addition formula, we then have

$$D = \begin{vmatrix} f_{p+1}f_{2n} + f_p f_{2n-1} & f_{p+n} & f_p \\ f_{q+1}f_{2n} + f_q f_{2n-1} & f_{q+n} & f_q \\ f_{r+1}f_{2n} + f_r f_{2n-1} & f_{r+n} & f_r \end{vmatrix}$$

$$= f_{2n}\begin{vmatrix} f_{p+1} & f_{p+n} & f_p \\ f_{q+1} & f_{q+n} & f_q \\ f_{r+1} & f_{r+n} & f_r \end{vmatrix} + f_{2n-1}\begin{vmatrix} f_p & f_{p+n} & f_p \\ f_q & f_{q+n} & f_q \\ f_r & f_{r+n} & f_r \end{vmatrix}.$$

By the addition formula, column 2 of the first determinant is a linear combination of columns 1 and 3. So $D = f_{2n} \cdot 0 + f_{2n-1} \cdot 0 = 0$.

13. It follows by the identity $l_{n+3k}l_{3k} - l_{n+6k} = (-1)^k l_n$ that the rows of the determinant are linearly dependent. So the determinant is zero.

15. It follows by the identity $f_{n+3k}l_{3k} - f_{n+6k} = (-1)^k f_n$ that the rows of the determinant are linearly dependent. So the determinant is zero.

17. Let D_n denote the given determinant. By Exercise 39.10, $D_n = (-1)^n D_0$. Since $D_0 = 2x^2(x^2+4)^3$, it follows that $D_n = 2(-1)^n x^2(x^2+4)^3$.

19. Let B_n denote the given determinant. First, add $-xR_2 - R_3$ to R_1. Now multiply C_2 by x, and then subtract $C_1 - C_3$ from the resulting column. This gives $xB_n = (-2x^2) \cdot 2B_{n-2}$. So $B_n = (-4x)B_{n-2}$, where $B_1 = 2$, $B_2 = 4 - x^2$, and $n \geq 3$. It then follows inductively that $B_{2n-1} = 2(-4x)^{n-1}$ and $B_{2n} = (4 - x^2)(-4x)^{n-1}$.

21. Let $|V_n|$ denote the given determinant. We claim that $|V_n| = x^n l_n$. It is true when $n = 1, 2$, and 3. Assume it is true for all positive integers $< n$, where $n \geq 3$. Expanding $|V_n|$ by row n, $|V_n| = x^2|V_{n-1}| - (ix)|W_{n-1}|$, where

$$W_{n-1} = \begin{vmatrix} x^2 & ix & 0 & 0 & & 0 & 0 & 0 \\ ix & x^2+1 & ix & 0 & & 0 & 0 & 0 \\ 0 & ix & x^2 & ix & \cdots & 0 & 0 & 0 \\ & & & & \vdots & & & \\ 0 & 0 & 0 & 0 & \cdots & ix & x^2 & 0 \\ 0 & 0 & 0 & 0 & & 0 & ix & ix \end{vmatrix}$$

$$= (ix)\begin{vmatrix} x^2 & ix & 0 & 0 & & 0 & 0 \\ ix & x^2+1 & ix & 0 & & 0 & 0 \\ 0 & ix & x^2 & ix & \cdots & 0 & 0 \\ & & & & \vdots & & \\ 0 & 0 & 0 & 0 & \cdots & ix & x^2 \end{vmatrix} = (ix)|V_{n-2}|.$$

Thus $|V_n| = x^2|V_{n-1}| - (ix)^2|V_{n-2}| = x^n(xl_{n-1} + l_{n-2}) = x^n l_n$.

So, by PMI, $|V_n| = x^n l_n$ for every positive integer n.

23. The general solution of the recurrence is $B_n = Au^n + Bv^n$, where $u = \dfrac{c + \sqrt{c^2 - 4ab}}{2}$, $v = \dfrac{c - \sqrt{c^2 - 4ab}}{2}$, $B_0 = 1$, and $B_1 = c$. The initial conditions imply that $A = \dfrac{u}{u - v}$ and $B = -\dfrac{v}{u - v}$. Thus $B_n = \dfrac{u^{n+1} - v^{n+1}}{u - v}$, where $n \geq 0$.

25. We have $h_{ii} = 2, h_{i,i+1} = -1$, and $h_{i,j} = 1$ when $i > j$. By recurrence (39.8),
$$|H_n| = 2|H_{n-1}| + \sum_{r=1}^{n-1}\left((-1)^{n-r}h_{n,r}\prod_{j=r}^{n-1} h_{j,j+1} \cdot |H_{r-1}|\right) = 2|H_{n-1}| + \sum_{r=1}^{n-1}|H_{r-1}|.$$
In particular, $|H_4| = 2|H_3| + \sum_{r=1}^{3}|H_{r-1}| = 2 \cdot 13 + 1 + 2 + 5 = 34 = F_9$.

27. $|T_4| = (x^2 + 1)T_3 + T_0 + T_1 + T_2 = (x^2 + 1)^4 + 3(x^2 + 1)^2 + 2(x^2 + 1) + 2$.

EXERCISES 40

1. It follows from equation (40.5) that $\sum_{k=1}^{n} \tan^{-1} \dfrac{x}{f_{2k+1}} = \tan^{-1} \dfrac{1}{x} - \tan^{-1} \dfrac{1}{f_{2n+2}}$. This yields the given result.

3. Δ^2 LHS $= (\alpha + \beta) + (\alpha^3 + \beta^3) = (\alpha + \beta)(\alpha^2 - \alpha\beta + \beta^2 + 1) = x(l_2 + 2)$
$= x\Delta^2$; so LHS $= x$.

5. By Exercise 40.2 and formula (40.7), we have
$$\sum_{n=1}^{m} \tan^{-1}\frac{1}{F_{2n+1}} = \sum_{n=1}^{m}\left(\tan^{-1}\frac{1}{L_{2n}} + \tan^{-1}\frac{1}{L_{2n+2}}\right)$$
$$= \tan^{-1}\frac{1}{3} + 2\sum_{n=2}^{m}\tan^{-1}\frac{1}{L_{2n}} + \tan^{-1}\frac{1}{L_{2m+2}}$$
$$\sum_{n=1}^{\infty} \tan^{-1}\frac{1}{F_{2n+1}} = \tan^{-1}\frac{1}{3} + 2\sum_{n=2}^{\infty}\tan^{-1}\frac{1}{L_{2n}} + 0$$
$$2\sum_{n=2}^{\infty}\tan^{-1}\frac{1}{L_{2n}} = \sum_{n=1}^{\infty}\tan^{-1}\frac{1}{F_{2n+1}} - \tan^{-1}\frac{1}{3} = \frac{\pi}{4} - \tan^{-1}\frac{1}{3}$$
$$\sum_{n=2}^{\infty}\tan^{-1}\frac{1}{L_{2n}} = \frac{1}{2}\left(\tan^{-1} 1 + \tan^{-1}\frac{1}{3}\right) = \frac{1}{2}\tan^{-1} 2 = \tan^{-1}(-\beta).$$

7. Follows from Exercise 40.6.

9. Follows from Exercise 40.7.

11. Using Exercise 40.6, $\sum_{n=1}^{m} \tan^{-1} \frac{xl_{2n+1}}{l_{2n+1}^2 + x^2 + 5} =$
$\sum_{n=1}^{m} \left(\tan^{-1} \frac{1}{l_{2n}} - \tan^{-1} \frac{1}{l_{2n+2}} \right) = \tan^{-1} \frac{1}{l_2} - \tan^{-1} \frac{1}{l_{2m+2}}$.
So $\sum_{n=1}^{\infty} \tan^{-1} \frac{xl_{2n+1}}{l_{2n+1}^2 + x^2 + 5} = \tan^{-1} \frac{1}{l_2} - 0 = \frac{1}{x^2 + 2}$.

13. Using the identity $\tan^{-1} x + \tan^{-1} 1/x = \pi/2$, identity (40.8) yields $\tan^{-1} 1/F_{2n+1} = (\pi/2 - \tan^{-1} L_{2n}) + (\pi/2 - \tan^{-1} L_{2n+2})$. This yields the desired result.

15. Using the identity $\tan^{-1} x + \tan^{-1} 1/x = \pi/2$, identity (40.14) yields
$\tan^{-1} \frac{(-1)^n}{f_{2n+2}} = \left(\frac{\pi}{2} - \tan^{-1} \frac{l_{n+1}}{l_n} \right) - \left(\frac{\pi}{2} - \tan^{-1} \frac{l_{n+2}}{l_{n+1}} \right) =$
$\tan^{-1} \frac{l_{n+2}}{l_{n+1}} - \tan^{-1} \frac{l_{n+1}}{l_n}$.

17. The result is true when $n = 1$. Assume it is true for an arbitrary positive integer k. Then, by Corollary 40.2, $\sum_{i=1}^{k+1} (-1)^{i+1} \tan^{-1} \frac{1}{f_{2i}} =$
$\left[\tan^{-1} \frac{f_{k+1}}{f_{k+2}} + \tan^{-1} \frac{(-1)^{k+1}}{f_{2k+2}} \right] + (-1)^{k+2} \tan^{-1} \frac{1}{f_{2k+2}} = \tan^{-1} \frac{f_{k+1}}{f_{k+2}}$.
Thus, by PMI, the result is true for every $n \geq 1$.

19. Since $\tan^{-1}$ is a continuous increasing function, $\tan^{-1} \frac{1}{f_{2n}} > \tan^{-1} \frac{1}{f_{2n+2}}$; and $\lim_{n \to \infty} \tan^{-1} \frac{1}{f_{2n}} = \tan^{-1} 0 = 0$. So the series converges; and
$\sum_{n=1}^{\infty} (-1)^{n+1} \tan^{-1} \frac{1}{f_{2n}} = \lim_{k \to \infty} \sum_{n=1}^{k} (-1)^{n+1} \tan^{-1} \frac{1}{f_{2n}} = \lim_{k \to \infty} \tan^{-1} \frac{f_k}{f_{k+1}} =$
$\tan^{-1} \left(\lim_{k \to \infty} \frac{f_k}{f_{k+1}} \right) = \tan^{-1} \frac{1}{\alpha(x)} = \tan^{-1}(-\beta(x))$.

21. Follows from Exercise 40.18.

23. Since $\beta(2x) = \delta(x)$, the result follows from Exercise 40.19.

25. Let $\theta_n =$ RHS. Then $\tan \theta_n = \frac{\frac{1}{f_{2n}} - \frac{1}{f_{2n+2}}}{1 + \frac{1}{f_{2n} f_{2n+2}}} = \frac{x f_{2n+1}}{f_{2n+1}^2} = \frac{x}{f_{2n+1}}$. The given result now follows.

27. Let $\theta_n =$ RHS. Then $\tan \theta_n = \frac{\frac{1}{l_{2n}} - \frac{1}{l_{2n+2}}}{1 + \frac{1}{l_{2n} l_{2n+2}}} = \frac{xl_{2n+1}}{(l_{2n+1}^2 + \Delta^2) + 1} =$
$\frac{xl_{2n+1}}{l_{2n+1}^2 + x^2 + 5}$. The given result now follows.

29. Using formula (40.19), $\sum_{k=1}^{n} \tanh^{-1} \frac{x}{f_{2k+2}} =$
$\sum_{k=1}^{n} \left(\tanh^{-1} \frac{1}{f_{2k+1}} - \tanh^{-1} \frac{1}{f_{2k+3}}\right) = \tanh^{-1} \frac{1}{f_3} - \tanh^{-1} \frac{1}{f_{2n+3}}$. So
$\sum_{k=1}^{\infty} \tanh^{-1} \frac{x}{f_{2k+2}} = \tanh^{-1} \frac{1}{f_3}$.

31. By formula (40.24), $\sum_{k=1}^{\infty} (-1)^{k-1} \tanh^{-1} \frac{2Q_{2k+3}}{P_{2k+3}^2} = \tanh^{-1} \frac{1}{12}$; that is,
$\sum_{k=1}^{\infty} \tanh^{-1} \frac{2(-1)^{k-1} Q_{2k+3}}{P_{2k+3}^2} = \tanh^{-1} \frac{1}{12}$.
Then $\frac{1}{2} \sum_{k=1}^{\infty} \ln \frac{P_{2k+3}^2 - 2(-1)^k Q_{2k+3}}{P_{2k+3}^2 + 2(-1)^k Q_{2k+3}} = \frac{1}{2} \ln \frac{13}{11}$. This gives the desired result.

33. Using the identities $f_{n+k} f_{n-k} - f_n^2 = (-1)^{n+k+1} f_k^2$ and

$$f_{n+k} - f_{n-k} = \begin{cases} f_n l_k & \text{if } k \text{ is odd} \\ f_k l_n & \text{otherwise,} \end{cases} \quad \text{we get LHS} = \tanh^{-1} \frac{\frac{f_n}{f_{n+k}} - \frac{f_{n-k}}{f_n}}{1 - \frac{f_n f_{n-k}}{f_n f_{n+k}}} =$$

$$\tanh^{-1} \begin{cases} \dfrac{(-1)^{n+k+2} f_k^2}{f_n \cdot f_n l_k} & \text{if } k \text{ is odd} \\[2ex] \dfrac{(-1)^{n+k+2} f_k^2}{f_n \cdot f_k l_n} & \text{otherwise} \end{cases} = \tanh^{-1} \begin{cases} \dfrac{(-1)^{n-1} f_k^2}{f_n^2 l_k} & \text{if } k \text{ is odd} \\[2ex] \dfrac{(-1)^n f_k}{f_{2n}} & \text{otherwise.} \end{cases}$$

35. By Theorem 40.8, $\sum_{i=1}^{n} \tanh^{-1} \frac{(-1)^{n-1} f_k^2}{f_{ik}^2 l_k} =$
$\sum_{i=1}^{n} \left[\tanh^{-1} \frac{f_{ik}}{f_{(i+1)k}} - \tanh^{-1} \frac{f_{(i-1)k}}{f_{ik}}\right] = \tanh^{-1} \frac{f_{nk}}{f_{(n+1)k}} - \tanh^{-1} 0 =$
$\tanh^{-1} \frac{f_{nk}}{f_{(n+1)k}}$.

EXERCISES 41

1. Since $a(ix/2) = i\alpha(x)$ and $b(ix/2) = i\beta(x)$, by formulas (41.20) and (41.21), $l_n = 2(-1)^n \cdot \frac{[i\alpha(x)]^n + [i\beta(x)]^n}{2} = \alpha^n(x) + \beta^n(x)$.
3. This follows by the Binet-like formula $2T_n(x) = (x + \sqrt{x^2 - 1})^n +$ $(x - \sqrt{x^2 - 1})^n$.

5. Since $a(-x) = -b(x)$ and $b(-x) = -a(x)$, $T_{2n}(-x) = \dfrac{a^{2n}(-x) + b^{2n}(-x)}{2} = \dfrac{a^{2n}(x) + b^{2n}(x)}{2} = T_{2n}(x)$. Similarly, $T_{2n+1}(-x) = -T_{2n+1}(x)$. Thus T_{2n} is even and T_{2n+1} is odd.

7. Follows by the Chebyshev recurrence.

9. LHS $= (2xT_{n+1} - T_n) - (2xT_{n-1} - T_n) = 2x(T_{n+1} + T_{n-1}) = 2x \cdot 2xT_n$ $=$ RHS.

11. Since $T_0 = 0$, the given result follows from the identity (41.22).

13. Using Exercise 41.11, LHS $= T_{2\cdot 2n} = 2T_{2n}^2 - 1 = 2(2T_n^2 - 1)^2 - 1 = 8T_n^4 - 8T_n^2 + 1 =$ RHS.

15. By a repeated application of the Chebyshev recurrence, $T_{2n+1} = 2xT_{2n} - T_{2n-1} = 2x(T_{2n} - T_{2n-2}) + T_{2n-3} = 2x(T_{2n} - T_{2n-2} + T_{2n-4}) - T_{2n-5} = \cdots$ $= 2x(T_{2n} - T_{2n-2} + T_{2n-4} - \cdots + \nu T_2) - \nu T_1 =$ RHS.

17. $2T_{m+n} + 2T_{m-n} = (a^{m+n} + b^{m+n}) + (a^{m-n} + b^{m-n}) = a^{m+n} + b^{m+n} + a^m b^n + a^n b^m = 4T_m T_n$. This yields the given result.

19. Follows by Exercise 41.17.

21. By Exercises 41.18, 41.19, and 41.42, $T_{2n+1} - T_{2n-1} = 2T_n(T_{n+1} - T_{n-1}) = 2T_n \cdot (x^2 - 1)U_{n-1} =$ RHS.

23. Since $a(-x) = -b(x)$ and $b(-x) = -a(x)$, it follows by formula (41.33) that

$$U_{2n}(-x) = \frac{a^{2n+1}(-x) - b^{2n+1}(-x)}{a(-x) - b(-x)} = \frac{-b^{2n+1}(x) + a^{2n+1}(x)}{-b(x) + a(x)} = U_{2n}(x).$$

Similarly, $U_{2n+1}(-x) = -U_{2n+1}(x)$. So U_{2n} is even and U_{2n+1} is odd.

25. By formula (41.34), $U_n(x) = (n+1)x^n + \sum_{k\geq 1} \binom{n+1}{2k+1}(x^2 - 1)^k x^{n-2k}$. So $U_n(1) = n + 1$.

27. $(a - b)U_{-n} = a^{-n} - b^{-n} = -\dfrac{a^n - b^n}{(ab)^n} = -(a^n - b^n)$. So $U_{-n} = -U_n$.

29. Follows by Exercise 41.28.

31. $(a - b)$ LHS $= (a^{n+2} - b^{n+2}) - (a^n - b^n) = a^{n+1}(a - b) - b^{n+1}(b - a) = (a - b)(a^{n+1} + b^{n+1})$. So LHS $= 2T_{n+1} =$ RHS.

33. Using Exercise 41.31, LHS $= (2xU_{n+1} - U_n) - (2xU_{n-1} - U_n) = 2x(U_{n+1} - U_{n-1}) = 4xT_{n+1} =$ RHS.

35. $(a - b)^2$ LHS $= (a^{n+2} - b^{n+2})^2 + (a^{n+1} - b^{n+1})^2 = a^{2n+3}(a + b) + b^{2n+3}(b + a) - 4 = 2x(a^{2n+3} + b^{2n+3}) - 4 = 2x \cdot 2T_{2n+3} - 4$. Thus $(x^2 - 1)(U_{n+1}^2 + U_n^2) = xT_{2n+3} - 1$.

37. RHS $= (a^n - b^n)^2 + 2 = a^{2n} + b^{2n} =$ LHS.

39. Since $f_{n+1} = (-i)^n U_n(ix/2)$, it follows by Exercise 41.38 that
$f_{n+1} = (-i)^n \sum_{k=0}^{\lfloor n/2 \rfloor} \binom{n+1}{2k+1} \left(-x^2/4 - 1\right)^k (ix/2)^{n-2k}$. This gives the desired formula.

41. Clearly, the formula works when $n = 0$. Assume it works for an arbitrary nonnegative integer n. Then, by identity (41.36),
$\frac{\sin(n+2)\theta}{\sin\theta} = \frac{\sin[(n+1)+1]\theta}{\sin\theta} = \frac{\sin(n+1)\theta\cos\theta + \cos(n+1)\theta\sin\theta}{\sin\theta}$
$= xU_n(x) + T_{n+1} = U_{n+1}$. Thus, by PMI, the formula works for every $n \geq 0$.

43. LHS $= (T_{n+1} + T_{n-1})(T_{n+1} - T_{n-1}) = (2xT_n)[(x^2-1)U_{n-1}] =$
$x(x^2-1)(2U_{n-1}T_n) = x(x^2-1)U_{2n-1} =$ RHS.

45. $-$(LHS) $= \cos(m-n)\theta - \cos(m+n)\theta = 2\sin m\theta \sin n\theta =$
$2\sin\theta\, U_{m-1} \cdot \sin\theta\, U_{n-1} = 2(1-x^2)U_{m-1}U_{n-1}$; so LHS $=$
$2(x^2-1)U_{m-1}U_{n-1} =$ RHS.

47. Using the product formula $2\sin A \sin B = \cos(A-B) - \cos(A+B)$, $\sin^2\theta$(RHS) $= \sin(m+1)\theta\sin(n+1)\theta - \sin m\theta \sin n\theta$. Then $2\sin^2\theta$(RHS) $= [\cos(m-n)\theta - \cos(m+n+2)\theta] - [\cos(m-n)\theta - \cos(m+n)\theta] =$
$[\cos(m+n)\theta - \cos(m+n+2)\theta] = 2\sin\theta\sin(m+n+1)\theta$. So
RHS $= \frac{\sin(m+n+1)\theta}{\sin\theta} = U_{m+n} =$ LHS.

49. (LHS)$\sin^2\theta = \sin(n+2)\theta\sin n\theta - \sin^2(n+1)\theta = \sin(n+2)\theta\sin n\theta -$
$[1 - \cos^2(n+1)\theta] = \frac{1}{2}[\cos 2\theta - \cos(2n+2)\theta] + \cos^2(n+1)\theta - 1 =$
$\frac{1}{2}\{(2\cos^2\theta - 1) - [2\cos^2(n+1)\theta - 1]\} + \cos^2(n+1)\theta - 1 = -\sin^2\theta$. So
LHS $= -1 =$ RHS.

51. $T_{n+1} = \cos(n+1)\theta = \cos n\theta\cos\theta - \sin n\theta\sin\theta = xT_n - \sin^2\theta\, U_{n-1} =$
$xT_n - (1-x^2)U_{n-1}$.

53. It follows from Exercise 41.45 that $T_{m+n}(ix/2) - T_{m-n}(ix/2) =$
$2[(ix/2)^2 - 1]U_{m-1}(ix/2)U_{n-1}(ix/2)$. This implies $\frac{l_{m+n}}{2(-i)^{m+n}} - \frac{l_{m-n}}{2(-i)^{m-n}} =$
$-\frac{x^2+4}{2} \cdot \frac{f_m}{(-i)^{m-1}} \cdot \frac{f_n}{(-i)^{n-1}}$; that is, $l_{m+n} - (-1)^n l_{m-n} = \Delta^2 f_m f_n$.

55. $U_n(x) = \frac{\sin(n+1)\theta}{\sin\theta}$; so $U_n' = \frac{d}{d\theta}\left[\frac{\sin(n+1)\theta}{\sin\theta}\right] \cdot \frac{d\theta}{dx}$
$= -\frac{1}{\sin\theta}\left[\frac{\sin\theta\cdot(n+1)\cos(n+1)\theta - \sin(n+1)\theta\cos\theta}{\sin^2\theta}\right]$
$= \frac{-(n+1)T_{n+1} + xU_n}{1-x^2}$. This gives the desired result.

57. Using Exercises 41.54 and 41.55, $T_n'' = nU_{n-1}' = n\left(\dfrac{nT_n - xU_{n-1}}{x^2-1}\right) = \dfrac{n}{x^2-1}\left[(n+1)T_n - (T_n + xU_{n-1})\right] = \dfrac{n}{x^2-1}\left[(n+1)T_n - U_n\right].$
This gives the desired result.

59. By Exercise 41.56, $\dfrac{1}{2n}T_{2n}' = 2T_{2n-1} + \dfrac{1}{2n-2}T_{2n-2}' = 2(T_{2n-1} + T_{2n-3}) + 2T_{2n-5} + \dfrac{1}{2n-6}T_{2n-6}' = 2(T_{2n-1} + T_{2n-3} + T_{2n-5}) + \dfrac{1}{2n-6}T_{2n-6}' = 2(T_{2n-1} + T_{2n-3} + T_{2n-5} + \cdots + T_1)$. This implies the given result.

61. Since $T_{n+1}T_{n-1} - T_n^2 = x^2 - 1$, $T_{n+1}(ix/2)T_{n-1}(ix/2) - T_n^2(ix/2) = -(x^2+4)/4$. This implies $i^{n+1}l_{n+1} \cdot i^{n-1}l_{n-1} - (i^n l_n)^2 = -(x^2+4)$; that is, $l_{n+1}l_{n-1} - l_n^2 = (-1)^{n+1}(x^2+4)$.

63. Using Exercise 41.22, $T_{n^{k+1}} = T_{n^k}(T_n)$, where $T_1(T_n) = T_n$ and $k \geq 0$.

65. Clearly, $R_0 = 1 = Q_0$, $R_1 = 1 = Q_1$, and $R_2 = 3 = Q_2$. Assume the formula works for all nonnegative integers $< n$, where $n \geq 3$. Then $R_n = \sum\limits_{k\geq 0} b_{n,k} = 2\sum\limits_{k\geq 0} b_{n-1,k} + \sum\limits_{k\geq 0} b_{n-2,k-1} = 2R_{n-1} + R_{n-2}$. This recurrence, coupled with the initial conditions, implies that $R_n = Q_n$.

67. Clearly, $z_0 = 1$. So let $n \geq 1$. Then by formula (41.28),
$$z_n = \sum_{k=0}^{n} |a_{n+k,k}| = \sum_{k=0}^{n} \frac{n+k}{n}\binom{n}{k}2^{n-k-1} = \sum_{k=0}^{n}\left[\binom{n}{k} + \binom{n-1}{k-1}\right]2^{n-k-1}$$
$$= \tfrac{1}{2}(2+1)^n + \tfrac{1}{2}(2+1)^{n-1} = 2 \cdot 3^{n-1}.$$

69. We have $z_n = 3z_{n-1}$, where $z_0 = 1$. Let $g = 1 + 2x + 6x^2 + \cdots$. Then $(1-3x)g = 1 - x$, so $g = \dfrac{1-x}{1-3x}$.

71. Since $d_{0,0} = 0 + 1$ and $d_{1,0} = 1 + 1$, the result is true when $n = 0$ and $n = 1$. Assume it is true for all nonnegative integers $< n$, where $n \geq 2$. Then, by recurrence (41.44), $\sum\limits_{k=0}^{\lfloor n/2 \rfloor} d_{n,k} = 2\sum\limits_{k\geq 0} d_{n-1,k} - \sum\limits_{k\geq 0} d_{n-2,k-1} = 2n - (n-1) = n + 1$. Thus, by PMI, the result is true for every $n \geq 0$.

73. We have $h_k(x) = \dfrac{-h_{k-1}}{1-2x}$, where $k \geq 1$. but $h_0 = \dfrac{1}{1-2x}$. So, by PMI,
$$h_k(x) = \frac{1}{1-2x}\left(\frac{-1}{1-2x}\right)^k, \text{ where } k \geq 0.$$

75. Let S_n denote the nth row sum. Clearly, $S_0 = 1$ and $S_1 = 2$. Now let $n \geq 2$. Then $S_n = \sum\limits_{k\geq 0} d_{n,k} = \sum\limits_{k\geq 0}(2d_{n-1,k} + d_{n-2,k-1}) = 2S_{n-1} + \sum\limits_{k\geq 0} d_{n-2,k} = 2S_{n-1} + S_{n-2}$. Thus S_n satisfies the Pell recurrence. Since $S_0 = P_1$ and $S_1 = P_2$, it follows that $S_n = P_{n+1}$, where $n \geq 0$.

77. This is true since $d_{n,k} = |c_{n,k}|$ for every $n, k \geq 0$.

EXERCISES 42

1. The sum of the weights of tilings of a $1 \times n$ board is U_n. Suppose an arbitrary tiling has k dominoes. Then it contains a total of $n - k$ tiles. So there are $\binom{n-k}{k}$ such tilings. The sum of the weights of such tilings is $(-1)^k\binom{n-k}{k}(2x)^{n-2k}$. Thus $U_n = \sum_{k=0}^{\lfloor n/2 \rfloor}(-1)^k\binom{n-k}{k}(2x)^{n-2k}$.

3. Sum $= -\sum_{k \text{ odd}}\binom{7-k}{k}(2x)^{7-2k} = -192x^5 - 8x$.

5. Answer $= \sum_{k \text{ even}}\binom{11-k}{k}(2x)^{11-2k} = 2048x^{11} + 4608x^7 + 280x^3$.

7. Consider a $1 \times (m + n)$ board. The sum of the weights of its tilings is U_{m+n}. Now consider an arbitrary tiling T. Suppose it is breakable at cell m. The sum of the weights of such tilings is $U_m U_n$. On the other hand, suppose it is *not* breakable at cell m. The sum of the weights of such tilings is $-U_{m-1}U_{n-1}$. Thus the sum of the weights of all tilings of the board is $U_m U_n - U_{m-1}U_{n-1} = U_{m+n}$.

9. Let S_n denote the sum of the weights of tilings of length n. Then $S_0 = 1$ and $S_1 = 2x$.
   Consider an arbitrary tiling T. Suppose it ends in a domino. The sum of the weights of such tilings is $-S_{n-2}$. On the other hand, suppose T ends in a square, black or white. The sum of the weights of such tilings is $2xS_{n-1}$. Thus $S_n = 2xS_{n-1} - S_{n-2}$, where $S_0 = 1$ and $S_1 = 2x$. So $S_n = U_n$.

11. Consider a $1 \times (m + n)$ board. By Theorem 42.6, the sum of the weights of its tilings using Model III is T_{m+n}.
    Let T be an arbitrary tiling. Suppose it is breakable at cell m. Then, by Theorems 42.1 and 42.6, the sum of the weights of such tilings is $T_m U_n$. On the other hand, suppose it is *not* breakable at cell m. The sum of the weights of such tiling is $-T_{m-1}U_{n-1}$. Thus the sum of the weights of all tilings of the board is $T_m U_n - T_{m-1}U_{n-1} = T_{m+n}$.

13. Let C_n denote the number of tilings of length n. Clearly, $C_0 = 1 = Q_0$ and $C_1 = 1 = Q_1$. Since $T_n = 2xT_{n-1} - T_{n-2}$, it follows that $C_n = 2C_{n-1} + C_{n-2}$, where $n \geq 2$. This recurrence, coupled with the initial conditions, implies that $C_n = Q_n$.

15. Using Exercise 42.14, $T_n = U_n - xU_{n-1} = (2xU_{n-1} - U_{n-2}) - xU_{n-1} = xU_{n-1} - U_{n-2}$.

17. Let S_n denote the sum of the weights of tilings of a circular board with n cells. Then $S_1 = 2x = 2T_1$ and $S_2 = 4x^2 - 2 = 2T_2$.
    Consider an arbitrary n-bracelet T, where $n \geq 3$. Suppose a square occupies cell 1. Then cells 2 through $n - 1$ can be glued to form an

$(n-1)$-bracelet; the sum of the weights of such $(n-1)$-bracelets is S_{n-1}. So the sum of the weights of such n-bracelets is $2xS_{n-1}$.
On the other hand, suppose *no* square occupies cell 1. Then a domino must occupy cells 1 and 2, or cells 1 and n. By removing the domino and gluing the resulting ends, we can form an $(n-2)$-bracelet. The sum of the weights of such $(n-2)$-bracelets is S_{n-2}. So the sum of the weights of such n-bracelets is $-S_{n-2}$.
Thus the sum of the weights of all n-bracelets is $S_n = 2xS_{n-1} - S_{n-2}$, where $S_1 = 2T_1$ and $S_2 = 2T_2$. Consequently, $S_n = 2T_n$.

EXERCISES 43

1. $f_{11} = xf_{10} + yf_9 = x^{10} + 9x^8y + 21x^6y^2 + 35x^4y^3 + 5x^2y^4 + y^5$. Similarly, $l_{11} = x^{11} + 11x^9y + 44x^7y^2 + 77x^5y^3 + 5x^3y^4 + 11xy^5$.
3. $f_6l_4 + yf_5l_3 = (x^5 + 4x^3y + 3xy^2)(x^4 + 4x^2y + 2y^2) + y(x^4 + 3x^2y + y^2)(x^3 + 3xy) = x^9 + 9x^7y + 27x^5y^2 + 30x^3y^3 + 9xy^4 = l_9$.
5. Adding the equations $u^n - v^n = \sqrt{x^2+4y}\,f_n$ and $u^n + v^n = l_n$, the result follows.
7. $(u-v)\text{LHS} = u^{-n} - v^{-n} = (-v/y)^n - (-u/y)^n = \dfrac{(-1)^{n+1}}{y^n}(u^n - v^n)$; so LHS = RHS.
9. $(u-v)f_{2n} = u^{2n} - v^{2n} = (u^n - v^n)(u^n + v^n)$; so $f_{2n} = f_nl_n$.
11. $(u-v)\text{LHS} = (u^{n+1} - v^{n+1}) + y(u^{n-1} - v^{n-1}) = u^n(u + y/u) - v^n(v + y/v) = (u^n + v^n)(u - v)$. So LHS = RHS.
13. $\text{LHS} = y(u^n + v^n)^2 + (u^{n+1} + v^{n+1})^2 = y(u^{2n} + v^{2n}) + (u^{2n+2} + v^{2n+2}) = u^{2n+1}(u + y/u) + v^{2n+1}(v + y/v) = (u-v)(u^{2n+1} - v^{2n+1}) = (u-v)^2f_{2n+1} = (x^2+4y)f_{2n+1}$.
15. $\text{LHS} = (u^{n+1} + v^{n+1})(u^{n-1} + v^{n-1}) - (u^n + v^n)^2 = (-y)^n(u/v + v/u) - 2(-y)^n$
$= (-y)^n\left(\dfrac{u^2+v^2}{uv}\right) - 2(-y)^n = (-y)^{n-1}(x^2 + 2y) - 2(-y)^n$
$= (x^2+4y)(-y)^{n-1} = \text{RHS}$.
17. $\text{LHS} = (xl_n + yl_{n-1}) + yl_{n-1} = l_{n+1} + yl_{n-1} = (x^2+4y)f_n = \text{RHS}$.
19. $\text{LHS} = xf_n + (xf_n + 2yf_{n-1}) = 2(xf_n + yf_{n-1}) = 2f_{n+1} = \text{RHS}$.
21. Since $u^2 + v^2 = x^2 + 2y$, $\text{LHS} = u^n(u^2 + y^2/u^2) + v^n(v^2 + y^2/v^2) = u^n(u^2 + v^2) + v^n(v^2 + u^2) = (u^n + v^n)(u^2 + v^2) = (x^2 + 2y)l_n = \text{RHS}$.
23. $\text{LHS} = (u^{n+2} + v^{n+2}) - y^2(u^{n-2} + v^{n-2}) = u^n(u^2 - y^2/u^2) + v^n(v^2 - y^2/v^2) = u^n(u^2 - v^2) + v^n(v^2 - u^2) = (u^n - v^n)(u^2 - v^2) = x(x^2+4y)f_n = \text{RHS}$.
25. $\text{LHS} = u^{2n} + v^{2n} = (u^n + v^n)^2 - 2(uv)^n = l_n^2 - 2(-y)^n = \text{RHS}$.

27. LHS $= (f_{n+1}(x,y) - yf_{n-1}(x,y))(f_{n+1}(x,y) + yf_{n-1}(x,y)) =$
$xf_n(x,y)\cdot l_n(x,y) = xf_{2n}(x,y)$.

29. LHS $= (l_{n+2}(x,y) + y^2 l_{n-2}(x,y))(l_{n+2}(x,y) - y^2 l_{n-2}(x,y)) =$
$(x^2+2y)l_n(x,y)\cdot x(x^2+4y)f_n(x,y) =$ RHS.

31. Follows from Exercise 43.30.

33. $(u-v)$LHS $= (u^{m+1} - v^{m+1})(u^n + v^n) + y(u^m - v^m)(u^{n-1} + v^{n-1}) =$
$u^{m+n}(u + y/u) - v^{m+n}(v + y/v) + u^m v^n(u + y/u) - u^n v^m(v + y/u) =$
$(u^{m+n} + v^{m+n})(u - v) + 0 - 0$. So LHS = RHS.

35. LHS $= \sum_{k=0}^{n}\binom{n}{k}(u^k + v^k)(u^{n-k} + v^{n-k}) = \sum_{k=0}^{n}\binom{n}{k}(u^n + v^n) +$
$\sum_{k=0}^{n}\binom{n}{k}(u^{n-k}v^k + v^{n-k}u^k) = 2^n l_n + 2(u+v)^n = 2^n l_n + 2x^n =$ RHS.

37. The formula works when $n = 1$ and $n = 2$. Suppose it works for all positive integers $\le k$, where $k \ge 2$. By Pascal's identity, $\sum_{j=0}^{\lfloor k/2\rfloor}\binom{k-j}{j}x^{k-2j}y^j =$
$\sum_{j=0}^{\lfloor k/2\rfloor}\binom{k-j-1}{j-1}x^{k-2j}y^j + \sum_{j=0}^{\lfloor k/2\rfloor}\binom{k-j-1}{j}x^{k-2j}y^j = A + B$ (say).

Suppose k is even. Then $A = \sum_{j=0}^{\lfloor (k-2)/2\rfloor}\binom{k-j-2}{j}x^{k-2j-2}y^{j+1} =$
$y\sum_{j=0}^{\lfloor (k-2)/2\rfloor}\binom{k-j-2}{j}x^{k-2j-2}y^j = yf_{k-1}$; $B = x\sum_{j=0}^{\lfloor (k-2)/2\rfloor}\binom{k-j-1}{j}x^{k-2j-1}y^j =$
$x\sum_{j=0}^{\lfloor (k-1)/2\rfloor}\binom{k-j-1}{j}x^{k-2j-1}y^j = xf_k$. Thus $A + B = xf_k + yf_{k-1} = f_{k+1}$.

So the formula works when $n = k+1$ and k is even.
Similarly, it works when $n = k+1$ and k is odd. Thus, by the strong version of PMI, it works for all positive integers n.

39. RHS $= \sum_{k=0}^{n}\binom{n}{k}x^k y^{n-k}(u^k + v^k) = \sum_{k=0}^{n}\binom{n}{k}(ux)^k y^{n-k} + \sum_{k=0}^{n}\binom{n}{k}(vx)^k y^{n-k}$
$= (ux+y)^n + (vx+y)^n = u^{2n} + v^{2n} = l_{2n} =$ LHS.

41. $f_5 = \sum_{j=0}^{2}\binom{4-j}{j}x^{4-2j}y^j = x^4 + 3x^2y + y^2$;
$f_8 = \sum_{j=0}^{3}\binom{7-j}{j}x^{7-2j}y^j = x^7 + 6x^5y + 10x^3y^2 + 4xy^3$.

43. $l_5 = \sum_{j=1}^{3}\frac{6-j}{3}\binom{3}{j}x^j y^{3-j} f_j = x^5 + 5x^3y + 5xy^2$;
$l_9 = \sum_{j=1}^{5}\frac{10-j}{5}\binom{5}{j}x^j y^{5-j} f_j = x^9 + 9x^7y + 27x^5y^2 + 30x^3y^3 + 9xy^4$.

45. $\sum_{k=0}^{4}(-1)^k\binom{4}{k}l_k l_{4-k} = 2(l_0l_4 - 4l_1l_3) + 6l_2^2 = 2(x^2+4y)^2$; similarly, $\sum_{k=0}^{5}(-1)^k\binom{5}{k}l_k l_{5-k} = 0$.

47. LHS $= \sum_{k=0}^{n}(-1)^k\binom{n}{k}(u^k - v^k)(u^{n-k} - v^{n-k}) = l_n\sum_{k=0}^{n}(-1)^k\binom{n}{k} - \sum_{k=0}^{n}(-1)^k\binom{n}{k}(u^{n-k}v^k + v^{n-k}u^k) = l_n \cdot 0 - [(u-v)^n + (v-u)^n] = -[1+(-1)^n](x^2+4y)^{n/2} =$ RHS.

49. $(u-v)$LHS $= \sum_{k=0}^{n}(-1)^k\binom{n}{k}(u^k - v^k)(u^{n-k} + v^{n-k}) = \sum_{k=0}^{n}(-1)^k\binom{n}{k}[(u^n - v^n) - u^{n-k}v^k - v^{n-k}u^k] = (u^n - v^n)\sum_{k=0}^{n}(-1)^k\binom{n}{k} - (u-v)^n - (v-u)^n = (u^n - v^n)\cdot 0 - [1+(-1)^n](x^2+4y)^{n/2} =$ RHS.

EXERCISES 44

1. The general solution is $J_n(x) = Au^n + Bv^n$, where $u = \dfrac{1+\sqrt{1+4x}}{2}$ and $v = \dfrac{1-\sqrt{1+4x}}{2}$. Using the initial conditions $J_0(x) = 0$ and $J_1(x) = 1$ we get $A = \dfrac{1}{u-v} = -B$. So $J_n(x) = \dfrac{u^n - v^n}{u-v}$.

3. Suppose n is odd. Then $\dfrac{2^n}{3} < J_n = \dfrac{2^n+1}{3}$; so $J_n = \lceil 2^n/3 \rceil$. When n is even, $\dfrac{2^n-1}{3} = J_n < \dfrac{2^n}{3}$; so $J_n = \lfloor 2^n/3 \rfloor$.

5. Number of digits in $J_{14} = \lfloor 2^{14}/3 \rfloor = 5{,}461$; number of digits in $J_{19} = \lceil 2^{19}/3 \rceil = 174{,}763$.

7. 3(LHS) $= [2^n - (-1)^n] + [2^{n+1} - (-1)^{n+1}] = 3\cdot 2^n$; so LHS $= 2^n =$ RHS.

9. It follows by Exercise 44.7 that every two consecutive Jacobsthal numbers have the same parity. Since J_1 and J_2 are odd, it follows that every J_n is odd.

11. $3(J_{n+2} - J_n) = [2^{n+2} - (-1)^{n+2}] - [2^n - (-1)^n] = 2^{n+2} - 2^n = 3\cdot 2^n$; so LHS = RHS.

13. 3(LHS) $= [2^n - (-1)^n] + [2^{n+3} - (-1)^{n+3}] = 9\cdot 2^n$, so LHS $= 3\cdot 2^n =$ RHS.

15. $3J_n = 2^n - (-1)^n = 4\cdot 2^{n-2} - (-1)^n$; so $(-1)J_n \equiv 0 - (-1)^n \pmod 4$ and hence $J_n \equiv (-1)^n \pmod 4$, where $n \geq 2$.

17. $9(\text{LHS}) = [2^{n+k} - (-1)^{n+k}][2^{n-k} - (-1)^{n-k}] - [2^n - (-1)^n]^2 =$ $(-1)^{n-k+1}2^{n-k}[2^{2k} + (-2)^{k+1} + 1] = (-1)^{n-k+1}2^{n-k} \cdot 9J_k^2$; so LHS = RHS.
19. Follows by PMI and the addition formula $J_{m+n} = J_{m+1}J_n + 2J_mJ_{n-1}$.
21. $3(\text{LHS}) = [2^{n+k} - (-1)^{n+k}] - [2^{n-k} - (-1)^{n-k}] = 2^{n-k}(2^{2k} - 1)$ $= 2^{n-k} \cdot 3J_{2k}$; so $\text{LHS} = 2^{n-k}J_{2k} = \text{RHS}$.
23. $\text{LHS} = (J_{n+1} + J_n)(J_{n+1} - J_n) = 2^n \cdot 2J_{n-1} = 2^{n+1}J_{n-1} = \text{RHS}$.
25. $\text{LHS} = 2J_{n-k} + (J_{n-k} + 2J_{n-k-1}) = 2J_{n-k} + J_{n-k+1} = J_{n-k+2} = \text{RHS}$.
27. Using Exercise 44.21 and the identity $2^nJ_{-n} = (-1)^{n-1}J_n$, we have
$$J_{m-n} = J_mJ_{-(n-1)} + 2J_{m-1}J_{-n} = \frac{(-1)^{n-2}J_mJ_{n-1}}{2^{n-1}} + \frac{(-1)^n 2J_{m-1}J_n}{2^n}$$
$$= \frac{(-1)^n}{2^{n-1}}(J_mJ_{n-1} - J_{m-1}J_n).$$ So $2^{n-1}J_{m-n} = \text{RHS}$.
29. $9(J_n^2 + J_{n+1}^2) = [2^n - (-1)^n]^2 + [2^{n+1} + (-1)^n]^2 = 2(2^{2n+1} + 1) + 2^{n+1}[2^{n-1} -$ $(-1)^{n-1}] = 6J_{2n+1} + 3 \cdot 2^{n+1}J_{n-1}$. This gives the desired result.
31. Using the Jacobsthal recurrence,
$$\text{LHS} = 2\sum_{i=0}^{n} J_i = \sum_{i=0}^{n}(J_{i+2} - J_{i+1}) = J_{n+2} - J_1 = \text{RHS}.$$
33. $$\text{LHS} = \sum_{i=0}^{n}[2^{2i+1} - (-1)^{2i+1}] = \frac{2^{2n+3} + 3n + 1}{3} = \frac{3J_{2n+3} + 3n}{3} = J_{2n+3} + n$$ $= \text{RHS}$.
35. $$3(\text{LHS}) = \sum_{i=0}^{n}\binom{n}{i}[2^{i+k} - (-1)^{i+k}] = 2^k\sum_{i=0}^{n}\binom{n}{i}2^i - (-1)^k\sum_{i=0}^{n}\binom{n}{i}(-1)^i$$ $= 2^k3^n - 0$. So $\text{LHS} = 2^k3^{n-1} = \text{RHS}$.
37. $$3J_n = 2^n - (-1)^n = (3-1)^n - (-1)^n = \sum_{i=0}^{n}(-1)^i\binom{n}{i}3^{n-i} - (-1)^n$$ $$= \sum_{i=0}^{n-1}(-1)^i\binom{n}{i}3^{n-i}.$$ This gives the desired result.
39. $$\text{LHS} = \sum_{i=0}^{n}(-1)^i\binom{n}{i}[2^{i+k} - (-1)^{i+k}]^2 = \sum_{i=0}^{n}(-1)^i\binom{n}{i}[4^{i+k} + 1 - 2(-2)^{i+k}]$$ $= 4^k(1-4)^n + 0 + (-2)^{k+1}3^n = [(-1)^n4^k + (-2)^{k+1}]3^n = \text{RHS}$.
41. $$\text{RHS} = (-1)^n\sum_{k=0}^{n}(-2)^k = (-1)^n \cdot \frac{1 - (-2)^{n+1}}{3} = \frac{1}{3}\left[2^{n+1} - (-1)^{n+1}\right]$$ $= J_{n+1} = \text{LHS}$.
43. $2^n = (3-1)^n = \sum_{k=0}^{n}(-1)^k\binom{n}{k}3^{n-k}$; so $2^n/3 = \sum_{k=0}^{n-1}(-1)^k\binom{n}{k}3^{n-k-1} +$ $(-1)^n\left(\frac{1}{3}\right)$. Let n be even. Then $\lceil 2^n/3\rceil + \lfloor 2^n/3\rfloor = \frac{2}{3}\sum_{k=0}^{n-1}(-1)^k\binom{n}{k}3^{n-k} +$ $1 = \frac{2}{3}\left[(3-1)^n - 1\right] + 1 = \frac{2^{n+1}+1}{3} = J_{n+1}$. The case n odd follows similarly.

45. Let n be odd. Then $2^n \equiv 2 \pmod 3$; so $2^n = 3a + 2$ and $2^{n+1} = 3b + 1$ for some positive integers a and b. Then $\lfloor 2^{n+1}/3 \rfloor - \lfloor 2^n/3 \rfloor = b - a = \dfrac{2^{n+1} - 1}{3} - \dfrac{2^n - 2}{3} = \dfrac{2^n + 1}{3} = J_n$. The other case follows similarly.

47. LHS $= \frac{8}{9}\left[2^{n-1} + (-1)^n\right]\left[2^{n-2} - (-1)^n\right] + 1 = \frac{1}{9}[2^n - (-1)^n]^2 = J_n^2 =$ RHS.

49. LHS $= \dfrac{2^n}{3}\left[2^{-n} - (-1)^{-n}\right] = -\dfrac{1}{3}\left[(-2)^n - 1\right] = \dfrac{(-1)^{n+1}}{3}\left[2^n - (-1)^n\right]$
$= (-1)^{n+1} J_n =$ RHS.

51. LHS $= \left[2^{n+1} - (-1)^n\right]^2 - 2\left[2^n + (-1)^n\right]^2 = 2^{2n+1} - 1 = j_{2n+1} =$ RHS.

53. LHS $= [2^n + (-1)^n] + [2^{n+1} - (-1)^n] = 3 \cdot 2^n =$ RHS.

55. $j_n = 2^n + (-1)^n = 4 \cdot 2^{n-2} + (-1)^n \equiv 0 + (-1)^n \equiv (-1)^n \pmod 4$, where $n \geq 2$.

57. Since $j_n + j_{n+1} = 3 \cdot 2^n$, LHS $= (j_n + j_{n+1}) + (j_{n+2} + j_{n+3}) = 3 \cdot 2^n + 3 \cdot 2^{n+2} = 15 \cdot 2^n =$ RHS.

59. Since $j_n = 2^n + (-1)^n$, $j_n - 2j_{n-1} = [2^n + (-1)^n] - 2[2^{n-1} + (-1)^{n-1}] = 3(-1)^n$. So $j_n = 2j_{n-1} + 3(-1)^n$.

61. LHS $= [2^{n+k} + (-1)^{n+k}][2^{n-k} + (-1)^{n-k}] - [2^n + (-1)^n]^2$
$= (-2)^{n-k}[2^{2k} + (-2)^{k+1} + 1] = 9(-2)^{n-k} J_k^2 =$ RHS.

63. RHS $= 2(2^{2n-1} - 1) + 3 = 2^{2n} + 1 = j_{2n} =$ LHS.

65. LHS $= (j_{n+1} - j_n)(j_{n+1} + j_n) = (3 \cdot 2^n)(2j_n - 1) = 3 \cdot 2^{n+1} j_{n-1} =$ RHS.

67. LHS $= [2^{n+1} - (-1)^n]^2 + 2[2^n + (-1)^n]^2 = 3(2^{2n+1} + 1) = 3(j_{2n+1} + 2)$
$=$ RHS.

69. Follows by the identity $[x^2 + y^2 + (x+y)^2]^2 = 2[x^4 + y^4 + (x+y)^4]$.

71. $2\sum_{i=0}^{n} j_i = \sum_{i=0}^{n}(j_{i+2} - j_{i+1}) = j_{n+2} - 1$, as desired.

73. LHS $= \sum_{i=0}^{n}\left(2^{2i+1} - 1\right) = \dfrac{2^{2n+3} - 2}{3} - (n+1)$

$$= \frac{2\left[2^{2n+2} - (-1)^{2n+2} - 3n - 3\right]}{3} = 2J_{2n+2} - n - 1 = \text{RHS}.$$

75. LHS $= \sum_{i=0}^{n}\binom{n}{i}[4^{i+k} + 2(-2)^{i+k} + 1] = 4^k 5^n + 2(-2)^k(-1)^n + 2^n =$ RHS.

77. 3(LHS) $= 3[2^n + (-1)^n] - [2^n - (-1)^n] = 2 \cdot 2^n + 4(-1)^n$
$= 4[2^{n-1} - (-1)^{n-1}] = 4 \cdot 3J_{n-1}$; so LHS = RHS.

79. 3(LHS) $= [2^{n+1} - (-1)^{n+1}] + 2[2^{n-1} - (-1)^{n-1}] = 3[2^n + (-1)^n] = 3j_n$. So LHS = RHS.

81. LHS $= (j_{n+1} - 2j_{n-1})(j_{n+1} + 2j_{n-1}) = j_n \cdot 9J_n = 9J_{2n} =$ RHS.

83. LHS $= (2^{2n} + 1) - 2(-2)^n =$ RHS.

85. Let $d = (J_n, j_n)$. Since $j_n = 3J_n + 2(-1)^n$, $d = (J_n, 3J_n + 2(-1)^n) = (J_n, 2(-1)^n))$. Since every J_n is odd, this implies d is odd. So d is odd and $d|2(-1)^n$. Consequently, $d = 1$.

87. LHS $= [2^n + (-1)^n]^2 - [2^n - (-1)^n]^2 = 4 \cdot 2^n(-1)^n = (-2)^{n+2} =$ RHS.

89. $3(\text{LHS}) = [2^m - (-1)^m][2^n + (-1)^n] + [2^n - (-1)^n][2^m + (-1)^m]$
$= 2[2^{m+n} - (-1)^{m+n} = 3 \cdot 2J_{m+n} = 3(\text{RHS})$. So LHS = RHS.

91. LHS $= [2^m + (-1)^m][2^n + (-1)^n] + [2^m - (-1)^m][2^n - (-1)^n]$
$= 2[2^{m+n} + (-1)^{m+n} = 2j_{m+n} =$ RHS.

93. $3(\text{RHS}) = [2^m + (-1)^m][2^{n+1} + (-1)^n] + 2[2^{m-1} - (-1)^m][2^n - (-1)^n]$
$= 3[2^{m+n} + (-1)^{m+n}] = 3j_{m+n}$. So RHS = LHS.

95. LHS $= [2^{m+n} - (-1)^{m+n}]^2 + [2^{m-n} + (-1)^{m-n}]^2 = 4^{m-n}(2^{4n} + 1) - 2(-2)^{m-n}(2^{2n} - 1) + 2 = 4^{m-n}j_{4n} - 6(-2)^{m-n}J_{2n} + 2 =$ RHS.

97. $\text{LHS} = \lim\limits_{n\to\infty} \dfrac{3[2^n + (-1)^n]}{[2^n - (-1)^n]} = \lim\limits_{n\to\infty} \dfrac{3[1 + (-1/2)^n]}{[1 - (-1/2)^n]} = 3 = \text{RHS}.$

99. Let n be odd. Then $\dfrac{2^{n+1} - 1}{3}$ is an integer, but $\dfrac{2^{n+1} - 2}{3}$ is not. So

$$\left\lceil \frac{2^{n+1} - 2}{3} \right\rceil = \left\lfloor \frac{2^{n+1} - 2}{3} \right\rfloor + 1 = \left\lfloor \frac{2^{n+1} + 1}{3} \right\rfloor = \left\lfloor \frac{2^{n+1} - 1}{3} + \frac{2}{3} \right\rfloor$$

$= \dfrac{2^{n+1} - 1}{3}$. When n is even, $\dfrac{2^{n+1} - 2}{3}$ is an integer; so

$\left\lceil \dfrac{2^{n+1} - 2}{3} \right\rceil = \dfrac{2^{n+1} - 2}{3}$. Thus the given formula works in both cases.

101. When n is even, both $\dfrac{2^n - 1}{3}$ and $\dfrac{2^{n+1} - 2}{3}$ are integers. So
$c_n = \lfloor (2^{n+1} - 1)/3 \rfloor - \lfloor (2^n - 1)/3 \rfloor = \lfloor (2^{n+1} - 2)/3 - 1/3 \rfloor - (2^n - 1)/3$
$= \dfrac{2^{n+1} - 2}{3} - \dfrac{2^n - 1}{3} = \dfrac{2^n - 1}{3} = J_n$.

103. The characteristic roots of the recurrence $x_n = x_{n-1} + 2x_{n-2}$ are 2 and -1. So the particular solution corresponding to the nonhomogeneous portion is of the form $Cn2^n$. Substituting $Cn2^n$ in the given recurrence yields $C = 1/12$. Thus the general solution of the given recurrence has the form $x_n = A \cdot 2^n + B(-1)^n + (n/12)2^n$. Using the initial conditions, this yields $A = 5/18$ and $B = 2/9$. Thus $x_n = (5/18)2^n + (2/9)(-1)^n + (n/12)2^n$, where $n \ge 2$.

105. The general solution of the homogeneous part is of the form $A \cdot 2^n + B(-1)^n$ and the particular solution of the form $Cn2^n$. Substituting $Cn2^n$ in the recurrence $x_n = x_{n-1} + 2x_{n-2} + 2^{n-2}$ yields $C = 1/6$. So $x_n = A \cdot 2^n + B(-1)^n + (n/6)2^n$. This, paired with the initial conditions, yields $A = -1/9 = -B$. Thus $x_n = -(1/9)2^n + (1/9)(-1)^n + (n/6)2^n$, where $n \ge 0$.

107. The general solution of the given recurrence is $z_n = A \cdot 2^n + B(-1)^n + Cn2^n$. Substituting $Cn2^n$ in the given recurrence yields $C = 1/3$. Then $z_n = A \cdot 2^n + B(-1)^n + (n/3)2^n$. Using the initial conditions, this yields $A = 5/18$ and $B = 2/9$. Thus $z_n = (5/18)2^n + (2/9)(-1)^n + (n/3)2^n$. Since $2^n = J_n + J_{n+1}$ and $(-1)^n = J_{n+1} - 2J_n$, this yields $z_n = \left(\frac{2n+3}{6}\right) J_{n+1} + \left(\frac{2n-1}{6}\right) J_n$.

109. The characteristic roots of the recurrence $x_n = 2x_{n-1} + 8x_{n-2}$ are -2 and 4. So the general solution of the given recurrence has the form $x_n = A \cdot 4^n + B(-2)^n + C \cdot 2^n + D(-1)^n$. Substituting $C \cdot 2^n$ in the recurrence $x_n = 2x_{n-1} + 8x_{n-2} + (4/3)2^{n-1}$ yields $C = -1/3$. Similarly, substituting $D(-1)^n$ yields $D = -4/15$. Then $x_n = A \cdot 4^n + B(-2)^n - (1/3)2^n - (4/15)(-1)^n$. Using the initial conditions, this yields $A = 4/15$ and $B = 1/3$. Thus $x_n = (4/15)4^n + (1/3)(-2)^n - (1/3)2^n - (4/15)(-1)^n$.

111. By Exercise 44.110, we have $e_n = e_{n-1} + 2e_{n-2} + J_{n-2} = e_{n-1} + 2e_{n-2} + \frac{1}{3}\left[2^{n-2} - (-1)^n\right]$. The general solution is of the form $e_n = A \cdot 2^n + B(-1)^n + Cn2^n + Dn(-1)^n$. Substituting $Cn2^n$ in the recurrence $e_n = e_{n-1} + 2e_{n-2} + (1/3)2^{n-2}$ yields $C = 1/18$, Likewise, $Dn(-1)^n$ yields $D = -1/9$. So $e_n = A \cdot 2^n + B(-1)^n + (n/18)2^n - (n/9)(-1)^n$. Using the initial conditions $e_1 = 0 = e_2$, we get $A = -2/27 = -B$. Since $2^n = J_n + J_{n+1}$ and $(-1)^n = J_{n+1} - 2J_n$, $e_n = -(2/27)2^n + (2/27)(-1)^n + (n/18)2^n - (n/9)(-1)^n = \left(\frac{5n-4}{18}\right) J_n - \left(\frac{n}{18}\right) J_{n+1}$.

113. Let x_n denote an arbitrary composition of n. When 2 is added to the first summand of x_n, the parity of x_{n+2} remains the same. So there are $starte_n$ such summands in x_{n+2}. Placing "2+" in front of x_n results in x_{n+2} with an even summand. Since there are $a_n = J_n$ such compositions x_{n+2}, it follows that $starte_{n+2} = starte_n + J_n$, where $starte_1 = 0 = starte_2$.

115. Using Exercise 44.114, $starto_n = a_n - starte_n$ $= J_n - \left[-\frac{1}{4} - \left(\frac{2n-3}{12}\right) J_{n+1} + \left(\frac{4n-2}{12}\right)\right] J_n = \frac{1}{4} + \left(\frac{2n-3}{12}\right) J_{n+1} - \left(\frac{2n-7}{6}\right) J_n$.

117. By Exercise 44.116, $es_{n+2} = es_{n+1} + 2es_n + (8/9)2^n - (7/18)(-1)^n - (n/3)(-1)^n - 1/2$, where $es_1 = 0 = es_2$. The general solution of the recurrence has the form $es_n = A2^n + B(-1)^n + Cn2^n + Dn(-1)^n + En^2(-1)^n + F$. Substitution of each particular solution in the corresponding recurrence yields $C = 4/27, D = -1/27, E = -1/18$, and $F = 1/4$. Then $es_n = A2^n + B(-1)^n + (4n/27)2^n - (n/27)(-1)^n - (n^2/18)(-1)^n + 1/4$. Using the initial conditions, this gives $A = -8/27$ and $B = 5/108$. Thus $es_n = -(8/27)2^n + (5/108)(-1)^n + (4n/27)2^n - (n/27)(-1)^n - (n^2/18)(-1)^n + 1/4 = \left(\frac{4n^2+8n-13}{36}\right) J_n - \left(\frac{2n^2-4n+9}{36}\right) J_{n+1} + 1/4$, where $n \geq 1$.

119. Clearly, $b_2 = 1 = b_3$. Let x_n denote an arbitrary composition of n with the last summand even, where $n \geq 4$.
Placing a "1+" in front of x_{n-1} gives a composition x_n; there are b_{n-1} such compositions. A "2+" in front of x_{n-2} gives x_n; there are b_{n-2} such compositions. Adding 2 to the first summand of x_{n-2} also gives b_{n-2} compositions x_n. Since these steps are reversible, $b_n = b_{n-1} + 2b_{n-2}$, where $b_2 = J_1$ and $b_3 = J_2$. Thus $b_n = J_{n-1}$, where $n \geq 2$.

121. The number of summands in x_n is one more than that of plus signs in it. Then, by Exercise 44.120, $s_n = p_n + b_n = p_n + J_{n-1} = \left(\frac{n-3}{12}\right) J_{n+1} + \left(\frac{n+1}{12}\right) J_n + J_{n-1}$.

123. By Exercise 44.122, $z_n = z_{n-1} + 2z_{n-2} + \frac{4}{3}\left[2^{n-4} - (-1)^{n-4}\right]$, where $z_1 = 1$, $z_2 = 0$, and $n \geq 3$. The general solution of this recurrence has the form $z_n = A2^n + B(-1)^n + Cn2^n + Dn(-1)^n$. Substituting $Cn2^n$ and $Dn(-1)^n$ in the corresponding recurrences gives $C = 1/18$ and $D = -4/9$. So $z_n = A2^n + B(-1)^n + (n/18)2^n - (4n/9)(-1)^n$. Using the initial conditions, this yields $A = 8/54 = -B$. Thus $z_n = (8/54)2^n - (8/54)(-1)^n + (n/18)2^n - (4n/9)(-1)^n = \left(\frac{17n+8}{18}\right) J_n - \left(\frac{7n}{18}\right) J_{n+1}$.

125. Let $m \equiv 1 \pmod 3$ and $i = \frac{m-1}{3}2^k + J_k$, where $k \geq 2$. Then

$$\left\lfloor \frac{3i+1}{2^k} \right\rfloor = \left\lfloor \frac{3\left(\frac{m-1}{3}2^k + J_k\right) + 1}{2^k} \right\rfloor = \left\lfloor \frac{(m-1)2^k + 2^k - (-1)^k + 1}{2^k} \right\rfloor = m,$$

and no smaller value of i yields m. The other two cases follow similarly.

127. Since $m \equiv k \pmod 2$, $3(J_{m+1} - J_{k+1}) = 2^{m+1} - (-1)^{m+1} - 2^{k+1} + (-1)^{k+1} = 2^k(2^{m-k+1} - 2) = 2^k\left[2^{m-k+1} - (-1)^{m-k+1} - 3\right] = 3 \cdot 2^k(J_{m-k+1} - 1)$. The given result now follows.

129. $2^{k+1}(\text{LHS}) = (J_{m+1} - 2^k + J_{k+1})\left[3(J_{m+1} - J_{k+1}) + 2^{k+1}\right]$. Then $3 \cdot 2^{k+1}(\text{LHS}) = \left[2^{m+1} - 2^k + 2(-1)^k\right] 2^{m+1}$, so LHS = RHS.

131. Using $q = J_{m-k+1}$, $r = J_k$, and Theorem 44.7, LHS =
$\frac{J_{m+1} - J_k}{2^{k+1}}\left[3(J_{m+1} - J_k) - 2^k\right] + 3J_{m-k+1}J_k$. Then 3(LHS) =
$\frac{1}{2^{k+1}}(2^{m+1} - 2^k)(2^{m+1} - 2^{k+1}) + (2^{m-k+1} - 1)\left[2^k - (-1)^k\right] =$
$2^{2m-k+1} - 2^m - (-1)^k 2^{m-k+1} + (-1)^k$, so LHS = RHS.

133. Let $n = 5 = 2^k q + r$, where $(k, q, r) = (1, 2, 1)$ or $(2,1,1)$. Then, by Theorems 44.8 and 44.9, $N_1^*(5) = 8 = D_1^*(5)$ and $N_2^*(5) = 7 = D_2^*(5)$. Since $N_k^*(5) = D_k^*(5)$ for every k, $A(5)$ is odd.

135. Let $n = 6 = 2^k q + r$, where $(k, q, r) = (1, 3, 0)$ or $(2,1,2)$. Then $N_1^*(6) = 24 = D_1^*(6)$, but $N_2^*(6) = 11 \neq 10 = D_2^*(6)$. Since $N_k^*(6) \neq D_k^*(6)$ for some k, $A(6)$ is even.

137. Let $g(t) = \sum_{n=0}^{\infty} J_n(x)t^n$. Then $(1 - t - xt^2)g(t) = t$. So $g(t) = \dfrac{t}{1 - t - xt^2}$.

139. It follows from Exercise 44.137 that $\sum_{n=0}^{\infty} J_n t^n = \dfrac{t}{1 - t - 2t^2}$.

141. Using the generating functions $f(x) = \sum_{n=0}^{\infty} F_n x^n = \dfrac{x}{1 - x - x^2}$ and $j(x) = \sum_{n=0}^{\infty} j_n x^n = \dfrac{2 - x}{1 - x - 2x^2}$, we have

$$\begin{aligned}
\sqrt{5}f(x)j(x) &= \left(\frac{1}{1-\alpha x} - \frac{1}{1-\beta x}\right)\left(\frac{1}{1-2x} + \frac{1}{1+x}\right)\\
&= \frac{1}{(1-\alpha x)(1-2x)} + \frac{1}{(1-\alpha x)(1+x)} - \frac{1}{(1-\beta x)(1-2x)}\\
&\quad - \frac{1}{(1-\beta x)(1+x)}\\
&= \frac{2\sqrt{5}}{1-2x} - \frac{\sqrt{5}}{1+x} - \frac{\sqrt{5}\alpha}{1-\alpha x} - \frac{\sqrt{5}\beta}{1-\beta x}.
\end{aligned}$$

Equating the coefficients of x^n from both sides, we get $\sum_{k=0}^{n} F_k j_{n-k} = [2^{n+1} + (-1)^{n+1}] - (\alpha^{n+1} + \beta^{n+1}) = j_{n+1} - L_{n+1}$.

143. With the generating functions $l(x) = \sum_{n=0}^{\infty} L_n x^n = \dfrac{2 - x}{1 - x - x^2}$ and $j(x) = \sum_{n=0}^{\infty} j_n x^n = \dfrac{2 - x}{1 - x - 2x^2}$, we have

$$\begin{aligned}
l(x)j(x) &= \left(\frac{1}{1-\alpha x} + \frac{1}{1-\beta x}\right)\left(\frac{1}{1-2x} + \frac{1}{1+x}\right)\\
&= \frac{1}{(1-\alpha x)(1-2x)} + \frac{1}{(1-\alpha x)(1+x)} + \frac{1}{(1-\beta x)(1-2x)}\\
&\quad + \frac{1}{(1-\beta x)(1+x)}\\
&= \frac{6}{1-2x} + \frac{3}{1+x} - \frac{\sqrt{5}\alpha}{1-\alpha x} + \frac{\sqrt{5}\beta}{1-\beta x}.
\end{aligned}$$

Equating the coefficients of x^n from both sides, we get $\sum_{k=0}^{n} L_k j_{n-k} = 3[2^{n+1} - (-1)^{n+1}] - \sqrt{5}(\alpha^{n+1} - \beta^{n+1}) = 9J_{n+1} - 5F_{n+1}$.

145. Using the generating functions $p(x) = \sum_{n=0}^{\infty} P_n x^n = \dfrac{x}{1-2x-x^2} = \dfrac{1}{2\sqrt{2}}\left(\dfrac{1}{1-\gamma x} - \dfrac{1}{1-\delta x}\right)$, and $j(x) = \sum_{n=0}^{\infty} j_n x^n = \dfrac{2-x}{1-x-2x^2} = \dfrac{1}{1-2x} + \dfrac{1}{1+x}$, we have

$$
\begin{aligned}
2\sqrt{2}p(x)j(x) &= \left(\frac{1}{1-\gamma x} - \frac{1}{1-\delta x}\right)\left(\frac{1}{1-2x} + \frac{1}{1+x}\right) \\
&= \frac{1}{(1-\gamma x)(1-2x)} + \frac{1}{(1-\gamma x)(1+x)} - \frac{1}{(1-\delta x)(1-2x)} \\
&\quad - \frac{1}{(1-\delta x)(1+x)} \\
&= \frac{1+\sqrt{2}\gamma^2}{\sqrt{2}(1-\gamma x)} + \frac{1-\sqrt{2}\delta^2}{\sqrt{2}(1-\delta x)} - \frac{4\sqrt{2}}{1-2x} + \frac{\sqrt{2}}{1+x}.
\end{aligned}
$$

Equating the coefficients of x^n from both sides, we get

$$
\begin{aligned}
\sum_{k=0}^{n} P_k j_{n-k} &= \frac{1}{4}(\gamma^n + \delta^n) + \frac{1}{2\sqrt{2}}(\gamma^{n+2} + \delta^{n+2}) - 2\cdot 2^n - \frac{1}{2}(-1)^n \\
&= \frac{1}{2}Q_n + P_{n+2} - \frac{1}{2}[2^{n+1} - (-1)^{n+1}] - 2^n \\
&= P_{n+2} + \frac{1}{2}(Q_n - 3J_{n+1}) - 2^n.
\end{aligned}
$$

147. With the generating functions $q(x) = \sum_{n=0}^{\infty} Q_n x^n = \dfrac{1-x}{1-2x-x^2} = \dfrac{1}{2}\left(\dfrac{1}{1-\gamma x} - \dfrac{1}{1-\delta x}\right)$, and $j(x) = \sum_{n=0}^{\infty} j_n x^n = \dfrac{2-x}{1-x-2x^2} = \dfrac{1}{1-2x} + \dfrac{1}{1+x}$, we have

$$
\begin{aligned}
2q(x)j(x) &= \left(\frac{1}{1-\gamma x} + \frac{1}{1-\delta x}\right)\left(\frac{1}{1-2x} + \frac{1}{1+x}\right) \\
&= \frac{1}{(1-\gamma x)(1-2x)} + \frac{1}{(1-\gamma x)(1+x)} + \frac{1}{(1-\delta x)(1-2x)} \\
&\quad + \frac{1}{(1-\delta x)(1+x)} \\
&= \frac{1+\sqrt{2}\gamma^2}{\sqrt{2}(1-\gamma x)} - \frac{1-\sqrt{2}\delta^2}{\sqrt{2}(1-\delta x)} - \frac{4}{1-2x} + \frac{2}{1+x}.
\end{aligned}
$$

Let S_n denote the given sum. Equating the coefficients of x^n from both sides, we get

$$\begin{aligned}2S_n &= \frac{1}{\sqrt{2}}(\gamma^n - \delta^n) + \left(\gamma^{n+2} + \delta^{n+2}\right) - 4\cdot 2^n + 2(-1)^n \\ &= 2P_n + 2Q_{n+2} - 2[2^{n+1} + (-1)^{n+1}] \\ S_n &= P_n + Q_{n+2} - j_{n+1}.\end{aligned}$$

149. The congruence is true when $k = 0$. Assume it works for an arbitrary $k \geq 0$. Then $4^{3^k} = a \cdot 3^{k+2} + 3^{k+1} + 1$ for some $a \geq 1$. So $4^{3^{k+1}} = (a \cdot 3^{k+2} + 3^{k+1} + 1)^3 = b \cdot 3^{k+3} + (3^{k+1} + 1)^3$ for some $b \geq 1$. It now follows that $4^{3^{k+1}} \equiv (3^{k+1} + 1)^3 \equiv 3^{3(k+1)} + 3^{2k+3} + 3^{k+2} + 1 \equiv 3^{k+2} + 1 \pmod{3^{k+3}}$. So the congruence works for $k + 1$ also. Thus, by PMI, the result follows.

EXERCISES 45

1. Since $uv = -x$, $(u - v)^2(\text{RHS}) = (u^{m+1} - v^{m+1})(u^{n+1} - v^{n+1}) + (-uv)(u^m - v^m)(u^n - v^n) = (u - v)(u^{m+n+1} - v^{m+n+1}) = (u - v)^2(\text{LHS})$. So LHS = RHS.
3. Since $uv = -x$, $(u - v)(\text{RHS}) = (u^{m+1} + v^{m+1})(u^{n+1} - v^{n+1}) + (-uv)(u^m + v^m)(u^n - v^n) = (u - v)(u^{m+n+1} + v^{m+n+1}) = (u - v)(\text{LHS})$. So LHS = RHS.
5. Follows by recurrence (45.6).
7. Since n 2×2 tiles cover a $2 \times n$ board in exactly one way, it follows that $a_{2n,n} = 1$.
9. By recurrence (45.6), $a_{2n,n-1} = a_{2n-1,n-1} + a_{2n-2,n-1} + a_{2n-2,n-2}$. Letting $x_n = a_{2n,n-1}$, this yields the recurrence $x_n = x_{n-1} + n + 1$, where $x_1 = a_{2,0} = 2$. The particular solution of this recurrence has the form $n(An + B)$. Substituting this in the recurrence and then equating the corresponding coefficients yields $A = 1/2$ and $B = 3/2$. Then $x_n = x_{n-1} + (1/2)n^2 + (3/2)n$. So the general solution of the recurrence has the form $x_n = C + (1/2)n^2 + (3/2)n$. Since $x_1 = 2$, this implies $C = 0$. Thus $x_n = a_{2n,n-1} = (1/2)n^2 + (3/2)n$.
11. The general solution has the form $b_n = A \cdot 2^n + B(-1)^n + C$. Substituting $b_n = C$ yields $C = 1$. Then $b_n = A \cdot 2^n + B(-1)^n + 1$. This, paired with the initial conditions, yields $A = 1$ and $B = 2$. Thus $b_n = 2^n + 2(-1)^n + 1$, where $n \geq 0$.
13. The result is true for $1 \leq n \leq 3$. Now assume it is true for all positive integers $< n$, where n is odd. Then, by recurrence (45.8), $b_{n,0} = b_{n-1,0} + b_{n-2,0} - 2 = (L_{n-1} + 2) + L_{n-2} - 2 = L_n$. On the other hand, let n be even. Then, by recurrence (45.8), RHS $= (F_n + 2F_{n-1}) + 2 = L_n + 2 = b_{n,0}$. So the formula works for n. Thus, by PMI, it works for all $n \geq 1$.

15. The result is true when $n = 1$ and $n = 2$. Assume it is true for all positive integers less than an arbitrary integer n. Then, by recurrence (45.10), $b_{n,1} = b_{n-1,1} + b_{n-2,1} + L_{n-2} = (n-1)F_{n-2} + (n-2)F_{n-3} + (F_{n-1} + F_{n-3}) = n(F_{n-2} + F_{n-3}) = nF_n$. So the formula works for n and hence for all $n \geq 1$.

17. The result is true when $n = 1$ and $n = 2$. Assume it is true for an arbitrary positive integer n. When n is odd,

$$A^{n+1} = \begin{bmatrix} 0 & 1 & 1 \\ 1 & 0 & 1 \\ 1 & 1 & 0 \end{bmatrix} \begin{bmatrix} J_n - 1 & J_n & J_n \\ J_n & J_n - 1 & J_n \\ J_n & J_n & J_n - 1 \end{bmatrix} = \begin{bmatrix} 2J_n & 2J_n - 1 & 2J_n - 1 \\ 2J_n - 1 & 2J_n & 2J_n - 1 \\ 2J_n - 1 & 2J_n - 1 & 2J_n \end{bmatrix}$$

$$= \begin{bmatrix} J_{n+1} + 1 & J_{n+1} & J_{n+1} \\ J_{n+1} & J_{n+1} + 1 & J_{n+1} \\ J_{n+1} & J_{n+1} & J_{n+1} + 1 \end{bmatrix}.$$

Similarly, when n is even, $A^{n+1} = \begin{bmatrix} J_{n+1} - 1 & J_{n+1} & J_{n+1} \\ J_{n+1} & J_{n+1} - 1 & J_{n+1} \\ J_{n+1} & J_{n+1} & J_{n+1} - 1 \end{bmatrix}$.

So the formula works for n in both cases, and hence for all $n \geq 1$.

19. Let n be odd. Then, by Theorem 45.5, $a_{11}^{(n)} = J_n - 1 = \frac{1}{3}(2^n + 1) - 1 = \frac{1}{3}(2^n - 2) = 2J_{n-1}$. Similarly, when n is even, $a_{11}^{(n)} = J_n + 1 = \frac{1}{3}(2^n - 1) + 1 = \frac{1}{3}(2^n + 2) = 2J_{n-1}$.

21. The result is true when $n = 1$ and $n = 2$. Assume it is true for an arbitrary positive integer $< n$. Then

$$A^n = \begin{bmatrix} 0 & 1 & 0 \\ 1 & 1 & 1 \\ 0 & 1 & 0 \end{bmatrix} \begin{bmatrix} J_{n-2} & J_{n-1} & J_{n-2} \\ J_{n-1} & J_n & J_{n-1} \\ J_{n-2} & J_{n-1} & J_{n-2} \end{bmatrix} = \begin{bmatrix} J_{n-1} & J_n & J_{n-1} \\ J_n & J_{n+1} & J_n \\ J_{n-1} & J_n & J_{n-1} \end{bmatrix}.$$

So the formula works for n, and hence for all $n \geq 1$.

23. $|A^n| = |A|^n = 0^n = 0$.

25. Since n is odd, $J_{n+1} = \frac{1}{3}\left(2^{n+1} - 1\right) = \frac{2}{3}(2^n + 1) - 1 = 2J_n - 1$.

27. 111111, 111211, 112111, 121111, 121211, and 123211

29. With $a = 0$, the statement is true for $n = 1$. Assume it is true for an arbitrary odd integer $n \geq 1$. Then

$$M^{n+2} = M^2 \cdot M^n = \begin{bmatrix} 2 & 1 & 1 \\ 1 & 2 & 1 \\ 1 & 1 & 2 \end{bmatrix} \begin{bmatrix} a+1 & a+1 & a \\ a+1 & a & a+1 \\ a & a+1 & a+1 \end{bmatrix}$$

$$= \begin{bmatrix} 4a+3 & 4a+3 & 4a+2 \\ 4a+3 & 4a+2 & 4a+3 \\ 4a+2 & 4a+3 & 4a+3 \end{bmatrix}.$$

So M^{n+2} also has the desired form. Thus, by PMI, M^n has the given form for all $n \geq 1$.

31. Answer $= J_{11} = 683$.

33. Answer $= J_{12} + 1 = 1{,}366$.

35. Since $M^{r+s} = M^r \cdot M^s$, it follows by Corollary 45.6 that

$$\begin{bmatrix} * & J_{r+s} & * \\ * & * & * \\ * & * & * \end{bmatrix} = \begin{bmatrix} J_r + 1 & J_r & J_r \\ * & * & * \\ * & * & * \end{bmatrix} \begin{bmatrix} * & J_s & * \\ * & J_s + 1 & * \\ * & J_s & * \end{bmatrix}$$
$$= \begin{bmatrix} * & (J_r + 1)J_s + J_r(J_s + 1) + J_rJ_s & * \\ * & * & * \\ * & * & * \end{bmatrix}.$$

This gives the desired identity.

37. Since $M^{r+s} = M^r \cdot M^s$, it follows by Corollary 45.7 that

$$3\begin{bmatrix} j_{r+s} + 1 & j_{r+s} - 2 & j_{r+s} - 2 \\ * & * & * \\ * & * & * \end{bmatrix} = \begin{bmatrix} j_r + 2 & j_r + 2 & j_r - 1 \\ * & * & * \\ * & * & * \end{bmatrix} \begin{bmatrix} j_s + 2 & * & * \\ j_s + 2 & * & * \\ j_s - 1 & * & * \end{bmatrix}.$$

The given identity follows from this matrix equation.

39. Since $M^{r+s} = M^r \cdot M^s$, it follows by Corollary 45.7 that

$$3\begin{bmatrix} * & * & j_{r+s} - 1 \\ * & * & * \\ * & * & * \end{bmatrix} = \begin{bmatrix} j_r + 2 & j_r + 2 & j_r - 1 \\ * & * & * \\ * & * & * \end{bmatrix} \begin{bmatrix} * & * & j_s - 2 \\ * & * & j_s - 2 \\ * & * & j_s + 1 \end{bmatrix}.$$

The given result follows from this matrix equation.

EXERCISES 46

1. Follows from Theorem 46.3 by replacing m with n and n with $n + 1$.

3. By Theorem 46.3, $f_{2n} = f_n(f_{n+1} + yf_{n-1}) = f_nl_n$.

5. Since $f_n(2x, y) = p_n(x, y)$, the given result follows from Exercise 46.1.

7. $f_{10}(x, y) = \sum_{i,j=0}^{4} \binom{4-j}{i}\binom{4-i}{j} x^{8-2i-2j+1}y^{i+j} = x^9 + 8x^7y + 21x^5y^2 + 20x^3y^3 + 5xy^4$.

9. Since $f_8(x, y) = x^7 + 6x^5y + 10x^3y^3 + 4xy^3$, $p_8(x, y) = 128x^7 + 192x^5y + 80x^3y^3 + 8xy^3$.

11. $J_{10} = f_{10}(1,2) = \sum_{i,j=0}^{4} \binom{4-j}{i}\binom{4-i}{j} 2^{i+j} = 341.$

13. Let $x = 1 = y$. Then weight(colored tiling) = 1. So the sum of the weights of colored tilings of length n = the number of colored tilings of length $n = p_{n+1}(1,1) = P_{n+1}$.

15. Let $S_n = S_n(x,y)$ denote the sum of the weights of tilings of a $1 \times n$ board. Clearly, $S_0 = 2 = l_0(x,y)$, $S_1 = x = l_1(x,y)$, and $S_2 = x^2 + 2y = l_2(x,y)$. Suppose $n \geq 3$. Let T be an arbitrary tiling of the board. Suppose T ends in a square. The sum of the weights of such tilings is xS_{n-1}. On the other hand, suppose T ends in a domino. The sum of the weights of such tilings is yS_{n-2}.
Thus the sum of the weights of all tilings of the board is $xS_{n-1} + yS_{n-2} = S_n$. This recurrence, coupled with the initial conditions, yields the desired result.

17. Let $S_n = S_n(x,y)$ denote the sum of the weights of tilings of an n-bracelet. Clearly, $S_0 = 2 = q_0(x,y)$, $S_1 = 2x = q_1(x,y)$, and $S_2 = 4x^2 + 2y = q_2(x,y)$. Consider an arbitrary tiling T of the bracelet, where $n \geq 3$. Suppose T begins with a square. The remaining $n-1$ cells can be used to form an $(n-1)$-bracelet. The sum of the weights of such tilings is $2xS_{n-1}$. On the other hand, suppose T begins with a domino. The sum of the weights of such tilings is yS_{n-2}.
Thus the sum of the weights of all tilings T is $S_n = 2xS_{n-1} + yS_{n-2}$. Since both S_n and $q_n(x,y)$ satisfy the same recursive definition, it follows that $S_n = q_n(x,y)$.

EXERCISES 47

1. Since $r(0) = i$ and $s(0) = -i$, $v_n(0) = i^n + (-i)^n = [1 + (-1)^n]i^n$.

3. Follows by changing x to $-ix$ in the formula $V_n(x) = i^{n-1} f_n(x)$.

5. $v_{-n}(x) = v^{-n} + w^{-n} = \dfrac{v^n + w^n}{(vw)^n} = v^n + w^n = v_n(x)$.

7. Since $v(-x) = -w(-x)$ and $w(-x) = -v(x)$, $v_n(-x) = [-v(-x)]^n + [-w(-x)]^n = (-1)^n v_n(x)$.

9. LHS $= (v^n - w^n)^2 + (v^{n+1} - w^{n+1})^2 = (v+w)(v^{2n+1} + w^{2n+1}) - 4$
$= xv_{2n+1} - 4 =$ RHS.

11. $(v - w)$LHS $= v^{2n} - w^{2n} = (v^n - w^n)(v^n + w^n) = (v-w)V_n v_n$; so LHS = RHS.

13. It follows from Exercise 31.49 that $l_{2n}(-ix) = -(x^2-4)f_n^2(-ix) + 2(-1)^n$; that is, $v_{2n}/i^{2n} = -(x^2-4)(V_n/i^{n-1})^2 + 2(-1)^n$. This yields the given identity.

15. LHS $= (V_{n+1} - V_{n-1})(V_{n+1} + V_{n-1}) = v_n \cdot xV_n =$ RHS.
17. LHS $= (v^{n+1} + w^{n+1}) - (v^{n-1} + w^{n-1}) = (v - w)(v^n - w^n) = (v - w)^2 V_n = (x^2 - 4)V_n =$ RHS.
19. LHS $= v_{n+1} + (v_{n+1} - xv_n) = v_{n+1} - v_{n-1} = (x^2 - 4)V_n =$ RHS.
21. Using the identity $l_{n+1}^2(x) - (x^2 + 4)f_n^2(x) = xl_{2n+1}(x)$, we have $l_{n+1}^2(-ix) + (x^2 - 4)f_n^2(-ix) = -ixl_{2n+1}(-ix)$; that is, $(v_{n+1}/i^{n+1})^2 + (x^2 - 4)(V_n/i^{n-1})^2 = -ix \cdot v_{2n+1}/i^{2n+1}$. This yields the given identity.
23. LHS $= i^{n+1}f_{n+2}(-ix) + i^{n-3}f_{n-2}(-ix) = i^{n+1}[f_{n+2}(-ix) + f_{n-2}(-ix)] = i^{n+1}[(-ix)^2 + 2]f_n(-ix) = (x^2 - 2)[i^{n-1}f_n(-ix)] = (x^2 - 2)V_n$.
25. LHS $= (V_{n+2} + V_{n-2})(V_{n+2} - V_{n-2}) = (x^2 - 2)V_n \cdot xv_n =$ RHS.
27. LHS $= i^{n+2}l_{n+2}(-ix) + i^{n-4}l_{n-2}(-ix) = i^{n+2}[l_{n+2}(-ix) + l_{n-2}(-ix)] = i^{n+2}[(-ix)^2 + 2]l_n(-ix) = (x^2 - 2) \cdot i^n l_n(-ix) = (x^2 - 2)v_n =$ RHS.
29. $(v_{n+2} + v_{n-2})(v_{n+2} - v_{n-2}) = (x^2 - 2)v_n \cdot x(x^2 - 4)V_n = x(x^2 - 2)(x^2 - 4)V_{2n}$.
31. Since $l_n(x)f_{n-1}(x) + f_n(x)l_{n-1}(x) = 2f_{2n-1}(x)$, we have $v_n/i^n \cdot V_{n-1}/i^{n-2} + V_n/i^{n-1} \cdot v_{n-1}/i^{n-1} = 2V_{2n-1}/i^{2n-2}$; that is, $v_nV_{n-1} + V_nv_{n-1} = 2V_{2n-1}$.
33. $(v - w)$RHS $= (v^{m+1} + w^{m+1})(v^n - w^n) - (v^m + w^m)(v^{n-1} - w^{n-1}) = (v - w)(v^{m+n} + w^{m+n})$. So RHS $= v_{m+n} =$ LHS.
35. $(v - w)$RHS $= (v^m - w^m)(v^n + w^n) + (v^n - w^n)(v^m + w^m) = 2(v^{m+n} - w^{m+n})$. So RHS $= 2V_{m+n} =$ LHS.
37. RHS $= (v^m + w^m)(v^n + w^n) + (v^m - w^m)(v^n - w^n) = 2(v^{m+n} + w^{m+n}) =$ LHS.
39. Using Exercise 31.111, $[(-ix)^2 + 4]f_n^3(-ix) = f_{3n}(-ix) - 3(-1)^n f_n(-ix)$; that is, $-(x^2 - 4)(V_n/i^{n-1})^3 = V_{3n}/i^{3n-1} - 3(-1)^n V_n/i^{n-1}$. This implies $(x^2 - 4)V_n^3 = V_{3n} - 3V_n$.
41. It follows from Exercise 31.73 that $v_m/i^m \cdot v_{m+n+k}/i^{m+n+k} - v_{m+k}/i^{m+k} \cdot v_{m+n}/i^{m+n} = (-1)^{m+1}(x^2 - 4)V_n/i^{n-1} \cdot V_k/i^{k-1}$. This yields the given identity.
43. Since $v_0 = 2$ and $v_1 = x$, the result is true when $n = 0$ and $n = 1$. Assume it is true for every nonnegative integer $< n$. Then, by the Vieta recurrence, LHS $= [xv_{n-1}(x) - v_{n-2}(x)]v_{n-1}(x) + v_{n-1}(x)[-xv_{n-1}(-x) - v_{n-2}(-x)] = -[v_{n-1}(x)v_{n-2}(-x) + v_{n-2}(x)v_{n-1}(-x)] = 0$. So the result is true for n. Thus, by PMI, it is true for every $n \geq 0$.
45. Clearly, the result is true when $n = 0$ and $n = 1$.
Assume it is true for every nonnegative integer $< n$. Since $l_{2n} = (x^2 + 2)l_{2n-2} - l_{2n-4}$, by the Vieta recurrence, $v_n(x^2 + 2) = (x^2 + 2)v_{n-1}(x^2 + 2) - v_{n-2}(x^2 + 2) = (x^2 + 2)l_{2n-2} - l_{2n-4} = l_{2n}$.
So the result is true for n. Thus, by PMI, it is true for every $n \geq 0$.

47. Using Exercise 47.45, $v_n(2) = l_{2n}(0) = 2$.
49. By identity (47.17), $p_n(1) = (-i)^{n-1}V_n(2i)$, so $P_n^2 = (-1)^{n-1}V_n^2(2i)$. Since $2Q_{2n} = v_n(6)$ and $v_n(-x) = (-1)^n v_n(x)$, letting $x = 2i$ in identity (47.21), we get $v_n(-6) - (-8)V_n^2(2i) = 2$; that is, $(-1)^n \cdot 2Q_{2n} + 8 \cdot (-1)^{n-1}P_n^2 = 2$. This yields the given result.
51. Follows by identity (47.27).
53. Replacing x with $i/\sqrt{x}$ in the identity $x^n j_n(-1/x^2) = i^n l_n(-ix)$, we get $(i/\sqrt{x})^n j_n(x) = i^n l_n(1/\sqrt{x})$; that is, $j_n(x) = x^{n/2} l_n(1/\sqrt{x})$.
55. It follows by identities (47.14) and (47.26) that $l_{2n} = v_n(x^2+2) = (x^2+2)^n j_n\left(-1/(x^2+2)^2\right)$.
57. This follows from the identity $U_{n-1}(x/2) = i^{n-1}f_n(-ix)$.
59. Follows by identity (47.43).
61. Since $v_n(x) = V_{n+1}(x) - V_{n-1}(x)$, it follows that $2T_n(x/2) = U_n(x/2) - U_{n-2}(x/2)$. This yields the given result.
63. By identity (47.19), $p_{2n}(x) = 2xV_n(4x^2+2) = 2xU_{n-1}(2x^2+1)$.
65. Follows by Exercise 47.63.
67. Using identity (47.21), $2T_n\left((x^2-2)/2\right) - (x^2-4)U_{n-1}^2(x/2) = 2$. The given result follows from this.
69. Using identity (47.43), $2T_n\left((x^2-2)/2\right) - 4T_n^2(x/2) = -2$. This yields the given result.
71. Letting $x = -i/2\sqrt{2}$ in identity (47.43), we get $U_n(-i/2\sqrt{2}) = (-i/\sqrt{2})^n J_{n+1}(2)$; that is, $J_{n+1} = (\sqrt{2}i)^n U_n(-i/2\sqrt{2})$.
73. Identity (47.49), coupled with identity (47.13), implies $\left[xV_{n+1}(x^2+2)\right]^3 - (x^2+2)\left[xV_n(x^2+2)\right]^3 + \left[xV_{n-1}(x^2+2)\right]^3 = x^2(x^2+2)[xV_{3n}(x^2+2)]$; that is, $f_{2n+2}^3 - (x^2+2)f_{2n}^3 + f_{2n-2}^3 = x^2(x^2+2)f_{6n}$.
75. Since $v_n = v^n + w^n$, $v_n' = nv^{n-1}\dfrac{dv}{dx} + nw^{n-1}\dfrac{dw}{dx} = \dfrac{n(v^n - w^n)}{\sqrt{x^2-4}} = nV_n$.
77. Since $v_n(3) = L_{2n}$, the result follows by Exercise 47.76.
79. Replace x with $-ix$ in the identity $l_{n+3}^2 = (x^2+1)l_{n+2}^2 + (x^2+1)l_{n+1}^2 - l_n^2$. Then multiply both sides by $(-i)^{2n+6}$. We then get $[(-i)^{n+3}v_{n+3}]^3 = -(x^2-1)[(-i)^{n+2}v_{n+2}]^2 - (x^2-1)[(-i)^{n+1}v_{n+1}]^2 - [(-i)^n v_n]^2$. This yields the given identity.
81. Since $v_n = x^n j_n(-1/x^2)$, Exercise 47.79 shows that $[x^{n+3}j_{n+3}(-1/x^2)]^2 = (x^2-1)[x^{n+2}j_{n+2}(-1/x^2)]^2 - (x^2-1)[x^{n+1}j_{n+1}(-1/x^2)]^2 + [x^n j_n(-1/x^2)]^2$. Replacing x with $i/\sqrt{x}$, we get
$$\frac{-(-1)^n}{x^{n+3}}j_{n+3}^2 = -\frac{x+1}{x}\cdot\frac{(-1)^n}{x^{n+2}}j_{n+2}^2 - \frac{x+1}{x}\cdot\frac{(-1)^n}{x^{n+1}}j_{n+1}^2 + \frac{(-1)^n}{x^n}j_n^2.$$
This gives the desired identity.

83. Since $2T_n(x) = i^n l_n(-2ix)$, replacing x with $-2ix$ and multiplying the resulting equation by i^{2n+6} yields the given identity.
85. Replace x with $-ix$ in the identity $l^3_{n+4} = (x^3 + 2x)l^3_{n+3} + (x^4 + 3x^2 + 2)l^3_{n+2} - (x^3 + 2x)l^3_{n+1} - l^3_n$. Then multiply both sides by i^{3n+12}. Since $v_n(x) = i^n l_n(-ix)$, the given result follows.
87. Replace x with $1/\sqrt{x}$ in the identity $l^3_{n+4} = (x^3 + 2x)l^3_{n+3} + (x^4 + 3x^2 + 2)l^3_{n+2} - (x^3 + 2x)l^3_{n+1} - l^3_n$. Multiply the resulting equation by $x^{(3n+12)/2}$. Since $j_n(x) = x^{n/2} l_n(1/\sqrt{x})$, the given result follows.
89. Replace x with $-2ix$ in the identity $l^3_{n+4} = (x^3 + 2x)l^3_{n+3} + (x^4 + 3x^2 + 2)l^3_{n+2} - (x^3 + 2x)l^3_{n+1} - l^3_n$. Multiply the resulting equation by i^{3n+12}. Since $2T_n(x) = i^n l_n(-2ix)$, the given result follows.
91. Consider a $1 \times n$ board. The sum of the weights of its tilings is V_{n+1}. Let T be an arbitrary tiling of the board. Suppose T contains exactly k dominoes. Then it contains $n - 2k$ squares and hence a total of $n - k$ tiles. The weight of such a tiling is $(-1)^k x^{n-2k}$. Since the k dominoes can be placed among the $n - k$ tiles in $\binom{n-k}{k}$ different ways, the sum of the weights of all tilings T equals $\sum_{k\ge0} (-1)^k \binom{n-k}{k} x^{n-2k} = V_{n+1}$.
93. Consider a $1 \times (2n + 1)$ board. The sum of the weights of its tilings is V_{2n+2}. Let T be an arbitrary tiling of the board. Since the length of the board is odd, T must contain a median square M:
$$\underbrace{i \text{ dominoes}, n-i-j \text{ squares}}_{n-j \text{ tiles}}\ \boxed{x}\ \underbrace{j \text{ dominoes}, n-i-j \text{ squares}}_{n-i \text{ tiles}}.$$
The weight of such a tiling T is $(-1)^{i+j}\binom{n-i}{j}\binom{n-j}{i}x^{2n-2i-2j+1}$. So the sum of the weights of all tilings T equals $\sum_{\substack{i,j\ge0\\ i+j\le n}} (-1)^{i+j}\binom{n-i}{j}\binom{n-j}{i}x^{2n-2i-2j+1} = V_{2n+2}$, as desired.
95. Let A, B, and C denote the sets of tilings of length $n - 1, n + 1$, and $n - 3$, respectively. The sums of the weights of such tilings are V_n, V_{n+2}, and V_{n-2}, respectively. Then the sum of the weights of tilings in $B \cup C$ is $V_{n+2} + V_{n-2}$.
We will now develop a three-step algorithm to construct a one-to-three correspondence between A and $B \cup C$. To this end, let $T \in A$.
Step 1. Append two squares at the end of T. The resulting tiling belongs to B and has weight $x^2 V_n$.
Step 2. Append a domino at the end of T. The resulting tiling also belongs to B and has weight $-V_n$.
These two steps do *not* create tilings in B that end in $\boxed{\;-1\;}\boxed{x}$.

Step 3A. Suppose T ends in a square. Insert a domino immediately to the left of the square; this produces a tiling in B that ends in $\boxed{\;-1\;}\boxed{x}$.

Step 3B. Suppose T ends in a domino. Deleting the domino yields a unique element in C.

Steps 1–3 produce a single element in $B \cup C$. The sum of the weights of elements in $B \cup C$ is $(x^2 - 1)V_n - (xV_{n-1} - V_{n-2}) = (x^2 - 2)V_n$.

The algorithm is reversible. So the sum of the weights of tilings in $B \cup C$ and A are equal; that is, $(x^2 - 2)V_n = V_{n+2} + V_{n-2}$, as desired.

97. Consider a $1 \times (m+n-1)$ board. The sum of the weights of its tilings is V_{m+n}.
Using the concept of breakability, the sum of the weights of tilings breakable at cell m is $V_{m+1}V_n$, and that of tilings unbreakable at cell m is $-V_mV_{n-1}$. So the sum of the weights of all tilings is $V_{m+1}V_n - V_mV_{n-1} = V_{m+n}$, as desired.

99. Consider a $2 \times (m+n-1)$ board. By Theorem 47.10, the sum of the weights of its domino tilings is V_{m+n}.
Using the concept of breakability, the sum of the weights of tilings breakable at cell m is $V_{m+1}V_n$, and that of tilings unbreakable at cell m is $V_m \cdot i^2 \cdot V_{n-1} = -V_mV_{n-1}$. So the sum of the weights of all tilings is $V_{m+1}V_n - V_mV_{n-1} = V_{m+n}$, as desired.

101. Consider a board of length $2n$. By Theorem 47.11, the sum of the weights of all its tilings is v_{2n}.
The length of the board is even, so every tiling must contain an even number of squares. Consequently, every tiling must contain at least n tiles. Consider an arbitrary tiling T. Suppose it contains k squares (and hence $n-k$ dominoes) among the first n tiles. There are $\binom{n}{k}$ such tiles. The sum of the weights of such tilings is $(-1)^{n-k}\binom{n}{k}x^k v_k$. So the sum of the weights of all tilings T is $\sum_{k=0}^{n}(-1)^{n-k}\binom{n}{k}x^k v_k = v_{2n}$.

103. Consider a circular board with $m+n$ cells. Using Model V, the sum of the weights of all its tilings is v_{m+n}. Now follow the argument in Example 33.12.

105. Consider a $1 \times n$ board. The sum of the weights of tilings of the board is V_{n+1}, where $w(\text{square}) = x$ and $w(\text{domino}) -1$. Now follow the technique in Example 33.14 to establish that $v_n + v_{n+2} = (x^2 - 4)V_{n+1}$. The given identity follows from this.

107. Let $g(t) = \sum_{n=0}^{\infty} g_n t^n$, where g_n satisfies the Vieta recurrence. Then $(1 - xt + t^2)g(t) = g_0 + (g_1 - xg_0)t$. When $g_0 = 2$ and $g_1 = x$, this yields
$$\frac{2 - xt}{1 - xt + t^2} = \sum_{n=0}^{\infty} v_n(x)t^n.$$

EXERCISES 48

1. By identity (48.3) and the Cassini-like formula for Lucas numbers, LHS = $2(L_{n-1}^2 + L_{n-1}L_n + L_n^2)^2 = 2(L_{n+1}L_{n-1} + L_n^2)^2 = 2[2L_n^2 - 5(-1)^n]^2 =$ RHS.
3. Using identity (48.4) and the Cassini-like formula for Pell–Lucas numbers, LHS $= 2(Q_{n-1}^2 + 2Q_{n-1}Q_n + 4Q_n^2)^2 = 2[Q_{n-1}(Q_{n-1} + 2Q_n) + 4Q_n^2]^2 = 2(Q_{n+1}Q_{n-1} + 4Q_n^2)^2 = 2[5Q_n^2 - 2(-1)^n]^2 =$ RHS.
5. Using identity (48.5), Jacobsthal recurrence, and the Cassini-like formula $j_{n+1}j_{n-1} - j_n^2 = 9(-2)^{n-1}$, LHS $= 2(4j_{n-1}^2 + 2j_{n-1}j_n + j_n^2)^2 = 2[2j_{n-1}(j_n + 2j_{n-1}) + j_n^2] = 2(2j_{n+1}j_{n-1} + j_n^2) = 2[3j_n^2 - 9(-2)^n]^2 =$ RHS.
7. By identity (48.7), LHS $= 2\{p_{n-1}(x)[2xp_n(x) + p_{n-1}(x)] + 4x^2p_n^2(x)\}^2 = 2[p_{n+1}(x)p_{n-1}(x) + 4x^2p_n^2(x)]^2 = 2\{[p_n^2(x) + (-1)^n] + 4x^2p_n^2(x)\}^2 =$ RHS.
9. Using identity (48.5), LHS $= 2\{2xJ_{n-1}(x)[J_n(x) + 2xJ_{n-1}(x)] + J_n^2(x)\}^2 = 2[2xJ_{n+1}(x)J_{n-1}(x) + J_n^2(x)]^2 = 2\{2x[J_n^2(x) - (-2x)^{n-1}] + J_n^2(x)\}^2 =$ RHS.
11. Follows from identity (48.2).
13. Follows from identity (48.2).
15. Follows from identity (48.16).
17. Follows from Exercise 48.14 when $S_n(2x, 1) = p_n(x)$.
19. Letting $x = 1$ in Exercise 48.16, LHS $= 5F_{n+1}F_nF_{n-1}(F_n^2 + F_nF_{n-1} + F_{n-1}^2) = 5F_{n+1}F_nF_{n-1}(F_n^2 + F_{n+1}F_{n-1}) = 5F_{n+1}F_nF_{n-1}[2F_n^2 + (-1)^n] =$ RHS.
21. Letting $x = 1$ in Exercise 48.17, LHS $= 10P_{n+1}P_nP_{n-1}(4P_n^2 + 2P_nP_{n-1} + P_{n-1}^2) = 10P_{n+1}P_nP_{n-1}(4P_n^2 + P_{n+1}P_{n-1}) = 10P_{n+1}P_nP_{n-1}[5P_n^2 + (-1)^n] =$ RHS.
23. Letting $x = 1$ in Exercise 48.18, LHS $= 10J_{n+1}J_nJ_{n-1}(J_n^2 + 2J_nJ_{n-1} + 4J_{n-1}^2) = 10J_{n+1}J_nJ_{n-1}(J_n^2 + 2J_{n+1}J_{n-1}) = 10J_{n+1}J_nJ_{n-1}[3J_n^2 + 2(-1)^n] =$ RHS.
25. Follows from Exercise 48.15 when $s_n(x, 1) = f_n(x)$.
27. Follows from Exercise 48.15 when $s_n(1, 2y) = J_n(y)$.
29. Letting $x = 1 = y$ and $s_n = L_n$ in Exercise 48.15, LHS $= 7L_{n+1}L_nL_{n-1}(L_n^2 + L_nL_{n-1} + L_{n-1}^2)^2 = 7L_{n+1}L_nL_{n-1}(L_n^2 + L_{n+1}L_{n-1})^2 = 7L_{n+1}L_nL_{n-1}[2L_n^2 - 5(-1)^n]^2 =$ RHS.
31. Letting $x = 2, y = 1$, and $s_n = Q_n$ in Exercise 48.15, LHS = $14Q_{n+1}Q_nQ_{n-1}(4Q_n^2 + 2Q_nQ_{n-1} + Q_{n-1}^2)^2 =$ $14Q_{n+1}Q_nQ_{n-1}(4Q_n^2 + Q_{n+1}Q_{n-1})^2 = 14Q_{n+1}Q_nQ_{n-1}[5Q_n^2 - 2(-1)^n]^2 =$ RHS.
33. Let $x = 1, y = 2$, and $s_n = j_n$ in Exercise 48.15. Then LHS = $14j_{n+1}j_nj_{n-1}(j_n^2 + 2j_nj_{n-1} + 4j_{n-1}^2)^2 = 14j_{n+1}j_nj_{n-1}(j_n^2 + 2j_{n+1}j_{n-1})^2 =$ $14j_{n+1}j_nj_{n-1}[3j_n^2 + 18(-2)^{n-1}]^2 = 126j_{n+1}j_nj_{n-1}[j_n^2 - 3(-2)^n]^2 =$ RHS.

35. Since $l_n(x) = a^n(x) + b^n(x)$, $l_n(x/\sqrt{y}) = \dfrac{u^n + v^n}{y^{n/2}}$. Then $y^{n/2} l_n(x/\sqrt{y}) = u^n + v^n = l_n(x, y)$.

37. Follows by Exercises 48.35 and 48.36.

39. Using Exercises 48.35 and 48.37, $l_{2n}(x, y) = y^n l_{2n}(x/\sqrt{y}) = y^n \sum_{k=0}^{n} \frac{2n}{2n-k} \binom{2n-k}{k} (-1)^k \left(\frac{x^2 + 4y}{y}\right)^{n-k} = \text{RHS}.$

41. Using equation (48.25), $\sum_{k=1}^{n} \binom{n}{k} kx^{k-1} y^{n-k} l_k(x, y) = \sum_{k=1}^{n} \binom{n}{k} kx^{k-1} y^{n-k} (u^k + v^k) = n[u(ux + y)^{n-1} + v(vx + y)^{n-1}] = n(u^{2n-1} + v^{2n-1}) = nl_{2n-1}(x, y) =$ RHS.

EXERCISES 49

1. 111111, 11121, 1131, 11211, 12111, 21111, 1311, 3111, 231, 321, 1221, 2121, 2211, 11112, 1122, 1212, 2112, 222, 132, 312, where we have avoided the plus signs for brevity.

3. When $n \geq 0$, $B(n, 0) = 1 = B(n, n)$. When $n \geq 2$ and $j \geq 1$, $B(n, j) = B(n-2, j-1) + B(n-1, j-1) + B(n-1, j)$.

5. 0

7. $$Q^2 = \begin{bmatrix} x^2 & 1 & 0 \\ x & 0 & 1 \\ 1 & 0 & 0 \end{bmatrix} \begin{bmatrix} x^2 & 1 & 0 \\ x & 0 & 1 \\ 1 & 0 & 0 \end{bmatrix} = \begin{bmatrix} x^4 + x & x^2 & 1 \\ x^3 + 1 & x & 1 \\ x^2 & 1 & 0 \end{bmatrix};$$

$$Q^3 = \begin{bmatrix} x^2 & 1 & 0 \\ x & 0 & 1 \\ 1 & 0 & 0 \end{bmatrix} \begin{bmatrix} x^4 + x & x^2 & 1 \\ x^3 + 1 & x & 1 \\ x^2 & 1 & 0 \end{bmatrix} = \begin{bmatrix} x^6 + 2x^3 + 1 & x^4 + x & x^2 \\ x^5 + 2x^2 & x^3 + 1 & x \\ x^2 & 1 & 0 \end{bmatrix}.$$

9. 8, 15, 29.

11. T_n^*

13. $Q^n =$

$$\begin{bmatrix} t_{n+1}^* & t_n^* & t_{n-1}^* & t_{n-2}^* \\ x^2 t_n^* + x t_{n-1}^* + t_{n-2}^* & x^2 t_{n-1}^* + x t_{n-2}^* + t_{n-3}^* & x^2 t_{n-2}^* + x t_{n-3}^* + t_{n-4}^* & x^2 t_{n-3}^* + x t_{n-4}^* + t_{n-5}^* \\ x t_n^* + t_{n-1}^* & x t_{n-1}^* + t_{n-2}^* & x t_{n-2}^* + t_{n-3}^* & x t_{n-3}^* + t_{n-4}^* \\ t_n^* & t_{n-1}^* & t_{n-2}^* & t_{n-3}^* \end{bmatrix}.$$

15. By Exercise 49.14, $|Q_n| = (-1)^{n+1}$. Since $t_n^*(1) = T_n^*$, it follows that the given determinant equals $(-1)^{n+1}$.

17. x

19. x^3

21. 121121

23. 12312311

25.

x	1	x^2

27.

x	1	x^2	x^2

29. There are two walks of length 4: 21112 and 23112, The sum of their weights is $x^5 + x^2$.

Index

C

D

E

F

M

N

O

P

Q

R

S

T

U

V

W

Y

Z

PURE AND APPLIED MATHEMATICS

A Wiley Series of Texts, Monographs, and Tracts

Founded by RICHARD COURANT

ADÁMEK, HERRLICH, and STRECKER—Abstract and Concrete Categories

ADAMOWICZ and ZBIERSKI—Logic of Mathematics

AINSWORTH and ODEN—A Posteriori Error Estimation in Finite Element Analysis

AKIVIS and GOLDBERG—Conformal Differential Geometry and Its Generalizations

ALLEN and ISAACSON—Numerical Analysis for Applied Science

*ARTIN—Geometric Algebra

ATKINSON, HAN, and STEWART—Numerical Solution of Ordinary Differential Equations

AUBIN—Applied Functional Analysis, Second Edition

AZIZOV and IOKHVIDOV—Linear Operators in Spaces with an Indefinite Metric

BASENER—Topology and Its Applications

BERG—The Fourier-Analytic Proof of Quadratic Reciprocity

BERKOVITZ—Convexity and Optimization in $\mathbb{R}^n$

BERMAN, NEUMANN, and STERN—Nonnegative Matrices in Dynamic Systems

BOYARINTSEV—Methods of Solving Singular Systems of Ordinary Differential Equations

BRIDGER—Real Analysis: A Constructive Approach

BURK—Lebesgue Measure and Integration: An Introduction

*CARTER—Finite Groups of Lie Type

CASTILLO, COBO, JUBETE, and PRUNEDA—Orthogonal Sets and Polar Methods in Linear Algebra: Applications to Matrix Calculations, Systems of Equations, Inequalities, and Linear Programming

CASTILLO, CONEJO, PEDREGAL, GARCÍA, and ALGUACIL—Building and Solving Mathematical Programming Models in Engineering and Science

CHATELIN—Eigenvalues of Matrices

*Now available in a lower priced paperback edition in the Wiley Classics Library

†Now available in paperback

CLARK—Mathematical Bioeconomics: The Mathematics of Conservation, Third Edition

COX—Galois Theory, Second Edition

COX—Primes of the Form $x^2 + ny^2$: Fermat, Class Field Theory, and Complex Multiplication, Second Edition

*CURTIS and REINER—Representation Theory of Finite Groups and Associative Algebras

*CURTIS and REINER—Methods of Representation Theory: With Applications to Finite Groups and Orders, Volume I

CURTIS and REINER—Methods of Representation Theory: With Applications to Finite Groups and Orders, Volume II

DINCULEANU—Vector Integration and Stochastic Integration in Banach Spaces

*DUNFORD and SCHWARTZ—Linear Operators

Part 1—General Theory

Part 2—Spectral Theory, Self Adjoint Operators in Hilbert Space

Part 3—Spectral Operators

FARINA and RINALDI—Positive Linear Systems: Theory and Applications

FATICONI—The Mathematics of Infinity: A Guide to Great Ideas, Second Edition

FOLLAND—Real Analysis: Modern Techniques and Their Applications

FRÖLICHER and KRIEGL—Linear Spaces and Differentiation Theory

GARDINER—Teichmüller Theory and Quadratic Differentials

GILBERT and NICHOLSON—Modern Algebra with Applications, Second Edition

*GRIFFITHS and HARRIS—Principles of Algebraic Geometry

GRILLET—Algebra

GROVE—Groups and Characters

GUSTAFSSON, KREISS, and OLIGER—Time-Dependent Problems and Difference Methods, Second Edition

HANNA and ROWLAND—Fourier Series, Transforms, and Boundary Value Problems, Second Edition

*HENRICI—Applied and Computational Complex Analysis

Volume 1, Power Series—Integration—Conformal Mapping—Location of Zeros

Volume 2, Special Functions—Integral Transforms—Asymptotics—Continued Fractions

Volume 3, Discrete Fourier Analysis—Cauchy Integrals—Construction of Conformal Maps—Univalent Functions

*Now available in a lower priced paperback edition in the Wiley Classics Library

†Now available in paperback

*HILTON and WU—A Course in Modern Algebra

*HOCHSTADT—Integral Equations

JOST—Two-Dimensional Geometric Variational Procedures

KHAMSI and KIRK—An Introduction to Metric Spaces and Fixed Point Theory

*KOBAYASHI and NOMIZU—Foundations of Differential Geometry, Volume I

*KOBAYASHI and NOMIZU—Foundations of Differential Geometry, Volume II

KOSHY—Fibonacci and Lucas Numbers with Applications

KOSHY—Fibonacci and Lucas Numbers with Applications, Volume One, Second Edition

KOSHY—Fibonacci and Lucas Numbers with Applications, Volume Two

LAX—Functional Analysis

LAX—Linear Algebra and Its Applications, Second Edition

LOGAN—An Introduction to Nonlinear Partial Differential Equations, Second Edition

LOGAN and WOLESENSKY—Mathematical Methods in Biology

LUI—Numerical Analysis of Partial Differential

MARKLEY—Principles of Differential Equations

MELNIK—Mathematical and Computational Modeling

MORRISON—Functional Analysis: An Introduction to Banach Space Theory

NAYFEH—Perturbation Methods

NAYFEH and MOOK—Nonlinear Oscillations

O'LEARY—Revolutions of Geometry

O'NEIL—Beginning Partial Differential Equations, Third Edition

O'NEIL—Solutions Manual to Accompany Beginning Partial Differential Equations, Third Edition

PANDEY—The Hilbert Transform of Schwartz Distributions and Applications

PETKOV—Geometry of Reflecting Rays and Inverse Spectral Problems

*PRENTER—Splines and Variational Methods

PROMISLOW—A First Course in Functional Analysis

RAO—Measure Theory and Integration

RASSIAS and SIMSA—Finite Sums Decompositions in Mathematical Analysis

RENELT—Elliptic Systems and Quasiconformal Mappings

RIVLIN—Chebyshev Polynomials: From Approximation Theory to Algebra and Number Theory, Second Edition

ROCKAFELLAR—Network Flows and Monotropic Optimization

*Now available in a lower priced paperback edition in the Wiley Classics Library

†Now available in paperback

ROITMAN—Introduction to Modern Set Theory

ROSSI—Theorems, Corollaries, Lemmas, and Methods of Proof

*RUDIN—Fourier Analysis on Groups

SENDOV—The Averaged Moduli of Smoothness: Applications in Numerical Methods and Approximations

SENDOV and POPOV—The Averaged Moduli of Smoothness

SEWELL—The Numerical Solution of Ordinary and Partial Differential Equations, Second Edition

SEWELL—Computational Methods of Linear Algebra, Second Edition

SHICK—Topology: Point-Set and Geometric

SHISKOWSKI and FRINKLE—Principles of Linear Algebra With *Maple*™

SHISKOWSKI and FRINKLE—Principles of Linear Algebra With *Mathematica*®

*SIEGEL—Topics in Complex Function Theory

Volume 1—Elliptic Functions and Uniformization Theory

Volume 2—Automorphic Functions and Abelian Integrals

Volume 3—Abelian Functions and Modular Functions of Several Variables

SMITH and ROMANOWSKA—Post-Modern Algebra

ŠOLÍN–Partial Differential Equations and the Finite Element Method

STADE—Fourier Analysis

STAHL and STENSON—Introduction to Topology and Geometry, Second Edition

STAHL—Real Analysis, Second Edition

STAKGOLD and HOLST—Green's Functions and Boundary Value Problems, Third Edition

STANOYEVITCH—Introduction to Numerical Ordinary and Partial Differential Equations Using MATLAB®

*STOKER—Differential Geometry

*STOKER—Nonlinear Vibrations in Mechanical and Electrical Systems

*STOKER—Water Waves: The Mathematical Theory with Applications

WATKINS—Fundamentals of Matrix Computations, Third Edition

WESSELING—An Introduction to Multigrid Methods

†WHITHAM—Linear and Nonlinear Waves

ZAUDERER—Partial Differential Equations of Applied Mathematics, Third Edition

*Now available in a lower priced paperback edition in the Wiley Classics Library

†Now available in paperback